Cells and Organelles

Third Edition

Eric Holtzman
Columbia University

Alex B. Novikoff
*Albert Einstein College of Medicine
Yeshiva University*

SAUNDERS COLLEGE PUBLISHING

Philadelphia   New York   Chicago
San Francisco   Montreal   Toronto
London   Sydney   Tokyo   Mexico City
Rio de Janeiro   Madrid

Address orders to:
383 Madison Avenue
New York, NY 10017

Address editorial correspondence to:
West Washington Square
Philadelphia, PA 19105

Text Typeface: Souvenir
Compositor: The Clarinda Company
Acquisitions Editor: Michael Brown
Project Editor: Sally Kusch
Copyeditor: Elizabeth Galbraith
Managing Editor & Art Director: Richard L. Moore
Art/Design Assistant: Virginia A. Bollard
Text Design: Caliber Design Planning, Inc.
Cover Design: Lawrence R. Didona
Text Artwork: J & R Technical Services, Inc.
Production Manager: Tim Frelick
Assistant Production Manager: Maureen Iannuzzi

Front cover credit: Rat cerebellum section. Immunocytochemical localization of lysosomal β-galactosidase. (*Journal of Cell Biology,* in press.) Micrograph by Phyllis M. Novikoff.
Back cover credit: Rat hepatocytes surrounding pancreatic islet cells. Red spheres indicate cytosolic lipid. (*Proc. Natl. Acad. Sci. USA,* 1979.) Micrograph by Phyllis M. Novikoff.

Library of Congress Cataloging in Publication Data

Holtzman, Eric, 1939–
    Cells and organelles.

    Rev. ed. of: Cells and organelles/Alex B.
Novikoff, Eric Holtzman. 2nd ed. c1976.
Includes bibliographies and index.

    1. Cytology. 2. Cell organelles. I. Novikoff, Alex
Benjamin, 1913– . II. Novikoff, Alex
Benjamin, 1913– . Cells and organelles. 2nd
ed. III. Title. [DNLM: 1. Cells. 2. Organoids. QH
581.2 N943c]
QH581.2.H64 1983   574.8'7   83–7683

ISBN 0–03–049461–3

Cells and Organelles (3/e)   ISBN 0–03–049–461–3

3456   032   987654321

CBS COLLEGE PUBLISHING
Saunders College Publishing
Holt, Rinehart and Winston
The Dryden Press

Preface to the Third Edition

The past few years have been eventful ones in cell biology. We have revised our text to take into account the many conceptual and technical advances that have occurred. Our general approach remains the same as in the previous editions: We have integrated structural, molecular, biochemical, and physiological information and have tried to provide a view of the experimental and observational groundwork on which current concepts rest. We think it important also to give students a feeling for present uncertainties and for the thrust of current research. Thus, we continue to point out loose ends and inadequacies in present information.

In addition to integrating structural and biochemical approaches, we have updated and expanded our summaries of *background* biochemistry and molecular biology. Our experience, however, indicates that biology students are being exposed to such material at increasingly earlier stages of their education and that the attempt both to incorporate an introduction to basic chemistry and to provide a comprehensive introduction to cell biology within a single book would lead either to too long a book or to problems with level, scope, or depth of treatment. Consequently, we have retained the focus implied by our book's title.

A number of organizational infelicities present in the last edition have been eliminated in the present one through reordering of the sequence of treatment of several topics. We are grateful to the students and colleagues who called our attention to these and other features of the book that needed improving.

Many colleagues and friends have contributed to this revision. Those who provided micrographs are acknowledged in the corresponding legends. Arthur Karlin read the manuscript in its entirety and made numerous valuable suggestions and criticisms. So did Catherine Fussell, Andrew Hamilton, and Joseph Scott. Sherman Beychok, Larry Chasin, Bill Cohen, Art Forer, Jack Harding, John Hildebrand, Nancy Lane, Myron Ledbetter, Alberto Mancinelli, Jim Manley, Sandra Masur, Debby Mowshowitz, Bob Pollack, Diane Robins, Dave Soifer, Cathy Squires, Alex Tzagoloff, and Maurice Zauderer commented helpfully on long segments of the manuscript.

Note also that many of the figure legends accompanying diagrams indicate the names of the principal contributors to the research summarized in the diagrams.

New York, N.Y.   *E. H.*
June 1983   *A. B. N.*

Preface to the Second Edition

Little need be added to what we have written above, regarding either the sense of excitement among students of the cell, or our general approach to presenting the material.

Note should be made that the 1974 Nobel Prize for Physiology or Medicine was awarded to three cell biologists "for their discoveries concerning the structural and functional organization of the cell": Albert Claude, Christian de Duve, and George E. Palade. Their contributions to cell fractionation procedures and the techniques of electron microscopy permitted them, and others, to lay the basis for much of the cell biology of today.

In planning the second edition we gave serious consideration to the suggestion from a number of teachers and students that we include references for the statements we make in the book. We resisted the temptation chiefly for two reasons: We did not wish to overly enlarge the book's size, or diminish its readability. The book is being utilized for courses at quite a variety of levels, and reference lists suitable for all would be unwieldy. In addition, in our experience as teachers we find reference lists to go out of date very rapidly. Even to support an "old" point, one tends to refer to newly appearing review articles and research reports in which recently appreciated nuances are covered. We do intend our Further Reading lists to provide access to the background literature, and we have revised and updated them accordingly.

We take this occasion to thank most sincerely those who read and reviewed our book in manuscript form, and those who have troubled to write us their views since publication of the first edition.

New York, N.Y.
February 1976

A. B. N.
E. H.

Preface to the First Edition

Cytologists study cells, all cells, and by many techniques. Because of their dependence on the microscope, they tend to analyze living systems in terms of visible structure, but in recent years cytologists have become increasingly concerned with biochemistry. *Cell biologists* start with a bias toward viewing cells in terms of molecules. Recently they have been much concerned with nucleic acids and proteins, the macromolecules which molecular biology has revealed to play primary roles in heredity. However, as studies progress, it is becoming more and more difficult to make a sharp differentiation among the various approaches to the cell—structural (classical cytology), physiological (biochemistry and biophysics), and molecular (cell biology). As distinctions between cell biology and cytology have become blurred, and perhaps outmoded, this book is about cytology and cell biology.

The book is divided into five parts. The first part introduces the major features of cells and the methods by which they are currently studied. In the second, we consider each of the organelles in turn, presenting structural and functional information. Then, in the third part, we discuss the diversity of cell types constructed from the same organelles and macromolecules. The fourth part presents major mechanisms by which cells reproduce, develop, and evolve. The final part is a brief look at the progress and the future of cell study.

Many of the cells used to illustrate the principles we discuss are from higher animals. This is not only because the authors have had more direct experience with such cells. It also reflects the historical development of cytology; the cells of higher animals have been best analyzed from the viewpoint of correlated structure and function. However, increasing attention is being focused on protozoa, higher plants, and bacteria and related cells. Wherever possible we have referred to studies of such organisms.

We begin the book and each of the parts with introductions outlining the major themes of the chapters that follow and commenting upon some topics that do not fit conveniently into one chapter. In the chapters we combine descriptive and experimental information to provide key portions of the evidence on which contemporary concepts are based. The illustrations have been obtained from leading students of the cell. The figure legends and suggested reading lists will familiarize the reader with the names of some of those who have contributed to the progress of cytology and cell biology.

We hope that we convey some of the excitement felt today by students of the cell. In addition, we hope that not only past achievements, but important problems awaiting future solution become evident to the reader.

New York, N.Y.   *A. B. N.*
January 1970   *E. H.*

Contents

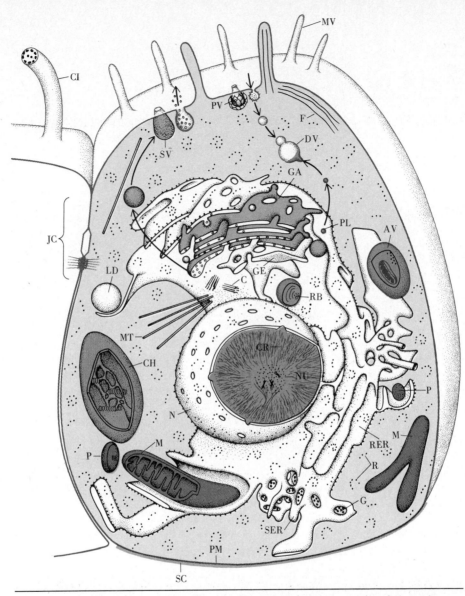

Schematic diagram of a cell and its organelles drawn to reveal their three-dimensional structure.

AV, autophagic vacuole; *C*, centriole; *CH*, chloroplast; *CI*, cilium; *CR*, chromatin; *DV*, digestion vacuole; *F*, filaments; *G*, glycogen; *GA*, Golgi apparatus; *GE*, GERL; *JC*, junctional complex; *LD*, lipid droplet; *M*, mitochondrion; *MT*, microtubules; *MV*, microvillus; *N*, nucleus; *NU*, nucleolus; *P*, peroxisome; *PL*, primary lysosome; *PM*, plasma membrane; *PV*, pinocytic vesicle; *R*, ribosomes and polysomes; *RB*, residual body; *RER*, rough endoplasmic reticulum; *SC*, extracellular coat (as drawn, "basal lamina"); *SER*, smooth endoplasmic reticulum; *SV*, secretion vacuole.

The organelles have been drawn only roughly to scale. Also, the sizes and relative amounts of different organelles can vary considerably from one cell type to another. For example, only plant cells show chloroplasts. A detailed enumeration of the organelle content of one cell type is presented in Chapter 2.12.

Introduction

The analogy between cells and atoms is a familiar one, and like many familiar comparisons it is both useful and limited. Cells and atoms are units. Each is composed of simpler components which are integrated into a whole that exhibits special properties not found in any of the parts or in random mixtures of the parts. Both exhibit considerable variation in properties, based on different arrangements of components; the number of variations far exceeds the number of major components. Both serve as basic building blocks for more complex structures.

However, the analogy cannot be pressed too far; cells can reproduce themselves, whereas atoms cannot. The ability to utilize the nonliving environment to make living matter is probably the most fundamental property of life, and cells are the simplest self-duplicating units. Duplication is based on DNA (deoxyribonucleic acid), which can be *replicated* to form perfect copies of itself. Thus the genetic information encoded in DNA is perpetuated from one cell generation to the next, sometimes without significant variation over vast periods of time. DNA is unique among macromolecules in its replication. Only in certain viruses has another macromolecule (RNA, ribonucleic acid) been shown to replace DNA in its central hereditary role; the replication of RNA in these viruses is based on the same principles as that of DNA.

Genetic information is expressed in cells by the mechanisms of *transcription* and *translation*. Transcription transfers the DNA-coded information to RNA molecules. Translation results in the formation of specific proteins whose properties are determined by the information carried by these RNA molecules. Among the proteins are *enzymes,* catalytic molecules that control most of the chemical reactions of cells. Enzymes differ in the kinds of molecules they affect (their *substrates*) and in the kinds of reactions they catalyze. Many are involved in the synthesis of the other cellular macromolecules, the nucleic acids (DNA and RNA), the lipids (fats and related compounds), and the polysaccharides (polymers made of linked sugar molecules). Through this chain of transcription, translation, and enzymatic activities, DNA directs its own replication and controls as well the rest of *metabolism,* the sum total of all the chemical reactions that take place in cells. The chain is universal, and thus all cells are made of the same classes of macromolecules (nucleic acids, proteins, lipids, polysaccharides) and smaller components such as water and salts. Duplication, and the presence in different cells of similar molecular and structural materials and mechanisms, are features of *cellular constancy,* one of the main themes of this book.

A second theme of the book is *cell diversity.* Cells may be classified into a large number of categories. *Eucaryotic* cells are distinguished from *procaryotic* cells, plant cells from animal cells, and muscle cells from gland cells. These distinctions derive from differences in morphology and metabolism. Eucaryotes differ from procaryotes in complexity of cellular organization. The unicellular protozoa, most algae, and the cells of multicellular plants and animals fall in the eucaryote category. In these cells, different specialized functions (such as respiration, photosynthesis, and DNA replication and transcription) are segregated into discrete cell structures, many of which are delimited from the rest of the cell by membranes. The cell's *organelles* reflect this segregation; they are subcellular structures of distinctive morphology and function. The most familiar of the organelles is the *nucleus,* which contains most of the DNA of the cell and enzymes involved in replication and transcription. The nucleus is separated by a surrounding membrane system from the rest of the cell, the cytoplasm. The cytoplasm contains many organelles including the *mitochondria,* the chief intracellular sites of respiratory enzymes; in plants, the cytoplasm contains *chloroplasts* in which are present the enzymes of *photosynthesis,* a metabolic process unique to plant cells. The mitochondria, chloroplasts, and a number of other cytoplasmic organelles are also delimited as discrete structures by surrounding membranes.

The procaryotes include the bacteria, the blue-green algae, and some other organisms. In contrast to the eucaryotes, they have relatively few membranes dividing the cell into separate compartments. This is not to say that all components are mixed together in a random fashion. The DNA, for example, does occupy a more or less separate nuclear region, but this is not delimited by a surrounding membrane. In fact, the traditional distinguishing feature of procaryotes is the absence of a membrane-enclosed nucleus (the suffix *caryote*—or *karyote,* as it is often spelled—refers to the nucleus). Respiratory and photosynthetic enzymes are not segregated into discrete mitochondria or chloroplasts, although, as will be seen, the enzymes are held in ordered arrangements within the cell.

The diversity of cell types owes its origin to evolution. By comparison with other macromolecules, DNA is remarkably stable. But *mutations* and other changes do occur at an appreciable, though low, frequency. Mutations alter the genetic information that is encoded in DNA and passed by a cell to its progeny; thus they can produce inherited changes in metabolism. Some result in a *selective advantage:* Roughly speaking, organisms with such alterations produce relatively more viable offspring and therefore will increase in relative frequency in the population. This can take place more or less rapidly in different circumstances. In the final analysis, however, the pattern of spread of a genetic change in a population depends on reproduction and therefore, ultimately, on division of cells.

Usually the daughter cells resulting from division of *unicellular organisms* are essentially similar to the parent cells; the daughters contain replicates of the parent cells' DNA, and the DNA establishes the range of potential responses to the environment by specifying the available range of metabolic possibilities. Given a similar environment, there is little difference between parent and daughter. If the environment changes, parents and daughters will change in similar fashion and within genetically imposed limits. Diversity of cell types rests upon mutation and attendant evolutionary phenomena.

In *multicellular organisms,* diversification of cell type without mutation is a regular feature of development. Most multicellular animals and plants start life as a single cell, a *zygote,* with a nucleus formed by the fusion of two parental nuclei. (Usually this results from fusion of sperm and egg or the equivalent.) The cell divides to produce daughter cells with identical DNA, but these *differentiate* into specialized cell types with different morphology and metabolism (for example, gland cells producing digestive enzymes or muscle cells rich in specially organized contractile proteins). In different cell types, different portions of the DNA are used in the transcription that underlies macromolecule synthesis. For a given cell type only a particular part of the total genetic information is responsible for the cell's characteristics. Thus constancy in DNA coexists with diversity in metabolism and morphology of the cells carrying that DNA.

Differentiation implies that cells are not mere aggregates of independent molecules or structures each "doing its own thing," and that DNA molecules are not autonomous rulers of subservient collections of other molecules. The expression of genetic information in a given cell depends largely on the interactions of DNA with other molecules—especially proteins—which control the production of RNA. Interactions among cells are key factors in determining the patterns of molecular interplay governing differential gene expression in a developing organism. The immediate environment of a given cell is strongly influenced by other cells, both neighboring and more distant; this environment can have major impact on the directions and timing of a cell's differentiation.

As in development, the normal functioning of adult multicellular organisms also depends upon the interaction of neighboring cells and upon long-range cell-to-cell interactions mediated, for example, by hormones or nerve impulses. Cells are integrated into tissues, tissues into organs, and organs into an organism. Similarly, cells are themselves highly organized: Molecules are built into structures in which they function in a coordinated and interrelated manner; they often show properties not found in a collection of the same molecules free in solution. The products of one organelle may be essential to the operation of another. Cell functions depend upon mutual interaction of parts. These interactions are elements of a network of mechanisms that regulate metabolism. *Cell organization* and the implications of organization for function are a third theme of the book. A fourth is the system of *regulatory* mechanisms by which cells—individually, and through their interactions—control the rates and directions of their activities: the timing of events; the nature and amounts of materials to be synthesized, broken down, or taken up from the outside; and cell shape and movement.

Reproduction and constancy, evolutionary and developmental diversity, the integration of cellular components into a functional whole—all are subjects for investigation in modern cell biology. The fifth theme of the book is the dependence of major biological findings upon the development of *new methods of study* and upon the *choice of the best organism* for the problem at hand. We shall illustrate the kinds of experiments and approaches currently used in cell biology. The microscope is a central tool, but (as outlined in the first Preface) microscopic, biochemical, and physiological approaches increasingly are interwoven: Present-day cell biologists merge the structural focus of classical "cytology" and the molecular focus with which "cell biology" originally developed. Similarly, descriptive and experimental approaches supplement each other. The great diversity of cells and organisms presents opportunity for choice of cell types especially well suited for analyses of new problems. Investigations of pathological material and of cells experimentally stressed by abnormal conditions provide valuable clues to normal functioning.

Study of the cell is progressing rapidly, and the solution of many problems presently unsolved may be anticipated with confidence. Some of the unsolved problems are of practical importance. As our understanding of cells increases, so does our ability to control and modify them. This ability is crucial for medicine and agriculture. It also raises important ethical and social questions.

PART 1

Cytology and Cell Biology of Today

About 60 years ago, the last edition of E. B. Wilson's great book, *The Cell in Development and Heredity,* was published. It was a summation and synthesis of a vast cytological literature. The work reflects the extraordinary ingenuity of early experimenters and the great excitement over what was then a recent appreciation of the roles of chromosomes in heredity. It is concerned mainly with eucaryotic cells. The nucleus had been extensively studied; some of the cytoplasmic organelles had been identified although clarification of their functions was only beginning. Biochemical analysis of cells was in its infancy.

Since that time, and especially in the last 25 years, there has been a remarkable development of techniques applied to the study of cells. The electron microscope has extended the cytologists' investigation of structure down to the level of macromolecules. Biochemists have separated and analyzed cell molecules and organelles and determined their metabolic functions. Cytology and biochemistry have been combined to the extent that modern cytology is often referred to as *biochemical cytology.*

To illustrate current views of cell *organization,* we will begin with the rat *hepatocyte,* the major cell type of the liver. This cell type has many important functions, ranging from the secretion of blood proteins and the storage of carbohydrates to the destruction of toxic material produced elsewhere in the body. This variety of physiological functions is one reason that hepatocytes are widely studied by biochemists. Biochemical study is facilitated by the relative homogeneity of the organ; hepatocytes constitute over 60 percent of the cells and 90 percent of the weight of the liver in the rat. (The remainder consists of cells of blood vessels, ducts, and supporting tissues and specialized *phagocytes,* cells that engulf and remove from the blood a variety of materials such as some damaged or aged red blood cells.) Thus, constituents isolated from the liver

5

come primarily from one cell type, the hepatocyte. Rats are readily available, and they have large livers (almost 12 grams in an adult rat) from which relatively great quantities of cell constituents may be obtained. The hepatocytes are easily disrupted, to provide isolated organelles that can be studied by biochemical techniques. In addition, the liver is relatively easy to prepare for both light microscopy and electron microscopy.

To illustrate current views of *cellular metabolism,* we have chosen the metabolic pathway responsible for the formation of most of the ATP (*adeno-*

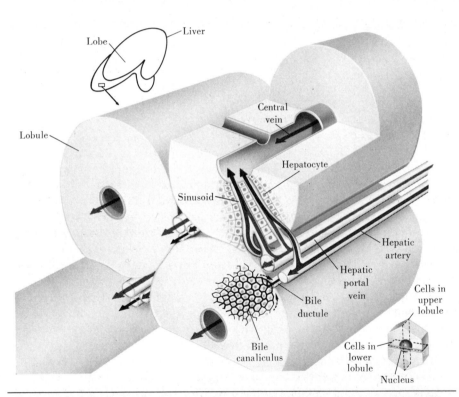

Figure I–1 *Diagram of hepatic lobules* (rat liver). The lobule at the upper right illustrates the relations of the liver cells (hepatocytes) to the blood; that at the lower right illustrates hepatocyte relations with the channels (bile canaliculi) into which the cells secrete bile. The two lobules diagram different views of the same cells as indicated by the dashed boxes at the lower right (these illustrate the planes along which the cells would be cut—Figure I–10—to generate the views shown). Nutrient-rich blood, carried from the intestine by the hepatic portal vein and oxygen-rich blood from the hepatic artery enter the sinusoids (modified capillaries) within each lobule. After exchanges have occurred with the hepatocytes arranged along the sinusoids, the blood enters the central veins of the lobules and is carried out of the liver.

sine triphosphate) of cells. ATP provides energy for virtually all cell functions: transport of substances into and out of the cell, chemical reactions within the cell, and integrated activities such as secretion, movement, and cell division.

To set the stage for subsequent consideration of the mechanisms by which the cell reproduces its structures (*organelle biogenesis*), this part of the book closes with a brief discussion of *self-assembly,* a group of processes basic to the building of complex structures from macromolecules.

Chapter 1.1 A Portrait of a Cell

The relations of hepatocytes to the architecture of the liver are shown schematically in Figure I–1. Liver, like any other organ, must be nourished by molecules that enter its cells from the blood (for example, small molecules such as sugars and water, salts, and some macromolecules). The protein *albumin,* *lipoproteins* (complexes of lipid and protein), and other macromolecules are secreted by the hepatocyte into the blood. Other materials released to the blood include sugars derived from the liver's carbohydrate stores, and wastes such as CO_2. Bile, which contains molecules formed in the breakdown of hemoglobin and of toxic substances along with detergent-like bile salts derived from cholesterol, is secreted into small extracellular channels, the *bile canaliculi.* These lead to the bile duct, which in turn empties into the intestine, where the bile salts help to solubilize fats from food, facilitating fat digestion.

The flow of bile past the hepatocytes is in the opposite direction from the flow of blood. The hepatocytes are aligned along the blood and bile spaces, and the arrangement of organelles within the cell reflects the polarization of functions; large, specialized areas of the cell surface are involved in the exchanges with the blood and other smaller areas in the secretion of bile. The key organelles are presented by a series of diagrams in Figures I–2 through I–5. The diagrams are based upon light microscopy (Fig. I–6), electron microscopy (Fig. I–7), and biochemical studies. To convey the three-dimensional structure of the organelles, a hypothetical "generalized" cell is also included (as the frontispiece), preceding the Introduction. Information on the number of different organelles is presented in Chapter 2.12. Two sets of definitions should be noted: *Intracellular* refers to things within the cell, *extracellular* to material outside the cell, and *intercellular* to components found between cells. *Vesicle* and *vacuole* are somewhat imprecise terms; both are used to designate a class of more or less spherical intracellular structures, each delimited by a membrane. A vesicle is a small vacuole.

The diagrams indicate the complexity of cell structure and function. Each of the different organelles has a distinctive morphology, biochemical composition, and function. Detailed discussions of each organelle are found later in this part and particularly in Part 2. We now turn our attention to the major techniques by which "portraits" of cells have become possible.

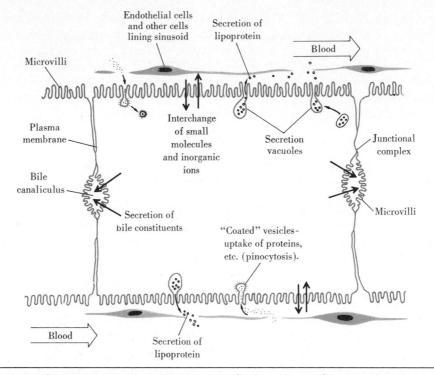

Figure I–2 Diagrammatic representation of a hepatocyte (from a rat) showing the *plasma membrane* that surrounds the cell. This membrane and structures associated with it control entry and exit of material. Arrows indicate interchanges of materials between cells and blood. Inorganic ions such as K^+ or Cl^-, and H_2O, O_2, CO_2, sugars, amino acids and other small molecules pass across the membrane. Larger molecules are transported in membrane-delimited vesicles and vacuoles that separate from or fuse with the plasma membrane. Macromolecules entering the cell are shown as small dots, although many are almost invisible by current techniques. The lipoprotein particles secreted by the cell are seen as small spheres by electron microscopy. Secretion of bile probably involves movement of some molecules across the membrane and release of others by fusion of vesicles.

The *junctional complex,* a specialized region of association between adjacent cells, restricts movement of material between hepatocytes: For example, it prevents most molecules from moving directly from the blood sinusoids into bile canaliculi. The complex also helps in holding adjacent cells together.

Chapter 1.2 Methods of Biochemical Cytology

The microscopic study of cells is limited both by the microscope and by the manner of preparing the specimen for observation. In general, living cells and tissues are difficult to study directly with the ordinary light microscope. Multicellular tissues are usually too thick to permit penetration of light; single living cells are often transparent, with little visible internal detail. Thus one line of

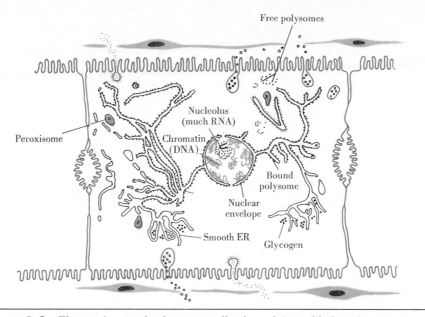

Figure I–3   The *nucleus* and other organelles have been added to the structures seen in Figure I–2. The *chromatin* of the nucleus contains most of the cell's DNA, whereas the nucleolus is particularly rich in RNA. The nuclear envelope is part of the *endoplasmic reticulum (ER)*, a complex interconnected system of membrane-delimited channels (large sacs and smaller tubules) in which certain types of proteins and other macromolecules are transported within the cell. Openings ("pores") in the nuclear envelope permit movement of macromolecules between nucleus and cytoplasm.

Ribosomes (granules composed of RNA and protein) are usually complexed with information-carrying *mRNA* (messenger RNA) to form *polysomes,* the sites of protein synthesis. Some polysomes are attached to part of the ER; other ("free") polysomes are not attached to ER. ER with ribosomes attached is called *rough ER. Proteins,* synthesized on the polysomes of rough ER, enter the ER channels to be carried to other parts of the cell. The *smooth ER* lacks ribosomes; it is involved, for example, in the synthesis of lipids such as the steroids. Stored carbohydrate, in the form of granules of *glycogen,* is found close to networks of smooth ER tubules.

Peroxisomes are membrane-delimited structures containing enzymes that catalyze reactions many of which involve *hydrogen peroxide;* one of their likely roles in hepatocytes is in the breakdown of fatty acids, components of lipids.

development of techniques for cell study is centered around improvements in microscopes and in methods for preparing and observing cells.

A second line of development is concerned with coordinating structural findings with biochemical information. Several methods that permit the direct use of the microscope in studying chemical and metabolic features of cells have been devised. These are best used in conjunction with the increasingly powerful, nonmicroscopic techniques available for isolation and chemical analysis of cell components.

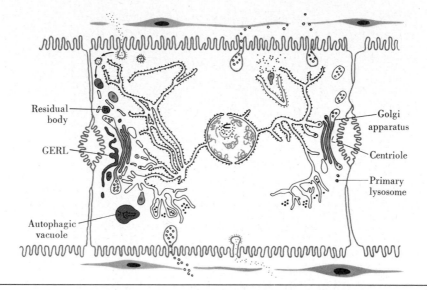

Figure I–4 The *Golgi apparatus*, which has been added to the structures seen in Figure I–3, consists of flat sacs and vesicles of varying sizes. Its functions include concentration of material produced in the ER, "packaging" of several classes of molecules into membrane-delimited vesicles, and roles in the synthesis of carbohydrate-containing macromolecules. The large Golgi vesicles (often called vacuoles) are filled with secretion material such as lipoproteins (Fig. I–2).

    Lysosomes are sites of intracellular digestion: *primary lysosomes,* probably including certain small vesicles of the Golgi region, transport the digestive enzymes; *autophagic vacuoles* and *residual bodies* are among the other types of lysosomes. *GERL* is a distinctive system of sacs, tubules and related structures associated with the Golgi apparatus; it may participate in lysosome formation or functioning. The roles of the pair of *centrioles* are poorly understood.

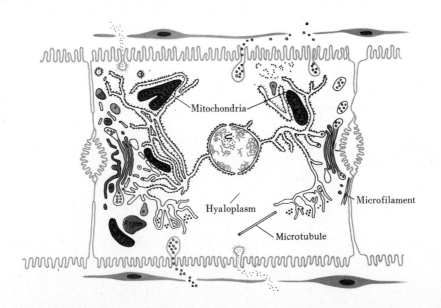

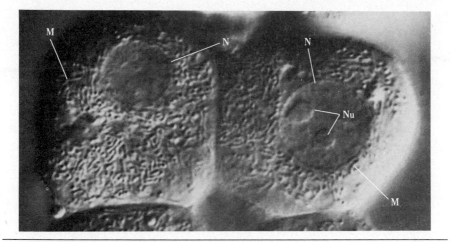

Fig. I-6 *Two hepatocytes,* still alive, isolated from the liver of a rat and viewed with a Nomarski differential interference microscope (Section 1.2.2 and Fig. I-12). This microscope produces an image that conveys a strong three-dimensional impression. *N.* indicates the cells' nuclei, *Nu* nucleoli, and *M* two of the many mitochondria visible in the cytoplasm. × 2,500. (From G.B. David, *Improved Isolation Separation and Cytochemistry of Living Cells.* Stuttgart: G. Fischer-Verlag, 1975).

Chapter 1.2A Microscopic Techniques

During the latter part of the nineteenth and early part of the twentieth centuries, the light microscope approached the theoretical and practical limits of its performance as an optical instrument. Improvements in lenses and design resulted, by the beginning of this century, in instruments fundamentally similar to those in use today. During this period, the basic techniques of preparation of material for microscopy were also developed:

1. *Fixing* of cells or tissue with agents that serve to kill and to stabilize structure with lifelike appearance, to immobilize the cells' macromolecules and to prevent postmortem disruptive changes known as *autolysis.*

◀ **Figure I–5** The diagram now shows a few of the cell's many *mitochondria.* The mitochondria are the chief sites of respiratory (oxygen-consuming) metabolism; they produce ATP (*adenosine triphosphate,* a key molecule in energy transfer and utilization) and are involved in many other of the cell's central metabolic processes.

Microtubules, microfilaments and other intracellular filaments participate in maintaining the cell's shape and in intracellular motion. The *hyaloplasm* ("cytosol" or "ground substance") is the "background" cytoplasm; by light microscopic techniques it appears structureless but some researchers think it is occupied in part by a network of organized material.

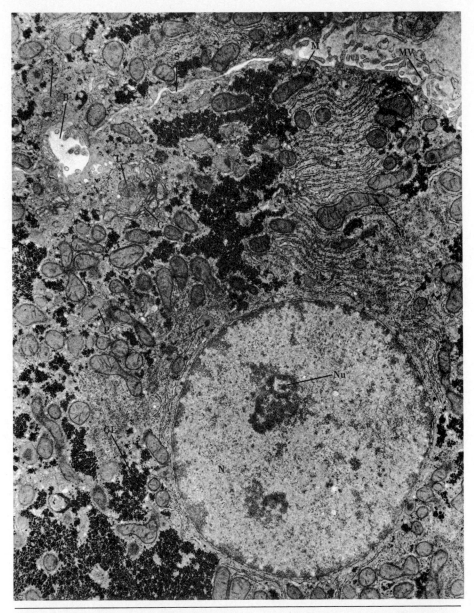

Figure I–7 *Major organelles of a rat hepatocyte.* This is a low-magnification electron micrograph that illustrates the overall appearance of the cell. *B* indicates a bile canaliculus, and *MV*, a microvillus at the sinusoid surface; many of the other microvilli near *MV* have been sectioned (Fig I–10) so as to appear unattached to the cell. The dense deposits at *GL* are masses of glycogen granules. *M* indicates mitochondria; *N*, the nucleus; *Nu*, a nucleolus; *P*, peroxisomes; *L*, lysosomes; *G*, the Golgi apparatus; and *S*, the intercellular space separating this cell from an adjacent hepatocyte. *E* indicates endoplasmic reticulum. × 25,000. (Courtesy of W.-Y. Shin.)

2. *Embedding* in hard materials that provide support of the tissue for *sectioning,* the preparation of thin slices. Sectioning tissue permits study of complex structures by providing an unimpeded view of deep layers.
3. *Staining* of cells with dyes that color only certain organelles, thus providing *contrast,* for example, between nucleus and cytoplasm or between mitochondria and other cytoplasmic structures.

At first glance, it might seem improbable that tissue put through such elaborate procedures would bear any resemblance to living material. Certainly there is reason for caution in interpretation. However, several generations of investigators have provided increasingly trustworthy methods for tissue preparation. When comparison is possible, the results often compare remarkably well with direct observation on living cells and with information obtained by various indirect means. A variety of preparative methods are available for study of specific cell features.

In the present century, development of cytological techniques has proceeded chiefly in four directions: (1) invention of microscopes based on newly understood physical principles, notably the *electron* microscope, which permits use of much higher magnifications; (2) development of new optical devices, such as the *phase-contrast and interference* microscopes and perfection of others, such as the *polarizing* microscope, that facilitate detailed study of living cells; (3) evolution of *cytochemical* methods for obtaining chemical information about microscopic preparations; and (4) development of techniques for the isolation of organelles and other components for biochemical study.

1.2.1 Electron Microscopy

Resolution The structures of interest to cytologists range widely in size. In terms of the units used for microscopic measurement (Fig. I–8), the diameter of a mitochondrion is about 0.5 μm, the lengths of many bacteria roughly 1

Unit	Symbol		Chief Use in Cytological Measurement
Centimeter	cm	= 0.4 inch	Macroscopic realm (naked eye) Giant egg cells
Millimeter	mm	= 0.1 cm	Macroscopic realm (naked eye) Very large cells
Micrometer (Micron)	μm (μ)	= 0.001 mm	Light microscopy Most cells and larger organelles
Nanometer (Millimicron)	nm (mμ)	= 0.001 μm	Electron microscopy Smaller organelles, largest macromolecules
Angstrom unit	Å	= 0.1 nm	Electron microscopy, X-ray methods Molecules and atoms

Figure I–8 **Units used in measuring the dimensions of cells and organelles.**

μm, and the diameters of most mammalian cells in the range of 5 to 50 μm. All microscopes are characterized by limits of *resolution,* which in turn determine the limits of useful magnification. The limit of resolution of a microscope determines how small an object can be adequately visualized with that instrument and thus establishes the fineness of detail accurately represented in the image produced.

That such limits exist is inherent in the interaction of light and other electromagnetic radiation with the specimen and with the system used to produce an image. No matter how perfect the microscope, an image can never be a perfect representation of the object (Fig. I–9). As a result, objects lying close to one another cannot be distinguished *(resolved)* as separate objects if they lie closer than approximately one half the wavelength of the light being used. The wavelengths of visible light determine color. Blue light has a wavelength of 475 nm and red light of 650 nm; the average wavelength of white light (for example, sunlight), which is a mixture of all colors, is about 550 nm. Thus for light microscopes, the limit of resolution is about 0.25 μm. If spaced apart by less than this limit, adjacent but separate objects appear in the light microscope as one object. This effect imposes an absolute limit on the details that can be

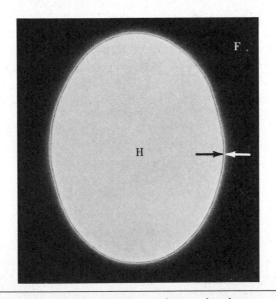

Figure I–9 *Diffraction, a factor limiting resolution.* An electron micrograph that illustrates one factor limiting resolution. A hole *(H)* in an electron-opaque film *(F)* was photographed under optical conditions that exaggerate the effect of diffraction, a phenomenon that occurs whenever light or electrons pass edges. The hole actually has a simple edge, but the image of the edge shows a "diffraction fringe." The fringe consists of concentric bands, one dark and one bright (arrows). Under usual conditions of microscopy, the distorting effects of diffraction and other comparable optical phenomena are less severe, but they cannot be totally eliminated. × 120,000. (Courtesy of L. Biempica.)

distinguished with the light microscope, irrespective of magnification. *Magnification* refers to the apparent size of objects; *resolution* refers to the clarity of the image. Microscope images can be enlarged almost indefinitely by optical and photographic means, but beyond a point the image appears increasingly blurred so that nothing is gained by further magnification. Thus light microscopes are rarely used at magnifications greater than 1000 to 1500 ×.

The development of the electron microscope (EM) as a practical instrument in the 1940s and 1950s made possible useful magnifications of 100,000 × or greater. Electrons can be regarded as an extremely short wavelength form of radiation. In practice, they can provide resolution of details separated by 1 to 5 Å, 1000-fold better than the light microscope. In the electron microscope, lenses focus electrons in a manner analogous to the way glass lenses of ordinary microscopes focus light. The magnified image is viewed on a fluorescent screen like a TV screen and is also recorded on photographic plates. Most biological macromolecules have dimensions on the order of 10 to 100 Å. The diameter of a DNA molecule (not the length, which is very much greater) is 20 Å; many protein molecules have diameters of 50 to 100 Å. Thus macromolecules cannot be distinguished in the light microscope but are well within the range of resolution of the electron microscope.

Electrons, however, cannot travel long distances in air. Thus both the specimen chamber and the lens systems in an electron microscope must be maintained under high-vacuum conditions by pumps that remove air and water vapor or other gases evaporating from the specimen. This requirement both complicates the instrument and limits the types of specimen that can be examined; living cells, for instance, cannot be studied with the instruments designed thus far.

Specimen Preparation   Fixation for light microscopy and for electron microscopy usually depends on the use of agents that precipitate macromolecules and keep them insoluble. *Formaldehyde* and related compounds such as *glutaraldehyde* are widely used. They react with protein molecules to produce linking of previously separate molecules (Section 2.3.2) and other changes promoting insolubility. *Osmium tetroxide* (osmic acid), commonly used in electron microscopy, probably also can link separate molecules; it appears to react with lipids and other components, but its reactions are incompletely understood (see Section 2.1.2).

For light microscopy, fixed tissue usually is embedded in paraffin wax. The tissue is dehydrated and soaked in molten wax, which penetrates the tissues and then is cooled to form a hard matrix. Embedded tissue is sectioned (on mechanical slicers called *microtomes*) with sharp steel knives to produce sections 2 to 10 μm or more in thickness. The sections are mounted on glass slides for study in the microscope. On occasion, tissue is not embedded but is instead frozen solid and sectioned while frozen. Freezing makes tissue hard enough for cutting reasonably thin sections but also tends to damage cells. The technique is used for quick preparation (during surgical operations, for example, when it is desirable to examine tissue samples for diagnostic purposes) and for other applications such as retention of certain cell lipids and enzymes. Several of the many staining procedures used for light microscopy are outlined in Sections 1.2.3 and 2.2.3.

For electron microscopy these techniques are inadequate. Electrons can pass only through very thin sections and are blocked entirely by glass. Thus to embed tissue, it is soaked in unpolymerized plastic, which is hardened by the addition of appropriate catalysts that promote polymerization. The plastics provide support for cutting sections 500 to 1000 Å or less in thickness. To accomplish this, sharp glass or diamond edges are used, as steel does not readily attain or keep an edge that is sharp enough. In place of glass slides, sections are spread out on a screenlike metal mesh (a "grid") and observed through the empty spaces of the screen (see Fig. I–10).

Sectioning imposes difficulties of interpretation. If the object being studied is, for example, a cylinder 1.0 μm long, ten to twenty sections 500 to 1000 Å thick can be cut in planes perpendicular to the cylinder axis. Each section cut from the cylinder will appear then as a flat disc (Fig. I–10). To ascertain the three-dimensional structure, it is best to cut *serial sections,* a series of sections made at the same angle. These can be used to reconstruct the shape of the object in three dimensions. In our example, piling a series of discs one on top of another (mentally or by using cut photographs) would reproduce the original cylinder. However, serial sectioning for electron microscopy is difficult. Thus for routine work, most microscopists take sections at random and compare the appearances at various angles to the axis (Fig. I–10).

The stains used in electron microscopy are not the colored dyes of light microscopy. Rather, they contain heavy metal atoms, usually lead or uranium, in forms that combine more or less selectively with chemical groups characteristic of specific types of structures in the cell. Since the presence of such atoms permits fewer electrons to pass through the specimen and produce an image, "stained" structures appear darker than their surroundings. Conventionally, a structure that appears dark in the electron microscope is referred to as *electron-dense* or *electron-opaque.*

Some specimens, particularly isolated structures of macromolecular dimensions, are suitable for viewing in the electron microscope without sectioning. Special techniques are used to make such structures stand out against the thin films of plastic and evaporated carbon used to support them in the microscope. Metals, such as platinum, can be evaporated onto the structures from an angle, which increases their electron density and can create a shadow-like effect that enhances the three-dimensional appearance (Fig. I–11). Structures in Figures II–57, III–20, III–36, and IV–10 were prepared by such "shadowing" techniques. Another method for preparing isolated macromolecules or small particles for electron microscopy, *negative staining,* is explained in Figure II–53.

Sometimes the electron microscope images obtained with various of these contrast-producing procedures are electronically processed through im-

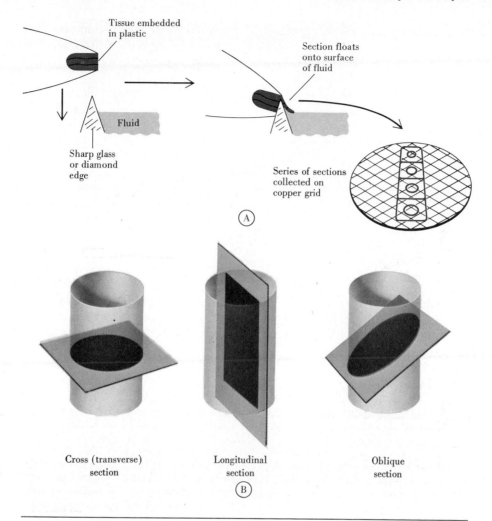

Tissue embedded in plastic

Section floats onto surface of fluid

Fluid

Sharp glass or diamond edge

Series of sections collected on copper grid

A

Cross (transverse) section

Longitudinal section

Oblique section

B

Figure I–10 *Sectioning tissue for microscopy.*

  A. To section material for electron microscopy, plastic-embedded tissue is brought down across a sharpened glass or diamond edge, using a device *(microtome)* that advances the tissue by a precisely controlled distance between successive passes across the edge. (In the illustration, the advance would be to the right.) The distance advanced determines the thickness of the slices. The thin slice (section) produced each time the tissue passes across the edge is floated away from the edge onto the surface of water or other liquid. This keeps the section flat, helps protect it from mechanical damage and facilitates collecting the sections onto the grid for viewing. Generally a series of sequential sections is collected onto a single grid, as illustrated. (In the diagram, the sections on the grid are shown in face view as they are viewed in the microscope, whereas the one being cut is shown in side view.)

  B. Sectioning a cylinder along different axes can produce the appearance of a circle, a rectangle or an ellipse. (The profiles shown are those obtained for a solid cylinder. See also Figs. III–45 B and C.)

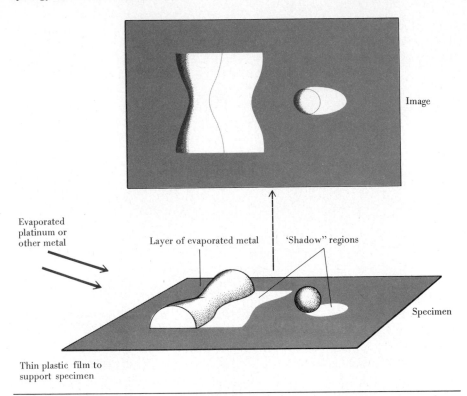

Evaporated platinum or other metal

Layer of evaporated metal

'Shadow" regions

Specimen

Image

Thin plastic film to support specimen

Figure I–11 *Shadowing of a specimen for electron microscopy.* **The specimen is supported on a thin film of plastic and carbon. Evaporated metal is directed at the specimen from an angle. The metal forms an electron-dense layer on the support film and on specimen objects that protrude above the support film. The metal deposits coat regions of an object on the side facing the source of the evaporated metal, but little or no metal accumulates in the "shadow" regions on the side facing away from the source. When the preparation is now viewed in an electron microscope, electrons pass through the "shadow" regions more readily than they do where the layer of metal is present; hence, in the final image produced, the shadows will appear lighter (less electron-dense) than the coated zones of the objects or the support film. The image obtained requires careful interpretation, both because of the presence of the shadow and because the image is a two-dimensional projection (on a photographic plate or microscope screen) of a three-dimensional object. Note also that sometimes the image is printed photographically in reverse contrast, giving a bright image of the object and a *dark* image of the shadow (Fig. IV–10A is printed in this way; Fig. IV–10B is in ordinary contrast.).**

In "rotary shadowing," used often for viewing DNA molecules, the specimen is rotated as the metal is evaporated, so that metal reaches the surface sequentially from many angles. This provides more adequate visualization of portions of the molecules that may be oriented unfavorably for shadowing at a given angle. The contrast produced depends partly on deposition of metal on the specimen and the shadow proper is somewhat diminished by the use of multiple angles. Figures II–12, IV–10 and IV–36 were prepared by methods including rotary shadowing or related procedures.

age-intensifying devices and procedures that further increase the contrast between specimen and background, so that fine details can be brought out. Image-enhancing techniques have proved particularly useful in analyzing structures with repetitive organization (see Figs. II–78 and II–84).

Freezing; Scanning Microscopes; High-voltage Microscopes Innovations in techniques and instruments are continually being introduced in electron microscopy. Impressive results have been obtained with methods of specimen preparation based on freezing of tissues at very low temperatures, attained with liquid nitrogen ($-196°$ C) or liquid helium ($-269°$ C). Such low temperatures can freeze cells quickly enough to catch them in different stages of very rapid physiological changes and also produce a noncrystalline form of ice, thus minimizing damage. Fixation, embedding, and sectioning can be avoided: In freeze-fracture and freeze-etch procedures, for example, the cells are quick-frozen and then broken open, exposing their interior structures (fuller descriptions are provided in Figs. II–6 and II–12).

Recently, *scanning* electron microscopes (SEMs) have come into widespread use as a means for viewing surfaces of specimens too thick for conventional ("transmission") microscopes (see Figs. II–95 and III–16). Rather than passing electrons through the specimen, as in the conventional instruments, these microscopes scan the surface with a narrow beam of electrons and construct an image based on the degree and direction of scattering of electrons; this scattering varies as the beam encounters details with different geometric features (size, shape, orientation) and electron opacity. Often the specimen is first coated with a thin layer of evaporated metal (as in the shadowing procedures described in Fig. I–11) to enhance electron scattering. Scanning instruments of moderate resolving power are widely available (they can distinguish objects spaced 50 to 100 Å apart). Higher-resolution instruments now coming into use combine scanning and transmission microscopy and are proving versatile for work at the macromolecular level. Electron beams tend to damage and distort molecules, limiting the detail that can be visualized; scanning of preparations with relatively low-intensity beams coupled to the use of electronic contrast-enhancement methods like the methods employed to improve pictures transmitted from satellites (see also Fig. II–78) is one of the means for minimizing such problems.

High-voltage electron microscopes utilize voltages of 1000 to 3000 kilovolts (1 kv = 1000 volts) to accelerate electrons, rather than the 40 to 100 kv conventionally employed. This permits the electrons to penetrate much thicker sections, up to several micrometers, so that aspects of organelle geometry and interrelations that are difficult to see in thinner sections may be made evident. Already, a better appreciation of the three-dimensional arrangement of microtubules, microfilaments, and other fine filaments has come from studying unsectioned, thin cultured cells (Fig. II–94). Possibly such microscopes will open the way for electron microscope observation of living cells, through use of special fluid-containing chambers.

1.2.2 Study of Living Cells

For the study of living cells, early light microscopists relied on a few large cell types (egg cells, some plant cells, and protozoa) and developed microsurgical techniques to introduce material into single living cells and to alter living cell structures. Aside from these efforts, cytology has been based predominantly on the study of fixed, sectioned, and stained material. The constant improvement of methods has enabled cytologists to obtain increasingly reliable information from fixed preparations, although it remains true that the preparation of cells for microscopy may introduce *artifacts* (distorted and misleading appearances; Fig. II–61 provides an example).

Living cells are dynamic entities, continually moving and changing; fixed and stained cells are stable and unchanging. Images from fixed material are at best only "snapshots" of dynamic cells. Often a sequence of events must be reconstructed by careful study of a large number of such "snapshots" taken at intervals.

Interference and Phase-contrast Microscopes Fortunately, ways are now available to the light microscopist for examining cells in action. The behavior and speed of light as it passes through a portion of a cell depend on the concentration, nature, and organization of the cellular constituents. For example, cell regions differ in *refractive index,* which can be thought of as a measure of the speed of light through matter. The refractive index depends in large part on the density of the region through which the light passes—that is, on the concentration of matter per unit volume. The higher the density, the higher the refractive index, and the slower the rate at which light passes. Thus light traveling through the nucleus will usually be affected differently from light traveling through the cytoplasm because of differences in the concentration of material. Differences in light paths due to factors of this type are not easily detectable in the ordinary microscope. The *phase-contrast* and *interference* microscopes are light microscopes modified to translate refractive index differences and related optical effects arising at boundaries between structures into visible contrast, either in brightness or in color. This is done by taking advantage of the phenomenon of *interference.* It is a basic physical principle that two identical light waves can, for example, be made to cancel each other, yielding darkness rather than brightness when they arrive together at a given point. This can result from relative shifts in the "phase" of vibration that occur when one wave passes through a medium of refractive index different from that encountered by the other wave (Fig. I–12). By using such effects to manipulate the light passing through the cell rather than manipulating the cell itself, an "optical staining" is achieved so that various organelles stand out in sharp contrast to their surroundings (see Figs. I–6, I–13, and IV–24). In addition, because of the influence of concentration on the optical properties of cells, the phase-contrast and interference microscopes can also be used to estimate the amounts of material present in different cells or cell regions; an example of such quantitative use is outlined in Section 2.11.1.

Darkfield and Polarizing Microscopes *Darkfield* microscopes are constructed so that of the light shone on the specimen only that scattered from edges or particles in the specimen can enter the lenses that produce the microscope image. Consequently, details such as small structures surrounded by a less organized solution show up as bright against a dark background (the effect is similar to the "shining" of dust particles in an intense light beam in a darkened room, such as a movie theatre). With such a microscope objects with dimensions smaller than the limit of resolution can be *seen*; the limit, however, severely restricts the accuracy with which the actual shape and size of such structures can be determined. In recent years darkfield microscopes have proved especially useful in observing patterns of movement of structures too small to be viewed effectively with other microscopes, such as bacteria or bundles of the fine filamentous structures (bacterial flagella; Fig. III–9) by which bacteria propel themselves. (See Fig. II–83 for an important example of such use.)

The *polarizing* microscope can detect regions in cells where constituents are disposed in highly ordered array. A *polarizer* built into the microscope produces *polarized light,* which is passed through the specimen. Ordinary light can be pictured as a beam in which wavelike vibrations are occurring in all possible directions perpendicular to the direction in which the beam is proceeding. In polarized light the vibrations are confined to a single set of directions (for example, left and right, or up and down). Such light represents a sort of oriented field of electromagnetic energy that can interact in a distinctive fashion with specimen areas in which macromolecules or subunits are also arranged in a regular and oriented manner (for example, like a stack of coins or in parallel rows). Built into the microscope is an *analyzer,* an optical device that translates the behavior of polarized light that has passed through a specimen into visible contrast in the image. Ordered areas can be made to appear darker or brighter than areas of more random arrangement (see Fig. IV–22); such areas are referred to as *birefringent.* The important point is that *order* can be detected and analyzed in living cells, even when the oriented array is not visible by ordinary light microscopy.

Culturing; Microcinematography In addition to improvements in optical techniques, the development of methods for growing cells in artificial conditions or *culture* (Fig. I–13) has opened broad possibilities for studying living cells, both unicellular organisms and cells separated from tissues. An increasing variety of cell types can be maintained in artificial well-defined growth media. Chapter 3.11 will discuss some of the important results obtained by cell biologists with cultured cells.

Some activities of organelles in living cells, particularly in unicellular organisms and in cells in culture, can be vividly demonstrated by taking motion pictures through a phase-contrast, polarizing, or interference microscope. Such *microcinematography* also *records* these activities and facilitates their detailed study. Processes thus recorded may be speeded up photographically or may

A. Interference

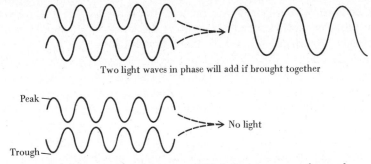

Two light waves in phase will add if brought together

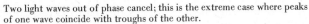

Two light waves out of phase cancel; this is the extreme case where peaks of one wave coincide with troughs of the other.

B. Principle of the differential contrast microscope

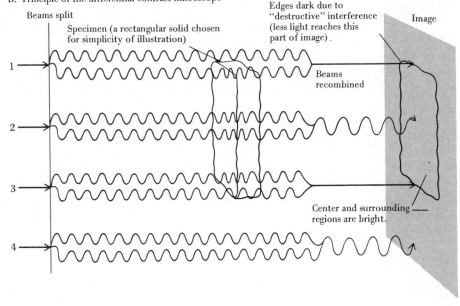

be viewed in slow motion. Section 2.10.3 refers to the study of the motion of cilia. Another striking example is the "degranulation" (merger of lysosomes and other granules with phagocytic vacuoles) of white blood cells when bacteria are engulfed by these cells (see Section 2.8.2). Videotaping is now being used for similar studies: it has proved particularly powerful when employed in conjunction with electronic image-enhancement methods that can increase the contrast among details in the TV image.

◄ **Figure I–12** *Principles of interference microscopy.*

A. Two light waves arriving together at a point in space interfere; the resulting intensity of light depends on the "phase relations" between the waves; that is, on the relative timing of arrival of the peaks and troughs of the two waves at the point of interest. Two extreme cases are illustrated: if the peaks of one wave coincide exactly with the peaks of the other, the two will add, giving high intensity; if the peaks of one coincide exactly with the troughs of the other, the two will cancel, and the light intensity will be zero. In the first situation the waves are referred to as "in phase"; in the second, they are "out of phase" (by a half wavelength or 180 degrees). Intermediate situations between these cases produce intermediate intensities; the waves are referred to as "out of phase" (by the corresponding fraction of a wavelength). For our purposes, the important point is that light intensity can be manipulated by altering phase relations between light waves.

B. Principles of the "differential contrast interference microscope" (Nomarski microscope). The essential feature of this microscope is the use of interference to make boundaries and regions of objects such as zones of sharp curvature stand out as darker (or lighter or differently colored) from their surroundings. Figs. I–6, IV–24 and IV–29 illustrate this. In more technical terms, as diagrammed here, a "beam-splitting" optical system is used to divide the illumination into two sets of beams before the light reaches the specimen. The resultant light can be thought of as consisting of pairs of beams, each pair being recombined into one beam after the light has passed through the specimen. The two members of each pair of beams are displaced with respect to one another so that they travel through slightly different points in the specimen. (This displacement is very slight; otherwise two overlapping images would be produced.) If both members of a given pair go through a similar region of the specimen (in the illustration, through the interior of the object being examined, or through the region outside the object . . beams 2 and 4) the two will simply recombine to give back the original intensity of the light from which they are produced. The difference at edges is that here there will be beam pairs for which one member goes *through* the object and the other *outside*. If, for instance, one beam goes through a cell and the other outside (or one through the nucleus and the other through the cytoplasm) the two will be out of phase when they recombine (beams 1 and 3). This occurs because of the refractive index differences mentioned in the text; these differences result in different speeds of light through different regions of a cell or its surroundings. Therefore, through interference, the images of cellular boundaries will be darker than those of adjacent regions. Similar phenomena occur when paired beams pass through a markedly curved body or through highly irregular structures because the two beams will go through different thicknesses of the same object. (Note also that the microscope can easily be adjusted so that edges stand out bright against a dark background or so as to provide contrasting colors.)

1.2.3 Cytochemistry

The procedures of cytochemistry aim at producing a color or special contrast, such as enhanced brightness or darkness, at the sites of specific constituents within cells. In cells photographed in ultraviolet light, for example, the regions where nucleic acids are concentrated appear darker than their surroundings, since nucleic acids selectively absorb light with wavelengths around 260 nm.

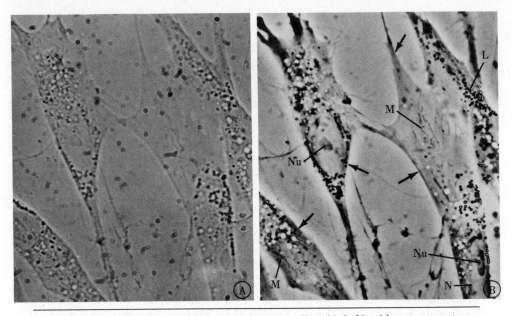

Figure I–13 *Phase-contrast.* The same living cells (chick fibroblasts grown in culture) photographed (A) through an ordinary light microscope and (B) through a phase-contrast microscope. The phase-contrast image clearly shows cell borders (arrows), nuclei *(N),* nucleoli *(Nu),* the elongate mitochondria *(M),* spherical lysosomes *(L),* and other details that are barely detectable in the ordinary microscope. × 800.

In the *Feulgen procedure,* sections are treated with dilute hydrochloric acid. The DNA, because of the unique sugar it contains (deoxyribose), is the only constituent changed in the right way (aldehyde groups form) to react with the *Schiff reagent,* a colorless form of the dye *basic fuchsin.* The modified DNA reacts with this reagent to restore the bright red color of the dye. This Feulgen reaction generally stains only the nuclei, where almost all cellular DNA is found. Figure I–14 shows a Feulgen-stained nucleus; the DNA is seen to be localized in the *chromatin,* of which chromosomes are composed. The small amount of DNA present in some cytoplasmic organelles usually does not show, because its concentration is below the limit needed for visibility.

If tissue sections are exposed to dyes that are soluble only in specific types of lipids, the cells will absorb the stain only in the areas that contain these lipids. Staining procedures that can be used to demonstrate RNA are also available. They show, for example, the localization of much RNA within a discrete intranuclear body, the *nucleolus* (Fig. I–15). Stain also appears in areas of the cytoplasm, which when viewed in the electron microscope are seen to contain many small RNA–protein granules, the *ribosomes.*

Such staining techniques can give quantitative information. For instance, the amount of dye taken up may be directly proportional to the amount of

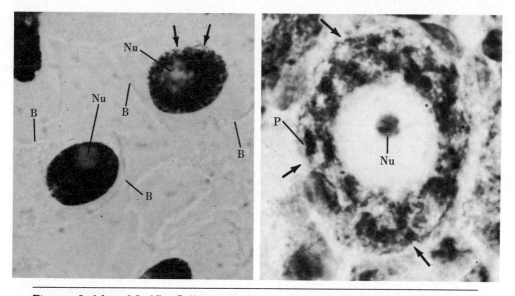

Figures I–14 and I–15 *Cells stained by specific procedures for localizing nucleic acids.* Figure I–14 shows two cells from a root tip of *Tulbaghia,* stained with the Feulgen method for DNA (see Section 1.2.3). The cytoplasm is unstained, therefore the cell borders *(B)* are barely visible. Most of the nuclear volume is occupied by intensely stained chromatin, and the arrows indicate points at which the tangled, threadlike character of the chromatin can be seen. Nucleoli *(Nu)* are unstained. × 1000. Figure I–15 shows a neuron (nerve cell) from a rat ganglion stained with the positively charged dye *pyronine,* which stains basophilic structures—those rich in negatively charged molecules—in this case RNA molecules (see Section 2.2.2). The red color of the dye shows as black when green light is used for photography, as it was in this micrograph. In the nucleus, the nucleolus *(Nu)* stains intensely, and in the cytoplasm, basophilic patches *(P)* are stained. Arrows indicate the cell's borders. × 1600.

stained component, and the amount of dye present at a given site within a cell can be measured through the microscope by determining how much light is absorbed by that site (*microspectrophotometry*). This quantitative approach using the Feulgen stain was important in establishing that the amount of DNA in cell nuclei is characteristic of the given species and constant from cell to cell in a given organism; these were historical steps in evaluating the role of DNA.

Tissues are sometimes studied by microscopes that exploit particular characteristics of cellular materials. *Acoustic* microscopes detect differences in "elasticity"—various materials reflect sound waves to varying degrees and show changes in such reflection under different conditions. *X-ray microscopes* detect concentrations of dense materials. With appropriately prepared specimens, *analytical* ("electron-probe") microscopes can reveal locations of certain important elements, such as calcium ions, at the electron microscope level;

these elements emit characteristic patterns of electromagnetic radiation, such as X-rays, when exposed to the microscope's electron beam.

Localizing Enzymes by Microscopic Means As mentioned in the Introduction, enzymes vary in the compounds they affect (their substrates) and in the nature of the reactions they catalyze. Under specific conditions the actions of some enzymes can generate insoluble products visible in the microscope. This permits direct microscopic study of enzyme distribution in cells. For example, *phosphatases,* enzymes that split the phosphate group from specific substrates, are localized by incubating the section with the appropriate substrate in the presence of a metal ion such as lead. The lead ions combine with the phosphates liberated at the sites of the enzyme. Lead phosphate is insoluble and precipitates as it forms. The resulting deposits at the enzyme sites are visible by electron microscopy (Figs. II–48 and II–72). For light microscopy, the lead phosphate is converted to lead sulfide, which has a deep brown color. In Figures I–16 through I–19 the plasma membrane, endoplasmic reticulum, Golgi apparatus, and lysosomes are shown to contain specific phosphatases that split different substrates. Figures I–20 and I–21 show two other organelles, mitochondria and peroxisomes, stained through effects of enzymes within the organelles. Thus far, only a few of the many enzymes known to biochemists have proved suitable for cytochemistry—these survive fixation and other steps in tissue preparation needed to keep the enzymes in place and are capable of yielding insoluble visible reaction products. Nevertheless, enzyme stains are useful in selectively visualizing organelles, as Figures I–16 through I–21 demonstrate.

Immunohistochemistry The locations of a variety of molecules have been determined by obtaining proteins that bind specifically to these molecules and preparing the proteins as stains. *Immunohistochemistry* is the most broadly applicable such approach. As we shall discuss in Chapter 3.6C, when higher vertebrate organisms are exposed naturally, or by injection, to microorganisms,

---

Figures I–16 to I–21 are light micrographs of preparations incubated to reveal sites of six different enzymes. In each case, "staining" is due to the deposition of a reaction product resulting from the action of the enzyme on a specific substrate. In Figures I–16 to I–19 the enzymes are phosphatases hydrolyzing different phosphate-containing substrates.

Figure I–16 Staining due to splitting of ATP by a plasma membrane enzyme (or enzymes) of cells in a liver tumor of a rat. Each light polygonal area is a cell outlined by the dark-reaction product on its surface. × 500.

Figure I–17 Staining due to splitting of the diphosphate of inosine (IDP) by enzymes at two sites in rat neurons (nerve cells). One site is the Golgi apparatus (G; see Fig. I–18). The other is the membrane systems of the endoplasmic reticulum; this results in staining of patches (arrows) in the cytoplasm comparable to the patches in Figure I–15 where it is the RNA of the ribosomes bound to the reticulum that is stained. *N* indicates nuclei. × 1000.

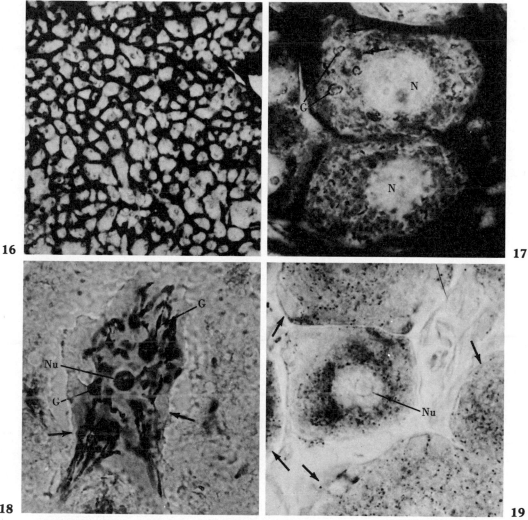

Figure I–18 Staining of the Golgi apparatus in a rat neuron due to the activity of an enzyme that splits thiamine pyrophosphate (TPP) and therefore is called thiamine pyrophosphatase (TPPase). The Golgi apparatus is the network indicated by *G.* Cell borders are seen at the arrows. *Nu* indicates a nucleolus. × 880.

Figure I–19 Staining of the lysosomes of rat neurons due to the presence of the enzyme acid phosphatase. Portions of several neurons are seen in this micrograph. *Nu* indicates the nucleolus within the nucleus of the one in the center of the field; the arrows indicate the edges of several others. The lysosomes are the dark granules scattered in the cytoplasm of the neurons. × 800.

(Figs. I–20 and I–21 on next page.)

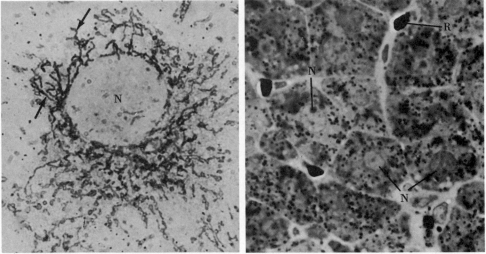

Figure I–20 Staining of the many mitochondria (two are indicated by arrows) in a mouse connective-tissue cell grown in tissue culture. The enzyme responsible for staining transfers hydrogens of NADH (Section 1.3.1) to a soluble "tetrazolium" dye and thus converts the tetrazolium to an insoluble blue "formazan." The nucleus *(N)* is unstained. × 1000. Compare this figure with the photograph of mitochondria in a living cell in Figure I–13.

Figure I–21 Staining of peroxisomes in rat hepatocytes owing to a reaction in which a soluble colorless compound, diaminobenzidine, is converted to an insoluble brown reaction product. The reaction is dependent on the enzyme *catalase* within peroxisomes, which are the dark cytoplasmic granules surrounding the unstained nuclei *(N)*. Hemoglobin can also catalyze the reaction with the result that the red blood cells *(R)* in the hepatic circulation also stain. × 800.

to proteins from other species, or to other foreign molecules, they synthesize *antibodies*. These are proteins that can bind selectively to the foreign materials that evoked their synthesis. Antibodies have proved to be versatile experimental tools for recognition and manipulation of cells and molecules since a wide variety, each able to recognize and bind tightly to a particular cellular component, can be obtained. For example, proteins that lack cytochemically demonstrable enzyme activity can be located by staining them with antibodies specific for them. For such use, the antibodies can be coupled with fluorescent dyes visible by light microscopy. Fluorescent dyes emit visible light when irradiated with ultraviolet light, which ordinarily is not itself directly visible. Hence structures to which the antibodies bind are differentially "stained": They stand out as bright against a darker background (Figs. II–86, II–92, II–93, and IV–23) or, with suitable fluorescent dyes, can be distinctively colored. Tracer molecules detectable in the electron microscope can also be coupled to the antibodies and used to reveal sites of antibody binding. (Fig. III–40 illustrates a variant of this approach.)

1.2.4 Autoradiography

A valuable technique for tracing dynamic events in cells is *autoradiography*. It is especially useful when employed quantitatively. Small *precursor* (metabolic forerunner) molecules that the cell uses as building blocks in the synthesis of macromolecules are made radioactive by incorporation of radioactive isotopes. The most widely used isotope is tritium (^{3}H), a radioactive form of hydrogen. Carbon-14, phosphorus-32, and sulfur-35 are also employed. Probably the most frequently used radioactive precursor is *tritiated thymidine* (^{3}H-thymidine); this is thymidine (used by the cell to synthesize DNA) made radioactive by the substitution of tritium atoms for some of its hydrogen atoms. Typically, radioactive precursors are presented to cells, which are then fixed at successive intervals and subsequently sectioned. The sections are coated with photographic emulsion similar to ordinary camera film (Fig. I–22). Exposure of camera film to light followed by photographic development leads to reduction of exposed silver salts in the film and to production of metallic silver grains. These grains form the image in the negative. Similarly, exposure to radioactivity and subsequent development produces grains in the autoradiographic emulsion. When sections prepared for autoradiography are examined microscopically, both the underlying structure and the small grains in the emulsion are seen simultaneously (see Figs. I–23 and II–43).

If, for example, radioactive precursors of RNA are given to cells that are fixed only a few minutes later, almost all the grains are seen over the nuclei, suggesting that it is here that RNA is made from precursor molecules. If the interval between the exposure to radioactive precursors and the fixation is lengthened to an hour or two, the percentage of grains over the nuclei diminishes, and more and more grains are seen over the cytoplasm (Figs. I–23 and I–24). These observations suggest that once RNA has been made in the nucleus, it moves out into the cytoplasm. Such use of "snapshots" taken at successive time intervals illustrates one of the most commonly used methods of studying dynamic biochemical events in cells.

The grains produced in the emulsion are similar, irrespective of the precursor used; that is, radioactive RNA, protein, and DNA all "look" alike. Interpretation of autoradiographs depends on the knowledge of biochemical pathways; precursors are carefully chosen so that they are used by the cell to build only one kind of molecule. Furthermore, "control" preparations are employed to verify the interpretation. For example, after exposure to RNA precursors and incorporation of radioactivity, tissue sections can be treated with a solution of purified RNase (ribonuclease; sometimes abbreviated RNAse). This enzyme specifically breaks down RNA and removes it from the tissues. If such treatment removes the radioactivity from a sample of the material being studied (so that no grains are produced in the autoradiographic emulsion), it is reasonable to conclude that the precursor has been incorporated into RNA.

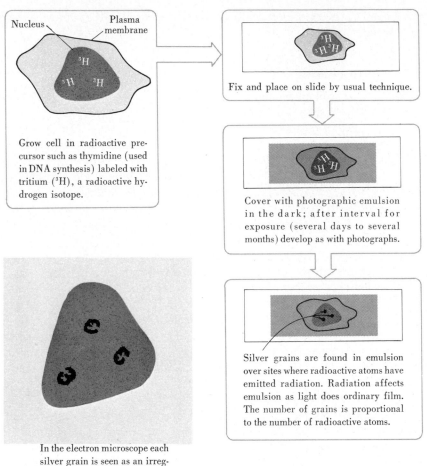

Nucleus

Plasma membrane

³H

³H   ³H

Grow cell in radioactive precursor such as thymidine (used in DNA synthesis) labeled with tritium (³H), a radioactive hydrogen isotope.

³H
³H ³H

Fix and place on slide by usual technique.

³H
H ³H

Cover with photographic emulsion in the dark; after interval for exposure (several days to several months) develop as with photographs.

Silver grains are found in emulsion over sites where radioactive atoms have emitted radiation. Radiation affects emulsion as light does ordinary film. The number of grains is proportional to the number of radioactive atoms.

In the electron microscope each silver grain is seen as an irregular metallic deposit.

Figure I–22 *Procedures of autoradiography.* The radiation emitted by the radioactive atoms in the specimen is emitted in random directions so not every emission enters the emulsion, which is present only on one surface of the specimen. Therefore, only a proportion of emissions generates grains. Nevertheless, the number of grains over a given region *is* proportional to the number of radioactive atoms in that region. This proportionality is the basis of quantitative use of the technique as in Fig. I–24. Those precursor molecules that have not been incorporated into macromolecules are washed out of the cells during fixation and therefore do not yield grains.

Chapter 1.2B Cell Fractionation

Cell fractionation is a central and versatile technique for studying cell chemistry. The different subcellular structures are separated by centrifugation of ho-

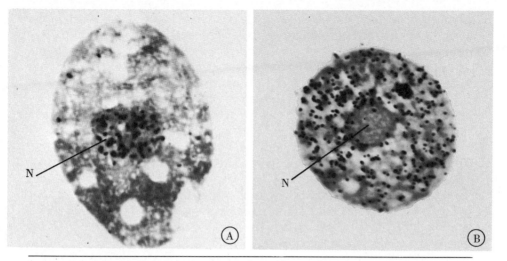

Figure I–23 *Light microscope autoradiographs illustrating a pulse-chase experiment on RNA metabolism in Tetrahymena,* a ciliated protozoan. The cells in A and B were both grown for approximately 10 minutes in tritiated cytidine (^{3}H-C), a radioactive precursor of RNA. The cell in A was fixed almost immediately thereafter. The cell in B was allowed to grow in nonradioactive solutions for 90 minutes after exposure to ^{3}H-C. In A almost all grains are seen over the nucleus *(N)*, whereas in B most are over the cytoplasm. Experiments of this type are known as pulse-chase experiments. A component is labeled during a brief exposure to radioactive precursor (a pulse), and the subsequent behavior of the radioactive molecules is followed in the chase, the period of growth in a nonradioactive medium. In the present case RNA molecules are labeled in the nucleus during the pulse. During the chase they migrate to the cytoplasm and are replaced in the nucleus by newly made (unlabeled) molecules (see also Fig. I–24). Approx. × 1000. (Courtesy of D. Prescott, G. Stone, and I. Cameron.)

mogenates. The latter are prepared by disrupting cells of liver or other organs in media that preserve the organelles. Solutions of sucrose or other sugars are generally used because they can be readily adjusted to maintain the integrity of organelles and to counteract the tendency of organelles freed from the cell to clump together. Some organelles, such as mitochondria and plastids, remain

	Percentage of grains over		
	Nucleolus	Rest of nucleus	Cytoplasm
30 minutes	54	41	5
30 minutes + 4 hours	14	19	67

Figure I–24 *RNA moves from nucleus to cytoplasm.* Mouse cells (L-strain fibroblasts) grown in culture were exposed for 30 minutes to radioactive RNA precursor (^{3}H-cytidine), then either fixed immediately or grown for an additional 4 hours in nonradioactive medium before fixation and autoradiographic study. "Rest of nucleus" refers to the nonnucleolar region of the nucleus. (From the work of R. P. Perry.)

essentially intact. On the other hand, the endoplasmic reticulum, which in the living cell is an interconnected network of membrane-bounded cavities, is broken into separate pieces that form rounded vesicles (Fig. I–25). Similarly, the plasma membrane is fragmented into pieces of variable sizes in most cells, but methods are available for obtaining plasma membranes or some large, specialized regions of plasma membranes almost intact from a few cell types. (Obviously, some disruption is inevitable when separating the plasma membrane from the cell's contents.)

Differential and Density Gradient Centrifugation The behavior of a particle in a centrifugal field depends chiefly upon its weight and upon the resistance it encounters in moving through the suspension medium. Its size, density, and shape influence the movement of a particle in a centrifugal field (see Fig. I–25, legend). The most widely used method for separating cell organelles is called *differential centrifugation*. It is based on differences in the speed with which structures sediment to the bottom of a centrifuge tube. At a given centrifugal force, structures that are relatively large and dense sediment most rapidly. Thus with comparatively low forces (slow speed of centrifuge), nu-

---

Figure I–25 *Cell fractionation.* Cells are disrupted by a variety of procedures. "Shearing forces" that break open the cells are produced by use of a blender such as the ones used in food preparation, or by movement of a close-fitting glass or plastic pestle within a tube as illustrated. The space between pestle and tube wall in such a "homogenizer" can be varied to achieve breakage of cells with minimum damage to organelles. Inevitably the plasma membrane is disrupted and the ER is extensively fragmented—the ER fragments seal off to form closed vesicles isolable as a major component of the "microsomes."

Modern ultracentrifuges are capable of rotating at 50,000 to 100,000 revolutions per minute. Such rotational speeds impose forces in excess of 250,000 times the force of gravity, permitting "sedimentation" (movement toward the bottom of the tube) of the smallest cell structures and even of macromolecules.

Useful equations and definitions: V = sa. The velocity *(V)* with which a given type of particle sediments when suspended in a particular medium depends on the "sedimentation coefficient" *(s)* characterizing the particle in the medium used for the sedimentation, and on the force imposed by the centrifuge *(a).*

$s = \dfrac{m(1 - v\rho)}{f}$: The sedimentation coefficient of a given structure ("particle") de-

pends on several factors, not only on *m,* the mass of the particle. The other terms in the equation relate to the influence of the medium and of the shape of the particle on sedimentation: *f* is a "frictional coefficient" that expresses, in quantitative terms, the effects of the shape of the particle and the resistance to movement (viscosity) of the medium; ρ is the density of the medium; and *v* is 1/ density of the particle in solution. Notice that when the density of the particle and the density of the medium surrounding it are identical, $v\rho = 1$. Hence $1 - v\rho = 0$ and thus from the equation above $s = 0$. In other words, when the medium's density matches that of the particle, the particle does not sediment.

By convention, sedimentation coefficients are measured in Svedberg units (S), named after a principal inventor of the ultracentrifuge. $1\ S = 10^{-13}$ second. Coefficients measured under a variety of actual conditions are referred by calculation to "standard conditions," usually sedimentation in water at 20° C.

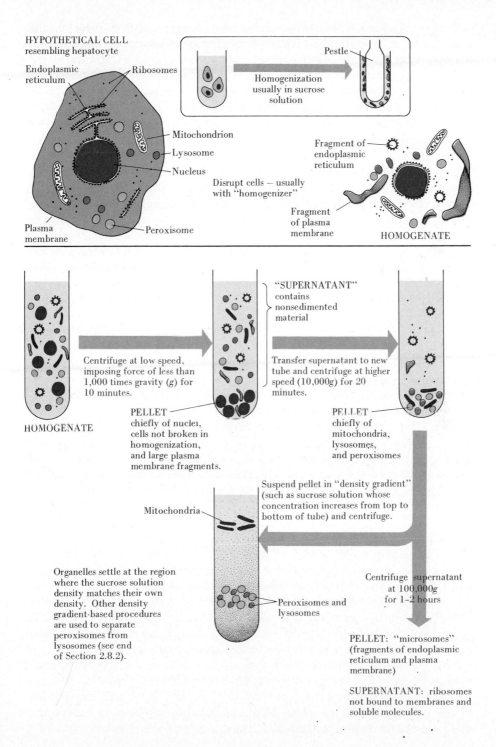

HYPOTHETICAL CELL
resembling hepatocyte

Endoplasmic reticulum

Ribosomes

Pestle

Homogenization usually in sucrose solution

Mitochondrion

Lysosome

Nucleus

Disrupt cells — usually with "homogenizer"

Fragment of endoplasmic reticulum

Plasma membrane

Peroxisome

Fragment of plasma membrane

HOMOGENATE

Centrifuge at low speed, imposing force of less than 1,000 times gravity (g) for 10 minutes.

"SUPERNATANT" contains nonsedimented material

Transfer supernatant to new tube and centrifuge at higher speed (10,000g) for 20 minutes.

HOMOGENATE

PELLET chiefly of nuclei, cells not broken in homogenization, and large plasma membrane fragments.

PELLET chiefly of mitochondria, lysosomes, and peroxisomes

Suspend pellet in "density gradient" (such as sucrose solution whose concentration increases from top to bottom of tube) and centrifuge.

Mitochondria

Organelles settle at the region where the sucrose solution density matches their own density. Other density gradient-based procedures are used to separate peroxisomes from lysosomes (see end of Section 2.8.2).

Centrifuge supernatant at 100,000g for 1–2 hours

Peroxisomes and lysosomes

PELLET: "microsomes" (fragments of endoplasmic reticulum and plasma membrane)

SUPERNATANT: ribosomes not bound to membranes and soluble molecules.

clei, large fragments of plasma membrane, and some mitochondria can be sedimented as a pellet (Fig. I–25), while the remaining cell structures remain in suspension. With somewhat higher centrifugal force, a second fraction containing most of the mitochondria is separated. Most lysosomes and peroxisomes are sedimented in the same fraction. The so-called "microsomes," which consist mainly of fragmented endoplasmic reticulum but also of some small fragments of plasma membrane, come down with still greater force. (Note that there are no microsomes *as such* in the cell; the fragments of ER seal off into closed vesicles much smaller than the sacs and tubules from which they arise.) Free polysomes and ribosomes (those not bound to membranes) often remain unsedimented, together with other small structures and soluble molecules. The ribosomes can be brought down by prolonged use of high centrifugal forces.

Centrifugation through gradients of increasing concentration of sucrose or other substances can yield separations on the basis of density. The higher the concentration of dissolved sucrose, the greater the density of the solution. Density gradients can be used to increase the differences in rates at which different structures sediment. Another approach is based on the choice of conditions of timing and concentration such that organelles situate themselves at the levels in the centrifuge tube where the densities of the solution match the densities of the organelles (Fig. I–25). The solution can then be removed carefully from the tube so that the different layers ("bands") are obtained separately; this can be done, for example, by punching a small hole in the bottom of the tube and slowly draining the contents, which emerge in the sequence with which they are found in the centrifuge tube. Density gradient centrifugation often enables separation of particles of roughly the same size but different density. Thus peroxisomes can be separated from lysosomes, or microsomes can be separated into those fractions in which the membranes are studded with ribosomes and those in which ribosomes are absent. Plasma membrane fragments also have been isolated in this manner. Modified procedures of gradient centrifugation can be used to achieve separations of macromolecules, such as RNA and DNA (see Fig. II–28).

The Cytosol, Hyaloplasm or Ground Substance The material that remains suspended under conditions sufficient to sediment even the ribosomes is referred to by biochemists as the *soluble fraction* or *cytosol*. This fraction usually contains some material lost from the organelles in the course of homogenization and centrifugation. However, it also contains other molecules, including some enzymes, that are not known to be part of the intracellular structures identifiable in the microscope. Such molecules are usually thought to derive from the *hyaloplasm* (also called ground substance, cell matrix, or cell sap), which was originally defined by microscopists as the apparently structureless medium that occupies the space between the visible organelles. As will be evident in subsequent chapters, with the improved techniques less and less of the cell appears structureless. Most investigators continue to believe that there does exist in the cytoplasm a "soluble phase"—a solution containing inorganic ions and small molecules, materials in transit from one cell region to another and dissolved or suspended enzymes that function as individual molecules or perhaps as parts of small groups of molecules rather than as components of large, microscopically recognizable organelles. But there is a lively debate going on over proposals that many components once thought to be suspended in this solution are actually bound, perhaps loosely, in structured arrangements that do not survive the disruption involved in cell fractionation (Section 2.11.3).

Analysis of Fractions; Coordinated Use of Centrifuges and Microscopes Isolated fractions are examined by electron microscopy to determine the integrity of the organelles and the purity of the fractions. This step is of importance since fractions thought to contain only one organelle may prove to be contaminated with others. For example, some lysosomes and peroxisomes may sediment with the mitochondria, and some mitochondria may sediment with the lysosomes and peroxisomes. Fractions that consist exclusively of one organelle are difficult to obtain, so that most work is done with fractions that are less than pure.

Fractions can be disrupted by chemical or physical means; subfractions can then be separated by additional centrifugation. Thus fragments of rough endoplasmic reticulum (isolated as microsomes) can be treated with detergents such as deoxycholate that solubilize the membranes and free the ribosomes.

Isolated fractions can be subjected to the full range of biochemical analysis in order to learn about the chemical composition, enzyme activities, and metabolic capabilities of subcellular structures. With the use of radioactive precursors and by isolating fractions at different time intervals, insights may be gained into temporal aspects of metabolic events. If cells are exposed, for instance, to radioactive amino acids, and fractions are isolated early after exposure, radioactive proteins are found only in the fractions that contain ribosomes, suggesting that ribosomes are the sites of protein synthesis. As the interval between exposure to radioactive amino acids and isolation of fractions is lengthened, other cell structures are found to contain radioactive protein. Thus it is possible to trace the transport of the protein, initially synthesized on the ribosomes, through its subsequent transport in the cell.

Biochemical approaches and microscope-based ones, such as autoradiography and cytochemistry, are most effectively used in conjunction with one another. The biochemical techniques on which much of the present understanding of cell chemistry and metabolism is based, such as cell fractionation, or the analysis of molecules with centrifuges (see Fig. II–28), columns (Section 2.3.5), or gels (see Fig. II–37), generally give information about the *average* properties of cells, organelles, and molecules. When the starting material is a complex mixture of cells, such as the brain, it can be quite difficult to decide which cell type is responsible for the properties of a cell fraction or the presence of a particular molecule. Even relatively homogeneous cell populations, such as a tissue culture, can show important variations from cell to cell. Microscopic techniques give less varied and detailed types of information about molecules or metabolism than do biochemical approaches, but they can focus on single cells or organelles or sometimes single macromolecules and thus reveal the distribution of properties among the individuals that comprise the population.

Chapter 1.3 A Major Metabolic Pathway

Figure I–26 outlines major metabolic sequences responsible for the breakdown of glucose. Full details are given in the biochemistry texts and in other references listed at the end of this part. Emphasis here will concern defining some major terms and summarizing certain features especially relevant to cell organization and useful for subsequent discussions.

Chapter 1.3A Metabolic Breakdown of Glucose

1.3.1 Aerobic and Anaerobic Pathways; Some Definitions

In hepatocytes, glucose (a 6-carbon sugar) is broken down first to two molecules of the 3-carbon compound, *pyruvic acid* (pyruvate). As seen in Figure I–26, this breakdown occurs in a series of steps, each step catalyzed by a specific enzyme. The product of each step serves as the substrate for the next enzyme; the products are referred to as *intermediates*.

The further metabolism of pyruvate to form carbon dioxide and water depends on a second series of reactions known as the *tricarboxylic acid cycle* or *Krebs cycle*. Each pyruvate molecule is first converted to a 2-carbon *acetate* molecule. The acetate undergoes the reactions shown in Figure I–26. Each molecule is combined with a 4-carbon molecule (oxaloacetate) to produce a 6-carbon product. This 6-carbon molecule is broken down by a series of steps that ultimately regenerate an oxaloacetate molecule, which can then combine with another molecule of acetate and go through another turn of the cycle. At two steps in the series of reactions, CO_2 is released; these CO_2 molecules plus the one CO_2 released in the pyruvate-to-acetate conversion account for all three carbons of the pyruvate molecule.

Oxidation and Reduction; Coenzymes Five of the Krebs-cycle reactions, starting with pyruvate, involve coupled *oxidations* and *reductions*. Several equivalent definitions applied to such reactions are often used. In chemical terms, oxidation–reduction reactions involve the transfer of electrons from one molecule to another, accompanied by a release or transfer of energy. The molecule losing electrons is oxidized; the one gaining electrons is reduced. In biological systems, many such electron-transfer reactions include the exchange of hydrogen atoms: Oxidation of a substrate takes place by loss of hydrogens;

reduction, by gain of hydrogens. The enzymes catalyzing such reactions are *dehydrogenases,* which are a large class of the general category of *oxidative* or *oxidation-reduction* enzymes. As is often the case with such enzymes, the dehydrogenases of the Krebs cycle transfer the electrons and some of the protons (H^+) of the hydrogens removed from substrates to *coenzymes.* These are a class of small, nonprotein molecules of several types that function together with enzymes in many metabolic reactions. In four of the steps we are considering, two hydrogen atoms are removed from substrates; from these, two electrons and one proton are transferred to *NAD* (nicotinamide adenine dinucleotide), with an additional H^+ being freed. The overall reaction for each step is

$$\text{Substrate}_{\text{reduced}} + \text{NAD}_{\text{oxidized}} \rightarrow \text{Subs}_{\text{oxi}} + \text{NAD}_{\text{red}} + H^+$$

Often NAD^+ and NADH are used to designate the oxidized and reduced forms of the coenzyme.

Aerobic and Anaerobic Pathways   The steps from glucose to pyruvate are *anaerobic*—they require no oxygen. Virtually all organisms are capable of degrading sugars by similar anaerobic pathways. In some microorganisms, the pyruvate molecules are further metabolized (anaerobically) to produce ethyl alcohol molecules; such production of alcohol from sugar is known as *fermentation.* Other cells instead convert pyruvate molecules anaerobically to lactic acid (lactate) molecules; the path from glucose to lactate is called *glycolysis.* Some bacteria, yeasts, and other microorganisms metabolize exclusively by fermentation, glycolysis, and other anaerobic pathways. Cells of most organisms have enzymes both for anaerobic pathways (glycolysis and others) and the Krebs cycle and related aerobic (oxygen-dependent) pathways. As Figure I–26 makes evident, the aerobic metabolism of glucose via the Krebs cycle is preceded by the glucose-to-pyruvate steps of glycolysis.

1.3.2 Glucose Metabolism and ATP

Energy Storage and Transfer Using ATP, Reduced Coenzymes, and Gradients   Generally, metabolism of glucose or related sugars and other carbohydrates is the major source of cellular ATP. ATP is usually formed by the addition of a phosphate group to ADP (adenosine diphosphate). The chemical bond formed in this reaction is called a *high-energy* bond; its formation requires an input of energy derived from metabolism. When such a bond is broken, energy is made available for release, or for transfer to other molecules. ATP and related molecules such as GTP (guanosine triphosphate) are major molecular forms in which energy is stored and transferred to sites of cell functions.

The mechanisms by which cells generate ATP will be discussed in greater detail in Part 2 (especially Sections 2.6.2 and 2.7.2). Points to be noted here are that oxidation-reduction reactions are central to cellular energy metabolism and that reduced coenzymes as well as ATP are carriers of metabolic energy.

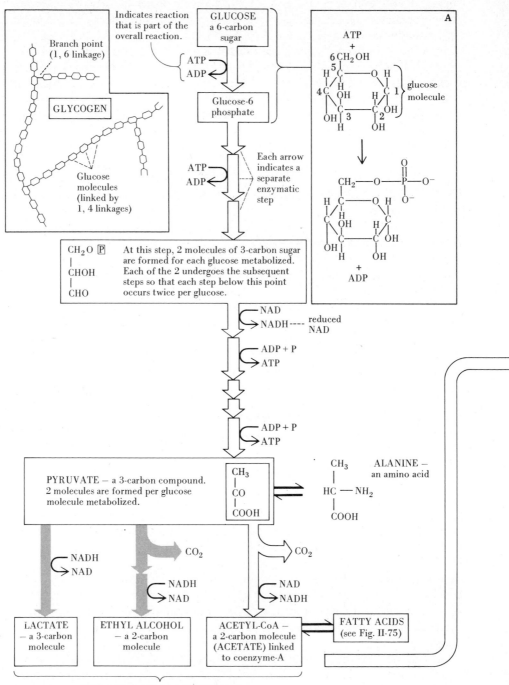

Branch point
(1, 6 linkage)

GLYCOGEN

Glucose
molecules
(linked by
1, 4 linkages)

Indicates reaction
that is part of the
overall reaction.

GLUCOSE
a 6-carbon
sugar

ATP
ADP

Glucose-6
phosphate

ATP
ADP

Each arrow
indicates a
separate
enzymatic
step

A

ATP
+
$6\,CH_2\,OH$

glucose
molecule

+
ADP

$CH_2\,O\; \boxed{P}$
|
CHOH
|
CHO

At this step, 2 molecules of 3-carbon sugar
are formed for each glucose metabolized.
Each of the 2 undergoes the subsequent
steps so that each step below this point
occurs twice per glucose.

NAD
NADH ---- reduced
NAD

ADP + P
ATP

ADP + P
ATP

PYRUVATE — a 3-carbon compound.
2 molecules are formed per glucose
molecule metabolized.

CH_3
|
CO
|
COOH

CH_3
|
HC —NH_2
|
COOH

ALANINE —
an amino acid

NADH
NAD

CO_2

NADH
NAD

CO_2

NAD
NADH

LACTATE
— a 3-carbon
molecule

ETHYL ALCOHOL
— a 2-carbon
molecule

ACETYL-CoA —
a 2-carbon molecule
(ACETATE) linked
to coenzyme-A

FATTY ACIDS
(see Fig. II-75)

ALTERNATE FATES OF PYRUVATE

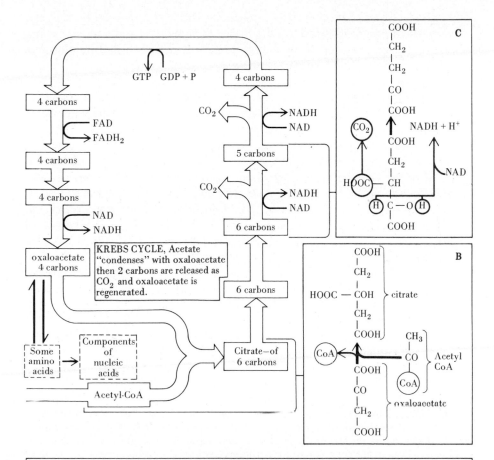

ENERGY STORAGE FROM GLUCOSE METABOLISM

1. In the steps that can occur anaerobically (glucose to pyruvate) 2 molecules of ATP are consumed and 4 ATPs are formed per glucose broken down.

2. In the Krebs cycle, starting with pyruvate, the equivalent of one ATP (a GTP; guanosine triphosphate, a compound similar to ATP) is formed per pyruvate (hence, 2 per glucose). Reduced NAD and reduced FAD also are formed. Through electron transport and oxidative phosphorylation — oxygen-consuming pathways — the reduced coenzymes can give rise to a total of 14 ATPs per pyruvate (28 per glucose). (Some investigators believe this is an overestimate; Section 2.6.2).

3. Two NAD molecules per glucose are reduced in the glucose-to-pyruvate steps. Their metabolic fate may differ somewhat from the coenzymes reduced in the Krebs cycle, but they represent energy available for metabolism equivalent to 6 ATPs that could be formed by electron transport and oxidative phosphorylation. (In anaerobic organisms and in certain normally aerobic cell types under anaerobic conditions, the reduced coenzymes produced in the glucose-to-pyruvate steps are used in reactions such such as the formation of *ethyl alcohol* or *lactic acid* (lactate).)

4. TOTALS: *Overall,* up to 38 molecules of ATP formed per glucose.
  "Aerobic" processes: 30 ATPs from Krebs cycle and associated electron transport
                6 ATPs from coenzymes reduced in glucose-to-pyruvate steps.
  Processes that can occur anaerobically: 2 ATPs from glucose-to-pyruvate steps (additional
                energy is stored in coenzymes; see 3 above).

(Legend on next page)

Through suitable oxidation-reduction reactions, the energy carried by reduced coenzymes can be transferred to other molecules, with the reduced coenzyme being reoxidized. Transfer of the energy to ATP is accomplished by a special series of processes; stated briefly:

1. The electrons from the coenzymes enter a sequence of reactions known as *electron transport*. This is carried out by a set of proteins and other components known as the *respiratory chain*.
2. Associated with electron transport is a sequence of events involving H^+ ions. This is still incompletely understood but is thought to include movement of the protons across mitochondrial membranes. (Cells can use energy to establish differences in concentration of electrically charged components—especially ions—across membranes and then draw upon the energy stored in the form of such *electrochemical gradients* in a variety of ways: see Sections 2.1.3, 2.6.2, 2.7.2, 3.2.4, and 3.7.3.) Eventually the electrons plus protons are transferred to oxygen, producing H_2O.
3. Much ATP is generated in association with electron transport. This process of formation of ATP is referred to as *oxidative phosphorylation*. It links oxygen consumption to the addition of phosphate to ADP.

Anaerobic and Aerobic Energy Storage   Two conclusions drawn from the information summarized in Figure I–26 should be noted: First, the aerobic

---

Figure I–26   *Some features of the reaction sequences in glucose metabolism.* The chemical structures of several key molecules are indicated. The conventional numbering scheme for the carbons of 6-carbon sugars is indicated in A. Parts A– C illustrate important biochemical reactions: A shows the *phosphorylation* of glucose; B, the *condensation* of a 2- and a 4-carbon molecule (acetate and oxaloacetate) to form a 6-carbon molecule (citrate); C, a *decarboxylation* (the removal of a carboxyl group (COOH) by its conversion to CO_2) plus the transfer of hydrogen atoms to coenzymes (the *oxidation* or *dehydrogenation* of a substrate molecule coupled with the *reduction* of a coenzyme molecule). Also illustrated are some of the many pathways that intersect with the glucose pathway.

   Technical notes: Glucose can also be polymerized into polysaccharides. Glycogen is the principal storage form of glucose in animals. It is composed of branched chains of glucose; in biochemical terms these chains are described as being of glucose units linked principally by 1,4 bonds (carbon #1 [C-1] of a given glucose linked, as shown, to #4 of the next). The chains form a multiple-branched structure by virtue of 1,6 linkages—between C1 of a glucose at the end of a chain and C-6 of a glucose within an adjacent chain, as illustrated. Such branch points occur, on the average, at every tenth glucose. (The 1,6 and 1,4 bonds are both of the type known as α—see Fig. II–44.)

   Additional terminology: NAD was once known as *diphosphopyridine nucleotide* and thus is still occasionally referred to as *DPN*. Sometimes, for convenience, the glucose-to-pyruvate steps are also referred to as *glycolysis* although strictly speaking, the term designates the path from glucose to lactate. "Aerobic" metabolism requires oxygen; "anaerobic" metabolism does not.

pathways generate much more ATP than that arising in the anaerobic pathways; this reflects chiefly the production of reduced coenzymes in the Krebs cycle. Some anaerobic organisms have evolved mechanisms that supplement sugar-based ATP production with other anaerobic pathways for energy storage; these facilitate survival despite the absence of aerobic pathways. Second, the linking of glucose metabolism to ATP production by aerobic cells involves three more-or-less readily distinguishable sequences of reactions: the formation of pyruvate; the degradation of pyruvate molecules in the Krebs cycle; electron transport and oxidative phosphorylation. The relations of this conceptual separation of glucose metabolism to actual cell organization will be discussed shortly.

Chapter 1.3B Metabolic Interrelations

1.3.3 Carbohydrates, Proteins, and Fats

As with many other metabolic sequences, the glucose pathways intersect with other pathways (Fig. I–26). Most obvious: reduced NAD is used in a variety of metabolic pathways as a reducing agent (a source of hydrogens or electrons); ATP participates in a very large number of reactions. In addition, however, pyruvate can be formed from the amino acid *alanine* by a reaction that is reversible, so that (if appropriate nitrogen sources are present) alanine can also be formed from pyruvate. By reactions of these sorts, carbohydrates can be used to synthesize various amino acids, which then combine to form proteins, or amino acids can be metabolized via the Krebs cycle to generate ATP. Fatty acids—components of fats (Fig. II–3)—can be converted to acetate and degraded by the Krebs cycle, or fatty acids can be built up from acetate units derived from carbohydrates. The many such intersections involving Krebs cycle intermediates contribute much to metabolic flexibility and diversity. They are one of the chief "benefits" of the fact that pathways such as the ones in Figure I–26 are based on sequences of relatively small changes. Another benefit is the efficient conversion of energy.

The controls that determine which sequence within the complex network of possible pathways an individual molecule will follow are actively being sought (see, for example, Sections 2.1.4 and 3.2.5). The overall metabolic pattern in a cell or tissue responds to the availability in the cells of ATP, NAD, oxygen, and various intermediates, and to external regulatory agents such as hormones. For example, in very active muscle, much glucose is rapidly broken down to pyruvate, but since the oxygen supply is inadequate for immediate breakdown of all the pyruvate to CO_2 and water, some pyruvate is converted anaerobically to lactate. This conversion is reversible, and the lactate is converted back to pyruvate either in the muscle itself during periods of rest, or in other tissues which the lactate reaches via the blood stream.

1.3.4 Metabolic Organization and Interactions of Organelles in Eucaryotic Cells

When hepatocytes or other eucaryotic cells are homogenized and cell fractions prepared by centrifugation, the following distribution of enzymes is observed.

1. Enzymes of the glycolytic pathway are present chiefly in the unsedimented supernatant fluid.
2. The enzymes and other proteins of the Krebs cycle, oxidative phosphorylation, and respiratory chain are largely confined to the mitochondria.
3. If the mitochondria are disrupted (for example, by exposure to *ultrasonic vibration,* very-high-frequency sound waves) most Krebs cycle enzymes are released as soluble material. Centrifugation of a preparation of disrupted mitochondria results in a pellet containing the membranous remnants of mitochondrial structure, to which are attached electron transport and oxidative phosphorylation enzymes. Most Krebs cycle enzymes remain in the supernatant.

The generally accepted conclusion from these observations is that the glycolytic enzymes are not firmly bound to any of the visible cellular organelles. Probably they are in the hyaloplasm, either suspended as free molecules or so loosely bound to intracellular structures that they are readily freed during homogenization (see Chap. 1.2B). In contrast, the enzymes and other proteins of the Krebs cycle, electron transport, and oxidative phosphorylation are located in the mitochondria. By centrifuging disrupted mitochondria, further compartmentalization is demonstrated; the proteins responsible for electron transport and oxidative phosphorylation are firmly attached to the mitochondrial membranes, whereas most Krebs cycle enzymes are not.

Such observations form the basis of Figure I–27, a final diagram of the hepatocyte. In many metabolic pathways there is interaction among several different organelles. Even our discussion of the glucose pathway could easily be extended to include other organelles. For example, in hepatocytes the breakdown of fatty acids generating acetyl-CoA for the Krebs cycle may sometimes involve sequential action of peroxisomes and mitochondria. Figure I–27 illustrates only a tiny fraction of the known metabolic interrelations.

Compartmentalization   The grouping of related enzymes in a common organelle facilitates cooperative or sequential action of the enzymes and coordinated control of their activities. The product of one enzyme need not diffuse far before it encounters the enzyme catalyzing the next step in its metabolism. In addition, the cell can maintain special environments within particular organelles. Lysosomal enzymes, for example, require acid conditions (low pH) to function effectively. The cell makes the interior of functioning lysosomes considerably more acid than the remainder of the cytoplasm, in which the pH is kept near neutrality (a pH of 7—that of pure water), as is required for the stability and operation of most cytoplasmic systems. Membranes are crucial elements in the compartmentalization of the cell; this will be the subject of much of Part 2.

Protein synthesis
involves polysomes
+ amino acids linked
to RNA.

Glucose breakdown—initial steps
in hyaloplasm, later ones
(Krebs cycle etc.) in mitochondria.

Golgi apparatus and
associated structures
have several roles
in packaging secretory
proteins and
lysosomal enzymes.

ATP

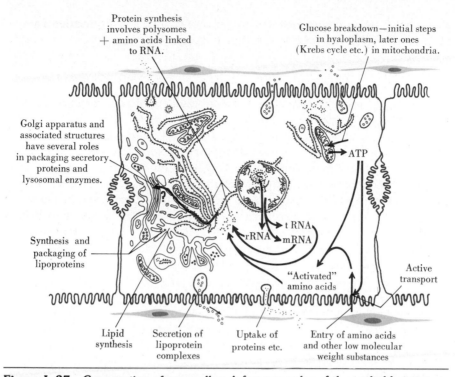

Synthesis and
packaging of
lipoproteins

t RNA
rRNA mRNA

"Activated"
amino acids

Active
transport

Lipid
synthesis

Secretion of
lipoprotein
complexes

Uptake of
proteins etc.

Entry of amino acids
and other low molecular
weight substances

Figure I–27 *Cooperation of organelles.* A few examples of the probable interrelations of organelle functions in important metabolic sequences—the synthesis, transport, and packaging of proteins and lipoproteins. Protein synthesis depends upon several categories of RNAs (Fig. II–23) produced in the nucleus, and on ATP synthesized by the mitochondria, and it utilizes amino acids that are transported through the plasma membrane. The proteins are made on polysomes (complexes of mRNA and ribosomes); those molecules synthesized on polysomes bound to the ER enter the ER and are transported to other cell regions. Some of the proteins synthesized by ribosomes on the ER, complex with lipids that also are made in the ER (partly in the smooth regions). The lipoprotein complexes formed in this way accumulate in membrane-delimited secretory packages; structures associated with the Golgi apparatus probably play a role in this process. The secretions are released by fusion of the membrane that surrounds them, with the plasma membrane.

Chapter 1.4 Chemistry and Structure

1.4.1 Bond Types

Two broad categories of chemical bonds are distinguished by biochemists: *covalent* and *noncovalent. Covalent* bonds, based on the sharing of electrons between two atoms, are responsible for the backbone structure of most biological molecules. They attach carbon atoms to one another and to various other atoms, such as the hydrogens and oxygens in sugars. Such bonds link amino acids to one another in proteins, sugars to bases and to phosphates in nucleic acids, and so forth. They are relatively strong; in almost all cases of cellular molecules, enzymes are needed to create or to disrupt covalent bonds at rates consonant with cellular phenomena.

The *noncovalent* interactions ("weak bonds") are of several sorts, which are not always sharply separable. In biological systems there are several of importance: *Ionic* ("electrostatic") *bonds* are based on mutual attraction of opposite charges. These are observed with inorganic ions (common ones in biological systems include sodium, potassium, calcium, and chloride: Na^+, K^+, Ca^{2+}, Cl^-) and with ionic forms of organic groups (such as carboxyl, amino, and phosphate groups: COO^-, NH_3^+, PO_4^{2-}) which carry discrete units of charge due to gain or loss of electrons and protons. In *hydrogen bonds,* a hydrogen covalently linked to one atom is attracted by another atom. (Usually a nitrogen or oxygen attracts a hydrogen, as is the case with the water molecules described in the next paragraph.) There are also *short-range interactions,* such as the *van der Waals attractive forces* in which, for example, atoms brought close enough to one another can induce fluctuating small changes in the distribution of charges in one another, resulting in weak attractions.

Hydrophobic interactions are a noncovalent type important for many biological structures and phenomena (see Section 2.1.2, Chap. 4.1, and the "thermodynamic digression" in Section 4.2.8). Interactions of this sort occur between groups with low affinity for water. The groups themselves bond only quite weakly to one another, but their association is stabilized by interactions among the water molecules that surround them. This reflects important properties of water. As is often true of oxygens and hydrogens linked covalently to one another, the electrons of the covalent bonds in a water molecule are not shared equally among the participating atoms, meaning that the oxygen tends to have some net negative charge and that the hydrogens tend to have net positive charges. In consequence, water molecules associate with charged components such as inorganic ions. For instance, in solution, a positive ion such as Na^+ is surrounded by a loosely held "shell" of water molecules oriented with their somewhat negative oxygens facing the ion. Moreover, water molecules can form hydrogen bonds with a variety of groups such as oxygen and nitrogen atoms in biological molecules, which often are somewhat charged on a basis similar to the oxygen in water. For our concerns here, note that water molecules also form hydrogen bonds with one another, so that the molecules of liquid water are involved in constantly shifting weak linkages with one another. Molecules or ions that can associate with water molecules (*hydrophilic* molecules) can dissolve or disperse in water. Components such as *hydrocarbon* molecules or portions of molecules (these are composed of only hydrogens and carbons) lack the groups needed to associate with water molecules. Consequently, when placed in water, they tend to associate with one another. The direct van der Waals attractions between hydrophobic groups are very weak. Around the associated hydrophobic groups, however, water molecules form their usual loosely structured arrays, linked by hydrogen bonds. The existence of these arrays helps to maintain the associations among the hydrophobic groups. In physical-chemical terms, the clustering of hydrophobic groups together is favored energetically because in this form they cause the least disruption of the interactions among water molecules. Less formally, one can think of the water arrays as a loose cage restricting the movements of the clustered hydrophobic groups.

The terms *polar* and *apolar* (nonpolar), though having a broader significance, are often used to refer to molecules or portions of molecules with hydrophilic and hydrophobic properties.

Molecular association via noncovalent bonds generally involves several or many such bonds. Therefore, the associations can be quite firm and stable even though the individual bonds are easily broken or changed. The formation and disruption of the bonds does not in general require enzymes, but it does require close association of the participating groups so that bonding is favored, for example, among molecules shaped so as to fit well together. The bonds respond to changes in temperature and to alterations in ionic environment such as in pH (recall that pH $= -\log [H^+$ concentration]; a low pH means a high concentration of H^+). Much of the importance of cells and organisms maintaining control over their ionic composition, temperature, and so forth arises from the need to maintain appropriate environments for noncovalent bonds.

1.4.2 Three Dimensions

Macromolecules are characterized by specific linear sequences of units attached by covalent bonds, such as the chains of amino acids (polypeptide chains) in proteins (Figs. I–28 and II–23) and of sugar–phosphate–base units (polynucleotide chains) in nucleic acids (see Fig. II–15). However, most also have characteristic three-dimensional shapes, stabilized chiefly by large numbers of weak bonds formed between different portions of the molecule. The DNA double helix is maintained by hydrogen bonds between the two polynucleotide chains (see Fig. II–15). The polypeptide chain of a given protein is folded into a specific secondary and tertiary structure (Fig. I–28); often, more than one folded chain is associated in the complete protein molecule. For most protein molecules, the three-dimensional pattern (*conformation*) produced is stabilized chiefly by hydrogen bonds and hydrophobic interactions, although certain covalent bonds such as disulfide bonds (see Fig. I–28) can also participate. The folded structures are characteristic for each type of protein and generally are crucial for that protein's biological role. A given enzyme, for example, possesses a specific "active" site, which is a three-dimensional distribution of particular amino acids that permit it to recognize, bind to, and act upon specific groups in the enzyme's substrate. Severe disruption of the enzyme's conformation ("denaturation" of the protein) abolishes its enzymatic activity.

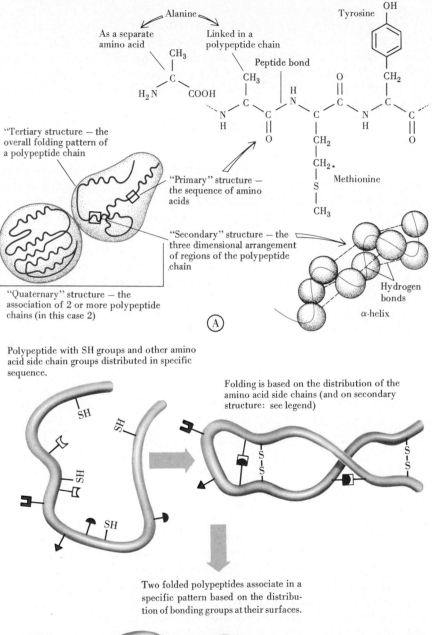

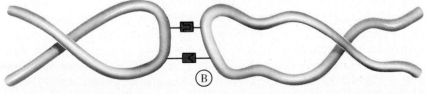

Alanine

As a separate amino acid

Linked in a polypeptide chain

Peptide bond

Tyrosine

Methionine

"Tertiary structure — the overall folding pattern of a polypeptide chain

"Primary" structure — the sequence of amino acids

"Secondary" structure — the three dimensional arrangement of regions of the polypeptide chain

Hydrogen bonds

α-helix

"Quaternary" structure — the association of 2 or more polypeptide chains (in this case 2)

Ⓐ

Polypeptide with SH groups and other amino acid side chain groups distributed in specific sequence.

Folding is based on the distribution of the amino acid side chains (and on secondary structure: see legend)

Two folded polypeptides associate in a specific pattern based on the distribution of bonding groups at their surfaces.

Ⓑ

The functioning of many biologically important proteins depends on their intimate association, through specific binding and interaction sites, with small nonprotein molecules with which they collaborate biochemically. The coenzymes discussed in Chapter 1.3 exemplify this cooperation. For a variety of proteins, coenzymes or comparable molecules are bound tightly to particular regions of the polypeptide chains and thus maintain long-term associations with the proteins. Such molecules are termed *prosthetic groups*; when the prosthetic groups are absent, the proteins are referred to as being in *apoprotein* form. The same type of prosthetic group can associate with different proteins, contributing its particular biochemical capacities (such as electron-carrying or oxygen-binding) to enzymes or other active proteins with diverse specificity and action. Familiar types include the *heme* groups, a class of iron-containing molecules (their structure is based on a *porphyrin ring* system similar to the

---

◀ **Figure I–28** *Proteins: structure and folding.*

A. Protein structure, summarizing terminology. The backbone structure of a protein is a polypeptide chain in which amino acids are linked to one another by the peptide bonds, formed by enzymatic reaction of the COOH and NH_2 groups all amino acids possess (see also Fig. II–23). The remainder of each amino acid protrudes from the backbone as a "side chain" (color); it is the sequence and three-dimensional arrangement of these side chains that confers its specific properties and capacities on the protein. The side chains characterizing three types of amino acid are illustrated here; additional ones are shown in Figures II–23, II–44, III–35 and elsewhere. The *secondary* structure shown is an α-helix, (α = *alpha*), a common form that often characterizes substantial lengths of the polypeptide chain: The spheres in the diagram indicate the approximate positions of successive amino acids; in the actual structure, the backbone is coiled into the helical form indicated by the helical line and the side chains are arranged around the outside of the helix. The regular pattern of hydrogen bonds (which form between the CO group of a given peptide bond and the NH group of a peptide bond at the next coil of the helix, four amino acids further on) makes the α-helical structure a stable one that resists disruption and deformation; thus α-helical regions are important frameworks affecting the overall conformation of the molecule. In the diagram of *tertiary* structure, the line indicates the backbone and the three-dimensional outline, the distribution of the side chains.

B. "Self-assembly" of protein structure. The folding of a protein into a specific three-dimensional conformation occurs spontaneously; the resultant conformation depends primarily on the amino acid sequence. For example, owing to the stability of the hydrogen bonded structure described in part A, portions of the polypeptide chain will spontaneously assume the α-helical conformation unless there are amino acid configurations present that interfere with this or favor some other arrangement. (The amino acid *proline,* for instance, has an unusual ring structure that prevents regions of the protein chains where it is present from twisting into the α-helical arrangement; see Fig. III–35.) Additional folding of the polypeptide is fostered by amino acid side chains that form non-covalent bonds with one another and by SH groups from the amino acid *cysteine* (Fig. II–23) which form S-S (disulfide) bonds that stabilize three-dimensional arrangements. Among other arrangements, amino acid side chains with hydrophobic properties tend to group together and separate α-helical regions of a given polypeptide chain can associate laterally with one another.

one illustrated in Fig. II–63 but containing an iron atom in place of the magnesium atom). A heme is associated with each of the four polypeptide chains of the *hemoglobin* molecule. *Myoglobin,* an oxygen-storing protein of muscle tissue, also utilizes heme as its prosthetic group and hemes are present as well in a variety of other proteins such as the electron transport proteins known as *cytochromes* (Chaps. 2.6 and 2.7), and the enzyme *catalase* (Chap. 2.9).

In subsequent chapters we shall have numerous occasions to refer to the regulation of the activities of enzymes and of other macromolecular properties through the action of regulatory agents of various kinds. For example, inorganic ions such as Ca^{2+} and small molecules such as the cyclic nucleotides affect the behavior of a variety of proteins (see Sections 2.1.4, 2.11.1, and 4.1.1, among others): The agents bind to the responding proteins; by so doing they inhibit or activate enzymatic capacities or alter the protein's abilities to interact with other protein molecules or with other types of macromolecules, such as nucleic acids. Comparable changes result from the action of regulatory enzymes, such as those that add or remove phosphate or methyl (CH_3) groups from specific target proteins (Sections 2.1.4 and 3.2.5) or those enzymes that activate other enzymes by cleaving specific bonds (Section 3.6.1). One of the insights achieved by biochemists and molecular biologists is the realization that many of these regulatory phenomena depend upon alterations in the three-dimensional structure of the responding molecules. Proteins sensitive to calcium or cyclic nucleotides, for example, possess specific sites to which these agents bind; the consequence of binding is a change in the conformation of the protein, leading to altered properties and activities. Effects of this sort, in which interactions with one part of a protein lead to more widespread changes in three-dimensional arrangement and in function, are referred to as *allosteric effects.*

Most of the regulatory phenomena with which we will deal are reversible: The cell can turn processes on and off or accelerate and then decelerate them. At the level of protein molecules, this reflects the reversibility of many of the conformational changes based on rearrangements of noncovalent bonds (or, when covalent addition of regulatory groups such as phosphates is involved, the ability of the cell to both add and remove such groups; Section 2.1.4).

1.4.3 Self-assembly

Certain of the three-dimensional characteristics of macromolecules derive relatively directly from the patterns of covalently linked subunits. The polysaccharides *glycogen* and *cellulose* are both composed of chains of glucose molecules, but the glucoses are linked in somewhat different patterns owing to differences in the enzyme systems that synthesize the two polysaccharides. The chains in glycogen (Fig. I–26) are extensively branched, facilitating its packing compactly into the small granules in which it is stored (Figs. I–3 and I–7).

Those of cellulose are unbranched and relatively flat (see Fig. III–21), permitting cellulose to form elongate thin fibrils of many parallel chains (Section 3.4.1).

The two strands of DNA can be separated by heating and are known to recombine on careful cooling to form again an ordered double helix (see Fig. IV–19). It has also been found that a number of proteins subjected carefully to treatments such as controlled heating, which disrupt their three-dimensional arrangement, will reassume the original ("native") structure when conditions are reversed. This can be observed, for example, with certain enzymes like RNase, which loses its enzymatic activity on heating but regains it on cooling. Too drastic a denaturation often is not reversible, and not all proteins are well-behaved in these respects (Section 4.1.1). However, findings of these sorts suggest that given proper conditions, the three-dimensional structure characteristic of a particular macromolecule can form spontaneously based upon the distribution of groups that associate by noncovalent bonds. Thus, for example, when a chain composed of different types of amino acids, such as a typical protein, is placed unfolded in water it will tend to fold into a form in which those amino acids whose side chains are *hydrophobic* will be inside, away from the water. The *hydrophilic* types of amino acids—those whose side chains have charges (see Fig. II–23) or other properties fostering their association with water—will be outside. This sort of concept is extremely illuminating because it obviates the need to postulate a special separate "structure-specifying" template to impose three-dimensional order on molecules such as proteins. The particular sequence with which amino acids appear in a given type of protein is determined genetically (that is, from nucleic acids; Chap. 2.3); this sequence itself determines the fundamentals of the protein's three-dimensional structure.

Molecules of fat (triglycerides; Section 2.1.2) are hydrophobic; thus in the cell they tend to aggregate together to form compact globular structures from which water is largely excluded (see Fig. II–41). This works out to be an efficient form in which to store metabolic reserves in minimal space, made use of evolutionarily in many cell types, including the specialized *adipose* cells (fat cells; Section 3.5.1) whose cytoplasm fills with such fat deposits.

Such concepts have been extended to explain the formation of larger cell structures made of many different macromolecules. As we shall see, the subunits of such structures will often associate spontaneously to reconstitute the structure: the correct association will *self-assemble* if the concentrations of the participant molecules are in the proper range and their environment is carefully controlled. This provides the basis of the ability of the cell to reproduce large and complex structures. The spontaneity, automatic aspects, and self-determined features of cellular assembly processes should not be overstated. Modifications of self-assembly are needed to explain the formation even of some fairly simple proteins (Chap. 4.1). But the basic thrust—that much of the formation of cellular structure depends on bonding potentials intrinsic to the participating molecules—has proved a powerful conceptual key.

Further Reading

General: Cell Biology and Biochemistry
There are many cell biology and biochemistry texts. Useful ones include the following.
Alberts, B., D. Bray, L. Lewis, M. Raff, K. Roberts, and J. D. Watson. *Molecular Biology of the Cell.* New York: Garland, 1983, 1200 pp. *Intended as a comprehensive introduction. A valuable reference, especially for topics in biochemistry and molecular biology of cells.*
Dyson, R. D. *Cell Biology: A Molecular Approach,* 2nd ed. Boston: Allyn and Bacon, 1978, 616 pp. *Helpful for biochemical background and for the appendix discussing methods.*
Karp, G. *Cell Biology.* New York: McGraw-Hill, 1979, 775 pp. *An extensive, detailed introductory text. References to the original cell biology literature.*
Lehninger, A. L. *Biochemistry,* 2nd ed. New York: Worth, 1975, 1104 pp. *An extensive and detailed treatment.*
Stryer, L. *Biochemistry,* 2nd ed. San Francisco: W. H. Freeman, 1981, 949 pp. *A well-written, up-to-date, general text.*
Wolfe, S. L. *Biology of the Cell,* 2nd ed. Belmont, California: Wadsworth, 1981, 544 pp. *A general text, whose special strengths are in its treatments of the nucleus and of some facets of cell motility. References to the original cell biology literature.*
Also useful:
Altman, P. L., and D. D. Katz. *Cell Biology.* Bethesda: Federation of American Societies for Experimental Biology, 1976, 454 pp. *This collection of data concerning various features of cells is somewhat out of date but is a convenient place to look for things like the chromosome numbers characterizing different cells or the composition of media employed to grow and maintain tissue cultures.*

Historical
Hughes, A. *A History of Cytology.* New York: Abelard Schuman, 1959, 158 pp. *A concise history of the early development of microscopic information about cells.*
Porter, K. R., and A. B. Novikoff. "The Nobel Prize for Physiology or Medicine." *Science* 186:516–520, 1974. *A brief summary of the main contributions of three Nobel laureates, Albert Claude, Christian de Duve, and George E. Palade, who pioneered in the application of electron microscopy and cell fractionation to cell biology. Each of these laureates also published a discussion of his own work and the pertinent background in* Science *during 1975.*
Wilson, E. B. *The Cell in Development and Heredity,* 3rd ed. New York: Macmillan, 1925, 1232 pp. *An extraordinary summary of the cytology of the period and of the preceding years, during which microscopic study of cells began to flower. Problems are discussed in a manner still of importance.*
"Discovery in Cell Biology." *J. Cell Biol.* 91 (3): part 2, Dec. 1981, 306 pp. *The entire issue is a collection of articles on most of the cell's organelles intended to trace the history of study of key problems up to the present. The last two chapters deal with the history of electron microscopy and of cell fractionation.*

On Techniques
Bradbury, S. *The Optical Microscope in Biology.* London: Edward Arnold, 1976, 76 pp. *A brief introduction to light microscopes.*

Gray, P. (ed.). *Encyclopedia of Microscopy and Microtechnique.* New York: Van Nostrand-Reinhold, 1973, 638 pp. *A collection of useful, compact discussions of major microscopic principles and techniques dealing both with electron microscopy and light microscopy. The level and quality of the entries varies, as with most encyclopedias.*

Griffith, O. M. *Techniques of Preparative, Zonal and Continuous Flow Ultracentrifugation,* 2nd ed. Palo Alto: Beckman Instruments Co., 1976, 44 pp. *A brief booklet outlining the theory and practice of ultracentrifugation with references to the background literature.*

Hayat, M. A. (ed.). *Principles and Techniques of Electron Microscopy.* New York: Van Nostrand-Reinhold. *Published at intervals since 1970, this series describes the variety of electron microscopes and preparative procedures in use.*

Methods in Cell Biology. New York: Academic Press. *Published at intervals since 1966, the series describes a vast variety of techniques and also recent advances in research.*

Bonds, Assembly, and Macromolecules

The biochemistry books listed above (especially Stryer's book) and the "assembly" references following Part 4 should be consulted for detailed discussion. A good starting point for reading is the section entitled "Protein Structure and Function" in the Molecules to Living Cells *collection of articles from* Scientific American *(see below), especially the article by D. E. Koshland, "Protein Shape and Molecular Control" from the Oct. 1973 issue of* Scientific American, *included in that collection.*

Articles and Reviews Concerning Recent Advances

Many clearly written articles dealing with progress in particular areas of cell biology appear in the monthly Scientific American. *Some of these are listed in the Further Reading sections following the next three parts of this book. The book* Molecules to Living Cells *(San Francisco: W. H. Freeman, 1980, 340 pp.) is a collection of about 20* Scientific American *articles dealing with topics covered in the text.*

Very current, short, readable discussions of recent research progress are published in the monthly Trends in Biochemical Sciences *and in the weeklies* Nature *and* Science. *Longer summaries of progress are presented in review periodicals; the ones of interest to cell biologists include* Annual Review of Biochemistry, International Reviews of Cytology, Annual Review of Plant Physiology, *and the* Cell Biology Monograph *series published at intervals by Springer-Verlag, Vienna. Chapman and Hall (London) and Halsted Press (New York) have been publishing a series of very short books, many on topics in cell biology, under the series title* Outline Studies in Biology.

The ongoing research literature in cell biology appears in a large variety of journals. Among many others, these include Journal of Cell Biology, Cell, Chromosoma, Proceedings of the National Academy of Sciences of the United States of America, Journal of Biological Chemistry, Journal of Histochemistry and Cytochemistry, Journal of Ultrastructure Research, Experimental Cell Research, Subcellular Biochemistry, Cell Motility, *and* Biochemical Journal. Hospital Practice *publishes many readable articles on cellular changes in disease.*

PART 2
Cell Organelles

This part discusses major features of the organelles of eucaryotic cells; Part 3 will consider the procaryotes and will include additional information on eucaryote organelles as specific cell types are discussed. Almost all of Part 4 is devoted to further consideration of the nucleus.

It is the membranes of eucaryotic cells that divide the cell into compartments distinctive in morphology and metabolism (Section 1.3.4). Membranes surround the cell as a whole and the *membrane-bounded* or *membrane-delimited* organelles: the nucleus, mitochondria, chloroplasts, lysosomes, peroxisomes, and the cavities of the endoplasmic reticulum and Golgi apparatus. The membranes are not simple mechanical barriers. They are highly ordered arrays of molecules, chiefly lipid and proteins, in which enzymes are integrated; they participate in diverse activities. Membranes provide selective barriers that control the amount and nature of substances that can pass between the cell and its environment and between intracellular compartments. Within several of the organelles, notably mitochondria and chloroplasts, enzymes and other molecules responsible for extensive sequences of reactions are built directly into membranes, providing both metabolic efficiency and functional capacities not possible for the same molecules free in solution.

But must a structure be membrane-bounded to qualify as an organelle? As techniques improve, less and less of the cell appears unstructured, but only some of the organization involves membranes. Nucleoli, chromosomes, ribosomes, centrioles, and microtubules all are distinctively structured and have specialized roles in the cell, but no membrane surrounds them. The non-membrane-bounded organelles grade down in size and complexity to protein filaments composed of a few hundred molecules. At the lower end of the size spectrum, the distinctions between organelle and macromolecule become difficult to define and perhaps meaningless. Should a multienzyme complex, in which a few or a few dozen enzymatically active protein molecules are complexed as a functional unit (Section 3.2.4), be called an organelle or a molecular aggregate? Are nucleoli to be considered organelles despite their being contained in other organelles (nuclei)? The decision is a matter of arbitrary definition.

The information that will be reviewed in this part suggests that cytologists and cell biologists may, before long, be able to specify the particular molecular arrangements in each organelle and the detailed chemical reactions that occur within it. The analysis of several organelles has passed from the stage of mere identification of the structure and of the molecular constituents to the point where the normal modes of formation and duplication of the organelles are being studied in detail, and functional parts of organelles are being reconstituted from simpler components in the test tube.

Turnover　Organelles are not static units of unchanging size and shape, with fixed relations to other organelles and with inflexible function. All show movement within most cells. Some can grow and produce duplicates. Most exhibit evidence of *turnover*. For example, if the mitochondria in a hepatocyte are labeled by the incorporation of radioactive amino acids into their proteins, by 5 to 8 days later, half of the label has been lost from the mitochondria. In number and mass, the population of mitochondria has remained the same, but despite this "steady state," half the organelles have "turned over." Does this mean that new mitochondria are continually arising as others are destroyed? Or does each mitochondrion add new material to its structure while simultaneously losing an equivalent amount of older constituents? Or do both processes take place? The answers are not yet known. The amount of a given component present in a cell at a given time depends on the rate of formation and on the rate of degradation. Both formation and degradation must be subject to precise cellular controls, and the relevant mechanisms are being sought.

Turnover is a general phenomenon affecting virtually all components of an organism; the major exception is the DNA of the nucleus. In evolutionary terms turnover may partly reflect the fact that enzymes and other macromolecules are subject to spontaneous changes that permanently damage or inactivate them. For example, some of the changes in conformation that proteins undergo (Section 1.4.2) are irreversible and abolish the molecules' enzyme activities or other functional capabilities. Hence, processes for degradation and replacement of macromolecules seem advantageous for the cell's economy.

Chapter 2.1　The Plasma Membrane

A plasma membrane is present at the surface of all cells. Although some other closely associated structures play important roles, this membrane is the primary barrier that determines what can enter or leave the cell. Its properties strongly influence the formation of multicellular aggregates (such as tissues), as well as the passage of material between closely associated cells. Specialized plasma membrane regions are often present. The surface structures *(junctions)* formed between cells associated in tissues, illustrated in Figures I–2, II–10, III–29, and III–30, are one important example. Figure I–2 also mentions the *microvilli;* these are tubular projections that increase the surface area of hepatocytes and of many other cell types. Some representative examples of specialized membrane structures will be discussed here. Additional examples are considered in Part 3. At this point our chief concern is with some general properties of plasma membrane structure and function.

2.1.1 Early Studies

Permeability; Osmosis Light microscopy cannot reveal the presence of the plasma membrane directly, since the thickness of the membrane is well below the resolving power of the light microscope. However, before the advent of electron microscopy, a great deal of indirect information about the membrane was accumulated from physiological experiments. The findings were interpreted in light of experience with experimental systems in which two solutions containing different concentrations of given substances dissolved in water were separated by artificial membranes that resembled the plasma membrane in being selectively permeable—that is, permitting only some types of molecules to pass. (Some of the membranes were "semipermeable"; these permit water to pass but not other components of solutions.) If a given component can pass readily through the membrane, a net movement or *diffusion* of the component will take place from the solution of higher concentration to that of lower concentration ("down the concentration gradient"), until the concentrations on both sides of the membrane are equal. If the membrane is impermeable or only slightly permeable to the dissolved components, but if, like the plasma membrane, it is moderately permeable to water, then water will pass rapidly into the more concentrated solution. This net movement of water in the direction that tends to equalize the concentrations of dissolved materials on each side of the membrane is referred to as *osmosis*. Overall, two compartments separated by a membrane tend to equilibrate with each other so that (1) the total concentration of dissolved material, in terms of the *number* of separate molecules and inorganic ions per unit volume, will be the same in both; (2) the concentration of each component that can cross the membrane will also be the same in both; and (3) in each compartment, the number of positively charged ions, molecules, or groups will equal the number of negatively charged ones. Figure II–1 includes an example of how such equilibria turn out when several components of different types are involved.

Early experiments on cells utilized eggs, blood cells, or plant tissues, which could be observed conveniently, alive in the microscope. The cells were suspended in solutions containing different molecules dissolved in water, and the volume changes of the cells were measured. The details of behavior of a cell in a given solution will depend on the total concentrations of all molecules dissolved inside the cell and in the surrounding solution, as well as on the relative rate of passage through the membrane of dissolved molecules as compared with water. Under appropriate experimental conditions (Fig. II–1), extensive volume changes can be interpreted in light of the work with artificial membranes as indicating relative impermeability of the plasma membrane to particular dissolved molecules and consequent osmotic influx or efflux of water; cells may shrink drastically or swell to bursting ("lysis"), depending on the concentration and composition of the solutions in which they are suspended. Such experiments, plus more sophisticated ones made possible by the evolution of techniques for direct measurement of concentrations of substances in cells, have made it clear that a plasma membrane exists and exhibits great selectivity about what can pass through. Gases move across the membrane with little difficulty. Water and other small molecules can pass through more readily than can larger molecules with comparable chemical properties. Material that is soluble in lipids generally enters the cell more rapidly than similarly sized substances not soluble in lipid. Some substances pass through by passive diffusion; others, only with the expenditure of energy by the cell.

Comparison of artificial systems and cellular ones soon made it clear that although cells certainly do show osmotic and permeability phenomena expected for membrane-bounded structures, they do not simply respond passively to their surroundings. It is now appreciated that the control of cellular volume and the regulation of the concentrations of dissolved materials ("solutes") in intracellular compartments and in the extracellular fluids of multicellular organisms are essential physiological activities. These activities depend on cellular structure (for example, the plasma membrane must be intact) and can require considerable expenditure of energy. They are responsive to regulatory and coordinating agents such as hormones (Sections 2.1.4 and 3.5.3). Major participating mechanisms include:

1. Control of the distribution of inorganic ions such as K^+, Na^+, and Cl^- across the plasma membrane and in extracellular spaces. The concentrations of these ions are important determinants of the osmotic relations of cells with their surroundings. The ions also have direct impact on cellular enzyme activities (Section 3.10.3) and on other processes. Regulation of these concentrations depends on activities of the membranes of each cell (Sections 2.1.3 and 3.7.3) and on organs, such as the kidney (Section 3.5.3).
2. Control of the macromolecular composition of circulating fluids such as blood and control of the interaction of these fluids with the tissues. This is based on regulation of the production and degradation of osmotically important circulating proteins and other components (Sections 2.8.2 and 3.5.3) and on maintenance of barriers and other factors regulating local circulation, such as the restrictive walls of the blood capillaries of some organs (Section 3.5.2).
3. Expulsion of water from cells, as in some algae, fungi, and protozoa, which possess specialized organelles for the purpose (Section 3.3.1). These organelles are lacking in most multicellular organisms.

4. The presence of rigid cell walls surrounding bacteria and plant cells (Sections 3.2.3 and 3.4.1). The walls prevent osmotically induced pressures—developed in fresh water, for example—from rupturing the cells.

Figure II–1 *Phenomena of osmosis and permeability.*   ▶

A. Osmotic behavior of a simplified cell whose membrane is impermeable to inorganic ions such as Na^+, K^+ or Cl^- or to sugars such as sucrose, but is permeable to small uncharged molecules such as glycerol (Section 2.1.2 and Fig. II–3). The concentration *(c)* of the cell's contents is initially equivalent to 0.2 M NaCl (for mammalian cells the actual value is about 0.15 M) and the cell's initial volume *(V)* is 1 cubic micrometer (μm^3). The effects of varying concentrations of solutes in the medium (boxed values) are illustrated. (The volume of the medium is sufficiently large that the medium is not affected by the movements to be described.) *Important note:* since salts dissociate in solution ($NaCl \rightarrow Na^+ + Cl^-$) a 0.2 M solution of NaCl or KCl is equivalent, *in terms of osmotically effective concentration* (number of separate molecules or ions per unit volume) to a 0.4 M concentration of a non-dissociating molecule such as sucrose or glycerol.

The osmotic flow of water between compartments is governed by the difference in "osmotic pressure" *(π)* between the two: $\pi = (n/V)RT$ where n/V is concentration given in terms of numbers *(n)* or moles of molecules or ions present in the volume *(V)* of the compartment. *T* is temperature and *R* is a constant (the "gas" constant).

(1) In an *isotonic* solution the cell volume is stable. (2) In a solution with solute concentration greater than that in the cell interior (*hypertonic* solution), water leaves the cell, decreasing the cell's volume. Since the same quantity of intracellular solutes is now present in a smaller volume, the effect is an increase in *c* inside the cell. Water continues to leave until the osmotic pressure inside the cell equals that outside. (3) In a solution more dilute than the cell's interior (*hypotonic* solution) the cell increases in volume owing to entry of water. This increase in volume decreases *c* in the cell. If the swelling is too great to be accommodated by the plasma membrane, the cell bursts. (4) In a solution containing a relatively high concentration of a solute of a type that can enter the cell (here, glycerol) there is an initial shrinkage due to osmotic movement of water out of the cell since, in general, water crosses the membrane more rapidly than do solutes. Subsequently, however, the solute enters the cell and since this tends to increase *c* in the cell, water re-enters and the cell swells. The influx of water and subsequent swelling will continue unless other factors intervene (Section 2.1.3) to control solute entry.

B. The term *Donnan equilibrium* as generally used in biology refers to the distribution of solutes and water attained where charged molecules and inorganic ions are present in compartments separated by a membrane that prevents passage of the larger molecules. In panel I a solution added on one side of a membrane contains KCl and negatively charged macromolecules *(mm)* whose (−) charges are accompanied, as they must be in ordinary solutions, by an equivalent number of (+) charges (in this case K^+ ions). Panel II shows the distribution that will result from diffusion of K^+ and Cl^- across the membrane with no net movement of water; bear in mind the requirement that the number of (+) charges in a solution equal that of (−) charges. The important point is that the presence of charged macromolecules in different concentrations on opposite sides of a membrane impermeable to those macromolecules, tends to produce concentration gradients in inorganic ions: In the illustration, the concentration of K^+ in the left compartment is greater than that in the right, while the gradient in Cl^- is in the opposite direction.

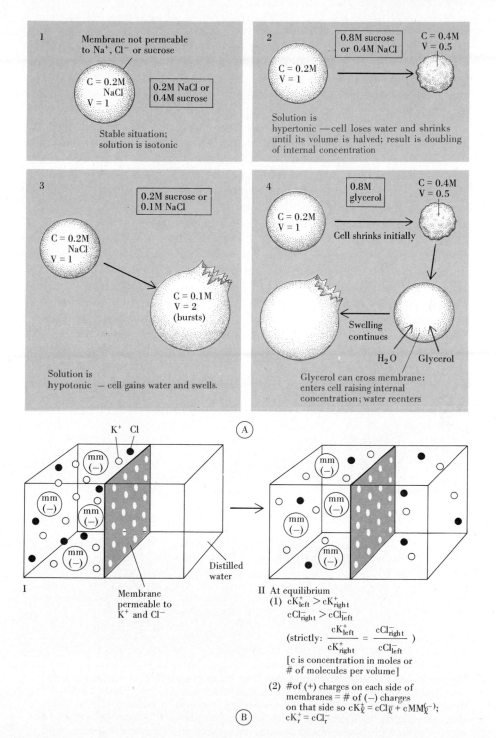

1 Membrane not permeable
to Na⁺, Cl⁻ or sucrose

C = 0.2M
NaCl
V = 1

0.2M NaCl or
0.4M sucrose

Stable situation;
solution is isotonic

2

0.8M sucrose
or 0.4M NaCl

C = 0.4M
V = 0.5

C = 0.2M
V = 1

Solution is
hypertonic —cell loses water and shrinks
until its volume is halved; result is doubling
of internal concentration

3

0.2M sucrose or
0.1M NaCl

C = 0.2M
NaCl
V = 1

C = 0.1M
V = 2
(bursts)

Solution is
hypotonic — cell gains water and swells.

4

0.8M
glycerol

C = 0.4M
V = 0.5

C = 0.2M
V = 1

Cell shrinks initially

Swelling
continues

H_2O Glycerol

Glycerol can cross membrane:
enters cell raising internal
concentration; water reenters

Ⓐ

K⁺   Cl

mm
(−)

mm
(−)

mm
(−)

mm
(−)

mm
(−)

mm
(−)

mm
(−)

mm
(−)

I

Membrane
permeable to
K⁺ and Cl⁻

Distilled
water

II At equilibrium

(1) $cK^+_{left} > cK^+_{right}$

$cCl^-_{right} > cCl^-_{left}$

(strictly: $\dfrac{cK^+_{left}}{cK^+_{right}} = \dfrac{cCl^-_{right}}{cCl^-_{left}}$)

[c is concentration in moles or
# of molecules per volume]

(2) # of (+) charges on each side of
membranes = # of (−) charges
on that side so $cK^+_\ell = cCl^-_\ell + cMM^{(-)}_\ell$;
$cK^+_r = cCl^-_r$

Ⓑ

"Unit Membrane" (Three-layered) Appearance The mechanisms thought to be involved in permeability and osmosis—especially the selectivity of the plasma membrane—plus evidence from other indirect observations suggest that the plasma membrane is a highly organized and structurally complex entity. However, the electron microscope usually shows a relatively simple structure; the architectural basis of complex membrane functions is still not well understood. In conventional electron microscope preparations the plasma membrane appears as two electron-dense layers enclosing a markedly less dense layer, with a total thickness of 7 to 10 nm (Fig. II–2). This three-layered structure has come to be called a *unit membrane*. In the history of its use this term has acquired a variety of connotations related to membrane origins and function, some of which proved misleading. As we and most others now use it, "unit membrane" refers simply to the three-layered microscopic appearance.

The microscope conveys little of the dynamic features of membrane structure—the movement of molecules within the membrane and of membranes within the cell. In these respects and others, it has become increasingly evident that the actual arrangement of molecules in many membranes is more complex than the unit-membrane structure seems to imply. In the first flush of excitement among microscopists at actually seeing membranes, it was widely believed that most cellular membranes were quite similar to one another, and that extensive interconnections existed among different membrane-delimited cell compartments. Diagrams from the early days of electron microscopy often show the plasma membrane continuous with the endoplasmic reticulum (ER)

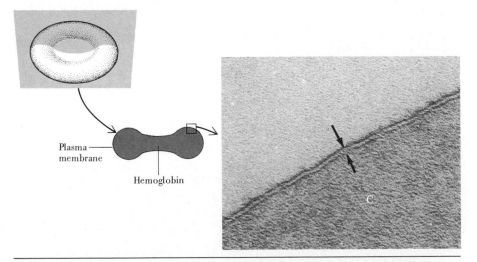

Plasma membrane

Hemoglobin

Figure II–2 *Plasma membrane.* An electron micrograph of a portion of a human red blood cell *(C).* The arrows indicate the three-layered (dense–light–dense) structure of the plasma membrane. × 250,000. (Courtesy of J.D. Robertson.) The sketch indicates the characteristic pinched-in ("biconcave") disc shape of the cell and the plane of section used to provide the view in the micrograph.

and include other membrane continuities among organelles so that the cell appears permeated by membrane-bounded channels continuous with the extracellular space. As work has progressed it has become clear that although membranes of different organelles do have key features in common, they also differ markedly in composition and in important facets of their organization. In addition, on closer examination with improved techniques, some apparent continuities—most notably those between the endoplasmic reticulum and the plasma membrane—turned out to represent a kind of optical artifact. The images of adjacent separate objects sectioned in the proper plane can overlap to create an impression of continuity. True structural continuity between the ER and the plasma membrane is observed only under most unusual circumstances. Some of the other seeming connections among organelles are probably better interpreted in terms of dynamic movement of membranes from one to the other than as static channels. Discussion of this problem continues below.

2.1.2 A Molecular Model of Membranes; Lipids

From a variety of correlated chemical and structural studies, a series of models thought to apply to the plasma membrane and to other cellular membranes has been proposed as conceptual frameworks for further investigation. Although it is evident that the models are oversimplified they have stimulated a great deal of experimentation and discussion.

Lipids   Cellular membranes are rich in lipids. This has been concluded both from direct chemical analysis and from the permeability characteristics of the plasma membrane, which provide indirect evidence for the presence of lipids. Because proteins are present as the other major membrane constituent, the membranes have frequently been referred to as *lipoprotein membranes.* Plasma membranes (and other membranes) of different cell types usually contain from one to four times as much protein as lipid, although a few cellular membranes contain more lipid than protein. This estimate is on a *weight* basis. Since lipid molecules are usually much smaller than protein molecules (lipids have molecular weights on the order of 1000, whereas those of proteins range from 10,000 to over 100,000) there are usually far more lipid than protein *molecules* in membranes. The ratio varies, ranging from 10 to 100 lipid molecules per protein molecule in different membranes. In very rare cases, such as the membranes surrounding the gas-containing vacuoles that influence buoyancy of certain procaryotes (Chap. 3.2.C), the membranous structures found seem to contain much protein but little lipid. These membranes lack the three-layered microscopic structure; they are still poorly understood.

Figure II–3 illustrates three of the major types of lipids found in nature: fats, phospholipids, and steroids. Fats consist of *fatty acids,* long hydrocarbon chains, linked to a "backbone" three carbons long. The backbone is formed from the compound *glycerol.* When all three glycerol carbons are attached to fatty acids, the resultant lipid is known as a *triglyceride,* the major component

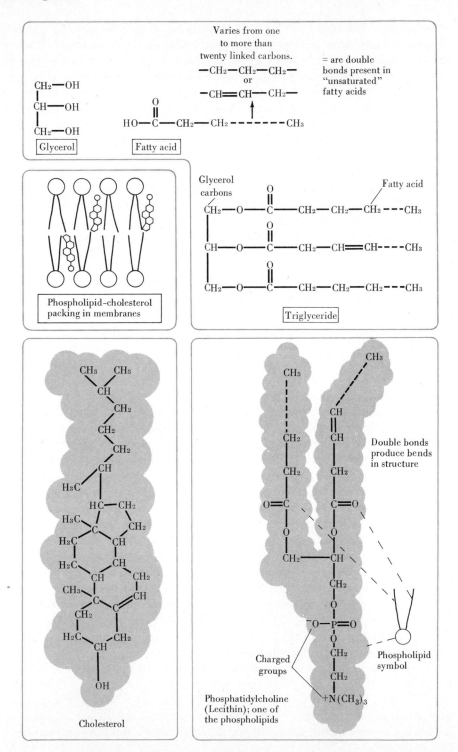

Varies from one
to more than
twenty linked carbons.

$-CH_2-CH_2-CH_2-$
or
$-CH=CH-CH_2-$

= are double
bonds present in
"unsaturated"
fatty acids

CH_2-OH
$CH-OH$
CH_2-OH

Glycerol

$HO-\overset{\overset{O}{\|}}{C}-CH_2-CH_2------CH_3$

Fatty acid

Phospholipid–cholesterol
packing in membranes

Glycerol
carbons

Fatty acid

$CH_2-O-\overset{\overset{O}{\|}}{C}-CH_2-CH_2-CH_2---CH_3$

$CH-O-\overset{\overset{O}{\|}}{C}-CH_2-CH=CH-CH_3$

$CH_2-O-\overset{\overset{O}{\|}}{C}-CH_2-CH_2-CH_2---CH_3$

Triglyceride

Cholesterol

Double bonds
produce bends
in structure

Charged
groups

Phospholipid
symbol

Phosphatidylcholine
(Lecithin); one of
the phospholipids

of most fats. (The term *triacylglyceride* is coming into use instead of triglyceride.) Mono- and diglycerides are constructed, as their names suggest, with one or two fatty acids linked to glycerol. Most phospholipids have a structure similar to that of triglycerides, except that in place of one of the fatty acids they have a more complex chain including phosphate and, often, nitrogen-containing groups. (Different chains in this position characterize different phospholipids. Fig. II–3 illustrates the widespread phospholipid *phosphatidyl choline;* in another common one, *phosphatidyl ethanolamine,* the group attached to the phosphate is —CH_2—CH_2—NH_2; and in *phosphatidyl serine* it is the amino acid *serine* [linked via its —OH group; see Fig. II–44B].)

Steroids have an architecture that is quite different from that of the fats and phospholipids; they are relatives of *cholesterol,* a molecule containing interconnected rings of carbon atoms.

An additional group of membrane lipids, known collectively as *sphingolipids* because they are based on the backbone molecule *sphingosine* (Fig. II–13), include phospholipids similar to the glycerol-based phospholipids. Common membrane *glycolipids* (lipids with sugars attached) also are included in this group (see Fig. II–13).

The Bilayer   The most abundant lipids in most cellular membranes of higher animals and plants are the glycerol-based phospholipids. These are accompanied by varying proportions of cholesterol and sphingolipids; some plasma

---

◄ **Figure II–3** *Structures of lipid molecules.* Common natural fatty acids have 16 to 20 carbons in the region indicated by the dotted lines. The nitrogen-containing compound attached to the phosphate on the third glycerol-derived carbon in lecithin is *choline;* hence this lipid type is now generally referred to as phosphatidyl choline.

The phospholipids of membranes are roughly 25 to 30 Å long; the steroids are somewhat shorter. These two lipid types associate in membranes as indicated at the middle left. The association with cholesterol and the "bends" in their tails where unsaturated fatty acids are present affect the closeness with which the tails of phospholipids pack together within membranes; this influences membrane properties (Section 2.1.3). (After diagrams by J.B. Finean, L. Stryer, S.L. Wolfe and others.)

Symbols: Several conventional symbols for phospholipids are in use for diagrams of lipid arrangements in membranes or other structures. The one in this diagram is the simplest, but ones where the circle has a short stem or the lines are wavy are also frequently employed, as in Figure II–4 and elsewhere in this text. These symbols are equivalent—the differences are matters of taste. The circle indicates the hydrophilic zone, usually called the "polar head" and the pair of lines, the hydrophobic "tails" (the hydrocarbon chains of the two fatty acids). Other lipids in membranes are sometimes schematized by a circle plus a line like the symbol used for a generalized polar lipid in Fig. II–5. For steroids the circle (hydrophilic region) corresponds to the end where the OH group is shown in the present figure and the line to the ring structure, which is predominantly hydrophobic. Another representation of cholesterol, presenting the skeletal structure of the molecule, is utilized in the present figure at the middle left; the small circle corresponds to the OH.

membranes contain almost as many molecules of the two latter lipids as of phospholipids, whereas some intracellular membranes contain very few. All three of these types of lipids are *amphipathic* ("amphiphilic"), meaning that different portions of the molecule show different affinities for water. The "hydrocarbon" regions composed of carbons and hydrogens, such as most of each fatty acid, are hydrophobic (Section 1.4.1). But groups such as the charged phosphate and nitrogen groups of phospholipids, the sugars of glycolipids, and the hydroxyl (OH) group found at one end of the cholesterol molecule (Figs. II–3 and II–44) confer hydrophilic properties on the regions of the molecules in which they are present. The models being discussed are based heavily on the amphipathic nature of most membrane molecules. They suggest that the central region of membranes consists of two layers of lipid molecules (Fig. II–4). The grounds for this suggestion are strong; they include the fact that when membrane lipids are placed in water, the lipids tend to orient spontaneously into such a *bimolecular layer* or *"bilayer"* (Fig. II–5). The thickness of such artificial bilayers, roughly 5 to 7 nm, is similar to that of cellular membranes. This is evident both in electron microscopy of fixed, embedded, and stained preparations and with physical techniques that avoid the need for such

---

Figure II–4 *Membranes.* ▶

    A. Relations of proteins and lipids envisaged in current models thought to apply both to the plasma membrane and to most other biological membranes. Where the proteins extend into the lipid bilayer, close lateral association of hydrophobic portions (Fig. II–13) with lipids may serve to "seal" the protein in, in the sense of eliminating "spaces" through which material could leak. The three-dimensional sketch illustrates the "fluid-mosaic" model, versions of which are almost universally accepted at present. The diagram stresses the concept that protein molecules occur as three-dimensional structures associated in various ways with the lipid bilayer. (After H.A. Davson, J.F. Danielli, S.J. Singer, G. Nicolson, V.T. Marchesi and many others.)

    B. Some possible transport mechanisms discussed in the text. The upper row illustrates carriers that might diffuse across the hydrophobic core of the bilayer, channels selective for particular components on the basis of size and charge, and channels that open in response to evoking agents, permitting ions or other components to pass through. The lower sketches indicate how changes in conformation of integral membrane proteins could conceivably contribute to transport. *I* shows a hypothetical carrier mechanism based on binding of the transported molecule to a specific site on the carrier protein, which then moves to the other side of the membrane. The movement involves changes in the local three-dimensional arrangement of portions of the protein; the protein itself remains in place in the membrane. Models of this type may more realistically apply to movements across a limited span of the membrane rather than to movements across the entire thickness: perhaps there are "barrier regions" within transmembrane proteins that are traversed by such phenomena. *II* illustrates the opening of a channel in response to the binding of the molecule to be transported. Both for *I* and *II*, there could be situations in which the conformational changes are induced by ATP, ions such as Na^+ or other agents, rather than by the molecule to be transported or in concert with the molecule; this could account for some of the phenomena discussed in Sections 2.1.3 and 2.1.4.

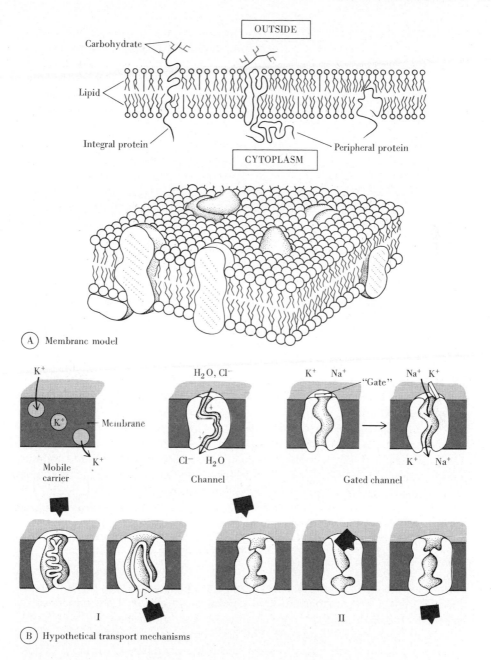

OUTSIDE

Carbohydrate

Lipid

Integral protein

Peripheral protein

CYTOPLASM

(A) Membrane model

K$^+$

K$^+$

Membrane

Mobile
carrier

K$^+$

Mobile carrier

H$_2$O, Cl$^-$

Cl$^-$ H$_2$O

Channel

K$^+$ Na$^+$

"Gate"

Na$^+$ K$^+$

K$^+$ Na$^+$

Gated channel

I

II

(B) Hypothetical transport mechanisms

extensive manipulation of the specimen: For example, X-ray crystallographic procedures for determining molecular ordering have been applied to certain highly ordered natural arrays of membranes such as the spiral stacks in myelin (Fig. III–43) or the piled sacs of photoreceptor cells (Fig. III–44).

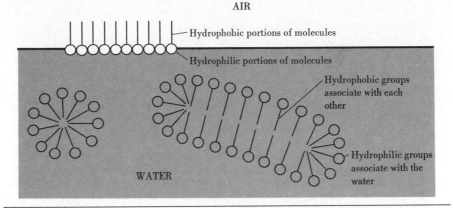

Figure II–5 *Lipid molecules associate spontaneously to form ordered arrays.*
**Lipids with polarized structure (the symbol used here refers to any sort of lipid
with a hydrophilic end and a hydrophobic one) form layers when placed at air-
water interfaces. When immersed in water they aggregate in various arrays in
which the hydrophilic groups face outward and the hydrophobic ones inward.
The diagram shows cross-sections of what, in three dimensions, would be a
sphere or a cylinder (left) and a flat disc, a ribbon or a sheet (right). The precise
form assumed depends on the nature of the lipids, the contents of the surrounding solution, the temperature, and other conditions.**

Crucial evidence favoring a bilayer structure came from studies on the
plasma membrane of red blood cells (see Figs. II–2 and II–7). This plasma
membrane has been studied more intensively than any other because the cells
are convenient to obtain and to study by microscopic, biochemical, and physiological means. Red blood cells generally lack intracellular membrane systems,
and in mammals they even lack the nucleus—the plasma membrane surrounds
a cytoplasm consisting chiefly of one protein, hemoglobin (see Fig. II–2). Thus
more than 50 years ago it proved possible to isolate selectively the lipids of red
blood cell plasma membranes and to demonstrate that the amounts present
were roughly those required to make a layer two molecules thick over the
surface area of the cell.

The view of the membrane as based on a lipid bilayer was especially
attractive to electron microscopists because it agrees well with the "unit membrane" image. Presumably the light central layer of the three would correspond to the hydrophobic "core" of the bilayer; the two dense lines would
represent the hydrophilic surfaces. This correspondence by itself is only conjectural, since there is still much to learn about the chemistry of osmium deposition during fixation, and of heavy metal staining, the phenomena upon which
electron density in most conventional electron microscopic preparations depends (Section 1.2.1). However, when artificial bilayers of lipids are examined
with electron microscopic methods of the sorts used for cells, they show a
three-layered appearance similar to that seen with cellular membranes.

Proteins   Membrane proteins show a more complex distribution than the lipids. Analysis of this distribution has improved as proteins in general have come to be better understood and with application of new techniques to membranes. An early proposal, held to for quite some time, was that proteins are spread out on the two hydrophilic surfaces of the bilayer, sandwiching the lipids. Proteins, however, generally fold into compact three-dimensional arrays (Section 1.4.2). Once it became evident that this is true of many membrane proteins, calculations showed that there are far too few protein molecules to cover the lipid surfaces. Furthermore, the different proteins in plasma membranes, and in other cellular membranes, vary markedly in the ease with which they can be detached: Some are relatively readily solubilized by changing concentrations of salts or pH. Others can be removed only by drastic treatments, such as exposure to detergents that grossly perturb the lipid arrangement and disrupt the membrane. These differences are explained by the proposition that different proteins have different associations with the lipids, as illustrated in Figure II–4. Proteins that are inserted into the hydrophobic core of the membrane have come to be called *integral* or *intrinsic* proteins. Those with more superficial association are referred to as *peripheral* or *extrinsic*. It is assumed that integral proteins generally have extensive hydrophobic regions, corresponding to the portions inserted in the hydrophobic zone of the membrane. For a few membrane proteins, the amino acid sequences have been determined and have been shown to include portions where amino acids with hydrophobic side chains predominate (Fig. II–13C) or regions that can fold into arrays whose exterior surfaces are predominantly occupied by such hydrophobic groups. (The arrays include helices like the ones mentioned in Fig. I–28.) The portions of integral proteins that extend out of the bilayer have amino acid sequences in which hydrophilic groups are relatively abundant.

Peripheral proteins can, in principle, associate either with the the hydrophilic portions of the lipids or with integral membrane proteins. Those that have been studied in detail seem to have the latter sort of association.

Still to be adequately evaluated is the extent to which areas of the lipid bilayer of membranes are truly "naked"—that is, not covered by membrane-associated proteins or carbohydrates. The model in Fig. II–4 of proteins surrounded by a "sea" of lipid implies that very extensive regions of this sort should exist, but there is contention about the degree to which such models overstate the case.   .

As Figure II–4 suggests, the specific proteins and groups (especially carbohydrates) attached to proteins differ at the two membrane surfaces. For plasma membranes, the carbohydrates face the extracellular surface. There also are differences in lipid composition between the cytoplasmic and extracellular faces of the plasma membrane.

The fact that certain proteins actually span the membrane, stretching from one surface to the other, helps to explain some of the functional properties of plasma membranes outlined below. Evidence for the existence of such *transmembrane* or *membrane-spanning* proteins comes from experiments in which membranes are exposed to agents that react with proteins but that are too large to penetrate the membranes. (For example, the enzyme *lactoperoxidase* can be used to catalyze the addition of radioactive iodine atoms to membrane proteins. Alternatively, antibodies that bind to specific portions of particular membrane components can be employed.) The idea is to label the proteins with agents that can interact with them only from one side or the other of the membrane and thus to determine the side(s)—cytoplasmic or extracellular—on which particular proteins or portions of proteins are located. Red blood cells are exposed to these agents either as intact cells or after suitable disruption that gives the agents access to the inner (cytoplasmic) surface of their plasma membranes. It is even possible with red blood cells to make membrane fragments that seal off into inside-out vesicles, whose outer surface is the plasma membrane surface that originally faced the cytoplasm. The differences in the populations of proteins that react under different conditions confirm, for example, that certain proteins are restricted to the cytoplasmic surface of the membrane; these react only in disrupted cells. Since some proteins are accessible from either side of the membrane, they evidently stretch across the width of the structure, exposing specific portions on one side and different portions on the other.

Freeze-fracture and Freeze-etching Microscopy   Present views of membrane organization have also been profoundly influenced by the advent of *freeze-fracture* and *freeze-etch* techniques (Figs. II–6 and II–7). These approaches avoid fixation and embedding (and hence minimize some possibilities for artifact) and provide face views of the membrane, rather than the cross-sectioned views that dominate conventional electron microscope images. (Compare Figs. II–2 and II–7.)

As Figure II–7 illustrates, freeze-fracture preparations of plasma membranes show numerous globular structures, often 6 to 9 nm in diameter, present in the central zone of the membrane—that is, evidently inserted in the hydrophobic core. These *intramembrane particles* ("IMPs" or "intramembranous particles") are big enough to span the membrane width. Most investigators believe they are proteins or at least are due to the presence of proteins. This belief is supported by observations such as those indicating that artificial membranes made only of phospholipids usually show no particles, whereas the particles are present if the protein *rhodopsin* (Section 3.8.1) is mixed with the lipids. Caution is still required, however, in generalizing from such findings, since under exceptional conditions artificial lipid mixtures lacking proteins can be made to show particles. From the relatively large size of many of the naturally occurring intramembrane particles it has been suggested that rather than corresponding to single protein molecules, some represent complexes of lipid with protein and that many correspond to large protein complexes made up of several subunits.

Molecular Mobility   The freeze-fracture procedure has permitted some initial demonstrations of differences in the molecular organization of different por-

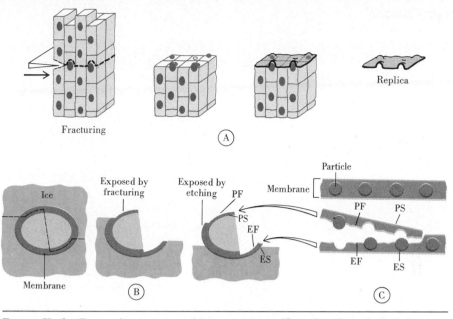

Figure II–6 *Freeze-fracturing and freeze-etching.* (See also Figs. II–7, II–12, II–20, and III–1.) A. Rapidly frozen tissue is fractured, usually with a sharp knife edge. A thin layer of carbon and heavy metal (often platinum) is evaporated onto the exposed surface. This produces a thin, detailed replica of the surface; it is this replica, rather than the surface itself, that is viewed in the electron microscope. The replicas provide face views of the exposed surfaces. Contrast is due to shadowing effects (as in Fig. I–11).

    B. Preparation of a simplified cell, bounded by a plasma membrane. Fracturing tends to expose the interior of a membrane, separating the two lipid layers (see Fig. II–4). After fracturing, the preparation can be "etched" before making the replica: In etching, additional ice is removed by evaporation (sublimation). This can expose the outer surface of the membrane (see Fig. II–7).

    C. Fracturing reveals particles within a membrane. (Based on work by D. Branton, H. Moor and others.) Often, most of the particles remain with one of the two surfaces produced by the fracture plane; the other surface shows what frequently seems to be a complementary pattern of pits or depressions. The two interior *faces* revealed by freeze-fracturing of the plasma membrane are called the *EF* (extracellular face) and *PF* (protoplasmic face), as indicated in the diagram. The actual *surfaces* of the original membrane are designated *ES* and *PS,* as also indicated. Similar conventions are used for intracellular structures: Thus, for example, the *ES* of the endoplasmic reticulum is the surface facing the lumen, and the *PS,* the surface facing the cytoplasm—the surface on which ribosomes attach (see also Figs. II–59 and II–64.) For the plasma membrane, the *PF* is the one that frequently retains more of the particles than does the corresponding *EF,* which has more of the pits. Note that C shows the faces at a given region of membrane, but in general, two different regions with different orientation must be examined to see the two faces, as in B. The *PS* often is difficult to make directly visible.

tions of the plasma membrane of a given cell. We shall see shortly (Section 2.1.5) that highly ordered arrangements of intramembrane particles characterize regions where adjacent cells form special junctional structures with one another. But although more or less stable specialized membrane regions do exist, the membrane is not a rigidly organized structure. Under varying experimental

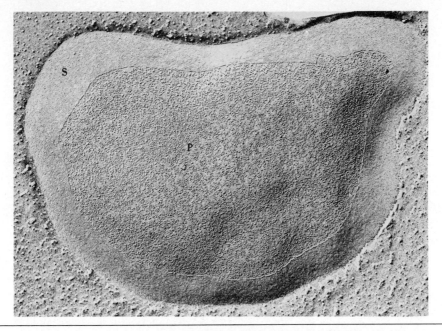

Figure II–7 *Surface of a human red blood cell prepared by the freeze-etch procedure* (Figure II–6). The area at *P* is the interior of the membrane (the *PF*, Figure II–6), exposed by fracturing. Numerous small (7 nm in diameter) particles are visible. The area at *S* is the outer surface of the membrane (the *ES*) exposed by etching. This surface lacks the particles. × 40,000. (From Marchesi, V.T., and T.W. Tillack. *J. Cell Biol.* 45:649, 1970. Copyright Rockefeller University Press.)

conditions, some of the intramembrane particles can show marked changes in distribution, suggesting that large molecular structures can move in the plane of the membrane. This possibility is also suggested by experiments in which cells are induced to fuse with one another (Chapter 3.11); the distribution of certain surface proteins is followed by use of immunohistochemical techniques (Section 1.2.3). The surface proteins of two fused cells rapidly intermingle. Additional studies using cytochemical procedures (Fig. II–8), or biophysical techniques such as fluorescence microscopy, electron spin resonance, and other methods for looking at molecular ordering and interactions, buttress the view that some of the proteins and many of the lipids can move laterally within membranes. In some membranes, lipid molecules, in diffusing, laterally traverse distances of many micrometers per minute. Proteins diffuse more slowly. In one series of experiments, electric fields were shown to induce oriented migration of certain charged proteins in plasma membranes—hinting that they may be circumstances under which biologically generated potentials (Sections 2.1.3 and 3.7.3) have similar effects.

By comparison with their lateral mobility, passive movement of proteins and lipids from one face of a membrane to the other is quite slow; specific

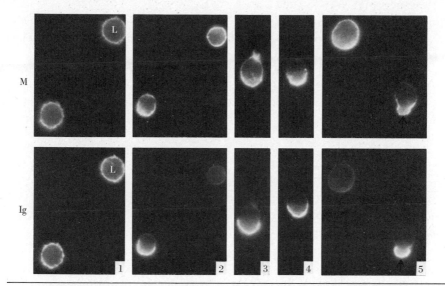

Figure II–8 *"Capping,"* an extreme form of movement of proteins in the plane of the plasma membrane. Lymphocytes (Chapter 3.6.C) isolated from mouse spleen were exposed to antibody molecules that bind specifically to certain proteins (immunoglobulins) of their plasma membrane. This binding induces "capping," the movement of these plasma membrane proteins (plus the inducing antibodies bound to them) to one pole of the cell. In these micrographs this movement has been made visible by virtue of the fact that the antibodies used to induce capping had the fluorescent molecule *fluorescein* attached to them. The lower series of panels (1 to 5) shows cells fixed at different stages in capping, photographed through a fluorescence microscope (see Fig. II–86 for discussion). *L* indicates one of the lymphocytes; note that the fluorescent material moves from a location surrounding the entire perimeter of the cell (panel 1) to a clustered distribution at one cell pole (arrow in panel 5).

The upper series of micrographs shows the same cells as the lower, but the bright fluorescence here is due to immunohistochemical staining (Section 1.2.3 and Fig. II–86) with antibodies specific for *myosin*. (These antibodies were linked to the fluorescent dye *rhodamine* which fluoresces red and thus can be distinguished from the yellow-green fluorescing fluorescein.) The fact that the fluorescence patterns are similar in the two series of micrographs demonstrates the occurrence of a migration of myosin in the cytoplasm underlying the plasma membrane; this migration parallels the capping of the plasma membrane molecules. As will be outlined in Section 2.11.4, it is suspected from such observations that myosin, actin and other cytoplasmic proteins participate in bringing about or controlling capping and act also in other related phenomena of anchoring or moving of plasma membrane molecules.

Technical note: In interpreting these micrographs, remember that they are views of spherical cells, somewhat flattened in preparation, and that each cell has a quite large, unstained nucleus, that occupies most of the cell's volume. The edges of the images of the cells are especially bright because the view here is down through a relatively great expanse of plasma membrane seen almost edge on. × 500 (approximately). (From G.F. Schreiner, K. Fujiwara, T.D. Pollard and E.R. Unanue, *J. Exp. Med.* **145**:1393, 1977. Copyright Rockefeller University Press.)

"transportation" systems are required to "aid" such passage. Proteins, especially, are impeded from crossing the bilayer by their size and hydrophilic zones.

The basic overall conception is that the lipid bilayer of the membrane normally is an appreciably fluid structure and that most proteins are *potentially* able to move laterally within this structure, although the actual movement of many is normally restricted through anchoring to one another or to molecules within or outside the cell. The molecular architecture of the cell surface is affected by, and reciprocally affects, cytoplasmic systems such as the "cytoskeleton" (Sections 2.11.3 and 2.11.4), intercellular junctions (Section 3.5.2) and extracellular material (Section 3.4.1 and Chap. 3.6B). Proteins found at the surfaces of the membrane, such as *spectrin* in the cytoplasm of red blood cells (Section 2.11.2) or *fibronectin,* at the extracellular surface of various cell types (Chap. 3.6B), help to mediate these interactions.

2.1.3 Transport Across the Plasma Membrane

How can molecules in aqueous solution (that is, dissolved in water) outside the cell pass through the plasma membrane and gain access to the aqueous solution inside the cell? One route is directly through the hydrophobic region of the bilayer: The small, somewhat lipid-soluble molecules that diffuse along this route are able to disrupt the relatively weak bonds by which they associate with water (Section 1.4.1), form appropriate interactions with the lipid layers, and then again form associations with water. The permeability properties of these molecules depend on their relative solubilities in water as compared to their solubilities in (affinities for) lipids. The direction of net movement is *down* concentration gradients: from regions of higher concentration to regions of lower concentration. This is always true of "passive transport"—movement across membranes based on diffusion that does not require direct intervention of membrane molecules: Since molecules diffuse in random directions, probabilities dictate that more will move from a region where they are in greater abundance to a region where they are less numerous than will move in the opposite direction.

Water itself enters the cell quite readily. Osmosis (Section 2.1.1) is, in fact, the rapid, passive movement of water down *its* concentration gradient. Where solute concentrations are high, the concentration of water molecules in the same volume of space is correspondingly diminished, and vice versa. The absolute concentration of water in biological solutions is generally very high, but it is the *difference* in concentration of water between different solutions that drives osmosis.

Work with artificial phospholipid bilayers has revealed the somewhat surprising fact that water diffuses through bilayers at rates roughly comparable to its diffusion across natural membranes. The hydrophobic core seems less of a barrier than might be thought. Evidently, even though water is only slightly soluble in the hydrophobic core of a bilayer, the high concentration of water in biological solutions permits appreciable amounts to penetrate.

Interestingly, the permeability of artificial membranes to water varies with the proportions of cholesterol added to the membrane, with the temperature, and with the relative abundance of unsaturated fatty acids in membrane lipids (see Fig. II–3). All of these affect the fluidity of the membrane (increased temperature and unsaturated fatty acids increase it; decreased temperature can actually "freeze" the membrane; cholesterol has more complex effects but can increase rigidity). The more fluid the membrane, the more readily water passes. This suggests that the movement of water across membranes may depend in part on constant small movements of lipid molecules, which may transiently open pathways or local discontinuities in the barrier. Such observations may also help to explain some of the variations in properties of natural membranes with different lipid composition. For example, the plasma membranes of cells that line the bladder of mammals are notably thicker (12 nm) than those of most cells. They show a number of other specialized structural features and are unusually rich in one type of lipid, the *cerebroside* variety of glycolipid (see Fig. II–13). These characteristics are accompanied by an unusually low permeability to water and to solutes of the types concentrated in urine.

Channels Though less than water's, fairly high rates of passive permeation into cells are found for a number of other substances not particularly soluble in lipids. Some features of the passage of water across membranes, and much of the transmembrane movement of inorganic ions like Na^+ and K^+, as well as of small hydrophilic molecules, could be explained if the membrane possessed special *channels* or *pores*. For the most part, these channels are still hypothetical, but estimates of their possible size can be made based upon the threshold sizes above which particular types of components are excluded. The results suggest diameters on the order of 5 to 10 Å, large enough for water molecules (the smallest of the molecules abundant in living systems). They are large enough also for common inorganic ions and small organic molecules *if* these are stripped of many of the water molecules that tend to associate with them in solution (Section 1.4.1; the water molecules make the effective size of the solute larger than its true size, but many of them are only loosely held and might well be lost during passage through a hydrophilic channel). The differences in selectivity of the several types of channels that may coexist in plasma membranes is thought to be based both on size and on other factors, such as charge: A positively charged ion, for example, would probably be precluded from passing through a narrow channel lined by positively charged groups, since like charges repel.

Channels need not be thought of literally as holes, nor as permanent. They may be organizations of membrane molecules or parts of molecules arranged so that a series of appropriate groups (like hydrophilic ones) or of loosely bound water molecules extends from one side of the membrane to another. Passage through the channel may involve the making and breaking of weak, noncovalent bonds between the permeating molecule and membrane components. Some channels may arise transiently, as suggested in the last subsection. Others may have a longer life span but have "gates" that open them and close them in response to agents that affect the cell's permeability (Section 3.7.3 and Fig. II–4).

The fact that they must span the membrane, together with the postulated selectivity and complexity of behavior of many of the suspected channels, suggests they may be comprised in part of proteins or groups of proteins. Protein-poor membranes, such as the plasma membranes that make up the myelin sheath surrounding certain of the axons of nerve cells, seem to carry out few transport or enzymatic activities. (Myelin's primary role is as a passive insulator [Section 3.7.4].) One of the transmembrane proteins isolated from red blood cells is thought to participate in the cell's passive permeability to anions (recall that "anions" are negatively charged ions, and "cations," positively charged ones). Perhaps by forming a selective channel, this protein contributes to the rapid exchanges of Cl^- and HCO_3^-, between the cells and the fluid in which they are suspended, that are important for the transport of CO_2 by the blood. Figure II–9 shows a multiprotein complex that contains a receptor-controlled channel responsible for changes in permeability to cations such as Na^+ and K^+ that take place in muscles and certain other cells responding to nervous stimulation.

Figure II–9 *The acetylcholine receptor.* ▶

A. Electron micrograph showing a face view of a fragment of plasma membrane from the electric organ of the fish *Torpedo*. The membrane is rich in receptors for the neurotransmitter acetylcholine (see C below). Each of the donut-shaped structures in the micrograph is such a receptor; the donuts are about 9.5 nm (95 Å) across. Upon stimulation by binding of acetylcholine molecules released from neurons (nerve cells), these receptors alter the permeability of the plasma membrane to Na^+ and K^+; this is the trigger for the electrical discharge produced by the organ. (The preparation was made by isolating the membrane, "negatively staining" it [Fig. II–53] and printing the electron micrograph in reverse photographic contrast.) × 400,000. (Courtesy of J. Hainfield, A. Karlin, D. Wise and the Brookhaven Scanning Transmission Electron Microscope Facility established by J. Wall.)

B. Cutaway view of one receptor like those in panel A, seen from the side. The receptor is composed of five, folded, polypeptide chains (portions of three are visible in this diagram) probably arranged around the perimeter of a narrow channel as shown. The present consensus is that on binding acetylcholine, the receptor undergoes conformational changes that open the channel, making it wide enough to permit passage of Na^+ and K^+ ions. Upon dissociation of acetylcholine from the receptor the channel closes. (After proposals by A. Karlin.)

C. Three molecules important in cell to cell communication. Acetylcholine and norepinephrine serve as "neurotransmitters"; they are released by neurons (Section 3.7.5). Norepinephrine also serves as a hormone; like epinephrine (adrenaline) it is secreted into the blood stream by cells of the adrenal gland. The diagram also indicates the molecular site of splitting of acetylcholine into acetate and choline by the enzyme acetylcholinesterase, which inactivates acetylcholine as part of a regulatory mechanism governing the effects of neurotransmission (See Section 3.7.5).

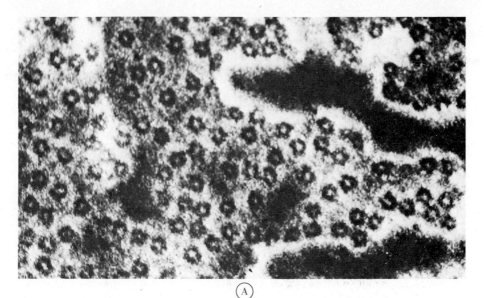

(A)

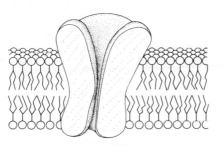

CYTOPLASM

(B)

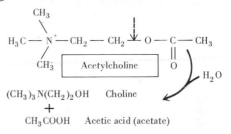

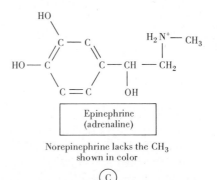

Epinephrine
(adrenaline)

Norepinephrine lacks the CH_3
shown in color

(C)

Carriers; Ionophores   Some membrane components may function as *carriers,* aiding the passage of ions or small molecules. Carriers are presumed to attach to the transported components at one surface of the membrane, to move across the membrane, and then to release the components at the other side. Some carriers might be portions of larger molecules such as proteins, which themselves remain in place but which change their shapes so that the carrier portions move (see Fig. II–4). Others might function by passing the transported material along a series of sites more or less fixed in the membrane (thus resembling specialized channels). Still others might be small enough and so constructed that a complex between the entire carrier and the molecule being transported can cross the lipid bilayer.

An experimental model of this last possibility is provided by the antibiotic *valinomycin.* Inorganic ions usually pass through artificial phospholipid bilayers far more slowly than they do through natural membranes. But when valinomycin is present there is a great increase in the rate at which Na^+ and especially K^+ cross such membranes. Valinomycin has the form of a small circle or donut (molecular weight 1100), with a hydrophobic exterior surrounding a hydrophilic interior in which it can selectively bind ions like K^+, substituting for the "hydration shell" of loosely associated water molecules that normally surrounds the ions in solution. The binding of ions within valinomycin molecules is reversible. Its hydrophobic exterior permits valinomycin to traverse the membrane's lipid layer and thus to act as a carrier. Transport by valinomycin across artificial membranes, or plasma membranes in which it has been inserted experimentally, is down concentration gradients. Ions can both associate with and dissociate from the carrier at each of the two surfaces of the bilayer. Net transport between solutions of different concentration results because the balances between association and dissociation depend on the concentration of ions in the solution to which the carrier molecules are exposed; the higher the concentration, the greater the proportion of carriers that are complexed with ions at a given instant. Thus, the population of valinomycin molecules inserted in a membrane tends to acquire relatively more K^+ ions at the surface of the membrane facing the solution with higher concentration of K^+. This produces a gradient of K^+–valinomycin complexes *within* the membrane—more such complexes are present near one membrane surface than near the other; the complexes tend to diffuse down this gradient, in the membrane, to the opposite membrane surface. Here they encounter the solution with lower K^+ concentration, and on balance, a proportion of the complexes loses K^+ ions to the solution. Since "empty" carriers will now be relatively more abundant at this membrane surface, there will be a gradient favoring diffusion of such carriers "back" to the other membrane surface where they can "reload" with K^+.

Valinomycin is one of a group of molecules known as *ionophores,* molecules that complex with specific inorganic ions and aid their passage across membranes. Other types of ionophores align in membranes, producing channels for particular ions. Most known ionophores are synthetic molecules or special products of microorganisms. They are useful experimental tools for controlling ionic permeabilities of membranes. They also give clues to the types of molecular arrangements that might profitably be sought among the molecules of membranes to explain natural permeability properties.

Facilitated (Mediated) Transport Glucose and a variety of other biologically important molecules enter some animal cells by "facilitated" or "mediated" transport mechanisms in which membrane proteins, occasionally referred to as "permeases," take part. Proteins are thought to be implicated in sugar transport because sugars with only slightly different chemical structure may enter cells at very different rates, even if they are similar in size, solubility properties, and other features that govern passive movement through membranes. This argues for highly selective sites that bind the sugar molecules as part of the transport mechanism. The transport itself could depend either on a carrier or on the transient opening of a channel as a result of the interaction of the sugar and protein.

Active Transport; Na^+ and K^+ Unlike the mechanisms discussed so far, energy-dependent transport—termed *active transport*—can operate against concentration gradients. It can, for example, result in concentrations within the cell that are considerably higher or lower than concentrations outside, for materials that would tend to be at equal concentrations inside and outside without the effects of active transport. Energy production by cells can be interrupted by exposure to low temperature or by treatment with metabolic poisons such as iodoacetic acid (which interrupts glycolysis) or cyanide (which disrupts cellular respiration). Treatments of these sorts strongly affect the movement into or out of the cell of several important types of ions and molecules.

Intracellular sodium ion concentration in most animal cells is kept much lower than extracellular concentration. In "tracer" experiments in which cells are placed in a medium containing radioactive sodium ions, some of the radioactive ions rapidly enter the cell and replace nonradioactive sodium ions that leave. These results indicate that Na^+ continually enters the cell and leaves the cell and that the normal low intracellular concentration is not due to an inability of the ions to cross the membrane. Rather, an active extrusion mechanism (a "pump") balances the entry of sodium ions and maintains the low intracellular level. Potassium ions have the reverse distribution; they are more concentrated within the cell than outside.

An ATPase that splits ATP at a much accelerated rate when Na^+ and K^+ are present in the medium has been found in many plasma membranes; this enzyme is considered to participate in Na^+ transport. Evidently the ATPase is a complex of protein molecules that picks up sodium ions from inside the cell and exchanges them for potassium ions outside; the K^+ ions are then carried to the inside and exchanged for additional Na^+ ions. In the course of these events, a phosphate derived from ATP is temporarily linked to the enzyme, somehow providing energy for the transport. That the ATPase is involved in ion transport is indicated by numerous observations including the following: (1) Both Na^+ and K^+ must be present for rapid activity of the enzyme in splitting ATP; (2) *ouabain,* a drug that inhibits the "pumping" of Na^+ out of the cell, also inhibits the splitting of ATP by the enzyme; (3) when the ATPase is prepared in the form of isolated molecules and then inserted into artificial membranes, the membranes acquire the capacity to transport Na^+ and K^+ actively. How the enzyme is able to act asymmetrically, treating Na^+ and K^+ differently on the two sides of the membrane, is not known. The two ions do, however, differ in size and in the strength of noncovalent association with water and other molecules. The Na^+–K^+ ATPase is made of two types of protein molecules, one of which spans the thickness of the plasma membrane, as might be expected from the pump's function. (In its functioning form the pump is of four molecules, two of each type.) Most current hypotheses assume that its actions depend on conformational changes in the proteins accompanying the addition and loss of the phosphate, which cyclically alter its affinity for Na^+ and for K^+ and accomplish the actual transfer of ions across the membrane (see Fig. II–4).

Electrochemical Gradients; Membrane Potentials   Ion gradients, such as the ones maintained by the Na–K active transport system, are an important form in which the cell "stores" energy, making it available for other functions. The Na^+ gradients described here as well as the gradients of H^+ that bacteria establish across their plasma membranes (Section 3.2.4) are used by the cells to drive the movement of other components such as sugars and amino acids across the plasma membranes. In such "cotransport" phenomena, Na^+, in many cells of animals, or H^+, in bacteria, moves down its concentration gradient into the cell; tied to this is movement of the other component into the cell ("symport," as with sugars or amino acids) or out of the cell ("antiport," as in one type of transport of Ca^{2+} ions; Section 2.1.4). Specific proteins responsible for some of these processes have been tentatively identified and in cases such as the transport of glucose it is believed that both Na^+ and the sugar are moved by the same protein. The Na^+–K^+ ATPase comes into play in pumping Na^+ back out. Hence ATP is required for the overall series of events; if the Na^+ gradient is not maintained, transport of the other component ceases.

The relevance of sodium and potassium ion distributions to nerve transmission will be discussed in Section 3.7.3. For the moment it should be noted that primarily as a result of the gradients of Na^+ and K^+ maintained by active transport, an electrical potential difference, comparable to the differences between poles of a battery, exists across the plasma membrane; the inside of the cell is normally electrically negative in comparison with the outside. Physical-chemical theory indicates that the existence of this potential reflects the asymmetries in ion distribution and the fact that some ions penetrate the membrane more readily than others.

Ionic gradients involving both differences in concentration of ions across membranes and an electrical potential are called *electrochemical gradients*. Both the concentration differences and the electrical potential can be harnessed by the cell to do work. The active transport of sugars and amino acids referred to above is one example. In mitochondria, the energy of an H^+ gradient is used both to drive movement of a variety of ions and molecules and, most importantly, to generate ATP (Chap. 2.6). Hypotheses about the mechanisms by which gradients are used to drive transport and other processes focus on several types of phenomena; these are closely interrelated, and more than one may be significant in a given case. The electrical potential, for example, represents an electric field that can influence the direction of movement of charged components: Negatively charged molecules entering membrane channels, or negatively charged complexes of molecules with their carriers, would, for instance, tend to move toward the more positive side of a membrane. Explanations along somewhat different lines start from the fact that interactions with ions (and also electrical potentials) can alter the conformation of proteins or of other membrane molecules. Thus the ionic gradients may control the opening or closing of membrane channels (see Section 3.7.3) or may lead to changes in conformation that favor *association* of a transported component with the transport system on one side of a membrane and *dissociation* on the other.

2.1.4 Communication Across the Plasma Membrane: Receptors

Cells respond physiologically and biochemically to diverse agents that reach the cells via the extracellular milieu. Regulatory *hormones* (such as insulin, which increases the uptake or storage of carbohydrates in liver, muscle, and fat cells) or the *neurotransmitters* produced by nerve cells, certain of which alter electrical properties of recipient cells, are examples of molecules released by one cell type to affect other cells. Most such agents act via specific *receptors,* molecules or complexes of molecules, that recognize and bind the agents and mediate their effects. Most of the known receptors are proteins; but in some receptor systems, specific lipids seem centrally important. Receptors for a number of agents are intracellular, and the effective agents first cross the plasma membrane. The steroid hormones, being lipid-soluble, can traverse the membrane: In the cytoplasm they bind to specific proteins, and the steroid-receptor complexes enter the nucleus. There, through interactions with the chromatin they influence the cell's genetic expression (Section 4.4.6). In many other cases the agent itself need not enter the cell. Rather, it interacts with receptors at the plasma membrane, and the receptors' responses control subsequent events.

Calcium One important type of response produces alterations in intracellular concentrations of inorganic ions. This can occur directly, through changes in plasma membrane permeability, as is the case with some receptors for neurotransmitters (Fig. II–9 and Section 3.9.1), or less directly as a result of release of ions from intracellular stores (Section 3.10.3). Even small changes in concentrations of ions like calcium ions (Ca^{2+}) can profoundly affect the cell. The "resting levels" of Ca^{2+} in the cytoplasm are quite low (10^{-6} or 10^{-7} M or less); small increases can alter activities of a number of enzymes, affect cell motility (Chap. 2.11), and promote fusions of membranes with one another (Sections 2.5.2 and 3.7.5). For a number of these actions, calcium itself has been found to act via other intermediaries. A principal one of these is the protein *calmodulin.* (Another, tropinin, will be considered in Section 2.11.1). When Ca^{2+} binds to calmodulin, the resultant complex can associate with other cellular proteins whose activities it changes (Sections 1.4.2, 2.11.4, and 3.9.2).

The rise in intracellular calcium concentration resulting, for example, from the interaction of a hormone or neurotransmitter with a cell-surface receptor is usually transitory. The hormone or transmitter dissociates from the receptors, leading to reversal of the permeability changes. Ca^{2+} dissociates from calmodulin, and calmodulin dissociates from its complexes with other proteins. The calcium free in the cytoplasm is rapidly removed by binding to various proteins and by active transport out of the cell or into storage compartments in the ER, mitochondria, or elsewhere (Section 3.9.1). These mechanisms enable the cell to limit the duration and intensity of its responses: It can turn them off once the evoking agents are gone and grade them according to the amounts and timing with which the agents reach the receptors.

An ionophore known as *A23187* can be used to increase plasma membrane permeability to Ca^{2+} (Section 2.1.3), permitting experimental introduction of higher-than-normal levels of the ion into the cell.

Cyclic AMP A somewhat different sequence of events from those involving Ca^{2+} and calmodulin is exemplified by the action of the hormone *epinephrine* (shown in Figure II–9) on hepatocytes. This hormone, released into the blood stream by cells of the adrenal gland, can bind to receptors located on the plasma membranes of hepatocytes and some other cell types. The result is the stimulation of the activities of a cell-surface enzyme system termed *adenyl cyclase.* This enzyme converts ATP into *cyclic AMP* (cyclic adenosine monophosphate; see Fig. II–16); and it acts with such orientation that the binding of hormone *outside* the cell is reflected in a rise of cyclic AMP *inside* the cell. Cyclic AMP can associate with a number of types of proteins, including members of a class of specific enzymes—*kinases.* The kinases are activated by cyclic AMP to catalyze the addition of phosphate groups (phosphorylation) to other proteins. Such phosphorylation can powerfully influence enzyme activities, interactions of proteins with one another, and other properties of the molecules. In the case of epinephrine, kinase-mediated phosphorylation produces *stimulation* of the enzyme *glycogen phosphorylase,* which contributes to the release of glucose molecules from the polymeric glycogen molecules in which they are stored (see Fig. I–26), and *inhibition* of a different enzyme (glycogen synthase) that catalyzes a step in the opposite direction—that is, in the polymerization of glucose into glycogen. The net effect is release of glucose by the hepatocyte, an important aspect of the physiological responses controlled by epinephrine.

Rises in cyclic AMP levels in the cell are counteracted by the enzyme *phosphodiesterase,* which converts cyclic AMP into an inactive, noncyclic form, permitting the cells to "turn off" their responses and eventually return to the prestimulation state.

Cyclic AMP and Ca^{2+} are examples of *second messengers,* agents that function intracellularly to mediate the influence of other agents, notably those that act at the cell surface. Release of many calcium ions or production of many molecules of cyclic AMP may be triggered by a very few molecules of hormone or neurotransmitter, amplifying the effect and facilitating the calling forth of multiple responses in a cell by a given agent. Calcium and cyclic AMP can interact—for example, Ca^{2+} can affect the activities of certain kinases and perhaps also of adenyl cyclases, or of phosphodiesterase—which helps to integrate the impact of the diverse agents that may simultaneously affect a given cell. Such integration, plus the specificity of response, is among the benefits arising from the apparent complexities of the series of events just described. Likewise, different cell types have different batteries of cell-surface receptors and so respond to different, though often overlapping, sets of hormones and transmitters; hence, a particular hormone can influence a selected set of cell types. Cell types also differ in the specificities and activities of the systems that are sensitive to cyclic AMP, Ca^{2+}, or other regulatory components, so that a given agent can evoke different responses in different cells.

Involvement of the Plasma Membrane   The changes in cell permeability produced by a number of types of receptors are based on the reversible opening of channels in the plasma membrane (an example is illustrated in Fig. II–9). Other receptors may function via more complex transport systems that move specific molecules into the cell. In several cases the *mobility* of membrane molecules seems central to the actions of agents binding to receptors. Some physiological effects of insulin, for example, are mimicked by molecules quite different from those of insulin, which engender clustering together of insulin-receptor proteins in the plane of the plasma membrane; antibodies that bind specifically to these receptors have this effect experimentally (and perhaps also in some rare diseases). Movements in the plane of the membrane might permit previously separate membrane molecules to interact with one another, yielding altered activities, changes in permeability, or changes in the relations of membrane components with molecules in the cytoplasm underlying the membrane (Section 2.11.4); these alterations in turn could stimulate or inhibit intracellular enzyme activities or modify interlinking of cell structures.

The plasma membrane itself can be a source of biologically important communication molecules. When suitably stimulated, a number of cell types, including the phagocytes important in the body's defenses (Section 2.8.2), release derivatives of the fatty acid *arachidonic acid.* Arachidonic acid can be converted by cellular enzymes to a variety of *prostaglandins* and *thromboxanes,* which have many effects on the cells that produce them and on other cells; some prostaglandins, for example, influence cyclic AMP levels. Cells' major stores of arachidonic acid are in membranes; presently popular theories suggest that they are released from these stores enzymatically, by *phospholi-*

pases. The drug *aspirin* appears to exert its several effects largely through inhibition of prostaglandin synthesis.

2.1.5 Direct Communication Between Cells: Gap Junctions

Even when closely packed in tissues such as in the liver, most cells remain separated by a space of at least 10 to 20 nm that is occupied by extracellular materials. Plant cells are separated by a distinct wall; they communicate with one another via small cell-to-cell continuities that traverse the wall, the *plasmodesmata* (Section 3.4.3). Only a few animal cells show such cytoplasmic continuities; thus, many cell interactions, even among neighboring cells, depend on release of metabolites or regulatory molecules by one cell into extracellular spaces and their receipt by another. For a variety of animal cell types, however, more direct communication is possible between neighbors. Exchange of materials without release to the extracellular environment can be demonstrated by using fine glass tubes (micropipettes) to inject dyes or fluorescently labeled molecules of various sorts into one cell and then observing their direct entry into adjacent cells. Similarly, microelectrodes can be used to induce and detect ready flow of currents carried by inorganic ions within networks of cells abutting one another. If unmodified, the plasma membranes should prevent such ionic flow, because their lipid layers have the properties of an electrical insulator.

In virtually all of the many cases where such movements of molecules and ions between cells of animals are demonstrable, the electron microscope detects the presence of *gap junctions,* one type of the specialized junctional structures that occur at the interfaces between adjacent animal cells. At local regions, the space between the cells is reduced to a gap of 20 to 40 Å (2 to 4 nm), and the array of structures illustrated in Figure II–10 is observed. The junctions have the appearance of a repeating series of units (sometimes called "connexons") forming roughly cylindrical bridges through which what seem, functionally, to be channels penetrate from one cell to the next. The extracellular space persists around each bridge, so that suitably small material in the extracellular space can percolate between the cells (Fig. II–10). Although gap junctions illustrate well the ability of cells to maintain long-term specialized local arrangements of proteins, they seem to form relatively readily when appropriate cells come in contact (formation in tissue culture takes place in minutes) and can rapidly seal off, isolating cells from one another. Sealing off may be controlled by Ca^{2+} and may be based on a rearrangement of the hexagonal array of particles bordering each unit, closing the channel. Movement of the units themselves within the plane of the membrane is known to alter the overall appearance of the junction reversibly between highly ordered crystal-like arrays and less ordered ones. According to some investigators, the more ordered arrays of the units characterize the closed-off state.

The protein and other membrane molecules responsible for the unique properties of gap junctions are not yet well characterized.

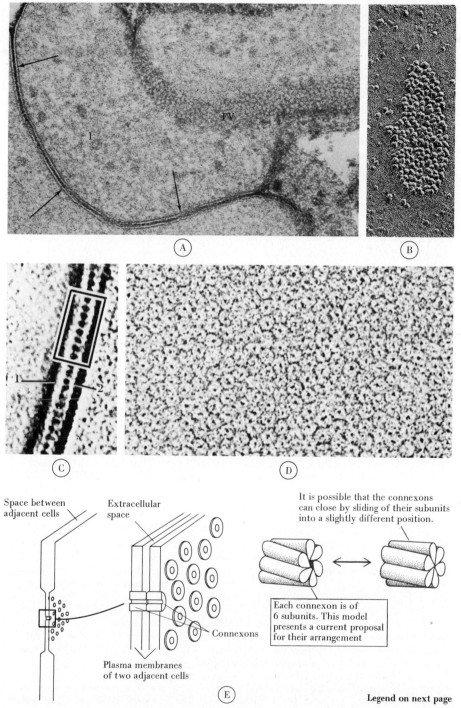

Space between adjacent cells

Extracellular space

It is possible that the connexons can close by sliding of their subunits into a slightly different position.

Connexons

Plasma membranes of two adjacent cells

Each connexon is of 6 subunits. This model presents a current proposal for their arrangement

E

Legend on next page

The diameter of each gap junction channel may be as large as 15 to 20 Å in some cases, permitting molecules of molecular weights somewhat greater than 1000 to pass through (there may be limited selectivity on the basis of charge and other properties). This means that many important metabolic products and regulatory molecules such as cyclic AMP can move from cell to cell via these junctions, although proteins and other macromolecules cannot. The significance of such passage for cell functioning and tissue organization has only begun to be investigated in detail, but there are strong reasons to suspect that gap junctions aid in coordinating metabolism of groups and layers of cells as well as conveying signals of several sorts (Sections 3.7.5, 3.9.2, and 4.4.1) and permitting sharing of nutrients.

When we consider tissues in Section 3.5.2 we shall describe other types of junctions between animal cells that help to maintain the functional architecture of multicellular arrays. These junctions include *desmosomes,* which help to anchor cells to one another, and *tight junctions,* which prevent passage of materials through the extracellular spaces of some tissues.

2.1.6 Bulk Transport: Pinocytosis and Phagocytosis (Endocytosis)

With a few special exceptions (Sections 2.8.4, 3.1.1, and 3.11.3), materials of macromolecular dimensions or larger cannot cross membranes. When they en-

---

Figure II–10 *Gap junctions.*
A. Two rat hepatocytes *(1,2)* associated by gap junctions. Each cell is bordered by a three-layered (unit) plasma membrane seen clearly at the arrows; the two adjacent membranes are separated by a small space or "gap." Electron-dense antimony salts (pyroantinomate) have been introduced into the extracellular space and have penetrated between the cells so that the gap appears dark. The area at *FV* is a gap junction so sectioned that a face view is seen; small particles or subunits are outlined by the tracer. × 130,000. (From D.S. Friend, and N.B. Gilula, J. Cell Biol. 53:758. Copyright Rockefeller University Press.)
B. Freeze-fracture preparation of a gap junction from a rat ovary cell. The membrane shows the characteristic array of closely packed intramembrane particles generally seen at gap junctions. × 100,000. (Courtesy of N.B. Gilula.)
C. and D. Edge-on view (C) and face views (D) of gap junctions isolated from rat hepatocytes and prepared for microscopy by "negative staining" (Fig. II–53) so that the membranes and junctional structures appear light against the dark matrix of stain. In face view the isolated junction is seen as a collection of more-or-less circular units 8 to 9 nm in diameter and packed together in "hexagonal" arrays. Each unit shows a dark center whose interpretation is still uncertain; it is generally presumed to reflect the presence of a channel through the unit, but the assumption that the dark area is simply a hole is probably an oversimplification. In edge view the two membranes *(1,2)* are seen to be connected by fine bridges that traverse the darkly stained extracellular space; the connections are most readily seen in the boxed area. Approximately × 500,000. (From Unwin, P.N.T., and G.Z. Zampighi, Nature **283**:545, 1980.)
E. Model of a gap junction as a group of paired *connexons,* each connexon being composed of 6 subunits. (Based on the Unwin and Zampighi reference in parts C and D and on work by N.B. Gilula, D. Goodenough, W.R. Lowenstein, J.P. Revel, and L. Staehelin.)

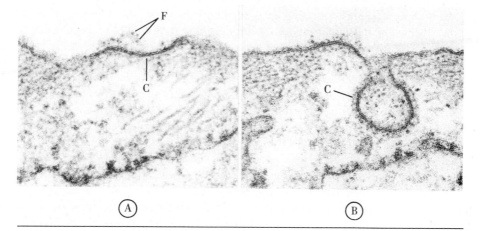

Figure II–11 *Receptor-mediated endocytosis.* Cultured human fibroblasts were exposed to low-density-lipoprotein particles *(LDL)* to which molecules of the iron-containing protein *ferritin* had been attached. The ferritin molecules are visible as small electron-dense (dark) "dots" *(F).* In panel A, LDL particles in the extracellular medium are grouped together because they have bound to LDL receptors that are clustered in a "coated pit"; the pit is the slight indentation of the plasma membrane whose cytoplasmic surface shows a faint band of fuzzy material *(C)* attached to the plasma membrane. (The plasma membrane exhibits the expected three-layered "unit" structure, though this is clearly visible only in places where the plane of section has intersected the membrane perpendicularly. The structure of the coat is seen better in panel B.) Panel B shows the beginning of the next stage in endocytosis: the invagination of the coated pit to form a "coated vesicle"; the coating at *C* appears to be made in part, of short projections from the cytoplasmic surface of the membrane. The vesicle in this micrograph is still attached to the cell surface; at the next stage of endocytosis the attachment will seal off, freeing the vesicle. × 80,000. (From Orci, L., J.-L. Carpentier, A. Perrelet, R.W. Anderson, J.L. Goldstein, and M.S. Brown, *Exp. Cell Res.* **113:** 1–13, 1978. Courtesy of Academic Press and the authors.)

ter cells, such materials generally do so in membrane-delimited compartments formed from the plasma membrane; the pertinent processes are referred to collectively as *endocytosis. Phagocytosis* of bacteria or other large structures, by protozoans or by the specialized phagocytic cells of higher animals, involves the incorporation of these particles into intracellular vacuoles that originate by the folding of the plasma membrane around the material being engulfed. *Pinocytosis* is a similar phenomenon, except that the particles taken up are of molecular or macromolecular dimensions: The membrane folds in to form small vacuoles or vesicles that move into the cell, carrying drops of the medium. This may be seen by the light microscope in living amebae because the vacuoles are quite large. In most cells of multicellular organisms, pinocytic vesicles are too small for light microscopy but can readily be seen by electron microscopy (see Figs. II–11 and III–31).

In many cases the occurrence of pinocytosis can be demonstrated by introducing tracer molecules into the organism or the media in which cells are grown. One such tracer is the iron-containing protein *ferritin,* which is visible in the electron microscope because the large number of iron atoms in each molecule scatter electrons much more effectively than the cell components do (Fig. II–11). Another tracer is the enzyme *peroxidase* isolated from horseradish; many cells will take up this protein by pinocytosis. If cells are then fixed and incubated with the enzyme's substrate, the sites of peroxidase activity will be marked by a dense product (see Fig. III–31). During the incubation, many molecules of the reaction product are formed by the action of each molecule of peroxidase taken up by the cell. Thus, the method is very sensitive; sites of only a few peroxidase molecules can be detected.

A "linguistic" note: Particles taken up by endocytosis are referred to by some investigators as *endocytized* and by others as *endocytosed;* similarly, different writers use *phagocytized* or *phagocytosed* and *pinocytized* or *pinocytosed.* We shall utilize the first member of each pair, but this is a matter of habit, not of special conviction.

"Coated" Vesicles   An interesting feature of many pinocytosis vesicles is the fuzzy "coating" they show on the membrane surface facing the cytoplasm. The coat is particularly evident during vesicle formation (Figs. II–11 and II–12). Similar coating is sometimes seen on the surface of vesicles not involved in pinocytosis (for example, some vesicles produced by membrane systems associated with the Golgi apparatus). The coat is made of an ordered array of protein molecules (Fig. II–12); the most prominent protein is *clathrin.* The possible functions of the coat are intriguing: Current hypotheses suggest it aids in the molecular rearrangements required for pinching off of small vesicles from larger membrane structures; or perhaps it takes part in controlling the composition of the membrane destined for the vesicle (see below), participates in interactions of the vesicle with the cytoplasmic filament systems responsible for intracellular movement (Section 2.11.4), or is responsible for responses to regulatory proteins such as calmodulin. The coat is not attached permanently to the membrane; it often disappears early in the life history of a vesicle, probably by disassembling into its subunits, which can then be reused.

Selectivity; Receptor-mediated Endocytosis   Although endocytosis can involve uptake of small "gulps" of the medium surrounding the cell and thus has nonselective features, both pinocytosis and phagocytosis often are selective to a quite significant degree. Rates of endocytosis respond to environmental signals such as the presence of particular inducing molecules in the medium. The major phagocytic cells of higher animals—the macrophages and polymorphonuclear leukocytes (Section 2.8.2 and Chap. 3.6C)—have specific surface receptors by which they recognize materials to be engulfed. In fact, the actual process of phagocytosis by these cells may depend upon a sort of "crawling" of their plasma membrane around the surface of the particle to be taken up,

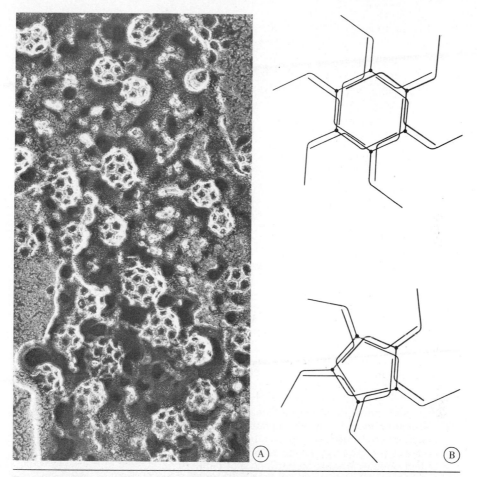

Figure II–12 *Coated vesicles*. Panel A shows coated vesicles, 0.1 μm in diameter (about half the diameter of the one in Figure II–11B) as isolated from brain. The micrograph demonstrates the coating surrounding the vesicles to be a cage-like "geodesic-dome" structure of hexagonal and pentagonal elements. The preparative procedures used (*"freeze-deep etching"*; a type of "freeze-drying") preserve delicate cell structures and provide striking and useful three-dimensional views (see also Fig. II–91). The material is frozen in liquid helium at −269° C (and, if appropriate, opened by fracturing as in Fig. II–6 or by other methods); much or all of the ice is then evaporated away (sublimed) in a high vacuum. A shadowed replica of the exposed surfaces, similar to the replicas made in freeze-fracturing, is examined in the microscope. (From J. Heuser, *Trends Biochem. Sci.* 6:64, 1981.)

Panel B diagrams how the proposed units of which vesicle coating is thought to be made could be fit together to give the hexagonal and pentagonal elements seen in panel A. The unit of structure has a geometric form referred to as a "triskelion" (one is shown in color); the triskelions are composed of clathrin and other proteins. (After work by Branton, D., R.A. Crowther, B.M.F. Pearse, E. Ungewickell, and others. See Cold Spring Harbor Symposium XLVI:703, 1982.)

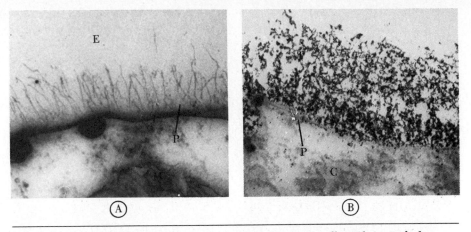

Figure II–13 *The extracellular surface of a cell is generally rich in carbohy-drate-containing materials, sometimes organized as a more or less elaborate layer.* A. Portion of the surface of an ameba. Numerous fine filaments are seen attached to the extracellular surface of the plasma membrane *(P)*. Like the "fuzz" in Figure III–28D, this is an exceptionally prominent form of the cell coat. The tip of a mitochondrion is present at M. E indicates the space outside the cell. × 55,000. B. Portion of a similar cell that had been exposed to thorium dioxide particles in suspension. The electron-dense metal-containing particles are adsorbed to the filaments *(T)*. Such adsorption is the first step in the uptake of material through pinocytosis. *C* indicates the cell cytoplasm. × 35,000. (Cour-tesy of G. Pappas and P. Brandt.)

      C. A glycoprotein, and D. glycolipids, of the plasma membrane. The car-bohydrate rich portions of these molecules are usually located at the extracellular surface of the membrane. (Also illustrated are the structure of certain membrane phospholipids related to the glycolipids.) Note that, like the lipids in Figure II–2, the glycolipids have a hydrophilic "head" (the carbohydrate-containing region) and two hydrophobic "tails" (a fatty acid and the sphingosine chain).

      Glycophorin illustrates well the presence of hydrophilic groups on portions of membrane proteins that extend into the cytoplasm or the extracellular space and of hydrophobic groups on portions that come to be located in the membrane interior. The cytoplasmic portion of glycophorin is about 40 amino acids long, the portion inserted in the membrane about 20 amino acids—mostly hydropho-bic—long and the extracellular portion, 70 amino acids long. More complex ar-rangements than that of glycophorin characterize some other membrane pro-teins; for instance there can be many hydrophilic groups in the portions within the membrane interior, especially if the protein's three-dimensional conformation is such that the surface of the protein where it faces the lipids (see Fig. II–3) is rich in hydrophobic groups—the hydrophilic groups can be concentrated in the "interior" of the three-dimensional structure, for example, lining channels like the ones in Figures II–3 and II–9. Protein and lipids with carbohydrates attached are sometimes referred to generically as *glycoconjugates*. Figures II–44, II–45 and III–35 present additional details.

---

with the membrane adhering to the surface through its receptors recognizing and binding to characteristic molecular configurations (such as the antibodies that attach to foreign particles; Chap. 3.6C). Protozoa also show selectivity in what they phagocytize.

      Pinocytosis was once thought of primarily in terms of the cell's "drinking" of a droplet of fluid. Such "fluid-phase" incorporation does occur, but it is now

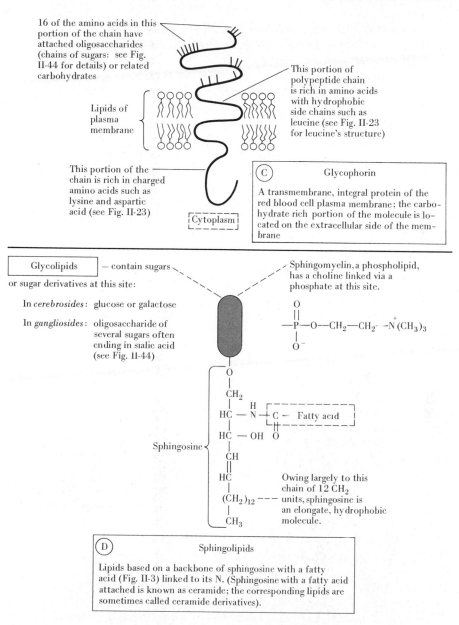

16 of the amino acids in this portion of the chain have attached oligosaccharides (chains of sugars: see Fig. II-44 for details) or related carbohydrates

This portion of polypeptide chain is rich in amino acids with hydrophobic side chains such as leucine (see Fig. II-23 for leucine's structure)

Lipids of plasma membrane

This portion of the chain is rich in charged amino acids such as lysine and aspartic acid (see Fig. II-23)

Cytoplasm

C Glycophorin

A transmembrane, integral protein of the red blood cell plasma membrane; the carbohydrate rich portion of the molecule is located on the extracellular side of the membrane

Glycolipids — contain sugars or sugar derivatives at this site:

In *cerebrosides*: glucose or galactose

In *gangliosides*: oligosaccharide of several sugars often ending in sialic acid (see Fig. II-44)

Sphingomyelin, a phospholipid, has a choline linked via a phosphate at this site.

$$-P-O-CH_2-CH_2- \overset{+}{N}(CH_3)_3$$

Sphingosine

O
|
CH_2
|
HC — N ⊢ C — Fatty acid
| ‖
HC — OH O
|
CH
‖
HC
|
(CH_2)_{12}
|
CH_3

Owing largely to this chain of 12 CH_2 units, sphingosine is an elongate, hydrophobic molecule.

D Sphingolipids

Lipids based on a backbone of sphingosine with a fatty acid (Fig. II-3) linked to its N. (Sphingosine with a fatty acid attached is known as ceramide; the corresponding lipids are sometimes called ceramide derivatives).

recognized that uptake of water is not the primary significance of pinocytosis, and that in many cases pinocytosis is preceded by adsorption of specific molecules to the plasma membrane so that the vesicle formed has a much higher concentration of particular molecules than that present in the bulk medium surrounding the cell. Figures II–11, II–13B, and III–49 illustrate examples of such "adsorptive endocytosis." Adsorption, unlike the actual internalization of endocytized materials, does not require cellular energy expenditure; it depends on noncovalent bonding and takes place even at temperatures as low as 4° C, which effectively block endocytosis proper.

Adsorption preceding pinocytosis often is to specific receptors, which may trigger endocytosis. An interesting feature frequently observed in receptor-mediated endocytosis (see Fig. II–11) is the clustering of specific receptors at sites of vesicle formation, so that the vesicle that forms is selectively enriched in the receptors and in their "ligands" (the molecules that bind to the receptors). This can occur before the binding of the ligands, or as a response to the binding. More extreme redistributions prior to endocytosis sometimes occur on a number of cell types whose receptors, when occupied by appropriate molecules, can move selectively to patchlike collections on the cell surface, or even to one pole of the cell (*capping;* Fig. II–8). For these longer-range redistributions at least, cellular energy is required (as it is also for endocytosis proper.) In these cases and some of those where redistribution is less extensive, endocytosis-related lateral redistribution of membrane molecules has tentatively been ascribed to actions of the coating of coated vesicles and to intervention of cytoplasmic anchoring and motility-generating systems (Sections 2.11.3 and 2.11.4). Redistributions are most often induced experimentally by agents possessing more than one binding site so that they can bind simultaneously to more than one receptor molecule ("bivalent or multivalent ligands," of which antibodies are an example; Chapter 3.6C); this tends to bring the receptors together or to hold them near one another once they have approached each other through lateral diffusion in the membrane.

For certain cases of receptor-mediated endocytosis, including that of low-density lipoprotein (LDL; Fig. II–11), the proposal is now being tested that some cell types continually internalize the receptors (and reinsert them in the cell surface; Section 2.8.4) whether or not they are occupied by the corresponding "ligands" (the molecules to be taken up). The notion is of a sort of "conveyor" belt kept running whether or not it is loaded. The studies on which this proposal is based were done on cells grown in culture, and the proposal applies only to a few receptors and ligands; more must be learned about the controls of endocytosis before the concept is accepted, particularly for cells "in situ" (in their normal locales with an organism).

Cellular Roles The materials taken up inside endocytic vacuoles and vesicles generally are degraded by enzymes to which they are exposed when the vacuoles and vesicles fuse with lysosomes (Chap. 2.8). For unicellular organisms, this permits feeding via endocytosis. In multicellular organisms, the digestive system has largely taken over the process of degrading nutrients to useable form, but even so, most cells retain capacities to form endocytic vesicles. Several functions have evolved for these capacities: (1) uptake of particular molecules too large to penetrate the plasma membrane (see Fig. II–70 and Section 3.10.2); (2) destruction of foreign organisms, tissue debris, and dead cells (Sections 2.8.2, 3.8.1, and 3.10.2 and Chap. 3.6C); (3) removal of macromolecules from extracellular spaces as part of the processes by which the composition of extracellular fluids is controlled (Section 2.8.2); (4) control of the composition of the plasma membrane by selective removal of specific molecules such as receptors; and (5) removal of ligands from the cell surface. (With "active" molecules such as hormones, these last two processes might contribute to turning off the cell's responses.) In addition, in certain cell types, endocytosis provides a means by which macromolecules can move from one extracellular surface of a cell to another, entering the cell in vesicles and leaving it when the vesicles fuse with another region of the plasma membrane. This is one of the ways in which specific molecules can pass across sheets of cells (Section 3.5.2 and Chap. 3.6C).

When phagocytosis and pinocytosis proceed rapidly, large amounts of plasma membrane are removed from the surface and taken into the cell as the membrane of vesicles and vacuoles. How the surface is replenished is still being investigated. Part of the answer probably lies in a process of *recycling* whereby vesicles release their *contents* to intracellular compartments by fusing with them, after which the vesicle *membrane* can return to the cell surface and reinsert in it (Sections 2.5.4 and 2.8.4).

2.1.7 Carbohydrates and Cell Surfaces; Polysaccharides, Glycoproteins, Glycolipids

Polysaccharides as a class are polymers of carbohydrate molecules. We have already mentioned glycogen, the major intracellular storage form of sugar in animals (see Fig. I–26). Another form of polymerized glucose, *starch,* is used in sugar storage by plants. We also have mentioned cellulose, a principal component of the walls that surround plant cells (see Fig. III–20). In animals as well, extracellular spaces often are occupied by polysaccharides specific for different tissue types, especially a category containing sulfates or other acidic groups and known as *glycosaminoglycans* or *mucopolysaccharides* (Chap. 3.6B). Glycosaminoglycans generally occur in company with proteins, to which they may be linked in large *proteoglycan* or *mucoprotein* arrays. Such proteoglycans are major components of the abundant extracellular materials that occupy most of the volume and are responsible for the major characteristics of certain tissues such as skeletal and connective tissues. Cartilage, for example, contains much of the polysaccharide *chondroitin sulfate,* along with proteins such as collagen. The polysaccharide *hyaluronic acid* and collagen and other proteins are major components of joints and the lower layer of the skin. The elaborate jelly coats surrounding many egg cells (Chap. 3.10) also are rich in polysaccharides and proteins.

Carbohydrates also are linked to many of the cell's own proteins and to some of its lipids (see Fig. II–13). Glycoproteins have carbohydrates attached directly to certain of their amino acids, either as "saccharides" (sugars and sugar derivatives) or as short chains, *oligosaccharides,* of several linked sugars or derivatives. This is true of most of the proteins secreted by cells and of many—probably most—proteins of the cell surface. The *cerebrosides* and *gangliosides* (glycosphingolipids) are the principal cellular glycolipids.

Later sections of the book will consider these various extracellular and cell surface molecules in more detail (see especially Sections 2.5.2 and 3.4.1, Chap. 3.6B, and Figs. II–44, III–21, and III–37). Our immediate concern is with the fact that the carbohydrate side chains of plasma membrane molecules help to determine significant features of the cell surface, even though such groups usually make up less than 10 percent (by weight) of the membrane. They are hydrophilic, and most if not all are attached to the portions of proteins and glycolipids located at the extracellular surface of the membrane. This location makes them potential participants in the interactions of cells with one another and with extracellular materials. The specificity of such interactions can determine how cells associate with one another (Sections 3.11.2 and 4.4.1) and even whether cells survive in certain situations. For instance, the substances determining blood group type include specific glycolipids of the red blood cell membrane. When blood is transfused between incorrectly matched donors and recipients, these groups trigger responses of the immune system, leading to death of the cells. Other membrane molecules generate similar events with tissue transplants (Chap. 3.6C).

The ability of several types of cell-surface receptors to bind the molecules that normally associate with them can be markedly altered by digestion of the cell surface with enzymes that remove carbohydrates. Digestion with the enzyme *neuraminidase,* which removes the negatively charged sialic acids often present at the ends of the carbohydrate chains (Fig. II–44), affects the overall surface charge; this can, for instance, reduce the binding of charged molecules at sites from which they normally are taken up into endocytic vesicles. Carbohydrate chains of membrane molecules may also influence the interaction of the molecules with one another and, along with the polar (hydrophilic; Section 1.4.2) amino acids of the proteins, help to limit the extent to which membrane molecules can be pulled back into the lipid bilayer. (Thus, perhaps the carbohydrates help membrane proteins to provide a sort of anchor for cytoplasmic structures attached to the membrane [Sections 2.11.4 and 3.5.2].)

The terms *glycocalyx* and *cell coat* have come into use as names for the carbohydrate-rich layer associated directly with the cell surface. Sometimes this forms a distinctive looking "fuzz" (Figs. II–13 and III–28) at "free" cell surfaces (those not abutting other cells or structures). Less obvious layers of glycocalyceal materials, however, are present also in the spaces between closely packed tissues, where they may contribute to "cementing" the cells together. Where the cell's own coat ends and extracellular materials begin is often not obvious, and in fact, much attention is now being given to specific molecules such as the protein *fibronectin* (Chap. 3.6B), one of whose probable roles is to attach the cell surface to extracellular materials. Still to be adequately explored is the extent to which molecules located at the cell's outer surface help to determine the architecture of deeper zones of the plasma membrane by, for example, helping to hold membrane molecules in place.

Lectins   The members of a diverse and widespread group of proteins known collectively as *lectins* can each recognize and bind to a specific type of carbohydrate or configuration of two or more linked carbohydrates, including those present on glycoproteins and glycolipids. They are presumed to function in cellular recognition and defense phenomena of several sorts (see the discussion of nitrogen fixation in Chap. 3.4), but for the most part, their natural roles are unknown. Nonetheless, lectins are valuable as experimental tools. A number evoke dramatic changes in cells (Section 4.2.3). Several have proved useful for determining the cell-surface and intracellular locations of macromolecules bearing particular sets of carbohydrates: Lectins can be tagged with fluorescent or electron-dense markers and employed more or less as antibodies are used in immunohistochemistry (Section 1.2.3).

2.1.8 Experimental Introduction of Materials into Living Cells

Understanding of the cell's receptors and other molecules in the plasma membrane permits experimental manipulation of cellular physiological and biochemical mechanisms through use of natural agents, such as hormones, or artificial ones constructed to evoke particular responses of the sorts discussed in the past few sections. Other strategies are more direct: Cell parts can be recombined in new combinations through microsurgical techniques such as the transplantation of nuclei from one cell to another (see Fig. II–14 and Section 4.4.3). Some macromolecules that seem at first glance too large to cross the plasma membrane (or the membrane bounding an endocytic vesicle) in order to enter the cytoplasm can nonetheless do so to a limited extent, especially when cells are exposed to them in high concentrations and under suitable conditions: This is true, for example, of DNA (Sections 3.2.6 and 3.11.3). Other molecules that do not cross the plasma membrane can be injected directly into the cytoplasm (Section 2.3.5). "Piggy-back" and cooperative systems, in which desired materials are introduced by incorporating them into another agent able to penetrate into the cell, have been devised. Viruses have been used as such "vectors" for nucleic acids experimentally incorporated into the viral chromosome. Artificial lipid vesicles called *liposomes* are being used to carry diverse types of molecules; these vesicles can be made with varying numbers of membranes surrounding interior compartments containing solutions of proteins, inorganic ions, drugs, or other molecules. Liposomes bounded by a single membrane can be made to fuse with the plasma membrane of a cell, emptying their contents directly into the cytoplasm. Other types, with multiple membranes, enter the cell via endocytosis and reach the lysosomes, where the molecules in the liposomes' interiors can be released when the membranes are degraded.

At present most such approaches are applicable to only certain cell types, and they are useful chiefly as experimental tools. For the future they offer promise of practical use (Chap. 5.2).

Chapter 2.2 The Nucleus

Cells without nuclei have very limited futures. The only common animal cell type without a nucleus, the mammalian red blood cell, lives only a few months; aside from its role in oxygen transport, it is extremely restricted in its metabolic activities. Egg cells from which nuclei have been experimentally removed may divide for a while, but the products of division never differentiate into specialized cell types, and eventually they die. Fragments without a nucleus, cut from such large unicellular organisms as amebae or the alga *Acetabularia* (Fig. II–14), survive temporarily, but ultimately they die unless nuclei from other cells are transplanted into them.

If a fragment containing the nucleus is cut from an *Acetabularia* of one species, characterized by a given morphology, the fragment will regenerate a whole cell of that species. This regenerative ability permits experiments of the type illustrated in Figure II–14, in which nuclei of one species are combined with cytoplasms from different species. The conclusion drawn from such experiments is that the nucleus produces material that enters the cytoplasm and participates in the control of cell growth and cell morphology. The crucial finding is that the morphology of the regenerated cells eventually becomes like that of the species from which the nucleus is taken. In the hybrid fragments with the nucleus from one species and most of the cytoplasm from the other species, old cytoplasmic material persists for a while and may influence cell form. Eventually, however, this is depleted and replaced by newly produced material from the nucleus.

Thus the nucleus is essential to long-term continuation of metabolism and to the ability of cells to alter significantly their structure and function (as in differentiation). In large part, this reflects the primary role of the nucleus in producing the RNA required for protein synthesis. When cells change, their new functions and structures require new proteins. Even cells that are constant in metabolism and structure show continual replacement (*turnover*) of macromolecules and probably of organelles, including portions of the cytoplasmic protein-synthesizing machinery.

Our concern in this chapter and the next will be with general aspects of nuclear structure and function, and with the roles of the nucleus in providing the nucleic acid molecules essential for protein synthesis. Parts 4 and 5 will deal with the details of chromosome architecture, cell division, and facets of the regulation of genetic activity. The "nucleoids" of procaryotes are described in Sections 3.2.1 and 3.2.3.

2.2.1 DNA: A Brief Review

The contribution of the nucleus to total cell mass ranges, in different cell types, from approximately 5 percent (some muscle cells) to 5 to 10 percent in hepatocytes, and up to 50 percent or more in cells (lymphocytes) of the thymus

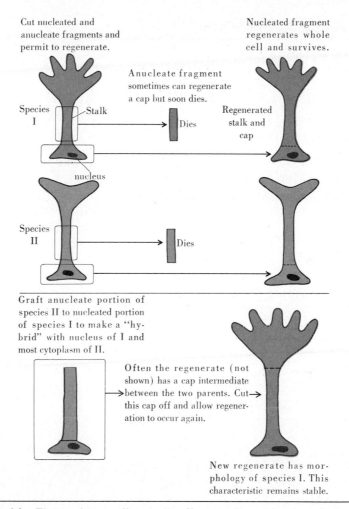

Cut nucleated and
anucleate fragments and
permit to regenerate.

Nucleated fragment
regenerates whole
cell and survives.

Anucleate fragment
sometimes can regenerate
a cap but soon dies.

Species
I

Stalk

Dies

Regenerated
stalk and
cap

nucleus

Species
II

Dies

Graft anucleate portion of
species II to nucleated portion
of species I to make a "hybrid" with nucleus of I and
most cytoplasm of II.

Often the regenerate (not
shown) has a cap intermediate
between the two parents. Cut
this cap off and allow regeneration to occur again.

New regenerate has mor-
phology of species I. This
characteristic remains stable.

Figure II–14 *The nucleus as "controlling" agent.* **Experiments with the large single-celled alga** *Acetabularia.* **The intermediate caps sometimes formed in the initial "hybrids" reflect the fact that some time elapses before "old" material in the cytoplasm is depleted and replaced by new material from the nucleus. The cytoplasmic materials responsible include RNA. (After work of Gibor, Hammerling, and others.)**

gland and in other rapidly dividing cells such as plant root-tip cells and cancer cells. Variations in the ratio of nuclear volume to cell volumes fall within the same range, and there are substantial differences in the dimensions and total mass of the nuclei of different cell types in a given organism.

The approximate composition of rat hepatocyte nuclei is 10 to 15 percent DNA, 80 percent protein, 5 percent RNA, and 3 percent lipid (based on dry

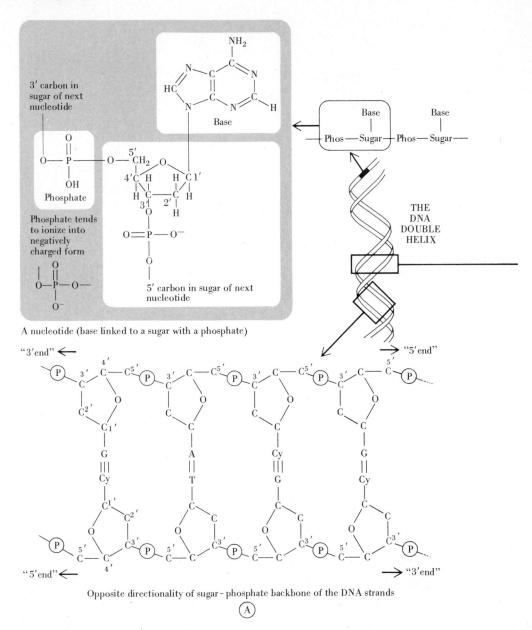

A nucleotide (base linked to a sugar with a phosphate)

Opposite directionality of sugar-phosphate backbone of the DNA strands

(A)

weight; water is a primary constituent of all cells and organelles and accounts for 70 percent of hepatocyte weight). An early cytochemical finding was the presence in nuclei of virtually all the DNA of cells; only later was the presence of small but significant amounts of DNA in some cytoplasmic organelles recognized. The DNA content of cells of different organisms varies greatly. How-

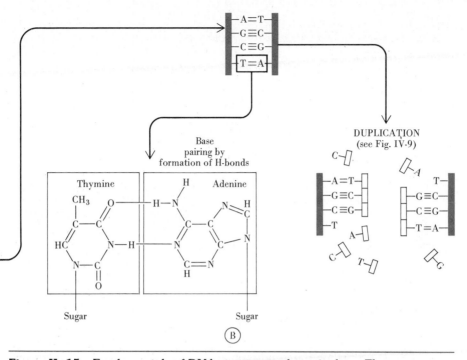

Figure II–15 *Fundamentals of DNA structure and terminology.* The conventional numbering of the 5 carbon atoms of the DNA sugar *deoxyribose* is indicated. The sugar's name derives from its lack of an OH group on the 2′ carbon, in contrast to ribose (see Fig. II–16B). The bonds, via phosphates, linking 3′ and 5′ carbons of successive sugars are called *phosphodiester* bonds. The base shown in panel A is *adenine*.

ever, quantitative cytochemical studies (Section 1.2.3) of cells in multicellular animals and plants established that despite the wide variations in nuclear size and total mass among cell types, the DNA content of most nuclei in a given organism is twice ($2 \times$) that of the sperm or egg cell nuclei (or, for some cells, a multiple of $2 \times$ based on continued doubling, such as $4 \times$, $8 \times$, or $16 \times$). This is as expected for the genetic material: Each gamete contributes an equal amount of DNA to the zygote nucleus, which by duplicating gives rise to all the nuclei in the cells of the organism. These considerations will be discussed further in Part 4.

DNA molecules are composed of two strands coiled together in a double helix (Fig. II–15). Each strand is a chain of nucleotides (a *polynucleotide* chain); all DNA nucleotides consist of a 5-carbon sugar (deoxyribose) with a phosphate group attached at one end and a nitrogen-containing ring compound (known as a *base*) at the other. Four different bases are present on different nucleotides: adenine (A), guanine (G), cytosine (C), and thymine (T). The two strands of nucleotides are aligned according to simple pairing rules:

Every A on one strand pairs with a T on the other; *every* G pairs with a C. The paired bases are held together by molecular interactions, chiefly *hydrogen bonds,* in which a hydrogen atom that is part of one base is also attracted to the second base (Section 1.4.1). The strands are referred to as *complementary,* since the sequence of bases on one strand exactly dictates that of the other. It is the base sequences that carry the hereditary information. The replication of DNA results in the duplication of base sequences (see Fig. II–15); this process underlies the role of DNA in the transmission of hereditary information, as will be discussed in Part 4.

Terminology The phosphate groups linking nucleotides attach the sugars by connecting the 3′ carbon of one to the 5′ carbon of the next (Fig. II–15). This defines a directionality of each polynucleotide chain that is important in nucleic acid replication and transcription (Sections 2.2.4 and 4.2.4). Note that in this sense the two strands of the DNA double helix run in opposite directions (Fig. II–15). Note also that the linkages of the sugars (deoxyriboses) located at the two ends of a strand differ. At one end, the 3′ carbon of the sugar will not be linked to another nucleotide; at the other, the 5′ carbon will similarly be "free." For convenience in referring to the directionality of polynucleotide chains, these ends are called the 3′ and 5′ ends, respectively. (As will be seen [Sections 2.6.3, 3.1.3, and 3.2.3], some DNA molecules are circular and may not have ends, except during their replication, but the differences in directionality of the two strands are universal.)

Another terminological convention that is now coming into wide use is the short-hand reference to DNA sizes in terms of "kilobases" (kb)—meaning 1000 nucleotides (nucleotide pairs for the double helix). A DNA double helix 3000 nucleotide pairs long (3 kb) is 1 μm in length. Corresponding molecular weights can be calculated from the fact that one nucleotide has a molecular weight of about 330, and one pair of nucleotides, 660. Section 3.1.3 summarizes additional information of this sort in terms of the genetic content of different nucleic acids.

A *gene* can be defined roughly as a sequence of DNA nucleotides that codes for an RNA molecule; many of the RNAs, in turn, specify the sequences of amino acids in protein molecules. There are, however, regulatory DNA sequences that do not fit this definition but often are referred to as genes or as parts of genes (Sections 3.2.5 and 4.4.6).

2.2.2 RNA: A Brief Review

The production of RNA is a most important aspect of DNA function. Autoradiographic evidence outlined earlier (Section 1.2.4) indicates that most of the RNA of cells is produced in the nucleus in close association with DNA. DNA and RNA are similar in that both are polynucleotide chains, but RNA differs from DNA in several ways: The RNA 5-carbon sugar is *ribose* instead of deoxyribose, and in RNA, the base *uracil* replaces the thymine of DNA (see Fig.

II–16). Most RNAs are single-stranded, except for those in certain viruses. However, in many RNA molecules, the nucleotide sequences are such that the molecule, by folding back on itself, can establish stretches of double-stranded association that follow the same base-pairing "rules" (A with U, G with C) as those for the DNA double helix. Such specific *secondary* and *tertiary* structure is well exemplified by the transfer RNAs (tRNAs; Section 2.3.1) in which the single chain is folded into a very compact looped structure as shown in Figure II–16. This arrangement orients different portions of the molecule with respect to one another and to the molecules with which tRNAs interact (Section 2.3.2) in ways essential for function (see Fig. II–23). Similar conclusions are now being drawn for other RNA molecules (Chap. 5.1). The folding of an RNA depends on the base sequence, imparted to it by DNA. This then reflects another kind of information—"assembly" information (Section 1.4.2)—additional to that required for synthesis of proteins, both types of information being carried by base sequences.

The actual specification of the base sequences of RNA molecules by DNA follows from base pairing. RNA nucleotides align along a DNA strand according to the base-pairing rules, and a transcription enzyme, RNA *polymerase* (Section 2.2.4), joins the nucleotides to form an RNA strand that is complementary to a DNA strand. By this mechanism, DNA acts as a *template* for RNA synthesis. The genetic code (Section 2.3.1) is such that for any given stretch of DNA, only one of the two strands of the double helix makes "sense" in terms of the specific information carried (that is, only one strand can specify a protein or a *functional* RNA). Nonetheless, under some test-tube conditions, either strand can be transcribed into an RNA molecule even though the RNA molecule from the second DNA strand will not correspond to a cellular product. This is one line of evidence indicating that there must be precise cellular control devices governing transcription—in this case, determining which DNA strand is to be copied. Sections 3.2.5 and 4.4.6 will consider some of these controls, which include the presence of DNA nucleotide sequences, recognized by RNA polymerases, that favor correct association of the enzymes with DNA.

The complementary relationship of RNA with DNA is the basis of an extremely valuable technique called *molecular hybridization,* or *RNA–DNA hybridization,* as shown in Figure II–17. The ability of separated *DNA* strands to reassociate can be exploited in somewhat similar ways (Section 4.2.6 and Fig. IV–19). The observation that test-tube "hybrids" can form between particular DNAs and RNAs is taken as good evidence that the two contain many complementary base sequences. Several different classes of RNA are present in cells; they differ in size and in role, as will be seen in the next chapter. With hybridization, it can be shown that all major types of RNA are complementary with DNA from the nucleus (with the general exception of the nucleic acids associated with mitochondria and plastids). This strongly supports the conclusion, drawn initially from autoradiographic and cytochemical studies, that all are synthesized in the nucleus.

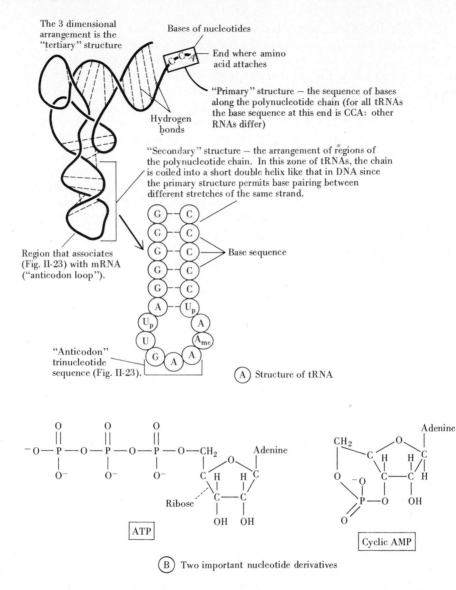

The 3 dimensional arrangement is the "tertiary" structure

Bases of nucleotides

End where amino acid attaches

"Primary" structure — the sequence of bases along the polynucleotide chain (for all tRNAs the base sequence at this end is CCA: other RNAs differ)

Hydrogen bonds

"Secondary" structure — the arrangement of regions of the polynucleotide chain. In this zone of tRNAs, the chain is coiled into a short double helix like that in DNA since the primary structure permits base pairing between different stretches of the same strand.

Region that associates (Fig. II-23) with mRNA ("anticodon loop").

Base sequence

"Anticodon" trinucleotide sequence (Fig. II-23).

(A) Structure of tRNA

ATP

Ribose

Adenine

Cyclic AMP

Adenine

(B) Two important nucleotide derivatives

Figure II–16 *RNA; nucleotide derivatives.*

A. Structure of a tRNA molecule (one from the bacterium *E. coli,* that is specific for the amino acid *phenylalanine*). At the upper left, the overall conformation of the polynucleotide chain is illustrated by the heavy line. Teh specific sequence of bases is shown for the nucleotides at the end of the molecule where amino acids attach, and (below) for the region where the tRNA binds to mRNAs during protein synthesis. The diagram outlines terminology useful in referring to RNAs and illustrates the presence of base-paired regions in RNA molecules (see also Fig. V–1). A_{me} and U_p (methyl adenine and pseudouridine) are modified bases present in tRNAs and certain other RNAs.

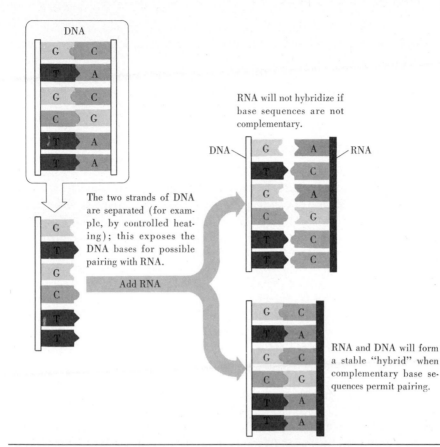

DNA

G — C
T — A
G — C
C — G
T — A
T — A

RNA will not hybridize if
base sequences are not
complementary.

DNA — G   A — RNA
T   C
G   A
C   G
T   C
T   C

The two strands of DNA
are separated (for exam-
ple, by controlled heat-
ing); this exposes the
DNA bases for possible
pairing with RNA.

G
T
G
C
T
T

Add RNA

RNA and DNA will form
a stable "hybrid" when
complementary base se-
quences permit pairing.

G — C
T — A
G — C
C — G
T — A
T — A

Figure II–17 *Hybridization of nucleic acids.* Purified nucleic acid molecules are
mixed in the test tube. The formation of hybrids can be detected in several ways.
For example, hybrids between radioactive single-stranded RNA and nonradioactive DNA will be double-stranded (one RNA strand, one DNA strand) and radioactive. Their radioactivity distinguishes them from the original double-stranded
DNA, while their double-strandedness produces differences (for example, in density) that permit them to be separated (by centrifugation, absorption to special
gels or filters, or other methods) from the original single-stranded RNA molecules.

     B. Two biologically important derivatives of adenine nucleotides. Both com-
pounds illustrated in this diagram are built of the same components as are RNA's
adenine nucleotides—the sugar is *ribose,* which differs from the deoxyribose of
DNA (Fig. II–15) in possessing an OH group on its 2' carbon, as illustrated. The
compounds are shown here to emphasize the fact that cells use the same building blocks in a variety of ways. ATP, for example, gives rise to the adenine nucleotides of RNAs (Section 2.2.4), serves as an energy source for an enormous
diversity of processes and reactions, may produce allosteric effects upon binding
to some proteins (see Sections 1.4.2, 2.10.5 and 2.11.1), and is the metabolic
precursor of cyclic AMP (Section 2.1.4) and of other molecules.

Once the RNA is made in the nucleus, much of it moves rapidly into the cytoplasm (see Figs. I–23 and I–24). In rat hepatocytes, less than 10 percent of the total cellular RNA is present in the nucleus at any given moment; much of this is in the process of synthesis or the maturation and processing events that follow transcription. Some of the RNA molecules are in transit to the cytoplasm.

2.2.3 Proteins of the Nucleus

In the nucleus, DNA is complexed with proteins, among which are the *histones.* The phosphate groups of the DNA are negatively charged. In cells stained with "basic dyes" (those dyes that are positively charged), the regions containing DNA (or RNA) stain because their negatively charged phosphate groups attract the oppositely charged dye molecules. Such regions are referred to as *basophilic.* The histones are *basic* proteins; that is, they have a high content of the basic amino acids lysine (see Fig. II–23), arginine, and histidine, which contain amino groups (NH_2) or derivatives. These amino groups can acquire an H^+ and assume a positively charged (NH_3^+) form: The proportion in this form increases with increasing H^+ concentration (decreasing pH), and at the pHs near neutrality (pH 7) that prevail in the cell, most of the amino groups in proteins occur as NH_3^+. Histones are *acidophilic;* their positively charged amino group derivatives attract acid (negatively charged) dyes. In nuclei, however, histones do not stain strongly with acid dyes unless the DNA is first removed by chemical or enzymatic means. This finding led to the early recognition that the histones and DNA in the nucleus are closely associated. The initial interpretation was simply that the DNA phosphate groups form ionic bonds (Section 1.4.1) with the histone amino groups, minimizing attraction of the histones for acid dye molecules. As more information about histone–DNA interactions accumulated (Section 4.2.5) this explanation was modified—some of the effect apparently is due to the facts that the histones are located within coils of DNA and that the dye molecules are impeded in penetrating to the proteins, both by physical barriers resulting from this arrangement and by the negatively charged DNA phosphates, which tend to repel the dyes.

DNA and histones are usually present in the nucleus in approximately equal amounts. There are only a few types of histones, typically five, present in a nucleus. These differ somewhat from one another in size and amino acid composition (Section 4.2.5), but they vary little from cell type to cell type. Much more variation both in kind and in amount is observed for the nonhistone proteins. The amounts differ among cell types, but generally there is more nonhistone protein per nucleus than histones. The nonhistone proteins are a heterogeneous collection about which surprisingly little is known in detail. Some are enzymes, and some may be very closely associated with the histones and DNA; evidence that these associations are important in the control of nuclear function will be developed in Section 4.4.6. Other proteins, associated with RNA molecules, include components of ribonucleoprotein particles such as the ribosomes that the nucleus "exports" to the cytoplasm, as well as proteins that may participate in the maturation of RNAs (Section 2.3.5 and Fig. IV–40). Of the proteins found in the nucleus that are not associated with nucleic acids, there are a number whose major known roles are cytoplasmic. Efforts are now being made to determine whether their presence simply reflects the fact that the barrier between the nucleus and the cytoplasm is not very restrictive, so that cytoplasmic proteins can "leak" in, or whether the proteins have distinctive nuclear functions. For example, *actin* and sometimes *myosin* are reported to be present in nuclei of at least some cell types. These proteins have increasingly well-understood cytoplasmic functions in generating motion and in maintaining cell shape and structure (Chaps. 2.10 and 2.11). Controversial proposals hold that they have comparable functions in the movement of nuclear components in cell division, in passage of material between nucleus and cytoplasm, or in the maintenance and alteration of nuclear structure.

Rough correlations can be drawn between the relative amounts of nonhistone proteins and RNA in nuclei and the activities of the nuclei, particularly those related to the support of protein synthesis in the cytoplasm. Sperm cell nuclei, for example, which are quite inactive, are small and compact (see Fig. III–48); they have virtually no RNA or nonbasic protein (DNA may account for one third to one half of the nuclear dry weight). In contrast, egg cell nuclei are relatively enormous and are rich in RNA and nonhistones, especially during the stages of maturation when the egg is growing rapidly and storing yolk and other materials. Given what is known now about the roles of macromolecules, such correlations are hardly startling, but historically they were important in pointing investigators in the directions that led to present knowledge of nuclear activities. The relative constancy of histone amounts and other properties generated the conviction that functions of this type of protein include ones that vary relatively little from cell to cell, such as roles in the basic structural organization of the nucleus. This now seems very likely (Section 4.2.5). The diversity and variation of the nonhistones suggests that they may contribute to the origin of cell differences (Section 4.4.6), although certain ones, especially some of the enzymes, have functions that are essentially universal.

2.2.4 Enzymes; RNA Polymerases

Quite a number of enzymatic activities and metabolic pathways have been attributed to nuclei, often on questionable grounds: Many, detected in what were thought to be pure preparations of isolated nuclei, turned out to reflect contamination of the cell fractions with cytoplasm. Others, such as the accumulation of glycogen occasionally seen, may characterize particular cell types under particular circumstances but appear not to be widespread. It is, however, generally agreed that most nuclei share a number of enzymatic activities, most notably those related to the production of DNA and RNA. Whether nuclei synthesize any of their own proteins or receive all of them from the cytoplasm has long been debated. Several laboratories have reported that preparations from some cell types containing isolated nuclei carefully freed of obvious cytoplasm can synthesize some proteins. However, as described in Section 2.2.6, the outer surface of the nucleus is itself part of the cytoplasm's endoplasmic reticulum and shares with it the organelles of protein synthesis. Whether proteins can be made in the interior of the nucleus must therefore still be regarded as an open question. At present, it seems unlikely, for most cell types, that extensive manufacture of proteins takes place in this compartment.

DNA replication will be considered in Section 4.2.4. As for transcription of RNA; the nuclei of eucaryotic cells have three distinct forms of polymerases, each responsible for producing different types of RNA molecules: Polymerase I, present in nucleoli, transcribes the large RNA molecules destined to be part of the structure of the ribosomes. Polymerase II specializes in producing messenger RNAs (mRNAs), the molecules that actually carry the information with which proteins are generated. Polymerase III transcribes certain of the smaller RNAs, including the tRNAs. This diversity is thought to facilitate cellular control over the amounts and types of RNAs to be produced (Section 4.4.6); experimentally, it is convenient because the different enzymes have somewhat different sensitivities to molecules that can inhibit RNA production. The drug α-amanitin, for instance, can be used to inhibit polymerase II selectively, thus permitting experimenters to shut off mRNA production differentially. In contrast, actinomycin D, as generally used, inhibits production of all types of RNA (see Fig. II–28).

Being polynucleotide chains, RNA molecules have the same sort of directionality as that of DNA strands. All three RNA polymerases copy the DNA sequentially, starting with the DNA nucleotide specifying the 5′ end of the forming RNA (the "5′ end of the gene" in current jargon) and adding nucleotides one at a time until the 3′ end of the RNA is reached (5′ to 3′ [5′ → 3′] synthesis direction). The forming RNA and the DNA strand that specifies it have the same sort of opposite directionality that characterizes the two strands of the DNA double helix. The nucleotides used for RNA synthesis are in triphosphate form. One is ATP; the others (GTP, UTP, CTP) have similar high-energy content. Energy for the polymerization is released by the splitting off of a pair of phosphates as the nucleotides are linked together.

The RNA polymerases elongate the RNA chains they produce, at rates of approximately 1000 to 3000 nucleotides per minute.

Figures II–31 and IV–40 illustrate transcriptions of RNAs from particularly active genes.

2.2.5 Structure

Chromatin As seen in the light microscope, the nucleus is bounded by a nuclear membrane. Within the nucleus there is a heterogeneous collection of fibrils and of dense areas that include *euchromatin, heterochromatin,* and *nucleoli* (Figs. II–18 and II–19; see also Figs. I–14 and I–15). "Chromatin" refers to the DNA-containing structures in the nucleus and corresponds to the cell's

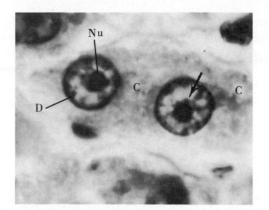

Figure II–18 Nuclear structure by conventional light microscopy. Two hepatocytes in a section of rat liver prepared by widely used routine procedures: fixation
in formaldehyde, embedding in paraffin, preparation of sections 5 to 10 μm
thick, and staining with hematoxylin (colors chromatin blue) and eosin (gives the
cytoplasm [C] a contrasting pink color). Prominent nucleoli *(Nu)* are present
within the nuclei. Much of the chromatin is dispersed and difficult to see in a
black and white photograph; the rest is in accumulations close to the nucleous
(arrow) or the nuclear envelope *(D)*. Although fixation can clump dispersed chromatin, producing "artifacts" (Section 1.2.3), dense accumulations such as these
are often seen in living cells. The term *heterochromatin* is used for the condensed
form of chromatin (see also Fig. II–19). × 1500.

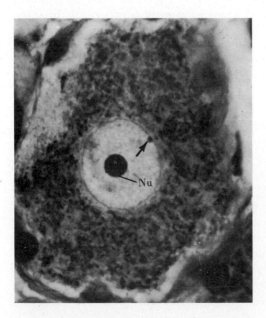

Figure II–19 *Sex heterochromatin.* Within the nucleus of the nerve
cell of a cat, a prominent nucleolus is visible *(Nu)*. Most of the
chromatin is in the form of fine
threads dispersed throughout the
nucleus. The arrow indicates the
sex heterochromatin or *Barr
body,* a characteristic accumulation of dense chromatin found in
several cell types in females of
many mammals. As in this cell,
the sex heterochromatin often is
associated closely with the nuclear
envelope. × 1000. (Courtesy of
M.L. Barr.)

chromosomes. During cell division each chromosome coils and folds into a compact configuration and is readily identifiable as an individual unit (Section 4.2.2). In cells not in division, chromosomes exist in a relatively unfolded state. In euchromatin, this unfolding is maximal; in heterochromatin, the chromosomes or parts of them are in a more compact folded and coiled array. Some nuclei show a stable, clearly nonrandom pattern of heterochromatin. For example, in many cells of female mammals, a small specific heterochromatic body is present. This is an X chromosome, one of the chromosomes that determines the sex of the animal (Fig. II–19 and Section 4.3.3). This heterochromatic region is absent in males, which have only one X chromosome as opposed to the two present in the cells of the females. Cells from males and females can be distinguished on the basis of this cytological characteristic. Evidence outlined later (Section 4.4.6) supports the conclusion that the heterochromatic X chromosome of female cells is genetically inactive in the sense that its DNA is not transcribed. This is one of the facts suggesting that a given region of chromatin is active when in the diffuse euchromatic state and inactive when converted to the condensed heterochromatic state.

The details of chromatin structure cannot readily be interpreted from conventional electron microscopy; all forms appear largely as collections of fine fibrils ranging down to 20 Å in diameter, the size of a DNA molecule. Only recently has it become clear, from studies to be described in Chap. 4.2, that chromatin structure is based on repetitive arrays of 100 Å–diameter beadlike structures (*nucleosomes*) strung out along DNA molecules. This ''string-of-beads'' appearance (see Fig. IV–17) results from the fact that at close intervals the DNA is wrapped tightly around small clusters of histone molecules so that each nucleosome corresponds to a well-defined complex of DNA and histone.

Nucleoli generally take the form of prominent spherical or ovoid bodies. The number per nucleus is characteristic of a given species of organism. Generally there are one to perhaps four per nucleus, although in some species, cell types, or circumstances, more can be formed. Nucleoli are composed of fine fibrils and granules packed in various arrangements (Fig. II–26 illustrates one); no membrane separates them from the rest of the nucleus. The fibrils and granules are primarily RNA molecules and RNA–protein complexes involved in the production of ribosomes for which nucleoli are responsible (Section 2.3.4).

The Nuclear Matrix Traditionally intranuclear structures were thought to be surrounded by a more or less fluid component, frequently called the nuclear ''sap.'' But as is true also for the cytoplasm (Section 2.11.3), it is now suspected that the ''sap'' may actually include a meshwork or ''matrix'' of loosely interlinked elements that helps to organize the nucleus, serving as a framework for nuclear shape, maintaining structural relations among the various nuclear components, anchoring some enzyme systems, and perhaps directing movements of materials within the nucleus and between the nucleus and cytoplasm. The nature of this system, and whether it indeed exists, is difficult to determine since as seen in the electron microscope the nucleus is full of a variety of fibrils and granules whose associations are difficult to discern and whose appearances change with different methods for preparing the material for microscopy. Some are parts of the chromosomes or nucleoli, and some are in transit between the nucleus and cytoplasm. In the zones adjacent to the chromatin there are seen characteristic sorts of small fibrils and also granules, 20 to 50 nm in diameter. Most of these structures are believed to contain RNA since they disappear in preparations digested with RNase; proteins are likely to be present as well. One class of granules, approximately 40 nm (400 Å) in diameter, is often referred to as the *perichromatin* granules. Other, smaller granules are sometimes called the *interchromatin* granules. Some of these granules represent stages in the maturation of RNA molecules (see Fig. IV–40), and others (see Fig. II–21) are on their way to the cytoplasm.

2.2.6 The Nuclear Envelope

The structure of the "nuclear membrane" is of great interest for interactions of the nucleus with the cytoplasm ("nucleocytoplasmic" interactions). The electron microscope has revealed it to be a flattened and expanded part of the endoplasmic reticulum that surrounds the nuclear material. The term *nuclear envelope* is used to convey the fact that the nucleus is surrounded by a flattened sac (Fig. II–20), rather than a simple membrane. The outer surface of the sac, the one exposed to the cytoplasm, has ribonucleoprotein granules attached to it. These are probably ribosomes similar to those of the cytoplasm. The inner surface of the envelope, exposed to the nuclear contents, lacks ribosomes. Often, much chromatin is aggregated along this surface. Specific attachment of chromosomes to the envelope is known to occur in a number of cases (e.g., in various species the paired chromosomes of the special *meiotic* cell divisions of gamete formation [Section 4.3.1] attach to the envelope by their ends, at certain stages; see also Fig. II–19). Some investigators are convinced that special associations between chromatin and the nuclear envelope, perhaps involving the nuclear pores discussed below, are present in many cell types. In addition, in a growing number of cell types the inner envelope surface has been found to be coated with a layer of fibrillar, electron-dense material 30 to 80 nm thick, the *inner fibrous lamina.* This structure varies in prominence in different cell types; it is the best-defined component of the supposed supportive nuclear framework mentioned in the last section. Its composition includes a specific set of proteins only recently obtained for detailed study.

In many cell types the nuclear envelope disintegrates during cell division and reappears afterward, in association with the newly separated sets of chromosomes (Section 4.2.10).

Cytochemical studies and work on the membranes of isolated nuclear envelopes have shown a general biochemical and enzymatic similarity to the rest of the endoplasmic reticulum. Thus far, there have been found no special striking biochemical features that might be significant for the functioning of the

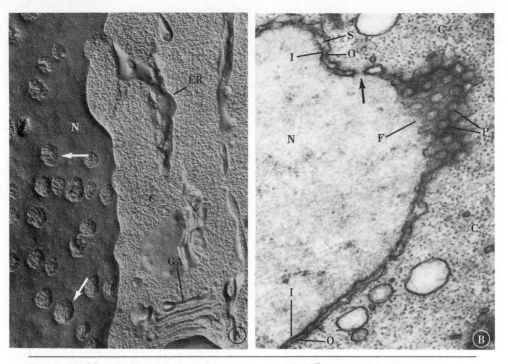

Figure II–20 *The nuclear envelope* as seen with different preparative proce-dures for electron microscopy. A. Portion of an onion root-tip cell prepared by freeze-etching (Fig. II–6). At *N* a portion of the nuclear surface is seen; the ar-rows indicate two of the pores; these appear as circular depressions or interrup-tions of the surface. In the cytoplasm *(C)* portions of the Golgi apparatus *(GA)* and endoplasmic reticulum *(ER)* are visible. × 30,000. (Courtesy of D. Branton.) B. Portion of an oöcyte of the toad, *Xenopus,* prepared by conventional methods of fixation, embedding, and sectioning. *N* indicates the nucleus; *C,* the cyto-plasm. In cross section, the nuclear envelope appears as a membrane-enclosed sac *(S).* The membrane at the surface of the sac facing the cytoplasm *(O)* and that at the surface facing the nucleus *(I)* are continuous at the edge of the pores; thus in cross section the pores appear (arrow) as gaps in the envelope. At *F* the envelope is twisted; the section provides a face view showing the pores *(P)* as circular openings. × 40,000. (Courtesy of J. Wiener, D. Spiro, and W.R. Loew-enstein.)

envelope, but such investigations are still at an early stage. The association with the inner fibrous lamina and the absence of ribosomes are the most obvious differences between the inner membrane and the outer, cytoplasmic-surface membrane. These two membranes are continuous with one another through the pores described in the next paragraphs; investigators are now following up initial evidence hinting that the membranes differ somewhat in the proteins present.

Pores and Pore Complexes Nuclear pores occur at intervals along the en-velope. These appear as roughly circular or polygonal areas, where inner and outer surfaces of the sac are fused, and the sac is thus interrupted (Figs. II–20 and II–21). The pores measure 50 to 80 nm across, and there may be as many as thousands of them, scattered across the nuclear surface. Commonly they occupy 10 to 30 percent or more of the nuclear surface area, and there are 20 to 50 per μm^2 of envelope. In a very few cell types, such as some mature sperm or the micronuclei of protozoa (Section 3.3.2), pores are much sparser than this (1 to 3 per μm^2), or even absent. This may reflect the relatively inactive states of such nuclei.

The pores serve as routes of nucleocytoplasmic interchange. Granules, or other bodies, are sometimes seen adjacent to both nuclear and cytoplasmic surfaces of the nuclear envelope and even inside the pores (Fig. II–21); some of the structures seen in these configurations are thought to contain RNA on its way to the cytoplasm. Experiments with marker substances indicate that material can also move in the other direction through the pores: When particles of gold or other tracers up to 10 nm in diameter are injected into the cytoplasm of amebae, some of the particles gain access to the nucleus. The electron opac-

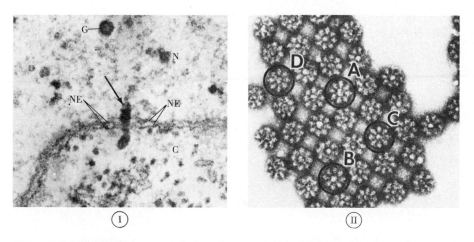

Figure II–21 *Nuclear pores.* I. Small portion of a cell from the salivary gland of the insect *Chironomus.* The arrow indicates a structure passing through a pore in the nuclear envelope *(NE).* The structure is a granule, similar to the one seen at *G* and containing RNA in passage from the nucleus *(N)* to the cytoplasm *(C).* These granules resemble ones produced by actively transcribing regions of the chromosomes (Balbiani rings; Section 4.4.4) and thus tentatively are thought to carry mRNAs complexed with proteins. × 100,000. (Courtesy of B.J. Stevens and H. Swift.)

II. Face views of pore complexes isolated from the nuclear envelope of a *Xenopus* oocyte by dissolving the membrane with detergent (Triton X100). The material was prepared by "negative staining" (Fig. II–53) so that the pore complex structures appear as light elements embedded in a dark matrix of stain. The four complexes at A to D provide particularly clear views; the structures seen are the spokes and central body diagrammed at *S* and *C* in Fig. II–22B. × 75,000. (From Unwin, P.N.T., and R.A. Milligan, *J. Cell Biol.* **93**:63, 1982. Copyright Rockefeller University Press.)

ity of the gold permits direct visualization of the particles in the electron microscope; some of the particles can be seen within the pores as if caught in the act of traversing the envelope.

However, the pores are not simply holes in the envelope. Much of the aperture is occupied by a cylindrical or ringlike arrangement (annulus) of moderately electron-dense granular or fibrillar material (Fig. II–22) that appears to be continuous with the inner fibrous lamina and remains attached to the lamina when the membranes are experimentally dissolved away with detergents. The annular material is organized as a *pore complex*. Investigators have yet to agree upon the finer details of the organization of the complex (Fig. II–22 illustrates two competing models), but most agree that the complex shows an overall octagonal structure when seen in face view, and that the central region of each complex exhibits different organization from that of the more peripheral portions.

Movement through the pores seems limited to molecules or particles less than 10 to 15 nm in diameter, or those that can distort to such a dimension (see Fig. II–21). This can be shown by introducing radioactively labeled or microscopically detectable molecules or particles of known size, such as the gold particles mentioned earlier, into the cytoplasm or into the medium sur-

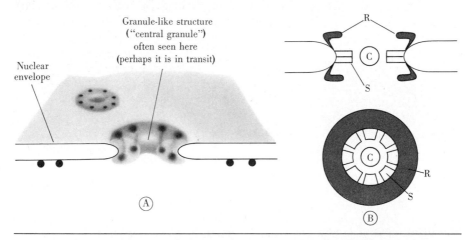

Figure II–22 *Models of nuclear pore complexes.* **Virtually all current models agree that pore complexes show some form of octagonal symmetry and most envisage this arrangement as including the presence of eight subunits more or less equally spaced around the perimeter. The nature and organization of the subunits is, however, still unclear. A. In this model the subunits are viewed as granules or coiled fibrils embedded, perhaps, in a ring of matrix materials. (After proposals by W. Franke, J. Gall, R.G. Kessel, G. Maul and others.) B. This model proposes that each complex is of a pair of rings *(R)* attached to a system of eight spokes *(S)*. A central "plug" *(C)* of some sort seems to be present. Particles sometimes seen at the cytoplasmic surface of the complex resemble ribosomes present elsewhere in the cytoplasm. (After Unwin and Milligan as in Fig. II–21, Part II). The lower diagram in B is a face view of the upper structure in the upper one.**

rounding isolated nuclei. The aperture in the membrane system is substantially larger than 15 nm; evidently movement is restricted to the central zone of the pore complex. How much selectivity there is beyond that based on size has proved difficult to determine. The nuclear interior and the cytoplasm clearly differ in composition, and the differences involve molecules small enough to pass through the pores; this extends even to inorganic ions like K^+. Nonetheless, differences need not depend on selectivity of the envelope. Molecules may accumulate on one side or the other, based upon their binding to structures for which they have particular affinities. Postively charged ions, for example, are attracted by negatively charged groups and thus would tend to accumulate in regions rich in nucleic acids, such as the nucleus (see Fig. II–1). Many molecules, both large and small, spend most of their time as parts of large complexes, which effectively restricts their movements, even though the associations are by relatively weak bonds. Thus, for example, it is possible that part of the reason why histones, made in the cytoplasm, accumulate in the nucleus is that only there do they encounter nucleic acids—specifically, replicating DNA—not already complexed with proteins. Once the histones associate with the DNA and with one another (Section 4.2.5) they no longer are free to leave the nucleus. Similarly, some RNAs confined to the nucleus may remain there owing to their complexing with nuclear proteins. Because of such considerations there is no general agreement about the extent of control exerted by the pore complexes on nucleocytoplasmic interactions. Some investigators believe that pore-associated enzymes, ATP-dependent transport processes, and reversible changes in the organization of the pore complexes (which might increase or decrease their effective aperture, transiently) participate in movement of the larger molecules and complexes, perhaps including ribonucleoprotein structures. This belief is based in part on experiments with isolated nuclei in which export of RNA seems enhanced by ATP. Such experiments are still difficult to interpret, however, and much more work is needed before firm conclusions can be drawn. Similar comments apply to the widely accepted idea that the ribosomes and messenger RNAs on the outer nuclear envelope are ones that have just emerged from the nucleus.

Annulate Lamellae   In a number of metabolically active types of cells, the nuclear envelope shows many infoldings which increase the surface available for nucleocytoplasmic exchange. In other cells, "annulate lamellae" are seen near the envelope. These are stacks of parallel flattened sacs with "pores" similar to the nuclear pores. From their structure and from observations of what seem to be annulate lamellae separating from the nuclear envelope, it has been suggested that these lamellae derive from the nuclear envelope of the cells and serve some function in the cytoplasm. Some investigators maintain that they give rise to endoplasmic reticulum; others believe that their chief role is in the transport of RNA from nucleus to cytoplasm or in the storage of RNA. There are also reports, yet to be fully verified, that annulate lamellae can form from the ER.

Chapter 2.3 Nucleoli, Ribosomes, and Polysomes

Ribosomes, the intracellular sites of protein synthesis, are present in virtually all cells. The few exceptions (for example, mature mammalian red blood cells and mature sperm cells have few, if any, ribosomes) show no protein synthesis. These exceptional cells have very restricted functions and relatively short life spans.

Ribosomes possess RNA of distinctive size and base composition known as *ribosomal RNA* (rRNA). Production of this RNA is a major function of nuclei. RNA–DNA hybridization experiments on many organisms indicate that well over a hundred (sometimes many hundreds) of apparently identical rRNA molecules can bind simultaneously to the DNA from one nucleus. This indicates that many duplicate copies of DNA sequences responsible for rRNA production are present in the *chromosomes*. The *nucleoli*, prominent in most cells of eucaryotes, are centers of rRNA synthesis.

2.3.1 Protein Synthesis: An Initial Overview

As protein synthesis is currently pictured (Fig. II–23 and Section 2.3.2), specific amino acids are "activated" by attachment in the cytoplasm to specific *transfer RNAs*. The tRNAs are relatively small molecules, 75 to 90 nucleotides long. They have some base sequences in common, and all show essentially similar overall folded compact conformations (see Fig. II–16), but each also has some distinctive base sequences (see Figs. II–16 and II–23). Certain of the features of each tRNA, perhaps specific base sequences and three-dimensional configurations, permit a specific enzyme to recognize it and to attach the proper amino acid to it. The amino acid–tRNA combinations ("aminoacyl–tRNAs") become aligned on a molecule of *messenger RNA* (mRNA), which has derived its base sequence from a specific region of DNA, the gene responsible for the synthesis of a particular protein. This mRNA base sequence can be pictured as a series of coding units *(codons)*, each a set of three nucleotide bases (a "trinucleotide sequence") complementary to a trinucleotide *anticodon* sequence of a tRNA. By base pairing, the nucleotide sequence of the mRNA specifies the alignment of the different tRNAs that associate with it. Since each tRNA carries a specific amino acid, amino acids are thus aligned according to the mRNA base sequence. The amino acids are linked together sequentially to form an elongating ("nascent") polypeptide chain, which, when complete, is released from the RNA (Fig. II–23). The amino acid sequence in a polypeptide chain results from an mRNA base sequence and thus, ultimately, from a particular base sequence in DNA. It is the sequence of amino acids that determines the specificity of the protein.

ATPs are split in amino acid activation, and GTPs are split during the initiation of polypeptide chain formation, the elongation of the chain, and the termination steps that lead to its release from the ribosome. Clearly, protein synthesis depends heavily on the cell's energy.

Polysomes; N- and C-termini of Proteins   Protein synthesis and the interaction of tRNA and mRNA take place on a ribosome–mRNA complex. A given mRNA molecule is simultaneously complexed at different points with a number of ribosomes to form a polyribosome or *polysome* (Figs. II–24 and II–41). The ribosomes and mRNA molecule move with respect to one another as successive portions of the mRNA molecule are "read" and "translated" into amino acid sequence. For each polypeptide chain, synthesis begins with the amino acids of the "N-terminal" (amino terminal) end and finishes with those of the "C-terminal" (carboxyl terminal) end. These designations refer to the fact that the amino acids at the two ends of the chain are each linked to only one neighbor, unlike those in the interior. Hence, at one end, an amino group is left unattached; at the other, a carboxyl group is similarly free (see Fig. II–23). This directionality of synthesis is, of course, dictated by the coding and control sequences in the RNAs and the specificities of the enzymes involved.

The mRNAs are translated in the same direction with which they are synthesized (Section 2.2.4). That is, the N-terminal end of the polypeptide chain is specified by nucleotides near the 5′ end of the mRNA.

Each ribosome of a polysome synthesizes the polypeptide chain coded for by the mRNA, so that at a given instant several chains of the same sort, but at different stages of elongation, are present along a given polysome. The number of ribosomes per polysome varies for different cells and proteins; polysomes with as many as 10 to 20 ribosomes, or rarely, many more, have been reported.

Many functioning proteins are composed of several polypetide chains, coded for by separate genes and translated from separate mRNAs. As Section 1.4.3 outlined, the chains often can associate with one another spontaneously in the test tube; assembly in the cell is sometimes more complex (see Chap. 4.1). Proteins may also undergo several types of modifications subsequent to their synthesis, such as addition of carbohydrates (Section 2.5.2) and limited enzymatic cleavage (Sections 2.4.3, 2.5.2, and 3.6.1).

"Universality"; Rates   For reasons of experimental convenience (Section 3.1.3 and Chap. 3.2B), more detailed information is available about nucleic acids and protein synthesis in bacterial (and viral) systems than that for eucaryotic systems. The bacterium *Escherichia coli (E. coli)* is the best studied of all cells in these regards. It is well established that the genetic code is essentially universal: In all organisms the same set of trinucleotide sequences can code for the same amino acids (for an "exception," see Section 2.6.3). The fundamentals of protein synthesis as well are pretty much the same in virtually all organisms. Therefore, we will use information derived both from procaryotes and eucaryotes in this chapter and several later ones. Bear in mind, however, that interesting differences in detail are emerging; these differences may be important for evolution, cellular organization, and the properties of specific cell types such as those of the immune system (Section 3.2.5 and Chap. 3.6C).

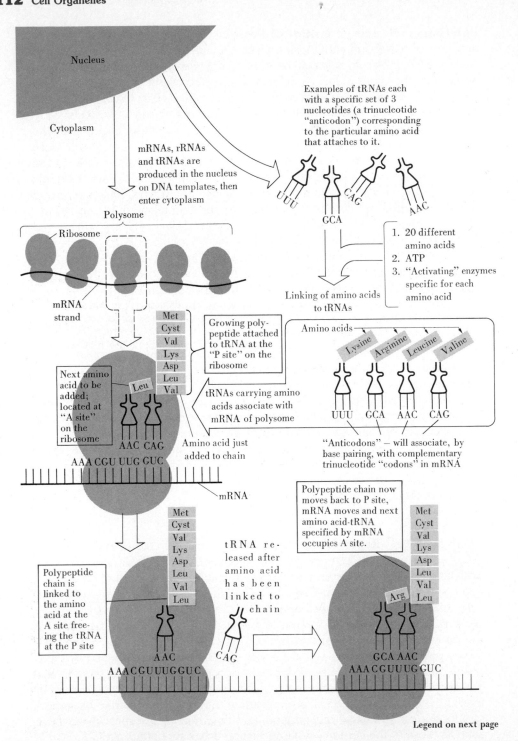

Nucleus

Cytoplasm

mRNAs, rRNAs and tRNAs are produced in the nucleus on DNA templates, then enter cytoplasm

Examples of tRNAs each with a specific set of 3 nucleotides (a trinucleotide "anticodon") corresponding to the particular amino acid that attaches to it.

UUU
GCA
CAG
AAC

1. 20 different amino acids
2. ATP
3. "Activating" enzymes specific for each amino acid

Linking of amino acids to tRNAs

Polysome

Ribosome

mRNA strand

Growing polypeptide attached to tRNA at the "P site" on the ribosome

Met
Cyst
Val
Lys
Asp
Leu
Val

Amino acids

Lysine Arginine Leucine Valine

UUU GCA AAC CAG

Next amino acid to be added; located at "A site" on the ribosome

Leu

tRNAs carrying amino acids associate with mRNA of polysome

AAC CAG

"Anticodons" — will associate, by base pairing, with complementary trinucleotide "codons" in mRNA

Amino acid just added to chain

AAA CGU UUG GUC

mRNA

Polypeptide chain now moves back to P site, mRNA moves and next amino acid-tRNA specified by mRNA occupies A site.

Met
Cyst
Val
Lys
Asp
Leu
Val
Leu

Polypeptide chain is linked to the amino acid at the A site freeing the tRNA at the P site

Met
Cyst
Val
Lys
Asp
Leu
Val
Leu

tRNA released after amino acid has been linked to chain

AAC

CAG

Met
Cyst
Val
Lys
Asp
Leu
Val
Leu

Arg

AAA CGU UUG GUC

GCA AAC
AAA CGU UUG GUC

Legend on next page

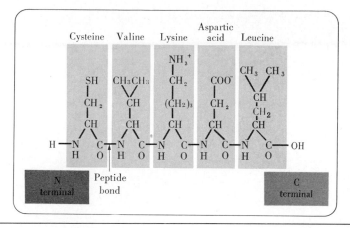

Figure II–23 *Major steps of protein synthesis.* The diagram on the opposite page shows only a few of the many different tRNAs and of the 20 amino acids. Figure I–28 illustrates additional amino acids. More details on protein synthesis are provided in Section 2.3.2. Elongation of the polypeptide chain (Pages 116–117) involves cycling of the chain between the P (peptidyl) and A (aminoacyl) sites on the ribosome as illustrated. The chain is attached, via the amino acid just added, to the tRNA that carried that amino acid. The polypeptide diagrammed above represents the first five amino acids that follow the methionine in the sequence whose synthesis is illustrated by the other drawings. The locations of the N- and C-terminals (*amino* and *carboxyl* terminals) are given as though this were the complete sequence of a finished polypeptide chain.

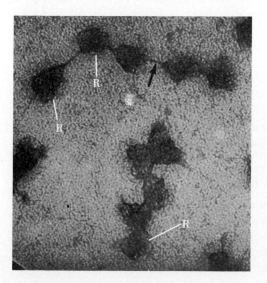

Figure II–24 Two *polysomes* isolated from human *reticulocytes* (progenitors of red blood cells). Each is composed of five ribosomes *(R).* The polysome at the top of the micrograph was stretched during preparation, and the thin strand (arrow) running between its ribosomes may be a strand of mRNA. × 350,000. See also Figure II–41. (Courtesy of A. Rich, J.R. Warner, and C.E. Hall.)

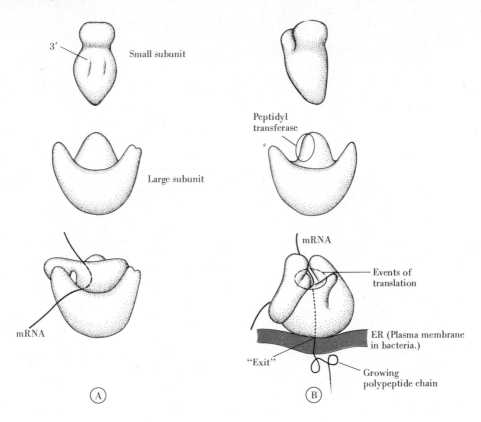

Even a relatively small bacterium contains 5000 to 10,000 ribosomes; a large eucaryotic cell may contain many million. Thus, many proteins can be made simultaneously by one cell. On the average, synthesis of a complete polypeptide chain may take from 10 to 20 seconds to a minute or two, depending on the cell type and on the length of the chain. For example, in mammalian cells producing the blood protein *hemoglobin*—a molecule made of four polypeptide ("globin") chains, each about 145 amino acids long—roughly one minute is required for completion of each chain. Bacterial cells can synthesize their proteins at a somewhat faster rate than this. Pertinent cellular control mechanisms are discussed in Sections 3.2.5 and 4.4.6.

2.3.2 Ribosomes, Polysomes, and Protein Synthesis

Ribosomes have been well characterized by a variety of techniques. They appear in the electron microscope as spherical to ellipsoidal bodies, roughly 150 to 300 Å (15 to 30 nm) in diameter, and consisting of two distinct subunits (Figs. II–25 and II–36). The subunits can be reversibly separated by various treatments—for example, by lowering the Mg^{2+} concentration of the medium.

◀ **Figure II–25** *Two proposed models of ribosome structure* based primarily on work with ribosomes of *E. coli* and other procaryotes.

A. The two subunits of the ribosome have distinctive shapes. The sites at which particular ribosomal proteins, portions of the RNAs and functional capacities are located are being mapped through, for example, experiments in which antibodies specific for particular components are bound to the ribosomes and the sites of binding determined by electron microscopy. Indicated in this illustration are the probable site of the 3′ end of 16S RNA (near the tip of the leader line, labeled 3′) and the association of the ribosome with mRNA in a polysome. (After G. Stoffler, p. 93 in *International Cell Biology, 1980–1981* [ed. H.G. Schweiger]. Vienna: Springer-Verlag, 1981.)

B. This model proposes a shape for the small subunit and an arrangement of the two subunits somewhat different from the one in part A; the differences emphasize the still-provisional nature of both models and the need for more evidence. Illustrated is the zone of probable location of the set of proteins that carry out peptidyl transferase activity (Section 2.3.2) and the association of the ribosome with membranes (the plasma membrane in procaryotes [Section 3.2.4] and the endoplasmic reticulum in eukaryotes). The model summarizes evidence that the polypeptide is formed through the translational events occurring in the region indicated, a region accessible to tRNAs and other necessary components from the cytoplasm. As it forms, the growing ("nascent") polypeptide chain remains in an unfolded configuration and maintains close association with the large ribosomal subunit as if traveling in a (hypothetical) tunnel or groove; when the chain reaches the opposite surface of the subunit from the "translational domain" it "exits" from the subunit and now can begin to fold. For membrane-bound ribosomes the "exit" region is close to the zone of association of the ribosome with the membrane so the nascent polypeptide can pass directly into the membrane (Section 2.4.3). (After J.A. Lake; see Bernabeu, C., and J.A. Lake, *Proc. Nat. Acad. Sci.* 79: 3114, 1982).

The two subunits differ in size. Generally, ribosome sizes are spoken of in terms of the speed with which they sediment in a centrifugal field. As pointed out in the legend to Figure I–25, the sedimentation coefficient is a rough measure of the mass of a particle or molecule but is influenced also by the density and shape of the particles and by the medium in which they are suspended. Intact ribosomes of eucaryotic cells have sedimentation coefficients of 80S, corresponding to a molecular weight of close to 5 million. These figures are derived from mammalian cells; values for other eucaryotes may differ slightly. The two ribosomal subunits sediment respectively at about 60S for the larger and 40S for the smaller; the larger subunit accounts for two thirds of the weight of the ribosome. Each subunit is a ribonucleoprotein particle with roughly equal amounts of RNA and protein. The larger subunit contains an RNA molecule (28S) 4000 to 5000 nucleotides long. The 18S RNA of the smaller subunit is about 2000 nucleotides long. (Each nucleotide contributes a molecular weight of 330; p. 96.) Two small RNA molecules (5S and 5.8S, about 150 nucleotides each) are associated with the large subunit. RNAs of ribosomes are referred to as rRNAs.

The ribosomes of *E. coli* are somewhat smaller than those of eucaryotic cells. Their sedimentation coefficient is 70S, which corresponds to a molecular weight of 2.8 million; the large subunit contains two RNAs, a 23S species (2900 nucleotides) and a 5S species (120 nucleotides); the small subunit has a 16S RNA (1540 nucleotides). Protein accounts for only one third of their weight, which is lower than in eucaryotes. The bacterial proteins have been more adequately characterized than those of eucaryotes. The large subunit of an *E. coli* ribosome contains 31 different polypeptide species (several copies of one of these may be present), and the small subunit, 21. In eucaryotes, the corresponding (approximate) numbers are 40 and 30.

Initiation of Translation   The several steps of protein synthesis require a shifting cast of characters whose interactions bespeak a precise and multifaceted organization of the ribosomes. In the initiation of synthesis, a small ribosomal subunit binds to an mRNA along with GTP and proteins known as initiation factors. A specific type of initiation aminoacyl–tRNA associates with this complex—one whose tRNA anticodon binds to the "start" sequence, AUG (or in some cases, GUG), on the mRNA. This mRNA codon, signaling the beginning of the information for a polypeptide, typically is located a number of bases after the actual beginning of the mRNA molecule (which is the 5′ end of the polynucleotide chain); the bases preceding this codon constitute a "leader" sequence, which helps the ribosome subunit to attach in the correct manner. The initiation aminoacyl–tRNA carries the amino acid *methionine* (see Fig. I–28). In bacteria (but not eucaryotes), the initiation methionine is modified by the presence of a formyl (CHO) group on its NH_2; hence this amino acid (formylmethionine) can link only to one other amino acid (via its COOH) and so can occur only at the N-terminal end of a polypeptide chain—the end synthesized first. (After synthesis is complete, the formyl group or the entire methionine can be removed, so that although growing proteins all begin with the same amino acid, mature ones need not.)

The complex of small subunit, GTP, initiation factors, mRNA, and initiation aminoacyl–tRNA associates with a large ribosomal subunit; the GTP is split, and the resultant GDP and the initiation factors are released in the process. In this way a functional polysome is formed.

Elongation   Once initiated, elongation of the chain takes place as follows for each amino acid to be added: First, the specific aminoacyl–tRNA bearing the next amino acid to be added associates with the polysome. Through a series of steps involving GTP, proteins known as "elongation" factors, and ribosomal components, this aminoacyl–tRNA comes to be positioned at a specialized region on the ribosome (the "A"—aminoacyl—site), where its anticodon sequence is bound to the mRNA codon specifying the amino acid. The three-dimensional conformation of tRNAs (Fig. II–16) is such that when a tRNA is bound at the A site, the amino acid carried by the tRNA is positioned in proper location to form a bond with the "nascent chain"—the growing polypeptide (Fig. II–23). The nascent chain itself is associated with the larger ribosomal subunit, but the amino acid most recently linked to the chain is still attached to its tRNA. This tRNA is still associated with the ribosome, and its anticodon is bound to the corresponding mRNA codon (Fig. II–23). The site of this growing end of the nascent polypeptide chain linked to a tRNA is called the "P" (peptidyl) site of the ribosome.

Ribosomal proteins, conventionally referred to collectively as the enzyme *peptidyl transferase,* now link the nascent chain to the newest amino acid, the one at the A site. This releases the chain from the P-site tRNA, leaving it attached to the A-site tRNA. Next, through reactions requiring GTP and "translocase activities" probably carried out by proteins, the now-free P-site tRNA leaves the polysome; the A-site tRNA with its attached polypeptide chain is then moved into the P site. The A site is now available to accept a new aminoacyl–tRNA once the mRNA moves with respect to the ribosome so that the next mRNA codon is in correct position to specify the entry of the next tRNA at the A site. When this has occurred the entire elongation sequence can be repeated.

Termination Termination of synthesis occurs when one of the mRNA "stop" sequences UAA, UAG, or UGA, which do not code for amino acids, moves into the A site. With GTP again required, protein molecules called "termination factors" help to release the newly made polypeptide chain from the ribosomes. They also free the ribosomal subunits from one another and from the mRNA; the ribosome separates from the mRNA while the polypeptide chain does, or soon after. The ribosomal subunits also separate from one another (see Fig. II–36); they move off and can now assemble into new polysomes. In the test tube the subunits can also assemble into ribosomes not associated with an mRNA, but the degree to which they normally do so in cells is not known.

The drug *puromycin,* an experimentally useful inhibitor of protein synthesis, has the effects of releasing forming polypeptides prematurely from the ribosomes and preventing new initiation. This drug, whose structure mimics aspects of an aminoacyl–tRNA, can associate with the ribosomal A sites and can be attached to the end of a polypeptide chain. But since puromycin lacks the groups needed for specific recognition of codons and for further elongation of the chain, when cells are exposed to it formation of proteins is interrupted.

Ribosome Architecture What specifically does the ribosome contribute to all these comings and goings, recognition events, enzyme actions, and transfers of large molecules? Proteins responsible for certain of the enzyme-like activities, such as that of peptidyl transferase, are built into the ribosome (Fig. II–25). In addition, the particles are large enough and appropriately organized to bind and hold the several participants in proper orientation. The small subunit seems more important in the initial interactions with mRNAs and aminoacyl–tRNAs; the large one, in the actual processes of linking of amino acids. More precise information is now emerging from investigations of the molecular anatomy of ribosomes (see Fig. II–25). The base sequences of all the rRNAs either are known or soon will be. That of the *E. coli* 16S molecule, for example, strongly suggests that the molecule has a well-defined, specific three-dimensional structure, based on loops and double-stranded regions (Section 2.2.2); this conformation is of the sort that might be expected for a molecule that interacts with ribosomal proteins to help bind them into a compact and precise structure like that of the small ribosomal subunit. In addition, at its 3′ end the 16S RNA has a base sequence more or less complementary to that found at the beginning of many *E. coli* mRNAs; by base pairing, this could help to establish the correct mRNA–ribosome association during initiation. (The comparable eucaryotic RNAs differ from this, at least in detail, although they still may be found to interact in an essentially similar manner.) Thus,

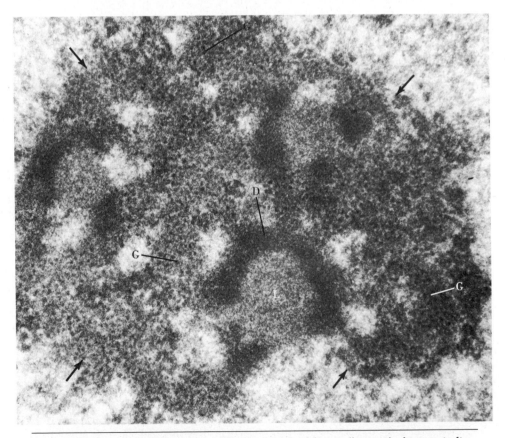

Figure II–26 *Nucleolus* from a cell in the vaginal lining (human). Arrows indicate the approximate edges of the nucleolus. The organelle contains numerous granule-like structures *(G)*, dense fibrillar regions *(D)*, and amorphous, somewhat less dense regions *(L)*. The pale irregular areas probably contain strands of chromatin. × 60,000. (Courtesy of J. Terzakis.)

although rRNAs are not translated into proteins, their DNA-derived base sequences still carry important information—for folding and for interaction with other RNAs.

Reconstitution   Ribosomes can be dissociated into their RNAs and proteins, and the various species of nucleic acids or proteins so obtained can be separated from one another. The molecules will reassemble into functional ribosomal subunits when mixed in the test tube under proper conditions of salt concentration, temperature, and so forth. Such experiments aid in the analysis of ribosome assembly (Sections 2.3.4 and 4.1.3). They also provide a useful approach for investigation of functional organization: Ribosomes lacking one or another of the proteins normally present can be constructed and studied to determine how the absence of specific proteins affects their function and molecular architecture. Most reconstitution experiments thus far have been done with bacterial ribosomes, especially those from *E. coli.*

The location of each of the ribosomal proteins is now being studied by several methods. The portions of the proteins at the surfaces of the particles are accessible enough to bind antibodies specific for them; the bound antibodies can be visualized by electron microscopy, thus locating the proteins. *Crosslinking* agents also are being employed. These are small molecules that can produce covalent bonds between neighboring molecules previously unlinked but close enough that the agents can bridge the space: When structures are treated with these agents and then disrupted, normally separable molecules often remain linked together, permitting the conclusion that they abut one another in the intact structure. One such cross-linking agent is glutaraldehyde (Section 1.2.1), which can form bonds between amino groups such as those on the side chains of lysines (see Fig. II–23) and thus cross-link proteins. Extensive cross-linking of this sort can render a structure quite resistant to disruption, which accounts for the usefulness of glutaraldehyde as a fixing agent for microscopic preparations (Section 1.2.1). (Section 3.6.5 and Fig. III–35 describe analogous cross-linking occurring in nature.)

2.3.3 Nucleoli

Autoradiographic evidence shows nucleoli to be the sites of extensive RNA synthesis (see Fig. I–24). Cytochemical and cell fractionation studies indicate that at least 5 to 10 percent of the nucleolus is RNA; the rest is mainly protein. Preparations of isolated nucleoli also contain some DNA, derived mainly from the *nucleolus-associated chromatin* that is attached to the nucleoli and sends strands deep into their substance (Fig. II–26). When discrete chromosomes become visible in cell division, nucleoli are often seen associated with specific *nucleolar organizer* regions of specific chromosomes (Fig. II–27). These include the same chromosome regions that contribute to the nucleolus-associated chromatin when the cells are not in division.

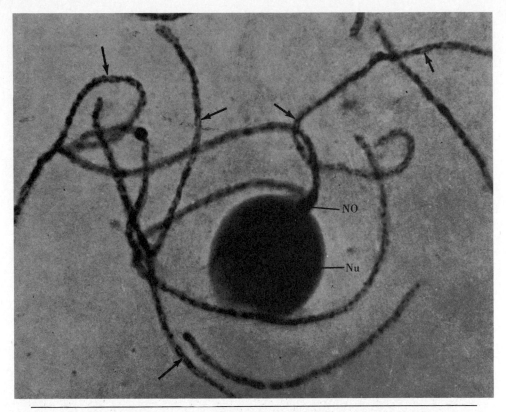

Figure II–27 *Nucleolar organizers as seen in meiotic chromosomes.* Preparation of a stage in male gamete formation in corn. The cell shown had been squashed prior to fixation, thus separating organelles and facilitating observation of some details. The threadlike structures are pairs (arrows) of closely associated chromosomes (pachytene stage of meiosis; see Chapter 4.3 and Figs. IV–30 and IV–33). Note the darker regions (chromomeres) present at intervals along the length of the threads (for example, just to the left of the arrow in the upper right hand corner). The nucleolus *(Nu)* is prominent. It is associated with a specific chromosomal site corresponding to the nucleolar organizers *(NO)* that are located on one pair of chromosomes. Approximately × 2000. (Courtesy of M.M. Rhoades and D.T. Morgan, Jr.)

As seen in the electron microscope, nucleoli are divided into zones of distinct structure. In addition to chromatin strands, there are areas rich in granules, roughly 150 Å in diameter, and regions that are primarily fibrillar or amorphous (Fig. II–26). The three-dimensional arrangement of these components varies among different cell types. They may be intermingled in apparently complex fashion (Fig. II–26), or, for example, as in hepatocytes and other cells in amphibia, the granules may be concentrated at the nucleolar periphery, surrounding a fibrillar core. Observations of the changes induced in nucleolar structure by digestion with RNase and proteolytic enzymes indicate that both the granular and the fibrillar zones contain RNA and proteins. A number of investigators have also suggested that DNA from the chromatin may extend into some of the fibrillar zones. A reasonable but still tentative interpretation of the structure in Figure II–26 is that zones of the type labeled L contain chromatin, those labeled D (for "dense fibrillar") are the sites where RNAs are actually being transcribed from specific DNA templates, and those labeled G contain products of this transcription that have complexed with proteins to form granules. This is suggested by autoradiographic evidence indicating that the sites of initial incorporation of radioactive RNA precursors in nucleoli are in portions of the nucleolus-associated chromatin and in fibrillar zones of the nucleoli. Subsequently the labeled molecules appear in the granular zones. The tentative explanation is that the fibrillar zones contain large rRNA precursor molecules, newly transcribed from rDNA, and that as these are modified, they pass into the granular zones. For one special case, that of amphibian oöcyte nucleoli, there has been a dramatic microscopic demonstration of the interrelations of DNA and RNA in the fibrillar core (see Fig. II–31), which probably corresponds to the D regions of Figure II–26.

Although no membrane delimits the nucleolus, contacts between nucleoli and nuclear envelopes are frequently seen; these might facilitate passage of nucleolar products into the cytoplasm.

2.3.4 Nucleoli and Ribosome Synthesis

Cytologists noted long ago that nucleoli are particularly prominent in cells that have high rates of protein synthesis. Information outlined in the preceding section suggests a role in RNA metabolism. A much more detailed analysis of nucleolar function is possible from several additional lines of evidence.

One important series of observations has been made on a mutant strain of the clawed toad *Xenopus*. It was initially noted that matings of certain *Xenopus* individuals produce offspring of which one quarter die. (The explanation for this percentage of nonviable offspring will be outlined in Section 4.3.2.) For the present purposes, the crucial fact is that cytological and genetic studies show that death can be attributed to an inheritable chromosomal defect affecting nucleolus formation. The nuclei of embryos carrying defective nucleolar chromosomes contain either no visible nucleoli or grossly abnormal ones. The embryos develop normally until just after hatching and then die. Death occurs shortly after the stage when normal embryos show a great acceleration in ribosome synthesis; before this, both normal and abnormal embryos rely on ribosomes stored in the egg during its maturation preceding fertilization (Sections 3.10.2 and 4.4.2). The embryos that die cannot make ribosomes, and the lack of nucleoli is a visible manifestation of this inability. Ribosomal RNA purified from tissues of normal *Xenopus* cannot hybridize with DNA from abnormal embryos, indicating that the DNA base sequences for rRNA formation are missing in the mutant chromosome. These observations demonstrate a central role for nucleoli in ribosome formation.

Initially it was believed that the small granules normally present in the nucleolus represented newly formed ribosomes that had accumulated there. However, the evidence now available does not support this idea. Properly isolated nucleoli contain few complete ribosomes. They do contain much RNA metabolically related to rRNA, but it seems to be in the form of extremely large molecules (up to 45S in some organisms; molecular weight about 4 million) that must be enzymatically fragmented and otherwise modified to become mature ribosomal constituents (see Fig. II–28). A single 45S molecule gives rise to one 18S and one 28S RNA plus one 5.8S RNA. The sum of the molecular weights of these three rRNAs is less than 2.5 million. This means that much of the precursor molecule is left over. Evidently, each precursor contains a seg-

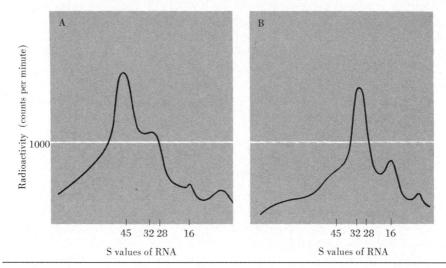

Figure II–28 *Ribosomal RNA synthesis in HeLa cells.* The cells were exposed for 25 minutes to radioactive RNA precursor (^{3}H-uridine). A. At this time a sample was taken; from it, RNA was extracted and centrifuged on a density gradient (see Chapter 1.2B) to separate RNA molecules of different size classes (measured here in terms of sedimentation coefficients; see Figure I–25 and Section 2.3.2). B. RNA extracted from cells that were exposed to radioactive precursor and then grown for an additional 20 minutes in nonradioactive medium in which the drug *actinomycin D* was included. This drug prevents the transcription of new RNA molecules on DNA templates. Thus changes in distribution of radioactivity during the 20-minute "chase" (see Fig. I–24, legend) reflect modifications of molecules made before the cells were placed in the drug-containing medium rather than synthesis of new RNA molecules. Comparison of part B with part A indicates that there are fewer radioactive 45S molecules and more 16-18S and 32S molecules. This is interpreted as meaning that 45S molecules give rise to 16-18S and 32S molecules; probably a given 45S RNA molecule is fragmented to produce one 16-18S and one 32S molecule. The 16-18S and 32S RNAs are precursors of the RNAs in the small and large ribosomal subunits. (From work of J.E. Darnell and J. Warner.)

ment corresponding to an 18S rRNA, one corresponding to a 28S RNA, and one corresponding to a 5.8S molecule, plus several "spacer" segments. These spacer segments may be important in the processing of the precursor into the rRNAs, but they are destined for removal from the maturing molecules and eventual degradation within the nucleus. Starting from its 5' end, the large precursor of rRNAs has the sequence

5' end, spacer, 18S segment, spacer, 5.8S, spacer, 28S, 3' end

The 5S RNA molecule of eucaryote ribosomes is coded for at a different chromosomal site from the other rRNAs and is not included in this precursor (Section 4.2.6). The processing steps converting the precursor to separate molecules are being sought (see Chap. 5.1).

The complexing of ribosomal proteins with the rRNA also begins in the nucleolus. The proteins are synthesized in the cytoplasm but migrate into the nucleus to participate in the initial steps of the assembly of ribosomes. Purified rRNAs and ribosomal proteins when mixed in the test tube will assemble through a form of self-assembly (Section 1.4.3) to reconstitute functional ribosomal subunits (Sections 2.3.2 and 4.1.3). Evidently, similar self-assembly occurs in the nucleolus.

Present evidence suggests that the large and small ribosomal subunits pass into the cytoplasm separately. At least in some cells, carefully timed studies with radioactive precursors show that a ribonucleoprotein particle containing the 18S RNA from a given precursor molecule appears in the cytoplasm a few minutes earlier than does one with the 28S RNA made from the same precursor.

Nucleolar Organizers; Repetitive rDNA The most obvious places to search for DNA template sequences coding for rRNA (the *rDNA* sequences) would be in the nucleolar chromatin.

Different strains of the fruit fly *Drosophila* can be bred so as to contain different numbers of nucleolar organizer regions in their chromosomes. The experiment shown in Figure II–29 demonstrates a direct proportionality between the number of organizers per nucleus and the number of DNA sites per nucleus that can form molecular hybrids with purified rRNA. In other words, it suggests strongly that the organizers, which originally were defined from microscopic appearances, are sites of rDNA. Less expected was another conclusion based on careful "bookkeeping" of the numbers of molecules involved. Several hundred molecules each of 18S and 28S RNA can hybridize with the DNA from one nucleus. Thus, each nucleus must contain several hundred copies of rDNA. In terms of the sequences coding for the rRNAs, these copies are indistinguishable from one another—they are very similar if not identical. (Subsequent studies have shown that *Drosophila* is unusual among organisms in that its rDNAs do exhibit detectable heterogeneity, particularly in sequences that are not included in the mature rRNAs: Specifically, in *Drosophila,* some of the rDNAs have "intervening sequences" [Section 2.3.5]; others lack them, as is

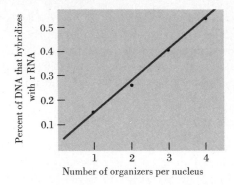

Figure II–29 *Nucleolar organizers and rRNA.* An experiment by F.M. Ritossa and S. Spiegelman on the relations between nucleolar organizers and rRNA in the fly *Drosophila.* The DNA purified from nuclei of individuals with different numbers of nucleolar organizer regions in the chromosomes is tested for the number of sites that can hybridize with purified rRNA (graphed as the percent of the DNA per nucleus that hybridizes).

generally found for rDNAs in most organisms. This situation, however, does not affect the central issues under consideration here.)

In *Drosophila,* a nucleus with two nucleolar organizer regions has about 400 copies of rDNA. Several thousand copies per nucleus are present in some

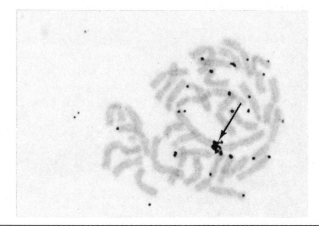

Figure II–30 *Nucleolar organizers.* Autoradiograph of chromosomes from *Xenopus* (mitotic metaphase; see Fig. IV–6) prepared by an *in situ* hybridization technique (also referred to as cytological hybridization). The chromosomes were exposed to alkaline conditions to "denature" the DNA, separating the two strands of the double helix. Then they were incubated with ribosomal RNA that was highly radioactive (it was synthesized in the test tube using highly radioactive precursors). The radioactive RNA has hybridized (Fig. II–17) with the DNA at the nucleolar organizer sites present on two chromosomes. These two sites are closely associated with one another in this preparation so that the autoradiograph shows only one cluster of grains (arrow). Individual grains are scattered through the remainder of the chromosomes, but this probably is due to nonspecific binding; such effects presently limit the usefulness of this technique to cases in which repetitive DNA sequences are studied and much RNA binds to a small region, resulting in a definitive cluster of grains (see Fig. IV–20 for another example). × 1500. (From Pardue, M.L., Cold Spring Harbor Symp. **38**: 475, 1974.)

other organisms. The rDNAs were the first well-defined cases of repetitive DNAs to be analyzed in detail. For them it seems likely that the repetition makes possible very rapid large-scale synthesis of ribosomal RNAs, as is required under at least some conditions of cell growth and metabolism. Repetition ("reiteration") of DNA sequences in perfectly ordinary nuclei has now been shown for a number of DNAs additional to the rDNAs (Section 4.2.6), though in many cases its significance is still unclear.

Their repetitiveness has also facilitated direct visualization of the chromosomal sites of rDNA by *in situ* hybridization. In this procedure, radioactive RNAs are hybridized with preparations of fixed cells rather than with purified DNAs. Figure II–30 describes the procedures and illustrates the binding of rRNA at the nucleolar organizer.

Certain strains of *Drosophila* bred so as to have only one nucleolar organizer undergo a "compensatory" increase in the number of rDNA copies. They show more than the number obtained with the strains used for the experiment in Figure II–29. This was an early illustration of the fact that DNA is not as staid and constant as was traditionally thought. Even more dramatic, though temporary, increases in rDNA content take place during the maturation of the egg cells of amphibia and other organisms. The cells make enormous numbers of ribosomes to be stored for use in embryonic development. Correspondingly, they form hundreds or even thousands of extra nucleoli, each equipped with a copy of the nucleolar-organizer DNA produced for the occasion by special variants of DNA duplication processes (Section 4.2.4). This situation has been exploited to permit microscopic examination of the functioning organizer since the nucleoli can be isolated and their contents spread on an electron microscope grid. Micrographs such as Figure II–31 permit several conclusions:

1. The multiple copies of rDNA sequences of the nucleolar organizer are arranged along a common DNA molecule in "tandem" ("head-to-tail") array—that is, one after another, with similar orientation of the transcribed sequences. They might instead have been more scattered, or clustered irregularly.
2. Many rRNA precursor molecules can be synthesized simultaneously along a single rDNA copy; an RNA polymerase associates with the beginning of the DNA sequence and moves along the sequence generating an RNA, but soon after it moves away from the beginning, another polymerase can attach and start to work (see Section 4.4.6 and Fig. IV–40).This permits very high rates of rRNA synthesis.
3. "Spacer DNA" regions are present between the DNA sequences that are transcribed. These are referred to as "nontranscribed" spacers, to distinguish them from the DNA sequences coding for the "spacer" portions of the rRNA precursor molecule described earlier; the latter DNA sequences are transcribed into the precursor though the transcripts are not included in the mature ribosomal RNAs. The word "spacer" carries no intended func-

tional connotation; spacer DNAs—transcribed and nontranscribed—may well have a variety of functions, though none are yet well understood (Section 4.2.6 and Chap. 5.1).

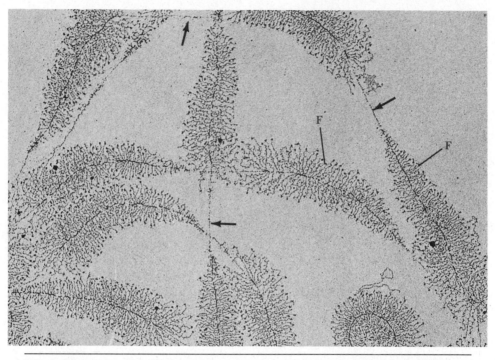

Figure II–31 *Transcription of ribosomal RNA.* Part of the material isolated from the fibrillar core of one of the extra nucleoli of a salamander *(Triturus)* oocyte (Section 3.10.1). In preparing the material for microscopy, the contents of isolated nuclei were dispersed in a suspension medium, then spread by centrifugation onto a support film (see Fig. I–11) on the surface of an electron microscope grid. Arrows point to the elongate DNA-containing fiber identifiable as such by its disruption in DNAse-treated preparations. Lateral filaments *(F)* protrude from the fiber; they are grouped in regions 2 to 3 micrometers in length, with each region showing about 100 filaments. Two to 3 micrometers of DNA contain enough information to specify one of the large (about 40S) rRNA precursor molecules, and each group of filaments is thought to represent many such RNA molecules being synthesized simultaneously by one gene. The filaments are known to be RNA from autoradiographic demonstration of the incorporation of appropriate precursors. The smaller filaments of a group are considered to be at an earlier stage in synthesis (see Fig. IV–40, part B, for more details). Each group of filaments is separated from the next by "spacer" DNA that apparently is not transcribed into RNA (see text). × 20,000. (Courtesy of O.L. Miller, Jr., and B.R. Beatty; see *J. Cell Physiol.* (Suppl. 1):225, 1969.)

(The term *transcriptional unit* [Section 3.2.5] is sometimes used for a stretch of DNA, like the several ones shown here, that is transcribed into a corresponding RNA molecule; this term helps as a reminder that the RNAs eventually produced often are much shorter than the initial transcripts due to the "processing" phenomena discussed in Chapter 2.3 and elsewhere.)

Some significant features of nucleolar behavior remain to be clarified. For example, during cell division the nucleolus usually disappears as a discrete entity, reappearing only in the terminal stages of division. Studies on plant cells indicate that the first sign of nucleolar reappearance is the occurrence of many small nucleolus-like bodies that eventually fuse to form the one or two nucleoli of the usual interphase cell. Although the origin of the small bodies is not yet known, such observations have led to the hypothesis that the nucleolar organizing region or the forming nucleolus collects either material synthesized at many points along the chromosomes or perhaps material that is dispersed among the chromosomes when the nucleolus disappears early in cell division (Section 4.2.10).

There are some organisms whose cells form many small nucleoli either normally or under experimental conditions; apparently the cells have many separate nucleolar organizer regions in their chromosomes rather than only a few.

Bacteria lack nucleoli probably because they have only a few copies of the DNA coding for rRNA, and these are not clustered together. Procaryotic rRNAs are, however, made from large precursor molecules much as in eucaryotes.

2.3.5 mRNA

Detailed analysis of mRNAs is made difficult by the fact that a given cell synthesizes many different mRNAs corresponding to its many different proteins. These mRNAs might be expected to vary considerably, not simply in base sequence but also in such properties as size: The proteins for which they code differ in length, and the relations of the information in mRNA to the length of a protein are set by the fact that three nucleotides are needed to specify one amino acid. Insulin, for example, contains about 50 amino acids; the pancreatic digestive enzyme precursor *chymotrypsinogen,* about 250; and individual chains of the contractile protein *myosin,* over 1500. These figures as such are slightly misleading, since it is now known that the number of amino acids in a mature protein may differ from that in the protein initially synthesized (Section 2.4.3 and Chap. 3.6A). Especially with such considerations it is easy to account for the fact that known mRNAs of mammalian cells range from a few hundred to several thousand nucleotides in length. In bacteria and viruses the range is potentially even greater: Not only do they make polypeptide chains as short as a few amino acids and as long as 1,000 or even more in exceptional cases, but in addition, a single mRNA molecule can carry information specifying two or more different polypeptide chains (see Section 3.2.5). (In eucaryotes a given mRNA generates a single species of polypeptide chain, though this may later be fragmented into more than one functional piece; Section 3.6.1.)

With bacteria or viruses, situations can readily be found or created in which one or a very few species of mRNA are transcribed in large amounts at particular times (enzyme induction or viral infection of cells, for example; Sections 3.2.5 and 3.1.2). Moreover, the information and techniques available for dealing with the genetics of bacteria and viruses are more advanced than is the case of eucaryotes (importantly, there is much more detailed understanding of the effects of mutations that modify aspects of transcription or translation). Hence, until recently the bulk of what was known of mRNA synthesis came from work on bacteria and some viruses. From this work it appears that production of mRNA begins "simply" by the association of an RNA polymerase with a "promoter" sequence in the DNA (Sections 3.2.5 and 4.4.6). The polymerase interacts with this site on the DNA molecule to position the enzyme at the correct place to start transcription. Comparison of bacterial and viral DNAs responsible for synthesis of a variety of proteins has revealed the presence of stretches of reasonably similar DNA base sequences preceding the information actually transcribed into RNA. Different genes might be expected to show such similarities for regions playing similar roles. The sequences in question could include sites such as the promoters, where proteins such as RNA polymerases interact with DNA, and other DNA sites responsible for regulation of gene activity (Sections 3.2.5 and 4.4.6). For many different viral and bacterial genes, the promoter regions include, among other nucleotide sequences, a short sequence in which Ts and As are prominent—TATAATG is one example—with its center located about 10 nucleotides "upstream" of the place where transcription is initiated. (Recall that transcription of DNA sequences starts with the 5' end of the RNA molecule.) Eucaryotic genes probably have comparable sequences in their promoter regions (Section 4.4.6).

As a bacterial mRNA molecule grows, its 5' end comes off the template (like the rRNAs of Fig. II–31 or the RNAs of Fig. IV–40); even before the molecules are complete, ribosomal subunits can attach and translation begin. It was early realized that such initiation of protein synthesis on a still-growing mRNA is much less likely in eucaryotes, since the sites of transcription (in the nucleus) are separated by the nuclear envelope from those of translation (in the cytoplasm). This difference between procaryotes and eucaryotes proved to be only the tip of the iceberg. It is now known that mRNA production in eucaryotes involves several types of "processing" steps not suspected from the work on bacteria.

Methods; PolyA Tails   Progress in studies of eucaryotic mRNAs came in part from careful choice of experimental material. To minimize the problem of mRNA heterogeneity, attention has been focused on cell types that specialize in the production of large quantities of specific proteins. Favored ones include *reticulocytes,* which are cells in process of maturing into red blood cells. Reticulocytes have lost their nuclei (in mammals) and therefore are no longer making RNA, but their cytoplasmic polysomes are still synthesizing proteins, making almost exclusively the globin chains that will complex with the small iron-containing *heme* groups to form hemoglobin. The cells of chick oviduct that secrete large amounts of the egg protein *ovalbumin* and those of silk glands in the larvae of certain insects are also quite useful. So are certain virally infected cells producing viral proteins via mRNAs whose basic features resemble those of normal eucaryotic cell mRNAs. Polysomes isolated from cells of these several sorts contain, predominantly, classes of mRNAs corresponding to the specialized protein products.

The size of a polysome depends largely on the length of the message (this determines how many ribosomes can bind). Therefore, polysomes enriched in particular mRNAs can be isolated by centrifugation. The globin-producing reticulocyte polysomes of Figure II–24, which typically have five ribosomes, can be purified in this way, and globin-specifying mRNA can be obtained from them. More direct isolation of mRNAs from complex mixtures has been greatly facilitated by the discovery that most mRNAs in the eucaryotic cytoplasm (those for histones are the best-known exceptions) have attached to their 3' end a chain of nucleotides, each of which carries the base adenine (Fig. II–15), a *polyA* "tail." The tail may be as much as 200 nucleotides long. The functional significance of the polyA is not understood, but its presence on a molecule is taken as presumptive evidence that the molecule is an mRNA or a relative, such as a precursor. For mRNA purification, artificial polynucleotides containing only T or U bases can be attached to filters or to inert matrices (modified cellulose, plastic beads, or others) packed in long columns. When mixtures of RNAs are passed through the filters, or percolated down the columns, those molecules with polyA tails will stick by base pairing to the polyT or polyU, while other RNAs will pass through; this provides a convenient way for separating out probable messengers.

Demonstration that a suspected mRNA is truly an mRNA requires showing that it can code for a protein. Such demonstration has been made easier by advances in design of experimental systems for translating suspected messengers. So-called *in vitro* (test-tube) protein-synthesizing preparations have been perfected: Purified ribosomes, tRNAs, and other components obtained from cells active in protein synthesis, such as reticulocytes, can be made to translate added mRNAs from foreign sources, generating the mRNA-specified product. A different approach with similar outcome is the injection of mRNAs into maturing egg cells at stages when the cells are making much protein (Section 3.10.2; *Xenopus* oöcytes are particularly useful).

Caps; Leading and Trailing Sequences Improved techniques of the sort just described have fostered rapid progress in identification of mRNAs, in the comparison of those from different cell types, and in analysis of their genesis. We now know, for example, that the polyA tail is added to an mRNA in the nucleus, soon after the mRNA is transcribed, by an enzyme that does not require a DNA template to do so. At the other end of the new molecule (5') a short "cap" is added by other enzymes: This consists of a nucleotide with guanine as its base, whose linkage to the mRNA chain is unusual (for example, unlike the 3' to 5' "phosphodiester" diester bond typical of nucleic acids [see Fig. II–15] the 5' carbon of this nucleotide is linked, via a pair of phosphate groups, to the 5' carbon of the first mRNA nucleotide). Methyl groups (CH_3)

are present characteristically on the guanine base of the cap nucleotide and often on the ribose sugars of the nucleotides at the adjacent 5′ end of the mRNA. From experiments comparing mRNAs with and without caps, it appears that with eucaryotic protein-synthesis machinery, the cap can speed the initiation of protein synthesis by contributing to the alignment of the ribosomes on the mRNA. Bacterial mRNAs lack caps. (They generally have been thought to lack polyA tails as well, but short tails may in fact sometimes be present, according to recent work.) Evidently then, with respect both to mRNA caps and to base pairing between mRNAs and rRNAs (Section 2.3.2), procaryotes and eucaryotes differ in at least the details of mechanisms by which polysomes form.

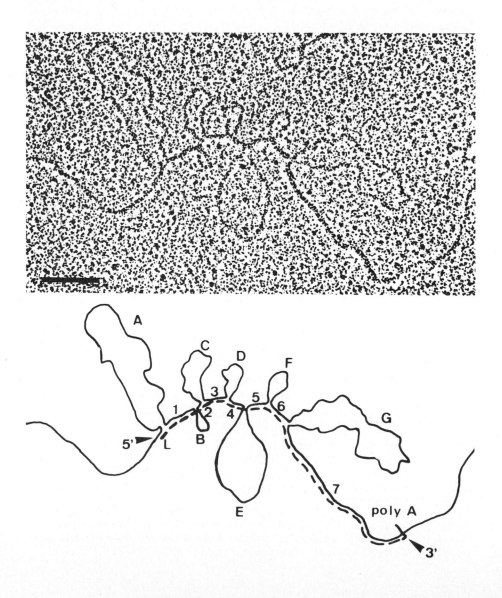

Both in procaryotic and eucaryotic mRNA molecules, there are short, somewhat variable, sequences at each end that are specified by DNA but do not code for portions of the polypeptide. These are the "leader sequences" at the 5′ end and the "trailer sequences" at the 3′ end; they are presumed to function in interactions with ribosomes or other components of the protein-synthesis machinery, aiding perhaps in polysome formation (Section 2.3.2) and in termination of translation (Chap. 5.1). (The starting and stopping points for actual *translation* of the mRNA are, of course, the trinucleotide sequences described in Section 2.3.2.)

Splicing; hnRNA   In labeling studies similar to the ones on rRNA illustrated in Figure II–28, cytoplasmic mRNAs are seen to arise from nuclear precursors that collectively comprise a fraction known as heterogenous nuclear RNA, or hnRNA, because of its wide range of molecular sizes and base compositions. The two processing steps already mentioned, the addition of a cap and of a polyA tail, occur when the mRNA is still in this form. The hnRNAs are, however, surprisingly large—some containing as many as 50,000 nucleotides—which led at first to the suspicion that they are shortened by steps analogous to the ones that generate rRNA. In general terms this suspicion proved true, but study of the details has revealed that the shortening involved not simply the degradation of spacers but also the splicing together of the mature message from several, sometimes many, small pieces. Figure II–32 illustrates one category of experiment from which this conclusion was drawn. Evidently the DNA coding for most eucaryotic mRNAs is organized as a series of "expressed se-

---

◄ **Figure II–32** *Splicing of mRNA;* an experimental demonstration. The upper panel is an electron micrograph showing a configuration of nucleic acids interpreted in the diagram below it. The preparation is of a chick DNA molecule (solid line) containing the sequence specifying the egg white protein ovalbumin; one strand from this molecule is seen after it was hybridized in the test tube (Fig. II–17) with an mRNA molecule (dotted line) specifying ovalbumin. The mRNA contains base sequences complementary to segments of the DNA sequence and thus forms a hybrid (RNA–DNA) double helix with the DNA in regions *L* and 1 to 7; (5′ and 3′ indicate the two ends of the RNA, poly A, the "tail" described in Section 2.3.5, *L* the RNA sequence specifying the signal sequence on the protein [Section 2.4.3] and 1 to 7, the mRNA sequences coding for the mature protein's amino acids.) The loops *A* to *G* are single-stranded regions of the DNA with no corresponding RNA; their existence indicates that between the DNA segments that do have complementary copies in the RNA there are DNA segments that do not have such copies. As now understood, this reflects the production of mRNAs by the splicing mechanisms described in Section 2.3.5. The initial RNA transcript that is formed would have hybridized with the DNA to give a continuous loop-free hybrid double helix, but once the intervening sequences have been removed from the RNA only discontinuous stretches of the DNA can "find partners" along the RNA, as in the micrograph. Approximately × 200,000; length of bar = 0.1 μm (approx.). The material was prepared for microscopy by a modified shadowing procedure (Fig. I–11). (Courtesy of P. Chambon and F. Perrin; see P. Chambon, *Scientific American* **244**(5):61, 1981, for more details and discussion.)

quences," or "exons," separated from one another by "intervening sequences," or "introns." The expressed sequences are those destined for eventual inclusion in the functional message, but the RNA transcript made initially is a single continuous molecule including both types of sequences. Soon after transcription, or perhaps even as it being completed, the intervening sequences are "spliced" out: They are selectively excised from the transcript, leaving only those nucleotide sequences that code for the protein plus the leader and trailer stretches at the 5' and 3' ends. The process is referred to as *splicing,* since the expressed sequences become joined end-to-end, yielding a single, continuous mRNA. (The mature eucaryotic "message," then, consists of a cap followed in turn by a leader sequence, the nucleotide sequence to be transcribed, a trailer sequence, and, usually, a polyA tail.)

For a number of genes the total length of intervening sequences is markedly greater than that of expressed sequence. Not uncommonly, an hnRNA is five to ten times or more the length of the mRNA generated from it. Splicing does account for the disparity in size between hnRNAs and their mRNA "offspring" and helps to explain why as much as 15 to 30 minutes may elapse between initiation of a particular transcript and the appearance of the corresponding mRNA in the cytoplasm. But why is splicing so common and so extensive? What is its significance for the cell's economy? Current speculation centers around the evolutionary "advantages" of constructing genes in pieces, with "new" genes being formed, perhaps, by reshuffling of the pieces (see Section 4.5.2). Chapter 3.6C will consider the rapid progress now being made in analysis of a related genetic phenomenon: the rearrangements of DNA occurring during the maturation of antibody-producing cells through which essentially the same genetic information can be reorganized in various combinations to produce a diversity of proteins. One conceptual problem that must be addressed is that splicing has yet to be detected in procaryotes, and thus, if it occurs at all, it must be much less extensive than it is in eucaryotes. It is, for example, difficult to envisage how such a splicing process could be reconciled with the ability of procaryotic mRNAs to undergo translation before transcription is complete.

Certain genes in eucaryotes, including many of the ones coding for histones, also seem to lack intervening sequences, indicating that splicing is not universally required for all eucaryotic mRNAs.

After excision, the intervening sequences are degraded, although speculative proposals suggest that before this occurs they carry out intranuclear functions still to be discerned. This breakdown, plus the degradation of rRNA precursor spacer segments, helps a great deal in explaining what was a puzzling finding: Despite the fact that most of the known functions of RNA are in the cytoplasm and that most cytoplasmic RNAs come from the nucleus, a large proportion of the RNA sequences transcribed in the nucleus do not reach the cytoplasm. In addition to the degradation of segments of molecules, there seem also to be RNAs that are degraded *in toto* within the nucleus (see also Section 4.4.6).

Splicing steps, as well as the removal of stretches at either end of the transcript, occur in the formation of some eucaryotic tRNAs and even some of the rRNAs of a few organisms. Much work is now focused on the splicing mechanisms both for the mRNAs and for these other RNAs. The base sequence CAGG, or a close relative such as AAGG, is present at the junctions between intervening and expressed sequences in the precursors of a variety of mRNAs, and virtually all intervening sequences studied thus far begin with GU and end with AG. These base sequences may be parts of the recognition system through which the splicing enzymes act. (Splicing must be a quite precise process, since an error would change the nucleotide sequence in the mRNA product with disastrous consequences.) Like almost all other RNAs, nuclear and cytoplasmic, precursor and mature, the hnRNAs seem to occur as complexes with proteins—they are believed to be present in one or another of the several types of particles detectable in the nucleus both by microscopic and by biochemical means (Section 2.2.5 and Fig. IV–40). The proteins of the particles may take part in splicing, perhaps as enzymes or perhaps as agents that help the RNAs to assume essential folded conformations. Certain of the small nuclear RNAs (snRNAs), a heterogeneous class of RNAs (100 to 300 nucleotides) that are confined to the nucleus, are also postulated to play regulatory, processing, or structural roles in RNA maturation. Chapter 5.1 will explore some of these possibilities in the context of the present belief that the processing of RNA may depend upon the three-dimensional conformation of the RNA molecules.

The cap is put onto a growing mRNA very early in its transcription. The polyA tail is added after transcription, probably during splicing. The nucleotide sequence AAUAAA has been found to be present about 15 to 20 nucleotides from the 3' end (to which the polyA is attached) of a wide variety of eucaryotic mRNAs. Perhaps this sequence helps to indicate to the relevant processing enzymes where the transcribed part of the mature mRNA molecule should end and where the polyA tail should be added. (Recent evidence hints that in some cases the initial transcription may continue well beyond the future 3' end of the mature transcript, with the "excess" being trimmed off before the polyA tail is added.) Production of a fully processed, mature mRNA molecule takes 3 to 30 minutes, depending on factors such as the length of the molecule and of its precursor.

mRNA Turnover Still to be determined is whether all hnRNA molecules made in the nucleus give rise to cytoplasmic messenger RNAs, or whether some either have other functions or else represent potential mRNAs that are, however, totally degraded within the nucleus, perhaps as one aspect of the regulation of the cell's genetic expression (Section 4.4.6). Much more information is also needed about the events through which a new mRNA is transported to the cytoplasm and about the mechanisms that determine its life span once it arrives. Some bacterial messages survive only a few minutes before being degraded by RNases; there actually seem to be circumstances in which degradation of the 5' end of an mRNA begins while translation and even transcription of younger parts of the molecule is still under way. Eucaryotic messages can be very long-lived, lasting many hours or even many days, but this varies for different messages, different cell types, and different circumstances (see Sections 4.4.2 and 4.4.6). Fast turnover (short mRNA life spans) facilitates metabolic flexibility—the cell being able to replace one set of mRNAs with a somewhat different set and thus to alter its metabolic direction rapidly, as occurs in bacteria when environmental changes induce the synthesis of particular enzymes (Section 3.2.5). Reticulocytes or hen oviduct cells, on the other hand, make the same product for long periods of time and do so by reusing the same mRNA molecules repeatedly. Experimental removal of the polyA tail seems to shorten the functional life span of mRNAs injected into *Xenopus* eggs. At present, however, this is more important as a demonstration that the nature of the control of mRNA turnover is open to experimental attack than as decisive evidence that the polyA tail normally regulates mRNA half-life. Similarly, the cap on eucaryotic mRNAs diminishes their susceptibility to attack by certain classes of RNases, but whether this is responsible for the relatively longer lives of eucaryotic mRNAs as compared to procaryotic ones, as some investigators assert, is a matter for future evaluation.

Chapter 2.4 Endoplasmic Reticulum (ER)

A great contribution of electron microscopy was the demonstration that in many cells of eucaryotes an extensive membranous system, the endoplasmic reticulum (ER), traverses the cytoplasm. (The resolving power of the light microscope is too low for identification or analysis of this system.) The lipoprotein membranes delimit interconnecting channels that take the form of flattened sacs (known as *cisternae*) and tubules. Rough ER is studded with ribosomes most of which are in the form of polysomes. Rough ER and smooth (ribosome-free) ER are part of one interconnected system, whose interior is closed off by the ER membrane, which prevents direct entry by the larger varieties of cytoplasmic molecules such as macromolecules. The proportions of the two ER types vary in different cell types. Proteins (including certain enzymes), lipids, and probably other materials are transported and distributed to various parts of the cell through the ER. In some cases these substances may accumulate and may be stored within the ER for considerable periods. In striated muscle the ER takes a special form, the *sarcoplasmic reticulum,* involved in coupling nerve excitation to muscle contraction (Section 3.9.1) through control of cytoplasmic Ca^{2+} concentrations.

The endoplasmic reticulum is much more than a passive channel for intracellular transport or an extensive surface on which to organize production of certain proteins. It contains a variety of enzymes playing important roles in metabolic sequences, for example, in synthesis of steroids.

As mentioned earlier (Chap. 1.2B), the ER as such cannot be isolated intact from cells. Fragments of ER are the major component of the *microsome* fraction. The fragments are in the form of membrane-bounded closed vesicles (see Fig. I–25). The vesicles form as pinched-off portions of the ER during homogenization. The interior of the vesicles corresponds to the interior of the cisternae and tubules of the ER. The microsomes from rough ER retain their ribosomes, which thus stud the external surface of the vesicles.

2.4.1 Free and Bound Polysomes

Two distinct populations of ribosomes are recognized: those *bound* to membranes (endoplasmic reticulum); and those that are *free,* that is, not visibly bound to membranes. The two types seem to draw their ribosomal subunits from the same cytoplasmic pool of subunits cycling onto and off mRNAs (Section 2.3.2). Both types form polysomes with mRNA, and both play similar roles in protein synthesis. But the free polysomes leave the newly synthesized proteins in the hyaloplasm (Chap. 1.2B), whereas bound polysomes transfer the protein into the membrane of the endoplasmic reticulum or, often, through the membrane into the ER interior. Evidence for this difference in function rests on comparative studies of different cell types.

Animal cells that secrete proteins such as digestive enzymes or hormones have a high proportion of bound polysomes. The proteins enter the extensively developed system of rough endoplasmic reticulum and, by mechanisms considered in this chapter and the next, are "packaged" into membrane-delimited granules. Eventually the granules are released at the cell surface.

Similar synthesis on bound polysomes and packaging in membrane-delimited granules probably occurs with lysosomal enzymes (Section 2.8.4). These "packages" are retained by the cell for its own use. Many cell types that maintain markedly extensive specialized membrane systems such as neurons (Chap. 3.7) or photoreceptors (Chap. 3.8) also have a particularly well-developed rough ER.

Free polysomes are especially abundant in cells synthesizing much protein (enzymes and so forth) used internally in rapid growth but not specially packaged inside membrane-delimited structures (for example, cancer cells and most cells of embryos). Evidently both the proteins of the hyaloplasm and those of non-membrane-delimited organelles are synthesized on free ribosomes. In *reticulocytes,* the cells that mature into red blood cells (Section 2.3.5), most of the ribosomes are of the free variety. The hemoglobin they synthesize is not included within special membranes and is not exported from the cell (see Fig. II–2).

Hepatocytes synthesize much protein (such as serum albumin) that is secreted into the blood and other proteins for internal use (for example, in replacing proteins lost in turnover). These cells contain many free and many bound polysomes.

As with most generalizations, the ones in this section must be treated cautiously. For example, it would be premature to rule out completely the synthesis of a few proteins of the hyaloplasm on bound polysomes, and there is good reason to believe that certain of the proteins of membrane-bounded

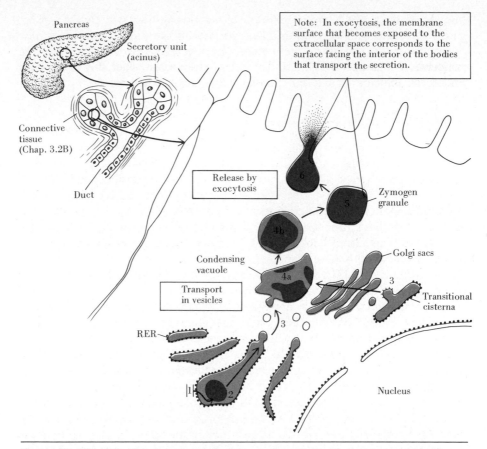

Note: In exocytosis, the membrane surface that becomes exposed to the extracellular space corresponds to the surface facing the interior of the bodies that transport the secretion.

Figure II–33 *ER, Golgi apparatus and secretion.* Path of secretory material from polysome *(1)* to lumen *(6)* in the exocrine gland cells (Section 3.6.1) of guinea pig pancreas. *RER* is rough endoplasmic reticulum. (After G.E. Palade, J. Jamieson, and others.) Condensing vacuoles are immature secretion granules that may arise from Golgi-associated sacs. They transform into secretion granules. Transitional ER cisternae are those that contribute material to the Golgi apparatus. Some of the vesicles involved in transport between ER and Golgi apparatus appear to be "coated" (Section 2.1.6). Each zymogen granule contains a mixture of the several types of enzymes produced by the cell.

The diagram illustrates a small portion of a secretory unit *(acinus),* as indicated. The gland is made up of many such units, each a more or less globular or flask-shaped group of cells that secrete into a lumen communicating with a duct system (see Chapters 3.5 and 3.6A for additional discussion of tissue organization). The units are surrounded by connective tissue.

compartments are made on free polysomes (Sections 2.5.4 and 2.6.3). Moreover, in the view of a few investigators, the simple classification of polysomes into "free" and "bound" is misleading. These investigators are impressed by the fact that, on the one hand, some proteins destined for insertion in ER membranes are made on what usually are thought to be free polysomes (Section 2.5.4) and that, on the other hand, bound polysomes may vary in the history of their association with the ER, depending on the protein being synthesized (Section 2.4.3). A "free polysome" making a membrane protein may conceivably establish a loose and transient relationship with the membrane, and there may be bound polysomes that remain "free" for an appreciable length of time as they initiate translation of their mRNAs. In other words, there could exist functionally important intergrades between strictly free and fully bound polysomes, whose existence will become evident with better techniques.

2.4.2 Protein-Secreting Cells

Autoradiographic Observations   In cells actively engaged in synthesizing and secreting proteins, *rough* ER is particularly elaborate. So is the Golgi apparatus, which led, early, to the suspicion that these two organelles collaborate in the secretory process.

The most intensively studied secretory cell is the type responsible for production of digestive enzymes in the guinea pig pancreas (Figs. II–33 and II–34). The fate of specific proteins (precursors of digestive enzymes made by the pancreas) has been followed by autoradiographic studies, combined with biochemical analysis of isolated organelles and of secretions released by the cells. In autoradiographs, labeled amino acids that are incorporated in these enzymes are found first over the rough ER, later in the region of the Golgi apparatus, still later in secretion granules, and finally in the extracellular space at one pole of the cell where secretions are released. (Upon release, the proteins enter ducts leading to the intestine; as released by the pancreas, the enzymes are called *zymogens* since they require activation in the intestine to function in digestion [Section 3.6.1]). From the polysomes where the proteins are manufactured, they enter the cisternae of the ER and move toward the Golgi apparatus. Then, apparently by transport in small vesicles, they pass from the reticulum either directly into "condensing" vacuoles, or into the sacs of the Golgi apparatus and from these into vacuoles derived from the sacs (Fig. II–33). "Packaged" into vacuoles, the secretory material undergoes concentration (condensation), which results in the formation of dense secretion granules. The vacuoles containing the secretion granules move to the cell surface, where the granules eventually are discharged by a kind of "reverse pinocytosis" (exocytosis), in which the vacuole membrane fuses with the plasma membrane (see Figs. II–33 and II–46).

Related Microscopic Observations   Though, as Chapter 3.6 will describe, there are differences in detail, routes and mechanisms basically similar to these are employed in a wide variety of secretory cells. In a few such cell types, the accumulation of considerable amounts of secretory proteins can be readily visualized with microscopic techniques. For instance, granules resembling secretion granules are found in the ER of secretory cells in dog pancreas and in the

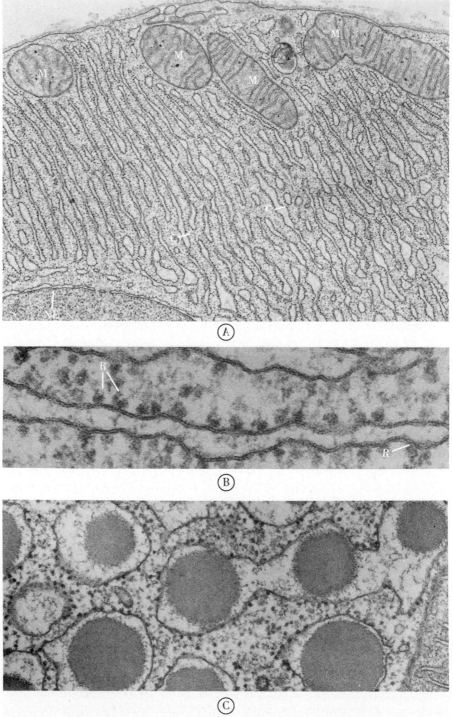

pancreas of fasting guinea pigs (Fig. II–34). This is direct confirmation of the autoradiographic evidence that secretory proteins are transported in the ER (see also Fig. III–40).

Cytochemical methods can be used to detect several specific types of proteins within the ER. For example, cells of the thyroid gland and some cells of salivary glands *synthesize* and *secrete* peroxidase enzymes. When preparations of these tissues are incubated in the proper cytochemical medium for demonstration of peroxidase activity (Section 2.1.6), the enzyme-produced reaction product is distributed as in Figure II–35. As with several other microscopically demonstrable enzymes, the nuclear envelope shows reaction product, as does the rest of the ER. Comparable cytochemical findings implicate structures in and associated with the Golgi apparatus, plus membrane-delimited secretory bodies that undergo exocytosis, in the intracellular transport and release of secretions. Immunohistochemical approaches also demonstrate involvement of ER and Golgi systems in cells secreting a variety of materials, including the pancreatic digestive enzymes. (Fig. III–40 illustrates the use of such methods in the study of secretion of antibody molecules themselves.)

2.4.3 Protein Entry into the ER

By what mechanisms does the cell determine which of its polysomes are to be free and which bound? How do proteins, which in their usual state are too large and too polar (Section 1.4.1) to cross membranes, pass into the interior of the ER? These two questions are fundamental to the understanding of secretion and of many other facets of cellular compartmentalization as well—especially the formation of membranes themselves (Section 2.5.4). Answers are coming from investigations with isolated polysomes, and with isolated microsomes derived from the rough ER of cells such as of the liver or pancreas. The picture that follows represents the results of recent work with such *in vitro* (test tube) experimental systems. Quite possibly some of the details will require revision as more information accumulates, and particularly as the interpretations are applied to situations *in vivo* (in the intact, living cell). But the basic outline is already strongly supported by diverse lines of evidence.

Polysomes' Associations with Microsomes When either naturally free polysomes or polysomes removed from microsomal membranes are made to

---

◄ **Figure II–34** *Features of the rough ER.* (Courtesy of G.E. Palade.) The three micrographs are from pancreas exocrine gland cells (Fig. II–33). A. Portion of a cell (rat pancreas) showing numerous flattened cisternae *(C)* of rough ER. Mitochondria are seen at *M* and the nuclear envelope at *NE*. × 15,000. B. Higher magnification micrograph of a single cisterna (guinea pig pancreas). The membrane surface is studded with numerous ribosomes *(R)*. × 150,000. C. Portion of a cell from a fasted guinea pig, in which large granules, probably of protein, have accumulated inside the cisternae. × 50,000.

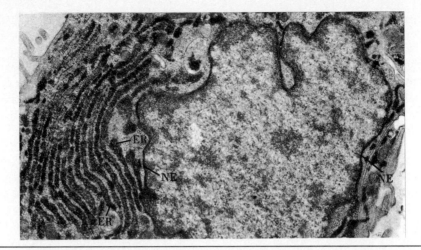

Figure II–35 *Newly synthesized proteins in the ER.* Portion of a rat thyroid epithelial cell from a section incubated with a cytochemical procedure that demonstrates sites of peroxidase enzymes. Deposits of electron-dense reaction product are present in the nuclear envelope *(NE)* and in the many cisternae of rough ER. The enzyme responsible for staining is one synthesized by these cells and thought to be ultimately released into the gland's lumen, where it participates in adding iodine to secreted protein (see Chapter 3.6A). × 15,000.

synthesize proteins in an experimental reaction mixture to which proteolytic enzymes are then added, the growing polypeptide chains can reach 30 to 40 amino acids in length before the enzymes begin to degrade them. Evidently, the association of the nascent chain with the large ribosomal subunit protects it for a time, but eventually the chain gets too long and begins to protrude out into the incubation medium where the proteases can get at it. When polysomes bound to microsomes are used, the protection is much more complete: The nascent polypeptide chain moves directly from the bound polysome into the interior of the microsomal vesicle where proteases cannot reach it (unless the microsomal membrane is first disrupted by the use of detergents such as Triton X-100 or deoxycholic acid [deoxycholate], which open the vesicles, exposing their interiors). The conclusion is that proteins move directly into the ER as they are synthesized (Fig. II–36), rather than first being completed and released from the ribosome to the cytosol and then taken up. Such "cotranslational" passage probably means that the protein enters the ER before its folding is completed. This makes the passage easier to conceive of (an unfolded polypeptide chain is roughly 10 Å in diameter, much narrower than a folded chain) and also helps to make passage unidirectional ("vectorial"), since once the chain has folded inside the ER it is too large to come back out.

That the large ribosomal subunit—the one with which the nascent polypeptide first associates—is attached to the ER membrane is suggested by images like the one in Figure II–36. Furthermore, when the level of Mg^{2+} ions in the medium surrounding microsomes is reduced by using agents like EDTA (ethylenediamine tetraacetate), which binds—"chelates"—Mg^{2+}, the ribosomal subunits separate, leaving the large subunits of membrane-associated polysomes still stuck to the membrane, while the small ones come free. The growing polypeptide chain inserted in the membrane also anchors the ribosome, so that removal of bound polysomes from microsomes with minimal disruption requires release of the chain from the polysomes; release can be induced, experimentally, with puromycin (Section 2.3.2).

Polysome-studded microsomes that transfer proteins to their interior can be isolated as such from the cell, but they also can be reconstituted *in vitro*. Successful reconstitution requires employment of microsomal vesicles originating from the ER; similar-sized vesicles from other cellular membrane systems will not do, indicating that specific features of the ER membrane are required. Polysomes originating from the cell's bound polysome population, or those that form *in vitro* when mRNAs coding for secretory proteins are added to appropriate reaction mixtures, can attach to these microsomes in proper fashion to transfer their products to the interior. Mixtures of heterogeneous origin—including mRNAs from one cell type or species, ribosomes, tRNAs, and enzymes from another, and membrane preparations from still another—will work; components from plant cells will even cooperate with those from animals. However, whereas *ribosomes* from many sources, including those from the free polysomes, will serve, free *polysomes* as such and *mRNAs of free polysome origin* cannot substitute for those from the bound population; they can form only loose associations not capable of protein transfer.

Thus, proper interaction between polysomes and membrane depends upon features of both. For the polysomes these features are determined by the mRNA rather than the ribosomes and are of a general sort—that is, bound polysomes producing diverse proteins can associate with the same ER sites.

Unifying Hypotheses   Is it the mRNA itself or the product it codes for that determines effective attachment to the ER membrane? One piece of evidence is the fact that when polysomes are induced to synthesize proteins before they are mixed with membranes competent to attach the polysomes, successful attachment, in many cases, takes place only when the mixing is carried out very soon after protein synthesis begins. Since the mRNA does not change during its translation, this experiment implicates the nascent polypeptide as a key participant in the formation of an effective association between polysome and ER membrane; the polypeptide, once long enough, would be expected to fold up, altering not only its effective size but also the distribution of groups that might interact with the membrane before folding became extensive. Figure II–37 illustrates a more startling variety of evidence: When the mRNA for a protein destined normally to enter the ER (or a rough microsome) is translated in a system lacking proper membrane, the product is *longer* than the protein normally produced. Experiments of this type led to the discovery that many secreted proteins are translated initially with a stretch of about 20 amino acids at

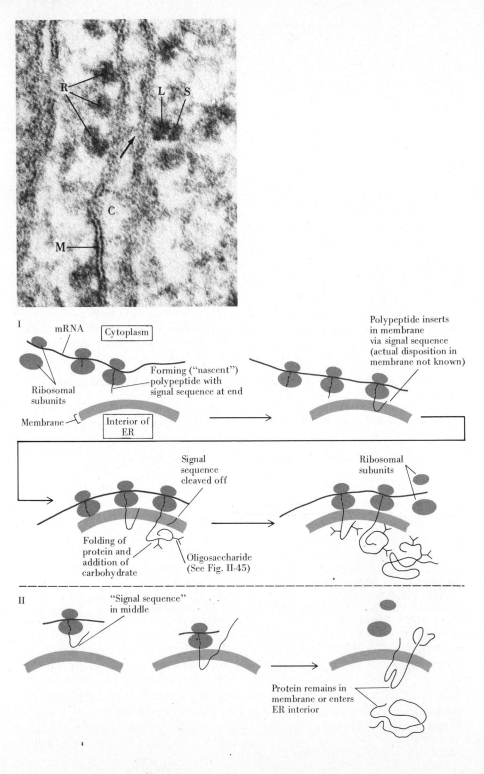

I

mRNA

Cytoplasm

Ribosomal
subunits

Membrane

Interior of
ER

Forming ("nascent")
polypeptide with
signal sequence at end

Polypeptide inserts
in membrane
via signal sequence
(actual disposition in
membrane not known)

Signal
sequence
cleaved off

Ribosomal
subunits

Folding of
protein and
addition of
carbohydrate

Oligosaccharide
(See Fig. II-45)

II   "Signal sequence"
        in middle

Protein remains in
membrane or enters
ER interior

the N-terminal (the earliest-made end). This segment of the polypeptide is removed very soon after the protein begins to enter the ER. The removal depends on the activity of a "peptidase" enzyme built into the ER and does not occur unless the protein crosses the ER membrane.

The segment of polypeptide chain exhibiting this behavior was designated the "signal" sequence. The amino acids present, and their arrangement, differ somewhat from one protein to another, but the features common to many proteins are thought to be more important: Most amino acids found in signal sequences have hydrophobic (apolar) side chains (like *leucine* and *valine* in Fig. II–23) and thus could establish conformations favoring penetration into the lipid bilayer. The signal sequence of amino acids is specified by mRNA;

◀ **Figure II–36** *"Bound" ribosomes; the "signal hypothesis."* Micrograph: Portion of a cisterna *(C)* of endoplasmic reticulum in guinea pig hepatocytes. The delimiting membranes show unit-membrane appearance *(M)* where they are sectioned in a plane perpendicular to the membrane; the more blurry appearance elsewhere results from curvature and irregularities in the membrane leading to oblique and face-on views of the membrane in the section. The division of the ribosomes into large and small subunits is seen at one of the ribosomes attached to the cisternae *(L* and *S)*. The large subunit is directly associated with the membrane, seen faintly at the arrow because it is sectioned obliquely. The other ribosomes included in the micrographs *(R)* are not as favorably oriented for observing their division into subunits. × 275,000. (Courtesy of D. Sabatini, Y. Tashiro, and G.E. Palade.)

Diagram: The "signal hypothesis" as formulated by G. Blobel and D. Sabatini. I. The polysome illustrated has three ribosomes. The one toward the right end of the mRNA strand was the first ribosome to associate and therefore has synthesized the greatest length of polypeptide (and is furthest along the message). As the "signal sequence" of amino acids at the beginning (N-terminal) of the polypeptide emerges from the large subunit of this ribosome, it mediates initial attachment of the polysome to the membrane. Subsequently, similar attachment occurs sequentially with the other two ribosomes of the polysomes. Not yet known are the details of signal entry and cleavage and of the passage of the polypeptide through the membrane. Nor is the ultimate fate of the signal sequence after it has been cleaved off understood. Once the polypeptide chain has been completed and has been released from the ribosome on which it was made, the ribosomal subunits are freed from the polysome. Note, however, that at the time this has occurred for the first ribosome, the other two are still in process of making their polypeptide chains, and thus the polysome remains attached to the membrane (see Section 2.4.3). Note also that glycosylation (Section 2.5.2) and folding of the polypeptide begins as it enters the ER cisterna and therefore can be well under way before the polypeptide chain has been completed. II. It is postulated that for some proteins, the signal sequence is located in the interior of the polypeptide chain rather than at the end (Section 2.4.3); these proteins may enter the membrane in "hairpin" configuration, as shown. Some types of proteins made in this way seem able to pass entirely through the membrane, with no cleavage of the signal, and to be released into the interior of the ER. Others may remain inserted in the membrane in various orientations specific for each species of protein: in the case shown, both ends of the polypeptide (the N-terminal and the C-terminal) remain on the cytoplasmic side of the ER membrane and a loop extends into the interior (Section 2.5.4).

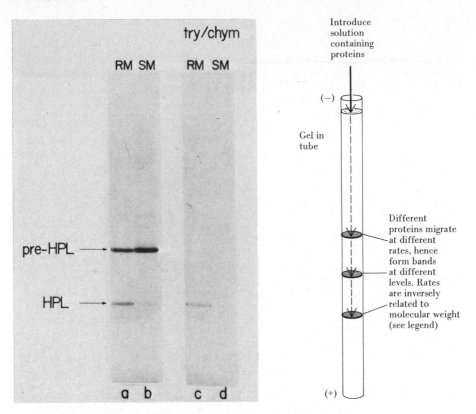

correspondingly, the differences in behavior between free and bound polysomes are due, in large part at least, to the absence of information for signal sequences in the mRNAs of free polysomes and the absence of signal sequences in the proteins they produce.

The proposal that a special hydrophobic portion of the polypeptide chain targets the polysomes to the ER and initiates the entry of the polypeptide into the reticulum—the "signal hypothesis" (Fig. II–36)—is now the reigning view and, as will be seen, has been the basis for fruitful thinking about other targeting and entry mechanisms for proteins of organelles. It should be noted, however, that not all proteins which enter the ER interior lose a segment on doing so, and that not all have a hydrophobic sequence located at their N-terminal end: For example, in the case of ovalbumin, a secretory product of the chicken oviduct (Section 2.3.5), the possibility is being explored that the signal sequence responsible for entry into the ER is a hydrophobic stretch of amino acids present in the middle of the protein that is never cleaved from the protein. In addition, for proteins that come to reside *in* the membrane rather than crossing it, the "signal hypothesis" has been supplemented or replaced by a

◄ **Figure II–37** *Entry of newly synthesized protein into the ER.* An *in vitro* (test tube) protein synthesizing system (Section 2.3.5) was prepared using radioactive amino acids (^{35}S-methionine; see Fig. I–28), ribosomes and other components from reticulocytes, and mRNA from human placenta. The predominant mRNA species present was one coding for human placental lactogen (HPL), a protein hormone (molecular weight 25,000) released by the placenta. To different samples of this mixture were added rat liver microsomes, either those derived from rough ER (RM) or those from the smooth membranes (SM). The columns ("lanes" or "tracks") marked a and b show the products made under these conditions, as analyzed by gel electrophoresis, a technique that separates proteins on the basis of molecular weight (see below). With smooth microsomes, the mixture produces unprocessed "preHPL" (the signal sequence is still present; Section 2.4.3). With rough microsomes the signal sequence is removed from some of the preHPL molecules generating HPL. Lanes c and d illustrate the effects of proteolytic enzymes (trypsin and chymotrypsin: Section 3.6.1) added to the reaction mixtures. The preHPL is degraded by the enzymes and so no longer is present, but the HPL is not degraded and so is still detectable on the gel. This "protection" of HPL indicates that the processed form of the protein that is produced in the presence of rough microsomes crosses the microsome membrane so that it is not accessible to the proteolytic enzymes. (Courtesy of D.D. Sabatini, G. Kreibich and colleagues.)

Right panel: *Polyacrylamide gel electrophoresis* (PAGE). In this technique for separating macromolecules from one another, the molecules are placed at the top of a tube or other system containing a synthetic gel (made of polyacrylamide) and subjected to an electric field. The rate of migration of a macromolecule through the gel depends on the molecule's size, shape and charge and the orientation of the electrical field. During a given time interval, molecules differing in these respects migrate to different extents and thus form "bands" at different levels in the gel. The bands can be visualized by staining or, if the molecules are radioactive as in the present case, by autoradiography. In studying proteins the technique is generally employed in such fashion that separation is based principally on differences in molecular weight of polypeptide chains; in the present case HPL is smaller than preHPL owing to the removal of the signal sequence. Before introduction onto the gel the proteins are exposed to a detergent such as sodium dodecyl sulfate (SDS) which unfolds polypeptide chains and separates the chains of multichain proteins. This "denaturation" minimizes the effects of differences in shape—secondary, tertiary and quaternary structure (Fig. I–28)—on proteins' behavior in the gel. In addition, SDS molecules bind to hydrophobic portions of the protein molecule; since the detergent molecules are negatively charged, this binding both masks the intrinsic charge differences among proteins and results in virtually all proteins migrating toward the (+) pole of the gel, as illustrated. The four "lanes" (a,b,c,d) at the left each represent a preparation like the one in the diagram at the right: timing and other conditions were kept uniform so in each lane a protein of a given molecular weight should appear at the same level. Because it is smaller than preHPL, HPL moves further along the gel in a given time interval than does preHPL.

number of other proposals needed to explain complex orientations of proteins in membranes (see Section 2.5.4). In part to emphasize the remaining uncertainties, some investigators refer to the signal sequence as the "leader" sequence (or "insertion" sequence), although this can produce a bit of confusion because "leader" sequences in nucleic acids have a different significance (Section 2.3.5).

Especially if "signals" do exist in the middle of some proteins (and if, as is likely, certain proteins associate with membranes via other types of hydrophobic zones—see Section 2.5.4), questions arise about the initial stages of association of a newly formed polysome with the ER membrane. If initial association depends solely on a segment of the nascent chain, there must be cases in which there is a quite appreciable delay between initiation of translation and association with ER—the time it takes between initiation of translation of the mRNA and the emergence of the requisite segment of the polypeptide from the large subunit of the first of the polysome's ribosomes (see Fig. II–36). Ovalbumin-synthesizing polysomes can associate with microsomal membranes *in vitro* at relatively later stages of translation than is true with polysomes synthesizing other proteins (see above), but the situation in the cell *(in vivo)* is yet to be explored.

Still being actively debated is whether proteins of the ER membrane help in the initial association of polysomes, and also whether a special channel is required for a nascent polypeptide to cross the ER membrane. Of the proteins built into the membranes of rough ER, only a very few are not also present in smooth ER. Much attention is focused on these in hypotheses to explain why polysomes attach only to the rough membranes. According to one hypothesis, two of the proteins ("ribophorins") serve as receptor sites that help to bind the ribosomes. Perhaps with additional ER components they also serve to recognize the "signal" and to establish a transmembrane channel or other translocating system for the polypeptide. Cross-linking studies (See page 119) do suggest that these proteins are located very close to the ribosomes in rough microsomes. But the channel function is more hypothetical; there still are those who argue that a special channel is not really required: Once inserted in a membrane by the hydrophobic "signal," the chain could, perhaps, be pushed through by the events intrinsic to the process of elongation of the nascent polypeptide, since the ribosome is anchored tightly to the membrane and the polypeptide chain has nowhere else to go.

As a hypothesis to account for coordination between polysomes and ER in timing their activities and interactions a current model suggests that once the polypeptide has grown long enough that the signal sequence emerges from the ribosome, this sequence is "recognized" by cytoplasmic proteins. These proteins interact with the polysome, setting off processes that inhibit continued translation of the mRNA. This supposed inhibition is reversed, it is proposed, by ER-associated proteins; the result is that growth of the polypeptide resumes only when the signal has encountered the ER in proper fashion and thus is in a position to enter the membrane. The signal-recognizing proteins are thought to function as components of distinctive signal-recognizing ribonucleoprotein particles. The particles, according to one recent study, consist of six different proteins complexed with a small molecule of RNA, 7S RNA. This RNA is about 260 nucleotides long, and is one of the "small cytoplasmic RNAs" whose functions are being sought (Section 5.1.2).

Once "delivered" to the ER as part of a bound polysome, how long can a particular mRNA molecule remain associated with the membrane? The answer is not known, but long-term association *might* result if the mRNA itself forms stabilizing attachments to the membrane. Alternatively, such association could result from phenomena of timing: A bound polysome will be linked to the membrane at each of its ribosomes (see Fig. II–36), and since each ribosome is at a different stage of protein synthesis, when a given polypeptide chain is completed the mRNA need not come free from the membrane; the "old" ribosome coming off one end may be replaced by a new one, initiating protein synthesis at the other end of the polysome—which itself is still attached by its other ribosomes. An intriguing suggestion is that the mRNA more or less "crawls" across portions of the membrane surface in this way, remaining attached by a renewable cast of ribosomes. Evaluation of this will require clear information about the degree to which ribosomes, mRNA and forming polypeptides are mobile in the plane of the membrane and about the significance of the spiral or hairpin configurations that often characterize ER-associated ribosomes (see Fig. II–41).

"Posttranslational" Processing For many proteins the removal of the "signal" sequence is only the first proteolytic step required for the production of the mature functional protein. The molecule that will give rise to the hormone insulin, for example, emerges from the ribosome as "pre-proinsulin"— the prefix "pre" indicating that a "signal" sequence is first removed (leaving "proinsulin"), and the prefix "pro" indicating that during its subsequent passage through the cell an additional sequence is excised (Sections 3.6.1 and 4.1.1). Other proteins, such as the digestive enzymes from the pancreas, have specific portions removed after their release from the cell (Section 3.6.1). The ER also is the site of other processing steps in the maturation of proteins, including addition of some of the carbohydrates of glycoproteins (see Section 2.5.2), formation of disulfide bonds (Fig. I–28), and complexing with lipids.

Different Proteins in Different Packages A given cell may produce quite a variety of different proteins destined for different membrane-delimited packages. Liver cells, for example, package lipoproteins, albumin, and other proteins and protein complexes for secretion and put lysosomal hydrolases into lysosomes, peroxisomal enzymes into peroxisomes, mitochondrial components into mitochondria, and so forth. At one point it was thought that most or all of these proteins came from the ER. This led to searches for specialized zones of rough ER within a given cell that might help to explain the sorting process on the basis of localized production of different proteins. There have been reports from cytochemical and cell fractionation studies that, especially when cells begin to make new proteins, one or another ER zone, such as the nuclear envelope, may transiently show different synthetic capacities from the rest. Well-documented cases of this sort are rare, but such behavior could reflect the pattern of spread of new mRNAs from the nucleus. A few investigators claim that there are portions of ER—those located close to mitochondria, for example—whose polysomes show stable differences in the relative frequencies of particular mRNAs from polysomes elsewhere in the cytoplasm. But these are infrequent and hotly debated findings, and although they may well point to important instances worth further study, by and large the search for distinct heterogeneity in the rough ER of most cells has failed thus far. The lack of specificity of polysome–ER interactions outlined above leads to the prediction—still a very tentative one—that cells may not readily be able to establish specialized subregions of rough ER.

Alternative explanations for the sorting of proteins gradually are finding convincing experimental support, as will be detailed in later sections. They include temporal mechanisms (the making of different products at different times; Section 2.5.1), "addressing" devices additional to the signal sequence (Section 2.8.4), and the surprising finding, for several categories of membrane-bounded organelles, that the proteins destined for the organelle interior may not be made on the ER (Sections 2.6.3, 2.7.4, and 2.9.3). Similarly, cell biologists are coming to appreciate that the ER itself may use a variety of mechanisms in the transport of materials that it makes, although very little is known of the fine details. Movement of proteins and other molecules inside the reticulum and movement of membrane plus contained material (by budding off of vesicles, for example) are the most obvious. But molecules may also move within the plane of the ER membrane, and certain types, particularly lipids, may leave the ER as individual molecules to be carried to other compartments (Section 2.5.4).

2.4.4 Smooth ER; ER Enzymes; Lipid Synthesis by ER

Smooth ER Whereas the rough ER is extensively developed in protein-secreting cells, it is the smooth ER that is extensive in cells secreting steroids, such as the hormone-secreting cells of the cortex of the adrenal gland and others (Fig. II–38). Upon homogenization of steroid-secreting cells, key enzymes of steroid synthesis are found in microsome fractions, which in these cells derive mainly from fragmented smooth ER.

Overall, the smooth ER seems capable of playing a number of roles. For the most part, particular functions have been studied only in one or a few cell types or circumstances where they are especially prominent, so that generalization would be premature.

Glycogen In hepatocytes, extensive deposits of glycogen are found intimately associated with the smooth ER. The polysaccharide is in the form of

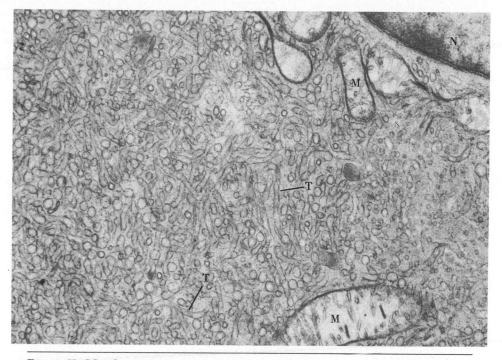

Figure II–38 *Smooth ER.* Portion of an interstitial cell in oppossum testis; these cells produce steroid hormones. Tubules of smooth ER *(T)* are abundant in the cytoplasm. An edge of the nucleus is seen at *N,* and mitochondria are indicated by *M.* × 23,000. (Courtesy of D.W. Fawcett.)

small granules that lie in the cytoplasm close to the tubules of the smooth ER meshwork (see Figs. I–3 and I–7). The significance of this striking relationship—found in several other cell types as well—is still being worked out. Presumably some of the enzymes associated with the ER membrane either play direct roles in glycogen breakdown (and synthesis?) or contribute to the control of this metabolism. One logical suspect is glucose-6-phosphatase, which removes phosphates linked to glucose and is found in the ER of hepatocytes and a variety of other cell types. Ways can be imagined in which this enzyme might participate in glycogen metabolism by converting phosphorylated forms of glucose into the nonphosphorylated forms; the latter forms, for instance, are specifically transported across plasma membranes and thus can leave the cell much more readily than the former. (The hepatocyte's roles include supplying glucose to the blood stream.) But at least two issues require more adequate confrontation. First, both rough ER and smooth ER have glucose-6-phosphatase activity, but only the smooth ER shows intimate associations with glycogen. Second, the portion of the glucose-6-phosphatase molecule responsible for its enzymatic activity is usually thought to be located on the inside of the

ER membrane, so that additional transport systems might be required to move substrates and products across the ER membrane. Very recent reports, however, argue that the latter systems are not required, and that the traditional views of the orientation of glucose-6-phosphatase in the membrane are in error. In general, the permeability properties, possible carrier systems, and other pertinent features of the ER membrane are only slowly coming under intensive investigation.

Oxidases; Drug Detoxification; Ca^{2+}   In hepatocytes, both rough and smooth ER can inactivate drugs such as *phenobarbital*. The enzymes responsible for this drug metabolism include enzymes, sometimes referred to as "mixed-function" oxidases, that take part in several metabolic pathways. In fact, the ER has short electron transport chains that differ in function and detail from the more familiar ones in mitochondria and chloroplasts (Chaps. 2.6 and 2.7) but, like them, serve in oxidation-reduction reactions (Section 1.3.1). As with the mitochondrial systems the ER chains include iron-containing electron-transporting proteins (ones called cytochrome b_5 and cytochrome P_{450} are present in hepatocyte ER) and coenzyme-linked enzymes with which these proteins interact (one characteristic ER enzyme of this sort, called NADPH–cytochrome c reductase, uses reduced NADP, a coenzyme closely related to NAD [Section 1.3.1]). In metabolizing drugs, the ER employs these systems to replace hydrogens in the drug molecules with hydroxyl (OH) groups. Study of these reactions has helped reveal the normal functions of the enzymes, which include catalysis of similar reactions in steroid metabolism and other pathways.

Following administration of phenobarbital to a rat, dramatic increases are seen in the amounts of hepatocyte smooth ER and of the drug-metabolizing enzyme. This interesting example of cellular changes affecting a particular organelle has proved useful in studying membrane changes as well as the enzymatic capacities of ER. It was once thought that the smooth ER in these circumstances was formed by the loss of ribosomes from the rough. The amounts of rough ER, however, remain fairly stable, and investigators are now studying proposals that the new proteins needed for the expansion of the smooth ER may instead pass from the rough to the smooth by moving within the plane of the membrane through the numerous direct continuities between the two systems. This would be consistent with the fact that, for the most part, rough and smooth ER membranes show similar composition. But it raises the interesting question: If some proteins can diffuse between rough and smooth ER membranes, why are the ribophorins or other components involved in ribosome binding restricted to rough ER?

Because of the many regulatory functions of calcium, there has been much recent interest in the possibility that one widespread role of smooth ER is in the control of cytoplasmic Ca^{2+} levels. This emerges from work on striated muscle cells, where such ER involvement is well established (Section 3.9.1) and seems likely for the axons of nerve cells (Section 3.7.5). But it is too early to know how many cell types make comparable use of their ER.

Lipid Synthesis   The ER is a major site of lipid synthesis, probably *the* major site in most cell types. Enzymes involved in synthesis of triglycerides, phospholipids, and steroids are present in isolated microsome fractions, and autoradiographic studies show that radioactive lipid precursors are incorporated rapidly into material in the ER. The electron transport systems referred to above take part in the conversion of saturated fatty acids to unsaturated ones (see Fig. II–3), another ER role in lipid metabolism.

Under some normal conditions, and some abnormal ones, lipids may accumulate as visible deposits within the endoplasmic reticulum. Thus, when fats are fed to a rat, triglycerides are broken down in the intestine into smaller molecules, chiefly monoglycerides, glycerol, and fatty acids. These then cross the plasma membrane of the absorptive cells lining the intestinal cavity. Within the absorptive cells the smaller molecules enter the ER, and enzymes in the ER convert them into triglycerides again. The triglycerides synthesized by the ER are visible as small droplets within the meshwork of smooth endoplasmic reticulum and vesicles that form from this ER (Fig. II–39). Triglyceride is also seen in the cavities of the rough ER, including the nuclear envelope. Eventually, the fat is secreted into the blood (Section 3.5.3).

Normally, much of the lipid synthesized in the hepatocyte ER complexes with proteins, also made in the ER, to form lipoprotein droplets that are secreted by the cell after passage through the Golgi apparatus. The lipoproteins secreted by the liver and the intestine are of great importance for the transport of fats and cholesterol in the blood (Section 3.5.3). The ones made by the liver include a category called "very-low-density lipoproteins" (VLDL) from their behavior in centrifugation, whose roles include the carrying of triglycerides, made by the liver, to fat-storing adipose tissue. These lipoproteins are readily visible as small particles in the smooth ER, as well as in Golgi-associated sacs and vacuoles (see Figs. II–41 and II–47), and it is widely assumed that they form from proteins synthesized in the rough ER complexed with lipids (these can include triglycerides, cholesterol, and phospholipids) made in the smooth. This assumption is part of a broader one: that in many cell types there is a general division of labor between rough and smooth ER, with the rough specializing in protein synthesis and the smooth in lipid synthesis. In the future this may be shown to be the case, but because both rough and smooth ER can synthesize lipids, it seems more likely to prove somewhat oversimplified. There may instead be *quantitative* differences between the two forms of ER, with the smooth making more of certain types of lipids, as seems obviously true in steroid-secreting cells.

Since the ER makes both lipids and proteins, it might be expected to participate centrally in the formation and maintenance of cellular membranes,

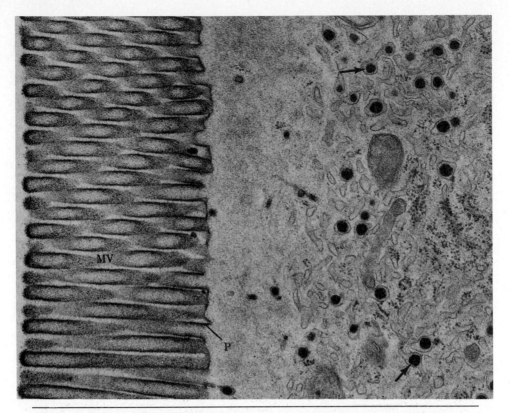

Figure II–39 *The ER synthesizes lipids.* Portion of an absorptive cell in the intestine (see Fig. III–27) of a rat fed a fat-rich meal. Many lipid droplets (arrows) are present inside the endoplasmic reticulum; they have been synthesized from components entering the cell from the intestinal lumen. At the lumen of the intestine, the plasma membrane *(P)* is seen to be extensively folded into closely packed microvilli *(MV)*. × 30,000. (Courtesy of S. Palay and J.P. Revel.)

its own included. It does but this role can be more comfortably discussed after we have considered the Golgi apparatus (see Section 2.5.4).

Chapter 2.5 The Golgi Apparatus

The history of our knowledge of the Golgi apparatus illustrates well the dependence of understanding of cell organelles upon the techniques available at a given time. The structure was discovered in 1898 by Camillo Golgi, who named it the "internal reticular apparatus." During the next few decades, Golgi's "metallic impregnation" method, involving long-term soaking in osmium tetroxide (later modified by using silver salt solutions), permitted light microscopists to demonstrate a similar system in many cell types and to obtain initial evidence linking the Golgi apparatus to secretion: The system was found to be prominent in gland cells, and secretory structures were found to arise in its vicinity. But Golgi's methods were difficult to control; despite several improvements by Golgi and others, they failed to give consistent results. Furthermore, the structure had not been seen in living cells. Thus a lively controversy developed between those who thought the Golgi apparatus to be a real cell structure and those who insisted it was simply an artifact of the fixing and staining procedure. A flood of publications failed to clarify the situation. With the development of electron microscopy, however, the controversy was settled unequivocally. All eucaryotic cells studied, with rare exceptions such as red blood cells, possess the Golgi apparatus. Wherever the classical cytologists had shown the Golgi apparatus in their drawings, electron microscopists found stacks of flattened sacs, each bounded by smooth-surfaced membrane. Located nearby, and sometimes connected to the flattened sacs, were vesicles of various sizes. It was this membrane system in which the osmium or silver was deposited during the classical impregnation procedures. Electron microscopy also showed that secretory materials in many types of gland cells are packaged within membrane-bounded bodies by the Golgi apparatus or related systems. Some other materials not to be secreted, such as components of pigment granules (see Fig. II–49) and possibly the hydrolytic enzymes grouped within lysosomes, also are packaged by Golgi-associated membrane systems.

A major recent advance in understanding Golgi apparatus functions is the development of methods for isolating cell fractions rich in fragments of Golgi apparatus; these procedures have been applied to a growing number of animal and plant cells. The fractions are found to be highly enriched in enzymes known as *glycosyl transferases*. These enzymes catalyze polymerization of sugars into polysaccharides that cells release to extracellular spaces, for example, certain components of plant cell walls (Section 3.4.1) and various secretions of animal cells. Glycosyl transferases also are responsible for the attachment of sugars to the glycoproteins that are abundant in many secretions and also in plasma membranes.

2.5.1 Functional Morphology

Terminological and conceptual problems abound in the literature on the Golgi apparatus. In part these reflect difficulties that sometimes arise in distinguishing the Golgi sacs and vesicles, from other vesicles or from smooth ER that may be present nearby. Moreover, with the usual thin sections used for electron microscopy, it is difficult to evaluate the three-dimensional arrangement of the Golgi apparatus. Thus in many animal cells, the apparatus has the form of a single continuous network, but a given electron microscope section will include only scattered portions of the network and thus may give the impression that the cell contains many separate stacks of Golgi sacs. The overall three-dimensional arrangement can better be appreciated by light microscopy of sections incubated in a cytochemical medium for demonstration of the enzyme *thiamine pyrophosphatase* (TPPase), which is present in high levels in the apparatus (see Figs. I–18 and II–40). Additional information has begun to come from use of relatively thick sections (0.5 to 5 μm) in the electron microscope; more of the apparatus is included in a given section than is true in the usual thin sections.

There is great diversity in the size and shape of the Golgi apparatus in different cell types. For instance, in neurons, an elaborate network surrounds the nucleus (see Fig. I–18), whereas in animal cells that secrete proteins or carbohydrate-rich materials, and in absorptive cells, the apparatus is more compact and is usually located between the nucleus and the cell surface where secretion or absorption takes place (Fig. II–40). In many cells of higher plants, and in a very few animal cells, the Golgi apparatus appears to consist of many unconnected units, called *dictyosomes*. Plant cells may contain dozens to hundreds (or thousands in some algae) of these dictyosomes, each a stack of Golgi sacs.

The number of sacs in the Golgi stack varies from three to seven in most animal and plant cell types (Figs. II–41 and II–42) up to ten to twenty in some cells such as the unicellular organism *Euglena*. Some investigators believe that certain lower plants utilize a single sac for their Golgi apparatus. In the stacks the sacs are separated from one another by a relatively constant distance of 200 to 300 Å. The nature of the material or forces holding them together as a stack is unknown, but in a few cells a thin layer of electron-opaque, sometimes fibrillar, material is seen in the middle of the 200 to 300 Å space. The sacs themselves are of variable width, sometimes quite flat and sometimes dilated by materials that accumulate within them. Quite often the rims show local dilations, which are stages in the budding off of vacuoles or vesicles.

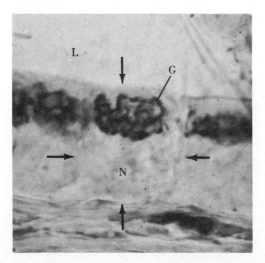

Figure II–40 *Golgi apparatus.* Thiamine pyrophosphatase preparation (see Fig. I–18) showing the Golgi apparatus *(G)* of a cell of rat epididymis, a male reproductive gland. As is usually the case for secretory and absorptive cells, the Golgi apparatus is situated between the nucleus (located at *N* but unstained) and the lumen *(L),* where secretions are released by secretory cells and materials taken up by absorptive cells. Arrows indicate the approximate locations of the borders of the cell. × 900.

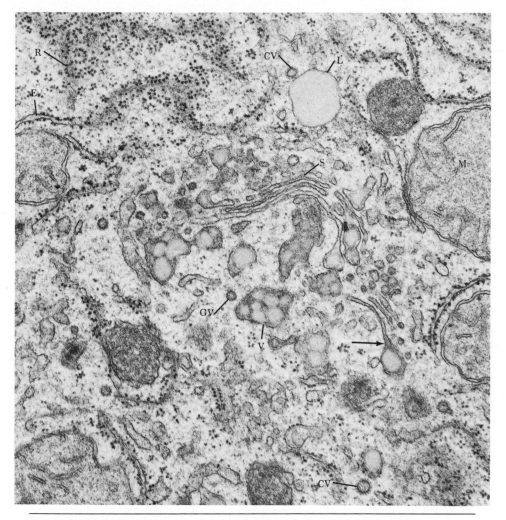

Figure II–41 *ER, Golgi, lipoproteins.* Portion of a rat hepatocyte. *M* indicates a mitochondrion; *E*, endoplasmic reticulum; *L*, a lipid droplet; *S*, Golgi sacs; *GV*, small vesicles of the Golgi apparatus; and *CV*, "coated vesicles" (Section 2.1.6). The structure at *V* is a large Golgi vesicle (vacuole) filled with lipoprotein droplets; the arrow indicates one such vesicle, probably in formation as a dilatation (swollen region) of a Golgi saccule. At *R* the endoplasmic reticulum has been sectioned so that the cisterna surface is seen in face view. The ribosomes are arranged in spiral, hairpin, and other patterns assumed by the polysomes bound to ER. A few tubules of smooth ER are seen at various places in the cytoplasm. × 30,000. (Courtesy of L. Biempica.)

Polarization; Cis and Trans Faces   In most cell types, plant and animal, the Golgi stack is morphologically polarized—one face looks different from the other (Fig. II–42). We shall use the terms *cis* and *trans* to refer to the two faces

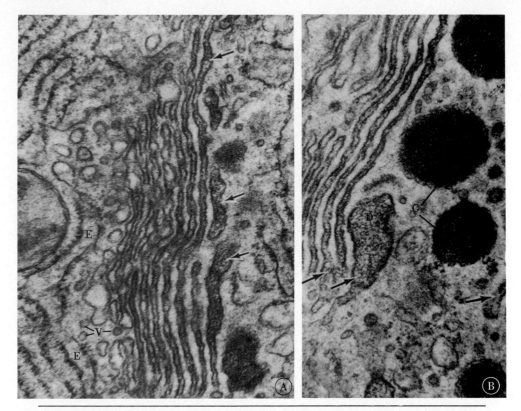

Figure II–42 *Golgi apparatus in a secretory cell.* **Small portions of cells in glands (Brunner's glands) of the intestine (mouse) showing the *Golgi apparatus*. Dense material is present within the Golgi sacs (arrows). This material appears more concentrated in the sacs toward one surface of the stack (the trans surface—toward the right) than in the sacs near the other surface (the cis one). As these micrographs suggest, secretory granules appear to form at least in part by accumulation of the dense material within dilated portions *(D)* of the Golgi sacs at the trans surface of the stack. Eventually the dilated regions pinch off and give rise to membrane-enclosed granules *(G)*. *E* indicates ER, and *V,* small Golgi vesicles. A, × 54,000. B, × 55,000. (Courtesy of D. Friend.)**

("convex" and "concave," "forming" and "maturing," "outer" and "inner," and "proximal" and "distal" have also been used). The sacs at the two faces may differ in a variety of features, few of which are understood in functional terms. For example, in many cell types it is only the sac at the cis face that shows osmium deposits with metallic impregnation methods, whereas only the ones toward the trans face show reaction product in tissues incubated to demonstrate TPPase activity. Biochemically the sacs also show heterogeneity. One model suggests that enzymes shared in common with the ER are particularly abundant on the cis side; while the trans side is richer in Golgi-specific glycosyl transferases. But in at least some cells, the enzyme distributions seem more complicated than this. In addition, there may be compositional and functional differences between the rims of a sac and its central regions, or other hetero-geneity within individual sacs.

Morphological, cytochemical, and enzymatic polarization of the Golgi apparatus is particularly interesting because the functions of the apparatus include what seem to be sequentially occurring steps in the modification or production of various macromolecules. It is easiest to envisage the ordering and regulation of such sequences if the responsible "machinery" is organized in separate compartments, each specializing in one or another step. This, however, may partly be a prejudice based on scientists' familiarity with the organization of industrial factories and assembly lines.

A striking feature of the TPPase-rich Golgi sac on the trans side of the apparatus in certain neurons and in hepatocytes is the presence of numerous interruptions or *fenestrations* (see frontispiece); indeed, the sac looks like a meshwork of tubules, interconnected in polygonal arrays. Fenestrations, generally less extensive, may be present in the other sacs of the stack and also in the endoplasmic reticulum near the Golgi apparatus. Their functional significance is uncertain. They could provide an increased surface area for interactions of the elements of the stack with adjacent structures. Alternatively, they might somehow relate to the fact that the functioning of the Golgi apparatus involves extensive rearrangements of membrane-delimited compartments, such as the fusion of vesicles with the sacs or the budding of vesicles from the sacs. Some investigators also suggest that, especially for cells with a very extensive Golgi apparatus, lateral continuity among different portions of the network is maintained in part by tubules that connect different regions of stacked sacs.

Distinct polarity in the Golgi stack occurs in the fungus *Pythium ultimum*, where the membranes of different sacs differ in thickness. At the cis face the sac membrane is relatively thin, similar to that of the ER. The closer the sac is to the trans face, the thicker its membrane. In many secretory cells there seems to be a similarly oriented gradient in the concentration of materials visible *within* the sacs (Fig. II–42). Such observations have been taken to indicate that materials enter the stack at this cis face and that the sacs move down the stack, from cis to trans, undergoing changes and concentrating their content as they move. Commonly it is at the trans face that the membrane-delimited "packages" produced by the Golgi apparatus form; the microscopic appearances suggest this occurs by budding of dilated regions from the trans-most sac (Fig. II–42).

Movements of sacs down the Golgi stack have long been thought likely, but techniques are still not available to establish unequivocally that they occur, and there is no basis for believing that passage from one surface of the stack to the other is an *invariable* feature of Golgi apparatus functioning. Some investigators argue that the sacs stay in place, transferring materials down the stack through continuities or by vesicles that bud from one sac and fuse with another. Each sac thus could retain functional and compositional individuality while processing secretions and other materials and passing them down the stack. Thus far, evidence for these views too has proved elusive: For instance, the requisite continuities between the sacs of a given stack are seen very rarely, if at all, and present methods could not demonstrate the postulated vesicle movements conclusively even if they do take place. Progress may soon come through employment of experimental agents that impede the movement of cellular products through the Golgi apparatus. Such an agent is *monensin,* which permits proteins to enter the Golgi apparatus from the ER while inhibiting their exit into secretory structures; it induces marked swelling of the Golgi sacs. (Monensin is an ionophore [Section 2.1.3] for H^+ and Na^+.)

Although the most prominent Golgi packages in many cell types separate from the sacs at or near the trans face, this is not invariable either. Membrane-delimited vesicles can also bud from the lateral margins of the sacs further up the stack. In the white blood cells known as polymorphonuclear leukocytes (Section 2.8.2), packaging of enzymes into lysosomes occurs at the *trans* face of the Golgi apparatus; later in the cells' maturation, a different set of enzymes is included in the "specific granules" that form at the Golgi face that apparently was the *cis* face earlier. Such observations hint at controls and complexities in the movement through the Golgi apparatus that merit detailed study before any route of movement into or through the Golgi apparatus is taken to be the "usual" one. The leukocyte, for example, seemingly achieves differential packaging of enzymes by synthesizing different ones at different times and by varying the sites in the Golgi apparatus where the packages are formed. That is, the cell is able to employ both temporal and spatial segregation mechanisms.

The cis–trans differences can be quite marked in secretory cells; by extension it often is assumed that the trans face, where secretory products are finally packaged, is simply oriented toward the cell surface where the products are slated to be released. Secretions then could move in more or less a straight line, from the apparatus into the stores of secretory bodies awaiting release (see Fig. II–33). But matters frequently are not this simple. In many of the cells with an elaborate Golgi apparatus, the stacked sacs either form cuplike arrays, with the trans face oriented toward the inside of the cup, or have a more complicated disposition. This can lead, for example, to a distribution of sacs like the one in the frontispiece, in which secretory granules form from trans-face structures that, over much of their extent, face "inward" toward the nucleus rather than "outward" toward the cell surface. The point is that passage of secretions from the Golgi apparatus toward the site of release from the cell sometimes may require orienting devices that start them out, at least, in the right direction. This and other oriented movements of membranes that seem required for Golgi apparatus functioning have been studied little as yet. The structures responsible for movements elsewhere in the cytoplasm—microtubules and filamentous material (Chaps. 2.10 and 2.11)—are present in the Golgi region as well and very probably play an essential part.

2.5.2 Packaging; Processing; Secretion

Autoradiography (ARG) has clarified important facets of the packaging and other roles of the Golgi apparatus. (In what follows, bear in mind that the

processes of tissue preparation for microscopy usually wash out small molecules, so that ARG shows only the labeled *macromolecular* population.)

When cells secreting proteins are administered amino acids labeled with
tritium (^{3}H; Section 1.2.4), the first organelles to show radioactivity are the
ribosomes and ER. In the mammalian pancreas (see Fig. II–33) this is observed
at 3 to 5 minutes after label administration. Only later (20 to 40 minutes) is
radioactive protein present in or near the Golgi apparatus (Fig. II–43). The
explanation is that proteins are synthesized by the ribosomes of the rough ER
and that some time is required for them to be transferred to the Golgi appa-

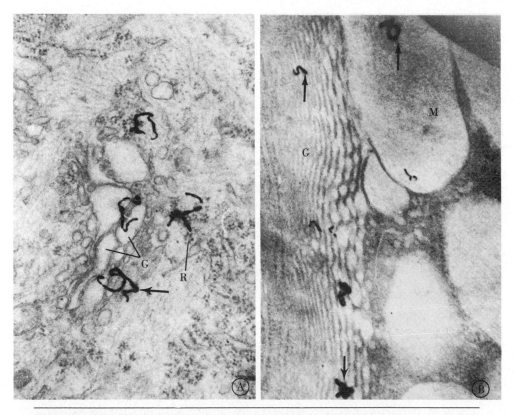

Figure II–43 *Electron microscope autoradiographs illustrating functions of the
Golgi apparatus.* Each irregular dense structure (arrows) would be seen as a single grain in the light microscope (Fig. I–22). A. Part of a neuron in a rat spinal
ganglion fixed 10 minutes after exposure to radioactive amino acid (^{3}H-leucine).
Two of the grains lie close to clusters of membrane-bound ribosomes *(R)*; the
other two lie over Golgi sacs *(G)*. Approx. × 40,000. (Courtesy of B. Droz.) B.
Portion of a mucus-secreting cell (a goblet cell; see Fig. III–27) in rat intestine
fixed 20 minutes after exposure to radioactive sugar (^{3}H-glucose). Grains are seen
over the Golgi saccules *(G)* and over a large Golgi vesicle (vacuole) containing
mucus *(M)*. × 45,000. (Courtesy of M. Neutra and C.P. Leblond.)

ratus. (The mechanism of such transfer will be discussed in the following section.)

With suitable tritium-labeled sugars, the ARG results can be dramatically different. Most plant cells secrete polysaccharide components of their surrounding extracellular walls (see Figs. III–22 and IV–28 and Section 3.4.1). Some animal cells also secrete polysaccharides; for example, cartilage cells release chondroitin sulfate, a major component of the stiff extracellular cartilage matrix (Chap. 3.6B), and goblet cells in the intestinal lining (Fig. II–43) secrete *mucin,* which contributes to the protective layer of mucus that lines the intestine. With these cell types radioactive sugars are seen to accumulate first in the Golgi apparatus, not in the ER. This holds also for radioactive sulfate in cells, such as those of cartilage, secreting sulfated polysaccharides. Apparently the Golgi apparatus is the site of synthesis of secretory polysaccharides, both the linking of the sugars to one another and the addition of groups such as sulfates (and phosphates; Section 2.8.4).

Glycoproteins; Sequential Actions of ER and Golgi Enzymes With cells secreting glycoproteins rather than polysaccharides, ARG results indicate that the linking of carbohydrates to the protein molecules ("glycosylation") begins in the ER and is completed in or near the Golgi apparatus. This is true, for example, of the *thyroglobulin* molecules synthesized and then stored extracellularly by the thyroid gland (Chap. 3.6A and Fig. II–70). Sugars are added in steps to the first ones, which are attached to amino acids of the proteins (Fig. II–44). Different radioactive sugars will give different labeling patterns, depending on where they occur in the "oligosaccharide" chains that are attached to the proteins (Fig. II–44). Galactose generally occurs near the end of the chain; this sugar is incorporated initially in the Golgi apparatus. Mannose typically occurs in the "core structure" near the beginning of the chain (see Fig. II–45) and is incorporated in the ER.

The discovery of glycosyl-transferase activities, notably that of galactosyl transferase, in Golgi-enriched fractions provided direct biochemical confirmation for the synthetic capacities suspected for the Golgi apparatus from the ARG data. These biochemical studies are being extended to yield a detailed picture of the cooperation of the ER and Golgi apparatus in the glycosylation of proteins. As Figure II–45 shows, for many of the chains the initial step in synthesis is the formation of a sequence of N-acetylglucosamines, mannoses, and glucoses linked to a lipid-related *dolichol* molecule associated with the lipid bilayer of the ER membrane. Transfer of this entire oligosaccharide sequence to the growing polypeptide occurs as the latter enters the ER interior—that is, during the elongation of the polypeptide. The oligosaccharide is linked by ER enzymes to the nitrogen in the side chain of the amino acid *asparagine* (Fig. II–44; the oligosaccharides are known as "N-linked" oligosaccharides). The enzymes recognize which of the protein's asparagines should receive oligosaccharides on the basis of the sequences of amino acids in the protein (evidently the presence of a threonine or a serine nearby is one "signal" that an asparagine is potentially to be glycosylated) and probably from the local three-dimensional conformation of the folding polypeptide. Next the oligosaccharide itself is modified both by the trimming off of some of the sugars initially present (*glucosidases* in the ER and *mannosidases* in the Golgi apparatus remove the sugars for which they are named—the mannosidases remove only specific ones of the several mannoses) and by the addition of new sugars (see Fig. II–45). Certain glycoproteins will remain in the ER to function there; for these, the oligosaccharides typically are modified to a final mannose-rich sequence. The glycoproteins that pass through the Golgi apparatus have a core structure that is the residue of the chain initially constructed in the ER but to which the Golgi apparatus adds sugars such as galactose, sialic acid, and fucose (see Fig. II–45).

The saccharide portions of glycolipids, as well as those glycoprotein oligosaccharide side chains that are linked to the amino acids threonine and serine (so-called *O*-linked oligosaccharides; Fig. II–44) are probably formed in a more direct way, by the sequential addition of dolichol-carried sugars to a growing chain. The final steps, at least, of this occur in the Golgi apparatus.

Note, then, that the sequence of sugars in oligosaccharides (and in polysaccharides) depends both on the specificities of the enzymes responsible for linking the sugars and trimming the chains and on the timing and location of exposure of the forming molecules to these enzymes. "Templates" such as those used for RNA and protein polymerization are not directly required. Instead, the cell segregates specific sets of glycosal transferases and "trimming" enzymes to the ER, and others to the Golgi apparatus, and transfers the maturing protein from one compartment to another so as to generate the proper sequence of events.

To have effective access to the molecules on which they act, the enzymes involved in most of these events must have their active sites located at the interior surface of the membranes to which they are attached—that is, facing the lumen of the ER cisterna or Golgi sac. There must also be transport systems for the passage of relatively large hydrophilic molecules, such as the sugars destined for the oligosaccharide chains, across the ER membranes. Linkage to dolichol phosphate, a molecule with a long hydrophobic portion (see Fig. II–45), may be key to the transport of certain sugars, including mannose. In fact at one time it was thought likely that the entire initial oligosaccharide sequence was constructed on the cytoplasm side of the membrane and then transferred as such to the interior via the dolichol carrier. Some investigators still adhere to this view, but others think it more probable that the actual linkage of the sugars occurs inside the ER, with the dolichol somehow serving to transport the sugars individually. Membrane-associated enzymes, perhaps including the ones responsible for linking the sugars into the forming oligosaccharides, may take part in such transport as well.

The sugar-adding glycosyl transferases under consideration are called *transferases* because they transfer sugars from an "activated" form, in which the sugars are linked to *nucleoside diphosphates* such as UDP (U is uracil as in nucleic acids), to polymers such as oligosaccharide chains and polysaccharides (see Fig. II–45). In so doing they release the nucleoside diphosphates.

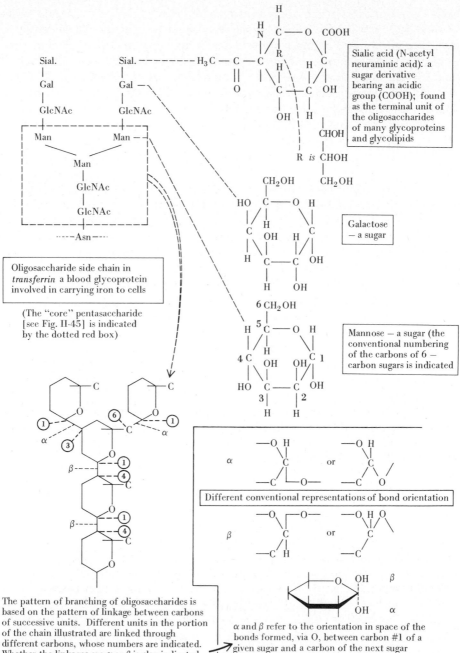

Sialic acid (N-acetyl neuraminic acid): a sugar derivative bearing an acidic group (COOH); found as the terminal unit of the oligosaccharides of many glycoproteins and glycolipids

Galactose — a sugar

Oligosaccharide side chain in *transferrin* a blood glycoprotein involved in carrying iron to cells

(The "core" pentasaccharide [see Fig. II-45] is indicated by the dotted red box)

Mannose — a sugar (the conventional numbering of the carbons of 6 — carbon sugars is indicated

Different conventional representations of bond orientation

The pattern of branching of oligosaccharides is based on the pattern of linkage between carbons of successive units. Different units in the portion of the chain illustrated are linked through different carbons, whose numbers are indicated. Whether the linkages are α or β is also indicated.

α and β refer to the orientation in space of the bonds formed, via O, between carbon #1 of a given sugar and a carbon of the next sugar (see legend)

Ⓐ

Legend on next page

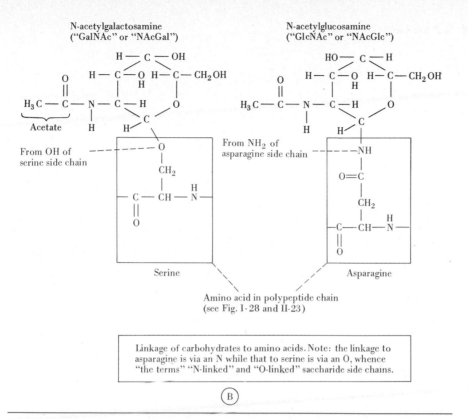

Ⓑ

Figure II–44 *Features of the oligosaccharide side chains of glycoproteins.* Linkage of the chains to proteins is through reactions with the OH or NH_2 groups at the ends of amino acid side chains, as illustrated; in the diagram the GlcNAc and GalNAc shown attached to polypeptides are intended to exemplify the initial units in chains of several to many saccharides (sugars and modified sugars).

Terminology related to linkages between saccharide units. As will be done in subsequent portions of this text, the linkages are referred to in terms such as α1,6 (or α1→6). The numbers specify the positions of carbons in the two units that are linked together. (Generally the linkages involve carbon 1 (C1) of one of the units and other carbons of the second unit.) α and β refer to the orientation of the linkages in space. As illustrated, various two-dimensional representations of saccharide units and chains are used to convey the three-dimensional structure. A useful way of thinking about the structure of a sugar is as a plane corresponding to the ring; the Hs, OHs and other groups that are attached to the ring, project up or down, that is, above or below the plane. The OH group on C1 of a given sugar can assume alternative α and β positions as shown, and these are responsible for the α and β orientations of linkages between sugars. The OH groups on the other carbons have a fixed orientation for a given sugar. Sugars often differ from one another simply on the basis of the orientation of particular OH groups—whether such a group is "up" or "down" can profoundly affect recognition of the sugar by proteins involved in transport or enzymatic reactions. (Compare galactose and mannose diagrammed here and see also Fig. III–21.) Correspondingly, the orientation of the linkages of saccharides to one another is an important specific feature of oligosaccharides and polysaccharides. The specific bonding pattern of a given sequence of saccharide is generated by the enzymes responsible for the synthesis of the chains and affects various metabolic behavior and other biological properties of oligosaccharides and polysaccharides. (Note: α and β = Greek *alpha* and *beta*.)

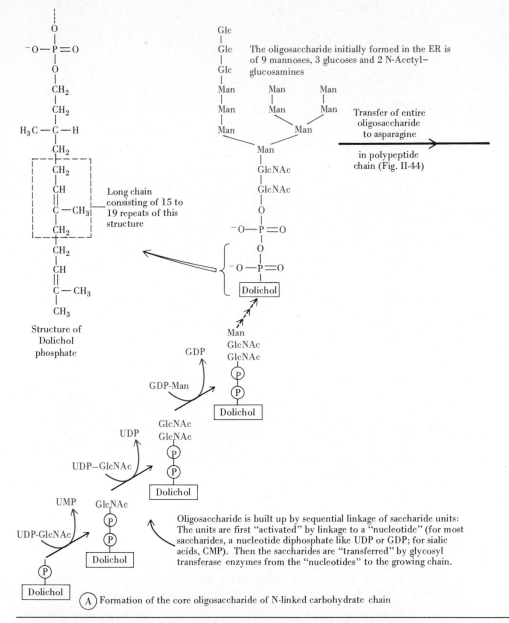

The oligosaccharide initially formed in the ER is of 9 mannoses, 3 glucoses and 2 N-Acetyl–glucosamines

Transfer of entire oligosaccharide to asparagine in polypeptide chain (Fig. II-44)

Structure of Dolichol phosphate

Long chain consisting of 15 to 19 repeats of this structure

Oligosaccharide is built up by sequential linkage of saccharide units: The units are first "activated" by linkage to a "nucleotide" (for most saccharides, a nucleotide diphosphate like UDP or GDP; for sialic acids, CMP). Then the saccharides are "transferred" by glycosyl transferase enzymes from the "nucleotides" to the growing chain.

Ⓐ Formation of the core oligosaccharide of N-linked carbohydrate chain

Figure II–45 *Biosynthesis of oligosaccharide chains of glycoproteins.* **The chain initially constituted is transferred intact from dolichol phosphate to the protein during the entry of the protein into the ER (Fig. II–36). Subsequently the chains are modified as indicated: Mannose-rich forms occur on proteins that remain in the ER or other intracellular sites. In proteins destined for the plasma membrane or for secretion, a number of sugars are removed from the chain initially constructed, leaving a core of five (often plus a GlcNAc added during the**

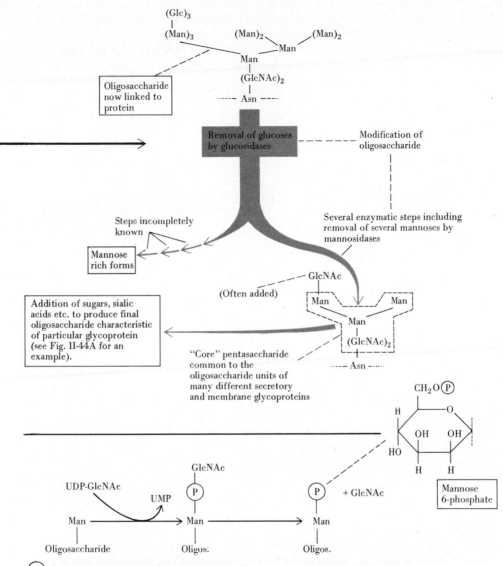

(B) In lysosomal hydrolases, phosphates are added to mannoses of N-linked oligosaccharides yielding the mannose-6-phosphate group that may "target" the hydrolases to the lysosomes (Section 2.8.4)

early stages of modification); additional saccharides are then added to the core, in sequences and patterns that vary among different proteins. In many cases the terminal saccharide of the completed chain is sialic acid (Fig. II–44). When the sugar *fucose* (Fig. III–21) is present, it is frequently linked to one of the initial GlcNAcs (the GlcNAcs directly attached to the polypeptide). (After presentations and work by S. Kornfeld, E. Neufeld, K. Van Figura, W. Sly, and many others.)

This is of interest here because the nucleoside diphosphates can inhibit glycosyl transferase (an example of "feedback inhibition"; Section 3.2.5) but will not do so if the phosphates are removed by other enzymes. The Golgi-associated TPPase (see Fig. II–40) may be one of these enzymes; other "nucleoside diphosphatases" are found in the ER and Golgi apparatus. Thus there is a possibility that TPPase aids or even regulates the transferases—a possibility especially worth investigating since the cellular roles of this enzyme have not been understood despite its utilization as a "marker" of the Golgi apparatus in cytochemical and cell fractionation studies.

Biological roles for the oligosaccharides of glycoproteins are still being sought. Given the early stage at which the oligosaccharides are first attached, it is conceivable that they affect the final folding of the polypeptide as well as its interactions with other molecules. For a time it was believed that the particular oligosaccharides present on glycoproteins served in sorting out the proteins destined for different locales. A somewhat more skeptical view prevails now, since although transport of some glycoproteins to their proper places is prevented by treatment of cells with the drug *tunicamycin* (this inhibits formation of *N*-linked oligosaccharides), transport of others is not. Still, there are cases—lysosomal hydrolases and proteins of the blood are examples (Sections 2.8.2 and 2.8.4)—in which carbohydrates can direct the proteins to which they are linked, along specific transport routes. Also potentially illuminating is that most of the proteins that pass through the ER do become glycosylated (that is, acquire saccharide side chains), whereas this is true of very few, if any, of the proteins made by free polysomes.

As outlined in Section 2.1.7, the oligosaccharide side chains of membrane glycoproteins and glycolipids contribute, perhaps, to controlling molecular orientation and to determining the specificities with which cells recognize molecules that bind to their surfaces (possibly including molecules by which cells recognize one another [Sections 3.11.2 and 4.4.1]).

Condensation and Release Most of the membrane-delimited bodies that leave the Golgi region contain mixtures of several different products: several types of proteins, or proteins and polysaccharides, or sometimes complexes of lipids and proteins. The contents of the bodies frequently are highly concentrated. How the proper mixtures are put together in the Golgi region is not known. As for the concentration or "condensation" process, some speculation suggests that there is an active expulsion of water from forming secretory bodies. However, an active process may not be necessary at least in some cases. In mixtures of proteins and polysaccharides, like some of the mixtures produced by the Golgi apparatus, molecules can associate to form large multimolecular aggregates. The associations depend on ionic bonds and other noncovalent interactions; inorganic ions such as Ca^{2+} may also take part. If aggregation occurs in a membrane-delimited structure such as a vacuole containing secretory materials, the result could be a spontaneous osmotic efflux of water; the osmotic properties of a solution depend upon the total number of independent molecules in that solution (Section 2.1.1), and aggregation reduces the number.

The secretions packaged by the Golgi apparatus and related structures are released from the cell by exocytosis (Figs. II–46 and II–47). Upon appropriate stimulation (Section 3.6.4) the membranes delimiting the secretory bodies fuse with the plasma membrane, releasing the contents to the extracellular space. In the pancreas of a fasted and refed guinea pig, radioactive secretions begin to be released into the secretory ducts within 1 to 2 hours after initial administration of labeled amino acids.

For many secretory cells the immediate trigger for exocytosis is a rise in cytoplasmic Ca^{2+} concentration, perhaps highly localized (Section 3.7.5), which seems to permit the membranes delimiting the secretory bodies to associate intimately with the plasma membrane. The actual fusion of the membranes requires rearrangements of the proteins and lipids still to be described satisfactorily. According to some hypotheses, specific proteins and lipids promote the fusion and ensure that the secretory bodies fuse specifically with the plasma membranes at the proper zones of the cell surface; the filament and

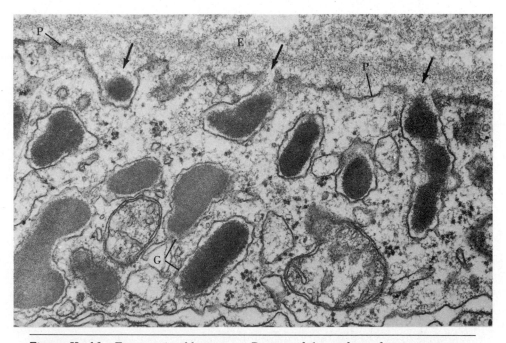

Figure II–46 *Exocytosis of hormones.* **Portion of the surface of a rat pituitary gland cell that produces the hormone prolactin. Secretion granules are seen at G. The arrows indicate granules in process of release from the cell by** *exocytosis,* **the fusion of the membranes enclosing the granules and the plasma membrane** *(P). E* **indicates the extracellular space. × 23,000. (Courtesy of M. Farquhar.)**

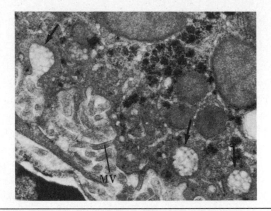

Figure II–47 *Exocytosis of lipoproteins.* Region of a rat hepatocyte near its sinusoidal surface (*see* Figs. I–1 and I–2). Arrows indicate vacuoles filled with lipoprotein droplets (*see* Fig. II–41). The vacuole at the upper left shows fusion of its delimiting membrane with the plasma membrane and presumably exocytotic release of its contents into the extracellular space (compare with Fig. II–46). One of the surface microvilli is seen at *MV.* × 20,000. (Courtesy of N. Quintana.)

microtubule systems that orient and generate cytoplasmic movement may contribute to this as well (Chaps. 2.10 and 2.11). Since metabolic poisons that inhibit energy metabolism also inhibit exocytosis, it appears that cellular energy is required to produce the fusion. A presently popular suggestion is that as fusion begins, permeability changes lead to osmotic movements of water into the secretory body, which generates the force that completes the process.

By conventional electron microscopy, exocytosis appears as the sequential fusion of the three microscopically visualized layers of the membrane delimiting a secretory body with the three layers of the plasma membrane. This could imply that fusion takes place in steps, with the two bilayers initially interacting closely through the hydrophilic surfaces facing the cytoplasm and then undergoing rearrangements involving their hydrophobic interiors and leading to full continuity between the bilayers. Freeze-fracture studies on protozoa have shown distinctive intramembrane-particle arrays to be present at sites of exocytosis (Section 3.3.1), but for most cells, such special arrays are not seen. Some investigators believe that exocytosis generally involves characteristic movements of the intramembrane particles in the fusing membranes away from the points of fusion; others think the observations suggesting this result from artifacts of tissue preparation. Decision between the two views is still impossible, since the actual fusion of given membrane regions can occur in a fraction of a second and thus is difficult to study.

After secretions are released, the membranes that carried them are retrieved from the cell surface. Retrieval is accomplished through endocytosis-like formation of tubules and vesicles, some of which are of the coated-vesicle variety (Section 2.1.6).

Consideration of other aspects of secretion and secretory cells will continue in Chapter 3.6.

2.5.3 ER and Golgi Apparatus: Routes of Interaction

Analysis of the transfer of materials from the ER to the Golgi apparatus has been hampered by the need to reconstruct the process from static electron micrographs and autoradiographs. Recently, progress has begun on reproducing this transfer and other steps in Golgi-apparatus functioning with test-tube mixtures of partly purified cell fractions. This should soon help to clear up the present confusion. The knowledge available now suggests that somewhat different ER-to-Golgi transfer mechanisms may operate in different cell types or circumstances. But in all cell types that have been studied in this regard, it appears that, for large molecules such as proteins, the materials transferred remain within membrane-delimited structures.

Occasionally, continuities between ER cisternae and Golgi sacs are found; whether they are long-lasting or intermittent is not known. Most often the ER and Golgi apparatus seem structurally separate, with transport dependent upon a kind of "membrane flow" or "shuttling." Vesicles (see Figs. II–33 and III–34), or sometimes sacs and tubules, with their contents separate from the ER and fuse with or transform into Golgi components. Frequently, this "flow" seems to converge on the cis Golgi face, with the sacs receiving their ER-derived contents there. This is why the cis face is often referred to as a "forming" face or as "proximal" to the ER. As outlined already, the cis sacs, or portions thereof, are widely supposed to pass down the stack, ultimately to give rise to vesicles or larger secretory structures at the trans face (Fig. II–33 summarizes the proposal that in the mammalian pancreas, at least some ER vesicles may also deliver their contents *directly* to forming secretory structures).

Sometimes, closely apposed to Golgi sacs, ER cisternae may be seen to have ribosomes only on the surface facing away from the Golgi apparatus (see frontispiece). A possible inference is that the ER sometimes gives rise directly to sacs of the Golgi apparatus, as well as generating vesicles that fuse with Golgi sacs. Findings in amebae support this possibility. When an ameba is deprived of its nucleus by microsurgical techniques, the Golgi apparatus is no longer detectable. If a new nucleus is implanted, the apparatus reappears, evidently by transformation of ER sacs.

Apparently the net passage of materials from the ER to the Golgi region is balanced in the steady state by passage of materials out of the region in the form of secretory bodies and other structures formed there. (Many investigators believe that vesicles can shuttle back to the ER to be reused in transporting proteins.) But when cells are stimulated to enhance their rate of production of secretions, they may show increases in the size of the Golgi stacks and in the prominence of the other structures of the region.

In the mammalian pancreas, when energy metabolism is interrupted, by exposing tissue slices to respiratory inhibitors such as cyanide, transport of proteins from the ER to the secretion granules forming in the Golgi region is blocked. The simplest explanation is that the vesicle-mediated transport from the ER that occurs in these cells (see Fig. II–33) is an energy-dependent process. Initial results of experiments with test-tube mixtures of cell components confirm that energy is required for transport steps that seem to occur at the cis face of the Golgi stack. Along with earlier microscopic observations, they also suggest that the vesicles participating in transport in the Golgi region have a coat for at least some of their life history, and that like the coat on endocytic vesicles (Section 2.1.6), this is made in part of clathrin molecules.

In some cell types, small structures resembling beads, 10 nm in diameter but of unknown composition, are found arrayed around vesicles budding from ER surfaces facing the Golgi apparatus. These have been described especially in insects but may have a broader distribution. Their location makes them prime suspects for agents or controlling devices in ER-to-Golgi transport.

GERL   In many cell types, a distinctive-looking collection of fenestrated sacs, tubules, and vesicles is found at the trans face of the Golgi apparatus. These structures are generally distinguished through cytochemical procedures for demonstrating certain enzymes (Figs. II–48 and II–49). The most widely observed of the enzymes is *acid phosphatase*. A similar enzyme is present in

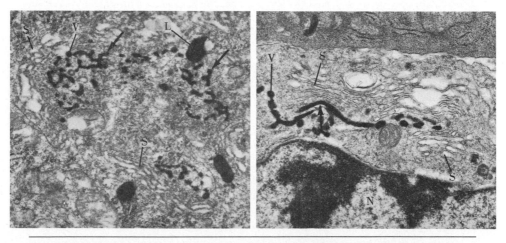

Figures II–48 and II–49   *Enzyme localizations in GERL,* a distinctive membrane system associated closely with the Golgi apparatus. Left, Figure II–48 shows the reaction product for acid phosphatase in *GERL* (arrows) in small vesicles *(V)* and in large lysosomes *(L)* in a rat neuron. Golgi sacs are indicated by *S.* × 20,000. Right, in Figure II–49 (showing a similar region from a cell of mouse *melanoma,* a pigmented cancer) *GERL* shows reaction product for *tyrosinase. N* indicates part of the nucleus. Once packaged in cytoplasmic granules, tyrosinase catalyzes the formation of the pigment *melanin,* which comes to fill the granules. × 20,000.

lysosomes, which are often clustered near the Golgi apparatus and sometimes appear as if budding from or fusing with the sacs and tubules in question. Thus the name *GERL* (*G*olgi-associated *ER* from which *L*ysosomes form) was coined to express the proposal that the sacs and tubules are especially concerned with lysosome formation or functioning. The term also summarized the hypothesis that the structures in question serve as a direct route, continuous with the ER, through which lysosomal enzymes might reach the lysosomes, perhaps without going through the Golgi stack. (In a variety of cell types ER cisternae are found closely associated with Golgi sacs at the trans face as well as at the cis face of the stack.) These roles and transport routes are still under investigation, as are the relations of GERL to the Golgi stack and the ER. There is disagreement about whether GERL is best thought of as ER or as part of the Golgi apparatus and about the nature of the material that moves through the system. Some secretory material may, for example, be processed or sorted there. Of likely significance are findings that GERL in some cell types can receive materials taken into the cell by endocytosis, including lysosomal hydrolases (Section 2.8.4). In this respect and others, GERL can resemble an extensive, interconnected network of lysosomal structures that may take part in degradative processes or in the cycling of membrane back and forth between the Golgi region and the cell surface (see below). Another fact to keep in mind is that there seems to be a special "addressing" system by which lysosomal hydrolases are sorted out from other proteins made in the ER; thus the hydrolases could very well follow a route different from that of other proteins (Section 2.8.4).

Even for vesicle-mediated transport of secretory products known to pass into the Golgi stack, the idea that the stack can sometimes be bypassed has arisen in the form of suggestions that some of the transport vesicles may fuse directly with forming secretion granules (see Fig. II–33). A few investigators have also made still-disputed suggestions that certain glycoproteins or other molecules may be completed in the Golgi apparatus and then move back to the ER. (Components of the specialized sarcoplasmic reticulum of muscle cells [Section 3.9.1] supposedly follow this route.) The essential conclusion to be drawn at present is that the structures at and near the trans Golgi face may be more complex and more heterogeneous than is evident in the transport and packaging schemes discussed thus far. More generally, this reemphasizes the need to retain a flexible viewpoint about the organization of the Golgi region, the transport that occurs there, and the interactions of the Golgi apparatus and the ER.

2.5.4 Membrane Formation, Maintenance, and Recycling

Cellular membranes arise from preexisting membranes. The maintenance of particular membranes, their expansion, and the elaboration of new types of membrane are based on the insertion of components into already existing membrane structures. (Removal of components also takes place, as will be considered when we discuss the degradative phases of intracellular turnover in Section 2.8.3.)

Membrane Proteins   Two otherwise puzzling features of many plasma membrane proteins can be accounted for by schemes proposing their synthesis on the ER, movement as part of the membranes bounding vesicles or other transport structures, glycosylation by the ER and Golgi apparatus, and insertion in the cell surface by exocytosis-like fusion of a Golgi-derived vesicle or vacuole. That proteins can span membranes would result simply from their insertion by bound ribosomes with only part of the polypeptide passing into the ER interior; a hydrophobic portion is left in the membrane, and a "tail" of different lengths for different proteins sticks out into the cytoplasm (see Fig. II–36). This also could readily explain the asymmetric orientation of the proteins—the fact, for example, that every transmembrane protein of a given type in the plasma membrane is oriented with the same groups (such as the oligosaccharide chains of glycoproteins) facing the extracellular space. Just like secretory glycoproteins released by the cell, those portions of membrane proteins inserted by the ribosomes to face the *interior* of the ER will be accessible to the glycosylating enzymes; these portions of the molecules will be exposed to the *extracellular* surface of the cell on exocytic fusion. (See Fig. II–33.)

Such mechanisms are not simply conjectural. That the carbohydrates of the cell surface arrive via the Golgi apparatus has been demonstrated by autoradiographic and cytochemical studies. In dividing plant cells (see Fig. IV–28), Golgi vesicles fuse to establish the boundaries—plasma membrane and cell wall—separating daughter cells. In other cell types, cell coats and even some plasma membrane regions have microscopically distinctive appearances—the fuzz on the ameba surface (Fig. II–13A) and the thickness of the bladder membrane (Section 2.1.3) are examples. In several such cases, Golgi vacuoles with contents or membranes appearing similar to the cell surface materials can be seen in exocytosis-like configurations at the cell surface. More recently, immunohistochemical techniques and cell fractionation at successive intervals after labeling (Chap. 1.2B) have been used to trace the movement of particular membrane proteins along the ER–Golgi–plasma membrane route. Results have been most convincing for viral proteins that come to be inserted in the plasma membrane of virally infected cells (Section 3.1.1), but evidence for movement along this route is accumulating for some normal cellular proteins as well. An interesting finding of still uncertain significance is that during their transport through the cell, certain of the protein molecules destined to enter membranes acquire one or a few fatty acids (see Fig. II–3) linked to the proteins by action of enzymes in the Golgi apparatus or ER. An obvious possibility is that these groups help to orient or anchor the proteins in the membranes.

Peripheral membrane proteins on the cytoplasmic surface of the membrane probably are added from the free ribosomes, associating with the membrane through interactions with the proteins already there. Those on the extracellular surface get there by exocytosis.

Schemes comparable to those just summarized for the plasma membrane can account for genesis of proteins of the ER itself, of the Golgi apparatus, and of other cellular compartments derived from the ER or Golgi apparatus. Oriented insertion, as occurs from bound ribosomes, can, for example, explain how enzymes come to be oriented with their active sites on the proper side of the membrane. There are, however, three instructive "complications":

1. Increasingly strong evidence suggests that certain of the integral proteins of the ER, such as cytochrome b_5, are synthesized on *free* rather than on bound ribosomes. The next few chapters will indicate that a number of proteins of mitochondria, plastids, and peroxisomes also originate on free ribosomes. Unlike the "cotranslational" insertion of proteins into membranes as they emerge from bound ribosomes, these proteins probably insert into or pass across the membranes "posttranslationally" (after translation and release from the ribosomes). Apparently, this is made possible by hydrophobic regions of the protein molecules. For certain cases it is hypothesized that insertion in the membrane involves an interaction of the proteins with preexisting membrane components that change the protein's three-dimensional conformation, exposing a hydrophobic portion that previously was "buried" within the molecule. Upon exposure this portion can mediate association with the hydrophobic zone of the lipid bilayer. Selectivity in such "triggering"—the involvement, for example, of specific membrane components—might explain how proteins made on free ribosomes associate only with some intracellular membranes rather than inserting indiscriminately everywhere.
2. The "signal hypothesis" (Section 2.4.3) in its simplest form predicts that the N-terminal end of a membrane protein should stick into the interior of the ER, and the C-terminal end, out into the cytoplasm (since the N-terminal is made first and the C last). Frequently this is true, but it certainly is not the case for many membrane proteins. Some loop back and forth across the membrane several times in a precisely determined pattern, and some have their C-terminal the "wrong" way: It protrudes into the interior of the ER. Hypotheses based on modifications of the "signal" hypothesis and on "posttranslational" events like the ones just described have been developed to explain these orientations and are now being tested. For instance, "signal" sequences in the middle of a protein (Section 2.3.4) could lead to the N-terminal end remaining on the cytoplasmic side of the membrane (see Fig. II–36). Clarification of mechanisms is also required for the fact that bound polysomes can send some types of proteins (secretory proteins, for instance) entirely through the ER membrane into the interior, whereas others are left with portions embedded in the membrane. Does some "stop-transfer" device operate to limit the movement of membrane proteins, or alternatively, is some special mechanism needed to permit the final (C-terminal) part of a secretory protein to complete passage into the ER interior?
3. The ER, Golgi apparatus, and plasma membrane show characteristic and major differences in protein composition; there may even be differences in

enzymatic capacities among different portions of the Golgi apparatus. In other words, membranes cannot be thought of as moving unchanged from the ER to the Golgi apparatus and thence to the cell surface. In theory at least, selective movement of particular membrane components from one compartment to another could be based on differential movements of specific molecules within the plane of a membrane and on transport from one compartment to another via vesicles enriched in or depleted of specific components of the parent membrane; there are precedents for both such possibilities (Sections 2.1.6 and 2.4.4). We also mentioned earlier in this chapter that in some cases the Golgi apparatus shows a polarity in membrane thickness from ER-like to plasma membrane–like. This has led to suggestions that, while passing through the Golgi apparatus, membranes undergo transformation so that membranes received from the ER are converted into forms suitable for insertion in the cell surface. (Secretory granules, for example, are known to be clad in relatively impermeable membranes, quite unlike the ER, evidently to insure that exocytosis does not severely disrupt the cell's permeability.) This transformation could occur by selective loss of some components, perhaps through shuttling back to the ER, and by chemical modification of others. Alternatively, it might reflect some still-to-be-appreciated subtleties about the properties of the Golgi stack.

Insights into these various problems and possibilities should come soon from ongoing investigations of known membrane transformations and maintenance phenomena. For example, at about the time of birth, the ER of hepatocytes in rats begins to make its glucose-6-phosphatase. The subsequent very rapid appearance of this enzyme throughout the entire expanse of the reticulum suggests that it is inserted into the preexisting ER, rather than segregated into a "new" ER zone that gradually grows to replace the old. Such insertion of protein molecules is most readily envisioned as occurring from polysomes bound to many, widely distributed sites on the rough ER or from free polysomes. The chief alternative possibility is that the enzyme is added to a few sites in the ER but diffuses very rapidly in the plane of the membrane to assume a uniform distribution. There is as yet no strong reason to postulate this latter process for the rough ER itself, but movement through interconnections followed by diffusion throughout the receiving membrane, does seem to be the simplest explanation for passage of integral membrane proteins synthesized by the rough ER, into the smooth ER (Section 2.4.4). On the other hand, despite the direct continuities between rough and smooth ER, the membranes of the two are distinct, at least insofar as proteins involved with ribosome binding are concerned. These proteins may be segregated to the rough zones by anchoring to one another or to the ribosomes, or by interactions with other membrane molecules or with molecules in the adjacent cytoplasm or ER lumen.

Membrane Lipids   Most membrane lipids have hydrophilic "heads" directed toward the surface of the membrane (as in Fig. II–4). This characteristic could be expected to prevent, or at least impede, the passage of the lipid molecules from one half of the bilayer to the other, since such passage requires crossing the hydrophobic "core"—the central zone—of the membrane. Some lipid molecules must, however, make this crossing to maintain or expand the bilayer as membranes grow or turn over: The ER enzymes that make phospholipids and other membrane lipids are thought to leave the newly synthesized molecules in the half-bilayer of the ER membrane that faces out toward the cytoplasm; a subsequent process is required to move some of the molecules into the other half-bilayer. Such "flipping" from one half of the bilayer to the other takes place at a relatively rapid rate in those biological membranes that have been studied (notably the plasma membrane of bacteria, which is a site of lipid synthesis). Since these rates are markedly more rapid than the spontaneous "flip-flop" rates obtained with artificial bilayers containing only lipids, it is presumed that biological systems use specific membrane proteins to aid the lipids in their passage from one side of the bilayer to the other. This might also help to explain how differences arise in the relative proportions of different lipids in the two halves of the same bilayer. Differential lateral distribution yielding different proportions of the several types of membrane lipids at different places in the plane of the bilayer probably results from binding and other interactions of specific lipids with membrane proteins and with other lipids.

The membrane-mediated transport between the ER and the Golgi apparatus and between the apparatus and the cell surface is one of the routes available for lipid movement. Lipids can, of course, also diffuse rapidly within the plane of the membrane. In addition, there are *lipid-exchange* proteins that can move individual phospholipid molecules from the membrane of one compartment to that of another, even in the absence of direct continuities or contacts between the membranes: The proteins apparently can carry lipids through the cytoplasm since they can transport lipid molecules through suspending media used in test-tube experiments. As with membrane proteins, some combination of transport and receipt mechanisms and targeting and recognition devices is needed to explain the fact that different compartments that seem to communicate with one another—via continuities, the movement of vesicles, or lipid exchange proteins—differ in their lipid composition. The ER, for example, has less cholesterol than that found in the plasma membrane or in some of the Golgi membranes.

Membrane Recycling In considering endocytosis, we pointed out that when endocytic vesicles withdraw membrane at rapid rates from the cell surface, there is a compensatory movement of membrane back into the surface, which may in part be accomplished by a return of the vesicles themselves (Section 2.1.6). Similarly, following exocytic release of secretions, membranes added to the cell surface from secretory bodies are returned to the cell interior by formation of vesicles and tubules that separate from the cell surface (Section 2.5.2): Some, at least, of these latter membranes move back to the Golgi region and fuse with Golgi structures to be reused for packaging. The transport of Golgi-destined material from the ER may be mediated by vesicles that shuttle back and forth between the compartments, fusing with one and then separating off and fusing with the other.

Too little is known yet for us to say much about the extent of such cycles or about their importance—seemingly substantial—in the cell's membrane economy. Intriguing speculation has been put forth: It is imagined, for example, that vesicles and vacuoles cycle selectively between the cell surface and the *rims* of Golgi sacs, serving to transport secretions produced in the more stable central portions of the Golgi sacs. The membranes of the latter portions of each sac, it is speculated, remain in place, carrying out the enzymatic activities and other processing functions of the Golgi apparatus; products are shipped from one sac to another and, ultimately, out of the apparatus, via the rims whose membrane supply is continually replenished by the return of vesicles from the cell surface. Whether this proposal proves accurate or not, the cycles known to occur can eliminate or reduce considerably the "drain" on metabolism that would result from the cell's having to make an entirely new piece of membrane for each endocytic vesicle or Golgi vacuole. How the membrane fusions and fissions that produce the vesicles and vacuoles that move back and forth are brought about and targeted, so that the cycling membranes move between the proper compartments, is an important aspect of more general unresolved questions about control of membrane budding (are coats involved?) and about mechanisms by which the cell moves its membrane-delimited structures (Chaps. 2.10 and 2.11).

Chapter 2.6 Mitochondria

Mitochondria are present, often in great abundance, in virtually all eucaryotic cells, both plant and animal. Their number has conventionally been thought to range from a few per cell to 1000 or more in a rat hepatocyte, and up to tens of thousands in one of the giant amebae. Although their sizes vary considerably in different cell types, diameters of 0.5 to 1.0 μm and lengths of 5 to 10 μm or more are common. Mitochondrial structure is striking and readily recognizable in the electron microscope.

Mitochondria are easily isolated from homogenates relatively free of other cytoplasmic components. Dry weight proportions are typically 25 to 35 percent lipid and 60 to 70 percent protein. It was learned early that mitochondria are the major sites of ATP production linked to oxygen consumption. Especially by virtue of their possession of the Krebs cycle enzymes, they play pivotal parts in many metabolic pathways, including the breakdown and synthesis of carbohydrates, fats, and amino acids (Chapter 1.3). They illustrate well the grouping of a number of sequentially acting enzymes within a membrane-limited organelle. This promotes efficiency, in part since the products of one reaction do not have far to travel before they are likely to encounter the enzymes catalyzing the next. Such effects can be especially striking when sequentially acting components are held in close relationship with one another by being actually incorporated within the same membrane, as is true in mitochondria. The profound functional implications of the arrangement of proteins and other molecules in mitochondrial membranes are becoming increasingly evident, largely through reconstitution experiments in which the membrane macromolecules are isolated and recombined.

Electron Transport: A Brief Review   Terminology important for this chapter was summarized in Section 1.3.1.

Chapter 1.3 outlined the major chemical steps in the Krebs cycle, including the reduction of coenzymes. ATP formation by mitochondria is based on the subsequent transport of electrons from the reduced coenzymes in an orderly manner by the "respiratory" or "electron transport" chain, and on related movements of protons. The electron transport system is outlined in Figure II–50; prominent among the participants are iron-containing proteins called cytochromes. The chain can be thought of as starting at a position of high energy and ending at one of low energy. Each step in the chain involves loss of electrons (oxidation) by one molecule and electron gain (reduction) by the next. The series of oxidation-reduction events results in a release of energy. Some of the energy made available in this way is used for the formation of ATP from ADP plus inorganic phosphate; this process overall is called *oxidative phosphorylation*.

2.6.1 Compartments Within the Mitochondria

Figure II–51 illustrates the basic structure of the mitochondrion. The organelle is bounded by an outer membrane, separated from the inner membrane, by a space 6 to 10 nm wide. The inner membrane is thrown into many folds, or *cristae* (Fig. II–52). Enclosed by the inner membrane is the *matrix*. Techniques are available for obtaining separate outer and inner membranes and for isolating some of the matrix substances. With these techniques, the segregation and organization of different enzymatic functions in the different membranes and compartments are being studied. As more is learned, the description presented here will undoubtedly require modification.

As Section 1.3.4 indicated, some mitochondrial enzymes are readily solubilized when the organelles are disrupted. These include soluble enzymes of the matrix and of the compartment between the inner and outer membranes. Perhaps some enzymes, loosely attached to one or the other membrane, are also detached during the disruption. Most Krebs cycle enzymes are in the readily solubilized category; the interpretation is that they come from the matrix in the inner compartment. Enzymes for fatty acid breakdown by a metabolic sequence known as the β-oxidation pathway (see Section 2.9.2 and Fig. II–75) also are present there. One Krebs cycle enzyme (succinic dehydrogenase) is known to be linked to the inner membrane.

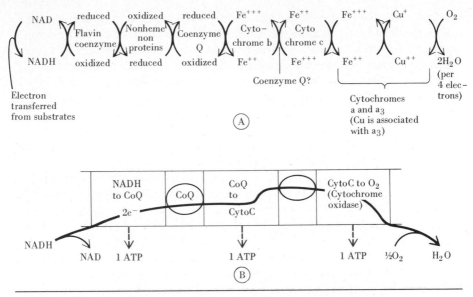

Figure II–50 *Major steps in mitochondrial electron transport.*

A. Probable sequence of transfer of electrons. Each component of the chain receives electrons from the previous component and passes them to the next. In the case of the cytochromes, the receipt of electrons reduces iron ions in the molecule to Fe^{2+}; the irons are present in *heme* prosthetic groups (Section 1.4.2). Subsequent loss of the electrons to the next component regenerates the oxidized, Fe^{3+} form. "Non-heme iron proteins" are sometimes referred to as *iron-sulfur* or *FeS proteins;* some uncertainty persists about details of their involvement in electron transport. Coenzyme Q is also known as *ubiquinone;* its role in electron and proton transport is discussed below and in Figure II–56 and Section 2.6.1. Prevailing working hypotheses suggest that coenzyme Q also serves to transfer electrons from cytochrome b to cytochrome c_1 through a cycle involving cytochrome b molecules located at different places in the membrane (as in Fig. II–56). The complex involving cytochromes a and a_3 is called "cytochrome oxidase."

B. *Functional complexes.* The electron transport components of part A are thought to be grouped into three functional complexes in the inner mitochondrial membrane, each of which can be isolated as a multiprotein complex. The segment of the transport of electrons carried out by each is indicated in the diagram. (A fourth inner-membrane multiprotein complex, that contributes to the electron transport chain by reducing coenzyme Q through oxidation of succinic acid, is not shown.)

It has generally been considered that the functioning of each of the three complexes results, ultimately, in the formation of one ATP molecule per pair of electrons transported. (One pair of electrons is needed for each H_2O produced.) This does not, however, mean that each complex is *directly* responsible for generating an ATP. Current views of the relations of electron transport to formation of ATP (Fig. II–56) are that the coupling of the two is somewhat "indirect" in the sense that what the electron transport chain does is to establish a proton gradient and that this in turn is used to generate ATP (see also Section 2.6.2 for amounts of ATP produced). (From work and formulations by A.L. Lehninger, E. Racker and many others.)

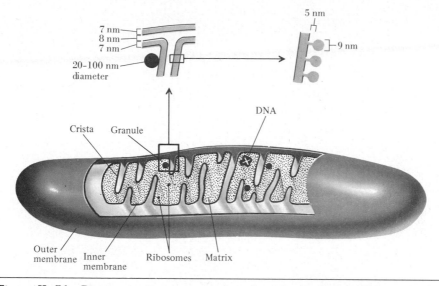

Figure II–51 **Diagrammatic representation of a mitochondrion. (After work of L. Ornstein, G.E. Palade, F. Sjostrand, H. Fernandez-Moran, and others.)**

Functional Arrays on and in the Inner Membrane   The major constituents of the inner membrane include lipids, especially phospholipids, and a heterogeneous group of proteins. Most of the proteins whose roles are known participate in electron transport, oxidative phosphorylation, or transfer of substrates, products, other essential metabolites, and ions into and out of the mitochondria. Functions of other proteins are not yet understood. Some have been referred to historically as "structural" proteins to convey the possibility that their primary responsibility is to maintain a framework, along with the lipids, that helps to organize the functioning of the other proteins. This function, however, was suggested because of a lack of specific knowledge and because of problems in preserving enzymatic activities during purification of membrane proteins. The present conception is that the membrane enzymes and electron transport proteins themselves fulfill the requisite structural roles and that very few, if any, specifically structural proteins are present.

The electron-transport components responsible for respiratory metabolism, and the systems that carry out oxidative phosphorylation, are both tightly bound to the inner membrane. This joint localization might be expected since, in the intact mitochondrion, electron transport and oxidative phosphorylation are *coupled*—as electron transport occurs, leading ultimately to oxygen consumption, phosphate groups are added to ADP, to produce ATP. Usually the coupling is "tight," meaning that neither ATP formation nor electron transport can occur alone; both must take place simultaneously. (The next section will take up these processes in detail.)

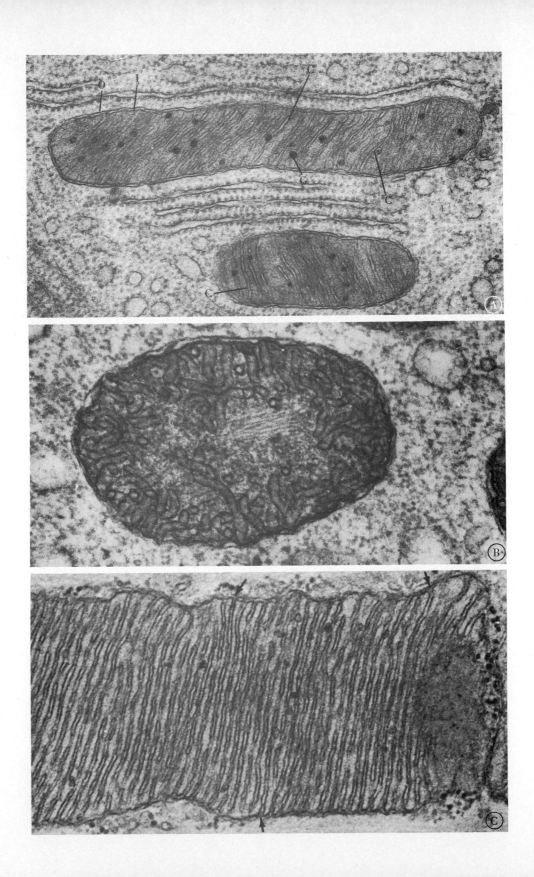

F_1; Negative Staining; Reconstitution In electron micrographs of sectioned cells fixed with the usual procedures, the inner membrane (and the outer one) has the three-layered appearance expected for a membrane of lipid and protein. It is thinner (5 to 7 nm) than most plasma membranes (7 to 11 nm). However, a distinctive structural feature of the inner membranes is evident from "negative staining" (Fig. II–53) of isolated mitochondria (Fig. II–54). The membrane shows a great many small, roughly spherical particles covering the surface of the inner membrane that faces the matrix. Each particle is attached by a stalk to the membrane. When these were first discovered, it was thought that each sphere might be a *respiratory assembly* containing all the enzymes responsible for electron transport and associated phosphorylation. This proved not to be the case; the spheres are too small to hold all the needed enzymes. Furthermore, when suitable experiment procedures are used to remove the spheres (see Fig. II–55), the stripped membranes still possess all the essential components of the respiratory chain. The spheres correspond to an enzyme called F_1. When extracted from the mitochondrion in soluble form, F_1 acts as an ATPase; that is, it splits ATP into ADP and phosphate. However, in the intact mitochondrion, F_1 acts in the opposite direction and is, in fact, responsible for coupling phosphorylation of ADP with electron transport. This role of F_1 has been elucidated by reconstitution experiments, illustrated in Figure II–55. Like those with the ribosomes (Section 2.3.2), such experiments permit investigators to determine the functions and interactions of different parts of a complex structure by comparing properties of individual components in isolation with properties of various combinations and of the intact assembly.

Respiratory Assemblies? As for the remainder of the inner membrane, several reasons led investigators to conclude that the agents of electron transport are arranged as repetitive arrays, in which proteins responsible for sequential steps are held in physical proximity to with one another: First, F_1 has an intimate functional relation to the enzymes of the respiratory chain, and it is visible in the form of a repetitive unit. Second, when mitochondrial membranes are fragmented into small vesicles (by ultrasonic vibration, for example, or by agitation with glass beads), the vesicles can carry out electron transport and oxidative phosphorylation (Fig. II–55). Even the smallest fragments of the mitochondrial membrane produced in such experiments have the molecules of the respiratory chain present in the same ratio as that in the intact mitochon-

◄ **Figure II–52** *Mitochondria* as seen by usual electron microscopic techniques. A. In a pancreas cell (bat). The outer *(O)* and inner *(I)* mitochondrial membranes and the flat cristae *(C)* are readily visible. Many intramitochondrial granules *(G)* are present; most cell types show fewer such granules. × 50,000. (Courtesy of D.W. Fawcett.) B. In the protozoan Epistylis. As in many protozoa and algae, the cristae are tubular. × 60,000. (Courtesy of P. Favard.) C. In a muscle (bat). The cristae are numerous and closely packed, correlating with the high state of activity of the muscle. At the arrows it is evident that the cristae are infoldings of the inner mitochondrial membrane. × 45,000. (Courtesy of S. Ito.)

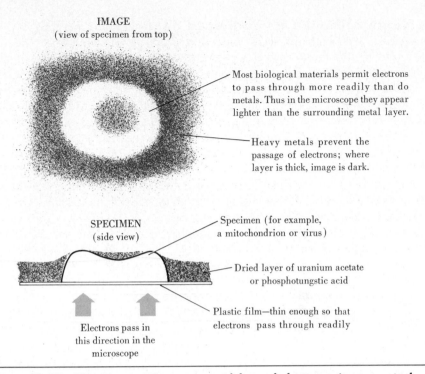

IMAGE
(view of specimen from top)

Most biological materials permit electrons to pass through more readily than do metals. Thus in the microscope they appear lighter than the surrounding metal layer.

Heavy metals prevent the passage of electrons; where layer is thick, image is dark.

SPECIMEN
(side view)

Specimen (for example, a mitochondrion or virus)

Dried layer of uranium acetate or phosphotungstic acid

Plastic film—thin enough so that electrons pass through readily

Electrons pass in this direction in the microscope

Figure II–53 *Negative staining,* a very widely used electron microscope technique for examining three-dimensional and surface aspects of cell structures. The specimen is not sectioned. It is placed on a thin plastic film and covered with a drop of solution containing heavy metal atoms (usually uranium or tungsten compounds such as uranyl acetate or phosphotungstic acid). The solution is allowed to dry, leaving the specimen in a thin layer of electron-dense material.

dria; this would not be expected if there were large specialized patches of membranes, each devoted to just one step in the sequence. Third, *multiprotein complexes* can be isolated from fragmented mitochondria. These are highly integrated assemblies of relatively few molecules that contain several sequentially acting components of the electron transport chain (see Fig. III–8 and Section 3.2.4). Segments of the respiratory chain can be separated as such discrete complexes (see Fig. II–50) and can be reconstituted as functional entities into artificial membranes.

One interpretation is that the respiratory chain is organized overall as distinct respiratory assemblies, in each of which *all* the necessary components maintain long-term structural associations with one another. Alternatively, multiprotein complexes, each carrying out a segment of the chain, might exist as separate, neighboring entities in the membrane, interacting with each other by virtue of their mobility in the membrane or through intermediate carriers, especially coenzyme Q (also called *ubiquinone*). Unlike the other components of

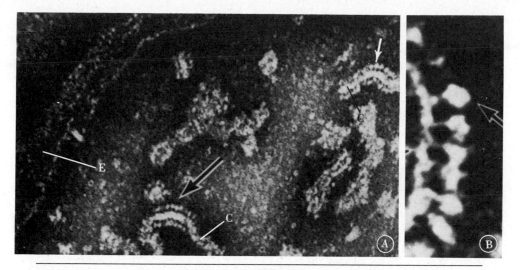

Figure II–54 F_1. Portion of a mitochondrion from heart muscle (cow). The organelle has been isolated, broken open, and then negatively stained (see Fig. II–53). A is a view ($\times$ 80,000) showing part of the edge of a mitochondrion *(E)* and several cristae *(C)*. Along the cristae surfaces facing the matrix (Fig. II–51) are arrays of spheres attached to the cristae by stalks (arrows). B is at higher magnification ($\times$ 600,000) and shows several spheres. These structures correspond to the F_1 component of the membrane discussed in the text. (Courtesy of H. Fernandez-Moran, T. Oda, P.B. Blair, and D.E. Green.)

the chain, which are mostly proteins, coenzyme Q is a small, highly lipid-soluble molecule that can diffuse in the hydrophobic interior of the membrane. There are many molecules of coenzyme Q for each of the multiprotein complexes with which it interacts. These molecules could represent a pool of mobile carriers available to transfer electrons between the complexes, and to transport protons, by migrating within the membrane. In any event, it has been estimated that a square micrometer of inner mitochondrial membrane surface has many hundreds of respiratory assemblies or their equivalent, providing an organized arrangement of the molecules responsible for electron transport and oxidative phosphorylation. Individual mitochondria of mammalian cells contain many thousands to tens of thousands of such assemblies, for which the folding of the membrane into cristae provides the necessary surface area.

Although most early models of the mitochondrial inner membrane envisaged close structural links between F_1 and the electron transport system, the somewhat indirect functional connection between the two discussed in the next section may not require direct, intimate association in the plane of the membrane (see also Section 2.7.3).

Cristae; Outer Membrane Consistent with the central role of the inner membrane in respiratory metabolism, the number of cristae per mitochondrion

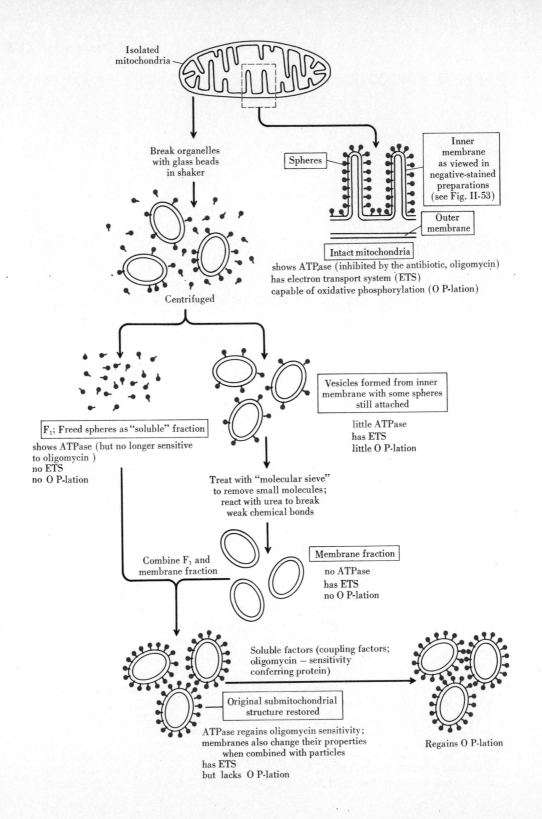

Isolated
mitochondria

Break organelles
with glass beads
in shaker

Spheres

Inner
membrane
as viewed in
negative-stained
preparations
(see Fig. II-53)

Outer
membrane

Intact mitochondria

shows ATPase (inhibited by the antibiotic, oligomycin)
has electron transport system (ETS)
capable of oxidative phosphorylation (O P-lation)

Centrifuged

Vesicles formed from inner
membrane with some spheres
still attached

little ATPase
has ETS
little O P-lation

F_1; Freed spheres as "soluble" fraction

shows ATPase (but no longer sensitive
to oligomycin)
no ETS
no O P-lation

Treat with "molecular sieve"
to remove small molecules;
react with urea to break
weak chemical bonds

Combine F_1 and
membrane fraction

Membrane fraction

no ATPase
has ETS
no O P-lation

Soluble factors (coupling factors;
oligomycin – sensitivity
conferring protein)

Original submitochondrial
structure restored

ATPase regains oligomycin sensitivity;
membranes also change their properties
when combined with particles
has ETS
but lacks O P-lation

Regains O P-lation

and their total surface area tend to be greater in cells with particularly intense respiratory activity (see Fig. II–52). The mitochondria themselves are often concentrated in intracellular regions of greatest metabolic activity—for example, close to the contractile fibrils in muscle cells, at the base of kidney tubule cells where extensive active transport of ions occurs (see Fig. III–32), or wrapped around the flagella that move sperm cells (see Fig. III–50).

The physiological significance of the various forms of cristae in different cell types is less clear. The cristae may appear (Fig. II–52) as simple plates (most cells of higher organisms) or as tubules (protozoa, algae, cortex of the mammalian adrenal gland); in some cases they may be organized into more complex three-dimensional structures.

The outer mitochondrial membrane has several oxidative enzymes different from those of the inner membrane, such as *monoamine oxidase*. The membrane is composed of lipids and proteins in roughly equal amounts, by weight. The proteins include ones believed to establish transmembrane channels that permit ready movement of components, essential for mitochondrial function and maintenance, across the membrane. Very little is well established about the space between the two membranes, though one or two cytochromes and enzymes have provisionally been identified there.

Permeability and Transport   The mechanisms by which mitochondria exchange molecules with their surroundings are varied and complex. The outer membrane appears fairly permeable to many small molecules and ions. The inner membrane is not; it serves as a barrier, regulating the entry and exit of many molecules. Some participants in mitochondrial metabolism, like oxygen, water or certain small fatty acids, enter the inner mitochondrial compartment readily while others, such as reduced NAD (NADH), do not (*electrons* from NADH can enter mitochondria, but this requires an indirect transfer to NADs already inside, via intermediate carriers and enzymatic reactions). For ADP, ATP, pyruvate, and some intermediates of the Krebs cycle, specific carrier systems mediate passage across the inner membrane. One of these exchanges an ATP from inside the mitochondrion for an ADP

◄ Figure II–55   *Reconstitution experiments with mitochondrial components.* Disruption of isolated beef-heart mitochondria and separation of the resulting fragments into membranous and "soluble" components, followed by reconstitution of some mitochondrial structure and function. The drug *oligomycin* is known to inhibit F_1-related functions in intact mitochondria through interaction with a protein complex, F_0, present in the inner membrane (Fig. II–56). Sensitivity to this drug is one measure of the extent to which reconstitution succeeds in reproducing conditions similar to those in mitochondria. The restoration of phosphorylation in the reconstituted particles depends both on the enzymatic activity of F_1 and on the reassociation of F_1 with F_0 (Fig. II–55 and Section 2.6.2).

from outside, obviously an important exchange if the cell is to use mitochondrial ATP and the mitochondrion is to acquire ADP for phosphorylation. This exchange and other transport across the mitochondrial membrane is accomplished by specific proteins. A small, probably mobile carrier, *carnitine* ($[CH_3]_3N^+$—CH_2—CHOH—COOH), is responsible for transferring long fatty acids, in cooperation with enzymes at the two surfaces of the membrane.

Isolated mitochondria can accumulate ions such as Ca^{2+} from the medium; this is an energy-dependent process that stops if mitochondrial respiration is inhibited (for example, by adding cyanide to the medium). Phosphate is taken up with the Ca^{2+}. As are other mitochondrial transport systems, the one for calcium is driven (as in Section 2.1.3) by the gradient in protons maintained across the inner mitochondrial membrane, to be discussed in the next section. When large amounts of calcium (or Ba^{2+} or Sr^{2+}) and phosphate have been taken up, prominent granular deposits of calcium phosphate are seen in the mitochondria. Somewhat similar though usually less electron-dense granules, 20 to 50 nm in diameter, are present normally in mitochondria (see Fig. II–52). Many workers infer that in the intact cell, these serve to bind calcium and phosphate and perhaps act as storage depots for these ions, but this is still unsettled.

2.6.2 Oxidative Phosphorylation

Present views about how the energy from electron transport is used to form ATP from ADP plus phosphate attribute central importance to specific features of the inner mitochondrial membrane; especially significant are the membrane's impermeability to protons and to other ions and the arrangement of its proteins.

For many years, the prevailing theory about energy storage was that at each of three steps in electron transport, energy is transferred from electron transport components to special "intermediate" molecules, not directly part of the electron transport chain. The energy is supposedly conserved in "high-energy bonds" (Section 1.3.2) in these intermediates, enabling them to engage in a series of reactions leading eventually to ATP formation. An alternative concept is that before being used to synthesize ATP, the energy is stored through alteration of the three-dimensional conformation of membrane molecules—a phenomenon conceivable from physical-chemical theory. That intermediate molecules and especially conformational changes play some roles in oxidative phosphorylation is not entirely excluded. Intense efforts, however, have failed to turn up the postulated chemical intermediates, and the conformational changes remain essentially theoretical matters not easy to test with available methods. The model for oxidative phosphorylation most widely accepted today has a quite different emphasis, one that was highly unorthodox at the time it was proposed 20 years ago: The *chemiosmotic* hypothesis suggests that the energy made available by electron transport is stored as an electrochemical gradient (Section 2.1.3) of H^+ across (or within) the inner mitochondrial membrane and that this gradient drives the formation of ATP.

Chemiosmotic Mechanisms   That electron transport *can* produce a "proton gradient" is shown by experiments in which oxidation of substrates by isolated mitochondria or submitochondrial vesicles, like those in Figure II–55, results in a change in the pH of the medium in which they are suspended, or of the vesicle's interior. The gradient reflects the fact that the electron transport components are organized in an asymmetric manner, which permits certain groups or components of the chain to acquire protons at one surface of the inner mitochondrial membrane (that facing the matrix) and to pass them to other members of the chain, which release the protons at the opposite surface of the membrane (that facing the exterior). Figure II–56 summarizes current proposals about this organization based on experiments like the ones done to determine the distribution of proteins in the plasma membrane (Section 2.1.2). The H^+ cannot simply diffuse back since the inner mitochondrial membrane is impermeable to protons (H^+); this permits the maintenance of a gradient. The processes leading to phosphorylation do, however, have the effect of moving H^+ down the gradient. Overall, therefore, the net distribution of H^+ across the membrane at a given moment represents a balance among electron transport, which establishes the gradient, and phosphorylation and other membrane activities, which tend to diminish the gradient as they make use of its energy.

The energy in the proton gradient is represented both by the differences in H^+ concentration across the inner mitochondrial membrane and by an electrical potential across the membrane analogous to that across plasma membranes (Section 2.1.3). Figure II–56 outlines a relatively simple hypothesis for how the proton gradient might bring about production of ATP. It is based on the known properties of F_1 and on the fact that when F_1 is removed from submitochondrial vesicles, they become "leaky" to protons: The F_0 proteins with which F_1 normally is associated are thought to be channels through which protons can move to F_1 from the opposite membrane surface; the removal of F_1 from the vesicles leaves these channels open. More complicated proposals for the harnessing of the energy are based on more sophisticated theories; one set concerns possible subtle effects of ions or potentials on membrane-associated enzymes, that might produce conformational changes to "energize" the enzymes.

Uncertainty also exists about the actual amounts of ATP that can be generated. The prevailing view has been that for each NADH that contributes its pair of electrons to the electron transport chain, three pairs of protons are transferred across the mitochondrial membrane, with the electrons ending up in a molecule of water (see Fig. II–50). Each proton pair represents enough chemiosmotic energy to support synthesis of one ATP. Thus the P : O ratio (molecules of ATP formed per atom of oxygen consumed) would be 3. These

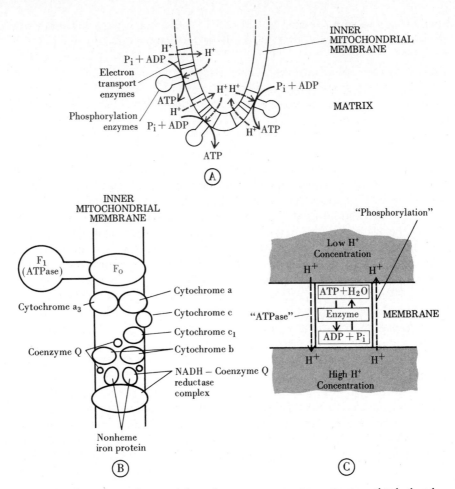

numerical relations are being debated anew—some investigators think that less ATP is actually made or that other numbers of protons are moved—and so it is still not certain, for example, precisely how much ATP is generated in the metabolism of a molecule of sugar (see Fig. I–26).

Evidence   Despite the remaining uncertainties, there is good evidence supporting the chemiosmotic model as the best available working hypothesis for mitochondrial oxidative phosphorylation:

1. When the impermeability of the inner mitochondrial membrane is compromised by disruption or by the addition of ionophores such as valinomycin (Section 2.1.3) transmembrane gradients cannot be maintained, and the mitochondria cease phosphorylating even though electron transport continues.

2. Experiments with a number of membrane-associated ion-transporting ATPases have shown that artificially imposed ion gradients can drive the

◀ **Figure II–56** *Chemiosmotic hypotheses of oxidative phosphorylation.*

A. Electron transport enzymes are thought to carry protons (H^+) from the matrix across the inner mitochondrial membrane. The electrochemical proton gradient then interacts with phosphorylation enzymes leading to the formation of ATP from ADP and phosphate (P_i).

B. Asymmetric arrangement of respiratory chain and phosphorylation enzymes as currently postulated. The roles and locations of the "non-heme-iron proteins" are unsettled, and there is still uncertainty about locations of cytochrome *b*. F_1, the cytochrome oxidase complex (*a* and a_3) and other components are each made of several subunits not shown.

The most widely accepted present version of the chemiosmotic hypothesis argues that as electrons are transported, the electron transport components transfer protons across the membrane as follows: Per pair of electrons entering the chain from NADH at the matrix surface of the inner mitochondrial membrane, it is thought that on net, three pairs of protons are transferred out of the inner mitochondrial compartment (see Fig. II–50). One pair of protons is transferred by the NADH-CoQ reductase; the transmembrane orientation of the complex presumably makes this transfer possible. Transfer of the other two pairs appears to involve cooperation of membrane proteins with coenzyme Q; the coenzyme seemingly helps to transport the protons across the membrane to be finally "ejected" with the intervention of appropriately located cytochromes, perhaps *b* for one pair and c_1 for the other.

C. One way in which an H^+ concentration gradient could be used to generate ATP. An enzyme system built into the membrane, through which H^+ cannot pass by passive diffusion, might be capable of coupling H^+ transport either to the hydrolysis of ATP ("ATPase" activity) or to the formation of ATP from ADP and P_i ("phosphorylation" activity). The ATPase activity can predominate when the enzyme transports H^+ from a region of low concentration to one of high concentration—an active transport process for which energy is required (Section 2.1.3). But when a concentration gradient has been established, for example, through action of electron transport enzymes, the enzyme can carry H^+ down the gradient and in so doing *produce* ATP. This can be thought of as reversing active transport and using energy represented by the concentration gradient for phosphorylation. (Based on work by P. Mitchell, E. Racker, and many others.)

ATPases in reverse—that is, make them produce ATP. This has been demonstrated for the mitochondrial F_1 ATPase system incorporated in artificial lipid vesicles.

3. Chemiosmotic mechanisms like the one suggested for mitochondria have provided convincing explanations for the functioning of other membrane systems with which mitochondria share some features, especially chloroplasts (next chapter) and the specialized "purple" zones of the plasma membrane of certain bacteria (Section 3.2.4).

Controls   The rate of operation of the respiratory chain is controlled by various factors, including the level of available ADP. Because the coupling of electron transport and phosphorylation normally requires that ADP be phosphorylated as electrons move down the chain, the rate of electron transport increases and decreases in accordance with increases and decreases in the quantities of ADP available for phosphorylation. The higher the rate at which a cell utilizes

ATP, the higher the ADP level, leading to higher rates of electron transport, oxygen consumption, and ATP production. Thus an effective mechanism is provided for balancing ATP production with its utilization.

Certain chemicals such as *dinitrophenol* (DNP) "uncouple" ATP production and permit the respiratory chain to function without simultaneous phosphorylation. DNP, like the ionophores mentioned earlier, affects mitochondrial permeability to protons, "collapsing" the electrochemical H^+ gradient. The extent to which similar effects occur naturally in cells is unknown, although such uncoupling has been proposed as a source of cell abnormality in one or two diseases, such as a fungus-induced blight of corn. In the tissue known as *brown fat* of hibernating mammals, uncoupling of mitochondria by a specific protein apparently permits the respiratory chain to release as heat the energy normally destined for ATP. (This is in addition to the heat released by ordinary metabolism as a result of the fact that even the more efficient of metabolic processes lose as heat half or more of the energy theoretically available to them.)

2.6.3 DNA, RNA, and Protein Synthesis: Mitochondrial Duplication

Nucleic Acids and Ribosomes The power of molecular biology and its rapid progress are evident in the success of recent experiments on mitochondrial nucleic acids. Morphological observations had shown the presence of fibrils 20 to 30 Å wide and of ribosome-like particles in mitochondria, and cytochemical investigations had studied the removal of the fibrils by DNase (which hydrolyzes DNA specifically) and of the ribosome-like granules by RNase. The results of these efforts were the first indicators that mitochondria contain traces of both DNA and RNA, but they were the subject of much controversy and could not clarify the nature or roles of these nucleic acids. When molecular biologists, armed with the techniques applied so successfully in the study of nucleic acids and protein synthesis in microorganisms, turned to this problem, it took only a year or two to establish the presence in purified mitochondria of DNA and RNA and of machinery for synthesizing protein. Mitochondrial nucleic acids and ribosomes were shown to differ in size from and to have properties unlike comparable components of the rest of the cell. Only small amounts are present in mitochondrial fractions isolated from cells, but the distinctive properties permitted the firm conclusion that the nucleic acids or ribosomes found in isolated mitochondria are not contaminants derived from other organelles during isolation.

The DNA of mitochondria (mtDNA) is a double helix, as it is in the cell nucleus. In almost all organisms studied, however, it has been shown to exist as a circular molecule (Fig. II–57). Most of the few exceptions apparently reflect technical problems, such as the breakage of the circles during DNA isolation. Mitochondria possess the entire machinery for replicating this DNA as part of their duplication. The RNAs present include a set of tRNAs, rRNAs, and

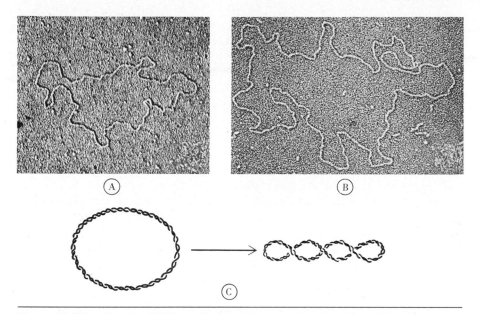

Figure II–57 *Circular DNA molecules from mitochondria.* The molecules shown in parts A and B were isolated from mitochondria of a leukemic human and prepared for electron microscopic examination by shadowing procedures (Fig. I–11). As is true of mitochondrial DNA molecules from cells of most higher animals, the molecule in part A is 5 μm long. (This is the length of the thin fiber, a DNA double helix, that delineates the perimeter of the distorted circle in the micrograph; see C.) The molecule in part B is 10 μm long and probably is a dimer made of two 5-μm molecules. Dimers and interlocked pairs of 5-μm circles have been found (in small numbers) in mitochondrial DNA preparations from several sources. × 35,000 (Courtesy of D.A. Clayton and J. Vinograd). C. As is true of most circular DNAs, those of mitochondria when isolated in their natural ("native") configuration are supercoiled (see Fig. III–11 and Section 5.1.2); the native configuration relaxes into an open circle as a result of artifactual nicking of the DNA that occurs during its isolation. In the cell, the supercoiling and untwisting of the DNA is controlled by enzymes (Section 4.2.4). (After A. Kornberg, *DNA Replication,* San Francisco: Freeman, 1980.) See Section 4.2.4 for replication of mtDNA.

presumed mRNAs: The RNAs hybridize with the DNA, reflecting their origin within the mitochondria.

Mitochondrial rRNAs, and the ribosomes of which they are a part, vary in size among different types of organisms (some are as small as 55S) but generally are markedly smaller than "cytoplasmic ribosomes" (the ordinary ribosomes in the cytoplasm outside the mitochondria). They also differ from cytoplasmic ribosomes in the use of formylmethionine (Section 2.3.2) as the initial amino acid of the nascent polypeptide and in their sensitivity to drugs (see below). In both these respects, mitochondrial ribosomes resemble bacterial ribosomes; such similarities have been much emphasized as evidence for the theory that mitochondria are evolutionarily related to procaryotic organisms

(Section 4.5.2). More recently, however, excitement has arisen over some striking features in which mitochondrial nucleic acids and protein synthesis differ from those both of procaryotes and of other eucaryotic systems. Mitochondria, for example, use a "stripped-down" translation system in which the tRNA populations are less varied than is true elsewhere. (For a number of amino acids there exist more than one tRNA species that can form the aminoacyl–tRNA complexes involved in cytoplasmic protein synthesis, so the variety of tRNAs in the cytoplasm is considerably greater than that of amino acids. This is less the case in mitochondria, which, in yeast at least, seem to have a total of only 24 species of tRNA to handle the 20 different amino acids.) The "coding" assignments of certain trinucleotides also are unusual: For instance, in mitochondria the mRNA codon UGA codes for the amino acid *tryptophan* rather than for the end of the polypeptide chain (Section 2.3.2). In other words, in the strict sense the genetic code is not universal—mitochondria can interpret particular trinucleotide sequences in unique ways. The significance of this has only begun to be explored. It may be primarily an "accident" of evolution, or it may provide a regulatory device by which mitochondrial protein synthesis is kept segregated from that in the rest of the cell.

The total amount of DNA in mitochondria generally is small except in cells with unusually abundant cytoplasm such as egg cells. In an L-cell (a cultured mouse fibroblast; Section 3.11.1), the 250 mitochondria contain a total amount of DNA equal to 0.1 to 0.2% of the DNA in the nucleus. There are 1000 mtDNA molecules per L-cell, and each mitochondrion has up to six of the 5-μm circular DNA molecules, but there is still little basis to believe that the six molecules (or, usually, the DNAs of the different mitochondria in a given cell) differ significantly from one another. For mitochondria of cultured cells of human origin (HeLa cells; Section 3.11.2), the base sequence of the 5-μm circle is known: There are about 16,500 base pairs (16,569 according to a recent determination). The information is arranged in unusually compact fashion—the usual noncoding DNA stretches found between typical genes of the eucaryotic nucleus (Section 4.2.6) are missing, and coding sequences for tRNAs are interspersed among those for mRNAs in closely packed arrangements. The DNA is transcribed into a large RNA molecule, which subsequently is fragmented into the separate, functional RNAs. Even with this compact arrangement, however, a single mitochondrion contains enough DNA to code for only a few of the hundred or more different proteins present in the organelle, since much of the DNA is required to specify rRNAs, tRNAs, and the like.

In unicellular organisms such as yeast, and in some higher plants, mitochondrial DNAs are longer than in cells of higher animals (up to 20 to 30 μm or more) and in yeast, where the individual molecules are about 75,000 nucleotide pairs long, the total DNA content of mitochondria may approach 10 percent of the nuclear content. In yeast and other fungi, however, the information is not as compactly organized as in higher mammals, and some of the DNA is made of stretches consisting largely of As and Ts, whose roles are not known. Certain of the genes in yeast mtDNA have intervening sequences (introns; Section 2.3.5. Section 4.2.6 outlines proposals for the functions of the introns.).

Both in yeast and in human mitochondria, the two organisms best studied, known polypeptides coded for by mtDNA include several subunits involved in the electron transport chain (such as three of the seven subunit proteins that comprise the "cytochrome oxidase" complex; Fig. II–50) as well as polypeptide chains that are part of the F_1–F_0 ATP-synthesizing system. Several polypeptides of still unknown function also are coded for by these mitochondrial DNAs, as are the two rRNAs present in the ribosomes and the two dozen tRNAs that the mitochondria use. Higher plants may make 20 or more distinct polypeptide chains in their mitochondria, in comparison with the total of 10 or so made by yeast. Plant mtDNA also codes for a third rRNA; this 5S RNA is similar to one of the small types of cytoplasmic ribosomal RNA but does not appear to have a counterpart in yeast or mammalian mitochondrial ribosomes.

Dual Origin of Components Yeast cells and the large *hyphae* (elongate structures with many nuclei in a common cytoplasm) that constitute the mold *Neurospora* have proved especially useful for experiments on the origins of mitochondrial components; they can be grown in bulk as large, fairly homogeneous populations into which drugs and radioactive precursors can conveniently be introduced and from which mitochondria can be isolated. Moreover, mutant strains with genetically based mitochondrial abnormalities can be obtained. The cells can survive even without functional mitochondria by relying on anaerobic metabolism (Section 1.3.1). (The *petite* mutants of yeast and the *poky* mutants of *Neurospora* are so named because of the alterations in growth reflecting mitochondrial abnormalities. See also Section 4.3.6.)

With such organisms, and with others, it has been shown that mitochondrial properties are under the control of two sets of genes, those in the nucleus and those in the cytoplasm. There is complementary biochemical evidence suggesting a dual origin of mitochondrial proteins. Especially illuminating results have been obtained by exposing cells to drugs: *Cycloheximide* (CHI) is one of a group of drugs that stop protein synthesis on cytoplasmic ribosomes but do not affect mitochondrial translation. In contrast, drugs like *chloramphenicol* (CAP) selectively affect the mitochondria. (Bacterial protein synthesis is also differentially sensitive to CAP and not to CHI.) The synthesis of many mitochondrial proteins, including such characteristic ones as cytochrome *c* (Fig. II–50), is not inhibited by CAP but is by CHI, indicating their origins on non-mitochondrial ribosomes. Only for the few proteins specified by mtDNA is the opposite pattern of inhibition seen. Even the proteins of mitochondrial ribosomes are made outside the mitochondria.

Duplication The question of the origin of mitochondrial proteins is part of the more general question: How are mitochondria duplicated? Early cytologists observed the division of certain cells containing only two or three mitochondria (for example, developing sperm cells of invertebrates). They showed that new mitochondria could arise by division of preexisting ones. Similar observations have been made recently in electron microscope studies of algae. In these cases mitochondrial division is coordinated with the division of the cells. Electron microscope images of cells such as hepatocytes suggest that mitochondria can also divide in nondividing cells.

When the protozoan *Tetrahymena* is grown in tritiated thymidine, the DNA of its mitochondria becomes labeled. If the cells are then transferred to a nonradioactive growth medium, the fate of this labeled DNA during cell growth can be followed by autoradiographic studies analogous to the pulse-chase experiment in Figure I–23. As the cells grow and divide, the mitochondria increase proportionally, so the number *per cell* remains fairly constant. Counts of the number of autoradiographic grains present over mitochondria at different times after labeling demonstrate that the DNA synthesized during the labeling period is randomly distributed throughout the entire mitochondrial population. In other words, the labeled DNA per mitochondrion is halved each time the total mitochondrial population doubles. This is as expected if mitochondria duplicate by some sort of division process in which a parental organelle contributes its molecules, including DNA, more or less equally to "daughter" mitochondria. Similar results have been obtained using labeled lipid precursors with *Neurospora*. Experiments of these types are not completely unambiguous since, especially with lipids, there are mechanisms other than growth and division through which radioactive macromolecules can spread from labeled organelles to initially unlabeled ones. Transient organelle fusions may occur, as well as "exchanges" such as the ones mediated by lipid exchange proteins (Section 2.4.4), in which individual macromolecules enter and leave otherwise intact structures. The following section raises the possibility that mitochondria can sometimes interconnect as large branched networks. If so, the spread even of DNA within the mitochondrial population might not always require growth and division. However, the results obtained have helped to convince most investigators that growth and division is the predominant mode of mitochondrial duplication under most circumstances.

Promitochondria   A variant of the growth-division pathway for mitochondrial origin is found in yeast, becoming evident when they shift from anaerobic to aerobic metabolism. When yeast are maintained in the absence of oxygen, they lack almost all cytochromes, and the electron microscope shows few if any typical mitochondria. They possess, instead, a few small membrane-bounded bodies with little internal structure. Upon addition of oxygen to the medium, there are both a marked increase in the number of such premitochondrial bodies ("promitochondria") and an extensive synthesis of cytochromes and other mitochondrial components; eventually the bodies become typical mitochondria. The precise manner of this transformation is variable, depending upon concentrations of glucose, lipid precursors, and other components of the growth medium. Apparently the premitochondrial bodies can divide and perpetuate themselves in the absence of oxygen.

From time to time it has been claimed that mitochondria arise in yeast or other cell types possessing no recognizable premitochondrial bodies. This led historically to suggestions that mitochondria can arise *de novo* (by assembly entirely from separate molecules), or from other membrane systems, such as the plasma membrane or nuclear envelope. However, most of these claims are now known to stem from difficulties in recognizing mitochondrial precursor structures, which sometimes lack a strongly distinctive appearance. There is not yet a fully documented and universally accepted case of mitochondrial origin in the absence of membrane-delimited primordia, presumed to contain mtDNA.

Entry of Proteins; Assembly How might a mitochondrial primordium or a fully formed mitochondrion grow? The proteins made inside the mitochondrion could insert directly into the inner membrane during or after their synthesis. But how do the proteins made outside, most of which are destined for the inner membrane or the inner compartment, traverse one or both of the mitochondrial membranes? Initial hypotheses were derived from similarities between the outer mitochondrial membrane and the endoplasmic reticulum (these share some enzymes but differ markedly in others) and from observations of close proximity between mitochondria and large sheets of rough ER. Ribosomes were also seen to stud the outer surface of yeast mitochondria under some conditions of growth. Can we conclude, then, that proteins move directly across the outer mitochondrial membrane from bound ribosomes or enter from the ER via continuities or by transport in vesicles? These possibilities remain open to an extent, particularly with regard to the mitochondria-associated ribosomes in yeast, but direct supporting evidence is scanty and controversial. Unambiguous ER–mitochondrial continuities are seen rarely, if at all, and there is no obvious population of ER-to-mitochondria transport vesicles. Instead, evidence has begun to accumulate that the mRNAs for at least certain mitochondrial proteins are present on *free* polysomes. This seems to be implied by findings indicating that entry of some mitochondrial proteins into mitochondria is initiated subsequent to completion of the polypeptide chains; entry, that is, may be "posttranslational" rather than "cotranslational" as in the ER. How protein molecules can pass across both mitochondrial membranes to enter the inner compartment is not known; some investigators propose that the routes of such entry are points at which the inner and outer mitochondrial membranes are closely associated with one another, but the reality of such associations is in dispute. The entry processes are currently being studied with the use of isolated mitochondria and proteins synthesized in well-defined *in vitro* systems (Section 2.4.3). Entry appears to require energy and sometimes results in the clipping off of a small chain of amino acids from the end of the entering polypeptide. For certain proteins, final folding is not completed until they are inside. Cytochrome *c* enters without its small iron-containing heme group, acquiring this group within the mitochondrion to form the complete, functional protein. Presumably, receptor systems on the mitochondrial outer membrane recognize features of the proteins to be taken up—sequences of amino acids or facets of their three-dimensional conformation—but this has only begun to be studied. Perhaps the outer membrane, which is relatively porous (Section 2.6.1) offers less of a barrier to proteins than might have been anticipated. In a number of organisms the outer membrane shows arrays of units reminiscent of the ones in Figure II–9; these are under investigation as possible sites of transmembrane channels.

How is the detailed molecular architecture of the inner membrane repro-duced as the mitochondrion grows? The reconstitution experiments described earlier suggest that once again, spontaneous assembly can provide much of the explanation. F_1 particles, for example, once formed seem able to associate correctly with the mitochondrial membrane in the absence of external guides. Interesting investigations are underway on the interaction of the mitochondri-ally synthesized proteins with those coming in from outside: To what extent do the ones made inside serve as guiding devices for the correct insertion of the ones made outside? (See also Chap. 4.1.) Are membrane proteins made out-side modified as they enter, so that, for example, they become more hydro-phobic and thus remain in the membrane?

One of the principal mitochondrial lipids, the phospholipid *cardiolipin* (di-phosphatidyl glycerol), is made locally; most other lipids are imported from the endoplasmic reticulum. This transfer might be facilitated by the close associa-tions observed between the two organelles.

2.6.4 Plasticity in the Living Cell

One or Many Mitochondria Per Cell? The study of living cells by phase-contrast microscopy and by microcinematography reveals that in many cells, mitochondria are continually in motion. They show dramatic changes in shape and volume, and they can probably fuse with one another or separate into several parts. The static images of the electron microscope cannot reveal such movement.

In some muscles of the rat and in a variety of other cells, highly branched mitochondria are seen in electron micrographs. One interpretation is that these are snapshots of mitochondria in the process of fusing or fragmenting. But it has also been reported that yeast grown in media rich in glucose contain only a single, quite large, multiply branched mitochondrion per cell and that this is a stable characteristic, persisting, for example, through cell division, in which each daughter receives part of the large mitochondrion. This and related ob-servations on other unicellular organisms have led a few microscopists to sug-gest that the usual impression of many separate mitochondria per cell may be misleading, based as it generally is on examination of very thin sections of cells. When serial sections (see Section 1.2.1) are used to reconstruct three-dimensional cell structure, separate profiles like the ones in Figure II–58 are seen to be interconnected. The conclusion occasionally advanced that mitochondria even of higher organisms *usually* exist as one or a very few large, irregular networks per cell must still be regarded as an overstatement; observations with other techniques (see Fig. I–20) and on living, intact cells (see Fig. I–13) do seem clearly to show many, individual, separate mitochondria. Even yeasts appear to contain many separate, spherical mitochondria when grown under some conditions. But the observations emphasize that the three-dimensional form of mitochondria, the plasticity of this form, and the abilities of mitochondria to divide and to fuse may be more complex than has usually been appreciated. In the development of sperm (Section 3.10.2) fusion of mitochondria into large, distinctive arrays is common.

Form Changes Related to Function   The thyroid hormone *thyroxine,* calcium ions, and other physiological substances can induce mitochondrial swelling in the test tube. Form alterations in isolated mitochondria are also observed upon addition of dinitrophenol and other nonphysiological agents that uncouple phosphorylation from oxidation.

Figure II–58 illustrates morphological changes related to mitochondrial function. Isolated mitochondria assume the *orthodox* configuration when the medium is depleted of ADP so that phosphorylation and the coupled respiration are stopped. If ADP is added, phosphorylation and respiration resume, and the mitochondria show the *condensed* configuration. Similar changes can be induced in cells (see Fig. II–58) by treatment with 2-deoxyglucose, which produces an increased level of intracellular ADP. Some investigators suspect that the normal state of mitochondria in cells falls between these extremes, and that mitochondria often appear in the orthodox configuration in electron micrographs owing to inactivation of metabolism during fixation of the cells for microscopic examination. Others suggest that the configurational changes reflect or parallel structural rearrangements, such as conformational changes in inner membrane proteins, that are directly important for ATP formation. The reduction of volume of the inner-compartment matrix in the condensed state could be imagined to pack the proteins there more tightly, promoting interactions of the enzymes and perhaps generating transient complexes capable of more efficient metabolism. This, however, is still essentially speculation. More prosaic views attribute condensation of the matrix to movements of ions, metabolites, and water.

Chapter 2.7 Chloroplasts

The plastids of plant cells are of various types, which contain different proportions of several pigments and of components such as starch. The most familiar and abundant plastids are the *chloroplasts,* characterized by their content of large quantities of the green pigment *chlorophyll.* They are responsible for the

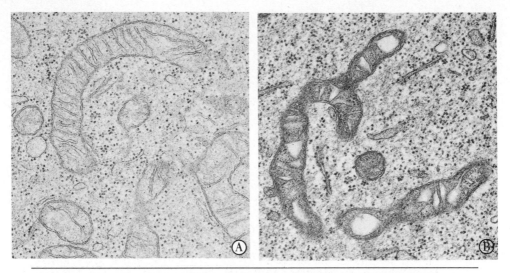

Figure II–58 *Changes in mitochondrial form associated with activity.* Mito-chondria in Ehrlich ascites tumor cells (Section 3.12.1). Panel A shows the "or-thodox" mitochondrial configuration in an untreated cell. Panel B shows a deoxy-glucose-treated cell (see text): The mitochondria have assumed the "condensed" configuration; the matrix is more electron-dense than in panel A, and the form of the cristae is altered. × 27,000. (From C.R. Hackenbrock, *J. Cell Biol.* **151**:123, 1971, Courtesy of the author and Rockefeller University Press.)

photosynthetic use of the energy of sunlight to effect the transformation of carbon dioxide and water into carbohydrates with the simultaneous release of oxygen.

Chloroplasts are large organelles that vary in size and shape from species to species. In some algae, one or two chloroplasts are present per cell and have the form of cups or elongate spirals that fill much of the cytoplasm (see Figs. II–61 and III–17). In other algae and in higher plants, a great many chlo-roplasts may be present in each cell in the form of ovoid or disclike bodies; leaf cells, for example, contain several dozen chloroplasts, each measuring 2 to 4 by 5 to 10 μm. Typical dry weight figures for chloroplast composition are 40 to 60 percent protein, 25 to 35 percent lipid (largely glycolipids), 5 to 10 percent chlorophyll, 1 percent pigments other than chlorophyll, and small amounts of DNA and RNA. Chlorophyll and the other pigments are lipid-sol-uble, chlorophyll being composed of a magnesium-containing *porphyrin ring* attached to a long lipid-soluble *phytol* tail (see Fig. II–63). (It is the porphyrin ring with its "conjugated" system of bonds—single bonds alternating with dou-ble bonds [C=C and C=N]—that is responsible for the absorption of light by a chlorophyll molecule.)

2.7.1 Structure

Chloroplasts are bounded by two membranes (see Fig. II–59). Within, they contain additional membranes surrounded by a matrix, the *stroma*. Granules containing starch often are present, scattered in the stroma. In many algae, however, starch accumulates near a special region known as the *pyrenoid* (see Fig. II–61 and Chap. 3.4A), in which the starch is apparently synthesized from glucose.

The internal membrane systems of the chloroplast are chiefly in the form of flattened sacs called *thylakoids.* In many algae the thylakoids are arranged in parallel arrays and run much of the length of the plastid. The details of arrangement differ in different types of algae. Often the thylakoids form groups or stacks over much of their length; in several classes of algae there tend to be three thylakoids per group (see Fig. III–17), but the number varies in different classes. In green algae such as *Chlamydomonas* (Figs. II–59 and II–61), the thylakoids are stacked into *grana;* these are arrays in which adjacent thylakoids are very closely apposed (''fused''). The precise arrangements—dimensions of the arrays, number of thylakoids per granum, and so forth—vary somewhat, even within a given chloroplast and, more appreciably, among different species of green algae. In higher plants the structure also varies somewhat in detail but usually resembles the arrangements shown in Figures II–59 and II–60. Each granum of a chloroplast in a higher plant consists of a stack of closely apposed thylakoids, resembling a pile of coins. The thylakoids of different grana are connected to each other by membranes running in the stroma.

2.7.2 Photosynthesis: A Brief Review

Photosynthesis involves two major sets of reactions, each set consisting of many steps (Fig. II–62). One set, the *light reactions,* transform (''transduce'') energy acquired through the absorption of light into usable chemical forms, namely ATP and the reduced form (NADPH) of the coenzyme NADP. This is accomplished via oxidation-reduction sequences that include the removal of electrons from water, releasing oxygen and protons (H^+). These protons and others transferred across the thylakoid membranes during the oxidation-reduction reactions contribute to an ATP-producing proton gradient like the one described for mitochondria in the last chapter.

The other reactions in photosynthesis, the *dark reactions,* occur both in the light and in the dark. They use the ATP and NADPH to build carbohydrates from CO_2. The most common pathway by which this is done is the ''carbon reduction cycle'' (the Calvin cycle), which starts with the reaction of CO_2 with the 5-carbon sugar *ribulose 1,5-diphosphate.* This generates two 3-carbon molecules that undergo the cycle schematized in Figure II–62. This initial CO_2 *fixation* is catalyzed by the enzyme *ribulose 1,5-diphosphate (''bisphosphate'') carboxylase,* often called RuDP carboxylase or carboxydismutase.

The Light Reactions The green color of chlorophyll is due to its selective absorption of light with wavelengths of 650 to 700 nm. (Light of these wavelengths appears to the eye as red; when red light is removed from white light, as by absorption, what is left appears green.) Chlorophyll also absorbs light at shorter wavelengths (400 to 475 nm, corresponding to the violet-blue region

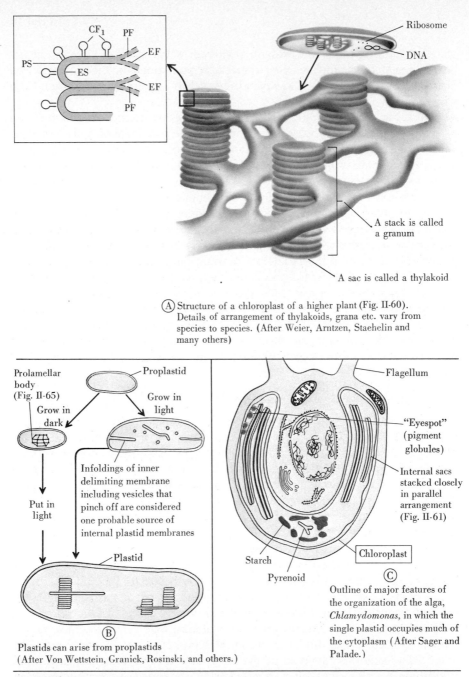

CF$_1$ PF

PS EF

ES EF

PF

Ribosome

DNA

A stack is called a granum

A sac is called a thylakoid

(A) Structure of a chloroplast of a higher plant (Fig. II-60). Details of arrangement of thylakoids, grana etc. vary from species to species. (After Weier, Arntzen, Staehelin and many others)

Prolamellar body (Fig. II-65)

Proplastid

Grow in light

Grow in dark

Put in light

Infoldings of inner delimiting membrane including vesicles that pinch off are considered one probable source of internal plastid membranes

Plastid

(B)

Plastids can arise from proplastids (After Von Wettstein, Granick, Rosinski, and others.)

Flagellum

"Eyespot" (pigment globules)

Internal sacs stacked closely in parallel arrangement (Fig. II-61)

Chloroplast

(C)

Starch

Pyrenoid

Outline of major features of the organization of the alga, *Chlamydomonas*, in which the single plastid occupies much of the cytoplasm (After Sager and Palade.)

Figure II–59 *Schematic representation of some structural and developmental features of chloroplasts.* The upper left sketch in part A indicates the nomenclature applied to the membrane surfaces and faces visualized by freeze-fracture procedures (Figs. II–6 and II–64), which separate membrane faces in the planes shown by dotted lines. It also illustrates the likelihood that the CF$_1$ particles of chloroplasts are present only on those portions of the thylakoids that are not stacked against an adjacent thylakoid.

of the spectrum). Two major chlorophyll types are present in eucaryotic plastids: Chlorophyll *a* shows *absorption maxima* (peak absorption) at about 670 nm and at about 490 nm. Chlorophyll *b* absorbs maximally at about 650 and 470 nm. Wavelength is an indicator of the energy carried by packets *(photons)* of light. The shorter the wavelength, the greater the energy. The absorption maxima of the chlorophylls and other photosynthetic pigments indicate the portion of the spectrum of sunlight that can be captured for energizing photosynthesis. When a chlorophyll molecule absorbs a photon, an electron of the chlorophyll's porphyrin ring acquires the photon's energy and thereby is *excited.* This is the basis of the chloroplast's ability to transduce light energy to chemical energy. The excited electron can enter a series of electron transport reactions involving iron-containing proteins (ferredoxins, cytochromes), a copper-containing protein *(plastocyanin),* lipid-soluble molecules *(plastoquinones)* resembling coenzyme Q (Section 2.6.1) and other participants, some yet to be identified. Two distinguishable "photochemical units", *photosystems,* cooperate in photosynthesis. Figure II–62 diagrams this cooperation as currently envisaged by a majority of investigators. The functional heart of each unit is a *reaction center* (action center) that carries out the essential processes by which energy derived from light is made available to the electron transport sequences and other phenomena of the light reactions. Each photosystem is characterized by a particular form of pigment molecule associated with the reaction center; the pigments almost certainly are chlorophylls. In photosystem I (PSI), the reaction center pigment absorbs light maximally at 700 nm and hence is called *P700.* Excited electrons from P700 pass from the PSI reaction center to the electron transport sequence that reduces NADP. The electrons "lost" from P700 in this process are replaced through electron transport from photosystem II (PSII); the cyclic pathway diagrammed in Figure II–62 represents an alternative replenishment mechanism, which some investigators believe is actually the predominant one, at least under some physiological circumstances. The reaction center pigment of PSII, P680, absorbs light preferentially at 680 nm. The PSII reaction center transfers electrons from this pigment to an electron transport chain that links PSII to PSI. The "loss" of electrons from P680 is made up by the removal of electrons from water molecules; it is this PSII-mediated oxidation ("splitting") of water that generates O_2.

In net effect the two photosystems collaborate via electron transport to transfer electrons from water to NADPH. Accompanying this transfer is the establishment of an electrochemical H^+ gradient across the thylakoid membranes: The thylakoid interior is maintained at low pH relative to the stroma. The H^+ ions freed in the "splitting" of water also contribute to this gradient (see Figs. II–62 and II–63). The proton gradient is used to generate ATP *(photosynthetic phosphorylation).* As with mitochondria, vesicles formed from fragmented thylakoids are capable of both electron transport and phosphorylation but can no longer phosphorylate if broken open or otherwise treated to disrupt the permeability barrier inherent in the membrane. Moreover, chloroplast vesicles can produce ATP for a brief period if simply soaked for a time in a medium of low pH, until their interior equilibrates with the medium, and then

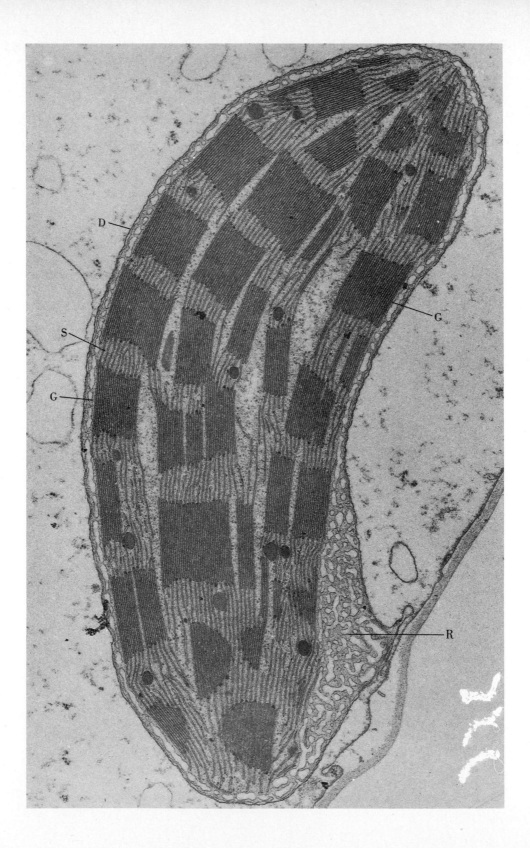

transferred into a medium of high pH, establishing a transient H^+ gradient across the vesicles' membranes. Experiments of this sort were crucial in verifying the plausibility, in general, of chemiosmotic mechanisms for production of ATP.

2.7.3 Structure and Function

When chloroplasts are broken open, many of the dark reactions' enzymes are released, while the components of the light reactions remain with the membranes. It is inferred that the dark reaction components are largely present in the stroma, whereas those of the light reactions are on and in the thylakoid membrane. The electron transport components show an asymmetric arrangement in the membrane related, presumably, to their production of the proton gradient (Fig. II–63). As might be anticipated from the fact that protons accumulate *within* the thylakoids, phosphorylation–coupling particles resembling the F_1 particles of mitochondria (Section 2.6.1) and called CF_1 are present on the *stromal* side of the thylakoid membranes. F_0-like complexes are present in the membranes, and the membranes show the permeability properties requisite for chemiosmotic ATP production. The membranes bounding the chloroplast, especially the innermost of the two, govern exchanges of substrates and metabolites with the cytoplasm through a variety of carrier and other transport mechanisms.

Light Harvesting   The reaction center pigments, P680 and P700, are tentatively identified as forms of chlorophyll *a* with absorption characteristics different from those of the average chlorophyll *a* owing to their particular associations with other membrane molecules. In addition to these pigments, chloroplasts contain several hundred chlorophyll molecules per reaction center. Of those chlorophylls, the vast majority serve as "antennae" or "light-harvesting" systems in the sense that when excited as a result of light absorption, they transfer the resultant excitation energy to a reaction center pigment (P680 or P700) rather than directly to an electron transport chain. Such transfer can involve the migration of the energy through several or even many chlorophylls associated closely in the thylakoid membrane, before it reaches the reaction center. Evidently an excited chlorophyll can transfer its energy directly to an adjacent one, exciting the recipient as it returns to the unexcited ("ground") state. When the energy reaches a reaction center, it is "trapped" by the fact that some is lost (eventually as heat) so that this chlorophyll no longer has enough energy to excite others. The fact that the absorption maximum of

---

◀ **Figure II–60**   *A chloroplast from a leaf cell of corn (Zea mays).* **The two delimiting membranes are seen at** *D*. **The grana** *(G)* **consist of stacks of saclike thylakoids. Membranes in the stroma connect the grana** *(S)*. **The reticulum seen at** *R* **is present only in some plants (Section 2.7.5). × 75,000. See Figure II–59 for a diagrammatic interpretation. (Courtesy of L.K. Shumway and T.E. Weir.)**

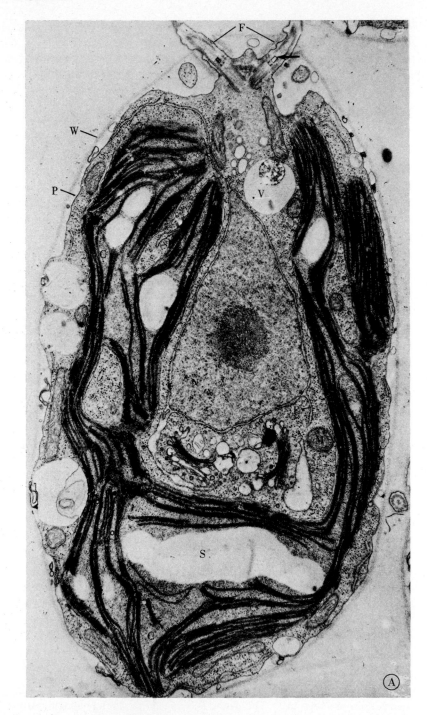

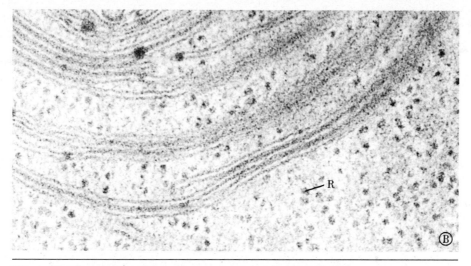

Figure II–61 *Chlamydomonas, an alga* (see also Fig. II–59).

A. (Left), Electron micrograph showing the major organelles: *C*, portions of the single large chloroplast; *E*, endoplasmic reticulum (note the continuity with the nuclear envelope—the nucleus is indicated by *N*); *G*, Golgi apparatus; *M*, mitochondria; *Nu*, the nucleolus; *P*, plasma membrane; *V*, a vacuole; *W*, cell wall. *F* indicates the two flagella (the arrow points to a poorly understood electron-dense structure present within the axoneme [Section 2.10.2] at the end of the central tubules at the point where the axoneme joins the basal body). The hole at *S* is an artifact introduced in preparing the cell for examination (Section 1.2.2); in the living cell, stored carbohydrate (starch) is present in this zone, but it is difficult to avoid extracting this material during fixation and embedding. Many free polysomes are present in the cytoplasm: their ribosomes are the numerous small dark granules some of which stud the ER. × 20,000.

B. (Above), Higher magnification view showing a small region within a chloroplast. The stacking of the thylakoids in closely appressed groups is readily evident. *R* indicates a choloroplast ribosome. Approximately × 100,000. (Micrographs courtesy of U. Goodenough.)

P700 is at a longer wavelength than other chlorophylls reflects this trapping. The absorption maxima measure the energy transitions in the molecule that are "permitted" by the laws of quantum physics. As implied above, absorption at a longer wavelength indicates an excited state of lower energy level. Energy transfer from one pigment molecule to another in the thylakoid membrane can only be to a molecule whose excited state is at a similar energy level or lower than that of the donor. The energy in an excited reaction center chlorophyll like P700 can, however, be passed further "downhill" via electron transport. Thus, energy funnels from the light-harvesting chlorophylls to the reaction centers. The net effect of the fact that P680 and P700 need not absorb light directly to be excited is an amplification of the light-gathering capacities of the

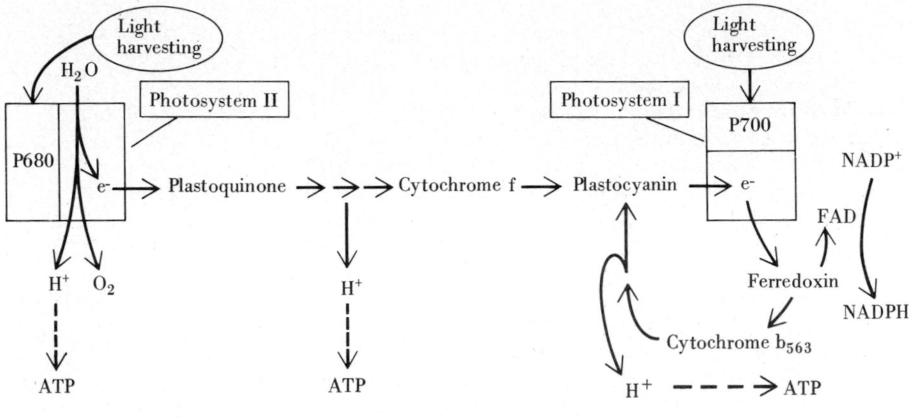

(A) Light reactions

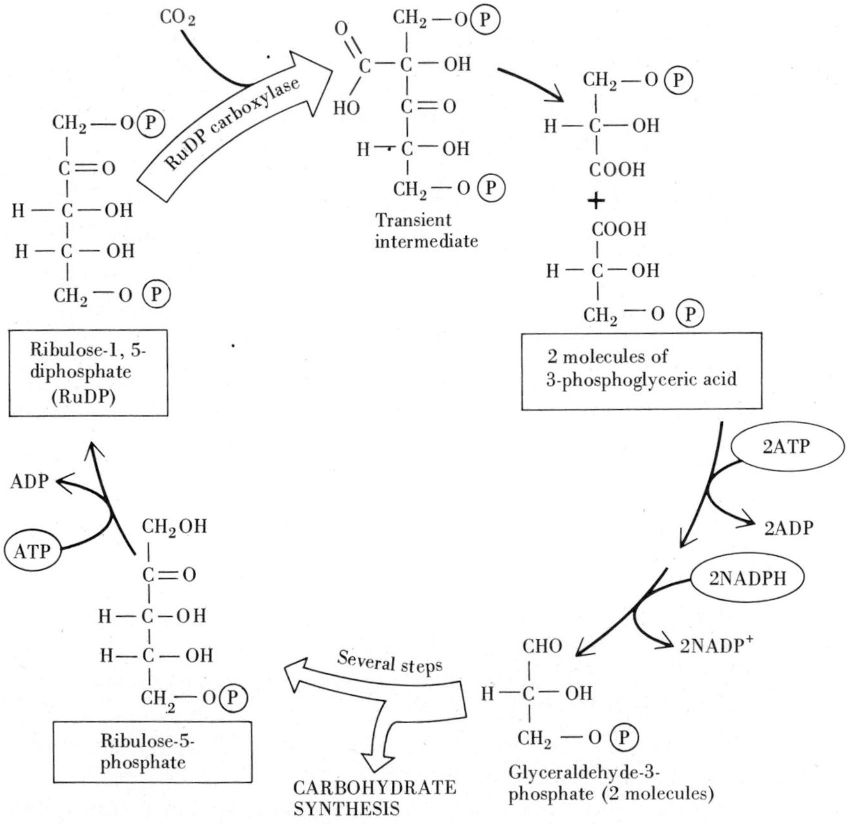

(B) The Calvin cycle

◀Figure II–62   *Outline of photosynthesis.* (After Arnon, Calvin, Clayton, Duysons, Emerson, Hill, Mitchell, Racker, Trebst and many others.)

A. Overall scheme of the light reactions according to the present consensus. Most investigators believe that under ordinary circumstances the flow of electrons is as diagrammed; light-excited electrons ($2e^-$ per H_2O split) pass from photosystem II (PS II) through an electron transport chain to photosystem I (PS I); in the course of this series of oxidation-reduction transfers, energy is released, much of which is made available in the form of an electrochemical proton gradient (as in Chapter 2.6). The electrons reaching PS I are re-excited by light, and the energy so provided is used to reduce NADP and contribute to the proton gradient. Note, however, that PS I can apparently function independently of PS II by employing a cyclic route (shown in red) whereby electrons from PS I are transferred back to PS I, probably by passage from cytochrome b_{563} into the plastoquinone–cytochrome *f*–plastocyanin chain. The extent to which this cyclic path operates in nature and the extent to which *cooperation* of the two photosystems normally is *obligatory* for photosynthesis need further investigation. Cytochrome *f* is also known as cytochrome c_{552}.

B. Major events of the Calvin cycle, the dark-reaction pathway by which CO_2 is "fixed" (converted into biologically metabolizable carbon). The ATP and NADPH molecules generated by the light reactions in the thylakoids are used, as indicated, in reactions that occur in the chloroplast's stroma. In essence, in each turn of the cycle one CO_2 and one RuDP (a molecule of 5 carbons [5-C]) are joined and the product split to produce two 3-C molecules. These can undergo a series of reactions which regenerate the 5-C molecule for use in another turn of the cycle. But since for each turn one *new* carbon enters from CO_2, for each three turns one "extra" 3-carbon molecule is produced—that is, there is a "surplus" beyond that needed simply to keep the cycle going, and this can be used to synthesize sugars and other carbohydrates.

---

photosynthetic apparatus. Further transfer of energy from absorbed light to the reaction centers occurs from another set of pigments in the thylakoids, the *accessory pigments,* of which the *carotenoids* are usually the primary types in higher plants. These absorb light at 400 to 500 nm (violet, blue, and green), thus extending the range of wavelengths whose energy can be used effectively by the chloroplasts.

The Thylakoid Membrane   How are the pigments and proteins of the thylakoid arranged? At one time it was thought that repetitive bundles or *quantasomes* in the chloroplast membranes each contained all the components of both photosystems plus a set of light-harvesting chlorophylls. Investigators now favor more complicated models (Fig. II–63).

On careful subfractionation of isolated thylakoids, multicomponent complexes carrying out the reactions of photosystem I can be separated from complexes containing photosystem II. The complexes include electron transport components, the pigments and proteins of reaction centers, and amounts of antenna chlorophyll that vary, depending on details of preparation. Most of the thylakoids' pigment molecules are isolable as "chlorophyll–protein" complexes ("light-harvesting complexes") containing chlorophylls and often carotenoids as well, associated with protein molecules. One such chlorophyll protein complex, containing both chlorophylls *a* and *b,* often appears in subfractions rich

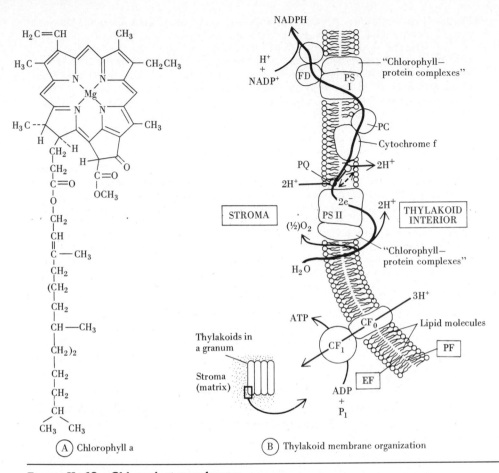

(A) Chlorophyll a

(B) Thylakoid membrane organization

Figure II–63 *Chloroplast membranes.*

A. A molecule of chlorophyll *a* showing the *porphyrin* ring system, which absorbs light, and the phytol "tail," an elongate hydrophobic, lipid-soluble chain, which permits the pigment molecules to integrate in photosynthetic membranes. The porphyrin system is in red.

B. Schematic model of *features of a thylakoid membrane* from a higher plant illustrating current proposals for the asymmetric distribution of major events and of the principal proteins and multiprotein complexes, and the path of the electrons. The lipid bilayer of the membrane is rich in galactose-containing glycolipids and is quite fluid. Photosystem I and photosystem II each include several polypeptide chains as well as pigments. The two photosystems differ in their lateral distribution in the membrane, to an extent still to be fully evaluated; the grana seem particularly rich in PS II and the stroma thylakoids in PS I. The implication of this difference for photosynthetic mechanisms and especially for the presumed collaboration between the two systems are being studied. As indicated in Figure II–59 and in the text, CF_1 and CF_0 are located at those surfaces of grana not stacked against adjacent membranes; they probably are found in stroma thylakoids as well.

Photosystems I and II transfer electrons from H_2O to $NADP^+$ yielding NADPH at the stromal side of the thylakoid membrane. As described in Figure in photosystem II and thus is thought to be structurally and functionally linked to this system. Other complexes are markedly richer in chlorophyll *a* than in *b;* certain of these may be associated with photosystem I. Although there is still some confusion about the details of organization of the light-harvesting systems, it is likely that the proteins help to hold the pigment molecules in orientations favoring the energy transfers that take place. Still uncertain is whether all of the pigment molecules are specifically associated with one or the other photosystem or whether there also is a "pool" of antennae pigments that can transfer energy to either system. Evidence favoring the latter possibility comes from biophysical analyses indicating that when the relative rates of operation of the two photosystems with respect to one another are experimentally altered, the thylakoid membrane can divert light energy from one to the other.

When chloroplast membranes are examined by negative staining and freeze-fracture microscopy (Fig. II–64), numerous particles are seen on the thylakoid surfaces and within the interior of the membranes. The particles on the surfaces that are exposed to the stroma include structures corresponding to CF_1; others, of RuDP carboxylase, are loosely bound to the membranes. A potentially significant observation is that CF_1 seems sparse or excluded from the regions where the membranes are closely apposed in stacks, being found chiefly at the exposed surfaces at the edges and top and bottom of the stack (see Figs. II–59 and II–63). If so, an apparent implication is that close linkage between particular CF_1s and particular electron transport systems (or photosystem units) is not required since the electron transport chains are not distributed similarly to CF_1 (see below). In chemiosmotic models this makes sense, since the gradient established by the sac as a whole or even the stack as a whole "drives" phosphorylation; the protons transferred across the membrane by

---

II–62, this transfer is mediated by components including plastocyanin (PC) and ferredoxin (FD). The splitting of water releases protons that accumulate within the thylakoid since the membrane is impermeable to H^+ except for movement through CF_0. Additional protons probably are transported to the thylakoid interior by plastoquinone molecules (PQ) some of which are thought to serve as a pool of mobile carriers that shuttle back and forth across the interior of the membrane (see Section 2.6.1) undergoing a cycle of reduction and oxidation in which they acquire and then release H^+ and e^-. CF_0 and CF_1, like their counterparts in mitochondria, collaborate to generate ATP (released into the stroma) while transferring protons from the interior of the thylakoid.

It is estimated that four quanta of light (photons), two per PS I and two per PS II, are required to release or transfer four H^+s to the interior of the thylakoid; this works out to eight quanta per O_2 molecule released. Two molecules of water are split per O_2 released, and this provides four electrons to the electron transport chain. Present estimates are that three protons are moved to the stromal space outside the thylakoid per ATP molecule generated.

PF and EF are the fracture faces exposed in freeze-fracture procedures (see Fig. II–6, Fig. II–59, and Fig. II–64). (After work by Andersson, Anderson, Arntzen, Miller, Staehelin, Stryer, and many others.)

Figure II–64 *Chloroplast membranes.* Freeze-fracture preparation of a thylakoid membrane from a granum of a barley chloroplast showing the prominent intramembrane particles. The zones labeled *EF* and *PF* correspond to the membrane faces similarly designated in Figure II–59. The asterisks [*] indicate regions where the membrane was not stacked directly against another membrane, as at the edges of the thylakoid stacks in grana or in the stroma thylakoids. The other faces are from stacked regions where a given thylakoid of a granum is closely appressed to the next thylakoid. Each of the zones shows a distinctive pattern of intramembrane particles sustaining the conclusion (see text and Fig. II–63) that there are differences between stacked and unstacked regions of the same membrane in terms of the relative abundance of different functional components. The very large particles characterizing the EF are the ones suspected to be locales of photosystem II (Section 2.7.2). × 100,000. (From Miller, K.R., *Scientific American,* **241**(4):102, 1979.)

a particular photosystem-electron transport cluster, can be utilized by any CF_0–CF_1 system in the same membrane, not just by an immediately neighboring one.

The particles visible in freeze-fractured thylakoid membranes fall into several size classes whose functional significance has yet to be worked out fully (Fig. II–64). Useful information is coming from comparison of the timing of appearance of particular particle classes with the timing of appearance of particular functional capacities during plastid development (see following section). Also valuable are observations of the structural effects of mutations that result in the loss of one or another chloroplast component, such as parts of particular photosystems. From such work it has been concluded provisionally that large (diameters of 15 to 18 nm) multi-subunit membrane-spanning particles visible in the E face of the thylakoids of some plants (Figs. II–63 and II–64) consist of an 8- to 10-nm core containing the photosystem II reaction center, surrounded by light-harvesting chlorophyll–protein complexes. A smaller set of particles (10 nm) may represent photosystem I with associated chlorophyll–protein complexes; these have provisionally been identified in the P face of the thylakoid membranes.

The structural organization underlying interaction of the two photosystems is poorly understood. A significant issue is whether the very close stacking of the thylakoids in grana permits collaboration of the active systems on the pairs of abutting membranes from neighboring sacs. Efforts are also being made to determine whether the stacked grana of higher plants differ from the stroma thylakoids with which they are directly continuous. The data thus far obtained from subfractionation suggest that the grana are much richer in photosystem II than are the stroma systems. Finally, there has been speculation that small movements of protein complexes in the plane of the membrane, as well as alterations in stacking of the thylakoids, are important regulating mechanisms for thylakoid functions. Changes in spacing of the various participants in the light reactions, involving movements of a nanometer or two, could profoundly affect their interplay and functioning.

2.7.4 Reproduction

Plastids contain their own DNA and RNA, distinct from both nuclear and other cytoplasmic nucleic acids. They are capable of protein and RNA synthesis, and they participate in lipid and pigment synthesis as well. Plastid ribosomes are somewhat smaller than other cytoplasmic ribosomes, generally resembling ribosomes of procaryotes in size (70S) and in molecular weights of their RNAs. Plastid protein synthesis is sensitive to chloramphenicol (and insensitive to cycloheximide), as is true also with mitochondria and bacteria (Section 2.6.3). The DNA isolated from chloroplasts of higher plants such as spinach or peas is in the form of 45-μm circles. Estimates of the numbers of circles present per chloroplast range from 10 to 50 for different plants, but there is little evidence that those in a given chloroplast vary significantly in their genetic content. The DNAs are capable of coding for a fair number of proteins, in addition to the chloroplast RNAs. As many as 50 or even more different polypeptide chains may be made in plastids including subunits of CF_1, cytochromes, and proteins of light-harvesting complexes. However, as with mitochondria, chloroplasts obtain some of their proteins and lipids from outside. In the case of the key dark reaction enzyme, ribulose diphosphate carboxylase, the smaller of the enzyme's two types of subunits is coded for by a nuclear gene and produced on cytoplasmic ribosomes, whereas the other subunit is coded for and produced in the chloroplast. Most Calvin cycle enzymes are made on the cytoplasmic ribosomes.

Both in the unicellular alga *Chlamydomonas* (Figs. II–59 and II–61) and in higher plants such as spinach, it appears that many of the proteins made outside the chloroplast are translated on free polysomes. These include proteins of the membranes and of the stroma. The behavior of the RuDP carboxylase small subunit has been studied most intensively. If made in a test-tube protein-synthesizing mixture, the small subunit can be taken up by isolated chloroplasts: In *Chlamydomonas,* the uptake is energy-dependent and involves cleavage of a sequence of 44 amino acids, mostly with apolar side chains, from the N-terminal end of the protein. In peas, the comparable cleaved-off sequence is of 57 amino acids and is less hydrophobic. Once inside, the small subunit complexes with the large subunit to form functional enzymes (RuDP carboxylase consists of eight copies of each subunit). The receptors on the membranes bounding the chloroplast that permit selective incorporation of proteins made outside of chloroplasts have not yet been identified.

As is true also with mitochondria, close associations of ER with the chloroplast surface are seen, especially in some algae (see Fig. III–17). For these algae it is possible that the ER contributes some proteins to the plastid, perhaps by budding of vesicles, but this is still a controversial suggestion. More generally, as with mitochondria it cannot yet be taken for granted that all or even most proteins made outside the chloroplast are translated on free polysomes, even in cells where intimate associations of ER and mitochondria or plastids are not evident. Only a few species of mitochondrial or plastid proteins have yet been adequately studied as to where, in the cytoplasm, they originate. Hence, any generalization must be of a preliminary sort.

Certain characteristic chloroplast glycolipids, notably those containing the sugar galactose, appear to be made by the membrane system that bounds the plastid and to move from there to the thylakoids.

Proplastids　In algae, plastids are seen to divide regularly as part of the cell division cycle and thus maintain a constant number. When a single large plastid is present, it is divided at the time the cell itself divides. In higher plants also, plastids can divide. In addition, plastids can develop from much simpler organelles, the proplastids. These appear to be self-duplicating but do not contain the elaborate patterns of internal membranes found in mature plastids. They are present in gametes and embryos.

Higher plants grown from seed in the dark provide a system for studying plastid formation from proplastids that, though "abnormal," has helped in the understanding of normal development. The tissues of such plants are usually colorless, containing no fully developed plastids and little chlorophyll. Modified plastids ("etioplasts") are present (Figs. II–59 and II–65); they contain arrays of membranous tubules, often in highly regular arrangements, called "prolamellar bodies." The latter contribute to the formation of normal internal membranes when the plants are illuminated. On exposure to light, the small amounts of the chlorophyll precursor *protochlorophyllide* present in the etioplasts are converted to chlorophyll, and additional chlorophyll synthesis begins;

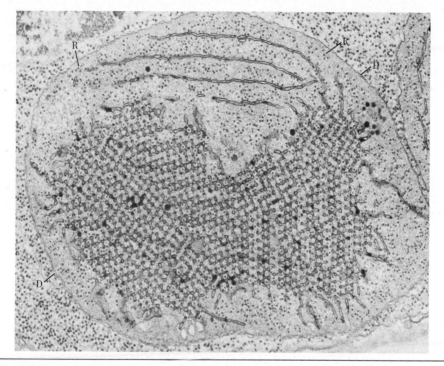

Figure II–65 *Plastid from an oat plant grown in the dark.* Plants grown in the dark are called etiolated, and the plastids like that shown here are etioplasts. Most of the organelle is occupied by a prolamellar body, a network of regularly arranged tubules. *D* indicates the pair of membranes that delimit the plastid, and *R*, a few of the many plastid ribosomes. Ribosomes are also numerous in the cytoplasm outside the plastid. × 35,000. (From Gunning, B.E.S., in Gunning and Stern (eds.) *Ultrastructure and Biology of Plant Cells.* London: Arnold Co., 1975.)

this is paralleled by the development of the usual internal chloroplast structure. During such "greening," the size of the intramembrane particles in the plastid membranes increases from 7 to 8 nm to the dimensions seen in mature plastids. This is one of the reasons some of the structures in the mature membrane are thought to be composite particles, each with a core containing a reaction center, surrounded by chlorophyll–protein complexes. It is assumed that the addition of the latter components to the particles in greening etioplasts is responsible for the observed increases in particle size.

Assembly   Mutants of *Chlamydomonas* show various alterations in details of chloroplast structure and development. One such strain is proving useful in attacking the question: How is the enzymatically complex membrane of chloroplasts formed? *Chlamydomonas* normally can synthesize chlorophyll in the dark, but the mutant cannot; thus when grown in the dark its plastids do not mature into chloroplasts. When dark-grown mutant cells are illuminated, they begin synthesis of chlorophylls and other chloroplast molecules, and chloroplast membranes form. This provides a very convenient experimental system. When the cells are illuminated in the presence of chloramphenicol, chloroplast membranes appear, but they lack some key proteins. Thus they are nonfunctional; they also do not form their normal stacked granal arrays. If the cells are now grown in cycloheximide, the chloroplasts develop into normal functioning organelles. Apparently the abnormal membranes formed initially are "cured" by the addition of proteins made in the chloroplast. This demonstrates that the membranes can be made in a multistep process—a framework can be built from some of the components and completed by the insertion of others. Presumably some of the subassemblies present in and on the mature membrane form as multimolecular arrays before adding to the membrane (Chap. 4.1). Eventually the observations should also provide insight into the mechanisms by which the overall three-dimensional structure of a complex organelle like a mature plastid arises (Section 4.1.4). More specifically, how does *Chlamydomonas* control stacking of its chloroplast membranes into their characteristic arrays, and what are the molecular requirements for membranes to form such intimate associations as are seen in the grana?

Invaginations of the inner delimiting membrane (see Fig. II–59) are frequently seen during maturation of proplastids into plastids. It has been claimed that thylakoid membranes form from these invaginations; some investigators also assert that extensive continuities persist between the inner membrane and the internal membrane systems of mature chloroplasts. However, definitive evidence is still being sought. In *Chlamydomonas* there are few, if any, invaginations of the delimiting membranes. For the mutants discussed above, it has been proposed that (1) thylakoids form by expansion of small sacs and other membranes that persist in the plastids of dark-grown cells; (2) this expansion involves continual addition of new macromolecules throughout the expanse of the growing structure (there may be no "new" and "old" membrane regions; Section 2.5.4); and (3) chloroplast ribosomes bound to the growing membranes insert new proteins directly, during translation.

The fact that chloroplast membrane formation, as well as synthesis of some pertinent macromolecules, is tied closely to the synthesis of chlorophyll represents an important cellular "control." Photosynthesis cannot proceed without chlorophyll, and evolution apparently has led to mechanisms that avoid formation of an elaborate photosynthetic apparatus lacking this pigment.

2.7.5 C$_4$ Metabolism

Photosynthesis is among the more ancient of integrated metabolic processes. It directly serves fundamental metabolic requirements of living systems in a considerable variety of environments. Some of the stages in the evolution of photosynthesis are represented among procaryotes and will be discussed later (Chap. 3.2C and Sections 3.2.4 and 4.5.2). The photosynthetic pathways described thus far in this chapter apply to many higher plants. Important variations do occur, however. Among the significant ones are those encountered in the group of plants known as C_4 *plants* because the initial products of CO_2 incorporation are 4-carbon molecules (see Fig. II–66) rather than the usual 3-carbon products. Not all aspects of this specialization are common to all C_4 plants, and not all are completely understood, but the situation seems substantially as follows: (We shall refer to plants using the pathways discussed so far as C_3 plants.)

Both in C_3 and in C_4 plants, *mesophyll* cells make up the *parenchyma* of the leaf—the bulk of the tissue located between the upper and lower leaf surfaces (the surfaces are composed of *epithelial cells* and their products). Frequently, some of the mesophyll cells form a layer beneath the upper surface of the leaf. Below this, in C_3 plants, the remainder of the mesophyll cells form a meshwork of irregularly shaped cells. Within the latter zone run the *vascular bundles*—branches of the familiar veins that supply the leaf. In C_4 plants the mesophyll tissue assumes the form of prominent cylindrical layers of cells oriented around the vascular bundles (Fig. II–66). Enclosed by these layers, immediately surrounding each vascular bundle is a cylinder of *bundle sheath cells* containing prominent chloroplasts. In C_3 plants the bundle sheath cells generally show few well-developed chloroplasts. Correspondingly, in leaves of C_3 plants most photosynthesis occurs in mesophyll cells. In contrast, the mesophyll cells of C_4 plants use CO_2, by a nonphotosynthetic pathway, to produce malic acid (see Fig. II–66). The pathway responsible involves both the chloroplasts and the cytoplasms outside them, with the malic acid then diffusing to the bundle sheath cells via fine cell-to-cell connections (plasmodesmata; Section 3.4.3). In the bundle sheath cells, the malic acid is enzymatically decarboxylated (see Fig. I–26), releasing CO_2 (Fig. II–66). The CO_2 now enters the Calvin cycle as in "ordinary" photosynthesis. (The pyruvate produced when CO_2 is liberated in the bundle sheath cells returns to the mesophyll cells for reuse.) In other words, unlike ordinary photosynthesis, C_4 metabolism depends on cooperation between two separate sets of cells. In effect, the mesophyll cells "funnel" CO_2 to the bundle sheath cells, which use it, photosynthetically.

The chloroplasts of C_4 plants show an unusual membranous reticulum (see Fig. II–60) of unknown significance. Moreover, in some C_4 species, the chloroplasts of the bundle sheath cells lack well-defined, stacked-coin-like grana (see Fig. II–61), thus somewhat resembling the plastids of many algae. The functional significance of this has yet to be ascertained, although there are some metabolic features of the chloroplasts that could be relevant. For instance, it is suspected that the bundle sheath cells make more use than usual of *cyclic phosphorylation* (see Fig. II–62 and below) to generate ATP in their chloroplasts; conceivably, this implies different balances between photosystems I and II.

C_4 metabolism is found in various plants of tropical or desert origin including corn, sugar cane, and a number of other important crop plants. It is evolutionarily adaptive in that it results in more efficient photosynthesis and water retention than C_3 metabolism, particularly under conditions of high O_2 levels, elevated temperature and light intensity, and low CO_2 levels. This stems from the fact that the enzymatic activity of RuDP carboxylase alters "direction" with different conditions. When O_2 is high and CO_2 low, RuDP carboxylase,

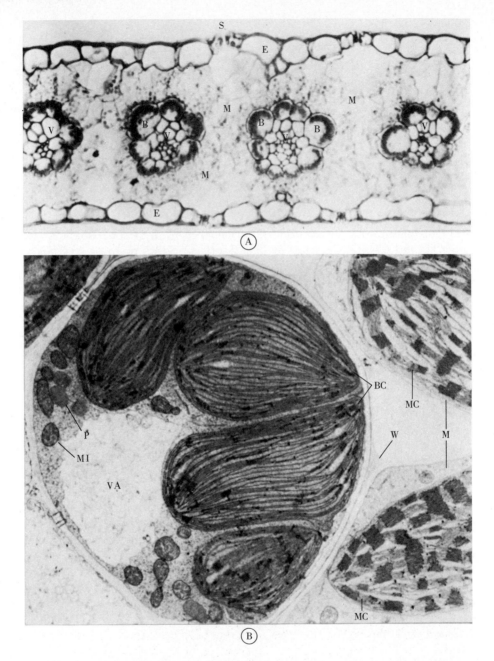

as well as fixing CO_2 for the Calvin cycle, can convert Calvin cycle intermediates to the 2-carbon molecule *glycolic acid (glycolate)*. The balance between these activities depends on conditions, with glycolate production being relatively favored when O_2 levels are high and CO_2 levels low (see Section 2.9.2). The glycolate enters the photorespiratory metabolic sequences described in

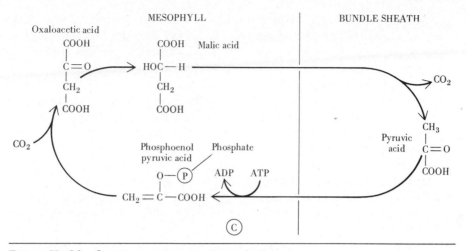

Figure II–66 *C_4 microanatomy and metabolism.*

A. (Left, top); Light micrograph showing a transverse section of a leaf of corn (*Zea mays;* see also Fig. II–60). *S* indicates one of the stomata that permit exchange of gases between the leaf interior and the surroundings (see also Fig. III–18). *E* indicates cells of the epidermis, the outermost layer of the leaf. The structures at *V* are cross sections of bundles of vascular tissue (Section 3.4.4); each bundle is encircled by a sheath of large, bundle sheath cells *(B)*. These in turn are surrounded by mesophyll tissue *(M)*. × 100.

B. (Left, below); Electron micrograph showing cells from a leaf of crabgrass *(Digitara)*. The large cell with a prominent vacuole *(VA)* is a bundle sheath cell. Note that its chloroplasts *(BC)* show elongate internal thylakoids that are not differentiated into the distinct granal and stromal arrays diagrammed in Figure II–59; they run much of the length of the plastid. This contrasts with the chloroplasts of the mesophyll cells, portions of which are seen at *M;* in these chloroplasts *(MC)* distinct grana are evident. *MI* indicates a mitochondrion, *P* a peroxisome and *W* part of a cell wall. × 12,000.

C. (Above); The fixation of CO_2 in a C_4 plant. The malic acid formed in the mesophyll cells moves to the bundle sheath cells where CO_2 is liberated to be used in the Calvin cycle.

(A and B from T.B. Ray and C.C. Black, *Encyclopedia of Plant Physiology,* Vol 6, Vienna: Springer-Verlag, 1979, pp. 79 and 80. C after M.D. Hatch, C.R. Slack, H.P. Kortschak and others.)

Section 2.9.2, which actually can *release* CO_2; from the viewpoint of photosynthesis the effect is a waste of carbons. C_4 metabolism and corresponding structure minimize this, since RuDP carboxylase is avoided in the initial steps (in the mesophyll cells), coming into play instead in the bundle sheath cells; here CO_2 "transport" via malic acid increases the local CO_2 concentration, which favors CO_2-fixing (Calvin cycle) role of RuDP carboxylase.

One consequence of these metabolic differences is that the leaves of C_4 plants can function efficiently with smaller openings of their *stomata*, the structures through which the leaf interior exchanges gases with the outside (see Fig. III–18). This retards water loss and favors survival under high-temperature, low-humidity conditions. High illumination and high temperature promote rapid photosynthesis, which consumes CO_2 and releases O_2. Exchange of these gases with the outside atmosphere can, however, be hindered under such circumstances by the narrowing of the stomatal openings in response to water-conserving regulatory mechanisms (Section 3.4.2). Thus O_2 levels build up, and CO_2 becomes depleted within the leaf, creating local conditions that favor diversion of CO_2 to photorespiration. In C_4 plants, evidently, the bundle sheath cells, which lie relatively deep within the leaf, are somewhat buffered against these effects by the volume of gas in more superficial layers of the leaf and, more importantly, by the delivery of CO_2 via malic acid, which maintains relatively high CO_2 concentrations locally, within the cells. In addition, if these bundle sheath cells do rely substantially on cyclic phosphorylation, they would generate markedly less O_2 than that produced by other photosynthesizing plant cells; low levels of O_2 favor CO_2 fixation.

Some possible evolutionary implications of matters raised in this section are outlined in Section 4.5.2.

Chapter 2.8 Lysosomes

Historically, the study of organelles usually begins with the accumulation of morphological observations and then passes to the isolation of the organelle in relatively pure fractions and to biochemical study. For lysosomes, however, this pattern was reversed. Their discovery began with an investigation of *hydrolytic enzymes,* enzymes catalyzing reactions of the type A_1—A_2 + H_2O → A_1—H + A_2—OH. Biochemical analyses of cell fractions separated from rat liver homogenates revealed that five such enzymes, all acting optimally in *acid* (low pH) media, sedimented together in centrifugation. Furthermore, it appeared that these enzymes were inactive toward their potential substrates if the fraction was carefully prepared to avoid disruption of the fragile organelles. This led to the hypothesis that the *acid hydrolases* were packaged together in an organelle previously undescribed. Since the enzymes it contained were hydro*lytic,* the organelle was called a *lysosome.* From their sedimentation characteristics it was predicted that hepatocyte lysosomes had a certain size, about 0.4 μm in diameter. From the observation of *latency,* that is, the fact that substrates added to isolated lysosomes were not split unless disruptive procedures were used, it was predicted that the organelle was surrounded by a membrane. When the acid hydrolase–containing fractions were examined by electron microscopy, morphologically distinctive cytoplasmic particles were indeed found, and the predictions concerning both size and outer delimiting membranes proved correct.

In subsequent years, biochemists found many acid hydrolases to be present in lysosomes (Fig. II–67). It also became evident that lysosomes are versatile organelles involved in a diversity of cell functions and showing corresponding variation in morphology and other properties. Unlike the situation with most mitochondria and chloroplasts, it often is impossible to ascertain that a structure is a lysosome simply from its microscopic appearances. Biochemical or cytochemical evidence is required.

2.8.1 Forms and Functions

Biochemical and cytochemical studies have already demonstrated lysosomes in protozoa, insects and some other invertebrates, amphibia, mammals, and a variety of other vertebrates. Evidently, with a few exceptions such as mammalian red blood cells, lysosomes are ubiquitous in animal cells. This is probably true of plant cells as well, but that matter is still unsettled. Some plant lysosomes will be described in Section 3.4.2.

    Tentative identification of a cytoplasmic particle as a lysosome is possible if (1) the electron microscope shows it to be membrane-delimited and (2) cytochemical studies show it to have one or more of the hydrolase activities found in lysosomes as studied biochemically. Identification by cytochemical techniques alone is equivocal, since reliable methods are available for only a few of the lysosomal hydrolases; thus the presence of other hydrolases can only be *assumed*. *Acid phosphatase* activity is the most widely used "marker" enzyme demonstrable by staining procedures (see Figs. I–19 and II–72). In the large majority of cases, bodies identified initially as probable lysosomes by acid phosphatase cytochemical procedures generally have turned out to be lysosomes or close relatives when more complete studies were possible; cytochemists now suspect, however, that more than one type of acid phosphatase exists

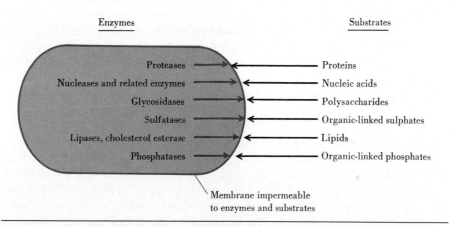

Figure II–67 *The biochemical concept of lysosomes* according to the work of C.deDuve, his collaborators, and others. There are now roughly 50 hydrolases known to occur in lysosomes, though not necessarily in all cell types. Several different enzymes hydrolyze each class of substrates shown to the right; the several enzymes within each group differ in specificity and details of their action.

in cells and that certain sites of this enzyme activity may not possess the full battery of lysosomal enzymes.

It would be of obvious advantage to be able to study all suspected lysosomes by biochemical methods and thus confirm their status. However, for one reason or another, many tissues are poorly suited for organelle isolation procedures (Chap. 1.2B). For example, some organs have large numbers of intermingled cell types, so that the source of isolated organelles cannot be easily identified as being cells of a given type. In other cases it is difficult to obtain amounts of tissue large enough or to achieve adequate purity of fractions. Cytochemical studies, with their direct access to single cells, can avoid many such problems. Ideally, cytochemical studies on cells and biochemical studies on isolated organelles are used together.

Many lysosomes are in the same size range as the first liver lysosomes studied, about 0.5 μm in diameter. The largest in animals cells, however, such as the "protein droplets" in cells of mammalian kidney (Section 3.5.3), are several micrometers in diameter. The smallest are the *Golgi vesicles;* these are only 50 to 75 nm in diameter. The identification of some of the vesicles of the Golgi region as probable lysosomes rests upon the presence of a delimiting membrane and upon cytochemical evidence that they possess several hydrolytic activities (including those of acid phosphatase, aryl sulfatase, and esterase). In addition, microscopic observations suggest that the vesicles fuse with endocytic structures, as is true of other lysosomes.

The size heterogeneity of lysosomes is paralleled by heterogeneity in form, origin, and function. All lysosomes are related, directly or indirectly, to *intracellular digestion* (Fig. II–68). The material to be digested may be of exogenous (extracellular) or endogenous (intracellular) origin. Collectively, the lysosomal enzymes are capable of hydrolyzing all the classes of macromolecules in cells (see Fig. II–67) and presumably would do so if they were not confined in structures delimited by membranes. There are a few known cases in which the hydrolases are secreted into extracellular spaces either normally or in pathological conditions. However, in most of the known functions of lysosomes, the material on which the acid hydrolases act must gain access to the interior of the lysosome, and the enzymes remain confined within the organelle.

Primary and Secondary Lysosomes; Residual Bodies There is much variation in the mechanisms by which material to be digested becomes enclosed inside a lysosome with the acid hydrolases. Figure II–68 is a diagrammatic representation of the major types of lysosomes and their possible modes of origin. The diagram illustrates the proposal that lysosomal enzymes, manufactured at the ribosomes, are transported by the endoplasmic reticulum to lysosomes, packaged by or near the Golgi apparatus. In some cells, *primary lysosomes* appear to bud from Golgi sacs (or GERL; Section 2.5.3). These lysosomes are thought to be packages that transport hydrolases to other membrane-delimited bodies with which the primary lysosomes fuse; the Golgi vesi-

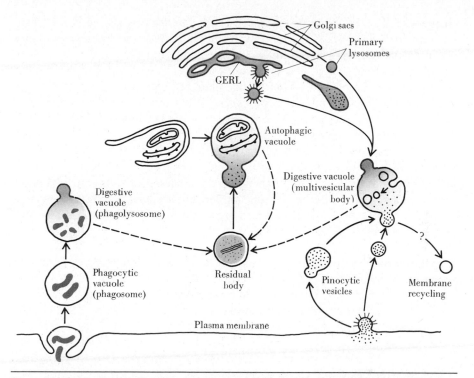

Figure II–68 *Possible origins and functions of lysosomes* of different types. Illustrated are (i) primary lysosomes, some of which are coated vesicles (Section 2.1.6), arising from sacs associated with the Golgi apparatus; (ii) fusion of lysosomes (both primary and secondary) with phagocytic, pinocytic, and autophagic vacuoles; (iii) formation of autophagic vacuoles by sequestration of cytoplasm within an enveloping sac; and (iv) formation of residual bodies as digestion proceeds in secondary lysosomes. Membrane recycling is discussed in Section 2.8.4.

cles referred to above probably fall into this category. When lysosomes contain both the enzymes and the material to be digested, being digested, or already digested, they are referred to as *secondary lysosomes*. Secondary lysosomes may accumulate large quantities of undigested or indigestible molecules; the resulting structures are known as *residual bodies*. The classification of lysosomes in these functional categories is useful in establishing relationships among structures that may differ considerably in morphology. This morphological variation reflects the variety of materials digested and the fact that lysosomes may form in several ways.

2.8.2 Hydrolysis of Exogenous Macromolecules (Heterophagy)

Many small molecules enter cells individually by passing through the plasma membrane. Macromolecules and larger particles, however, generally enter by one of the endocytic bulk transport processes, phagocytosis or pinocytosis

(Section 2.1.6). The endocytic vacuoles formed at the plasma membrane usually fuse with lysosomes, contributing their content to the lysosome interior where digestion takes place. The lysosomal membrane permits the passage of amino acids, small sugars, and other small molecules released through the hydrolysis of macromolecules. Thus the products of digestion can move out of the lysosomes into the adjacent cytoplasm. There are only a few special situations, such as the penetration of viruses or certain toxins from bacteria into cells (see below and Section 3.1.1), in which macromolecules or larger structures can enter the cytoplasm substantially intact. In protozoa, hydrolysis of material taken up by endocytosis is a central mechanism in feeding. Most multicellular organisms do not rely heavily on endocytosis for nutrition. However, endocytosis serves a number of functions in multicellular organisms (Section 2.1.6), so that most of their cell types are capable of pinocytosis, and a few, such as white blood cells, are phagocytes.

"Professional Phagocytes"   The phagocytic white blood cells of mammals are essential to the body's defense against bacterial and other invading organisms or toxic materials. They are characterized by distinctive cytoplasmic granules. These have been extensively studied by microscopy and cytochemistry and through biochemical work on isolated fractions.

Polymorphonuclear leukocytes ("neutrophils") possess two classes of granules, called *specific* granules and *azurophilic* granules (Fig. II–69). Both arise from the Golgi apparatus (Section 2.5.2). The azurophilic granules are primary lysosomes. When the phagocytes engulf bacteria or other materials, there is a rapid movement of granules to the phagocytic vacuoles. The membranes of the granules and vacuoles fuse, and the granule contents are emptied into the vacuoles. Within a few minutes the phagocytic vacuoles have acquired acid hydrolases from the azurophilic granules and thus are converted into (secondary) lysosomes. Through the actions of the hydrolases and of other enzymes and nonenzymatic material contributed by the two granule populations, most bacteria are destroyed (see Fig. II–69, legend). (A very few microorganisms, such as bacteria responsible for leprosy and tuberculosis, survive phagocytosis and multiply inside mammalian phagocytes: Some have coats that resist the lysosomal hydrolases; others can somehow inhibit fusion of lysosomes with the vacuoles in which they enter the cell. The bacteria responsible for whooping cough are believed to avoid destruction in part by producing large amounts of cyclic AMP, which inhibit defensive activities by the phagocytes.)

Phagocytes known as *macrophages* are numerous in the liver, spleen, lymph nodes, and other tissues of higher animals. In addition, macrophage precursor cells (monocytes) are present in the circulation; these accumulate at sites of injury or infection and develop into macrophages. Monocytes and macrophages produce primary lysosomes in their Golgi apparatus. Along with the white blood cells, macrophages participate in defense mechanisms such as those controlled by the immune system (Chap. 3.6C). In addition, they act as scavengers of cellular debris released in the course of tissue destruction follow-

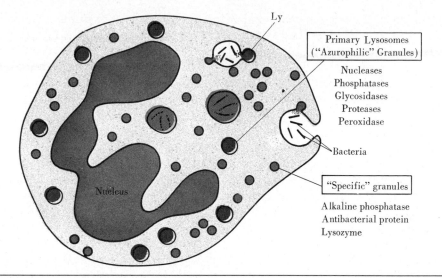

Figure II–69 *A "professional phagocyte."* Diagram of a rabbit white blood cell (of the type called a *polymorphonuclear leukocyte* or *neutrophil*) ingesting bacteria. Two classes of granules fuse with the phagocytosis vacuoles and contribute digestive enzymes and other components; the *"azurophilic"* granules are lysosomes. Some of the contents of the granules are listed. The names azurophilic and *specific* derive from distinctive staining reactions of the granule types. The presence of peroxidase in the lysosomes of these cells is a specialization related to the cells' ability to kill microbes: Peroxide (H_2O_2) is generated by an oxidase of the plasma membrane, which is internalized along with the bacteria. The peroxidase uses the peroxide in reactions ("halogenations") that help to kill the bacteria. In these reactions the peroxide is broken down to H_2O and O_2 so the cell itself is protected from potentially toxic effects (Section 2.9.2). The *lysozyme* in the specific granules aids in disrupting bacterial cell walls (Section 3.2.3); other proteins in these granules inhibit growth and survival of bacteria.

ing injury or infection, and in normal ("programmed") cell death. Extensive tissue breakdown occurs, for example, in the development of digits in the hand and foot and the resorption of the frog's tail during metamorphosis (see Section 4.4.7).

Normal Turnover of Cells and Cell Parts   Endocytosis and subsequent lysosomal degradation is the degradative route for turnover of the cells of blood. Thus in humans, red blood cells are produced and released by the bone marrow, circulate for 120 days, and then are destroyed. Normal degradation of "aged" red blood cells involves macrophages of the spleen and, to a lesser degree, of other organs. How the phagocytes discriminate among aged and younger cells is not known. Recognition of aged red blood cells may depend in part upon changes that occur in the red blood cell surface and upon alterations in mechanical properties, such as flexibility, that retard the movement of cells through special narrow spaces in the circulatory system of the spleen. (For recognition of foreign materials the phagocytes rely on the organism's antibodies; see Chap. 3.6C.) Phagocytic degradation is also responsible for the destruction of certain cell components released from the cells as part of normal events of maintenance or differentiation. This is seen, for example, in the cyclical shedding of membranes by retinal photoreceptors (Section 3.8.1) and in the maturation of sperm (Section 3.10.2).

The Thyroid Gland   In the thyroid gland, endocytosis and partial hydrolysis in lysosomes have considerable physiological importance (see Fig. II–70). The glycoprotein *thyroglobulin* is secreted into the extracellular lumen of the gland and stored there (Fig. II–70; also Section 3.6.1). When the gland is stimulated to release thyroid hormone (for example, by hormones from the pituitary gland) the stored thyroglobulin is engulfed by the secretory cells in large pinocytosis vacuoles. Electron microscopic and cytochemical observations show that lysosomes fuse with these pinocytic vacuoles. The thyroglobulin is partially hydrolyzed within the vacuole; the thyroid hormone *thyroxine* is one of the digestion products. Thyroxine, an iodine-containing derivative of the amino acid *tyrosine* (see Fig. I–28), is a relatively small molecule (Fig. II–70). Precisely how it leaves the lysosomes and crosses the plasma membrane to enter the blood capillaries at the base of the cell (the opposite pole of the cell from where thyroglobulin is released) is still unclear.

Turnover and Processing of Circulating Components   The plasma suspending the cells in the blood is rich in proteins with numerous transport and protective roles. In their normal, steady-state replacement (see introduction to this part), different plasma proteins turn over at different rates. For various human plasma proteins, average life spans before replacement range from a number of hours to a number of days. But plasma proteins that have been purified, treated to alter irreversibly their three-dimensional structure (denatured; Section 1.4.2), and then reintroduced into the circulation are very rapidly endocytized and degraded by the liver macrophages known as Kupffer cells. In addition, hepatocytes possess receptors that enable them to recognize, bind, endocytize, and deliver to the lysosomes a variety of blood glycoproteins that have lost the sialic acid that normally terminates their oligosaccharide side chains. The receptors specifically recognize the galactoses that occur next in line along the chains. These capacities enable the liver to clear the blood of abnormal or altered proteins and very likely account for much of the normal turnover of blood proteins. That is to say: Although the factors governing the differences in rates of normal turnover of different proteins are not understood, it is known that rapid destruction of proteins is promoted by the types of changes just described, and these changes are of sorts that could occur under natural conditions.

  The normal delivery of some substances from the circulation for use in the cell's cytoplasm also depends on heterophagy. Lipids are transported in the blood largely in the form of lipoprotein particles in which a lipid core is

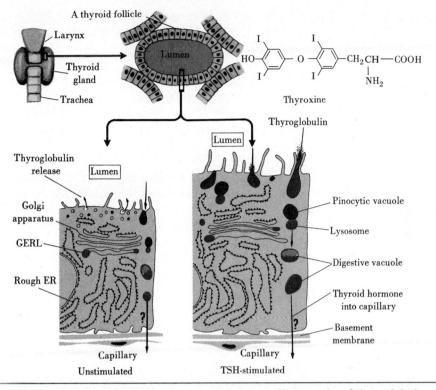

Figure II–70 *Lysosomal roles in thyroid function.* The uptake of thyroglobulin from its storage sites (colloid in the lumen) has been studied under normal conditions, but is more readily observed when the thyroid gland cells are stimulated with *thyrotropin* (TSH), a hormone produced by the pituitary gland. Thyrotropin increases the uptake of thyroglobulin from the lumen by pinocytosis and the consequent release of thyroid hormone to the blood capillaries at the base of the cell. The diagram illustrates the role of lysosomes whose enzymes (proteases) are responsible for the partial hydrolysis of thyroglobulin, which generates the thyroid hormones. The structure of the thyroid hormone, *thyroxine,* is also illustrated: the two rings are each of six carbon atoms; (compare with the structure of the amino acid *tyrosine* [Fig. I–28] from which thyroxine is produced).

surrounded by specific proteins. Different types are formed by the intestine from newly absorbed lipids, and by the liver from stored and circulating materials (Section 3.5.3 and Figs. II–39 and II–44). *Low-density lipoproteins* (LDL) are the main circulating form of cholesterol in humans. They are taken up by a number of cell types through endocytosis mediated by receptors that recognize their surface proteins (see Fig. II–11). Once in the lysosomes, the proteins are degraded, and the lipids are modified into forms in which they can exit to the cytoplasm. Lysosomal enzymes, for example, free cholesterol from the *cholesterol ester* forms in which it circulates. Cholesterol is used in the cytoplasm for synthesis of other steroids or insertion in membranes. When much cholesterol is coming in through heterophagy, the cell's own synthesis of cholesterol is suppressed; eventually, further endocytic uptake is also suppressed. These regulatory mechanisms apparently serve normally to prevent the cell from becoming "overloaded" with cholesterol.

We have already mentioned the likelihood that heterophagy helps to determine the duration of stay of hormones and other active agents at the cell surface (Section 2.1.6).

The incorporation of nondigestible materials in lysosomes has proved useful in preparing lysosome fractions that are largely free of the extensive contamination with mitochondria and peroxisomes present with conventional techniques. Thus, large amounts of the detergent Triton WR 1339 accumulate in the lysosomes of liver cells in rats administered this compound—Triton-loaded lysosomes are much less dense than normal and can be purified by centrifugation in a density gradient (see Fig. I–25). Beads of synthetic polymers ("latex") taken up by phagocytes can similarly facilitate purification of phagocytic structures.

2.8.3 Hydrolysis of Endogenous Macromolecules: Intracellular Turnover

Autophagy Most or all eucaryotic cells have a mechanism through which bits of their own cytoplasm are surrounded by a membrane and subsequently degraded. This process has been named *autophagy* to suggest self-*(auto)* phagocytosis. "Autophagic" vacuoles, a type of secondary lysosome, have been described in protozoa, cells of vertebrates and invertebrates, and plant cells. Autophagic vacuoles are identified by their content of one or more recognizable cytoplasmic structures such as mitochondria, plastids, bits of ER, ribosomes, glycogen, or peroxisomes (Fig. II–71).

At first glance, autophagy often appears to be an unselective process. The vacuoles seen in a given cell frequently appear almost to be random gulps of cytoplasm whose contents depend largely on the relative abundance of different structures in the cell. However, there are cases in which autophagy seems quite selective indeed. For example, during development of the larvae of the butterfly *Calpodes* the cells of the "fat body" undergo waves of sequential selective organelle destruction in which first many peroxisomes, later mitochondria, and then ER are degraded through autophagy and replaced by newly formed organelles. Less dramatic autophagic selectivity may be a more widespread phenomenon than presently recognized.

Although autophagy is a normal physiological phenomenon proceeding, probably at low rates, in normal cells, many more autophagic vacuoles are found when cells are under metabolic stress (when they are undergoing extensive developmental remodeling or responding to hormonal stimulation or to injury). Autophagy may be a means by which cells digest bits of their own

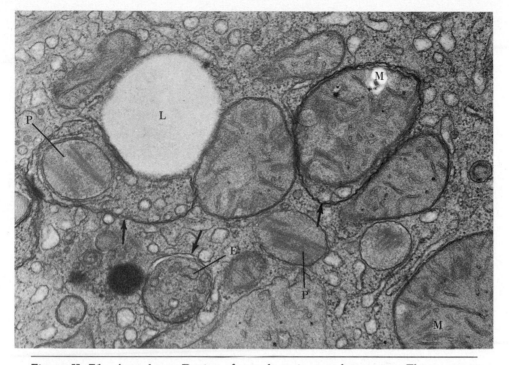

Figure II–71 *Autophagy*. Region of cytoplasm in a rat hepatocyte. Three autophagic vacuoles are present; their delimiting membranes are indicated by the arrows. Within one vacuole a mitochondrion is present *(M)*, a second contains a
peroxisome *(P)*, and the third, fragments of ER *(E)*. Other organelles, not included in autophagic vacuoles, lie nearby *(M, P)*. *L* indicates a lipid droplet in
the cytoplasm near one of the vacuoles. × 30,000. (Courtesy of L. Biempica.)

cytoplasm, to provide metabolites with which they can survive periods of starvation, stress, or metabolic urgency without self-destruction. It is not uncommon for cells that are soon to die to show an abundance of autophagic vacuoles. Such is the situation in many cases of cell death associated with the normal modeling of organs and tissues during development. But this should not
be construed to mean that autophagic vacuoles lead to the death of the cells.
Indeed, it may be that the cells die, under these circumstances, *despite* a "defense" mechanism involving autophagy.

How do portions of the cytoplasm come to be enclosed within the membranes of autophagic vacuoles, and how do acid hydrolases enter the vacuoles? Figures II–68 and II–72 outline the leading hypothesis, which suggests
that the endoplasmic reticulum, particularly in the Golgi zone, is involved in
forming the delimiting membrane of the vacuole; how the ER sac becomes
converted to the single membrane bounding the vacuole is not known. It also
has been proposed that the delimiting membranes of some autophagic vacu-

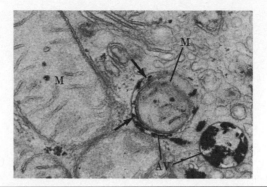

Figure II–72 *Two autophagic vacuoles (AV)* in a hepatocyte from a section of rat liver that was incubated to show sites of *acid phosphatase* activity. The reaction product appears black. Its presence in the two autophagic vacuoles confirms that they are lysosomes. The vacuole at the left contains a mitochondrion *(M)*. Arrows indicate appearances suggesting that the membranes surrounding this mitochondrion may be forming by the flattening and transformation of an acid phosphatase–containing sac as proposed in Figure II–68. A second mitochondrion *(M)* at the left of the micrograph is not within an autophagic vacuole. × 30,000.

oles derive from the Golgi apparatus. Fusion of other lysosomes with autophagic vacuoles has been demonstrated in hepatocytes; presumably this contributes hydrolases to the vacuoles. Hydrolases might also be provided to the vacuoles by the ER or Golgi-associated structures that participate in forming the delimiting membranes.

In cells of some endocrine glands, the membranes surrounding secretion granules can fuse with lysosomes, leading to a degradation of the granule contents. This autophagic process, called *crinophagy,* increases markedly when exocytic release of the secretions is suddenly decreased, relieving the consequent temporary imbalance between synthesis of the secretions and their export from the cell. This raises the question: What features of lysosomes or secretion granules are altered so that crinophagic fusion is promoted and excess secretory materials do not accumulate?

Autophagy in the Turnover of the Cell's Molecules and Organelles Lysosomes are the major intracellular sites of most of the known common enzymes capable of extensively degrading macromolecules; thus they might be expected to participate centrally in the degradation phases of intracellular turnover. The *half-life* (time required for half the molecules of a population to be replaced) of hepatocytic mitochondria is about 5 or 6 days. For peroxisomes the figure is about 1 to 2 days, and hepatocyte ribosomal RNA has a half-life of about 5 days. The formation in each hepatocyte of one autophagic vacuole containing one mitochondrion every 10 minutes or so could account for the known rates of mitochondrial turnover. This seems not to be excessively rapid, although the actual rates of autophagy are not known accurately.

As might be expected if partially nonselective autophagic processes participate in turnover, the turnover of intracellular macromolecules is at random with respect to age. Although each species of macromolecule has a characteristic *average* life span measured by its half-life, at any given time, a given recently made molecule has much the same probability of being degraded as an older one of the same type. This is seen, for example, by giving cells brief exposure to radioactive amino acids and following the subsequent fate of the labeled proteins made during that period. Radioactivity disappears from various populations of proteins with an exponential (random) time course. (The curves of experimental data have the same shape as the ones illustrating probabilistic events in Fig. IV–7B.) The same is true for the plasma proteins degraded by heterophagy considered in the preceding section. This contrasts with the turnover of red blood cells, which seem to have more determined individual life spans. The likely explanation is that for the cells, many events must occur (perhaps many surface molecules must undergo a change) before a threshold is passed and the cell is targeted for destruction. For molecules, conceivably only one critical "event" is required. Usually it is felt that the determining changes are of the sort that render proteins inactive or otherwise defective: irreversible denaturation, attack by another enzyme, or perhaps alteration of an oligosaccharide side chain. Such alterations can occur at random with respect to a molecule's age. But strictly speaking, it has not been shown that normal turnover selects altered proteins, even though, as outlined in the preceding section for the plasma proteins, denaturation or modification of oligosaccharide side chains can predispose proteins to particularly rapid destruction. It has not been ruled out that normal, still-functional molecules are broken down in the course of turnover, nor has it been demonstrated that *organelles* destroyed by autophagy are necessarily abnormal or nonfunctional at the time they meet their doom.

The degradation of an organelle within an autophagic vacuole involves essentially simultaneous destruction of all visible components of the organelle. Therefore, if turnover depended simply upon autophagy of whole organelles, all macromolecules of a given type of organelle should show essentially the same turnover rates. This does appear to be the case for several proteins of peroxisomes. In contrast, the turnover rate of mitochondrial outer membrane proteins is higher than that of inner membrane proteins, although the various proteins of the inner membrane and matrix do have half-lives similar to one another. Membrane lipids of ER and other organelles turn over at different rates from the proteins of the same structures. Various soluble enzymes of the hyaloplasm turn over at very different rates.

Differences in turnover rates among components of a given organelle are tentatively considered to reflect dynamic exchanges and replacement mechanisms by which a large structure can gain and lose individual macromolecules or arrays of molecules without extensive disruption of the organelle. This ties in with the information about membrane formation and changes discussed in earlier sections (Sections 2.5.4 and 2.7.4). There is no contradiction between molecule-by-molecule replacement and autophagy; both could occur simultaneously and contribute to turnover. It should also be noted that if molecule-by-molecule turnover of an organelle occurs, there still must be some explanation for the eventual degradation of the molecules themselves. Is there some kind of "microautophagic" process by which individual molecules are taken into lysosomes and degraded? Or are the nonlysosomal degradative mechanisms outlined below responsible for this facet of turnover? The same questions arise with regard to the supposedly soluble enzymes of the hyaloplasm. Various investigators have speculated that microautophagy might occur by a pinocytosis-like incorporation of cytoplasm during events in which vesicles bud from the lysosome surface into the interior (see below and Fig. II–68). Present techniques, however, have yet to provide a decisive test of the concept.

Nonlysosomal Degradative Mechanisms in Turnover Rough correlations have been noted between turnover rates and specific features of proteins such as their susceptibility to attack by proteolytic enzymes in the test tube, or even their size. Might such features determine the rate at which proteins encounter a hypothetical signaling system whose effect is to promote their uptake by lysosomes? Work on the plasma proteins (Section 2.8.2) suggests that enzymes that alter the oligosaccharide side chains on proteins, or enzymes that introduce one or a few cuts in the polypeptide chain, could be part of such a system involved in heterophagic phenomena. For autophagy, hypothetical lysosome surface receptors for altered proteins would recognize the altered proteins and promote their uptake.

Thus far the discussion has stressed lysosomes. A different perspective emerges from findings that some cells can recognize and selectively degrade abnormal proteins injected within them or made by the cell as the result of genetic mutations or of the presence of puromycin (Section 2.3.2) or artificial abnormal amino acids. The first such observations were made on bacteria; this is particularly significant because bacteria lack lysosomes. More recently, some other cell types have been used, including reticulocytes (Section 2.3.5) studied at stages where few if any lysosomes are present or active. Such observations emphasize the strong likelihood that nonlysosomal degradative mechanisms coexist with the lysosomal ones. The tentative conclusion that such mechanisms contribute to normal turnover is supported by studies in which lysosomal activities are inhibited (by "feeding" the lysosomes inhibitors of proteolytic enzyme activity via endocytosis or by raising the intralysosomal pH [Section 2.8.4]). Under such conditions intracellular turnover *is* markedly inhibited but it is *not* completely abolished. The experiments involved are complex and subject to several alternative explanations, but they point increasingly towards a cooperative enterprise in which the lysosomes play a major role in intracellular turnover, but not the only one or even always the predominant one.

If proteolytic and other lytic enzymes not confined within membranous barriers do contribute to cytoplasmic turnover, there must be mechanisms through which they are prevented from damaging the cell. Perhaps the enzymes can attack only substrates that have undergone some characteristic change. For instance, one of several competing models for turnover in reticulocytes proposes that the nonlysosomal system selectively degrades those proteins to which a polypeptide called *ubiquitin* has been linked through an ATP-dependent enzymatic reaction. This is by no means proved; it is mentioned here solely as an example of a plausible type of control mechanism currently being studied.

At any given time, the amount of a particular molecule present in the cell depends on the balance between its rate of formation and the rate at which it is being destroyed. Except for a few special cases (Section 4.4.3), nuclear DNA seems to escape destruction unless the cell as a whole is destroyed. RNA does turn over (Section 2.3.5), but little is known of the details of its degradation in eucaryotic cells. For RNA and for the other molecules discussed in this section, the absence of information about the mechanisms of the degradative phase of turnover is one of the largest gaps in the current picture of cellular metabolism.

2.8.4 Lysosomal Life History

The Sorting of Hydrolases Lysosomal enzymes are made by polysomes bound to the rough ER and must be segregated out for separate packaging from other rough ER products. Striking recent findings show that a number of cell types possess selective receptors with which they can bind recently synthesized lysosomal hydrolases selectively. These receptors were first discerned on the plasma membrane, where they mediate high-efficiency endocytosis of hydrolases experimentally introduced into the medium surrounding the cell (Section 2.8.5). The receptors recognize mannose sugars with phosphates attached (mannose-6-phosphates; Fig. II–45). These *phosphomannoses* are detectable in the oligosaccharide side chains of the hydrolases soon after translation of the proteins is completed (see Fig. II–45). The phosphomannose groups are removed, along with other portions of the hydrolase molecules once the enzymes reach the lysosomes. (It is suspected, from still ambiguous evidence, that certain of the hydrolases, especially the proteases [cathepsins], are enzymatically inactive as initially made and remain so until they reach the lysosomes, where their activation involves removal of short segments of the polypeptide chain; if so, this would protect the cell from damage by the enzymes during their transit to the lysosomes.)

The discovery of the receptor system led initially to the provocative suggestion that the cell packages hydrolases by first secreting them to the extracellular space and then taking them back up in endocytic structures that fuse with lysosomal or prelysosomal bodies. The hydrolases need never really be free in the extracellular milieu, because their binding to the receptors could keep them from moving away from the cell surface. Since uptake mediated by the receptors is highly selective for lysosomal hydrolases, the "secretion–recapture" cycle would accomplish differential packaging of the enzymes while other secreted proteins remained outside the cell. At present, however, most investigators favor the alternative proposal that the major transport of lysosomal hydrolases in most cells is by entirely intracellular routes. Phosphomannose receptors, like the ones at the cell surface, are found on intracellular membrane systems. These receptors have been suggested to serve inside the cell somehow to direct the proteins from the ER to the appropriate packages formed by the Golgi apparatus or the Golgi-associated structures of Section 2.5.3. Is it possible that some of the coated vesicles known to bud from Golgi-associated sacs transfer the hydrolases by virtue of their enrichment in the requisite receptors, much as coated endocytic vesicles (Section 2.1.6) selectively carry proteins from the cell surface?

Even if notions of these sorts prove accurate, they still leave unexplained the presence of the cell-surface receptors (do they minimize accidental "spillage" of enzyme to the outside?), and there are many other matters that will require clarification. For instance, how does the cell determine which of the many types of protein it makes in the rough ER should have phosphomannoses attached? Perhaps the polypeptide chains of the lysosomal hydrolases share unique conformational or other features recognized by the enzyme that generates these groups. That specific, testable hypotheses about such matters can now be framed reflects the enormous advantage of having known, identifiable, sorting devices—phosphomannose and the corresponding receptors—on which to focus study of the mechanisms by which proteins are put in the right package.

Recycling Phenomena Primary lysosomes are frequently invoked as carriers of hydrolases, but definite identification of a class of lysosomes of this sort has been accomplished only in a few cases, notably the phagocytic cells discussed in Section 2.8.2. In fact, it is well established that primary lysosomes are not the only source of the acid hydrolases that enter newly forming secondary lysosomes. Fusion of preexisting secondary lysosomes with new pinocytic or phagocytic vacuoles apparently is common in many cells. The experiment illustrated in Figure II–73 demonstrates such fusions. Tissue-culture cells (mouse fibroblasts) were permitted to engulf an electron-opaque material, finely dispersed iron particles. Like other engulfed material, these iron particles accumulate in secondary lysosomes. Having "marked" these lysosomes with iron, the cells were then permitted to engulf a second "marker," finely dispersed gold particles in a mixture of DNA and protein. Since the gold particles are readily distinguishable from the iron particles used in the first "feeding," the results were unequivocal. The presence of both iron and gold in the same lysosome demonstrates that fusions occur between old secondary lysosomes and newly formed digestive vacuoles. The same lysosomal enzymes apparently may be used for more than one round of digestion.

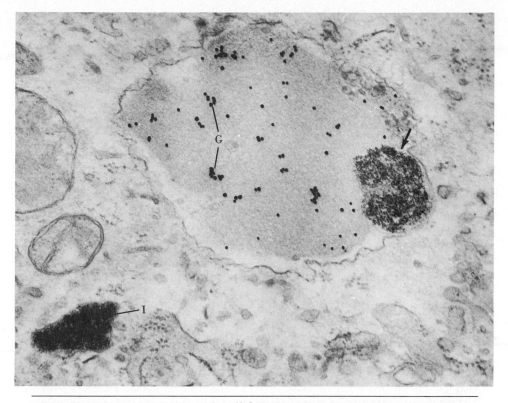

Figure II–73 *Lysosome "recycling."* Small portion of a cell (mouse fibroblast grown in culture) that was exposed first to colloidal iron and several hours later to colloidal gold (see text). A residual body *(I)* has accumulated a mass of electron-dense iron particles. The larger gold particles are seen within a digestive vacuole *(G)*. The presence of iron particles (arrow) in the vacuole containing gold strongly suggests that a lysosome formed as a digestive vacuole during the exposure to iron has fused with the more recently formed gold-containing vacuole. × 50,000. (From Gordon, G.B., L.R. Miller, and K. Bensch, *J. Cell Biol.* **25**:41–55, 1965. Copyright, Rockefeller University Press.)

Lysosomes tend to diminish in size as digestion proceeds and digestion products diffuse out. Water leaves osmotically. Decrease in surface area can occur by the budding off of small tubules and vesicles from the lysosome surface. Lysosomes known as *multivesicular bodies* result from such budding directed toward the interior of the organelle (Fig. II–68). This category of lysosome is often the depot to which pinocytized materials are delivered. When the membrane taken into the lysosome in this way originates as part of the plasma membrane (reaching the lysosome surface with the endocytic bodies it delimits), the lysosomes can probably participate in the turnover of cell-surface components. But although such turnover does occur, it is relatively slow: The half-lives of proteins and lipids in plasma membranes are often many hours to several days or more. In contrast, rapidly endocytizing cells, or secretory cells retrieving membrane from their surfaces (Section 2.1.6), can take into their cytoplasm very large amounts of their surfaces in very short times. The rates sometimes are such that within one hour or less the cell internalizes an area of membrane of the same order as the entire area of its plasma membrane. Were all of this to be degraded and replaced by new membrane, the molecules of the plasma membrane would turn over much faster than they are observed to do. It is precisely for this reason that the processes of membrane recycling outlined in Sections 2.1.6 and 2.5.4 were sought.

Membrane recycling seemingly implies that endocytic vesicles can fuse with lysosomes, delivering their *contents* to the lysosome interior but remaining only transiently inserted in the lysosome surface. Much of the vesicle membrane (or an equivalent area of membrane present previously in the lysosome surface) apparently "escapes" to return to the cytoplasm, moving perhaps to the Golgi apparatus or back to the cell surface. Electron microscopists examining food vacuoles undergoing shrinkage in protozoa have seen what very likely are vesicles budding from lysosome surfaces and migrating along pathways leading back to the plasma membrane.

Growing numbers of investigators have recently stressed their beliefs that material taken into the cell by receptor-mediated endocytosis (Section 2.1.6) is transferred from the vesicle responsible for initial uptake to an intermediate "carrier" vacuole ("endosome"; "receptosome"). This intermediate then goes on to interact with lysosomes or to be itself transformed into a lysosome. Before this occurs, the receptors return to the cell surface, perhaps traveling with vesicles recycling to the surface. The status of this concept is still unclear. It has long been known that endocytized materials do enter large prelysosomal structures by fusion of endocytic vesicles, but the implications for receptor behavior are still to be worked out. There are cases in which receptors responsible for endocytosis are known to reach the lysosomes; sometimes the receptors then undergo degradation, but in other cases it is seemingly from the lysosomes that receptors return to the cell surface. Discussion of these matters will continue below.

Why the membrane bounding the lysosome itself is not degraded by the enzymes within is not understood. Is there a barrier interposed at the inner surface of the membrane that limits access by the hydrolases?

Proper pH   Since lysosomal hydrolases function best at low pH, the lysosomal interior would be expected to be acidic. When phagocytes are made to engulf microorganisms stained with dyes whose color is pH-dependent (like the familiar dyes of litmus paper or other indicator devices), the interior of the phagocytic vacuole can be observed to become quite acid, rapidly reaching pHs below 5. Similar acidification occurs in compartments that receive pinocytized molecules, as can be demonstrated through use of fluorescent dyes whose fluorescence varies with pH. This acidity is the major basis for selective staining of lysosomes with a number of "vital dyes," stains usable even with living cells. Dyes such as *neutral red* and also other compounds, notably the antimalarial drug *chloroquine,* accumulate in lysosomes because they are so-called "weak bases." This means that at neutral pH (7), such as that in the cytoplasm or extracellular fluids, an appreciable proportion of the molecules is uncharged, whereas at low pH, NH_2 or comparable groups in the molecules acquire H^+ ions, becoming charged thereby (NH_3^+). In their uncharged forms the molecules can cross membranes and thus can enter the cell and lysosomes, but once inside a low-pH compartment, the molecules convert to the charged forms, which cannot cross membranes. The result is that the molecules accumulate within lysosomes, reaching concentrations that depend on the pH difference between the interior of the low-pH compartments (lysosomes) and the exterior (cytoplasm). Accumulation of high concentrations of molecules like chloroquine or neutral red within the lysosomes is prevented by metabolic inhibitors that deplete the cells' energy stores. This suggests that the low lysosomal pH is maintained in part by an active energy-dependent process; the "proton pumps" that might move H^+ into the lysosomes have eluded definitive characterization thus far but are presumed to resemble ion transport systems elsewhere in the cell. If the lysosomal pH is raised, as can occur when lysosomes are overloaded with chloroquine or with ammonium ions (NH_4^+), degradation within the organelles is inhibited.

Donnan equilibrium effects (see Fig. II–1) may contribute to the establishment or maintenance of a low pH inside lysosomes since charged macromolecules, such as components of cell-surface coats, accumulate, at least transiently, inside the organelles.

The pH of the cytoplasm outside the lysosomes is maintained near neutrality by a cluster of factors including buffering by both inorganic ions and organic groups and active control of the cell's internal composition by the many mechanisms discussed throughout this text. Unlike the lysosomal enzymes, most metabolic systems function poorly, if at all, at low pH.

Agents that accumulate selectively within lysosomes, through mechanisms like the ones just described, or through endocytosis, are frequently referred to as being *lysosomotropic.*

Acidification of Endocytic Structures; Toxin Penetration   A few recent studies on the timing of acidification of lysosome-related structures have been tentatively interpreted as showing that the pH inside structures of endocytic origin may begin to drop while they are still "prelysosomal"—that is, before they have fused with lysosomes. How widespread this pattern is remains to be seen, but it may point toward interesting control mechanisms. Experiments still in progress have been interpreted provisionally as indicating that the pH in prelysosomal endocytic structures—vesicles, vacuoles, or the intermediate "carriers" (see above) falls from 7 to 6 soon after the structures form. This might engender conformational changes in proteins, which could be of substantial biological importance. For instance, many of the ligands bound to cell-surface receptors (Section 2.1.6) dissociate from the receptors at pH 6; there is now much interest in the possibility that such dissociation frees the receptors in endocytic structures to recycle to the cell surface, as above, before the endocytic structures fuse with lysosomes. Such phenomena could avoid damage to the receptors that might occur were they to pass through the lysosomes during their cycling. The ligands released into the interior of the prelysosomal endocytic structure would go on to meet their degradative fate once fusion with lysosomes takes place; or some may pass into the cytoplasm.

Certain destructive materials, such as *diphtheria toxin,* a protein produced by diphtheria-causing bacteria, seem able to penetrate cellular membranes at pHs of 6 or below. The protein is believed to adsorb to specific cell-surface molecules, to undergo endocytosis, and then to enter the bilayer of the endocytic vesicle's membrane, perhaps by virtue of conformational changes caused by the low pH that expose hydrophobic zones in the protein's structure. Part of the protein can subsequently escape into the cytoplasm through incompletely understood events involving hydrolysis of covalent bonds in the protein by cellular enzymes; once in the cytoplasm, this portion of the toxin interferes with protein synthesis, accounting for the destructive effects of the bacteria.

Some possible implications of the acidification of endocytic structures for viral entry into cells are considered in Section 3.1.1.

Movement; Fusions Movement of lysosomes in cells presumably is accomplished by the system of filaments, microtubules, and other agents to be discussed extensively later on (Chaps. 2.10 and 2.11). Yet to be described are the devices that permit lysosomes to fuse selectively with incoming endocytic structures or with autophagic vacuoles, but not with mitochondria, the nucleus, or other organelles. The cases in which microorganisms inhibit lysosome fusion with endocytic structures in which they are present (Section 2.8.2) are being studied intensively for clues to fusion control mechanisms. The lysosome surface is also being investigated in the hope of turning up molecules or groups that might account for specificity of interaction with other membranes. Fusions have been induced in "cell-free" systems—mixtures of partially purified components—offering an advantageous system for studying membrane interaction, energy requirements, and other features of the process.

Fate; Lipofuscin Residua of digestion can be released from some cells through exocytic fusion of lysosomes with the cell surface. This is quite prominent with protozoa and some cell types of multicellular organisms. For other cells, however, the fate of many lysosomes is different. An important role of macrophages is to remove potentially dangerous materials from extracellular spaces; when the materials are indigestible, this role would be defeated were the cells simply to spit out their lysosome contents. Thus macrophages can retain lysosome contents for the many weeks or months the cells survive. (The cells themselves can be disposed of outside the body by a number of routes, but when they die inside the body, their contents are taken up anew by fresh phagocytes.) The recycling of secondary lysosomes through fusions with newly forming ones means that lysosome number need not increase as endocytosis continues, even if lysosome contents are retained by the cells.

For a number of cell types, indigestible substances experimentally administered, such as iron-containing particles, finely dispersed gold, and large carbohydrate polymers (dextrans) accumulate within lysosomes and can remain in the cell for prolonged periods. Among the most interesting *natural* accumulations are those that occur with aging in several tissues of humans and other animals. The number of lysosomes called *lipofuscin pigment granules* increases with age in human nerve, heart, and liver cells. These granules contain insoluble and indigestible forms of altered lipids and other residua of digestive events. Perhaps they form through the slow accumulation of indigestible materials in the lysosome population as it functions and intermingles its contents over the long lives of the cells. (Nerve cells in higher organisms never divide once mature and thus can survive as long as the individual.)

2.8.5 Storage Diseases; Cell Injury; Cell Death

Storage Diseases In many congenital "storage diseases," the lysosomes are abnormally large. Most of these diseases thus far studied appear to result from the inherited defect of a gene required for the synthesis of a specific hydrolase normally present in lysosomes. The diseases often are fatal early in childhood, but fortunately they are rare.

Among the first studied of these "lysosomal diseases" was a generalized *glycogen* storage disease, called Pompe's disease, in which the missing enzyme, *acid maltase,* normally found in lysosomes of various tissues, is involved in the degradation of glycogen. In the absence of this enzyme the liver lysosomes become engorged with glycogen. About 50 storage diseases are now known. Each reflects a deficiency in a specific enzyme, and each shows a characteristic pattern of accumulation of *mucopolysaccharides, lipids,* or related compounds in tissues such as nerve cells, muscle, spleen, or liver. Interestingly, when cells (fibroblasts) from a patient with a storage disease are grown in tissue culture along with normal cells (or cells from a patient with a different storage disease) the abnormal deposits may disappear. Evidently the normal cells release hydrolases into the medium, and the lysosomes of the abnormal cells obtain the missing enzyme through endocytosis. Chapter 5.2 will discuss some diagnostic and therapeutic aspects of storage diseases.

Not all abnormally "stuffed" lysosomes, however, need arise from enzyme deficiency. Diseases are known in which the lysosomes apparently contain a normal complement of enzymes but cannot cope with materials that enter them. This may result, for example, from the presence of abnormally large amounts of hard-to-digest molecules in the cell. A suggestion along these lines has been advanced for one of the factors that may contribute to the lipid deposits found in blood vessel walls in *atherosclerosis.*

Escape of Hydrolases? It was suggested soon after the discovery of lysosomes that abnormal conditions such as the uptake of injurious materials could lead to the leakage of lysosomal enzymes into the cell cytoplasm and to consequent cell injury or death. Although unequivocal evidence for this suggestion is very difficult to obtain, it remains a potentially important mechanism that may aid in understanding cell pathology. For example, some investigators believe that in the miners' disease, *silicosis,* silica particles, taken up by macrophages of the lung into phagocytic vacuoles that fuse with lysosomes, act upon the vacuoles' membranes to make them "leaky." Along with effects of the silica particles at the plasma membrane this may contribute to killing the cells. There is little reason to doubt that after a cell has died, its lysosomal hydrolases are released and participate in destroying cellular and extracellular materials.

When certain compounds, including drugs and hormones, are administered to cells, changes are observed in the membranes of lysosomes subsequently isolated. Some agents, such as vitamin A, appear to *labilize* the lysosomes (that is, make them less resistant to disruption), and others, such as cortisone, seem to stabilize the organelles. It is occasionally speculated that in the cell as well, labilizers promote leakage of lysosomal hydrolases and that stabilizers have the opposite effect. Alternatively, or in addition, such substances might affect the fusion of lysosomes with endocytic vacuoles.

In inflammations and diseases such as arthritis and "autoimmune" disorders (Chap. 3.6C), hydrolytic enzymes are released from phagocytes and other cells into surrounding tissues, where they may do damage. Sometimes this release results from cell death, but hydrolases may sometimes be released from lysosomes of living cells, probably through processes akin to exocytosis, provoked by abnormal conditions. The normal developmental remodeling of cartilage and bone involves extensive degradation of extracellular materials, also carried out, in part, by hydrolases secreted from the cells involved.

Chapter 2.9 Peroxisomes (Microbodies, Glyoxysomes)

Peroxisomes were seen in rodent kidney and liver in the early 1950s, when electron microscopy of sectioned biological materials was in its infancy; they were then called "microbodies." One of the peroxisome enzymes, urate oxidase, was among the first enzymes to be studied in isolated fractions of rat liver cells. The distribution of this enzyme in cell fractions was markedly similar to, but not identical with, that of the lysosomal enzyme acid phosphatase. It proved to be situated in peroxisomes rather than lysosomes.

A number of procedures have been developed to separate peroxisomes from other organelles, and cytochemical methods are now available for light microscope and electron microscope visualization of peroxisomes (see Figs. I–21 and II–74). It has been established that peroxisomes are distinctive organelles of widespread occurrence both in plants and animals.

2.9.1 Occurrence and Morphology

The peroxisomes of rodent liver and kidney, and those of various plant tissues, have diameters ranging from 0.5 to more than 1 μm. They are delimited by a single membrane, about 6.5 to 8 nm thick, and they contain a finely granular matrix. In a number of tissues, including mammalian liver (humans are an exception) and diverse plant cell types (Fig. II–74), a core or nucleoid is present. The structure of the nucleoid is characteristic of the cell type and species: For example, in rat liver the nucleoids consist of straight tubules arranged in parallel in a crystal-like array; in mouse and hamster liver, they have more the appearance of twisted strands.

Work with cell fractions from rat liver demonstrated the presence, in "microbody"-rich fractions, of enzymes catalyzing reactions that involve hydrogen peroxide; hence the name *peroxisomes* came into use in place of "microbodies." It was originally thought that peroxisomes were present only in liver and kidney of mammals, in a few other animal cell types (such as cells of the fat body of the butterfly *Calpodes*), and in certain plant cells (Fig. II–74 A and B).

The advent of a cytochemical method for the demonstration of catalase (Figs. I–21 and II–74E), one of the peroxisomal enzymes, changed this view. Catalase-containing bodies were found in virtually all cell types of mammals; the red blood cells are the chief exception known at present. For example, in the absorptive cells lining the small intestine (see Fig. III–27) such organelles are present in large numbers (Fig. II–74 D and E). They had been overlooked by microscopists largely because they lack nucleoids or other immediately striking morphological features. Peroxisomes similar to those of the intestine are present with greatly varying frequencies in most mammalian cell types. Such peroxisomes are often referred to as *microperoxisomes*. Usually they are quite small, measuring 150 to 250 nm in length.

Yeast grown in media rich in alcohols such as methanol form large amounts of a peroxisomal enzyme that oxidizes alcohol. Correspondingly their peroxisome population increases in size and number: These organelles come to occupy as much as 50 percent of the cytoplasmic volume.

2.9.2 Enzyme Activities and Functional Significance

A by-now familiar pattern of cell organization, the location of several metabolically related enzymes within one organelle, is well illustrated by peroxisomes (Fig. II–75). The enzymes initially identified in rat liver fractions included several that produce hydrogen peroxide (urate oxidase, D-amino acid oxidase, α-hydroxy acid oxidase) and catalase, which decomposes H_2O_2. Peroxisomes in different plant and animal cells can vary considerably in their enzymatic makeup, but they do tend to contain at least some peroxide-producing enzymes, and all (with one or two possible exceptions that still require verification) contain catalase. Urate oxidase is found only in the nucleoid-containing peroxisomes. It is possible that in some cases, such as rat liver, the nucleoids are actually crystals of this enzyme.

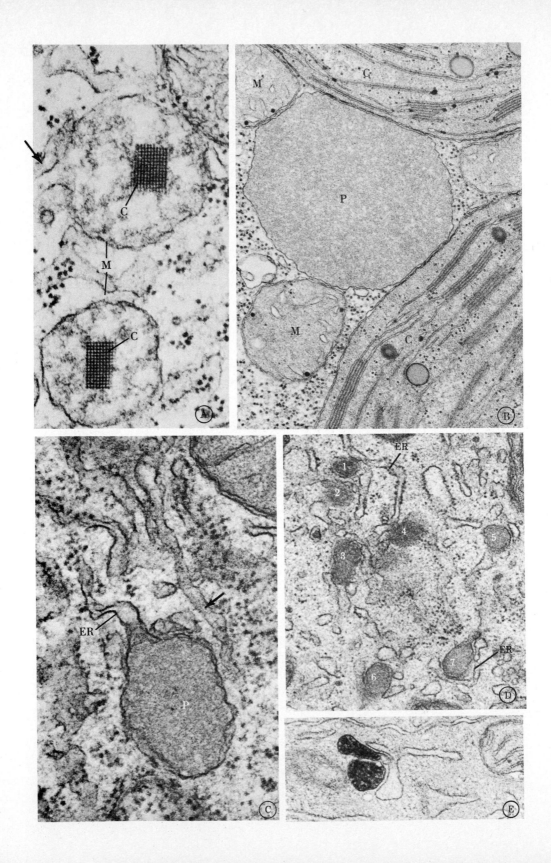

The enzymatic capabilities of microperoxisomes have been studied only in a few cell types. In some, oxidases are found along with the catalase, but for most, this has not yet been demonstrated. Like the larger peroxisomes, microperoxisomes vary from cell type to cell type in the spectrum of enzymes present.

From time to time investigators have proposed the presence of one or another enzyme or carrier system in the peroxisomal membrane. However, the history of peroxisomes is still a short one; most attention has been directed to their contents so that little is established about their membranes. Almost all of the known enzymes seem to be located in the interior of the organelles—in the matrix and, when present, the nucleoids.

Photorespiration; Glyoxysomes Our knowledge of the metabolic participation of peroxisomes is far more complete for plant cells than for animal cells. In many green leaves there is a light-dependent pathway resulting in oxygen consumption and CO_2 release known as *photorespiration*. This involves the interplay of chloroplasts, mitochondria, and peroxisomes. As mentioned in the discussion of C_4 metabolism (Section 2.7.5), the rate of photorespiration is increased by high light intensity, elevated temperature, and high O_2 concentrations; under these conditions the chloroplasts generate enhanced quantities of the 2-carbon compound glycolic acid, ("glycolate"). Glycolate is produced from the Calvin cycle intermediate RuDP (see Fig. II–62) through reactions initiated by the enzyme RuDP carboxylase; when levels of CO_2 are low and those of O_2 are high this enzyme tends to act as an "oxygenase," catalyzing production of glycolate with concomitant consumption of oxygen. Glycolate passes out of chloroplasts and gains access to the adjacent peroxisomes. Iso-

---

◀ **Figure II–74** *Peroxisomes.*
A. Two peroxisomes from a cell of the grass plant *Avena* show crystal-like nucleoids. The delimiting membranes *(M)* are continuous with those of smooth-surfaced sacs (arrow). × 75,000.

B. Portion of a cell of tobacco leaf showing a peroxisome *(P)* closely associated with chloroplasts *(C)* and mitochondria *(M)*. × 40,000.

C. Portion of a hepatocyte in rat liver. The peroxisome *(P)* was sectioned in a plane that did not include its nucleoid. Continuity of the peroxisome membrane with a sac of what appears to be ER is evident. Arrow: smooth ER. × 60,000.

D. Portion of an absorptive cell in guinea pig small intestine. A cluster of microperoxisomes *(1 to 7)* is present in this small area. ER, mostly smooth, is present nearby. The ER is quite tortuous, and the delimiting membranes of the microperoxisomes are irregular, so that continuities between the ER and peroxisomes, if they exist (Section 2.9.3), would be difficult to see. They are being sought through use of a device that permits rotation and tilt of a section in an electron microscope, allowing the selection of the most favorable angle from which to view a structure. × 37,000.

E. A portion of a cell like that in part D, but incubated by a cytochemical method utilizing diaminobenzidine (DAB) at alkaline pH (see Fig. I–21). The density of the peroxisomes is due to the reaction product (oxidized DAB) resulting from the action of catalase. × 44,000. (Parts A and B, courtesy of S.E. Frederick, E.H. Newcomb, E. Vigil, and F. Wergin; Part C, courtesy of L. Biempica.)

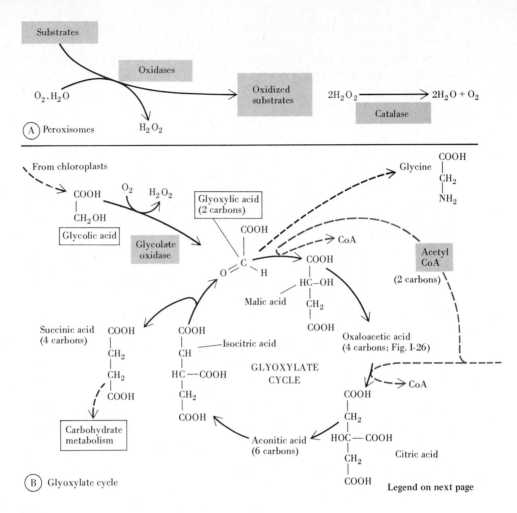

A. Peroxisomes

B. Glyoxylate cycle

Legend on next page

lated leaf-cell peroxisomes have been shown to possess the enzymes of the "glycolate pathway," among others. Through reactions in which oxygen is again consumed, these enzymes convert glycolate to glyoxylic acid, (glyoxylate; see Fig. II–75 and below). Glyoxylate, in turn, can be converted, by peroxisomes, to the amino acid glycine (see Fig. III–75), which passes out of the peroxisomes and enters the cytoplasm, where it can be used for protein synthesis or be transported into the mitochondria for further metabolism. The close associations often observed among chloroplasts, mitochondria, and peroxisomes (see Fig. II–74B) are thus apparently correlated with metabolic interaction among the organelles and the interweaving of amino acid and carbohydrate metabolism.

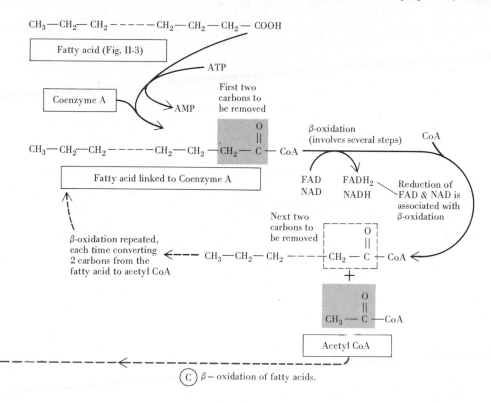

$$CH_3—CH_2—CH_2————CH_2—CH_2—CH_2—COOH$$

Fatty acid (Fig. II-3)

ATP

Coenzyme A

AMP

First two
carbons to
be removed

$$CH_3—CH_2—CH_2—————CH_2—CH_2—CH_2—C—CoA$$

β-oxidation
(involves several steps)

CoA

Fatty acid linked to Coenzyme A

FAD   FADH₂
NAD   NADH

Reduction of
FAD & NAD is
associated with
β-oxidation

Next two
carbons to
be removed

β-oxidation repeated,
each time converting
2 carbons from the
fatty acid to acetyl CoA

$$CH_3—CH_2—CH_2————CH_2—C—CoA$$

+

$$CH_3—C—CoA$$

Acetyl CoA

C β – oxidation of fatty acids.

Figure II–75 *Mctabolic pathways in peroxisomes.*

A. Peroxisomal oxidation results in *production* of hydrogen peroxide (H_2O_2). Peroxide is *decomposed* by the catalase present in peroxisomes.

B. Some enzymes of the *glyoxylate cycle,* though not necessarily the entire cycle, are found in a variety of plant and animal tissues. In a number of plant tissues the entire cycle is present in peroxisomes of a type which are often called *glyoxysomes.* The cycle links the two-carbon molecule glyoxylate (glyoxylic acid) to four more carbons by attaching to it the two carbons from an acetyl-CoA and then adding a second pair of carbons from another acetyl-CoA. (Acetyl-CoAs arise in many metabolic pathways including the oxidation of fatty acids [Part C]; see also Fig. I–26.) In subsequent steps glyoxylate is regenerated to continue the cycle. As indicated, products of the glyoxylate cycle can enter carbohydrate metabolism and amino acid metabolism (the latter, via the amino acid glycine). In plant cells, one of the peroxisomal oxidases, as shown, converts glycolic acid (glycolate) to glyoxylate. This provides the linking of chloroplast and peroxisomal metabolism in *photorespiration.*

C. In mitochondria and in peroxisomes, *fatty acids are oxidized by the β-oxidation pathway.* Carbons are removed, two at a time, by linkage to coenzyme A and by oxidative reactions that lead to the formation of acetyl-CoA molecules. These can enter other metabolic pathways, as indicated. Note also the coenzymes (FAD and NAD) are reduced during the oxidation so that, in mitochondria, fatty acids can be used to fuel ATP production. (From work by C. DeDuve, H. Beevers, N. Tolbert, P. Lazarow, and others.)

Seeds of castor beans and other plants store fats in tissues known as endosperm. During early development of the embryos, this fat is transformed into sugars (gluconeogenesis), which are used for ATP-generating metabolism. The transformation depends upon a sequence of enzyme reactions known as the *glyoxylate cycle* (Fig. II–75). The responsible enzymes are localized within peroxisomes that also possess catalase as well as enzymes that oxidize fatty acids (Fig. II–75C) with the production of H_2O_2; some plant scientists prefer to call these organelles *glyoxysomes*. The involvement of endosperm peroxisomes in lipid metabolism probably accounts for two morphological features: First, the peroxisomes are found quite close to the surfaces of lipid storage droplets in the cytoplasm, and second, the number of peroxisomes increases dramatically as lipid breakdown begins.

Lipid Metabolism in Animal Cells   Much remains uncertain about the functions of peroxisomes in animal cells, largely because the metabolic roles of the peroxisomal oxidases are unclear. Uric acid oxidase may participate in degrading purines (the class of nitrogen-containing compounds, of which the adenine and guanine bases of nucleotides are the most common examples). In frogs and chickens, two other enzymes of purine degradation accompany this oxidase in hepatic peroxisomes. Additional oxidases may participate in the metabolism of alcohols, though the primary systems (dehydrogenases and other enzymes) by which the liver deals with beverage alcohol (ethanol) seem to be in the mitochondria and endoplasmic reticulum. The roles of D-amino acid oxidase, an enzyme abundant in liver, kidney, and the cerebellar region of the brain, are mysterious. Except for some present in the walls of bacteria (Fig. III–7), the amino acids occurring in nature are of the L form (D and L, of course, refer to the mirror-image configurations of amino acids). It is not obvious why some cells possess an oxidase that acts on what would be a very rare substrate. Probably the enzyme has some other class of substrate yet to be discovered.

Recently a strong line of evidence implicating peroxisomes in lipid metabolism has developed. Peroxisomes, often of the microperoxisomal variety, are particularly abundant in cell types where lipid synthesis, storage, absorption, or breakdown is prominent. The intestinal cells shown in Figure II–74 D and E are examples. Others include the gland cells that synthesize steroid hormones and brown fat cells, which store and then degrade fats to provide heat for hibernating animals (Section 2.6.2). The abundance of peroxisomes in hepatocytes increases markedly in livers of rodents administered *hypolipidemic* drugs—drugs that decrease levels of cholesterol and other lipids in the blood.

This last observation was put to use for the demonstration that hepatocyte peroxisomes contain enzymes of the β-*oxidation* pathway for fatty acid breakdown. The "beta" (β) prefix indicates that these enzymes break fatty acids down into 2-carbon fragments available for further metabolic use (Fig. II–75). Similar enzymes are present in mitochondria, which historically had been thought to be the sole sites of fatty acid oxidation in animal cells. The first unequivocal clue that this sole localization is not the case was the finding that in plant cells possessing peroxisomes of the glyoxysome type, β-oxidation is chiefly a peroxisomal function. Subsequently, it was discovered that in liver responding to the hypolipidemic drugs, the levels of peroxisomal fatty acid–metabolizing enzymes are sufficiently increased to facilitate their discrimination from the more abundant mitochondrial enzymes. The peroxisomal enzymes have recently been found in normal liver and in some other tissues as well.

Fatty acid oxidation is an essential link between fat and carbohydrate metabolism. (In mitochondria, the acetyl-CoA produced can be used to generate ATP.) It is too early to know precisely how mitochondria and peroxisomes collaborate in this function within animal cells. The best present guess is that the two organelles favor somewhat different classes of fatty acids and that some types of fatty acids may be metabolized by initially undergoing partial breakdown in one of the two organelles and then moving to the other, where breakdown is completed.

Enzymes that may contribute to lipid synthesis also have been detected in peroxisomes.

Peroxide Hydrogen peroxide is a highly reactive molecule, capable of damaging cellular components. The catalase in peroxisomes undoubtedly serves to decompose the H_2O_2 produced by the peroxisomal oxidases, yielding O_2 and H_2O.

Though needing "protection" from them, the cell probably also uses peroxides both for defensive activities (Section 2.8.2 and Fig. II–69) and for reactions like the addition of iodine atoms to the precursors of thyroid hormones (Fig. II–70 and Chapter 3.6A). H_2O_2 can be produced metabolically, for instance, through action of enzymes that generate "superoxide" (O_2^-) and of *superoxide dismutase,* an enzyme that converts superoxide to peroxide. These enzymes are located outside the peroxisomes, but peroxisomal catalase might help to decompose the peroxide they produce, cooperating with other enzymes such as peroxidases (these generally are not peroxisomal enzymes, being located, probably, in the hyaloplasm) and some nonperoxisomal catalases.

2.9.3 Formation

Peroxisomes maintain close relations with the ER, often being found very close to tubules of the smooth ER. Moreover, as illustrated in Figure II–74, the membranes delimiting peroxisomes commonly show tubular extensions. When originally seen it was generally assumed that such extensions correspond to direct connections of peroxisomes with the endoplasmic reticulum. Thus it seemed likely that peroxisomal enzymes are synthesized on the rough ER and enter the peroxisomes directly from the ER. Serious doubt that this is the case came first from the finding that newly made catalase in hepatocytes does not turn up in microsome fractions isolated from the cells soon after an appropriate radioactive label is given. Rather, the catalase is found in the "cytosol" fraction

(Chap. 1.2B). (The enzyme is found there in the form of individual enzyme subunits [apocatalase]: Catalase is made of four such units that associate with one another and with an iron-containing *heme* group [Section 1.4.2] once inside the peroxisomes.) Subsequently, mRNAs for catalase and uric acid oxidase were detected in the free polysome populations of liver cells, suggesting that these enzymes are not made by the rough ER. Soon comparable findings were reported for enzymes of plant cell peroxisomes, although there is still dispute about certain of the enzymes. With plant material the first steps have been taken toward study of uptake of the enzymes by partially purified peroxisome fractions. Apparently, as with mitochondria and chloroplasts, the cell has "solved" the problem of sorting out the proteins destined for packaging in peroxisomes (Section 2.4.3) by using a route not dependent on the ER. Whether peroxisomes actually grow and divide is now being studied; in yeast, it is claimed that small peroxisomes can separate from large ones, but this has still to be studied thoroughly. As yet, there is no evidence that peroxisomes synthesize any of their own proteins or other major macromolecules.

What, then, of the continuities with the ER? Many investigators remain convinced that some of the structures observed by microscopy, like the one in Figure II–74C, do in fact represent such continuities. They point out that we still do not know how the *membranes* of peroxisomes are assembled and suggest that the ER may be responsible for this. Some even argue that peroxisome genesis involves the budding of membrane from the ER, with the enzymes entering the bud on the basis of some recognition system built into the membrane. The little that is known about the peroxisomal membrane does not yet permit a decisive test. For hepatocytes, studies very recently done suggest that its proteins may differ significantly from the proteins of the ER, but its lipid composition is roughly similar to that of the ER.

An alternative viewpoint is that most of the structures interpreted originally as attachments to the ER may instead correspond to connections between peroxisomes. Peroxisomes thus would be completely separate from the ER and would receive their components from elsewhere. Not all peroxisomes need remain forever linked together; still-to-be-detected budding processes could separate individual organelles from the interconnected network. Before this occurs they might, however, exchange contents with one another through the connections. This last possibility bears on the absence of age-dependency of turnover of peroxisomes (Section 2.8.3), since it would imply that there may be no "old" and "new" peroxisomes. Newly made peroxisomal proteins would intermingle with older ones in the same organelle, and if the organelles are degraded by autophagy, both newer and older proteins would meet the same fate.

These two points of view are not necessarily in conflict, and both may prove to contain elements of truth. In any event, there is no question but that peroxisomes are found in close proximity to the ER. Whether or not this serves in genesis of peroxisomes, it could contribute toward metabolic cooperation between the two organelles.

Chapter 2.10 Microtubules; Centrioles; Cilia and Flagella

This chapter and the next deal with elongate tubular and filamentous structures, each a polymer of one or a few types of proteins. These *microtubules* and *cytoplasmic filaments* participate in the establishment and maintenance of cell architecture, helping to anchor or support other structures and helping to determine the three-dimensional shape of the cell as a whole, and the shape of specialized cell portions. Thus, collectively they sometimes are referred to as the *cytoskeleton*. Both the microtubules and certain of the filamentous structures can assemble rapidly from soluble stores ("pools") of their protein components; most can also readily be disassembled. This provides flexibility for the cell, enabling it to change in shape and to modify the mechanical properties of local regions.

The other major roles of microtubules and filaments are in *cellular motility:* the movement of cells and the movement of structures within the cytoplasm. The two types of roles are closely interrelated, so that whether to call a particular function cytoskeletal or motility-related often almost is an arbitrary matter. Similarly, although for convenience and clarity we shall deal with the microtubules and filaments in separate chapters, these structures cooperate and interact in many, probably in most, of their functions.

2.10.1 Microtubules

As preparative methods for electron microscopy improved, microtubules became evident as a regular component of the cytoplasm of most cells. Microtubules are best preserved by fixatives such as glutaraldehyde or formaldehyde; only in the past 20 years have electron microscopists used these agents routinely to treat tissue before it is exposed to osmium tetroxide (Section 1.2.1), which when used alone often disrupts the tubules. Microtubules are thin cylinders, approximately 25 to 30 nm in diameter (Fig. II–76); it is not known whether there are any limits to possible lengths, since beyond several micrometers the measurements are very difficult to make in the thin sections usually employed for electron microscopy. They are not membrane-delimited but look hollow, with a central "channel" 15 nm across. From images such as Figure II–76A it has been concluded that the microtubule wall is made of some form of repeated basic structural building block. In cross section a microtubule has the appearance of a circular array of "subunits." It was early recognized that roughly a dozen such "subunits" are present, and as techniques improved, 13 were counted in the cross sections of walls of almost all naturally occurring microtubules. (In a few organisms other numbers, such as 11 or 15, are characteristically present.) In longitudinal view, particularly with negative staining, the walls appear to be made of parallel filaments ("protofilaments"), again generally numbering 13 (Fig. II–77).

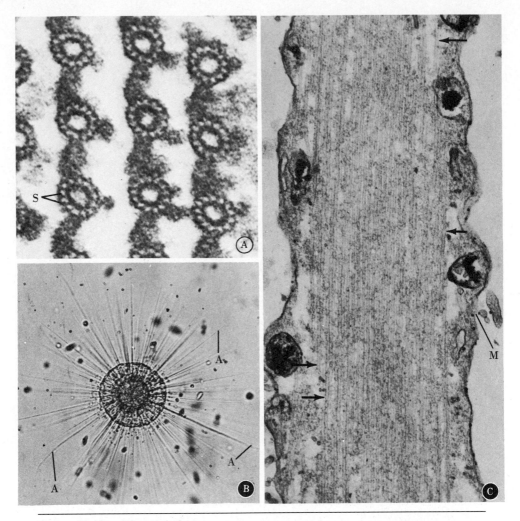

Figure II–76 *Microtubules.*

A. Several microtubules from a protozoan, viewed in transverse section after staining with tannic acid. Tannic acid brings out details of the tubule substructure; each tubule appears as a ring of 13 subunits (S). × 450,000.

B. The protozoan *Actinospherium* shows many modified pseudopods (axopods, *A*) that extend radially from the body of the cell. × 100.

C. A longitudinal section of an axopod like the ones in Part B. Its membrane *(M)* is part of the plasma membrane. Numerous microtubules (arrows indicate four) are found longitudinally arranged inside the axopod. Transverse sections would show many circular outlines (see Figs. I–10 and III–48). × 45,000. (From Tilney, L.G., and K.R. Porter; Part A is from *J. Cell Biol.* **59:267,** 1973. Copyright, Rockefeller University Press.)

A Molecular Model Microtubules are constructed chiefly of proteins called *tubulins.* Some other proteins, the microtubule-associated proteins (MAPs), also are present but have yet to be studied intensively. (One subclass of MAPs is known as the Tau proteins.) Tubulins have molecular weights of approxi-

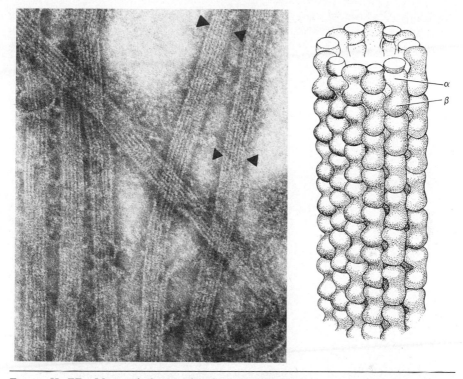

Figure II–77 *Microtubules: molecular organization.*

Micrograph: A half-dozen negatively stained (Fig. II–53) microtubules; the paired arrowheads indicate the lateral borders of two of them. The walls of each microtubule appear as if made of several longitudinal "protofilaments." In transverse section each protofilament would appear as one of the small circles seen in the microtubule walls in Figure II–76A; not all of the 13 units are visible in the present figure because the views are of intact, though somewhat flattened, microtubules; therefore some of the protofilaments lie underneath the ones that are visible. × 225,000. (Courtesy of O. Behnke and T. Zelander. See Dustin, P. *Scientific American* **243**(2):66–70, 1980; the present figure is a portion of a micrograph originally published in *J. Ultrastruct. Res.*)

Diagram: Interpretation of the structure of a microtubule as an assembly of subunits representing dimers of tubulin. Most investigators have concluded that each dimer is of one α and one β tubulin molecule but some now also postulate that there is more than one type of α or β tubulin present in a given microtubule so that not all of the α-β dimers need be identical. In line with present views (see Fig. II–78) microtubule structure has been drawn as what is called by mathematicians a *left-handed, three-start helix.* It is, however, still uncertain whether the relative positions of α and β indicated in the diagram are correct—conceivably the position indicated for α could actually be that of β and vice versa. (After L. Amos, P. Erickson, A. Klug and J. Randall.)

mately 55,000. Purified tubulin preparations from several sources consist of mixtures of equal amounts of two proteins (α and β tubulins), differing slightly in size and composition. The present consensus is that in the cell, the two tubulins are associated as "heterodimers" of two tubulin molecules, one of each type. The model diagrammed in Figure II–77 illustrates the structure of microtubules proposed from microscopy (Fig. II–78). The microtubule wall is thought to be made of globular subunits 4 to 5 nm in diameter (this is the size of a tubulin molecule). The subunits, in pairs corresponding to the heterodimers, are helically arranged around the wall and also fall into rows running

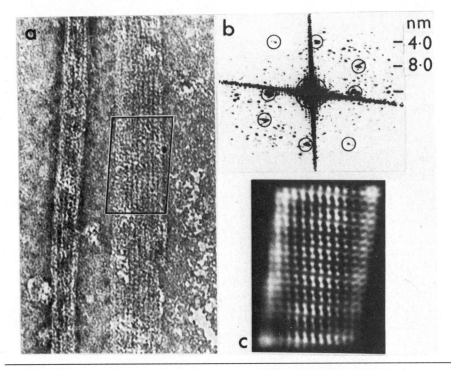

Figure II–78   *Three views of a microtubule wall.* In panel (a) portions of two microtubules run from top to bottom of the field (the two are the A [left] and B [right] tubule of a ciliary doublet; see Fig. II–80). The B-tubule has opened and flattened down on the support film during negative staining (Fig. II–53) so that all the protofilaments of its wall are visible and it appears wider than the intact microtubule at the left. Panel (b) shows the pattern obtained when the area of the B-tubule indicated by the box in panel (a) is subjected to *optical diffraction,* an analytical method that evaluates the spacing of regularly arranged units in a structure. The short lines indicated by enclosure in circles are the crucial details of the diffraction pattern recorded on a photographic film. The locations and arrangements of these lines indicate that the tubule wall has regularly spaced subunits of 4 nm and 8 nm in diameter; presumably these correspond to tubulin monomers and dimers. Panel (c) shows an *optically filtered* image of the region in the box in panel (a). Optical filtration "cleans up" the image by optical processing and photographic means that emphasize repetitive arrays (using the information from diffraction as in [b]), and diminish the contribution of irregularities to the image. (From Amos, L.A., and A. Klug, *J. Cell Science* **14:**523, 1974.)

parallel to the tubule's long axis. The longitudinal rows of dimers are the structures that appear as filaments in the wall.

There are hints that variations in composition or structural details sometimes occur among microtubules of different cells or at different sites in a cell, but the general organizational principles usually are quite similar in widely disparate species and cell types. An interesting possibility, still unverified, is that differences in microtubule functions in different cells or regions of cells reflect, in part, differences in the MAPs. The MAPs are believed to include proteins that extend laterally from the wall of the tubule, accounting for the existence of a "clear" zone of 5 to 10 nm or more from which other large structures are excluded. Fine filamentous material seen in this clear zone in some negatively stained preparations presumably corresponds to these MAPs. This material is under suspicion as a possible participant in lateral interactions of microtubules with one another and with other structures, as discussed later.

Cytoskeletal Distributions Microtubules frequently are distributed as might be expected for supportive, cytoskeletal structures. For example, microtubules are longitudinally disposed along the elongate processes of cells such as nerve cells (see Fig. III–41) or the rigid cytoplasmic extensions of some protozoa (Fig. II–76). The red blood cells of nonmammalian vertebrates (fish, amphibians, reptiles, and birds) as well as the *platelets* of mammals (cell fragments involved in blood clotting; Chap. 3.6B) have a flattened oval or disclike shape. Microtubules are arranged in a circumferential bundle ("marginal band"), like a hoop, at the margin of the disc, just below the plasma membrane. (The rather different "biconcave" shape of mammalian red blood cells [see Fig. II–2] is stabilized by other elements [Section 2.11.2].) During their maturation, sperm cells of many species change dramatically in shape and, as they do so, show microtubule arrangements that seem to generate or support the shapes produced (see Fig. III–48; Fig. III-34 illustrates another, comparable example).

2.10.2 Structure of Cilia and Flagella

The best understood participation of microtubules in motility is in cilia and flagella. Typical ciliated cells possess large numbers of cilia about 5 to 20 μm long (see Fig. III–16). Flagellated cells usually show only one or two flagella per cell (see Figs. II–61, II–79, and III–50); they are often as long as 100 to 200 μm. The diameter of a cilium or flagellum usually is less than 0.5 μm.

Axoneme Cilia and flagella are of similar structure, called a "9 + 2" arrangement, since it consists of *nine* microtubule "doublets" arranged as a cylinder 0.15 to 0.2 μm in diameter with an additional *two* tubules in the center of the cylinder (Figs. II–79 and II–80). As described in Figure II–80, one of the microtubules of each peripheral doublet (the "A"-tubule) is a complete microtubule, whereas the "B"-tubule is not—part of its wall is shared with the

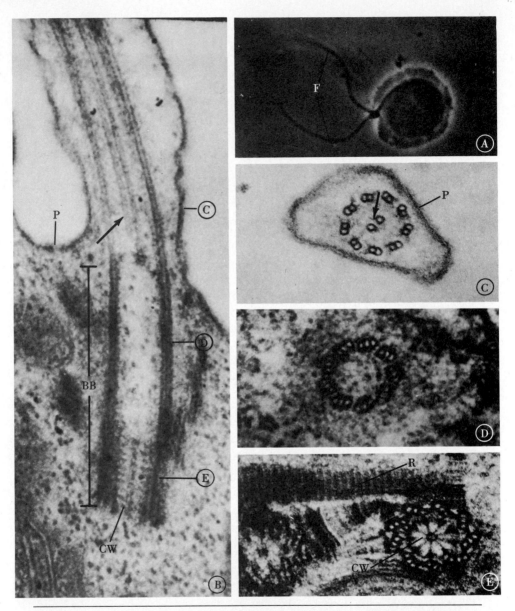

Figure II–79 *Flagella and basal bodies* of the unicellular organism *Naeglaria* during its flagellate phase. Part A is a phase-contrast micrograph; the two flagella are seen at *F*. Parts B to E are electron micrographs: B is a longitudinal section; C, D, and E are transverse sections at the levels indicated by the corresponding letters in B. *P* indicates the plasma membrane; *CW*, the "cartwheel" structure at one end of the basal body *(BB)*. The arrows show the central pair of tubules of the flagellum. In Part E a cross-striated rootlet *(R)* is seen. Such structures are often found extending from basal bodies into the cytoplasm; perhaps they help to anchor the basal body and to control the orientation of the flagella with respect to the cell. A, × 2000; B-E, approximately × 90,000. (Courtesy of A.D. Dingle and C. Fulton.)

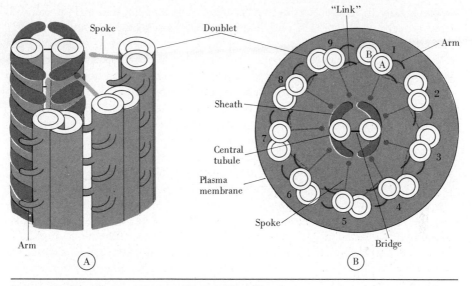

Figure II–80 *Structure of a cilium or flagellum*
    A. An interpretation of the arrangement of tubules and associated material based on transverse (cross) sections like those in Figure II–79 and on longitudinal views like those in Figures II–79, II–83 and II–84.
    B. The structures seen in a cross section of cilium or flagellum as it would be seen looking up from the base—that is, looking outward from the cell. Each doublet consists of an A- and a B-tubule; two arms are attached to the A-tubule (see Fig. II–84B for more details). Note that the A-tubule is a complete circle in cross section (it has the 13 protofilaments characteristic of most microtubules), whereas part of the wall of the B-tubule (which has only 10 or 11 "protofilaments") is shared with the A-tubule. Spokes occur at intervals along the length of each doublet; current hypotheses suggest they cyclically attach to and detach from the sheath surrounding the central tubules. The dynein arms of one doublet cyclically attach to and detach from the next doublet as described in the text. The fine links, made of the protein *"nexin,"* attaching adjacent doublets to one another, can be extensively stretched without breaking; there may also be bridges between the two central tubules occurring at intervals along the tubules' length. (After I. Gibbons, P. Satir, and many others. For diagrams used in preparing the present ones see Gibbons, I., *J. Cell Biol.* **91**:110S, 1981, and Satir, P., *Scientific American* **231**(4):44–52, 1974.)

A-tubule. Arms and other structures extend from the doublets as shown in Figure II–80. The arms are spaced regularly along the doublets, occurring every 24 nm, and the other longitudinal structures also are regularly spaced. The entire array of tubules is referred to as the *axoneme*. At its base in the cell, the axoneme of each cilium or flagellum is associated with a *basal body* (Fig. II–79 and Section 2.10.6). At their tips, the tubules in some cilia or flagella are inserted into electron-dense "cap" structures of various types.

Membrane   Cilia and flagella are bounded by a membrane that is an extension of the plasma membrane. This membrane is currently attracting increased attention because it represents a stable, specialized region of the cell surface of interest, both in relation to the functioning of cilia and flagella and for study of how the cell can establish and maintain such specialized membrane zones. Already established is that the base of the cilium generally is encircled by a *ciliary necklace,* a thin girdle made up of rows of intramembrane particles visible when the membrane is viewed in freeze-fracture preparations. *Plaques* of clustered particles frequently are present here as well. These particles may help to anchor structures inside the cilium or to control the permeability and transport properties of the membrane, perhaps regulating the entry or exit of Ca^{2+} (see Sections 2.10.3, 2.11.4, and especially 3.3.4). Attachments of axonemal structures to the membrane seem to exist along the length of the organelle. In some algae and other organisms, fine, hairlike projections ("mastigonemes") project from the surface of the membrane. These projections are thought to amplify the interaction between the moving flagella and the medium surrounding the organism. In *Chlamydomonas* (see Figs. II–59 and II–61), particles experimentally attached externally to the flagellar surface can move rapidly in both directions along the length of the flagellum, suggesting that membrane components can undergo similar movements. Also in *Chlamydomonas,* the membrane at the tip of the flagellum carries recognition molecules whereby the two "sexes" associate to initiate events that lead to cell fusion during sexual reproduction.

2.10.3 Movement of Cilia and Flagella

Cilia beat by producing an *effective stroke* (*power stroke*) followed by recovery—that is, a return to the original position (Fig. II–81). The rate of motion is rapid enough that the tip may attain velocities of several thousand micrometers per minute; a given cilium can beat 10 to 50 times a second. Flagella move with an undulating snakelike motion (Fig. II–81). For both cilia and flagella the overall movement pattern, as seen in three dimensions, sometimes has helical or circular aspects (Figs. II–81 and III–16). Movement of cilia and flagella enables cells to move through a fluid medium or to move the medium past themselves. For many protozoa and sperm, cilia or flagella endow the cells with motility. Movement of cilia of the cells lining the gills of molluscs induces circulation of water past the gills. In ciliated epithelia such as those lining the trachea of mammals, collective movement of several hundred cilia on each cell provides a strong upward current that helps carry mucus and trapped dust out of the lung. (Cigarette smoking inhibits normal ciliary action in the respiratory tract and thus affects the normal mechanisms for removal of foreign matter.)

Cilia typically beat in coordinated waves, so that at any given moment some cilia of a cell are in the effective positions of the cycle and others are in the recovery positions (see Fig. III–16). This assures a steady flow of fluid past a surface. The direction of the effective stroke, the frequency of beating, and the form of the overall motion of each cilium are controlled by cellular mechanisms; through changes in these parameters, unicellular organisms can back away from obstacles or orient with respect to environmental cues. Fragmentary

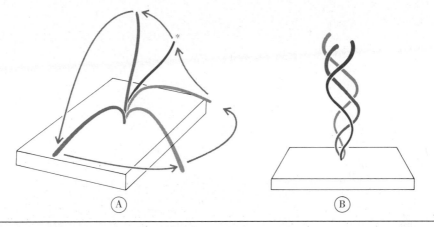

Figure II–81 *Successive stages in the motion of a cilium (A) and a flagellum (B).* The effective stroke of the cilium begins at *. For many cilia, especially in protozoa, recovery of the cilium occurs in a different plane from that of the effective stroke. As Figure III–16 shows, ciliary motion is often somewhat more complex than indicated here. (After B. Parducz and P. Satir.)

information (Section 3.3.4) indicates that mechanical factors, responses of the plasma membrane, and Ca^{2+} ions contribute to the control and coordination mechanisms.

Sliding Doublets   Much progress has been made in recent years toward relating the structure of cilia and flagella to their motion, and it is to be anticipated that a detailed molecular picture will be available before long. Cilia or flagella that have been detached from the cell and treated to make their membranes permeable to molecules in the medium can beat if ATP is added to the suspension medium. This indicates that cilia and flagella contain the machinery for motion, rather than being passively moved by activities elsewhere in the cell. If the tubules of the axoneme contracted to shorten by about 10 percent of their length, much of the motion could be accounted for; however, there is little evidence for tubule contraction. Instead, careful electron microscopic studies of cilia during bending indicate that there is a cycle of sliding (Fig. II–82) in which the doublets move longitudinally with respect to one another. Thus the mechanism resembles, to some extent, the contraction of striated muscle by the sliding of fine filaments along one another (discussed in the next chapter). To generate bending of the magnitude and form observed, the doublets need slide by less than one micrometer with respect to one another; the total relative movements of adjacent doublets may be less than 0.5 μm.

It was, however, necessary to demonstrate that the sliding of ciliary tubules is not simply the *result* of bending of the cilium: Isolated flagella were fragmented, their membranes were removed, and the axonemal fragments ob-

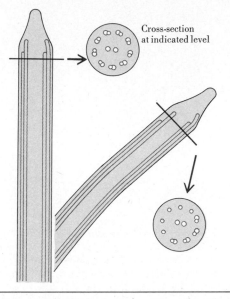

Cross-section
at indicated level

Figure II–82 *The sliding of doublets* during motion of a cilium in the gills of a mussel. One of the two tubules (the A-tubule) of each doublet is slightly longer than the other so that at the doublet's tip there is a short single tubule as shown. Consequently, cross sections at appropriate distances from the tip show a cyclical change in tubule pattern during bending, from all doublets to a mixture of doublets and single tubules. The entire doublets slide with respect to one another but the two members of each doublet do not—they retain their positions relative to each other. (After P. Satir.)

tained in this way were briefly digested with proteolytic enzymes. They were then exposed to ATP and studied by microcinematography using a darkfield microscope (Section 1.2.2). The structures elongated greatly (Fig. II–83), and when subsequently examined by electron microscopy, they showed a striking diminution in the overlap between adjacent doublets. These observations showed that sliding had taken place without bending, which strongly suggests that sliding is a primary event. Presumably sliding normally is harnessed to generate bending through intervention of the other ciliary components found to be disrupted in the experiment just described; these include the "spokes" and "links" shown in Figure II–80. Observations on changes in ciliary motion induced genetically—that is, by mutations that alter the proteins and visible structure of cilia—support this conclusion. A popular view is that the spokes and sheath engage each other cyclically, attaching and detaching, and that this helps to orient the movement induced by sliding into the observed bending.

Another factor likely to contribute to the conversion of sliding into bending is the attachment of one end of the axoneme to a basal body (see Fig. II–79 and below). This would appear to prevent sliding of the basal ends of the doublets with respect to one another.

Central Tubules The roles of the central pair of tubules require elucidation. Because certain nonmotile cilia serving sensory functions lack the central tubule pair (see, for example, Sections 3.6.3 and 3.8.1), it has sometimes been suggested that they are essential for motility. However, sperm and other cells of a few species and mutants of some unicellular organisms have cilia and flagella in which the central tubule arrangement is very different from that shown in Figure II–80: A single modified tubule may be present, there may be more than two, or the tubules may be absent. Many of the deviant forms are nonmotile, but some do bend or show other movements, indicating that the pattern of paired central tubules is not an absolute requirement for motion. Nonetheless, since a pair of central tubules is usually present, special roles for them in motility are being sought. Early studies indicated that the effective stroke of many cilia is oriented at right angles to the plane of the central pair (the beat would be up and down for the cilium in Fig. II–80B). Although matters may not always be quite this simple, the observations led to the suggestion that the central tubules establish mechanical properties of the axoneme related to the pattern of the ciliary beat. For some cilia or flagella, the central pair itself may not be fixed in space but may actually rotate as the cilium or flagellum moves; in fact, the entire axoneme may twist, accounting for the three-dimensional pattern of the beat referred to earlier. There have been proposals that the central pair might help to coordinate the sliding of the doublets or have some other nonmechanical function, but these are difficult to evaluate with present methods.

Dynein The molecular morphology of the axoneme is under intensive investigation. The tubules themselves are made of proteins of the *tubulin* class, similar to those of other cellular microtubules.

In a pioneering experiment, axonemes isolated from the protozoan *Tetrahymena* were treated with salt solutions that selectively extract the central tubules and the arms on the peripheral doublets. Residual axonemes comprised of armless peripheral doublets were left behind. From the salt extract the protein *dynein* was isolated. When purified dynein was added (with Mg^{2+} ions) to the armless doublets, the arms reappeared, demonstrating that dynein is a principal component of the arms (see also Fig. II–84). Further study has made clear that the arms are moderately complex structures, composed of several protein subunits. There are indications that the two arms of each doublet differ from one another in composition and in details of organization.

Dynein is an ATPase; this property and its location are strong evidence for its involvement in making energy available for sliding. Evidently ATP binds to the arm, producing changes (in protein conformation?) which lead to the arm's ability to form a bridge between the A-tubule of a given doublet and the B-tubule of the next. Each bridge permits a doublet to exert force on the adjacent doublet, so that the adjacent doublet is moved a short distance in the direction of the tip of the cilium or flagellum. During the cycle of bridge formation and movement the ATP is split, and ADP is subsequently released. The

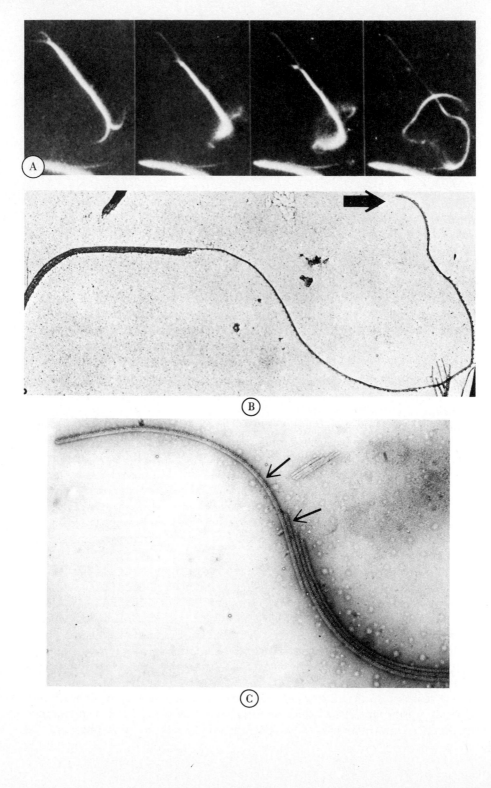

◀ **Figure II–83** *Doublet sliding in cilia and flagella*

    A. Frames from a movie taken through a darkfield microscope (Section 1.2.2). An axoneme from a sea urchin sperm flagellum was demembranated with the detergent *Triton* and digested briefly with the protease *trypsin*. ATP was then added. The micrographs show the structure at successive times separated by intervals of 10 to 30 sec. They illustrate the longitudinal sliding induced by ATP; the structure lengthens and becomes thinner. By the last panel a group of doublets has slid down toward the bottom, then curled back and curved across the remainder of the axoneme; this accounts for the narrow crescent structure crossing the field. (A connection of this crescent to the remainder of the axoneme is still present toward the lower left, but is barely seen since it has curled too far out of the plane of focus). × 1200. (From Gibbons, I.R., and K.E. Summers, *J. Cell Biol.* 91:117S, 1981.)

    Parts B and C are electron micrographs of preparations comparable to that in Part A. Part B shows an axoneme of *Tetrahymena* (Figs. II–59 and 61); the arrow indicates a doublet that has slid out from the remainder of the structure. (From W. Sale, *J. Protozool.* 24:498, 1977. Courtesy of the author and of P. Satir and S. Lebduska). Part C shows a segment of an axoneme from the *Chlamyclomonas* (Fig. II–59). The displacement of successive doublets owing to sliding is clearly evident; the arrows point to two of the doublets that have slid substantial distances. (From King, S.M., J.S. Hyams, and A. Luba, *J. Cell Biol.* 94:341, 1982. Copyright Rockefeller University Press.)

---

arm then detaches from the B-tubule ready for another ATP and another round of attachment and movement. The overall sliding pattern represents the cumulative effect of a number of such cycles occurring throughout the length of the structure. The force might be generated by a change in conformation or orientation of the dynein arms: Electron microscopic observations hint that there is a change in the angle of the arms with respect to the tubules, as though they were swiveling or cyclically bending and straightening as they function.

    The notion of sliding produced by bridges derived initially from the model of filament sliding in muscles (Section 2.11.1) but now has accumulated much independent support. For instance, mutations are known in several organisms that result in the absence of arms on cilia and flagella as a consequence of abnormalities in dynein. The cilia and flagella are rendered immobile. In humans, such a mutation causes sterility due to sperm immobility, and other problems.

2.10.4 Microtubules and Other Cellular Motion

Microtubules often are arranged along pathways of oriented motion of other structures in the cell and along directions of cell elongation. Rapid changes in the color of fish, by which they blend with their surroundings, result from responses of pigment cells such as *melanophores* and *erythrophores*. Within these cells, pigment granules (see Fig. II–94) migrate along paths delineated by microtubules, spreading through the cell or concentrating in a small area. The corresponding changes in light absorption by the cells alter the appearance of the fish's surface. But are the microtubules themselves responsible for the motion, or do they serve instead to *orient* motion generated by some other system such as the filaments or networks to be discussed in the next chapter? This is a major unresolved issue. Many investigators believe that among the microtubule-associated proteins, some at the tubule surface permit tubules to interact with one another, with the cell's filament systems, and with other structures. This interaction may be signaled by a fuzzy layer at the tubule surface or by the presence of fine bridges sometimes seen between adjacent microtubules or, more rarely, between tubules and the membranes of structures that may be moving nearby. The bridges could simply serve to stabilize arrays of microtubules, linking them together. But on the basis of analogies with cilia and flagella, it is tempting to think of the bridges as armlike devices by which a tubule can exert force on an adjacent structure, pulling or pushing it a short distance along the tubule length and perhaps leaving it in a position where the next bridge can form and continue the propulsion.

Other prominent cases of oriented movement along pathways marked out by microtubules include the passage of endocytic bodies toward the Golgi region, the accumulation of Golgi vacuoles at the forming cell wall of dividing plant cells (see Figs. IV–28 and IV–29), and the extensive transport of materials down the axons of nerve cells (Section 3.7.2). The most universal, and biologically the most important, is the movement of chromosomes to the poles of the *spindle* during cell division. Despite a century of study the mechanism of this movement is not known. The spindle is constructed largely of microtubules, and hypotheses presently in favor suggest that part of the force moving the chromosomes may originate by sliding or by similar lateral interplay of microtubules, and part by processes in which the tubules come to disassemble into their tubulin subunits. Sections 4.2.7 and 4.2.8 will present evidence concerning these possibilities.

The rapid and reversible lengthening of protozoa like *Stentor* (see Fig. III–13) is accompanied by sliding of tubule arrays past one another; this sliding might produce the lengthening or, alternatively, control and orient it. In other, slower elongations, such as the extension of processes by growing nerve cells or the elongation of maturing sperm, growth of microtubules or their involvement in transport of other materials may participate in bringing about or orienting the lengthening.

The possibility has been entertained that the interior of the microtubule is truly hollow and that small molecules move within the tubules like fluids in a soda straw. Evidence for this would be very hard to obtain, and virtually nothing is known of the organization of the tubule interior. Whether there really is freely movable material there is an open question.

Motion of cytoplasmic structures along tracks marked by microtubules often has a discontinuous appearance: Particles may move briefly at rates up to 10 μm per second, then stop and then move again. Such "jumping" is called *saltatory movement*. More continuous movements, at rates of 1 to 10 μm per second, also are observed, particularly for small structures, such as vesicles.

Dynein (Section 2.10.3) obtained by extraction from isolated cilia or flagella can be made experimentally to associate with microtubules from elsewhere in the cytoplasm, helping to convince some investigators (far from all) that a dynein-like protein is present on many tubules, although apparently not with the same obvious arrangement as is true in cilia and flagella. However, microtubule-related functions other than those in cilia and flagella seem normal in mutants lacking normal ciliary dynein, suggesting that such a dynein-like protein, if it exists, may be distinct from the ciliary protein.

Vanadate ions (VO_4^{3-}) and EHNA (erythrohydroxynonyladenine) are agents known to inhibit the ATPase activity of dynein. Recent studies (on permeabilized preparations; see introduction to Chap. 2.11) indicate that several types of microtubule-related movements, in addition to ciliary motility, are inhibited by these agents. Included are the saltatory movements of particles in tissue-culture fibroblasts and the spreading of pigment granules in erythrophores of fish. These findings are taken as presumptive evidence for involvement of a dynein-like protein in a variety of sorts of motility, though this is only provisional because the agents do affect enzymes other than dynein.

2.10.5 Assembly

Microtubules can be reversibly disassembled by exposing cells to high pressure, low temperature, or drugs such as *colchicine, vinblastine,* and *podophyllotoxin.* (Cilia and flagella are exceptional in this regard; their tubules resist such perturbations.) When the pressure is elevated or temperature lowered, the microtubules of thin pseudopods such as those shown in Figure II–76 disappear, and the pseudopods retract into the cell. With the return of the cell to normal conditions, both microtubules and pseudopods reappear. Tubule-disruption experiments of this type confirm the importance of the microtubules for aspects of cell shape and motion. Section 4.2.8 will consider such matters in more detail: The functioning of the mitotic spindle depends partly on controlled disassembly of microtubules.

Paralleling their behavior in cells, microtubules can disassemble and reassemble readily in the test tube. Assembly is induced in solutions of tubulin dimers if the temperature and ionic environment are appropriately adjusted (for example, Ca^{2+} must be low and Mg^{2+} present) and guanosine triphosphate (GTP) is added. Tubulin molecules themselves have GTP molecules or derivatives attached, but how GTP contributes to assembly is not known. GTP, like ATP, is a potential energy source, but GTP, like ATP, can have other effects on interactions of proteins. Both can affect protein conformation when bound to specific sites (Section 1.4.2) so that their association with proteins and their subsequent hydrolysis and release could help to regulate a variety of allosteric and other conformation-dependent processes, including the fitting together of proteins during assembly of biological structures.

Low temperature promotes microtubule disassembly, a fact that has been put to use in the purification of tubulin and which has theoretical interest as well (see Section 4.2.8).

Polarity; Action of Drugs   Microtubules are capable of adding tubulin dimers to both their ends, and disassembly ("depolymerization"), freeing dimers, can also occur at both ends. However, in test-tube assembly at least, microtubules show an intrinsic polarity of growth, adding new tubulin dimers preferentially at one end. Evidently the organization of the microtubule wall itself favors addition of new subunits at one end when tubulin concentrations and other conditions in the surrounding medium are favorable for tubule assembly. This "fast-growing" end is designated the (+) end. Slower growth of microtubules can take place from the (−) end as well; this can even account for most of the test-tube elongation of microtubules under conditions where the (+) end is blocked. When circumstances sharply favor disassembly, the (+) end again seems to predominate, so some call this end the "fast-assembly, fast-disassembly" end. Experimental conditions of tubulin concentration and other factors have been found in which microtubules simultaneously add subunits at their (+) ends and lose subunits at their (−) ends. This leads to the test-tube phenomenon known loosely as "treadmilling": Tubulin dimers add to the (+) end, become incorporated in the body of the microtubule as more dimers add, and eventually come off the tubule at the (−) end when the disassembly proceeding from that end reaches them. There are insufficient data to ascertain whether similar phenomena occur in the cell. Still, speculative schemes are being pursued in which treadmilling is supposed to participate in movement of the tubules themselves or of materials attached to the tubules. There are difficulties in envisaging this, since in the strict sense, the tubulins themselves are not really moving along the wall of the microtubule. Nevertheless, the notion that some microtubules, at least, are continuously undergoing balanced, polarized, simultaneous gain and loss of tubulins, is very widely accepted (see below and Section 4.2.8).

Colchicine and certain of the other microtubule-disruptive drugs bind to tubulin dimers in the nonassembled ("free") state. It is presently believed that colchicine–tubulin complexes associate with the (+) ends of microtubules preventing continued assembly, but not preventing disassembly, at least from the (−) ends, and perhaps from both ends. Colchicine's ability to depolymerize microtubules in the cell is taken as evidence that whether or not treadmilling takes place, microtubules in the cell do continually gain and lose subunits. The drug, by blocking the gain, shifts the system from equilibrium to net loss. The insensitivity of cilia and flagella to colchicine suggests that the axonemal microtubules may be in a stable state, but colchicine does block new assembly or growth of cilia and flagella.

The polarity of microtubules is visualized microscopically by observing the orientation of structures that assemble on the tubules from added tubulin or dynein (Fig. II–84). Section 2.11.1 and Figure II–91 describe analogous but longer-established methods for determining the polarity of microfilaments.

The drug *taxol* can slow or block depolymerization of microtubules and induce their assembly.

Microtubule Organizing Centers   Assembly and disassembly of microtubules provides for flexible control of cell shape and oriented motion. What factors influence the assembly process and microtubule distribution and functioning in the cell? Circumstantial evidence implicates ions such as Ca^{2+}, cyclic nucleotides such as cyclic AMP or cyclic GMP, and perhaps phosphorylation of proteins. In test-tube assembly experiments microtubules do not form if Ca^{2+} levels are too high, and in the cell, conditions that produce changes in Ca^{2+} or cyclic nucleotides often alter microtubule distribution or assembly. The Ca^{2+}-dependent regulatory protein calmodulin is associated with some microtubule-rich structures such as the mitotic spindle. But calcium and nucleotides have so many effects on cells it is still difficult to disentangle those that are specific for microtubules.

Although purified tubulin can assemble by itself, assembly is more rapid and extensive if certain of the microtubule-associated proteins are added and particularly if already assembled microtubules or fragments of tubules are present. It is speculated that microtubule-associated proteins stabilize the tubulin assemblies and thus help to regulate microtubule formation. In the test tube at least, MAPs can induce tubulins to form ringlike assemblies that subsequently open up to convert into segments of microtubule wall protofilaments. Preexisting tubules serve as "seeds" that grow by the oriented addition of tubulin (see Fig. II–87). Most investigators now believe that the assembly of microtubules in the cell is governed in part by structures known collectively as *microtubule organizing centers (MTOCs)*. Such structures would serve to initiate assembly at particular places in the cell and probably would remain attached to one end of the tubule, where they would also stabilize the microtubules by slowing or preventing disassembly. It is easiest to envisage situations in which the (−) end of a growing microtubule is the one associated with the MTOC, with the tubule extending away by elongation of its (+) end, but since slow assembly can occur at the (−) end of a microtubule there could be exceptions to this pattern.

A finding supporting the concept of MTOCs is that when tubules are depolymerized experimentally in the cell and then allowed to form anew, they tend to reappear near where they originally were, as though some governing structure or organization had persisted (see Fig. IV–25). A variety of candidates for MTOCs have turned up, and it could well be that there is more than one type. Included are aggregates of amorphous material in which microtubules are seen to terminate (see following section), as well as better-defined cellular structures such as the chromosomes' kinetochores, which can organize microtubules at least under experimental conditions (Section 4.2.7). Once again the case of cilia and flagella has been most thoroughly analyzed, as the next section will document.

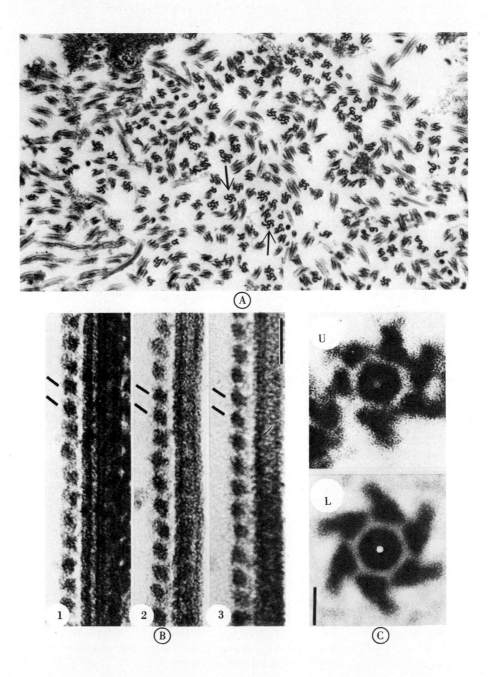

◀ **Figure II–84** *Microtubule polarity, decoration, and dynein.*

A. Microtubules in the spindle of a cultured rat-kangaroo cell (PtK$_1$) that had been "permeabilized," to permit subsequent entry of large molecules, by treatment with detergents, then exposed to high concentrations of tubulin extracted from brain. Under the conditions used, the tubulins bind laterally to the pre-existing microtubules, "decorating" the microtubules with curved sheets; the sheets appear as "hooks" in transverse (cross) sections of microtubules. The orientation of the hooks indicates the polarity of the tubules: the hooks point to the right (clockwise) when viewed from the (+) end (Section 2.10.5) of a microtubule and to the left when viewed from the (−) end. In the present micrograph most of the microtubules show clockwise hooks (see the ones at the arrows): from the orientation of the tissue section from which this micrograph was taken, this observation is interpreted as meaning that most microtubules in the spindle are oriented with their (+) end "distal to" (pointing away from) the spindle poles. (The possible implication that the (+) end—the "fast assembly end"—of kinetochore-associated tubules is oriented toward the kinetochore is a matter of great interest to investigators of cell division [Section 4.2.7]; some conclude, very tentatively, that kinetochores either attach to microtubules originating at the spindle poles or nucleate microtubules that then grow out by extension of their (−) ends.) × 48,000. (From Euteneur, U., and J.R. McIntosh, *J. Cell Biol.* **89**:338, 1981; see also *Nature* **286**:517, 1980.)

B. Longitudinal sections of microtubules associated with dynein (Section 2.10.3). In each panel, a tubule runs from top to bottom at the right, and dynein projects toward the left as roughly globular units attached at 24-nm intervals to the tubules. Panel 1 is a microtubule from an axoneme (Section 2.10.2) of *Chlamydomonas* showing the dynein arms normally present: each arm projects at an angle of 55° as indicated by the pair of lines. The axoneme is oriented so that its base—where it anchors in the cell—would be at the top of the figure. (The arms have the form of "lollipops" in that a thin—2- to 3-nm—stalk, barely visible in this preparation, connects the more globular regions—10 nm in diameter—to the microtubule.) The microtubules in panels 2 and 3 were reconstituted in the test tube from purified tubulin and then were exposed to dynein, which binds to them; again the dynein structures protrude at 55°. The angle made by dynein is an indication of microtubule polarity; in the axoneme, at least, the tilt of the arms is toward the (−) end of the tubule, the end at the basal body. The bar at the upper right corresponds to 50 nm. × 250,000. (From Haimo, L.T., B.R. Telzer, and J.L. Rosenbaum, *Proc. Nat. Acad. Sci.* USA **76**:5759, 1979.)

C. In this preparation, dynein from the protozoan *Tetrahymena* has been associated with microtubules in spindles isolated from the clam *Spisula*. Panel U is a microtubule viewed in cross section showing a counterclockwise arrangement of the associated dynein structures. Panel L is the same section subjected to one of the several techniques available for clarifying images of structures with repetitive or periodic organization. Since inspection of panel U suggests the presence of six more-or-less equally spaced "arms" projecting from the tubule, the image was rotated around its center ("Markham rotation") in six steps of 60° each (⅙ of a complete rotation of 360°). At each position the image was rephotographed. When the photographs were then superimposed, the "reinforced" image in L was obtained. As in part A, dynein "decoration" as illustrated here shows that most of the microtubules in a given region of the spindle have the same polarity. The bar at the lower left is equal to 25 nm. × 250,000. (From Telzer, B.R., and L.T. Haimo, *J. Cell Biol.* **89**:373, 1981. Copyright Rockefeller University Press.)

2.10.6 Centrioles and Basal Bodies; the Formation of Cilia and Flagella

Long before microtubules were discovered, microscopists were aware of the existence of a category of organelles known as *centrioles*. Centrioles usually occur in pairs. In typical nondividing cells there is one pair per cell (Fig. II–85). Frequently this pair is located close to the Golgi apparatus. The two members of the pair often lie with their long axes perpendicular to each other (Fig. II–85A). Characteristically, the centrioles are cylindrical structures, approximately 0.15 μm in diameter and 0.3 to 0.5 μm long. These dimensions are just at the limit of resolution of the light microscope, so that little was learned of their detailed structure until electron microscopy developed.

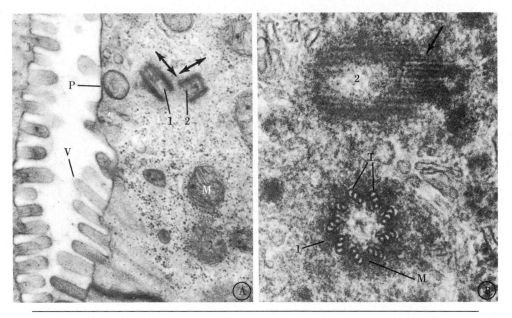

Figure II–85 *Centrioles.*
  A. Portion of the absorptive surface of a cell of the intestine (chick embryo) showing centrioles. Both centrioles of the pair *(1,2)* have been cut longitudinally. The long axes of the centrioles (arrows) are almost perpendicular to one another. *M* indicates a mitochondrion; *P*, the plasma membrane; and *V*, microvilli. × 35,000. (Courtesy of S. Sorokin and D.W. Fawcett.)
  B. Portion of a cultured cell from the Chinese hamster showing a centriole pair sectioned at a different angle from that in Part A. Centriole *1* is cut transversely; it shows the characteristic arrangement of tubules *(T)* as nine triplets embedded in a cylinder of dense matrix *(M)* surrounding a less dense central region. The "A" tubule of each triplet is the one closest to the center of the cylinder. (The middle member of the triplet is conventionally designated *B* and the third member, [to which the T's point in this picture] *C*). Centriole *2* is cut obliquely, thus obscuring the pattern of tubules. Part of one tubule cut in longitudinal section is seen at the arrow. Approximately × 70,000. (Courtesy of B.R. Brinkley and E. Stubblefield.)

Centrioles have a characteristic "9 + 0" appearance in the electron microscope. Each is made of nine microtubular units, arranged as the walls of a cylinder; each unit is a *triplet* composed of three associated tubular elements (Fig. II–85). There are no central tubules within the cylinder. In each triplet, the tubule closest to the center of the centriole (the A-tubule) is a complete microtubule; the other two (B-tubule and C-tubule) share portions of walls. The triplets are embedded in an amorphous matrix. At one end of the centriole a "cartwheel" structure is often seen (Fig. II–79). Granular or fibrillar material of uncertain arrangement may also be present elsewhere within the cylinder.

Centrioles, Centrosomes, Pericentriolar Satellites, and Microtubules The cytoplasmic zone in which the centrioles are located has been called the *centrosome, centrosphere,* or *cell center*. Characteristically, microtubules radiate from this region (Figs. II–86 and IV–25). Some authors now use the term "centrosome" to refer to centrioles plus associated "pericentriolar materials" (see below).

The fact that microtubules often radiate from cell regions where centrioles are present, and may even be anchored there, led initially to the suggestion that centrioles as such were microtubule-organizing centers, perhaps the major ones in the cell. This view has been modified as it became clear that some cells possess microtubules but lack centrioles (Section 4.2.7). Moreover, with improvement in microscopic techniques, the microtubules associated with centrioles were found to terminate not on the centrioles proper but on accumulations of amorphous material close to the centrioles. These accumulations collectively are known as "pericentriolar material." In many cases, the material appears as moderately electron-dense masses called "satellites." Satellites lack a well-defined structure and thus resemble the equally ill-defined bodies found at the ends of some of the microtubules that are not associated with centrioles. When centrioles initiate microtubule formation in test-tube assembly experiments, newly assembled tubules are seen terminating at the pericentriolar material. Neither the composition of pericentriolar material nor its relations to centrioles have yet been worked out, nor is it known yet why in their other guise, as basal bodies, organelles with the same structure as centrioles can initiate growth of tubules from the ends of their own tubules.

Basal Bodies A basal body is present at the base of each cilium or flagellum (Fig. II–79). This is true from the onset of formation of a given cilium or flagellum. Often fibers *(rootlets)* extend from the basal body deep into the cytoplasm, perhaps as part of a system that anchors the basal bodies and orients the cilium or flagellum (see Fig. II–79). In some cell types, especially in protozoa, numerous microtubules run between the basal bodies, serving apparently to maintain a relatively rigid supportive framework below the cell surface.

Each basal body has the same 9 + 0 structure as that of a centriole. The nine peripheral doublets of the associated cilium or flagellum appear as extensions of two of the three tubules of the basal body's triplets (the A-tubule and

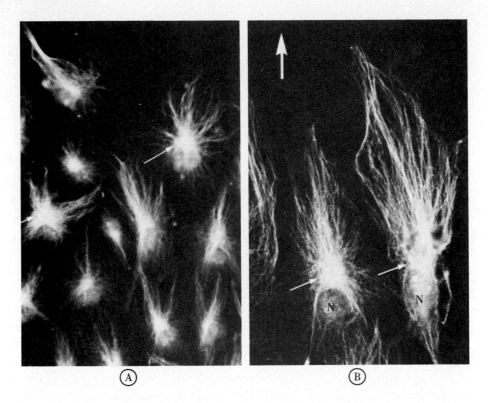

B-tubule). That basal bodies and centrioles are closely related, probably as alternate forms of the same organelle, is indicated by several lines of evidence; the distinction between the two is functional and derives from the pre–electron microscope era when the identity in structure could not be known. For example, in many nonciliated cell types, one of the two centrioles often has a rudimentary "primary cilium," with $9 + 0$ structure (page 257) attached. In some developing sperm cells a flagellum grows in association with a basal body during division, while the basal body is still in the position customary for centrioles of dividing cells—at the poles of the mitotic spindle (Section 4.2.7).

Formation of Cilia and Flagella Most of the growth of cilia and flagella depends upon the assembly of structural subunits that are "pre-formed," in the sense that subunit synthesis is a separable process from subsequent assembly into structures. If the two flagella of *Chlamydomonas* (Figs. II–59 and II–61) are broken off by mechanical agitation or other means, new ones regenerate. Preventing protein synthesis by the addition of inhibitors (such as cycloheximide) after amputation does not prevent regeneration, although the regenerates do not reach full length. If only one flagellum is removed, the other shortens for a while as regeneration occurs; then both return to normal length. When the numerous cilia at the surface of sea urchin embryos are removed, regeneration occurs even in the presence of inhibitors of protein synthesis.

◄ **Figure II–86** *Radiation of microtubules from the centrosome.* Microtubule distribution in cultured cells from the lining of pig aorta (endothelial cells; Section 3.5.2) as visualized immunohistochemically with fluorescent antitubulin antibodies (Section 1.2.3); (the method is outlined in detail below to provide an example of such techniques). *N* indicates the nuclei of several of the cells, and the small arrows point to the central zones near the nuclei, from which microtubules, the bright lines, radiate. This radiating pattern is thought to reflect the presence of centriole-associated microtubule organizing centers *(MTOCs)* near the nuclei. Of interest, but still uncertain significance, is that as the cells crawl along the surface on which they are grown, these MTOCs appear to be oriented between the nucleus and the leading edge of the cells (the direction of motion is indicated by the large arrow). A, × 300; B, × 700. (From Gotleib, A.I., L. McBarnie-May, L. Subrahmanyan, and V.I. Kalnins, *J. Cell Biol.* 91:591, 1981.)

Method: ("Immunofluorescence") The cells were fixed briefly with methyl alcohol to immobilize their contents then treated with acetone and air-dried to disrupt their plasma membranes, rendering the cell interior accessible to large molecules. They were then exposed to antibodies specific for tubulin, obtained from a tubulin-injected rabbit. Next, the preparations were exposed to fluorescent goat-antirabbit antibodies—antibodies obtained from a suitably injected goat that bind to rabbit antibodies and that also have the fluorescent dye *fluorescein* linked to them. The fluorescent antibodies reveal the locations of tubulin because they bind to the antitubulin antibodies and the latter are bound to cellular sites of tubulin. Such an "indirect" method—the use of labeled antibodies to detect other antibodies bound specifically to sites of interest—is frequently employed for several technical reasons that often make this approach preferable in terms of specificity and conservation of scarce antibodies to "direct" methods. (A "direct" method in the present case would be the use of fluorescein-labeled antitubulin antibodies; comparable direct methods were used for Fig. II–9). The cells were viewed in a fluorescence microscope: the light shone upon them from the light source was ultraviolet light (UV) with a wavelength specifically absorbed by fluorescein molecules. The fluoresceins then *fluoresce;* that is they re-emit most of the energy obtained from the UV light in the form of visible light. Since UV light is not visible to the eye (and is also blocked to an extent by glass lenses), where no fluorescent molecules are present the specimen appears dark. Stained structures, therefore, appear bright against a dark background, a situation quite favorable for visualizing even low concentrations of stain.

Presumably free subunits—tubulins and other proteins—are usually present in the cells studied in these experiments or are made available for assembly into cilia or flagella by disassembly of other microtubular structures. Thus, newly made proteins are not needed for regeneration of considerable lengths of cilia and flagella, and the inhibition of protein synthesis does not prevent regeneration. The experiment in which one flagellum is removed is interpreted as indicating that subunits can be free from one flagellum to support the growth of another.

This is not to say, however, that the cell lacks regulatory mechanisms enabling it normally to replenish its tubulin pools by new synthesis. Under ordinary circumstances, when regeneration or new formation of cilia or flagella is evoked, so are transcription and translation of the requisite mRNAs. Evidence obtained with a number of cell types in tissue culture indicates that the transcription and turnover of mRNAs coding for tubulin may be controlled by a system that responds to changes in the levels of free tubulin in the cytoplasm. The mechanisms underlying this, however, are still unknown. Furthermore, the research was done chiefly on cells that lack cilia or flagella, so it is not established that this type of regulatory system applies to all situations. In fact, the detailed relations between tubulins of cilia and flagella and those of other microtubules have yet to be worked out. Obvious differences have not been found, and experiments like the one in Figure II–87 demonstrate that tubulins from other microtubules can assemble onto axonemal microtubules. Still, the axonemal tubules do differ from the others in such respects as their sensitivity to colchicine and other drugs (Section 2.10.5). Although this may be due in part to "mechanical factors" such as the presence of basal bodies and the ensheathing of the axoneme in an extension of the plasma membrane, the existence of subtle differences among the tubulins has not been ruled out. A possibility arising in recent research is that the tubulins of cilia and flagella are drawn from the same pool as those of other microtubule structures but undergo modifications during their assembly into cilia and flagella that change their properties, perhaps irreversibly. On the other hand, a number of organisms have been shown to possess multiple copies of the genes coding for tubulins (see Section 4.2.6), and the possibility that different categories of microtubules sometimes contain tubulins of different genetic origin is not ruled out. Recently, there have been reports that the α form of tubulin reversibly undergoes addition of the amino acid tyrosine under various conditions; a few investigators have begun to construct hypotheses wherein this *tyrosinylation*, or subsequent phosphorylation of the tyrosines, is supposed to regulate microtubule assembly, disassembly or functioning.

The basal bodies in some way control assembly of ciliary and flagellar subunits. Of interest is the fact that the 9 + 0 structure of the basal body controls the formation of the somewhat different 9 + 2 structure. In one investigation cells were fed radioactive amino acids at different times during regeneration of flagella and were then studied by autoradiography. Results of the study indicated that as a flagellum grows it adds much of its new protein (identifiable by radioactivity) at the tip rather than at the base. Apparently the basal body initiates formation of the flagellum's doublets, and the tubules then grow by continued addition of subunits at the end opposite the basal body. Such oriented assembly takes place in the test tube when purified tubule proteins are added to isolated basal bodies or to portions of cilia (Fig. II–87). Experiments like this show that the end of the basal body on which the axoneme normally terminates (the "distal" end, opposite the "proximal," cartwheel end in Fig. II–79) is the more effective end in supporting tubule growth in the test tube. The differential assembly capacities of the two basal body ends offers an obvious though still hypothetical explanation for the orientation of ciliary growth in the cell. It is simply necessary to postulate that the processes by

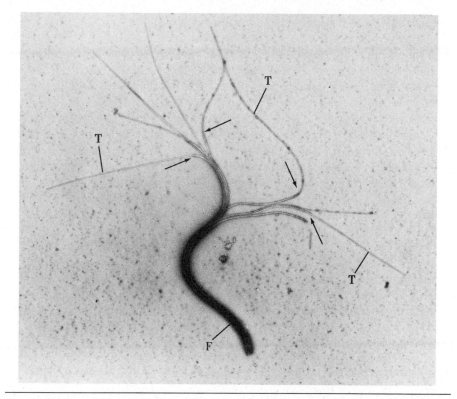

Figure II–87 *Growth of microtubules* by oriented addition of tubulin molecules. A flagellum *(F)* from *Chlamydomonas* (see Figs. II–59 and 61) was detached from the cell, and its membrane removed. It was then placed in a solution containing tubulin (purified from pig brain) under conditions favoring microtubule formation. The original ends of the doublets of the flagellum are indicated by arrows. Continuous with the A-tubules (Fig. II–80) of the doublets are the microtubules *(T)* formed from the pig brain tubulin. Negatively stained preparation (Fig. II–53). × 7000. (From Allen, C., and G. Borisy, *J. Mol. Biol.* **90:**381, 1974. Courtesy of the authors.)

which basal bodies themselves are duplicated (see following section) or those by which they approach the cell surface to initiate growth of a cilium or flagellum are so regulated as to ensure that the proper end of the basal body faces in the correct direction.

Formation of the central microtubule pair in the axoneme obviously differs from that of the doublets in that no corresponding initiating sites with microtubule structure are present in the basal body. One suggestion is that growth of these tubules is initiated by the membrane-associated electron-dense cap material that often is seen where the tubules end at the *tip* of the cilium or flagellum: In other words, these tubules might grow in the opposite direction from the rest, starting at what will become the end of the structure furthest from the cell. However, in some cases, plates or granules of somewhat more highly organized material are present where the central tubules abut on the basal body (see Fig. II–61), and the issues of the direction and control of central tubule assembly are not yet resolved.

The formation of the other structures of the cilium or flagellum—the arms, spokes, sheath, and so forth—has been little studied. From the fact that they can be reversibly dissociated (Section 2.10.3), it appears that the arms link to sites built into the tubule walls, but what is it that determines the locations of these sites? And how are the specializations of the ciliary membrane maintained despite its continuity with the remainder of the plasma membrane?

Analysis of the molecular details of basal body roles must await more adequate information about their composition (see the following section). Recent studies on isolated basal bodies suggest that they are rich in tubulins resembling those that constitute the ciliary tubules.

2.10.7 Centriole Duplication

Study of the duplication of centrioles and basal bodies faces two main difficulties: First, centriole chemistry is poorly understood; the organelles are too small for adequate cytochemical study, and they have only recently been isolated in a form pure enough to permit biochemical study. Second, the mode of formation may vary from cell to cell. When ciliated protozoa, such as *Paramecium*, divide by binary fission, the precise pattern of basal bodies near the surface (see Fig. III–15) is transmitted to both daughter cells. This involves the duplication of each basal body in coordination with cell division. In other cases, however, basal bodies appear at one point in the cell life cycle despite the previous absence of microscopically visible centrioles or basal bodies. This is true of the unicellular organisms *Naeglaria* (Fig. II–79). This organism can exist in a nonflagellated form, in which centrioles, basal bodies, or flagella have not been found, or it can become a flagellated cell with prominent flagella and basal bodies. This sudden appearance of basal bodies in cells that previously lacked visible centrioles or basal bodies also occurs in the sperm of some plants. It remains to be determined whether some self-duplicating structure is present that gives rise to basal bodies; it is possible that such structures have not been recognized in the microscope because of their small size or lack of distinctive appearance. Still, it is clear that a preexisting fully formed centriole or basal body is not needed for centriole formation.

Nucleic Acids?   There have been reports of the presence of nucleic acids in centrioles and basal bodies or in their immediate environs. For a while it was believed that DNA is present, but the evidence supporting this is presently considered unconvincing. (For example, DNA isolated in basal body–rich fractions turned out probably to be a contaminant from mitochondria or from bacteria previously phagocytized by the protozoa from which the organelles were obtained.) RNase treatment removes a bit of electron-dense material from within the cylinder interior of basal bodies of protozoa, and some other cytochemical evidence also suggests RNA may be present in or near basal bodies. Centriole-associated capacities to organize microtubules in cell division can be altered if the region of the centrioles is irradiated with highly localized, intense laser light under conditions expected to affect RNA (selective stains are used to sensitize the RNA). In addition, there is a report claiming that RNase treatment of isolated basal bodies inhibits their ability to function as microtubule-organizing centers in test-tube assembly of microtubules. However, these several observations and experiments are complex and subject to alternative interpretations. Although the evidence for centriole-associated RNAs is less widely doubted than that for DNAs, the final word has yet to be said about whether there really are nucleic acids in centrioles or basal bodies.

Mutations affecting the structure of basal bodies have been reported, but it is not yet known whether they occur in cytoplasmic nucleic acids or in those of the nucleus. Remember also that the work on ribosomal RNAs (Section 2.3.2) shows that RNAs can have roles in the assembly of complex structures independent of their specifying the sequence of amino acids in a protein (see also Chap. 4.1). If, therefore, RNAs are shown to be present in centrioles or basal bodies, it need not follow that these organelles synthesize proteins. Moreover, since the known RNAs of eucaryotic cells cannot self-duplicate, the presence of RNA need not be a sign of independent genetic information.

Procentrioles   In many cell types centrioles duplicate by the growth of a new *procentriole* near the "old" centriole. The procentriole generally contains the 9 + 0 pattern but is much shorter than the old centriole. It is characteristically oriented at right angles to the old centriole (Fig. II–88) and eventually grows into a complete centriole. This occurs in cells where centriole duplication is part of cell division, as well as in some cells where many basal bodies form. In ciliated protozoa preparing to divide, each basal body forms a procentriole near its "cartwheel" end, and a cartwheel is present early in procentriole development. In some other cells forming basal bodies, several procentrioles ap-

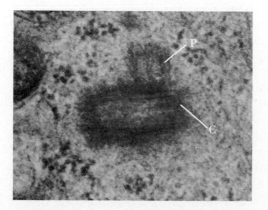

Figure II–88 *Procentriole.* Portion of a human cultured cell (HeLa; see Chapter 3.11) showing a centriole *(C)* and an associated procentriole *(P).* The procentriole grows at right angles to the centriole, eventually producing a configuration resembling that seen in Figure II–85. × 64,000. (Courtesy of E. Robbins.)

pear more or less simultaneously, radially arranged around the old centriole. Procentriole formation appears to involve neither fission nor direct budding from preexisting centrioles; how it does occur remains to be explained.

In cells of the respiratory and reproductive tracts of mammals, only two centrioles are present early in development; eventually large numbers of basal bodies are formed and produce the cilia that line the tracts. In the respiratory system, some of the basal bodies have been seen to form near the centrioles, but many arise from small aggregations of amorphous or finely fibrillar material far from the centrioles and with no obvious resemblances to them. Although these aggregates are considered by some to be "generative" structures, originally formed by centrioles, it is possible that they arise from some unrecognizable precursor material that originates elsewhere. In the cells that develop into the sperm of certain ferns and cycads, large compound bodies are formed from an unknown source and then fragment to produce numerous procentrioles, which ultimately give rise to the basal bodies of the many flagella found on each cell.

"Self-duplication," in the restricted sense that an old structure participates in the formation of a new one, more or less directly, may well be involved in the instances where procentrioles form near old centrioles. In the other cases, especially those where centrioles or basal bodies form in cells initially without recognizable organelles of this type, the existence of a self-duplicating structure or template yet to be identified, or the occurrence of unaided self-assembly (Section 1.4.3) of the component proteins into a centriole must be postulated.

Chapter 2.11 Filaments

No subspecialty in cytology or cell biology has undergone a more explosive recent development than the investigation of the protein filaments that abound in the cytoplasm of most eucaryotic cells. The filaments of striated muscle were the first to be analyzed in detail. Therefore, when others were found it was natural to postulate roles in motility, though it is now quite apparent that the mechanisms of such participation differ significantly among cell types. Involvement of filaments in the establishment and maintenance of three-dimensional cellular architecture is also evident. As already emphasized, both in motility and in cytoskeletal functions the filaments cooperate with the microtubules.

The advances in analysis of filament roles have come from technical progress on several fronts. Immunohistochemical approaches have been exploited impressively; antibodies labeled with visible tags have made possible detailed analysis of the intracellular geometry of what often are many kinds of proteins arrayed in several coexisting filament systems in the same cell. With fluorescent antibodies (antibodies with fluorescent dyes linked to them; Section 1.2.3) the light microscope can be used to view the entire cell and thus to observe patterns that would be very difficult to discern in the thin sections used for electron microscopy (see Figs. II–86, II–92, II–93, and IV–29). A related approach has recently been extended for studies of the dynamics of the filament systems: Labeled filament proteins are injected into living cells, and their distribution is followed by microscopy as they mingle with the cells' own proteins and function, apparently normally, along with them.

"Permeabilized Models"; Cytoplasmic Extracts "Simplified" model systems of several sorts are providing exciting information. Cells can be treated gently with membrane-disrupting detergents or glycerol, which renders them permeable and extracts small molecules but leaves major portions of the cytoskeletal and motility apparatus intact and functional. The absence of the permeability barrier facilitates experimental manipulation of the cell contents. At the other extreme, cytoplasmic systems are being reconstructed by mixing purified and partially purified components. With some mixtures, for example, it is possible to mimic the transitions from soluble, easily flowing cytoplasm to stiffer—more viscous or gelated—cytoplasm that are observed in cells. Through these experiments insight is gained into the factors that control such "sol–gel" transitions.

2.11.1 Filaments and Motion: Striated Muscle

The most thoroughly studied filaments are those of striated muscle (Figs. II–89 and II–90). The striated muscle fiber contains within it numerous elongate myofibrils (= muscle fibrils), which are made of regularly arranged fine filaments and which form the basis of muscle contraction. Each myofibril may be regarded as being composed of repeating contractile units or *sarcomeres* (Fig. II–89). Within the sarcomere there is a regular pattern of thick (11 to 14 nm) and thin (5 to 7 nm) filaments, as shown in Figure II–90. This alignment of filaments gives these muscle fibers their cross-banded (striated) appearance. The zones of thin filaments (I bands; Fig. II–89) can readily be distinguished from those of thick filaments (A bands) by light microscopy. (The individual filaments, however, are below the limit of resolution of the light microscope.) In three dimensions, the "Z-lines," which are taken as the end-boundaries of the sarcomere, are thin discs.

The polarizing microscope (Section 1.2.2) shows that a regular organization of longitudinal elements is present in living muscle as well as in fixed preparations. The interference and phase-contrast microscopes can be used for measuring the amount of material present in different regions of the muscle cell. The comparison of muscle fibers in their normal state with fibers that have been treated with solutions that extract the muscle protein *myosin* shows that only the A bands lose mass. This study and immunohistochemical ones demonstrate that myosin is found in the thick filaments and that another muscle protein, *actin,* is present in the thin filaments.

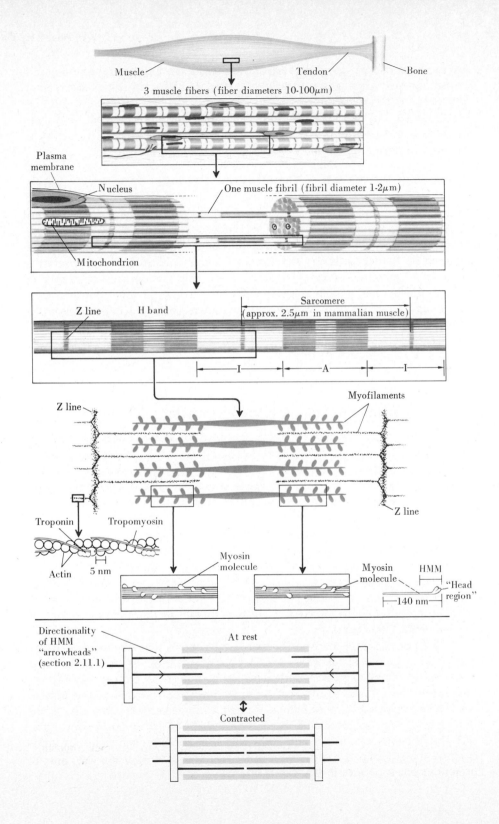

Muscle   Tendon   Bone

3 muscle fibers (fiber diameters 10-100μm)

Plasma
membrane

Nucleus   One muscle fibril (fibril diameter 1-2μm)

Mitochondrion

Z line   H band   Sarcomere
(approx. 2.5μm in mammalian muscle)

I   A   I

Z line   Myofilaments

Z line

Troponin   Tropomyosin

Actin   5 nm   Myosin
molecule   Myosin
molecule   HMM   "Head
region"

140 nm

Directionality
of HMM
"arrowheads"
(section 2.11.1)   At rest

Contracted

◀ **Figure II–89** *The structure of vertebrate striated muscle* (see also Fig. III–47). The unit comparable to a more conventional cell is a *fiber*. It is surrounded by a single plasma membrane but contains several to many nuclei. Within the fiber, *myofibrils* are present. These consist of *filaments* that are ordered in such fashion as to produce the appearance of alternating A and I bands along the fibrils. Each myofilament is of several hundred protein molecules. The diagram illustrates the probable arrangement of *myosin* molecules in the thick filaments and of *actin* molecules in the thin filaments. *HMM* indicates the approximate position of the "heavy meromyosin" fragment that can be prepared from myosin and used to evaluate polarity of the thin filaments (Section 2.11.1). Also shown are the locations of the elongate (40 nm) *tropomyosin* molecules that run along the "groove" between the two actin chains of the thin filament, and of troponin molecules associated with the tropomyosins. Each tropomyosin extends the length of seven actins and has one *troponin* unit (made of three subunits) bound to it. Many mitochondria are present throughout the fiber; they are found between adjacent myofibrils.

The region between two successive Z-lines (actually, *Z discs*) along a myofibril is referred to as a *sarcomere*. The sarcomere can be thought of as a repeating unit of function. The sliding of filaments within a sarcomere during contraction is illustrated at the bottom of the diagram. (Based on diagrams by A. Loewy, P. Siekevitz, S.L. Wolfe and others; after the work of H. Huxley, S. Ebashi, and many others.)

Sliding Filaments Figure II–89 diagrams the generally accepted theory about the arrangement and functioning of protein molecules in the filaments. According to this theory, muscle contraction takes place when the two sets of thin filaments of each sarcomere slide toward the center, past the central set of thick filaments. The sliding of filaments shortens the sarcomere. If the muscle is stretched and held at fixed length, the extent of the overlap of thick and thin

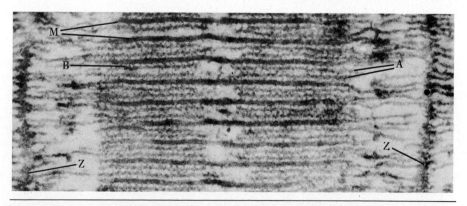

Figure II–90 *A sarcomere* from a rabbit muscle. The thin, actin, filaments *(A)* extend inward from the Z-line; in the center they overlap with the thick, myosin, filaments *(M)*. Indications of fine bridges between thin and thick filaments are seen in the zone of overlap *(B)*. × 100,000. (Courtesy of H. Huxley.)

filaments is reduced, and the force the muscle can generate is correspondingly reduced. This indicates that the interaction of the two sets of filaments is responsible for the force of contraction.

Each sarcomere in a muscle like that diagrammed in Figure II–89 shortens by less than a micrometer, but the summation of such changes in the thousands of sacromeres comprising a fiber yields muscle contractions of millimeters or centimeters. Thus individual filaments or sets of filaments need slide only short distances to contribute to large-scale effects.

Bridges; "Decoration" and Polarity of Actin Filaments The sliding of muscle filaments is ascribed to the action of bridges between thick and thin filaments (Fig. II–90), which pull filaments past one another. As Figure II–89 indicates, the myosin molecules in one half of the thick filament are oriented with polarity opposite to the molecules in the other half. In their interactions with myosin, the thin filaments also reveal a polarity that permits the thin filaments associated with the two ends of a thick filament to be "pulled" in opposite directions (both sets of thin filaments move "in," toward the center of the sarcomere). This polarity can be demonstrated by exploiting the fact that the bridges between thick and thin filaments correspond to portions of myosin molecules (the "head regions"; Fig. II–89); these portions can be cleaved from the molecules by gentle treatment with proteases. One type of actin-binding, ATP-splitting fragment of myosin prepared in this way is referred to as HMM— "heavy meromyosin." S_1, a somewhat shorter myosin "head" preparation, including less of the "tail" of the original myosin than that seen in heavy meromyosin, can be prepared by more extensive proteolysis. The "arrowheads" produced by "decoration" of striated muscle thin filaments with heavy meromyosin (as in Fig. II–91) all point toward the center of the sarcomere owing to the oppositely polarized organization of the two sets of thin filaments (Fig. II–89). S_1 gives a similar pattern.

Precisely how the bridges convert the chemical energy of ATP into the mechanical energy of contraction is not known. Virtually all hypotheses start with the presumption that in each cycle of bridge formation and release, ATP binds to myosin and is split. This, it is believed, generates conformational changes in the myosin that permit the myosin to bind to an actin in the thin filament and then allow the bridge portion (the "head") to swivel (or to contract), moving the thin filament roughly 10 nm along. ADP and phosphate now dissociate from the myosin, a new ATP replaces them, and the bridge releases and returns to its original conformation so that the myosin is now ready to undergo the cycle again. The cycles can repeat dozens of times a second, and many repeats obviously are required to account for overall sarcomere shortening, which may be to two thirds or less of the resting length. The *binding* of ATP to myosin apparently affects myosin's ability to bind to actin, whereas the *splitting* of ATP energizes conformational changes that generate movement.

Ca^{2+}; Troponin and Tropomyosin; α-Actinin From immunohistochemical staining and other procedures, two proteins in addition to actin have been

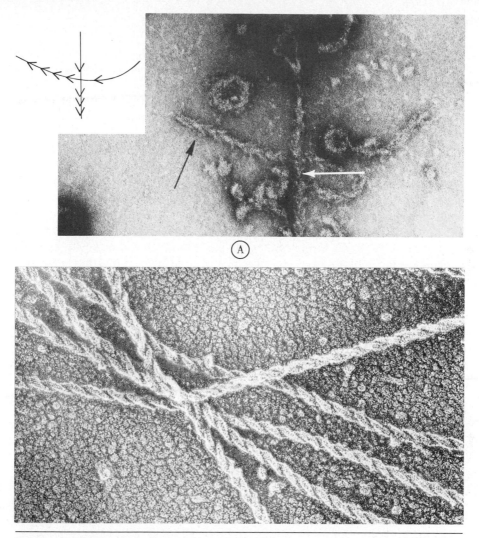

Figure II–91 *Decorated actin filaments.* A. Negatively stained preparation (Fig. II–53) showing actin microfilaments (arrows point to two) isolated from cultured hamster cells ("BHK-21 cells") and exposed to a solution containing "heavy meromyosin" (HMM) prepared from muscle myosin (Section 2.11.1). Numerous HMMs (red in the sketch) have bound to each filament, producing an "arrowhead" or "herringbone" pattern along the length of the filament; this is seen most clearly at the left-hand arrow. Note that the arrowheads all point in the same direction on a given microfilament, indicating a polarity in arrangement of the actin molecules. This polarity determines the direction in which force is exerted when actin and myosin filaments interact (Fig. II–89). × 112,000. (From R. Goldman, *Exp. Cell Res.* **90**:333, 1975.)

 B. Several actin filaments from vertebrate muscle prepared for microscopy by a freeze-drying method like the one described in the legend to Figure II–12B. S_1 fragments of myosin (Section 2.11.1) had been bound to the filaments before freezing and their presence helps make visible the double helical construction of the filaments (see Fig. II–89). From Heuser, J., *Trends Biochem Sci.* **6**:64, 1981.)

identified in the thin filaments, *tropomyosin* and *troponin* (Fig. II–89). These proteins participate in the control of muscle contraction by Ca^{2+} ions. The triggering of muscle contraction by a nerve impulse is mediated by the specialized endoplasmic reticulum *(sarcoplasmic reticulum)* of the muscle fiber. This membrane system will be considered in detail when we discuss the organization of different types of muscle fibers and cells (Chap. 3.9). In brief, when nerve impulses or other stimuli trigger changes in the electrical potential across the muscle fiber's plasma membrane, the sarcoplasmic reticulum responds by releasing Ca^{2+} to produce contraction. Subsequently the reticulum reaccumulates Ca^{2+}, and the muscle relaxes. It is thought that when Ca^{2+} levels are low (below 10^{-7} or 10^{-8} M), the troponin and tropomyosin molecules are so disposed on the thin filament (Fig. II–89) that they block interaction of actin and myosin. When Ca^{2+} concentrations rise, the Ca^{2+} interacts with troponin to change the conformation and positions of tropomyosin and troponin, permitting myosin and actin to form bridges, split ATP, and so forth. A slight shift in the position of the tropomyosin appears to be critical in exposing the sites needed for actin and myosin to interact.

Additional proteins associated with the fibrils apparently serve as mechanical linking and integrating agents. One of these, α-*actinin,* is concentrated at the Z-line, where it contributes to cross-linking and anchoring the actin filaments. Others will be discussed later (Section 2.11.3).

2.11.2 Microfilaments; Intermediate Filaments

As electron microscopic techniques improved it became apparent that cytoplasmic filaments are present in many cell types other than striated muscle. Sometimes they form distinctive and well-organized bundles, as is true in the microvilli of the cells lining the intestine (see Fig. III–28). In epithelial cells (Section 3.5.1) such as those of the skin, bundles of "tonofilaments" terminate near and probably anchor on *desmosomes,* junctional structures by which the cells adhere to one another (Section 3.5.2 and Fig. III–29). Many cells grown in tissue culture show prominent filament bundles, traditionally called *stress fibers,* running parallel to and near the surface of the cell attached to the culture vessel or slide on which the cell is growing (Fig. II–92).

Commonly, however, the filaments are less tightly ordered. They can be distributed through the cytoplasm as scattered filaments or loose bundles or meshes, and form meshworks or networks just below the plasma membrane. In this latter location they are responsible for the gel-like subsurface "cortex" detected in many cells when the consistency of different regions of the cytoplasm is tested by, for example, prodding with a glass microneedle or observing the movement and distribution of cytoplasmic particles. Most cellular organelles are excluded from this region.

Microfilaments; Nonmuscle Actin, Myosin, and Tropomyosin The filaments of different cells and cell regions range in thickness from about 5 nm

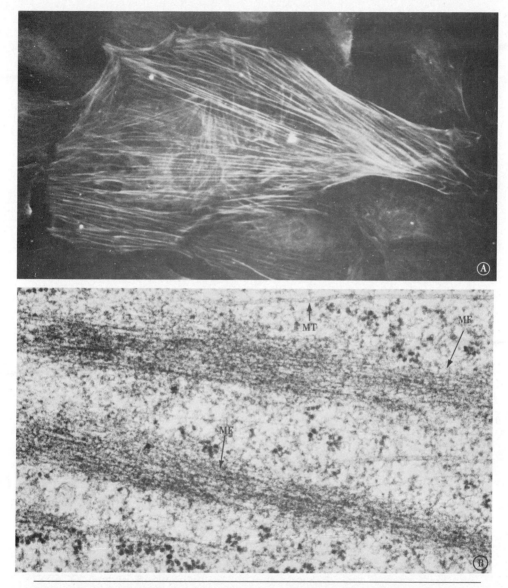

Figure II–92 *Microfilaments* in cultured mouse fibroblasts ("3T3 cells") A. This cell was photographed in the fluorescence microscope following immunohisto-chemical "staining" with fluorescent antibodies against muscle actin (Sections 1.2.3 and Fig. II–86). This image reveals the presence of actin in numerous elongate cytoplasmic fibers. B. Electron micrograph of a 3T3 cell showing bundles of microfilaments *(MF)*. These bundles correspond to the "fibers" in part A. *MT* indicates a microtubule. × 64,000. (Part A Courtesy of E. Lazarides and K. Weber. Part B, from Goldman, R., *Exp. Cell Res.* **90:**333, 1975.)

to more than 10 nm. This variation suggested early on that they are of more than one category. An initial key generalization emerged from two types of findings: First, immunohistochemical staining and incubation of fixed, glycerol-extracted cells with heavy meromyosin demonstrated that the most abundant filament type—the 5 to 6-nm category—stain as if the filaments were of actin. They bind antibodies specific for actin, and they form arrowheads with heavy meromyosin (Figs. II–91 and II–92). Second, proteins of the size and amino acid composition characteristic of actin were identified by biochemical procedures in nonmuscle cells and shown to comprise as much as 5 to 10% or more of the total protein content of some cells. As this work was extended it was found that the actins of nonmuscle cells differ slightly in amino acid sequence and perhaps in other minor ways from muscle actin. Evidently they are coded for by different genes (Section 4.2.6). But insofar as is known, they are functionally very similar to, if not identical to, the muscle protein. The term *micro-filament,* originally used for many types of cytoplasmic filament, has increasingly come to be used solely for the 5- to 6-nm actin filaments.

Microfilaments have been identified in a great variety of cell types. Somewhat surprising at first encounter was the discovery that they are present in plants (despite the manifest absence of muscles) where they show at least the major features characteristic of animal microfilaments: They bind heavy meromyosin and probably are made of a protein similar to animal cell actin.

The filaments in intestinal microvilli and stress fibers described above are chiefly of actin; that is, they are microfilaments, as are many of those in the cytoplasmic networks.

Proteins resembling myosin also are present in many cell types, although often they differ more from muscle proteins in size and other characteristics than does nonmuscle actin. In most cell types, far less myosin than actin is present, and distinct thick filaments are seen only rarely. This may mean that the myosin is in arrays or small aggregates distinctly different from muscle's thick filaments, or in some cases, that nonmuscle thick filaments are present in small numbers and are difficult to preserve and thus cannot readily be recognized. In either case, the relative scarcity of myosin and the absence of well-ordered sets of thick filaments provided an early warning against assuming too rapidly that the "actomyosin" system of nonmuscle cells functions just like the actomyosin system of striated muscles.

Both myosin and tropomyosin can be demonstrated in stress fibers through the use of fluorescent antibodies and also are found in the regions of some of the cytoplasmic actin networks in a variety of nonmuscle cells. Neither protein, however, is ubiquitously associated with actin: Both are absent from such prominent actin bundles as those of the intestinal microvilli. Troponin is not found in many of the nonmuscle cell types whose microfilaments are nonetheless affected by calcium; general opinion is that where troponin is absent, the effects of Ca^{2+} are mediated through the intervention of calmodulin. Calmodulin and troponin bear some structural similarities to one another and conceivably are evolutionarily related (as in Section 4.5.2).

Stress fibers have been studied chiefly in cultured cells. Their relevance to cells in their normal locales in tissues (in situ) is not certain, but their prominence and accessibility have made them favored objects for study of microfilament behavior and organization.

As seen with fluorescent antibodies the distribution of myosin in stress fibers is discontinuous. In contrast to antiactin antibodies, which stain the entire length of the filament bundle, antimyosin antibodies produce a series of bright spots spaced fairly regularly along the bundle. A few investigators interpret this as reflecting an organization of myosin somewhat akin to that in striated muscle, at least in the sense that discrete regions of organized myosin might pull on the microfilaments. This could contract the bundles or exert force transmitted to the bundle ends; these effects might move the plasma membrane zones to which the fibers attach. Stress fibers can in fact be made to contract in extracted and permeabilized cell preparations, but their role in the cell is not known. Proponents of a prominent role of stress fibers in motility find it disconcerting, however, that the regions of most active movement in cells possessing the fibers are the lamellipodia (Section 2.11.4) and other areas where the fibers are not present; these areas show instead a much less well-defined organization of their filaments and contractile proteins. In fact, in some tissue cultures, the cells with many stress fibers seem relatively nonmotile compared to those with fewer, suggesting the major role of the fibers may be in anchoring and orienting (and perhaps in producing limited, specific types of movement).

Intermediate Filaments   Filaments known as intermediate filaments, with thickness of 7 to 11 nm (10 is taken as typical), have recently been recognized as widespread. Despite similarities in dimensions, these are heterogeneous in composition and in other properties. A given cell may have more than one type of intermediate filament, though frequently a single type predominates. Often the filaments form loose bundles that course through the cytoplasm, sometimes as elaborate networks (Fig. II–93).

In many cell types, prominent intermediate filaments are made of proteins known as vimentin and desmin (either one or a mixture of the two may be present, depending on the cell type). The tonofilaments of the epithelial cells mentioned above are made of keratins, proteins related to the keratins of hair, feathers, and the skin of many vertebrates. (The tough, waterproof covering of the skin is provided by epidermal cells that fill with keratin filaments and die as they migrate to the body surface from the lower levels of the skin where the cells are generated.) In nerve cells, neurofilaments made of three distinguishable kinds of proteins are present throughout much of the cytoplasm and run longitudinally down the axon, along with microtubules. Certain of the supporting cells of the nervous system (glial cells; Section 3.7.4) possess filaments made of still another type of protein. These various intermediate filament proteins differ from one another in size and composition but do show some similarities in requirements for assembly into filaments and other properties. At present there is no simple way to make sense of the diversity, al-

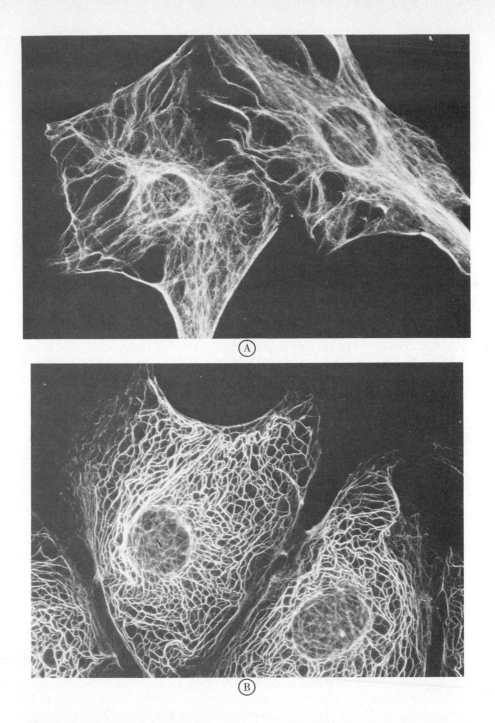

though to some extent cells with common embryological ancestry tend to use similar proteins for their intermediate filaments. Proposals now undergoing testing include the idea that despite their differences in overall amino acid composition, different intermediate filament proteins have regions where they are

◀ **Figure II–93** *Intermediate filament systems.* A. Cultured cell that originated as rat smooth muscle, stained with "immunofluorescence" procedures (Section 1.2.3 and Fig. II–86) to demonstrate the intermediate filament protein *vimentin*. N indicates the location of the nucleus. Surrounding the nucleus are brightly stained bundles of intermediate filaments. × 770. B. Cultured epithelial cell (Chapter 3.5) of rat kangaroo (PtK₂ cells) showing the immunofluorescence pattern obtained with antibodies to the intermediate filament protein *prekeratin*. A network of filament bundles spreads throughout the cytoplasm. N indicates the location of the nucleus. × 860. (Courtesy of W.W. Franke, E. Schmid, M. Osborn, and K. Weber.)

functionally and structurally similar. This could account for the observation that proteins from different types of intermediate filaments seem capable of collaborative assembly in the test tube, combining to form filaments. It seems likely to many investigators that in cells as well, there may be some filaments constructed of more than one type of intermediate filament protein.

Assembly and Disassembly; Actin-binding Proteins If concentrations of ions such as Mg^{2+} and K^+ are appropriately adjusted, actin filaments—the "F" form of actin, *F-actin*—can form in the test tube by polymerization of the more-or-less globular (actually, bean-shaped) actin molecules—*G-actin*. ATP bound to G-actin is split as the polymerization occurs. Actin filaments in turn can be linked into three-dimensional networks or associated into bundles. Various proteins promote such bundling or cross-linking or the polymerization, fragmentation, depolymerization, and perhaps the end-to-end merging of actin filaments; some of these proteins confer sensitivity to Ca^{2+} on the interlinking and bundling of the filaments or on their assembly and breakdown. Most of what is known directly about these assembly and cross-linking processes comes from work on test-tube mixtures of proteins (see also Section 4.1.2). From this work well over 25 proteins that control the gelation or solation of actin preparations, by influencing the assembly, length, and association of actin filaments, have so far been identified; these proteins are referred to collectively as *actin-binding proteins*. Cells may have several types, and different cells may differ somewhat in the types present. (The field is in a state of rapid growth; even the names used below for specific proteins are subject to change.) Precisely how the proteins work in the cell is still to be determined, but their existence undoubtedly explains the cell's ability to change the state of its actin rapidly and in controlled manner. Some of the proteins, like "profilin," which is widely present, bind to G-actin and inhibit polymerization; in the cell, presumably, this one or similar proteins maintain part of the actin supply in depolymerized form, ready for rapid deployment in response to ionic changes or other signals (Section 2.11.4). "Gelsolin," present in macrophages, loosens actin filament networks when Ca^{2+} is present, perhaps by severing filaments or by preventing their lengthening; such effects may permit the cell to adjust the fluidity and rigidity of local regions of its cytoplasm during the cell's ameboid motion (Section 2.11.4) and phagocytic activities.

Assembly of F-actin is hastened by the presence of preexisting filaments, suggesting that as with microtubules, initiation of growth by preexisting "seed" or nucleating sites may be one of the ways in which filament distribution can be controlled. Certain of the actin-binding proteins promote nucleation in the test tube and thus are good candidates for cellular controlling agents. There also are actin-binding proteins that seem to "cap" the filaments, binding to one end or the other, and preventing addition of G-actin there; this could help to regulate microfilament length and could influence the direction and polarity of growth. There is evidence that, under test-tube conditions at least, actin filaments have a preferred end for growth, although they can grow—more slowly—from the other end as well. The preferred end appears to be the one opposite the end toward which experimentally attached heavy meromyosin "arrowheads" point. Consequently, it is called the "barbed" end (in contrast to the "pointed" end), though of course normally no "arrowheads" or "barbs" are present on filaments in the cell. Somewhat perplexing are recent findings that growth from the barbed end seemingly can occur even when this end is associated closely with a membrane. Evidently such association need not "cap" the filament, in the sense of preventing any addition of new G-actins.

Agents such as *cytochalasins B and D* and *phalloidin* interfere with filament-related functions. Like colchicine for microtubules, they often are used as "diagnostic" tools—if a process is affected by them, this is taken as initial evidence that microfilaments may be involved. Circumspection is required in drawing such a conclusion since the reagents have direct effects on membranes and other systems—cytochalasin B, for example, inhibits sugar entry into the cell. Also, the agents are not always predictable in their impact on filaments. Probably the effects of cytochalasins and phalloidin on the filaments result in large part from their influencing actin assembly and interlinking. In the test tube, the cytochalasins bind to the barbed end of actin filaments, thus impeding further growth. They also inhibit gelation of actin-containing mixtures by preventing formation of networks. Phalloidin, when injected into cells, appears to make microfilament arrays excessively stable.

The intermediate filaments normally seem less "labile"—less prone to very rapid change—than are the microfilament systems, although their distribution in the cell does alter under varying circumstances, such as during the extensive rearrangements of the "cytoskeleton" occurring in cell division (Sections 4.2.7 and 4.2.8). The proposal has been made that the dismantling of intermediate filaments in nerve axons and elsewhere is accomplished by proteolytic enzymes activated by calcium.

Linkage to Other Structures; Vinculin; Spectrin Linkage of filaments to other structures is through additional specific proteins. One of the proposed roles for certain microtubule-associated proteins is in interactions with cytoplasmic filaments. Immunohistochemical studies have demonstrated α-actinin—like that present at the muscle Z-line—at some of the places where actin microfilaments are very closely associated with or linked to membranes. This protein may cross-link microfilaments near sites of membrane attachment, as it does at other sites. For a number of cell types a different protein called *vinculin* is detectable immunohistochemically at sites of microfilament–membrane association. Perhaps more than one protein is required for effective binding of actin to membranes, with different ones recruited for different roles, sites, or circumstances. In mammalian red blood cells, actin is linked to the protein *spectrin,* forming a cytoskeletal protein network that lies just below the cell surface. Via *ankyrin* and other proteins, the actin–spectrin network is bound to some of the integral proteins of the plasma membrane. This, at least, is a currently favored hypothesis developed in part to explain how the "biconcave disc" shape (see Fig. II–2) of the red blood cell is maintained. (Microtubule bands like the ones described in Section 2.10.1 are not present in mammalian red blood cells.) Spectrin-like proteins are now being detected in cell types additional to the red blood cells.

2.11.3 Cytoplasmic Architecture

In looking at pictures like Figures II–91 and II–93 it is important to keep in mind that a typical cell possesses a system of microfilaments, a system of microtubules, and a system of intermediate filaments, not just one of the three. Sometimes one system is preeminent, as in striated muscles, but quite often the three are extensively intermingled. For example, had the cells in Figures II–92 and II–93 been stained to locate tubulins instead of actin, vimentin, or keratin, they would have shown numerous microtubules radiating from the zone of the centrosome near the nucleus (Fig. II–86) and extending toward the cell borders and around the nucleus. When cells are treated with agents that depolymerize the tubules, the intermediate filament networks sometimes rearrange into large coils or bundles, suggesting that the tubules and filaments mutually influence one another's distribution.

These findings, in combination with what was said about microtubules in the last chapter and images such as Figures II–93 and III–28, make it easy to understand why investigators generally assume that along with the tubules, the filament systems serve as architectural or supportive elements. These systems are plausibly thought to help to maintain cell shape, to determine the position of the nucleus in relation to the rest of the cell, and to provide an anchoring and organizing framework for systems that produce intracytoplasmic movement. The cells shown in Figures II–86, II–92, and II–93 are tissue-culture cells, relatively thin and flat and therefore easy to study with the techniques used. The overall distribution of filaments and tubules is more difficult to study in other types of cells, but there is fragmentary evidence suggesting comparable interweaving of different filament and microtubular systems. In striated muscle cells, for example, desmin and vimentin intermediate filaments appear to run among the myofibrils, linking to them at the Z-lines. These filaments could help to integrate the muscle fiber, transmitting forces generated by the sliding of the actin and myosin filaments and keeping the sarcomeres of adjacent myofibrils aligned with respect to each other.

In the multicellular arrays constituting tissues, a cell's geometry is, of course, strongly affected by its neighbors' shapes and properties, and by extracellular materials (Chap. 3.6B). In addition, there are intracellular controls capable of influencing the cell's shape and cytoplasmic organization. As mentioned in the preceding chapter, microtubule-organizing centers enable cells recovering from conditions promoting microtubule disassembly to form new tubules at or near the old sites. The possibility is now being pursued that analogous "centers" affect assembly or distribution of other cytoskeletal elements such as intermediate filaments and microfilaments. The observations on microtubule reappearance are paralleled by other findings indicating that cells can "remember" both the broad outline and many details of their architecture and can reconstruct them after disruption. Sister cells in tissue culture, crawling away from one another after the cell division that produced them, may follow routes that are strongly correlated (for example, they may move along paths that are mirror images). They may also be of similar shape, even though during the division proper the cells round up (Section 4.2.7) and reorganize their microtubule and filament arrays. Since patterns of cellular movement as well as cell shape are influenced heavily by the filaments and microtubules (Section 2.11.4), these findings are evidence that some kind of information about overall architectural arrangement is passed on through division. For dividing protozoa (p. 354) some of this depends on duplication of precisely ordered arrays of structures such as basal bodies. Much less is known about other cells, although centrioles are among the organelles under investigation in these connections.

The Cytoplasmic "Matrix"; the "Microtrabecular Network"   We now turn to the discussion of the structure of the "structureless" background cytoplasm—the *hyaloplasm, cytosol, or ground substance*—deferred from earlier in the book (Chap. 1.2B). It is abundantly evident that the regions that showed little structure in the light microscope, or with early electron microscopic techniques, are in fact full of filaments and microtubules. These are intermeshed in various patterns and very likely can form transitory or more long-lasting linkages to one another. Suspicion is also growing that many of the enzyme molecules once thought of as freely soluble, individually functioning proteins may actually be capable of associating with one another and with cytoplasmic structures. Such associations, though easily disrupted and perhaps only temporary, could influence the conformations, functioning, and regulation of the enzymes; for instance, concentrations of ions or other components important for the enzymes' activities can be quite different in the immediate vicinity of a membrane or other structure than they are in the solution bathing that structure. Biophysical studies suggest that even much of the water in the cytoplasm may not exist as a freely mobile phase but may instead be organized owing to its binding to structures and surfaces (see Sections 1.4.1, 3.4.4, and 4.2.8).

A dramatic proposal currently undergoing scrutiny suggests that most cytoplasmic components, including the membrane-delimited organelles (perhaps excluding mitochondria), free polysomes, and plasma membrane, are linked to one another and especially to the filaments and microtubules. The linking is by short strands—"trabeculae"—of filamentous material, as thin as 2 to 4 nm in diameter, but of still uncertain nature. The overall effect is the integration of the entire cytoplasm into a loosely knit *microtrabecular network* capable of rapid change and rapid attachment and detachment of the structures linked in it. The spaces within the meshes contain soluble materials; it is within these spaces that soluble molecules and inorganic ions diffuse. Many of the larger molecules and structures traditionally thought of as free in the "cytosol" are actually linked loosely to the meshwork. The "structuring" of water in the cell could be by loose association with this meshwork as well as with surfaces of organelles. The cytoplasmic network communicates with the proposed nuclear matrix via the pore complexes and lamina (Section 2.2.6).

Needless to say, such a proposal has not met universal acceptance. It is supported chiefly by electron micrographs of the sort shown in Figure II–94. In addition, when cells are treated carefully with detergents that dissolve away their membranes' lipids, a cytoskeletal "ghost" may be left that retains much of the overall architecture of the cell. Polysomes remain attached to this structure, as do some plasma membrane proteins that apparently were linked to cytoplasmic elements adjacent to the membrane and remained linked when the membrane's lipids were removed. There are those who argue that both the microscopic preparations and the cytoskeletal ghosts exaggerate the degree of interlinking of cellular components due to changes induced by the preparative methods used ("artifacts"; Section 1.2.2); these investigators dispute the concept of a special network additional to the well-defined microtubules and filaments discussed in previous sections. The proposal does, however, emphatically codify the general agreement that in structural terms, as well as in the functional terms considered in Section 1.3.4, the cytoplasm cannot be thought of simply as a collection of organelles each separately "doing its own thing."

2.11.4 Motility

Filament arrays are assembled and assume suggestive orientations in all sorts of motility phenomena, ranging from the extension of pseudopods around material to be phagocytized to the bulk streaming of cytoplasm seen in some cells and the crawling of cells along surfaces. The hunt is now on for the mechanisms by which the filaments and their constituent proteins act. How far the field has yet to go is indicated by the fact that even for some types of muscle,

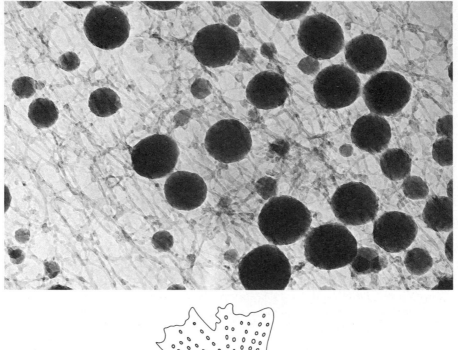

Figure II–94 *The "microtrabecular" network.* The red pigment granules of squirrel fish erythrophores as seen by high-voltage electron microscopy (Section 1.2.1). The granules move along linear paths as they disperse through the cytoplasm (see text); these paths are delineated by microtubules. When relatively thick sections of the cells (up to a micrometer or more) are examined as in the present micrograph the granules are seen as if embedded in a latticework of fine filamentous material, the presumed "microtrabecular network." (Courtesy of Porter, K.R., J.B. Tucker, and K.J. Luby-Phelps, *Scientific American* **244**(3):56, 1981.) Sketch: erythrophore seen from above showing nucleus and granules.

notably the nonstriated "smooth" muscle cells (Section 3.9.2), unambiguous information about filament organization and contractile processes is lacking.

Assembly and Disassembly Newly formed phagocytic vacuoles in polymorphonuclear leukocytes are surrounded by a meshwork of filaments, presumed to participate in the formation of the vacuole. Subsequently, for lysosomes to fuse with the vacuole membrane, at least part of the filament system must be dismantled or moved aside so that the membranes can approach one another. Very likely, in many other circumstances as well, local assembly or disassembly of filament networks can open or close pathways for movement and interaction of other structures.

Cells change their shape as they crawl along surfaces (Fig. II–95) or pass through narrow spaces as blood-borne phagocytes must do to squeeze through the blood vessel wall and enter tissue spaces. This requires control of deformability of cell regions. Movements of portions of cells with respect to one another also are influenced by balances of fluidity and rigidity. Pushes or pulls exerted on one part of a relatively stiff structure can be transmitted through the entire structure or even move it, whereas the same forces may be dissipated in a more deformable structure. These and other cellular motility phenomena are affected by the reversible interlinking of actin and associated proteins into gel-like networks or cable-like bundles, which have major impact on the mechanical properties of the cytoplasm.

Sperm cells preparing to fuse with eggs extend membrane-delimited processes from their surfaces that can be many tens of micrometers long and can grow to these lengths in seconds (Section 3.10.3). For the sperm of some organisms this depends upon the assembly of a bundle of microfilaments that, by its growth, pushes the sperm membrane out. The sudden filament assembly seems to result from changes in ion concentration (Section 3.10.3) that dissociate profilin-like proteins from actin, making the actin available for filament formation. The filaments form an elongate oriented bundle, probably because their growth is initiated at a yet-to-be-characterized mass of electron-dense material found at the base of the growing bundle.

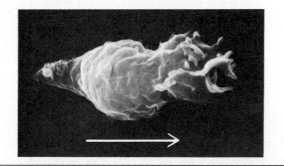

Figure II–95 *Cell "crawling."* Human polymorphonuclear leukocyte (see Fig. II–69) fixed as it was moving in the direction indicated by the arrow, then shadowed (Fig. I–11) with a mixture of gold and palladium and viewed in a scanning electron microscope (Section 1.2.1). At its front end the locomoting cell extends thin flattened "lamellipodia" that make up a "ruffled membrane." At the cell's rear, a tapering "uropod" (tail) is seen. Approximately × 6,000. The total cell length is approximately 10 μm. (From Zigmund, S., *J. Cell Biol.* **89**:588, 1981. Copyright Rockefeller University Press.) The view is from above.

Movement of Membrane Molecules and of Membrane-bounded Organelles The mobility of plasma membrane proteins in the plane of the membrane can be influenced by the actin and myosin systems in the cytoplasm nearby and probably by microtubules. In "capping," for example, the movement of receptor molecules to one cell pole is accompanied by accumulation of actin and myosin at that pole (see Fig. II–8). From such observations it has been concluded on the one hand that some plasma membrane molecules are kept in place by anchoring to cytoplasmic systems, and on the other that oriented movements in the plane of the membrane can be produced by the proteins, filaments, and tubules we have been discussing. At the same time, linkage to plasma membrane molecules may help to anchor the cytoplasmic systems, particularly when the plasma membrane is held firmly to another cell (Section 3.5.2) or to extracellular materials (Chap. 3.6B). Association of plasma membrane receptor molecules with cytoskeletal and motility-producing systems in the underlying cytoplasm could also be the basis of mechanisms through which some receptors affect intracellular events.

The inpocketing of the plasma membrane and pinching off of vesicles during endocytosis, as well as the highly oriented movement of pigment granules in melanophores (Section 2.10.4 and Fig. II–94), are examples of common varieties of intracellular movement that may also depend on linking of membranes to cytoplasmic filament and tubule systems. One line of current investigation is focused on the possibility that microfilaments pull on endocytic vesicles (cooperating, perhaps with the clathrin coating present on some of the vesicles [Section 2.1.6], which might help in the pinching-off of the vesicles from larger surfaces). The pinching-in (furrowing) of the cell surface during cell division is an important case in which an organized filament system of actin and myosin seems to pull on the plasma membrane (Section 4.2.9).

For the movement of pigment granules and other membrane-delimited structures in the cell, a thought-provoking hypothesis is that small amounts of actin or myosin or of other proteins that can interact with actin or myosin are attached to the membranes and produce movement by interacting with filaments: For example, the end of a motile filament might be anchored to the membrane; or a few membrane-associated molecules of myosin or some other protein that can bind to actin might suffice for repeated bridge formation with microfilaments, so that a membrane-delimited structure "crawls" along the filament in ways similar to those suggested for microtubules in Section 2.10.4. Some consider these ideas too fanciful or too vague. An alternative is that the proposed "microtrabecular network," if in fact it is linked to membrane-bounded organelles, could move the organelles. For the pigment granules it has been suggested that movement generated by the network is oriented along the tracts of microtubules, which determine the direction of movement, by virtue of the simultaneous association of the microtrabecular system with both the tubules and the granules (Fig. II–94).

Because of the presence of such a well-defined core of microfilaments, there is much interest in whether the microvilli of intestine show movements that might be controlled by the filaments. Initially it was thought that, under the influence of Ca^{2+} and ATP, the tips of the microvilli are pulled back toward the rest of the cell through interaction of their microfilaments with myosin molecules. The microvilli themselves lack myosins, but myosin is present in the filament system at the base of the microvilli (the "terminal web"; Fig. III–28). At the tips of the microvilli and perhaps along their length, the microfilaments are attached to the plasma membrane. The shortening of microvilli was frequently cited as a dramatic case in which interactions of actin with limited amounts of myosin in one region of the cell (the microvillar base) can exert force on structures located at some distance from this region (the microvillar tip). Doubt has recently been cast on this concept; it is now felt that the shortening induced experimentally by ATP and Ca^{2+} may have involved lateral contraction of the terminal web and disassembly of the microfilaments rather than direct pulling on the microvillar core through actin–myosin "sliding." In fact, a recent hypothesis suggests that changes in length of the microfilaments by assembly and disassembly is the way in which the cell normally adjusts the length of its microvilli.

Cytoplasmic Streaming   With more complex movements we enter realms of even greater uncertainty. In plant cells, the cytoplasm shows oriented streaming, referred to as "rotational streaming" or "cyclosis" (Fig. II–96). This has been studied most thoroughly in large algal cells such as those of *Nitella*. Such cells show a stationary gelated outer cortex (Fig. II–96) within which the

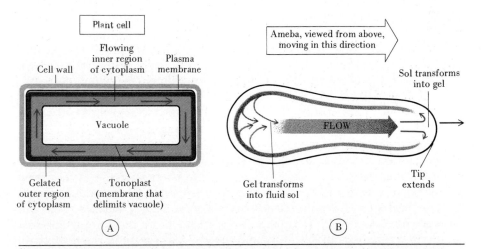

Figure II–96   *Two important types of cytoplasmic motion.* A. Rotational streaming ("cyclosis") in a plant cell—the rapid streaming of cytoplasm around a central vacuole. B. One type of *ameboid motion.* The extension of a pseudopod by an ameba is accompanied by constant flowing of cytoplasm in the direction of extension and probably by continual transformation from gelated *ectoplasm* or *cortex* (outer region of cytoplasm) to fluid *endoplasm* at the posterior end with the reverse transformation at the anterior end. (After conceptions summarized by Kamiya, Allen, and many others.)

"endoplasm" moves around the large central vacuole at rates of 25 to 50 μm per second or more. Bundles of microfilaments are found throughout the endoplasm, many seeming to anchor in the cortex. Most extensively studied have been the filaments located at the cortex–endoplasm interface. When "decorated" with HMM (Section 2.11.1) these show "arrowheads," all of which point "upstream," *opposite* the direction of motion. In striated muscle, the arrowheads point *in* the direction of motion (see Fig. II–89). Myosin is present in the algal cells, but its distribution is not known. An attractive hypothesis suggests that the myosin molecules are linked to mobile endoplasmic structures or networks; bridge formation between these myosins and the microfilaments could move the endoplasmic materials along, with the microfilaments remaining in place. In other words, the proposal is that the microfilaments pull myosin-linked structures past themselves—a reversal of the situation in striated muscle, where the myosin filaments pull on the actin filaments. (The basic mechanisms of actin–myosin interaction might well be the same in both situations, with the differences in motion arising from the ways in which the participating filaments and proteins are organized and anchored.) With permeabilized algal cells (see introduction to Chap. 2.11) addition of ATP does generate movement of cytoplasmic particles along the inner surface of the cortex, as this hypothesis predicts. There may, however, be other aspects to rotational streaming. The filament bundles extending deep into the endoplasm have been seen, in living cells, to exhibit undulatory (wavelike) movements, as well as movement of particles along their lengths. Whereas the hypothesis just outlined suggests that the motive force for movement is exerted chiefly at the interface between the cortex and endoplasm, these undulations could, by still-speculative mechanisms, directly propel cytoplasm deep within the endoplasm. Evaluation of the possibilities could come from further studies of cytoplasmic extracts; in such extracts addition of ATP results in movement of the filament bundles—which no longer are anchored in place, as they may be in the cell— through the medium.

A different type of cytoplasmic streaming, with velocities up to 1000 μm per second, is seen in the slime mold *Physarum*. This organism is organized as a multinucleate mass of cytoplasm (see page 307), rather than as separate uninucleate cells, and the cytoplasm surges rhythmically, reversing direction every few minutes. Such "shuttle" streaming has been studied in detail, using thin elongate strands of *Physarum* for experimental and observational convenience. Once again the movement is of a fluid endoplasm within a more gel-like cortical "ectoplasm." The motive force seems to be pressure, exerted by contraction of the ectoplasm. Actin and myosin are present in the ectoplasm. A widely adopted working model suggests that the contraction involves the aggregation of actin filaments into a meshlike network through linking by the myosins. In the subsequent relaxation the filaments return to a more parallel disposition. Some investigators believe there also is a cycle of polymerization and depolymerization of the actin filaments.

Evidence has begun to accumulate for oscillations in the intracellular levels of free Ca^{2+} in *Physarum;* these phenomena could contribute to regulating the contractions and imposing the observed rhythm. Ca^{2+} is also thought to participate in the regulation of rotational streaming. Calcium might exert control by influencing the actin–myosin interactions underlying contraction or filament sliding, by regulating assembly and disassembly processes, or by some combination of such effects.

Cell "Crawling" (Ameboid Motion) Amebae and other unicellular organisms move at rates of up to 100 μm per minute by extending pseudopods. As pseudopods form, a marked bulk flow of cytoplasm occurs in the direction of the ameba's motion. This involves a continual formation and dissolution of a gel-like outer cortex of the cytoplasm as diagrammed in Figure II–96. The ameba's cytoplasm contains much actin and some myosin. Both thick and thin filaments, reminiscent of the filaments of striated muscle, have been observed, although these are not arranged in sarcomere-like arrays. When ATP and Ca^{2+} are added to isolated masses of ameba cytoplasm, movements and filament interlinkings are observed that could be based on actin–myosin sliding. One theory has it that the flow of cytoplasm is motivated by contraction at the rear of the ameba (due to actin–myosin interactions in the cortex), so that cytoplasm is squeezed forward. Another suggests that contractions in the cortex at the front of the ameba pull the cytoplasm forward: The pull might be generated by actin-myosin interactions in the cortex and transmitted through loosely interlinked cytoplasmic elements—perhaps involving actin meshworks or the proposed "microtrabecular network."

Mammalian phagocytes and the tips of growing nerve cell processes, as well as a variety of cells grown in culture, crawl at speeds of a micrometer or two a minute by extending a fan or "ruffle" of flattened "lamellipodia," thin, elongate "filopodia," and smaller, spikelike protrusions (see Fig. II–95). This advancing "ruffled membrane" forms plaquelike attachments—called "adhesion plaques" or "focal contacts"—to the surface on which the cell is moving. The cell seems to pull itself forward over such attachments, leaving a narrow protrusion behind, which eventually is jerked forward. Many actively moving tissue-culture cells contain few if any "stress fibers"; fluorescent antibody staining shows a "diffuse" distribution of actin, and the electron microscope reveals microfilaments and meshworks but no prominent bundles. When "stress fibers"—actin microfilament bundles—are prominent in the body of a moving tissue-culture cell, they are oriented along the direction of movement, and many end on the plaques by which the cell attaches to the substrate; the actin and myosin in the extending lamellipodia, however, are in a much less obviously organized network. Overall, a variety of elements seemingly combine to produce the crawling movement: attachment of filaments to the plasma membrane and of the membrane to extracellular surfaces; changes in the state of assembly and aggregation of cytoplasmic filament systems; and still unknown mechanisms that actually move the cell forward. Mass flow of cytoplasm, as in protozoan ameboid motion, is not readily detectable in this form of crawling, although the bulk of the cytoplasm obviously must be moved forward.

Coiling of Filaments   In Section 2.10.4 we mentioned that the protozoan *Stentor* exhibits cell elongation that involves sliding of microtubules past one another. The corresponding contractions, which permit the cell to change its shape reversibly, seem to be generated by bundles of 3- to 4-nm filaments, longitudinally arranged in the cytoplasm. These filaments are neither actin nor myosin; their chemistry is still unclear. Quite unlike the systems we have been studying, the filaments shorten by coiling or folding up into thicker structures, evidently under the influence of Ca^{2+} (see Section 4.1.1 for one of the several possible mechanisms). A few other similar cases have been described.

Controls   As has been alluded to at several points already, governance of cell motility seemingly is accomplished through familiar agents: responses at the cell surface, calcium, cyclic nucleotides, phosphorylation of proteins, and the other factors that also influence cell architecture. The situation in muscle provides a good model for some of the controls, but once again differences are encountered. For example, muscle cells rely on their endoplasmic reticulum to adjust levels of Ca^{2+} in the cytoplasm. Other cell types probably do so as well, but some may also utilize plasma membrane Ca^{2+} "pumps" to a greater extent than that possible in large muscle fibers. In cells lacking troponin, interactions between actin and myosin may derive their Ca^{2+} sensitivity through calmodulin-controlled enzymes such as kinases (Section 2.1.4) that link phosphates to myosin: Phosphorylation of some types of myosin can activate ATPase activity and hence promote the energizing of filament interactions. Phosphorylation of microtubule-associated proteins and of intermediate filament proteins may also be involved in the mechanisms that regulate the functioning or assembly (or both) of microtubules and intermediate filaments. It is important to bear in mind that control of assembly and disassembly of filaments, microtubules, cross-linked meshworks of actin, and other structures is an essential aspect of the control of motility. Calcium ions and actin-binding proteins are the best established of the agents likely to govern assembly and disassembly, but there probably are many others that come into play in one or another cell type or circumstance.

Cell motility phenomena respond to cues of many types. The environmentally responsive pigment granule migrations in melanophores are controlled by neuronal or hormonal signals probably transmitted intracellularly by cyclic AMP. Left undisturbed a cell in tissue culture tends to crawl in the same direction for prolonged periods, but it will change directions to follow wrinkles, folds, fibers, and other mechanical "guides" it encounters. When tissue-culture cells collide on a surface, they reorient the direction of ruffled membrane extension formation and crawl off in new directions (Section 3.1.1). "Tactic" responses are common in many motile cell types and unicellular organisms—cells can orient their movement with respect to distant sources of chemical signals ("chemotaxis") or of light ("phototaxis"; Section 3.8.2).

Chemotaxis   Slime molds spend part of their life history as separate ameboid cells; later these cells come together to form elaborate reproductive multicellular aggregates. The chemotactic agent that brings the cells together is cyclic AMP, which the cells both respond to and release into their surroundings. A slime mold ameba exposed experimentally to a drop of cyclic AMP deposited in the growth medium nearby responds by extending a pseudopod in the direction from which the cyclic AMP is coming; this is the first step in the reorganization of the cell's motility apparatus that leads eventually to its moving toward the source of cyclic AMP. Evidently the cells can attract one another by comparable mechanisms. Mammalian phagocytes are attracted to sites of foreign bodies by chemotactic agents released into the blood (Chaps. 3.2B and 3.2C). Though specific receptors undoubtedly are involved, how cells sense the *gradients* of such agents—the direction in which the agents spread from the sites of their release or formation—is an open question. One model suggests that *time* is the important factor: Cells may sample their environment periodically, then move for a while and sample again. If more of a chemoattractive agent is encountered the second time than the first, the cell's receptors trigger continued movement in the same direction. If less is encountered, movement reorients, perhaps at random, and continues reorienting with repeated sampling until a direction is reached in which increases in the agent are encountered in successive samples. This model shows promise for explaining chemotaxis in bacteria (Section 3.2.4).

An alternative proposal, more likely with larger cells such as the phagocytes, is that the cell's receptor systems are sensitive enough to actually detect differences in concentration of materials between the medium at the front of the cell and the medium at the rear. At cellular dimensions such concentration gradients would generally be very small. However, minute differences can conceivably be amplified. For example, if the receptors involved are of the type that alter the cell surface's permeability to ions, slight differences in the extent of the permeability changes occurring at different regions of the surface could lead to an oriented flow of ionic current within the cell (see Section 3.7.3). This, in turn, could engender oriented changes in the organization and interaction of cytoplasmic filaments or other cytoskeletal and motility-related elements.

Chapter 2.12 The Hepatocyte: A Morphometric Census

This is a convenient point to return to the hepatocyte for a summary of its content of major cytoplasmic organelles.

It is difficult to obtain quantitative information about the numbers and volumes of organelles in a cell as large and complex as the hepatocyte. Estimates of the relative numbers of different organelles can be made from light microscope preparations such as those shown in Figures I–14 through I–21, but they are limited by several factors, such as the resolving power of the light microscope (Section 1.2.1). Sections for the electron microscope are thin, and a given section often includes only a small portion of a given structure. Organelles rarely have simple shapes; thus treatment by the usual geometric formulas is imprecise. For organelles such as the lysosomes, the variety of morphological types complicates estimates: Some types may be difficult to identify in the microscope for census purposes.

Isolated cell fractions are of limited use in estimating relative amounts of cell components unless extreme care is taken to ensure that the fraction is reasonably pure, and unless it is possible to determine what proportion of the total cell content of the organelle is present in the fraction studied. Sometimes this proportion can be approximated by comparing the fraction with the un-

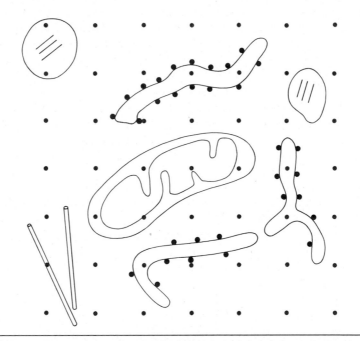

Figure II–97 *An example of morphometric approaches.* To estimate the relative volumes and surface areas of organelles whose shape is too irregular for simple geometric formulas, the cell is photographed, and a transparent overlay with a regular pattern of dots or lines is placed over the image. In the situation illustrated, the proportion of dots that fall over structures of a given kind gives a direct estimate of the percentage of the total area of the photograph occupied by those structures. (In this example, about 10 percent of the dots (5/49) fall over mitochondria, 10 percent over the ER, 2 percent over microtubules, and 4 percent over peroxisomes.) The percentages of *area* in the *photograph* are, in fact, estimates of the percentage of the *volume* of *tissue* that is occupied by each type of structure. The estimates are quite accurate, if an extensive enough sample of the tissue is examined. The technique permits much more rapid assessment of large numbers of cells than do methods for direct measurements of areas.

fractionated cell in terms of the amounts of characteristic components, such as the mitochondrial respiratory enzymes or nuclear DNA.

The most successful approach to quantitation, called *morphometry* or *sterology* (Fig. II–97), is based on the fact that an electron microscope thin section can be regarded as a statistical sample of the cell; the contents of the sample depend on the size, number, shape, and distribution of organelles and on the section thickness and angle. Statistical formulas are used to relate the information from sections to the cell as a whole.

Morphometric estimates indicate that a rat hepatocyte averaging 20 μm in diameter has a surface area on the order of 3000 square micrometers (μm^2) and a volume of roughly 5000 cubic micrometers (μm^3). The nucleus occupies 5 to 10 percent of the cell's volume, mitochondria occupy 15 to 20 percent, peroxisomes 1 to 2 percent, and lysosomes less than 1 percent. (The figure for lysosomes includes residual bodies, some pinocytic digestion vacuoles, and a small number of autophagic vacuoles; the vesicles that may be primary lysosomes are excluded, since their number and volume are exceedingly difficult to estimate.) Glycogen varies greatly in amount but may occupy 5 to 10 percent of the cytoplasmic volume. The ER occupies roughly 15 percent of the cytoplasmic volume and has a surface area of 30,000 to 60,000 μm^2; rough ER and smooth ER make approximately equal contributions, with smooth ER somewhat more extensive (estimates vary, depending in part on details of tissue preparation). Perhaps 300 ribosomes are present per square micrometer of rough ER surface. The Golgi apparatus is said to occupy 5 to 10 percent as much of the cell as the ER, but the diversity of components of the apparatus makes such estimates uncertain. On the order of half the volume of the cytoplasm is accounted for by hyaloplasm, free ribosomes, microtubules, and filaments.

The number of mitochondria per cell is usually thought to be about 1000 to 2000 (but see Section 2.6.4). A few hundred peroxisomes (perhaps 400 to 500) are present per cell, and excluding small vesicles, there are possibly one half as many lysosomes as there are peroxisomes. There are several million ribosomes, of which two thirds to three quarters are bound to the ER, one pair of centrioles, and a large but unknown number of microtubules and filaments.

Further Reading

General and Multiple Topics

Cold Spring Harbor Symposia. Proceedings of annual and occasional symposia on a variety of topics are published by the Cold Spring Harbor Laboratory. Volume 49 of the Cold Spring Harbor Symposium *series is an extensive set of papers drawn from the 1981 symposium that covered a wealth of topics relating to cytoplasmic structure and function.*

Fawcett, D. W. *The Cell*, 2nd ed. Philadelphia: W. B. Saunders, 1981, 861 pp. *An exceptional atlas of electron micrographs illustrating the organelles of animal cells, plus compact explanatory material.*

Gall, J. G., K. R. Porter, and P. Siekevitz (eds.). *Discovery in Cell Biology.* New York: Rockefeller University Press, 1981, 306 pp. *A special issue of the* Journal of Cell Biology *devoted to reviews of the history and current status of work on the major cellular organelles.*

Kornberg, W. (ed.). *Mosaic 12 (4): 49 July/Aug. 1981. An issue of the National Science Foundation magazine summarizing recent progress in several areas of cell research. At times elementary, but provides an interesting overview of cell interactions, motility, and other matters.*

Prescott, D. M., and L. G. Goldstein (eds.). *Cell Biology, A Comprehensive Treatise.* New York: Academic Press. *An ongoing series of review articles, initiated in 1977, discussing a variety of topics in cell biology, including many of the functions of organelles.*

Roos, A., and W. F. Boron. Intracellular pH. *Physiol. Rev.* 61: 296, 1981. *An extensive review on regulation of pH in the various cellular compartments. Valuable as an introduction to the literature.*

Schweiger, H. G. (ed.). *International Cell Biology, 1980–1981.* Berlin: Springer-Verlag, 1981, 1033 pp. *A compilation of short articles covering many topics of cell biology in up-to-date discussions. (International Cell Biology, 1976–1977, edited by Brinkley and Porter and published by Rockefeller University Press, is somewhat out of date but has many still-useful articles.)*

Consult also the cell biology and biochemistry texts in Further Reading, Part 1. *The review periodicals in that list should also be consulted, especially* International Review of Cytology, Journal of Cell Biology, Nature, Science, Cell, *and* Trends in Biochemical Science. *The* Journal of Supramolecular Structure *publishes the proceedings of many conferences of interest to cell biologists.*

Additional Books and Articles Dealing with Specific Areas
Plasma Membrane and Related Topics Such as Transmembrane Signaling, Membrane Genesis, and Cycling:

Cheung, W. Y. Calmodulin. *Scientific American* 246 (6): 62–70, 1982.

Evans, W. H. Communication between cells. *Nature* 283: 521–522, 1980. *A minireview on gap junctions.*

Finean, J. B., R. Coleman, and R. H. Michell. *Membranes and Their Cellular Functions.* New York: John Wiley & Sons, 1978, 157 pp. *An excellent concise paperback, discussing the structure and function of plasma membranes, mitochondrial, plastid, and bacterial membrane, and many other topics.*

Geisow, M. J. Intracellular membrane traffic. *Nature* 295: 649–650, 1982. *A brief summary of recent work on membrane cycling between intracellular compartments and the cell surface.*

Harrison, R., and G. Lunt. *Biological Membranes, Their Structure and Function,* 2nd ed. New York: John Wiley & Sons, 1980, 288 pp. *An intermediate-level paperback, with a useful appendix on methods for studying membranes.*

Jain, M., and R. C. Wagner. *Introduction to Biological Membranes.* New York: John Wiley & Sons, 1980, 382 pp. *Somewhat uneven, but has clear discussions of membrane biochemistry, biophysics and functions.*

Kuhl, F. A., and R. W. Egan. Prostaglandins, arachidonic acid and inflammation. *Science* 210: 978–984, 1980.

Lodish, H. F., and J. E. Rothman. The assembly of cell membranes. *Scientific American* 240 (1): 48–63, 1979.

O'Malley, B. W., and W. T. Schrader. The receptors of steroid hormones. *Scientific American* 234 (2): 32–43, 1976.

Pearse, B. Coated vesicles. *Trends Biochem. Sci.* 5: 131–134, 1980.

Silverstein, S., R. Steinman, and Z. Cohn. Endocytosis. *Ann. Rev. Biochem.* 46: 669–722, 1977.

Staehelin, L. A., and B. E. Hull. Junctions between living cells. *Scientific American* 238 (5): 140–152, 1978.

Nucleus and Nucleolus
(See also Further Reading, Part 4.)

Busch, H. (ed.). *The Cell Nucleus.* New York: Academic Press. *A series of books published at intervals since the mid-1970s, collecting articles on many features, functional and structural, of nuclei and nucleoli.*

Chambon, P. Split genes. *Scientific American* 244 (5): 60–71, 1981.

Gilbert, W. DNA sequencing and gene structure. *Science* 214: 1305–1312, 1981.

Lewin, B. *Gene Expression 2,* 2nd ed. New York: Wiley-Interscience, 1980, 1160 pp. *A comprehensive summary of the molecular biology of the eucaryotic nucleus that considers recent advances in work on topics such as the synthesis of RNAs and the organization of chromatin.*

Miller, O. L. The visualization of genes in action. *Scientific American* 228 (3): 34–42, 1979.

Paule, M. R. Comparative subunit composition of the eukaryotic nuclear RNA polymerases. *Trends Biochem. Sci.* 6: 128–131, 1981.

Watson, J. D. *The Molecular Biology of the Gene,* 3rd ed. New York: Benjamin, 1976, 739 pp. *A comprehensive introduction to molecular biology that also covers much modern biochemistry. The out-of-date segments are updated in* Molecular Biology of the Cell *by Alberts, et al. (see Further Reading, Part 1).*

Ribosomes, Protein Synthesis, and the Like

Chambliss, G., G. R. Craven, J. Davies, K. Davis, L. Kahan, and M. Nomura (eds.). *Ribosomes: Structure, Function and Genetics.* Baltimore: University Park Press, 1980, 1008 pp. *An extensive collection of articles on ribosomes and protein synthesis.*

Hunt, T., T. Caskey, and B. Clark. *Trends Biochem. Sci.* 5: 178–181, 207–209, and 234–237, 1980. *The authors each provided one of a three-part series on the mechanisms of protein synthesis.*

Lake, J. A. The ribosome. *Scientific American* 245 (2): 84–97, 1981.

ER and Golgi Apparatus

Apps, D. K. The uptake of amines by secretory granules. *Trends Biochem. Sci.* 7: 153–156, 1982. *An interesting mechanism by which a Golgi product functions.*

Hand, A. R., and C. Oliver (eds.). *Basic Mechanisms of Cellular Secretion.* In *Methods in Cell Biology,* Vol. 23. New York: Academic Press, 1982, 587 pp. *An extensive collection of articles dealing with the participation of the ER and Golgi apparatus in secretion and with exocytosis and related mechanisms.*

Tartakoff, A. M. Simplifying the complex Golgi. *Trends Biochem. Sci.* 7: 174–176, 1982. *Summarizes some of the present controversies.*

Zimmerman, M., R. A. Mumford, and D. F. Steiner (eds.). Precursor processing in the biosynthesis of proteins. *Ann. N.Y. Acad. Sci.* 343: 1980, 449 pp. *A collection of articles discussing processing of proteins by the ER and Golgi apparatus.*

Mitochondria and Plastids
(See also Higher Plant Cells in Further Reading, Part 3.)

Attardi, G. Organization and expression of the mammalian mitochondrial genome: a lesson in economy. *Trends Biochem. Sci.* 6: 100–103, 1981.

Bjorkman, O., and J. Berry. High efficiency photosynthesis. *Scientific American* 229 (4): 80, 1973.

Clayton, R. K. *Photosynthesis: Physical Mechanisms and Chemical Patterns.* Cambridge: Cambridge University Press, 1980, 281 pp. *An excellent, compact, but comprehensive discussion of photosynthesis in eucaryotes and procaryotes, including consideration of plastid structure and function.*

Govindjee, and R. Govindjee. The primary events of photosynthesis. *Scientific American* 231 (6): 68–82, 1974.

Hinkle, P. C., and R. E. McCarty. How cells make ATP. *Scientific American* 238 (3): 104–123, 1978.

Kirk, J. T. O, and R. A. E. Tilney-Bassett. *The Plastids: Their Chemistry, Structure, Growth and Inheritance.* New York: Elsevier-North Holland, 1978, 960 pp. *A comprehensive, useful reference work.*

Laetch, W. M. The C_4 syndrome: a structural analysis. *Ann. Rev. Plant Physiol.* 25: 27, 1974.

Miller, K. R. The photosynthetic membrane. *Scientific American* 241 (4): 102–113, 1979.

Tzagoloff, A. *Mitochondria.* New York: Plenum Press, 1982, 342 pp. *A compact, clear treatment, especially of the biochemistry and molecular biology of mitochondria.*

Lysosomes and Peroxisomes

Bock, P., R. Kramar, and M. Pavelka. *Peroxisomes and Related Particles in Animal Tissues.* Vienna: Springer Verlag, 1980, 239 pp. *Somewhat uneven in places, but contains a wealth of information, including a survey of methods for study of peroxisomes.*

Hasilik, A. Biosynthesis of lysosomal enzymes. *Trends Biochem. Sci.* 5: 237–240, 1980.

Holtzman, E. *Lysosomes, A Survey.* Vienna: Springer Verlag, 1976, 298 pp. *An effort at a comprehensive survey.*

Kindl, H., and P. B. Lazarow (eds.). Peroxisomes and glyoxysomes. *Ann. N.Y. Acad. Sci.* 386: 1982, 500 pp. *An up-to-date collection of research and review articles discussing many facets of peroxisomes in plants, animals, and microorganisms.*

Matile, Ph. *The Lytic Compartment of Plant Cells.* Vienna: Springer Verlag, 1975, 183 pp. *Lysosomes and lysosome functions in plants.*

Turnover at the intracellular level: *Examples of efforts underway to analyze the relative contributions of lysosomes and nonlysosomal systems are described in articles by Shaw and Dean (Biochem. J. 186: 385, 1980), by Voellmy and Goldberg (Nature 290: 419, 1981), and by Etlinger and Goldberg (J. Biol. Chem. 255: 4563, 1980).*

Cell Motility and Cytoskeleton: Microtubules, Microfilaments, and So Forth

Albrecht-Buehler, G. The tracks of moving cells. *Scientific American* 238 (4): 68–76, 1978.

Cohen, C. The protein switch of muscle contractions. *Scientific American* 233 (5): 36, 1975.

Cohen, C. M., and D. Branton. The normal and abnormal red cell cytoskeleton, a renewed search for molecular defects. *Trends Biochem. Sci.* 6: 266–268, 1981.

Craig, S. W., and T. D. Pollard. Actin-binding proteins. *Trends Biochem. Sci.* 7: 88–92, 1982.

Dustin, P. Microtubules. *Scientific American* 243 (2): 66–76, 1980.

Goldman, R., T. Pollard, and J. Rosenbaum (eds.). *Cell Motility.* Cold Spring Harbor, N.Y.: Cold Spring Harbor Laboratories, 1976. *A three-volume collection of articles discussing many aspects of cell motility and of microtubules, filaments, and so forth. Some of the discussions are now out of date, but many are still very useful introductions to the subjects, especially if read in conjunction with more recent reviews like those in the* International Cell Biology *and* Cold Spring Harbor Symposium 49 *volumes mentioned at the beginning of this section.*

Kamiya, N. Physical and chemical basis of cytoplasmic streaming. *Ann. Rev. Plant Physiol.* 32: 205–236, 1981.

Lackie, J. M., and P. C. Wilkinson (eds.). *The Biology of the Chemotactic Response.* Cambridge, Cambridge University Press, 1982, 177 pp. *A collection of articles dealing with leukocytes, unicellular organisms, and slime molds.*

Lazarides, E., and J. P. Revel. The molecular basis of cell movement. *Scientific American* 240 (5): 100–113, 1979.

Porter, K. R., and J. B. Tucker. The ground substance of the living cell. *Scientific American* 244 (3): 56–67, 1981.

Satir, P. How cilia move. *Scientific American* 231 (4): 42–44, 1974.

Snyderman, R., and E. J. Goetzl. Molecular and cellular mechanisms of leukocyte chemotaxis. *Science* 213: 830–837, 1981.

Wheatlley, D. N. *The Centriole: A Central Enigma of Cell Biology.* New York, Amsterdam: Elsevier, 1982, 232 pp. *Discusses the still-uncertain roles and origins of centrioles.*

PART 3

Cell Types: Constancy and Diversity

The diameters of known cells range from 0.2 μm for the simplest procaryotes up to 1 to 100 μm for most protozoa, algae, and cells of multicellular organisms; occasional specialized cells, such as some egg cells, have diameters in the millimeter or centimeter range. Correspondingly, procaryotic cells tend to have volumes of a fraction of a cubic micrometer (μm^3); "simpler" eucaryotic cells like yeast (Fig. III–1), volumes on the order of a few tens of μm^3; and most eucaryotic cells, volumes of many hundreds to many thousands of μm^3. This diversity in size is matched by a diversity in morphology and metabolism: The simplest cells are not much more complicated in structure than a mitochondrion or a plastid; the most complex show intricate patterns of hundreds of thousands of organelles.

Each cell type is characterized by a particular organization of its organelles and macromolecules. Structural differences among cell types are related

Figure III–1 *Yeast cell (Schizosaccharomyces pombe)* prepared by freeze-etching (Fig. II–6) for examination in the electron microscope. (The contrast has been reversed, photographically, so that "shadows" produced in shadowing (Fig. I–11) the replica appear dark.) The cell is surrounded by a cell wall *(W)* and plasma membrane *(P).* The nucleus is seen at *N.* Visible in the cytoplasm are mitochondria *(M),* Golgi apparatus *(G),* a few cisternae of ER *(E),* lipid droplets *(L),* and vacuoles *(V).* Polysomes also abound in yeast cell cytoplasm but are not readily seen in freeze-fracture preparations. × 20,000. (From Koop F., in *Methods in Cell Biology*, Vol. XI. (ed.), Prescott, D.M. New York: Academic Press, 1975.) ▶

to differences in function. Sometimes this relation is obvious; for example, chloroplasts are present in plant cells, which are capable of photosynthesis, and absent in animal cells, which are incapable of photosynthesis. Often, however, the relations between structure and function are more subtle. Thus in some insect flight muscle fibers, the thick filaments extend throughout the length of

the entire sarcomere, almost from Z-line to Z-line. The sliding of thick and thin filaments during contraction involves only short, probably oscillatory, movements producing length changes of less than 5 percent. (These muscles control the extraordinarily rapid wing motion.) By contrast, in the skeletal muscle of vertebrates where movements are slower, much greater length changes occur, often exceeding 20 percent. The thick filaments occupy only part of the sarcomere length, near the center. This is associated with the extensive filament sliding that takes place (see Section 2.11.1).

Although different organization characterizes cells doing different things, cells doing similar things generally show similar organization. For example, most animal cells engaging in the extensive synthesis of *proteins* that are secreted at the cell surface have large nucleoli and extensive rough ER. The Golgi apparatus in these cells is large and usually is located between the nucleus and the region of the cell surface where the secretion is released. In contrast, a variety of cells producing steroids have abundant smooth ER.

These similarities permit one to speak of cell types (protozoa, algae, absorptive cells, gland cells, and so forth). Findings made in the study of one cell type can often be extrapolated to others of similar structure or function, but this must be done with caution. Although similarities often result from common evolutionary origin, evolutionary changes do not stop once a particular organization develops; consequently, cells with general features in common often differ in detail. Also, comparable organization can be achieved from very different evolutionary starting points *(convergent evolution).*

The great diversity of cell types with different organization and functions has provided favorable experimental materials. Experiments on nuclear function described earlier (Fig. II–14) were possible because the alga *Acetabularia* is a large unicellular organism, relatively simple in structure, easy to maintain in the laboratory, and capable of surviving drastic microsurgery. The chloroplast biogenesis experiments on *Chlamydomonas* (Section 2.7.4) were facilitated by the availability of mutants in which chlorophyll synthesis is dependent on light. Studies of the extraordinarily large neurons of lobsters and squids have supplied much information about nerve impulse conduction. Bacteria have been used extensively in studies of heredity and biochemistry because of their relative simplicity, their rapid growth in media of simple and readily controlled composition, and the availability of many mutants.

Partial catalogs of cell types exist. Thus histology textbooks (*histology:* the study of tissues) may describe virtually all cell types of humans or of other organisms; a few such books are listed at the end of this part. This text will not attempt any such catalog. Rather, a number of cell types have been selected to illustrate major differences in the organization and features of procaryotes and eucaryotes, animal cells and plant cells, and specialized cells of multicellular animals. In the last group, mammalian cells have been chosen more often than others because they have been most extensively studied. Some discussion of specialized cells of plants and invertebrates has been included, as well as a brief consideration of the viruses. The latter are not cells, and their evolutionary status in the biological world is still uncertain. There has been much controversy about whether viruses are truly "living," for they depend on cells for their reproduction, synthesis of macromolecules, and most other activities. However, because of their simplicity and the variety of interesting and experimentally useful properties provided them by evolution, viruses have been of incalculable value in the study of many important biological problems. As with bacteria, they have been used to provide much of the present insight into the molecular mechanisms of heredity.

Classification and Terminology Schemes for classification of organisms generally attempt to group species with common characteristics and also to point up evolutionary relationships. Inevitably, the categories are somewhat arbitrary, and different classification systems have been used by different biologists and at different stages in the understanding of evolution. Thus, for instance, the protozoa usually are included in a kingdom referred to as *Protista*. Sometimes this term has been used to include all unicellular organisms, procaryotic and eucaryotic, but nowadays procaryotes are usually considered to constitute a kingdom of their own, sometimes referred to as the *Monera*. Still, different classification schemes for eucaryotes include different groups along with the protozoa as protists. We shall adhere to the widely used scheme that categorizes living organisms in five kingdoms: *Monera, Protista, Fungi, Plants,* and *Animals*. In this system, *Protista* refers to eucaryotic organisms that are unicellular for much or all of their life spans—primarily the protozoa plus certain unicellular chloroplast-containing organisms such as *Euglena* (Section 3.8.2), diatoms and related *Chrysophyte* algae, and dinoflagellates; many of the chloroplast-containing protists have flagella, thus mixing what had been traditionally thought of as "animal-like" and "plant-like" traits.

Other classification systems emphasizing the absence of appreciable cell differentiation refer to some primitive colonial or multicellular organisms as protists. Yeast and other fungi are sometimes so designated, although for example fungi such as *Neurospora* (Section 2.6.3) are coenocytic—that is, they form elongate hyphae with many nuclei in a common cytoplasm. In still other schemes fungi have been classified instead as plants, though they lack chloroplasts and have cell walls based on polysaccharides such as chitin, rather than on cellulose as in most plants.

Sometimes multicellular algae have been included in Protista even though some show stable differentiation of several cell types. The classification we shall use places most of the algae in the kingdom of Plants; the simplest plants, such as the unicellular alga *Chlamydomonas* (Fig. II–61), thus resemble some of the protists.

Chapter 3.1 Viruses

Viruses have none of the organelles discussed in Part 2, and strictly speaking, they have no metabolism. From an evolutionary point of view they represent the ultimate in specialization. For their duplication they depend entirely on their ability to enter cells of living organisms and to redirect the metabolism of the cells (by substituting their nucleic acid for the cell's DNA as the controlling element). Different viruses enter different cells, from bacterial to human, and make use of different host cell molecules and properties. In extreme cases, the host cell very rapidly becomes diverted from its normal functions to the production of new viruses; it ultimately dies, releasing the viral progeny to infect other cells.

Their dependence on cells for reproduction argues against the consideration of viruses as *primitive* in the evolutionary sense, that is, as the ancestors of cellular life. Perhaps the simplest notion is that they arose as a sort of parasite at the same time as, or subsequent to, the origin of cellular life; possibly some are degenerate forms that have evolved from structures once capable of independent life. Some may have originated from portions of the cellular genetic apparatus.

3.1.1 Structure; Entry into Cells; Release and Escape

A great variety of viruses are known (Fig. III–2). Many are extremely small, about the size of a ribosome; the largest have maximum dimensions of 0.1 to 0.3 μm.

Basically, viruses consist of a nucleic acid core (DNA or RNA) wrapped in a coat made of fewer than a hundred to several thousand protein molecules (see Figs. III–3 and IV–2). In the more complex viruses some lipid and oligosaccharide components are also present: For a number of the viruses that infect animal cells, a lipoprotein membrane surrounds the core composed of the viral nucleic acid complexed with proteins. The RNA viruses (for example, poliovirus and type A influenza virus) are the only biological systems known in which RNA and not DNA is the hereditary material. Often the RNA is present in its usual single-stranded form, but some viruses have a core of double-stranded RNA similar in properties to DNA. Reovirus has ten different double-stranded RNAs in its core, each coding for a different protein. (This may be an evolutionary reflection of the fact that it infects eucaryotic cells, and these generally do not translate mRNA molecules containing information for more than one polypeptide chain; see Sections 2.3.5 and 3.2.5.)

The viral protein coat protects the nucleic acid during the extracellular phase of the virus life cycle. Also, some of the proteins in the coat bind the virus to the cell surface prior to the entry of viruses into cells. Some viral coat proteins include enzymes that enable the virus to modify macromolecules of the host and possibly to release itself from molecules to which it may bind as it is penetrating to and through the surfaces of the cells. For example, a number of viruses, such as influenza viruses, have *neuraminidases* at their surfaces; these enzymes split off sialic acids and related groups from oligosaccharides (see Figs. II–44 and II–45). Roles proposed for the neuraminidases range from aiding the virus to pass through extracellular materials such as the mucus in

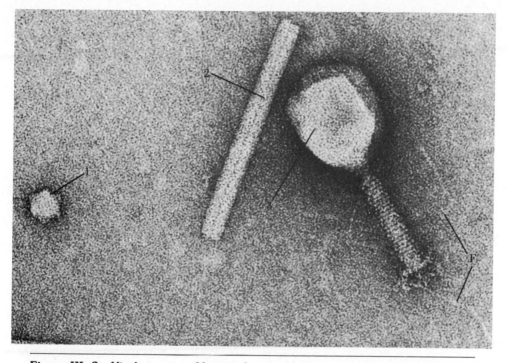

Figure III–2 *Viral structure.* Negatively stained preparation (Fig. II–53) showing three different viruses. *1* is ϕX-174, a single-stranded DNA virus infecting bacteria. *2* is a tobacco mosaic virus that infects tobacco plants. *3* is the bacteriophage T_4; the head and tail are clearly visible, but since the tail fibers are difficult to preserve, only some of them are seen (F; see Fig. IV–2). × 250,000. (Courtesy of F. Eiserling, W. Wood, and R.S. Edgar.)

the respiratory tract to participation in the release of newly matured viruses from cells in which they have replicated. The viral core may also include a few enzyme molecules; for example, some RNA viruses carry enzymes for RNA transcription, and others ("retroviruses") possess enzymes that generate DNA copies of RNAs (Section 3.12.3).

Figure III–2 illustrates the diversity of viral form. In the simpler viruses the coat (*capsid*) consists of a number of protein units, of one or a few types, geometrically arranged around the nucleic acid. In many small viruses the protein subunits surround a DNA (or RNA) core to form a polyhedron, commonly an icosahedron (20 faces) (Fig. III–3). In tobacco mosaic virus (TMV), which has the overall shape of a rod, the core of RNA is arranged as an elongate spiral and is surrounded by protein units in a helical pattern (see Fig. IV–2).

Entry into Cells Some viruses have a much more complex structure involving several different kinds of protein. For example, T_4, one of the *bacteriophages* (viruses that attack bacteria), has a head composed of DNA sur-

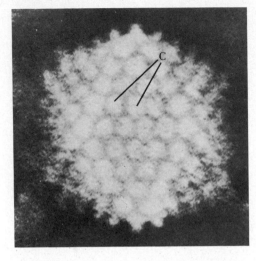

Figure III–3 *Adenovirus structure.* Negatively stained (Fig. II–53) adenovirus, a DNA virus that infects human and other animal cells. The micrograph shows the capsomeres *(C)*, protein subunits composing the "coat"; these have a polyhedral arrangement. (Careful study indicates that the coat consists of 252 capsomeres arranged as an icosahedron.) × 650,000. (Courtesy of R.A. Valentine and H.G. Pereira.)

rounded by protein subunits; in addition, it has a "tail" made up of several sets of other types of subunits including a set of fibers at the end (see Fig. IV–2). The tail attaches the virus to the cell surface; enzymes (viral *lysozyme*; see Section 3.2.3) in the tail help to digest the bacterial wall. The DNA passes through the tail and then through the hole in the bacterial wall produced by the tail enzyme. A contraction of the tail, apparently accomplished by rearrangement of the tail subunits yielding a shorter thicker array, aids the entry of the DNA into the cell. The rest of the virus is left outside. How the DNA actually crosses the plasma membrane is still undescribed. Nor has it been established whether ATP, carried by the virus from the cell where the virus formed, energizes tail contraction, though this is frequently assumed to be true.

For other types of viruses such as those infecting animal cells, the initial step of entry into the cells also is adsorption of the virus to the cell surface. Different viral strains have evolved coats with different specificities determining the cells to which they can attach and the particular surface components ("viral receptors") to which they bind. For many viruses infecting animal cells, adsorption is by means of filamentous protein "spikes" that protrude from the viral surface. Sometimes these spikes show neuraminidase or other enzyme activity as well as binding specificity.

For most viruses, the events subsequent to adsorption to the host cell surface are not as obvious as with T_4; the steps probably vary considerably for different viral strains. Eventually, however, viral nucleic acids are uncoated and appear in the cytoplasm, sometimes along with proteins that accompanied them in the viral core. Such penetration requires passage across at least one cellular membrane—the plasma membrane or a derivative thereof surrounding an endocytic vesicle. Penetration may require mechanisms of greater complexity when proteins must accompany the viral nucleic acid into the cytoplasm than those involved when only nucleic acids need enter. This, at least, is a possible implication of the fact that viral nucleic acids themselves often are "infective"—they can be taken up into cells in naked form if cells are exposed to them under suitable conditions (Section 3.11.3); but how this occurs is unknown.

"Escape" from Endocytic Structures   Commonly, with animal cells, viruses are seen to be endocytized after attachment to the cell surface. However, this observation alone does not establish that endocytosis is a *necessary* step in viral penetration. It is possible that a few of the viruses enter by a different, less readily detectable route, and that these are the ones that go on to multiply. Thus, many investigators believe that the coats of certain viruses can interact with the plasma membrane to permit direct penetration of the core across the plasma membrane. For other strains of viruses, membrane penetration seemingly does occur after the virus has been taken up in an endocytic vesicle. For membrane-enveloped viruses, fusion of the viral membrane with the plasma membrane at the cell surface or with the membrane bounding an endocytic structure is generally thought to permit passage of the viral core into the cytoplasm. For other types, proteins in the coat may interact with the cells' membranes to create a route for the viral core.

Lysosomal enzymes can degrade many varieties of viruses, but certain viruses may "escape" into the cytoplasm even after fusion of the endocytic vacuoles in which they enter the cell with lysosomes. *Reovirus* is taken into lysosomes by endocytosis but is not degraded; its components have evolved in such a fashion that when they are acted on by the lysosomal hydrolases the viral core is uncoated, and viral proteins apparently become modified so that their enzyme activity is activated. The core itself (double-stranded RNA plus some proteins) is resistant to hydrolysis. Evidently, while still within the lysosome, the core initiates the sequence of macromolecular replication and synthesis that leads, somehow, to the formation of viral progeny outside the lysosomes. For several other types of viruses, recent experiments show that a low pH facilitates the penetration of the viral core across membranes; probably the low pH induces conformational changes in coat proteins that expose hydrophobic regions or other groups favoring the appropriate interactions of the viral coat with the membranes. The low pH in lysosomes or in prelysosomal endocytic structures (Section 2.8.4) is thought to allow such viruses to cross into the cytoplasm.

When, for example, viruses are coated with antibodies before encountering cells, multiplication of the viruses is often inhibited as a result of the antibodies' binding to viral surface components. This can block access by the viral coat to the viral "receptors" on the plasma membrane inhibiting adsorption to the cells. Those viruses that are endocytized are degraded in lysosomes since the antibodies prevent the interactions of the viral coat with membranes needed for escape.

Release from Cells; Membrane Formation   Release of some types of viruses occurs with the death and rupture of host cells (*lysis*). In other cases, host cell destruction is much less rapid or even absent, and the virus is released by inclusion in a bud pinched off at the surface of the still-living cell. The bud remains around the virus: In this way most membrane-enveloped viruses obtain components of their membrane directly from the host cell surface. For a variety of viruses (such as *vesicular stomatitis virus*), proteins, including glycoproteins, in the membrane surrounding the virus are made only in infected cells, being coded for by the virus. These proteins are incorporated in the host's membrane prior to viral budding. The processes of incorporation are similar to those used for the cell's own membrane proteins. For instance, in eucaryotic cells viral glycoproteins that will become integral membrane proteins are made in the rough ER and pass to the plasma membrane via the Golgi apparatus. Thus, viruses are being used as experimentally convenient models for plasma membrane genesis. Matters of particular interest are how the viral core interacts with the membrane to produce budding at those places where viral proteins have been incorporated and why budding of different types of viruses from the same cell can occur at different specific regions of the cell surface. Answers to the latter question may help to explain how the cell itself maintains specialized plasma membrane regions.

The insertion of previously soluble proteins into membranes—a matter of considerable importance for membrane genesis (Section 2.5.4)—is also being studied with the aid of viruses. Section 2.8.4 mentioned the ability of diphtheria toxin to enter plasma membranes; this protein is actually coded for by a gene of a virus resident in the bacterium responsible for diphtheria (see following section). One group of investigators also claims that proteins of certain bacteriophages maturing in bacteria insert in the cell's plasma membrane posttranslationally; they fold initially into soluble forms that subsequently interact with the membrane and undergo conformational changes permitting the proteins to enter the lipid bilayer. Other investigators disagree, asserting instead that the proteins pass into the membrane directly from bound polysomes (Section 3.2.4).

It is suspected that some viruses that enter the ER obtain their membranes from the reticulum. Conceivably, others, such as *Vaccinia* virus, assemble an entirely new membrane inside the cell, using lipid molecules carried individually by proteins (Section 2.5.4). These processes of membrane acquisition, however, have been studied much less than has been that of budding from the plasma membrane.

3.1.2 Effects on the Host; Expression of Viral Information

Within the cell, different viruses concentrate in different regions. Some types are found in the nucleus, some in the endoplasmic reticulum (Fig. III–4), and others elsewhere in the cytoplasm. By one means or another the viruses evade or overwhelm mechanisms that tend to protect the cell. The resistance of reo-

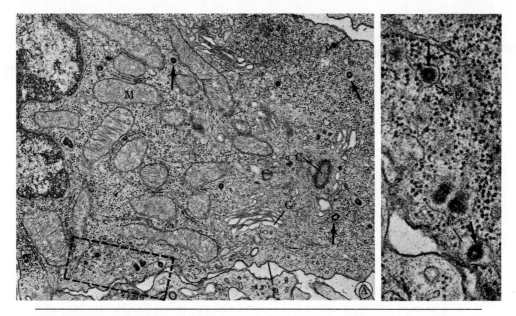

Figure III–4 *Viruses in the ER.* Portion of a cell in the spleen of a leukemic mouse. In A (× 18,000) *N* indicates the nucleus; *P,* the plasma membrane; *C,* a centriole; and *G,* a part of the Golgi apparatus. The electron-dense bodies within the endoplasmic reticulum (arrows) are enlarged in B (× 56,000). The delimiting membrane of the ER is seen at *E.* The bodies within the reticulum are viruses. (Courtesy of E. de Harven.)

virus RNA to lysosomal hydrolysis was mentioned in the preceding section. The DNA of some bacteriophages is modified in such a manner that they resist degradation by the nucleases in the host bacteria that normally restrict entry of foreign nucleic acids (Section 3.2.6); unusual forms of DNA bases, including types with glucose molecules attached, are sometimes present. In animal cells, viral infection can evoke production of a class of cellular proteins known as *interferons,* which in turn mobilize cellular antiviral activities. Interferons apparently inhibit viral multiplication through a number of effects that eventuate in inhibition of protein synthesis in the cells and activation of nucleases that degrade RNAs. The proteins are now under intensive study as potential therapeutic agents.

Effects of viruses on cell metabolism vary greatly. Certain viruses, such as some bacteriophages and the viruses causing polio and influenza, drastically alter cell metabolism, often eventually killing the cells. The binding of viruses to plasma membranes or the insertion of viral proteins into the membranes can lead to fusions of cells into abnormal, multinucleate giant cells and to clumping and lysis of red blood cells. Host cell functions may be entirely diverted to the production of new viral components, for which the nucleic acid of the infecting virus acts as a template. In the case of some viruses with single stranded RNA as their genetic material, a special "replicating form" of RNA is synthesized. The virus first brings about formation of a complementary copy of its RNA and then uses this in turn to produce new RNAs that are complementary to this copy. This results in the production of RNAs identical to the original RNA. Virus-infected cells also often synthesize a variety of enzymes not normally present in cells but involved in the production of specific viral constituents. For example, normal cells contain no enzymes capable of using RNA templates rather than DNA templates for synthesis of RNA, but cells infected with the RNA viruses just described do have such enzymes. The viral nucleic acid specifies production of these enzymes as well as production of the viral coat proteins. Host cell ribosomes and other components are used in the synthesis.

With poliovirus and others the viral RNA is translated into one very large polypeptide chain, which subsequently is cleaved by proteases of the host cell into several smaller, functional protein molecules. Among the cleavage products are protein molecules destined for incorporation in the viral coat and, probably, enzymes for RNA replication. This is one of many examples of the very efficient "packaging" of information in the viral genome. Another is the use of alternative patterns of splicing of RNA molecules (Section 2.3.5) to generate different mRNAs coding for different proteins from transcripts of the same stretch of DNA (see also Section 3.6.10).

Assembly; Coexistence   As viral components accumulate, assembly of viruses begins in the cell. Reconstitution experiments with some of the simpler viruses suggest that the nucleic acids and coat proteins can simply associate spontaneously to form the viruses. However, as Sections 4.1.2 and 4.1.3 will discuss, more complex processes may be involved even for viruses that do not acquire a lipoprotein membrane.

In some cases, rather than forming complete new particles and "escaping" from the cell, viruses enter into a coexistence with the host cell and remain reproducing in long-term residency more or less in synchrony with the cell. The extreme form of this is found in some bacteria where certain (lysogenic) viruses become closely associated with the bacterial DNA. In such "provirus" form they behave very much like part of the host chromosome; viral DNA and bacterial DNA are integrated in the same molecule and replicate at the same time. Both the insertion of viral DNAs into host cell DNAs and their excision occur by processes closely akin to those that normally produce genetic recombination ("crossing-over") between DNA molecules (Section 4.3.4). Long-term virus–host cell relationships break down under some circumstances. The virus becomes virulent, leaves the bacterial chromosome, duplicates much more rapidly than the host cell, and soon kills the cell. It is speculated that some diseases of higher organisms are due to such activation of latent viruses.

As discussed later (Section 3.12.3), certain viruses are known to cause tumors when injected into animals or to bring about rapid and abnormal growth when added to cell cultures.

Other Agents That Infect or Live in Cells A number of plant diseases have been traced to the presence of *viroids,* which are naked, infectious RNA molecules. The molecules are single-stranded but show a great deal of internal base pairing (see Fig. II–16 and Chap. 5.1), giving them stable, well-defined structures. Viroids are only a few hundred nucleotides long; although this is enough to code for a small protein, most speculation is that they act by simulating a cellular regulatory factor or in some other way not dependent on their being translated. Too little is known at present to be sure about this or about the manner(s) in which they disorder metabolism.

Viroid-like causative agents are now being sought in diseased tissues of animals including humans. *Scrapie* is a disease of sheep that can be transmitted by passage of what seems to be an infectious agent and leads slowly to degenerative changes of nervous tissue; it shows similarities to certain human disorders. Particular interest in scrapie has recently been aroused by controversial observations suggesting that the scrapie agent—which has yet to be fully purified and characterized—may have very little nucleic acid, if any. The smallest forms may be no bigger than a modest-sized protein molecule. How such a tiny agent perpetuates itself and exerts its efforts is mysterious; imaginative schemes attribute its powers to effects on the host cell genome or other regulatory systems.

Bacteria are the best known of the infectious.*cells.* Infectious eucaryotic cells include the fungi and protozoa such as *Plasmodia.* The latter enter red blood cells of humans by inducing infolding of the plasma membrane; within the resulting vacuoles, they digest hemoglobin, giving rise to the manifestations of *malaria.*

A variety of other interesting cases are known in which cells of quite different species enter into short-term or long-term symbiotic *(mutual benefit)* or parasitic relationships; one cell may live inside another *(endosymbiosis)* or at the surface of another. Infective or symbiotic particles—bacteria and algae—are found in the cytoplasm of some strains of *Paramecium* and other protozoa (see also Sections 3.3.5 and 4.5.2). In one strain of the fruitfly *Drosophila,* the viable offspring are almost entirely female; males die as embryos. This trait is due to the presence of an infectious procaryotic microorganism, a *spirochete,* transmitted through the ova. (In other *Drosophila* strains a virus, also transmitted by the ova, confers abnormal sensitivity to carbon dioxide on the flies.) We have already mentioned one of the wealth of effects that microorganisms can have on cells they enter: the inhibition of fusion between lysosomes and phagocytic vacuoles (Section 2.8.2). A different kind of effect, of profound significance, is the provision of usable nitrogen by bacteria to plant cells, which forms the basis of the nitrogen cycle in the food chain of higher organisms (Chap. 3.4).

Study of infectious and symbiotic situations promises much of value for understanding normal cellular processes, for better analysis of disease, and for untangling the evolutionary history of living systems (Section 4.5.2).

3.1.3 Viruses as Experimental Tools

Relative Genetic Complexity of Viruses, Procaryotes, and Eucaryotes Polypeptide chains in typical proteins consist of a few dozen to a few hundred amino acids; 200 to 400 is often used as the figure for an "average" chain although, of course, this is somewhat arbitrary. Three RNA bases are required to code for a single amino acid; since RNAs are transcribed from only one of the two complementary DNA strands, three DNA base *pairs* are needed to code for an amino acid. Thus, a *very* rough measure of the number of different proteins for which a cell or virus might carry information is obtained by dividing the number of base pairs in its DNA by three. (The comparable figure for single-stranded RNA viruses is the number of bases divided by three.)

Cells of vertebrates and higher plants each contain hundreds of millions to billions of base pairs (10^8 to 10^9); yeast, somewhat more than 10^7; bacteria, on the order of 10^5 to 10^7; mitochondria and plastids, 10^4 to 10^5; and viruses, between 10^3 and 10^5. In TMV the RNA is a single strand with slightly more than 6000 bases, and poliovirus RNA has 7500 bases. T_4 bacteriophage has roughly 200,000 base pairs specifying up to 50 different proteins. Several viruses are known to contain only a few thousand base pairs, enough to specify only three or four proteins, including the coat proteins and those required for function in a host cell. The DNA of *polyoma* virus (Section 3.12.3), whose coat is of 72 similar protein subunits, is a circular molecule 1.6 μm long. Since there are 3000 base pairs per μm (Section 2.2.1) this viral DNA contains about 5000 base pairs. By contrast, the circular DNA of the bacterium *Escherichia coli* is about 1 mm long, the total DNA content of a yeast cell nucleus is about 4.5 mm, and single chromosomes in many eucaryotic cells contain enough DNA to make a double helix that is millimeters or centimeters in length. Mitochondria of higher animals contain circular molecules 5 μm long (Fig. II–57). (To convert these numbers into approximate molecular weights multiply by 660 per nucleotide pair; Section 2.2.1.)

Exploiting Simplicity To an extent, comparisons of numbers of base pairs exaggerate the differences in genetic potential among different forms. Not all DNAs carry information for proteins or for RNAs such as the tRNAs and rRNAs. Some portions of DNA molecules are regulatory, some code for "spliced"-out regions of mRNA precursors or "spacers" destined for degradation, and some have no known function (Section 4.2.6). In eucaryotes especially, there is much repetitiveness of DNAs: We encountered a first instance in discussing the nucleolar organizer (Section 2.3.4), and even more dramatic repetition of sequences will be discussed later (Section 4.2.6). In the preceding section we noted that viruses make unusually efficient use of their DNA, cramming much information into short stretches. The same is true in mitochondria of some organisms (Section 2.6.3). Still, the comparisons do dramatize the relative simplicity of viruses. Researchers can now hope to identify the function of every gene on the chromosome (nucleic acid molecule) even in relatively complex viruses and to identify every protein for which the nucleic acid codes—a perspective that is presently unrealistic for chromosomes of cells.

As simple, "self-reproducing" structures that can be manipulated genetically and that use cellular organelles to produce easily recognizable molecules, viruses have been of great value as experimental tools. The demonstration that only the DNA of some bacteriophages enters bacteria was an important piece of evidence in establishing the primacy of nucleic acids in heredity. More recently, *adenovirus*—a DNA virus that infects mammalian cells, causing respiratory ailments in humans and other species—has been used to study mRNA production. This virus (10^5 base pairs) codes for about 20 proteins. Analysis of its growth cycle was instrumental in the discovery of RNA splicing: Its mRNAs are produced in very much the same manner as are the cell's own and make use of the cell's machinery. Section 3.1.1 mentioned the use of viruses to investigate membrane formation.

Reconstitution of certain viruses from their isolated components is relatively easy. This is of advantage in unraveling the mode of formation oi .nore complex biological structures (Chap. 4.1). A better understanding of how nucleic acid molecules many thousands of angstroms long are packed into spaces a few thousand angstroms or less in length within viruses will provide valuable clues to the rules governing nucleic acid arrangement in plant and animal cells. Another viral property of experimental use is that DNAs of viruses infecting eucaryotic cells often show the same sort of association with histones as is seen in the normal nuclear DNA of the cell.

Telling Time Viruses show evidence of temporal control mechanisms. Certain viral mRNAs and proteins are synthesized early in infection; others, only later. Viral core nucleic acids are replicated according to a schedule related to the synthesis of other molecules and the assembly of mature viruses. Elucidation of the molecular bases for these controls should shed light on the more complex temporal controls of DNA duplication and gene expression in cells and multicellular organisms (Section 4.2.3 and Chap. 4.4). Some of the regulation is relatively straightforward—certain products cannot be made until others, such as enzymes or "replicating forms" (Section 3.1.2) needed for nucleic acid replication, have been provided. Sequential appearance of some products reflects sequential arrangement of the corresponding genes along the DNA, these genes being transcribed in the order in which they are arranged. More subtle controls also come into play. For example, proteins made early may have conformations permitting some of the molecules to bind to DNA sites very near the ones specifying their own synthesis. This can block access of the RNA polymerases to portions of the DNA (the polymerases are provided by the hosts). In consequence, as "early" proteins accumulate, they differentially "turn off" their own synthesis by inhibiting continued transcription of the corresponding genes. Transcription of "late" regions of the chromosome is not inhibited so the "late" proteins are made as required. Binding of viral proteins near the portions of the viral chromosome where DNA duplication begins (the "origin of replication") is also one of the possible mechanisms for control of the timing of production of the replicates of the chromosome destined for inclusion in the progeny viruses.

The transformation of normal cells into cancer cells by viruses is an essential experimental tool in current work on the mechanisms governing rates of cell growth and division (Sections 3.12.3 and 4.2.3).

Chapter 3.2 Procaryotes

The small size of procaryotes and, until recently, difficulties in preserving their structure for electron microscopy have hampered extensive analysis by microscopy. Today the structure of some bacteria and blue-green algae is known in fair detail, and beginnings have been made in the study of procaryotes such as the *mycoplasmas*. The situation is quite different with respect to biochemistry and genetics; some bacteria, *Escherichia coli* (*E. coli*) and related forms, are probably the most intensively studied and best-understood organisms. Because of the wealth of such biochemical information, many questions about the relations of structure and function may be studied to advantage in bacteria.

The overall cellular dimensions of procaryotes are usually on the order of a fraction of a micrometer to a micrometer or two. Evidence points to a higher degree of intracellular organization than that once thought to be present in procaryotes, but this organization is simpler than that of eucaryotic cells. In some cases the cells contain moderately elaborate arrangements of membranes, but these usually are not organized as discrete organelles with a special surrounding membrane separating them from the rest of the cell.

Chapter 3.2A Mycoplasmas

Mycoplasmas are the simplest known cellular organisms. The diameters of the smallest mycoplasmas are 0.2 to 0.4 μm; the largest nonfilamentous forms measure almost a micrometer. Filamentous forms, consisting of chains of attached cells, reach more than 40 μm in length.

Thus in size, some mycoplasmas overlap with the largest viruses, and some with the smallest bacteria. Mycoplasmas produce diseases in animals and plants, including respiratory and other diseases in humans. Consequently at one time they were called pleuropneumonia-like organisms (PPLO). The possible pathogenetic (disease-causing) roles of recently described viruses found inside mycoplasmas need to be clarified. Vaccines have been produced against *Mycoplasma pneumoniae;* large-scale testing of their effectiveness in reducing human respiratory illness caused by the organism is underway. Special pains need to be taken to prevent contamination by mycoplasmas of cells grown in culture (Chap. 3.11).

3.2.1 Structure and Function

Figure III–5 is based primarily on the best-studied species of *Mycoplasma*. In all mycoplasmas studied thus far, a circular DNA molecule is contained in a nuclear region that is not separated from the remainder of the cell by a membrane. Ribosomes are the major structures visible in the "cytoplasm." A plasma

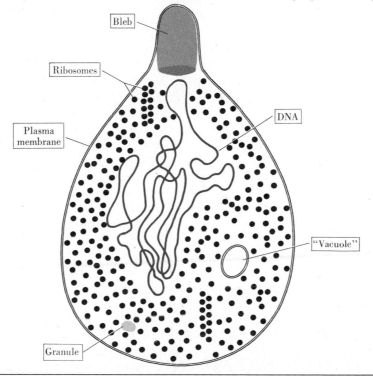

Figure III–5 *Diagram of a mycoplasma.* (After the work of D.R. Anderson and of J. Maniloff and H.J. Morowitz.) Of the great variety of mycoplasmas, only a few have been studied adequately by electron microscopy. These differ in cell shape, arrangement of DNA fibrils, patterns of ribosome distribution, and other features. Some of these differences may reflect intrinsic differences among different mycoplasma types. Others probably result from differences in growth conditions and in methods of preparation of cells for microscopy (the cells are readily deformable).

The mycoplasmas that have been studied all show the presence of a delimiting plasma membrane (with a "unit-membrane" appearance as diagrammed) and of ribosomes and DNA fibrils (a single circular double helix of DNA per cell is thought to be present). They have no cell wall and no nuclear envelope or other elaborate intracellular membrane systems. The *granule* and *vacuole* represent structures seen in some types and are of uncertain significance. For example, in some cases at least, the vacuole is actually a deep infolding of the plasma membrane that appears free from the plasma membrane in electron microscopic preparations only because connections were not included in the thin section being studied. Blebs also have been described clearly only in some types of mycoplasmas. Some researchers believe the DNA anchors to a disc-like structure at the base of the bleb.

membrane of the usual thickness and three-layered appearance delimits the cell. As in other cells, the membrane consists of lipids and proteins; some of the proteins including ones identified as glycoproteins have part of their polypeptide chain exposed at the external surface of the plasma membrane. Mycoplasmas do not synthesize their own cholesterol; their plasma membrane acquires it from the medium in which the organisms are growing. Some forms require such steroids for their survival, whereas others do not. (Steroids are generally not present in appreciable amounts in most types of procaryotic organisms other than mycoplasmas.) Unlike the situation in most other procaryotes, no elaborate cell wall surrounds the plasma membrane. This makes the cells somewhat fragile and deformable but also accounts for the fact that penicillin has little effect on them (see Section 3.2.4). Whether the DNA molecule is attached to a part of the plasma membrane, as in bacteria (see Fig. III–6), is not fully settled, but some investigators think that DNA replication involves sites on the membrane. Some mycoplasmas appear to be motile, but the basis of their movement is not understood. The possible presence of cytoplasmic filaments and of proteins resembling actin is a matter of current dispute.

The enzymes present in the cell include, among others, those required for DNA replication, for the transcription of RNA and the translation of proteins, and for the generation of ATP by anaerobic breakdown of sugar along pathways similar to those described earlier (Chap. 1.3). Mycoplasmas can live and grow by themselves in artificial growth media; unlike viruses, they do not require host cells for duplication. Unlike many bacteria, however, most known mycoplasmas require growth media of complex composition since they have adapted, evolutionarily, to growth in tissues and tissue fluids of higher organisms or in soil or sewage.

3.2.2 Macromolecules and Duplication

The smaller of the mycoplasmas contain DNA with about 750,000 base pairs and a few hundred ribosomes—most, presumably, associated with mRNAs to form polysomes. (By comparison, complex bacteria contain roughly 10 times as many DNA base pairs and about 50 to 100 times as many ribosomes.) Such cells probably can make no more than some 500 to 1000 different kinds of proteins. The largest mycoplasmas can make about twice as many.

Because of their small size and other difficulties involved in observing mycoplasmas, little is firmly established even about matters such as cell division. In most cases binary fission appears to be the mechanism of cell division, although according to reports not yet fully substantiated, budding occurs in some species, whereas growth into multicellular filaments that subsequently fragment may characterize other species. In *Mycoplasma gallisepticum* a bleb is present at one end of the cell (Fig. III–5). Preparatory to cell division a second bleb forms; during division, the two blebs come to lie at opposite poles of the cell. The cell then pinches into daughter cells, each with nuclear region, ribosomes, and bleb. An attractive hypothesis suggests the DNA is anchored to disclike structures at the blebs, which, it is speculated, control separation of daughter DNA molecules during division.

Mycoplasmas are probably far more complex than the primitive self-duplicating systems from which life arose. Yet some may be close to the minimum size and complexity required for independent cellular life and reproduction under present-day conditions on Earth. Knowledge of their metabolism and reproduction should shed light on control mechanisms and other aspects of macromolecular synthesis and structural duplication in more complex cells.

Chapter 3.2B Bacteria

There are a great many different types and species of bacteria, but only a few have been thoroughly studied with modern techniques. *E. coli,* the favorite experimental organism for geneticists, biochemists, and molecular biologists, is normally found in the mammalian digestive tract. An *E. coli* cell measuring 1 by 2 μm contains roughly 5000 distinguishable types of components, ranging from water to DNA and other macromolecules. Its genetic information is carried by a single DNA molecule; enough information is present (3×10^6 base pairs) to code for several thousand different proteins; on the order of 20,000 ribosomes are present. A single cell, by doubling at rates of more than once every 15 to 30 minutes, can give rise to millions of essentially identical cells in a short time. Growth and reproduction can occur in a simple, chemically defined growth medium containing only glucose and inorganic salts. This indicates that the cell contains all the necessary enzymes to synthesize both metabolic precursors and macromolecules from very simple molecules. The combination of relative genetic simplicity, ease of experimental manipulation (for example, production and isolation of mutants), metabolic versatility, and the ability to grow on simple media and to produce large uniform populations makes *E. coli* an outstanding experimental tool. This is exemplified by work with *E. coli* on the synthesis of the enzyme β-*galactosidase*. As outlined below, study of the production of this enzyme has progressed to the point where specific proteins controlling the transcription of the pertinent mRNAs have been identified. The specific sequences of nucleotides in the DNA regions with which these proteins interact have also been mapped.

Metabolism; Nitrogen Fixation   Bacterial metabolism is extraordinarily diverse. This variety has provided opportunity for insight into metabolic controls of many kinds. Many bacteria have pathways that differ only in some details from those of higher organisms. Aerobic and anaerobic pathways for sugar breakdown and mechanisms of electron transport and ATP formation show many features in common with comparable processes in eucaryotic plant and animal cells. Some metabolic pathways, however, are essentially unique to certain bacteria and related procaryotes. The conversion of atmospheric nitrogen to biologically usable forms is one of these. This is accomplished through an enzyme system, *nitrogenase,* which uses ATP and a short electron-transport chain, based on iron-containing *ferredoxin* proteins, to reduce nitrogen to ammonia ($N_2 \rightarrow NH_3$). Nitrogenase is irreversibly inhibited by oxygen. Therefore nitrogen-fixing strains live in habitats where oxygen levels are low—deep in soils, or sometimes within masses of other cells (Chap. 3.4).

The varied biochemistry of different bacterial strains is correlated with their survival in virtually every type of environment, often in circumstances where only very simple molecules are available on which to base metabolism. *E. coli* needs only water, NH_4Cl, glucose, and phosphates and other salts and inorganic ions to survive. Photosynthetic bacteria can use CO_2 in place of glucose and hence are capable of surviving in media consisting entirely of inorganic molecules and ions.

Archaebacteria The unique metabolic capacities of bacterial strains often reflect their origins relatively early in evolutionary history when the Earth's atmosphere and other resources were quite different from today (Section 4.5.2): It has, for example, recently been suggested that one set of unusual procaryotes represents a distinct line of evolution that diverged early from the ancestors of other modern-day procaryotes and from the evolutionary lines that led to eucaryotic cells. These "Archaebacteria" show varying distinctive metabolic properties but share common features of their ribosomal RNAs and other nucleic acids and of their cell walls (they lack peptidoglycan—see below). Many cannot survive where much oxygen is present and thus inhabit anaerobic environments. Some thrive in extreme conditions of temperature and acidity (thermoacidophiles) or extreme salinity (halophiles), and some generate unusual metabolic products such as the methane (CH_4) made by "methanogens." If the still-controversial suggestion that they originated as a distinct evolutionary line early in the history of life on Earth comes to be generally accepted, the *Archaebacteria* will join the "true" bacteria, the blue-green algae, and the mycoplasmas as a major conceptual division of the procaryotic world.

3.2.3 Structure

Electron microscopy shows that bacterial cells possess a plasma membrane under a cell wall, ribosomes, and one or several nuclear regions (Fig. III–6). Under most growth conditions, *E. coli* contains two or more nuclear regions (nucleoids), each containing apparently identical copies of the one chromosome—a circular structure made of a thin fiber about 1 mm long. As far as is known, the fiber is a single continuous double-stranded DNA molecule. In con trast to eucaryotic DNA, histones (Section 2.2.3) are not associated with bacterial DNA, although recent reports assert that some basic proteins are present, and *polyamines* (small organic molecules that contain amino groups) may be bound to phosphate groups of the bacterial DNAs and RNAs. As has been stressed before, the procaryote "nucleus" is not separated from the "cytoplasm" by a special membrane system. This does not imply, however, that the DNA is in a random or disorganized configuration. In fact, the nucleoid can be

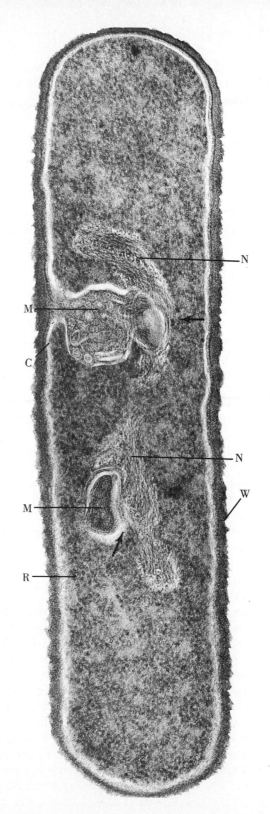

Figure III–6 *A Gram-positive bacterium, Bacillus subtilis,* showing the prominent wall *(W)* surrounding the cell, mesosomes *(M)*, and nuclear areas *(N)*. The mesosomes are infoldings of the plasma membrane *(C)* and also are closely associated (arrows) with the nuclear region. (Connection of the lower of the two mesosomes to the plasma membrane is not seen simply because the plane of section [Fig. I–10] did not include the appropriate region.) The small dense granules *(R)* in the "cytoplasm" are ribosomes.

As is true of procaryotes in general, the arrangement of DNA in the nuclear region cannot readily be analyzed by microscopy of sectioned cells: Many short fibrils are seen in thin sections; it is not possible to determine visually whether they are part of one folded DNA molecule or represent some other, unknown, arrangement. Study of the structure of procaryote chromosomes has thus depended heavily on investigation of material isolated from the cell. Approximately × 70,000. (Courtesy of A. Ryter.)

isolated by cell fractionation procedures (Chap. 1.2B) in the form of a compact, coiled, looped, and folded array with RNA polymerase molecules and growing (nascent) RNAs associated. Work has now begun on the functional geometry of the nucleoid and the manner in which enzymes, ribosomes, and other components interact with it. Section 2.3.5 pointed out that polysome formation and translation of a bacterial mRNA may commence as the RNA molecule is being transcribed. Correspondingly, nascent mRNA molecules with attached ribosomes are found associated with the DNA.

Bacterial ribosomes are notably smaller than the ribosomes of the cytoplasm of eucaryotes (they sediment at 70S as opposed to 80S; Section 2.3.2) and contain more RNA (over 60 percent) than protein (eucaryotic cell ribosomes contain 40 to 50 percent RNA). A single small (5S) ribosomal RNA is present rather than the two found in eucaryotes. The bacterial chromosome has only a few repeats of the rDNA information for rRNAs, and no nucleoli are formed. Intervening sequences (introns; Section 2.3.5) may not be present at all in bacterial genes; if any are, they are much less evident than in eucaryotes. Bacteria use only one type of RNA polymerase; eucaryotes, three (Section 2.2.4). No endoplasmic reticulum or Golgi apparatus is present in bacterial "cytoplasm"; nor are there lysosomes or peroxisomes. Cytoplasmic filament arrays or structures somewhat resembling microtubules have been observed in certain bacteria and other procaryotes and are now being examined to determine how closely they resemble eucaryotic systems and how widely they are distributed. At present, most seem to be restricted to a relatively few species.

Cell Walls   The extracellular wall that surrounds the cell is usually a moderately rigid structure responsible for the maintenance of the distinctive cell shapes of different bacterial strains (the spherical *cocci,* elongate *bacilli,* and so forth). Its removal leads to alteration of the cell into a fragile "protoplast," which may burst from water influx unless osmotic conditions in the surrounding medium are carefully adjusted (see Section 2.1.1).

In classifying bacteria, bacteriologists have long used *Gram* staining procedures based on treatment of cells with the dye *crystal violet,* followed by exposure to iodine, and extraction with alcohols or other solvents. Apparently those cells that retain the stain do so because their walls prevent its extraction. These procedures enabled the distinction of two different major bacterial classes by light microscopy. Many bacteria, such as the pneumonia-producing organism *Pneumococcus,* are colored by the Gram stains and are referred to as *Gram-positive.* Other bacteria, such as *E. coli,* are *Gram-negative.*

The cell walls of Gram-positive bacteria contain a *peptidoglycan* ("murein") network. This is a polymer of repeating units, each composed of two modified sugars with an attached "tetrapeptide" that frequently includes unusual types of amino acids (see Fig. III–7). The units are linked to one another into chains up to 50 (or more) units long; the chains are cross-linked to one another as illustrated in Figure III–7. The extensive linking produces a continuous, strong network surrounding the cell. The network is associated with other polymers producing a wall up to tens of nanometers thick, which, however, is sufficiently open in structure that macromolecules can pass across it.

The walls of Gram-negative bacteria contain a peptidoglycan network a few nm thick but have in addition an outer lipid bilayer often appearing as a three-layered membrane in the electron microscope. Present in this "outer membrane" are proteins and lipopolysaccharides—complex chains of carbohydrates with fatty acids attached. The space between this layer and the plasma membrane is the "periplasmic space." The outer membranous layer of the Gram-negative cell wall is a barrier to proteins and other large macromolecules. Sugars, amino acids, and the like can cross this layer and enter the periplasmic space though channels involving proteins sometimes called *porins;* evidently these form membrane-spanning pores in the outer membrane. Some components in Gram-negative cell walls are toxic to animals and account for certain effects of bacterial infection. The lipopolysaccharides are among the major components against which higher vertebrates responding to infection make antibodies through which their defensive systems can initiate antibacterial processes (Chapter 3.6C).

The differences in cell wall chemistry between Gram-positive and Gram-negative bacteria are associated with differences in sensitivity to important antibacterial agents. For example, Gram-positive bacteria are usually more sensitive to penicillin, probably because this agent cannot pass the outer layer of the Gram-negative wall. Penicillin interferes with the interlinking of the peptidoglycan during cell growth and thus reduces the rigidity of the wall and the resistance of the cell to osmotic rupture. The enzyme *lysozyme* present in phagocytic cells, in tears, and at other sites hydrolyzes the sugar backbone of the peptidoglycan, exposing the cells to disruption. When white blood cells phagocytize bacteria, cytoplasmic granules rich in lysozyme fuse with the phagocytic vacuoles, as do the lysosomes (Fig. II–69). Lysozyme also affects Gram-positive species more readily than Gram-negative ones; again, the lipid-containing layer probably acts as a barrier to penetration of the enzyme into the wall. Some bacteria survive uptake by white blood cells and other phagocytes, probably because their cell walls remain intact and protect the cell itself from osmotic and enzymatic disruption (Section 2.8.2).

In many bacteria, additional layers of protein or polysaccharides are present outside the wall layers already described. Occasionally this material appears as a mosaic of repeating tilelike units. These layers may aid in protecting the cells or in helping them adhere to surfaces. Wall components of some types of bacteria inhibit uptake by phagocytes.

Spores Certain bacteria can transform into dormant spores that are metabolically inactive and extraordinarily resistant to extremes of temperature, dehydration, and other drastic environmental conditions. The spores contain a single nuclear region and relatively little cytoplasm. Each is enclosed in a special thick extracellular wall. The control mechanisms of spore formation and germination are being actively studied. These processes show some analogies

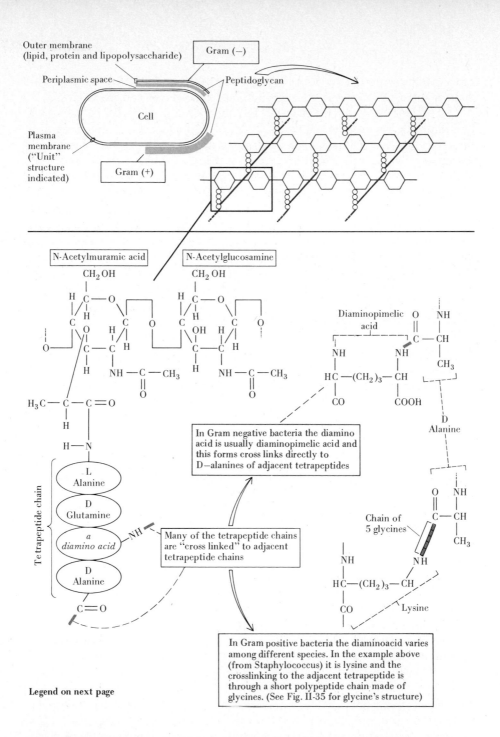

Outer membrane
(lipid, protein and lipopolysaccharide)

Gram (−)

Periplasmic space

Peptidoglycan

Cell

Plasma membrane ("Unit" structure indicated)

Gram (+)

N-Acetylmuramic acid

N-Acetylglucosamine

CH_2OH

CH_2OH

Diaminopimelic acid

D Alanine

Tetrapeptide chain

L Alanine

D Glutamine

a diamino acid

D Alanine

In Gram negative bacteria the diamino acid is usually diaminopimelic acid and this forms cross links directly to D−alanines of adjacent tetrapeptides

Many of the tetrapeptide chains are "cross linked" to adjacent tetrapeptide chains

Chain of 5 glycines

Lysine

In Gram positive bacteria the diaminoacid varies among different species. In the example above (from Staphylococcus) it is lysine and the crosslinking to the adjacent tetrapeptide is through a short polypeptide chain made of glycines. (See Fig. II-35 for glycine's structure)

Legend on next page

to the differentiation of specialized cells of the multicellular eucaryotes: During spore formation new cell products are made. During germination, synthetic machinery that previously was inactive is turned on; a similar process occurs

◀ **Figure III–7** *Bacterial cell walls.* As outlined in the text, both Gram-negative and Gram-positive bacteria have a layer of peptidoglycan in their walls. This layer is made of a continuous molecular network composed of units like the ones illustrated here (there is some variation in detail among different strains of bacteria, but the basic structural principles are constant.) In essence, the wall consists of long, cross-linked systems of modified sugars; in these "saccharide" chains, *N*-acetylmuramic acids (NAMs) alternate with *N*-acetylglucosamines (NAGs) so the chains often are described as consisting of repeating disaccharide units. Attached to the saccharide chains are "tetrapeptides," peptide chains (Figs. I–28 and II–23) four amino acids long. The latter chains are unusual in that they include D-amino acids as well as L-amino acids. (D and L refer to the alternative mirror-image geometrical forms—"optical isomers"—of the amino acids that can be constructed from the same atoms linked in the same sequence; the usual amino acids occurring in natural proteins are all of the L variety.) The third amino acid in the tetrapeptide chain is a "diamino acid"; the importance of this is that the side chains in diamino acids, as the name implies, terminate in an NH_2 group. In the cell walls, this group permits the amino acid to form a second peptide bond in addition to the one in the tetrapeptide chain. The second bond is key in cross-linking the tetrapeptides and thus in establishing the overall continuity of the peptidoglycan structure. The crosslinks attach the diamino acid of one tetrapeptide to the fourth amino acid of another. Several different cross-linking structures accomplish this in different bacterial species; two are illustrated. (The heavy lines used here for the cross-links are solely for diagrammatic emphasis; the actual links are peptide bonds.) (After N. Sharon, L. Stryer, R.Y. Stanier, et al., and many others.)

upon fertilization of egg cells (Chap. 4.4). Problems being investigated include the identification of the control mechanisms that operate to produce orderly synthesis of the right amounts of the right components at the right time, ensuring bacterial transformation into spores under some conditions and conversion into active cells under other conditions.

3.2.4 Features of Functional Organization

Transport; Respiration Figure III–8 shows a multienzyme complex that has been isolated from *E. coli.* Several dozen protein subunits are associated in a structure that carries out a number of sequential steps in pyruvate metabolism (as in Chapter 1.3). The complex can be dissociated into three distinct enzymes (and can then be reconstituted in the test tube.) The existence of multienzyme complexes in bacteria is one indication of the high degree of organization that underlies the apparent morphological simplicity.

The bacterial plasma membrane and its derivatives participate in a diversity of functions including several segregated to intracellular membranes in eucaryotic cells. The plasma membrane of *E. coli* has transport systems for sugars, amino acids, Na^+, and other molecules and ions. "Binding proteins" in the cell wall can accumulate specific components from the medium; plasma membrane proteins *(permeases)* then help to pass the components into the cell. Active transport of some components is driven by ATP. Much active trans-

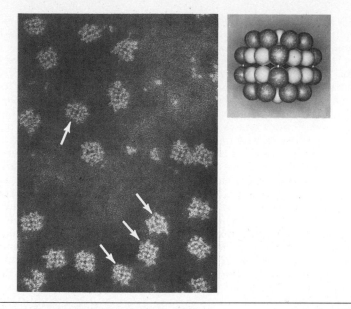

Figure III–8 *Pyruvate dehydrogenase multienzyme complexes* isolated from *E. coli* and negatively stained. The diameter of the complex is 300 Å, roughly twice that of a ribosome (see Fig. III–6). Each complex contains several molecules of three types of enzymes, which in turn are made of subunits (a total of 72) arranged as indicated in the model at the upper right. × 300,000. (Courtesy of L.J. Reed, R.M. Oliver, and D.J. Cox.)

port, however, draws energy (Section 2.1.3) from the H^+ gradient that the bacterium establishes across its plasma membrane through membrane-associated electron transport systems. (In aerobic bacteria O_2 is the final "acceptor" of the transferred electrons, yielding H_2O as in mitochondria. In anaerobic strains, groups such as nitrate $[NO_3]$ or sulfate $[SO_4]$ are used instead of O_2.) Probably by chemiosmotic mechanisms, the H^+ gradient can also be used to generate ATP.

Photosynthesis There are several types of photosynthetic bacteria. *Bacteriochlorophyll* and other photosynthetic pigments differ somewhat from the comparable pigments of eucaryotes. Bacterial photosynthesis operates on general principles basically similar to those described in Section 2.7.2 but differing in many details. Most strikingly, H_2S, H_2, or other molecules are used as electron sources in place of H_2O, and rather than O_2, the bacteria generate products such as sulfur. The dark reactions are based on the Calvin cycle as in eucaryotes.

In some photosynthetic bacteria, the photosynthetic pigments and associated components of electron transport are arranged in structures within the cell. In different species, these structures have the form of flattened sacs, tubules, or vesicles, but they are not separated from the rest of the cell as a discrete chloroplast by a delimiting membrane (see Fig. III–12 for a cell with essentially comparable organization). In the "purple" bacteria the internal vesicles have typical membrane structure and probably derive from the plasma membrane, perhaps by pinching off. In the "green" bacteria, some of the bacteriochlorophyll is found in the walls of internal vesicles that appear to be bounded by a single-layered (protein?) "membrane" rather than the usual three-layered lipoprotein membrane. These unusual structures may collaborate with pigments and other components of photosynthetic light reactions located in the cell's plasma membrane. Certain of the purple bacteria contain discrete intracellular bodies, not membrane-bounded, made largely of RuDP carboxylyase, the enzyme responsible for photosynthetic CO_2 fixation (Fig. II–62).

Purple Membranes   Some of the *Halobacteria* form specialized patches in their plasma membranes when O_2 levels are low and an alternative source of energy is required. These regions are known as "purple" membranes since their composition is dominated by "bacteriorhodopsin." (The color of the purple photosynthetic bacteria mentioned above is due to a different pigment; the purple bacteria do not have the purple membranes.) Bacteriorhodopsin is a protein pigment similar to the rhodopsin of vertebrate retinal photoreceptors (Section 3.8.1) in that its light absorption depends on a vitamin A derivative known as *retinal*. The protein molecules are arranged in regular arrays, each molecule folded so that its polypeptide chain crosses the membrane seven times. Illumination of the membrane results in the production of an H^+ gradient across the membrane caused by "pumping" of H^+ out of the cell by light-energized bacteriorhodopsin. This gradient can be used by the cell to generate ATP and for other energy-requiring processes. Each bacteriorhodopsin molecule can pump several hundred protons per second.

Purple membranes are of exceptional interest to investigators of chemiosmotic mechanisms, as well as to students of light reception in photosynthesis and in vision. In comparison with a chloroplast thylakoid or mitochondrial crista, the structure of purple membrane regions is relatively simple. A detailed map showing how each of bacteriorhodopsin's amino acids is placed with respect to the membrane is rapidly nearing completion; specific models of how H^+ may be transported should soon be available. (One of the several possibilities is that light liberates a proton from retinal and that these H^+s move through the protein by sequentially associating with certain of the many sites capable of forming hydrogen bonds [Section 1.4.1] within the molecule.)

The use of H^+ gradients to generate ATP by bacteria involves F_0 and F_1 components comparable to those of mitochondria (Fig. II–56), located on the cytoplasmic side of the plasma membrane. Interestingly, when artificial vesicles are created in which bacteriorhodopsin is incorporated together with *mitochondrial* F_1 and other coupling factors, ATP can be generated when the vesicles are illuminated and consequently establish an H^+ gradient. Such reconstituted "hybrids" offer strong support for the chemiosmotic hypothesis as applied to mitochondrial phosphorylation.

Mesosomes; Cell Division Interest has centered in another membranous structure, the *mesosome,* formed by infolding of the plasma membrane (see Fig. III–6). These are found chiefly in Gram-positive forms, although some Gram-negative forms show similar but simpler infoldings of the plasma membrane. They are generally held to be among the sites where membrane-associated respiratory enzymes are located.

Mesosomes, or similar special regions of plasma membrane, have been ascribed still-conjectural roles in bacterial duplication and division. Such division involves DNA replication, growth, and separation into two cells by formation of a *septum* across the cell. The septum grows in from the surface; it consists of a plasma membrane and cell wall. The central question is: How is DNA behavior controlled so that daughter cells receive the proper share of nuclear regions? By analogy with eucaryotic cells (Chapter 4.2) it might be expected that the chromosome is anchored to another structure that controls the chromosome's position during division. In electron micrographs the DNA of each nuclear region appears to be attached at one point to a mesosome, and the mesosomes are often seen near the forming septum and attached to daughter nucleoids, as if controlling separation. Evidence that is similarly circumstantial suggests that attachment of the nucleoid to the plasma membrane is important in DNA replication. For example, when *E. coli* nucleoids are isolated by gentle methods, cells actively replicating DNA yield nucleoids with some membrane fragments attached, whereas cells not replicating their DNA yield membrane-free nucleoids. Presumably this reflects corresponding alterations in nucleoid–membrane associations in the cell. *(E. coli,* being Gram-negative, lacks well-defined mesosomes.)

Exoenzymes and Other Exported Proteins Bacteria must secrete ("export") the binding proteins and other proteins that form part of their cell walls since these proteins are made in the bacterial "cytoplasm." Bacteria also produce "exoenzymes" that are secreted to the extracellular environment, or into the periplasmic space of Gram-negative forms. Many of the exoenzymes break down large molecules, such as macromolecules, into small units that can enter the cell. By degrading components of potential barriers, such enzymes can facilitate passage of bacteria into or within the tissues of organisms they infect, as well as providing nutrients for the bacteria.

Controlled breakdown and resynthesis of the extracellular wall is a necessary step in bacterial growth and division. Such remodeling is required to permit increases in total cell volume and the formation of a new wall between the daughter cells. The units of which peptidoglycan is formed are synthesized in the cytoplasm and, by carrier-mediated transport, moved to the extracellular space where linking into the network takes place. The enzymes that do the linking and also those that open up portions of the network, permitting expansion, must also be exported from the cell.

The release of extracellular enzymes by a bacterium has been likened to the emptying of lysosomal enzymes into a phagocytic vacuole of an animal cell: A phagocytic vacuole is, after all, a bit of the external milieu that has been enclosed within a membrane and taken into the cell. This analogy cannot be pushed far, however, because the export of proteins from the bacterial cytoplasm is not based on the fusion of membrane-delimited compartments. Neither endocytosis nor exocytosis is observed in bacteria (nor are lysosomes present). Rather, proteins are thought to pass across the bacterial plasma membrane directly from the polysomes on which they are made, by mechanisms essentially similar to those used for the entry of proteins into the endoplasmic reticulum. "Signal" sequences (Section 2.4.3) have been identified for some types of exported proteins; conceivably, other types cross the membrane posttranslationally by one or another of the mechanisms we have discussed (Sections 2.6.3, 2.7.4, 2.8.4, and 2.9.3).

Proteins that are constituents of the bacterial plasma membrane presumably pass *into* the membrane by mechanisms like those used for exported proteins (see Section 2.5.4). How proteins of the *outer* membrane of Gram-negative strains become inserted in this membrane is not known. Some researchers believe that local regions of special close associations ("adhesion" points or zones) exist between the plasma membrane and the membrane of the cell wall, permitting direct transfer. Elucidation of the transfer mechanisms may help to explain how proteins imported into mitochondria or chloroplasts can cross the outer membranes of these organelles to insert in the inner membranes or the interior compartments.

The lipids of bacterial membranes are made by enzymes associated with the plasma membrane.

There is considerable controversy about whether growth of the plasma membrane and cell wall is accomplished through localized addition of material at special growing points or whether new molecules are inserted directly throughout much of the expanse of the structure. This is difficult to resolve, in part because molecules added locally might subsequently spread from their sites of addition and eventually assume a random distribution in the wall or membrane. For some bacteria at least, localized growth of the cell wall has been demonstrated for particular circumstances. Examples are encountered during cell division, when new wall materials are added selectively at the equator (where the septum forms, as described above). *Caulobacter* cells extend elongate stalklike structures by which they attach to surfaces; the wall material of the growing stalk is added at the base, where the stalk joins the wall surrounding the remainder of the cell. In other cases, however, walls thicken by addition of material at many points. Similarly, although many experiments seem to show that lipids and proteins are added throughout the expanse of the plasma membrane, a few suggest a more localized addition of particular enzymes or other proteins, perhaps occurring at the cell equator during division.

Motility   Many bacteria have flagella (Figs. III–9 and III–10), but these are quite unlike the flagella of eucaryotic cells. There may be few or many flagella

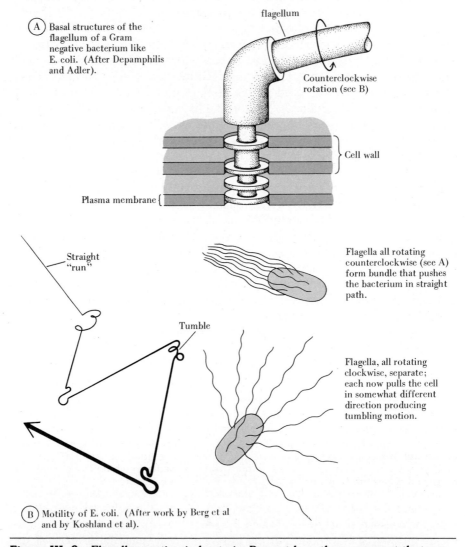

Figure III–9   *Flagellar motion in bacteria.* **Present hypotheses suggest that certain of the ringlike structures at the flagellum base are anchored in the wall or plasma membrane and that the remainder of the flagellum can rotate within these rings. (Diagrams based on Berg, M. *Scientific American* 232(2):36, 1975, and Adler, J., *Scientific American* 234(4):40, 1976.)**

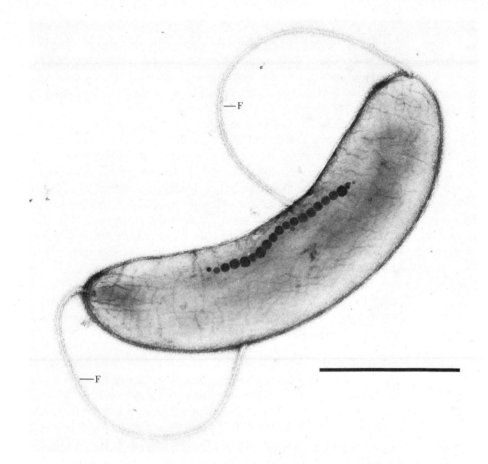

Figure III–10 *A bacterium that responds to magnetic fields.* Electron micrograph of a negatively stained (Fig. II–53) preparation showing a bacterium of a strain capable of orienting its direction of motion along the lines of force of the earth's magnetic field. It is theorized that this effect directs the bacteria towards the bottom of the bodies of water in which they live; the organisms survive best in anaerobic or low-oxygen conditions such as those occurring in sediments at the bottom of bogs and marshes. The chain of dark particles ("magnetosomes") within the cell are of Fe_3O_4—the ion oxide known as *magnetite*. The particles probably are formed by extraction of iron from the media in which the organisms grow and appear to be surrounded by a membrane-like sheath; the sheath may be associated with the plasma membrane and is thought to help keep the particles anchored in place. Magnetic fields exert an orienting force on the particles; the effect is to orient the bacteria so that they move along the lines of force. Flagella *(F)* present at both ends of the cell provide motility. (In this specimen the ends of the two flagella are folded back under the bacterium near its middle, but in life they are attached only at the ends of the cell, one flagellum at each end). The bar length represents 1 μm. × 30,000. (From Blakemore, R.P., and R.B. Frankel, *Scientific American* **245**: No. 6, 59, 1981. Courtesy of the authors, publishers, and Mrs. N. Blakemore.)

per cell (*E. coli* has a half-dozen), each a fiber 10 to 20 nm thick and up to 10 μm long protruding through the cell wall from a basal structure anchored in the plasma membrane. The fiber is made primarily of molecules of the protein *flagellin* in helical arrangement; it appears to form by growth from the tip, using flagellin molecules exported from the cell, which some believe reach the tip of the growing flagellum by passage through its hollow core.

The flagella do not bend; rather, the entire flagellum rotates around its axis at rates of 5 to 10 or more rotations per second. These rotatory movements move the bacteria at rates of tens of μm per second. The energy for this movement probably is supplied by the H^+ gradients across the plasma membrane—movement can be induced by artificially imposed H^+ gradients and seems not to require ATP. A leading, though unorthodox, theory to explain the motion proposes that the ions associate with the structures present at the flagellum's insertion in the plasma membrane; the actual movement would be generated by repulsions or attractions between the ions on the base of the flagellum itself and those on the surrounding structures, which cause the flagellum to rotate much like the shaft of an electrical motor (Fig. III–9). A more conventional alternative suggests that the ionic gradients set off conformational changes in proteins that propel the flagellum.

Chemotaxis Bacterial chemotaxis (Section 2.11.4), as studied in *E. coli*, is based on control of movement of the flagella. The sensory system depends in part on binding of attractants (for example, sugars and disaccharides such as maltose or amino acids such as serine) to specific receptor systems that include binding proteins in the cell wall. One consequence of the binding is the addition of methyl groups to other proteins. Since this is a well-defined chemical change that can, for example, alter proteins' conformations, binding properties, and enzymatic activities, it should provide a central clue to chemotactic mechanisms, but as yet what may lead from this methylation to changes in motion is still unknown. The mechanisms could include alterations in ionic permeabilities or other properties of the plasma membrane.

The net orientation of movement in chemotaxis results from the fact that when the flagella rotate in one direction (counterclockwise as looked at from the tip), the bacterium swims in relatively straight lines, whereas clockwise rotation results in a rapid nonoriented tumbling (Fig. III–9). In unstimulated state, the bacterium alternates these motions, moving in a straight run for a while, then tumbling so that when straight motion resumes it is in a random, new direction. When the bacterium moves into a zone of higher concentration in a concentration gradient of chemoattractant, the straight motion is prolonged, and tumbling is less frequent; this pattern favors, on balance, motion in the direction of higher concentration ("up the gradient"). Tumbling is more frequent when the bacterium finds itself going down the gradient; tumbling continues to reorient the direction of motion, probably at random, until the organism starts moving up the gradient, and once again, straight motion becomes relatively prolonged. Each period of tumbling or straight movement lasts for roughly one to a few seconds.

The ability of the bacterium to move in the direction in which concentrations of attractants are higher seems to depend on a "temporal" (timing-based; Section 2.11.4) mechanism. The pattern of flagellary movement by the bacterium responds to the concentration of attractant in which the bacterium is present at a given time. This response continues for a period but eventually "decays." The bacterium then again "samples" its environs and grades its next response according to whether the concentration now encountered is higher or lower than that in the prior "sample." Possible evidence for such a mechanism is the fact that the pattern of flagella rotation in *E. coli* responds to a sudden change in the overall concentration of attractant in the medium, much as it does to a concentration gradient—in other words, a temporal change in concentration can mimic the effects of concentration differences in space. Since methylation of proteins is part of the initial sensory response, it has been speculated that demethylation may be part of the "decay."

Other Movements and Tactic Responses; Spirochetes   Figure III–10 illustrates a strain of bacterium capable of orienting its movement with respect to magnetic fields, including that of the Earth. Such organisms exemplify the extraordinary range of evolutionary adaptations by bacteria.

Forms of motion other than that based on flagella are found in bacteria and related organisms. They include "gliding" over solid surfaces by a variety of organisms and the sinuous swimming motion of *spirochetes*. The latter organisms have bundles of flagella-like fibrils running in the periplasmic space, but how these generate motion is still obscure, as is the mechanism of gliding.

Pili   *Pili* are another class of cylindrical structures made of protein molecules; they extend from the surface of bacteria, sometimes in large numbers. Certain of them in Gram-negative strains are required for bacterial *conjugation,* a form of sexual reproduction (Section 4.3.3). Beyond this, however, not much is known about the roles of the various types of pili found on different bacteria.

3.2.5 Metabolic Controls: A Brief Review

Study of *enzyme induction* and *repression* in bacteria has elucidated mechanisms by which genes can be turned on or off. Many *E. coli* enzymes are *constitutive;* that is, they are synthesized irrespective of the growth medium. *Inducible* and *repressible* enzymes are produced in some growth media but not in others. The most intensive work on induction has been done on an *E. coli* metabolic pathway responsible for steps in the metabolism of some sugars and referred to as the β-*galactosidase system*. In the absence of the glucose–galactose disaccharide *lactose* (or other "inducing" compounds of similar molecular structure), the bacteria make very few molecules of the enzymes that metabolize lactose. Inducing compounds added to the growth medium initiate extensive and coordinated production of three related enzymes: β-*galactosi-*

dase, which splits lactose into galactose and glucose; a *permease* responsible for entry of lactose into the cell; and an *acetylase* involved in certain reactions of sugars. The proteins showing these enzymatic activities are coded for by DNA nucleotide sequences that are adjacent to one another along the chromosome; a single mRNA molecule transcribed from this DNA contains the information for all three species of protein. Polysomes with the mRNA translate the three proteins as separate molecules, using the "punctuation" trinucleotide sequences (the codons signaling the starting and stopping sites; Section 2.3.2) to determine where to begin and to end each polypeptide. Several thousand β-galactosidase molecules rapidly accumulate in the cell following exposure to lactose.

Promoters, Operators, and the Like The explanation of these findings resides in the mechanisms that control transcription of RNAs. RNA polymerases attach to DNA at DNA sequences called promoter sequences. Typically in bacteria these are located adjacent to the sequences to be transcribed (certain of the promoters in eucaryotes may overlap the transcribed sequences). As in the system we are considering, stretches of as many as 50 to 100 or more base pairs may participate in initiating binding of RNA polymerases. These long stretches include much shorter sequences that have been shown to be essential for a polymerase molecule to attach firmly to the DNA at the proper place and to begin to unwind the DNA so that the bases can be exposed for complementary matching during RNA transcription. For the β-galactosidase genes, a key short sequence in the promoter is TATGTTG spaced by 5 additional nucleotides from the place where RNA transcription actually begins. The sequence given is of the "noncoding" DNA strand, the nontranscribed strand. The corresponding complementary sequence on the DNA strand that actually serves as the template for the mRNA is, of course, ATACAAC—remember that the DNA information is transcribed in its 3'–5' direction, generating RNA that grows in the 5'–3' direction starting at the RNA's 5' end (Sections 2.2.4 and 2.3.5). (The convention of presenting the sequence for DNAs in this fashion stems from the fact that in the regions of genes that are transcribed into RNAs, the noncoding strand has the same base sequence as that of the RNAs so the information content of the DNA is conveniently indicated. [RNAs, of course, have *U*s in place of the DNAs' *T*s; Chap. 2.2.] Very likely in promoter regions both DNA strands actually are involved in interactions with RNA polymerase.)

For constitutive enzymes, it has been proposed that the relative rates of transcription of different genes reflect, in part, differences in the associated promoters leading to different affinities for RNA polymerase molecules and hence to different rates at which the polymerases attach and initiate transcription. For inducible enzymes such as β-galactosidase there are additional controls: An *operator* DNA sequence is present near the beginning of the information to be transcribed. (This sequence overlaps with part of the promoter and includes also some nucleotides that code for the initial, "leader" part of the mRNA; see Section 2.3.5.) To the operator there can bind a *repressor* protein that is the product of a separate, regulatory gene. Such binding blocks transcription, probably by blocking access of the RNA polymerase molecules to the necessary promoter sequences. When inducers are present they interact with the repressor protein, binding to it and changing its conformation in ways that infere with its binding to DNA. Thus the operator sequence is freed, permitting RNA polymerase molecules to bind to the promoter and in this way "depressing" transcription. In other words, regulation depends upon specific affinities of proteins—repressors and polymerases—for specific DNA sequences, the interference of one protein bound to DNA with the binding of another, and the effects of inducer on the affinity of the repressor protein for the specific DNA sequences to which it binds.

Another regulatory DNA sequence of the β-galactosidase system, located about 60 base pairs before the site where transcription starts, is recognized by a "catabolite gene-activating protein" (CAP). This protein is sensitive to cyclic AMP. When cyclic AMP is bound to the CAP, the protein interacts with the DNA to permit transcription. When cyclic AMP levels in the cell drop, the cyclic nucleotides tend to dissociate from the CAP, the protein comes off the DNA, and transcription rates fall. The evolution of such a mechanism reflects the fact that when E. coli cells are supplied with glucose, rendering metabolism of other sugars metabolically superfluous, intracellular cyclic AMP levels drop; the CAP system then permits the cell to avoid making β-galactosidase and certain other unneeded enzymes, even in the presence of inducers. Synthesis of the other enzymes' mRNAs is also controlled in part by the CAP's binding to corresponding regulatory DNA sites. Overall, CAP affects transcription of a number of genes located at considerable distances from one another in the chromosome and in so doing can help coordinate and regulate broad metabolic patterns.

Since all three of the β-galactosidase genes share the same promoter and are under common control, the three together are referred to as an *operon*. (Alternatively they are called a *transcriptional unit,* the general term now coming into use to designate a stretch of DNA transcribed into a continuous RNA molecule—in this case, one containing information for three different polypeptide chains.) Note that the β-galactosidase operon (or *lac* operon, as it is usually called) is under both "positive" and "negative" control: The CAP system is needed to *stimulate* it; the repressor system turns it *off*.

In Chapter 5.1 we shall consider some of the properties of nucleic sequences and of nucleic acid–protein interactions that may bear on the regulatory phenomena discussed in the past few paragraphs.

Several other metabolic pathways have also been shown to be under comparable controls. For the β-galactosidase system, which is responsible for *breaking down* components of the growth medium, enzyme synthesis depends on the presence of inducers. For other pathways, such as one responsible for *synthesizing* the amino acid histidine, production of the enzymes is repressed if the end-product molecule resulting from operation of the pathway is provided in the growth medium. The evolution of such mechanisms has provided flexibility in metabolism; the synthesis of certain enzymes is adjusted to the particular environmental conditions. Since β-galactosidase may contribute as much as 3 percent (3000 molecules) of the total protein in a bacterium, it is clearly advantageous that synthesis is halted when appropriate substrates are not available or if better alternatives such as glucose are available.

Feedback Inhibition; Regulation of Ribosome Synthesis   Feedback inhibition mechanisms have been most studied for cases in which the products of an enzymatically mediated metabolic pathway inhibit the further operation of the pathway. Thus, if enzyme E catalyzes the transformation of molecules of a into molecules of b and if b is then further metabolized by other enzymes, eventually to produce $d,$ in a variety of cases it is found that d inhibits the enzymatic activity of E. In this way, the quantities of d and of b present in the cell are regulated: As excess d builds up, further synthesis is slowed down. The underlying mechanisms generally are thought to depend on the binding of products to enzymes (in our example, of d to E) with consequent allosteric (Section 1.4.2) influences on the enzyme's activity.

Conceptually similar feedback regulation operating through other molecular mechanisms probably comes into play in a variety of ways. For example, the rate at which new ribosomes are formed by *Escherichia coli,* from newly synthesized rRNAs and proteins, alters substantially with rates of cellular growth: in rapidly growing cells where many ribosomes are required to sustain continued growth, the ribosomes may represent as much as one quarter of the cell's mass and ribosomal proteins, over 10 percent of the total cellular protein. One aspect of the relevant controls is the regulation of transcription of pertinent RNAs. Notice that the production of the several rRNAs from a common precursor molecule (Section 2.3.4) obviates the need for separate controls to ensure that the different rRNAs are all produced at the same rate. The cell, however, must also regulate the synthesis of the more than 50 different proteins of the ribosomes; a recent hypothesis now attracting much interest is that feedback inhibition plays a key part in this regulation: When ribosomal proteins are accumulating at rates greater than those at which the cell is assembling the proteins into new ribosomes, the excess molecules may specifically inhibit further *translation* of ribosomal proteins. Most likely this inhibition depends on binding of certain of the proteins to the corresponding mRNAs, making these mRNAs unavailable for protein synthesis until the excess of ribosomal proteins already made is used up. (Since individual mRNAs in *E. coli* carry information for several different ribosomal proteins, a given type of ribosomal protein probably can inhibit further synthesis of several others.)

Protein Breakdown   There have been some noteworthy observations on regulation of the degradative phase of macromolecule turnover in bacteria. For instance, in rapidly growing bacteria there is very little, if any, breakdown of the cells' own proteins. When, however, the cells are starved for carbon or nitrogen by placement in a deficient medium (where they stop growing), they increase intracellular protein degradation. This increase makes amino acids and other small molecules available for essential energy metabolism, and for other activities, including, at times, the synthesis of new induced enzymes that may help the cell to exploit new "resources." Another adaptive mechanism is the ability of bacteria selectively to degrade severely abnormal proteins, synthesized as a result of mutations or of the presence of abnormal metabolites (such as certain "amino acid analogs," compounds that substitute for normal amino acids). How the proteolytic systems are organized and regulated and how specific proteins or types of proteins are selected for breakdown, while the rest of the cell is spared, are issues of broad concern since special lysosome-like degradative structures have not been found in bacteria.

Abnormally folded proteins, such as those resulting from mutations or amino acid analogues, are more readily attacked by proteases than are the comparable normal forms. Thus, to some extent, selective breakdown of such proteins could occur simply on the basis of differential susceptibility of the molecules to attack. There are correlations between relative rates of degradation of different proteins in the bacterial cell and the relative susceptibility of these proteins to attack by proteases in the test tube. However, when starved, bacteria also degrade normal proteins. Moreover, although most known proteolytic enzymes do not require an input of metabolic energy to accomplish the actual breakdown of polypeptide chains, the overall process of degradation of a bacterium's proteins is ATP-dependent; this finding suggests the involvement of phenomena more complicated than simply the operation of indiscriminate proteases that select targets at random or on the basis of abnormal folding. Three lines of investigation are actively being pursued: First, since the best-known bacterial proteases are the ones released as exoenzymes, the possibility is being studied that some of the proteins to be degraded are broken down outside the plasma membrane after being released from the cell by hypothetical energy-requiring events. Second, several groups of investigators postulate that the ATP-requiring step in protein degradation produces an initial modification of the protein destined for degradation. One group believes the modification to be the selective linking to the target protein molecules of some "marker," perhaps a phosphate group or perhaps even a polypeptide known as *ubiquitin;* this, they assert, makes the protein more prone to subsequent recognition and destruction by proteolytic enzymes and explains how the proteases can distinguish molecules to be degraded from those to be left alone. Another group believes the initial modification corresponds to a conformational change. Third, there are recent observations that certain bacterial proteases are stimulated, in the test tube at least, both by ATP and by DNA. This last finding is one of several observations raising the still-speculative possibility that selective degradation of DNA-associated proteins could be involved in the regulation of genetic activity. The stimulation by ATP could indicate that the proteases require phosphorylation or an energy-dependent conformational change to be active.

A Mild Warning Feedback inhibition, allosteric effects, and the binding of regulatory proteins are general mechanisms applying very broadly to procaryotes and eucaryotes. It does not follow, however, that all cells use the same regulatory mechanisms or that they use a given one in the same way. Procaryotes and eucaryotes differ in details of genetic apparatus and protein-synthesizing machinery (Sections 2.3.2, 2.3.5, and 4.4.6). Eucaryotes, as already mentioned, seem not to make mRNAs with information for more than one polypeptide chain (though they sometimes fragment a given chain into several different components; Section 3.6.1). Thus, operons of the sort described for the β-galactosidase system, where a single promoter controls several adjacent genes, may not exist as such in eucaryotes. Programmed differentiation of varied cell types from a common progenitor is a central feature of the normal development of multicellular eucaryotes but occurs only in a very restricted sense in procaryotes. Knowledge of the regulatory systems governing differentiation is scant (Chap. 4.4), but recent studies indicate that these systems involve features of chromosome organization and structure that seem unique to eucaryotes. The point is not that bacteria have nothing to teach us about higher forms but only that caution is essential before assuming that little fundamental has changed in the evolution of genetic mechanisms. As the next section will show, this has practical as well as theoretical impact.

3.2.6 Plasmids; Recombinant DNA

Many bacteria possess *episomes*. These are DNA structures, usually circular (see Fig. III–11), with the ability to exist either free in the cytoplasm or inserted in the bacterial chromosome. As with viruses, the integration with and exit of episomes from the bacterial chromosome seem to occur by processes related to those responsible for genetic recombination between DNA molecules. One of the first episomes discovered was the bacterial sex factor known as *F*. The insertion of F into the bacterial chromosome results in the ability of the bacterium to transfer part or all of its chromosome linked to F to another cell, a mechanism of bacterial sexual reproduction. Such transfer takes place during conjugation when a bridge is established between cells, as will be briefly outlined in Section 4.3.3. The point to keep in mind here is that conjugation is a process controlled by the F factor in order to bring about transfer of F from cell to cell; transfer of parts of the *bacterial chromosome* occurs when F has been integrated into the chromosome but is not an essential aspect of the process.

Episomes are a category of *plasmids,* DNA molecules, generally circular and containing as many as 10^5 base pairs, that can replicate and maintain an independent existence in the bacterial cytoplasm, separate from the chromosome. Different types of plasmids are present in characteristic numbers, from 1 to 100; the number of copies and other facets of plasmid replication are controlled in large measure by the plasmid's own genes. Plasmids can be transferred in the laboratory from one bacterium to another by conjugation, which several types promote. Other mechanisms of transfer quite useful in the laboratory include *transduction*—the transfer of cellular DNA by viruses—and, especially, *transformation*—the entry of purified DNA into a bacterium, where the DNA can replicate or integrate in the recipient's chromosome. Transfer of plasmids by these three routes probably can take place in nature as well; conjugation is generally judged the most important in quantitative terms.

Among the known plasmids, a number carry genes whose effect is to confer resistance to various antibiotics on the bacteria they inhabit; these plasmids thus are of medical importance. One type, for example, has genes for *penicillinase,* an enzyme that inactivates penicillin. The general notion is that plasmids can contribute to the functioning of the cells in which they are present but that their contributions are not essential to normal metabolic processes. Their relationship with bacteria can be regarded as a form of molecular symbiosis, the plasmids obtaining machinery needed for replication and transcription and the bacteria acquiring abilities to conjugate and other capacities.

The abilities of certain plasmids and viruses to enter and leave the host cell chromosome, sometimes carrying host cell genes with them, and to pass from cell to cell, introduce mechanisms for the transfer and shuffling around of DNA. The effects of these movements on the genetics and evolution of cells have only begun to be explored (Sections 4.2.6 and 4.5.1).

"Genetic Engineering" Test-tube transformation of bacteria by plasmids is one of the foundations of the techniques known variously as "DNA cloning," "recombinant DNA technology," and "genetic engineering," which have received much attention of late. Virtually any type of DNA that is of interest can be inserted into purified plasmids, which can then be used to transform a recipient bacterium. The inserted DNA will replicate as part of the plasmid so that with the growth of large populations of the transformed bacteria, virtually limitless numbers of copies of the DNA being investigated can be produced.

The insertion itself is based on the use of *restriction enzymes* produced by bacteria. Quite a number of such enzymes have been isolated from different bacteria: Each recognizes a different DNA nucleotide sequence (usually involving 4 to 6 base pairs; see Fig. III–11) and cleaves the DNA at that sequence. In nature they probably function to control the entry of foreign DNAs into the bacteria. For laboratory use, their advantage is that they cleave DNAs at the usually quite limited number of sites where the proper sequence is present, and their variety is such that an enzyme that will permit cutting intact segments of interest out of larger molecules can generally be found. It is possible, for example, to isolate particular genes or parts of genes from chromosomes; once isolated, the DNAs are inserted into plasmids by the operations shown in Figure III–11. Note that with the illustrated procedures, the same enzyme used originally to obtain the DNA fragment of interest can later be used to isolate it from the plasmids, since it is joined to the plasmid DNA by the original restriction enzyme–susceptible nucleotide sequence.

The availability of abundant copies of copies of particular DNAs has facilitated analysis of a variety of molecular biological issues. Progress is now also being made in achieving the *expression* (transcription and translation) of eu-

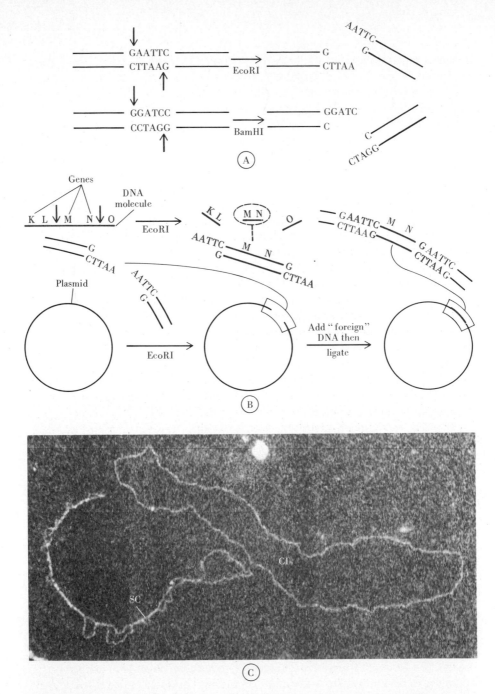

caryotic genes inserted in bacteria. The plasmid's own genes normally can be transcribed and translated in recipient cells, but for inserts of DNA from eucaryotes there are problems: These arise from the fact, emphasized earlier, that RNA production differs between eucaryotes and procaryotes (procaryotes, for example, lack splicing machinery; see Section 2.3.5), and there is a strong

◄ **Figure III–11** *Cloning of DNA through use of restriction enzymes and plasmids.*

A. *Restriction enzymes.* A variety of such enzymes can be isolated from different bacteria. Each type cleaves DNA at a different specific sequence of nucleotides, usually six (sometimes four) pairs long. The enzymes are named for the bacteria from which they are isolated; thus EcoRI comes from *E. coli* and BamHI from *Bacillus amyloliquifaciens.* The sites they cleave show symmetrical arrangements of bases; generally they are a type of *palindrome* (an arrangement of words, letters, etc. that reads the same forward and backward) in that the sequence on one strand "read" in the left to right direction is the same as that on its complementary partner, read right to left (see also Section 5.1.2). The bonds hydrolyzed by the enzymes are located at the sites indicated by the arrows, and the effect of the enzymes is to produce "staggered" cuts leaving complementary sets of unpaired bases ("sticky" ends) at the ends as illustrated.

In the bacterium, the enzymes hydrolyze foreign DNAs (when potentially susceptible sequences are present in the bacterium's own DNA, they are protected by addition of methyl groups to certain of the bases which prevents the enzymes from attacking). The experimental utility of the enzymes derives from the fact that it is generally possible to find a type of enzyme whose specificity is such that it cleaves DNA molecules of interest into fragments suitable for planned uses, as in B.

B. *Introduction of a foreign DNA into a plasmid.* A variety of plasmids are known with DNA sequences such that they are cut only at a single place by a given restriction enzyme such as EcoRI. This cut opens the circle but leaves the plasmid otherwise intact. To clone a DNA sequence from another source, such as genes *M* and *N* in this example, the source DNA is exposed to a type of restriction enzyme selected so as to release a fragment with the sequence included in it; the same sort of enzyme is used to open a suitable plasmid. This procedure yields fragment ends and plasmid ends with complementary base sequences so that the ends will base pair ("anneal") under proper conditions. The integrity of the plasmid circle can then be restored enzymatically (using DNA ligase; Section 4.2.4) producing a plasmid with a "foreign" DNA insert; this can now be cloned by growth in bacteria, as outlined in the text. Note that copies of the DNA of interest (here, *MN*) can subsequently be retrieved from the plasmid progeny through use of the same type of restriction enzyme employed to produce the combination in the first place—the insertion procedure results in the inserted DNA segment being bounded by base sequences susceptible to the enzyme.

C. Two DNA molecules, each of the *E. coli* plasmid "PSM1," as seen in an electron microscope (the scanning-transmission EM developed by J. Wall and J. Hainfeld at the Brookhaven National Laboratories). One of the molecules *(S)* is in its native, supercoiled form (see Fig. II–57C); the other has been enzymatically "nicked" during isolation and so the supercoil has relaxed, revealing the circular profile of the molecule *(CI).* (Courtesy of J. Hainfeld and S. Lippard.)

likelihood that somewhat different control mechanisms govern gene expression (Section 4.4.6). The hope, of course, is eventually to employ bacteria as factories producing proteins difficult to obtain in quantity by other means, such as some hormones, and to create bacterial strains with combinations of genetic capacities useful for particular agricultural, medical, scientific, or commercial purposes. Efforts are also under way to use "recombinant DNA techniques" to insert genes into eucaryotic cells, such as yeast, plant, or animal tissue-culture cells. This is being done by using transformation procedures, or viruses, to introduce the DNAs (Section 3.11.3). The first steps are also being taken toward introduction of DNAs, or of cells "engineered" by adding DNAs, into multicellular organisms, with the long-term goal of dealing with genetic abnormalities or constructing plants with agriculturally desirable traits.

At publication of the last edition of this book the scientific community and the public were engaged in a spirited debate over the safety and ethical issues raised by recombinant DNA technology. Now this new technology has engendered a great deal of commercial excitement. Arguments are arising over patent rights and profits, and over the relations of universities, which carried out most of the research with public funding, to the enterprises that will market products based on this research.

Chapter 3.2C Blue-Green Algae (Cyanobacteria)

Blue-green algae, the cyanobacteria, occupy a wide variety of habitats, and some are of considerable ecological or economic significance. In the context of this book, their most interesting characteristic is that the blue-green algae are *procaryotes,* resembling bacteria in many regards. A few grow as single separate cells, and many form filamentous multicellular colonies. Occasionally they are referred to as "blue-green bacteria" or "cyanophytes."

The metabolism of blue-green algae is based on photosynthesis. These algae are the most primitive plants to possess chlorophyll and in which photosynthesis produces oxygen. In addition to chlorophyll (the form present resembles chlorophyll *a* of eucaryotes; see Section 2.7.2), the algae contain unique pigments, collectively called *phycobilin;* one of these pigments *(phycocyanin)* is blue, and another *(phycoerythrin)* is red. The variability in color of different species of blue-green algae usually results from differing amounts of green (chlorophyll), blue, and red pigments.

The algae show an adaptive response to light. When grown in red light they form more pigment that absorbs red light; consequently they look blue. When grown in blue light they produce more blue-absorbing pigment and thus look red. These pigments are combined with polypeptides in the cell to form biliproteins, which can account for as much as 25 percent of the cell's dry weight. The energy they absorb (light wavelengths of 550 to 670 nm) can be transferred to chlorophyll for use in photosynthesis (Section 2.7.3). Thus, the cell can efficiently utilize a broader spectrum of wavelengths than that possible with chlorophyll alone. The phycobilin pigments along with other special metabolic features, as well as the ability to form spores that are quite resistant to the environment, facilitate growth of the algae under conditions of temperature, water salinity, dryness, and light intensity that would preclude survival of most higher plants. Many blue-green algae can fix nitrogen (Chaps. 3.2B and 3.4) and thus are capable of growing in exceedingly simple media with CO_2 as a source of carbon for photosynthesis, N_2 as a source of nitrogen, and inorganic components.

As in all procaryotes the DNA (as usual, a double-stranded helix, probably circular) is not segregated into a nucleus separated from the cytoplasm by a nuclear envelope (Fig. III–12). No endoplasmic reticulum, Golgi apparatus, or mitochondria have been observed. The polysomes are "free" and are similar to those of bacteria. Gas vacuoles often are present and permit the cells to float at water levels advantageous in terms of light and temperature. The vacuoles accumulate gases dissolved in the growth medium, by unknown mechanisms. They are bounded by protein-rich layers that seem not to be typical lipoprotein membranes (see Section 2.1.2).

The cell walls resemble the walls of bacteria: They contain lipoproteins, lipopolysaccharides, and extensively interlinked polymers involving sugars and amino acids. These polymers and their arrangement are like those of some Gram-negative bacteria (Section 3.2.3), and like the walls of many bacteria, under some circumstances the algal walls can be dissolved by the enzyme lysozyme to produce "protoplasts." (It is generally thought that bacteria and blue-green algae evolved from a common ancestral form.) An additional gelatinous sheath often surrounds the cell wall of the algae.

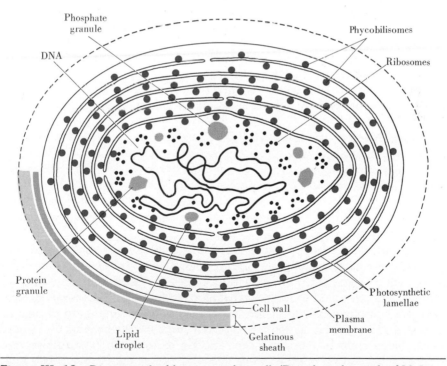

Figure III–12 *Diagram of a blue-green alga cell.* (Based on the work of M. Lefort-Tran and others.) The protein granules are suspected to include bodies composed of the dark-reaction enzyme ribulose-diphosphate carboxylase (Fig. II–62). Filamentous blue-green algae consist of chains of cells like this one, with occasional specialized *heterocysts* (see text) interspersed.

Photosynthetic Apparatus   The photosynthetic apparatus of blue-green algae is somewhat more highly organized than that of photosynthetic bacteria. The pigments are present in flattened sacs *(thylakoids* or *lamellae)* arranged in parallel array. In usual preparations their membranes show the typical three-layered appearance. In some cases the thylakoid closest to the periphery of the cell appears to be continuous with the plasma membrane; hence the plasma membrane is considered the likely source of the thylakoids.

Granules, about 10 by 30 nm in size, are observed between the thylakoids; most if not all are attached to the thylakoid membrane surfaces. These granules can be seen by electron microscopy of thin sections and of freeze-etch preparations. The granules are referred to as *phycobilisomes* or *cyanosomes*. The cells can be disrupted and fractions prepared by density gradient centrifugation. With this technique, the phycobilisomes are separated into a fraction found to contain phycobilins, while the thylakoid membranes are found in the fraction containing chlorophyll and other pigments (such as yellow *carotenoids;* Section 2.7.3). Negative staining shows the phycobilisomes to consist of subunits. To what extent the variation in subunit number and size reported by different investigators is due to different procedures and different species used is uncertain. Still unclear also are the functions of the subunits. However, a given granule almost certainly contains an ensemble of pigments in an arrangement that facilitates an orderly "downhill" transfer of energy from one to another and to the thylakoid membrane pigments (as in Section 2.7.3). One model suggests that the pigments are in layered arrays, with those absorbing at lowest wavelength (highest energy) placed furthest from the membrane, so that energy passes from layer to layer until it reaches the membrane.

Division; Motion; Heterocysts   The cells divide by inward growth of plasma membrane and wall, producing two cells of equal size as in bacteria. Mesosomes or similar structures have yet to be observed, and the mechanisms for separation of daughter chromosomes are not known.

Although they have no obvious specialized motion-related organelles such as cilia or flagella, blue-green algae are capable of gliding and rotary motions.

Heterocysts formed in filamentous blue-green algae are specialized cells occurring intermittently along the chain of cells of which the filaments are composed. They appear when the growth medium lacks usable forms of nitrogen (nitrates, ammonia, and so forth) so that nitrogen fixation is necessary. Heterocysts' walls are unusually thick; this apparently helps to limit entry of oxygen so that the nitrogenase system is protected from inactivation by O_2 (Chap. 3.2B). Since entry of nitrogen might also be expected to be impeded, the impact of the wall on entry of gases needs further study. No phycobilisomes are present in heterocysts, and the photosynthetic apparatus is biochemically simplified so that whereas ATP can be made by photosystem I activity (Section 2.7.2), no CO_2 fixation and, importantly, no O_2 production occurs (photosystem II is not present). Heterocysts communicate with the other cells of the filaments by narrow pore-like openings in the walls at the cell's ends. Probably through direct cell-to-cell continuities present in these zones (as in Section 3.4.3), nitrogen compounds pass out of the heterocysts, and metabolic products needed for heterocyst survival and function pass in from the neighboring cells.

Like spores in bacteria, heterocysts reflect the ability of procaryotes to differentiate into specialized forms upon proper environmental stimulus. Since a nucleoid is lacking, however, they cannot duplicate; new ones are produced instead by differentiation of ordinary blue-green algal cells.

Prochlorophytes   Prochlorophytes are a group of procaryotic algae that, like the blue-green algae, depend on chlorophyll-based photosynthesis. They differ from the typical blue-green algae but resemble chloroplasts of eucaryotic cells in possessing both chlorophyll *a* and chlorophyll *b;* in lacking phycobilisomes and phycobiliproteins; and in the fact that their thylakoids stack closely in pairs or larger aggregates. This group's feature of particular interest is that they occur as symbionts in association with marine invertebrate animals (tunicates); this association may shed light on the presumed evolutionary origins of chloroplasts through symbiosis (Section 4.5.2).

Only a single genus of prochlorophytes *(Prochloron)* has thus far been described. This was originally classified as a variant genus of blue-green algae but increasingly is now thought to represent a distinct evolutionary line.

Chapter 3.3 Protozoa

Protozoa have traditionally been thought of as primitive "animal-like" organisms chiefly because most are motile and because they are heterotrophic. (*Heterotrophic* organisms cannot use solely inorganic materials to make organic compounds and thus require external sources of organic materials on which to base metabolism. In contrast, the bacteria described in the introduction to Chapter 3.2B are *autotrophic,* as photosynthetic organisms generally are, in that they manufacture biological molecules from raw materials such as CO_2). Less emphasis is placed currently on trying to classify all unicellular organisms as plants or animals since the distinctions often are problematic (witness the motile chloroplast-containing protists; see introduction to Part 3) or can obscure important evolutionary relationships (Section 4.5.2).

Protozoa are single-celled eucaryotic organisms with the same basic functions and with the same organelles as those in cells of "higher" organisms. Many different types occur (amebae, ciliates, flagellates, and so forth), each with particular characteristics associated with its own mode of life. Unlike the cells of multicellular organisms, most protozoans must individually carry out all the functions required for both survival and duplication of the whole organism. For this reason, although they usually are regarded as low on the evolutionary scale, the protozoa include the most complicated and diversified types of known cells. Some are found as parasites or symbionts in higher organisms. For example, flagellates that inhabit the digestive tract of termites provide the enzymatic capacities *(cellulases)* for digestion of the cellulose in wood. The rumen of the cow's digestive system has a population of protozoa and bacteria with similar function.

We cannot begin to do justice to the protozoa in the space available. Instead, some selected features particularly illuminating for our considerations of organelle functioning and cellular controls will be discussed. The exaggerations of form and function in specialized protozoa yield excellent experimental material for a variety of concerns. Some protozoa can be conveniently grown as large populations in a medium of controlled and well-defined composition. Many are large enough for microsurgical techniques.

3.3.1 Specialized Structures

Pinocytosis and phagocytosis are used by most protozoa in feeding. Amebae often have elaborate surface coats (see Fig. II–13) that adsorb material to be taken in by phagocytosis and pinocytosis. Many other protozoans have special "mouth" regions where food—bacteria, algae, other protozoa—is accumulated, by virtue of the currents created by specialized arrangements of cilia, and is ingested by endocytosis (Figs. III–13 and III–15). The vacuoles and vesicles that form acquire lysosomal enzymes, which digest the contents. Finally, egestion, or defecation, of indigestible residues occurs by fusion of the vacuole membrane with the plasma membrane, sometimes at distinct "cytoproct" regions of the cell surface.

Secretory Phenomena   Protozoa of some species secrete a hard extracellular shell, often containing calcium or silicon. Others possess *trichocysts,* cytoplasmic bodies that fuse with the plasma membrane and extrude filamentous structures used in defense and trapping of food. (Similar but more complex bodies found in *Hydra,* called nematocysts, will be discussed in Section 3.6.3.) Most freshwater forms and some that live in salt water contain *contractile vacuoles* (water expulsion vacuoles). Especially in fresh water, the organisms are not in osmotic equilibrium with their environment, and they constantly tend to take in water. In bacteria and higher plants living in water, this tendency is counteracted by the presence of a rigid cell wall, which prevents swelling. In protozoa the water that comes in ultimately is removed by the contractile vacuoles, which can associate intimately with or fuse with the plasma membrane to expel the water at the cell surface. In some organisms, for example, *Paramecium* (see Figs. III–15 and III–16), the vacuoles are large, round, membrane-delimited bodies surrounded by an elaborate arrangement of small vesicles and branched, canal-like tubules that carry fluids to the vacuoles. Contractile vacuole functions are believed to be energy-dependent since they can be inhibited by cyanide or other inhibitors of respiration; this presumably

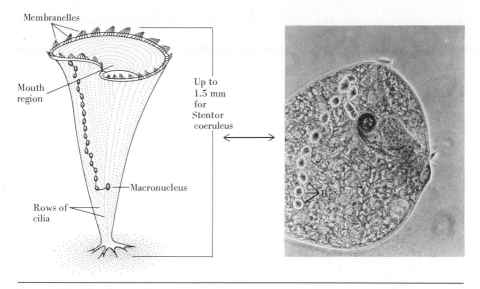

Figure III–13 *Stentor, a ciliated protozoan.*

As outlined in Sections 2.10.4 and 2.11.4, through operation of its microtubules and filament systems, this organism can assume a "trumpet" shape as illustrated in the diagram, or a shortened rounded appearance as in the micrograph. The macronucleus has the form of a chain of beads *(B)*. The micronuclei are too small to be well-demonstrated in the micrograph. Each membranelle is a row of coordinated cilia; the rows in turn are metachronally coordinated (Section 3.3.4) with respect to one another so as to produce currents that move food particles into the mouth zone. Additional cilia are aligned along the length of the body. (Diagram after M. Sleigh. Micrograph × 100, courtesy of L. Margulis.)

is why mitochondria frequently abound near contractile vacuoles. Perhaps ion pumps or other transport systems control osmotic water flow into the vesicles and tubules that feed the vacuoles; the pumps may maintain particularly low solute concentrations within these structures. The vacuole filling–expulsion ("diastole–systole") cycle can be completed in less than a minute; in an hour an organism can expel a volume of water several times its own volume.

The cell surface of ciliates is especially highly organized. Among the structures present are circular arrays of intramembrane particles found in the plasma membrane at sites where organelles such as trichocytes or *mucocysts* (these seem to contribute to maintenance of the surface) are "docked" awaiting the signal for exocytosis. The particle arrays may take part in the membrane interactions needed for exocytosis or in the triggering of exocytosis; currently, experimenters are trying to determine whether, for example, they control the plasma membrane's permeability to calcium, permitting a very localized influx of Ca^{2+} upon proper stimulation (Section 3.6.4).

Respiratory Organelles   Most protozoa possess mitochondria like the ones shown in Figure II–52B. Certain flagellate protozoa, however, contain *kinetoplasts,* unique structures having sufficient DNA for detection by light microscope cytochemistry, but having as well the typical mitochondrial membrane configurations. As the cells multiply, these structures divide. Study of kinetoplast DNA should produce insights into the replication and other properties of nucleic acids in more typical mitochondria. Much of it is in the form of quite small circles, but more typical mitochondrial DNAs are present as well. Other protozoa, such as certain of the specialized parasitic flagellates *(trichomonads)* found in vertebrates, lack functional mitochondria. The trichomonads instead possess unique "hydrogenosomes," membrane-delimited organelles that can oxidize pyruvate under anaerobic conditions and generate H_2 by reducing H^+, a most unusual adaptation to the anaerobic lifestyle of the organisms.

Cytoskeletal Structures   A few species of protozoa have giant "centrioles," rodlike structures tens of micrometers long that, like ordinary centrioles, appear to be involved in formation of the spindle for cell division. The still-limited information available suggests that the giant centrioles lack the $9 + 0$ tubule structure and otherwise differ in organization from ordinary centrioles, although their function is apparently similar. Further study may therefore help to establish which aspects of centriole structure are essential for function.

In certain of the flagellates inhabiting termites, the entire end region of the cell (plasma membrane plus contents) can rotate with respect to the rest. (The effect is somewhat as if the bleb in Fig. III–5 could rotate around its long axis once every second while the rest of the cell stays stationary.) An elongate bundle of microtubules running through the cell, with one end in the rotating region, seems to produce the rotation, but the force responsible is thought to be exerted on the microtubules by a sheath of microfilaments that encircles the microtubule bundle. For one end of the cell to rotate differentially, the plasma membrane where it joins the rest of the cell must be quite fluid; otherwise it is difficult to see how the integrity of the cell surface could be maintained.

3.3.2 Macronucleus and Micronucleus

In ciliates and some other protozoa, two types of nuclei are regularly present in the same cell. One of these often is quite large (Figs. III–13 and III–14), contains much DNA, and shows the presence of nucleoli. These *macronuclei* participate actively in producing RNA used in cell metabolism. The smaller *micronuclei* contain much less DNA; their most evident role is in sexual reproduction, when they participate in meiosis (Section 4.3.1). During sexual reproduction (*conjugation* in protozoa), two individuals exchange micronuclei through temporary cytoplasmic bridges; meanwhile the macronucleus of each degenerates, to be replaced later by alteration and growth of a micronucleus. This micronuclear transformation involves the synthesis of much DNA. In some species of protozoa, amicronucleate strains are known that are asexual but

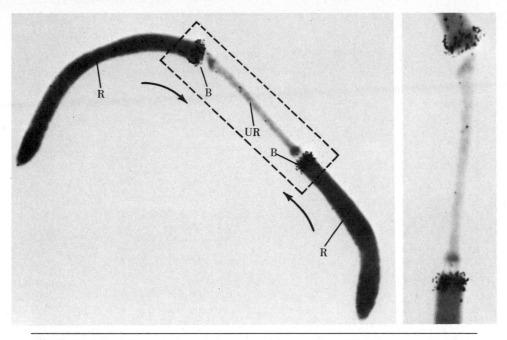

Figure III–14 *Macronuclear replication in a ciliated protozoan.* Autoradiogram of the macronucleus of the ciliated protozoan *Euplotes*. The cell had been exposed to tritiated thymidine for 20 minutes before fixation; the DNA synthesized during this interval is therefore radioactive. Grains are found only over two short lengths of the nucleus; these are the replication bands *(B)*. The bands reflect a wave of replication passing toward the center of the nucleus (arrows); the wave has not yet reached the zone at *UR*. The chromatin in the regions between the bands and the ends of the nucleus has already been replicated *(R)*. × 600. The enlargement at right is of the area in dotted lines. × 1000. (Courtesy of D. Prescott.)

survive and function with only a macronucleus. In other strains with both types of nuclei, removal of the macronucleus causes cell death even if a micronucleus is present.

Gene Amplification   The explanation for these facts is that the macronucleus contains many copies of the genetic information needed for the day-to-day functions of the cell. It seems to represent an "amplification" mechanism whereby numerous replicates of DNA, derived initially from the micronucleus, are made available for functions such as RNA production. This might explain the observations made on macronuclear regeneration in the ciliate *Stentor*. As Figure III–13 shows, the macronucleus of this organism has the form of a string of beads. All but one of the beads can be removed without killing the cell. Fragments of cells provided with a single "bead" can regenerate complete cells. If only one or two copies of given genes were contained in the macronucleus, it would be difficult to see how a complete nucleus, capable of sustaining cell life, could be regenerated from a small fragment that could not contain more than a small fraction of the original macronuclear DNA.

Direct analysis of the DNA of macronuclei and micronuclei has begun, using techniques such as DNA–RNA hydridization (Fig. II–17) and DNA reannealing (Section 4.2.6 and Fig. IV–19). One interesting finding is that in some ciliates, such as *Stylonychia,* many of the DNA sequences present in micronuclei are not found in macronuclei. As the macronucleus forms, extensive fragmentation and selective breakdown of DNA occur, as well as repeated duplication of some portions of the DNA. (During these processes the chromosomes of some ciliates transiently assume a banded polytene form like that described in Section 4.4.4.) Although differences in DNA sequences between macronucleus and micronucleus are absent or much less dramatic in other ciliates, these observations are of great potential interest, given the differences in roles of the two nuclear types. Seemingly, they imply that the macronucleus can be selectively constructed to contain many copies of the DNA-encoded information needed for its function and little or no superfluous information.

Macronuclei appear to be one device whereby the metabolic capacities of the nucleus are expanded to serve a large volume of metabolically active cytoplasm (see the following section). Polyploidy (Section 4.5.1), polyteny (Section 4.4.4), and selective amplification of specific genes as with the rDNA of amphibian oöcytes (Section 2.3.4), as well as DNA repetitiveness (Section 4.2.6), have similar effects.

Replication Asexual reproduction in protozoa is usually by division of the cell into two ("binary fission"). Informative studies of the macronuclear replication that precedes such division have been performed with the ciliate *Euplotes* (Fig. III–14). The macronucleus of this species is a long, sausage-shaped structure. Replication of the contents occurs in a progressive fashion, starting at both ends and moving toward the center. The actual points of replication can be distinguished in the microscope as a pair of bands whose appearance differs from the rest of the chromatin. This feature provides a most convenient system for studying duplication of nuclear material. The chromatin in the bands appears to transform from clumps and patches into dispersed fibrils (individual chromosomes have not been distinguished). Autoradiographic studies show that DNA synthesis occurs at the bands (Fig. III–14). On the other hand, RNA synthesis takes place everywhere but at the bands. Thus it is held that no synthesis of new RNA takes place on chromatin as it duplicates but that very soon after a particular portion of DNA is replicated it can again sustain transcription. From other evidence, this seems true of other cell types as well (Section 4.2.2). Newly synthesized proteins accumulate at the bands, suggesting that new chromosomal proteins become associated with DNA as it replicates (see Section 4.2.5). As the amount of DNA in the bands doubles, the histone content doubles. Newly labeled proteins of other sorts are also found in the replicating nucleus, away from the bands. The total duplication time for the *Euplotes* macronucleus is several hours. After duplication of its contents, the nucleus is "simply" pinched in two across its long axis as the cell divides. This is distinctly different from the behavior of the micronucleus, which undergoes ordinary mitotic divisions (as in Section 4.2.2).

3.3.3 Nucleus and Cytoplasm

The ability of amebae to survive microsurgical experiments well has facilitated a diversity of experiments. We referred in Section 2.5.3 to the influence, on the Golgi apparatus, of removing the nucleus and subsequently replacing it. Another line of investigation is based on labeling the nuclear macromolecules of one ameba by growing it in appropriate radioactive precursors, then transplanting the nucleus into another (host) ameba that has not previously been exposed to radioactive molecules. It can be shown by autoradiographic techniques that proteins and RNAs of the transplanted nucleus pass into the host's cytoplasm; some of these molecules subsequently turn up in the host's nucleus. Although the experimental system is a quite artificial one, the results suggest that there are a class of proteins and perhaps also some RNA molecules that can pass back and forth between nucleus and cytoplasm. Could such molecules serve regulatory or other functions in the interplay of nucleus and cytoplasm?

Surface-to-Volume Ratios   Other interesting experiments on the relationships of nucleus and cytoplasm have explored the hypothesis that the initiation of cell division depends on the cell's reaching a certain size. One form of this hypothesis suggests that division mechanisms have evolved so that a rapidly growing cell tends to divide before it exceeds a critical volume. For a sphere, the surface is given by $4\pi r^2$; the volume is $(4/3)\pi r^3$. For a given increase in r, the volume increases proportionally more than the surface, since cubes of numbers increase faster than squares. Although cells are not usually spheres, this sort of relationship of surface increasing at a slower rate than volume is a general one; as cells get larger, there is a *relatively* smaller surface available for exchange of material with the environment. Similarly, a fixed amount of DNA services an increasing mass of cytoplasm.

If portions of cytoplasm from growing amebae are repeatedly cut off so that the cells are prevented from attaining a certain size, the amebae do not divide. This finding suggests that one set of factors controlling cell division is related to cell size. However, the relationship is not a simple one. The growth of amebae may be curtailed by placing the cells in a nutrient-poor medium. Division is slowed but does occur despite the fact that the dividing cells are smaller than normal. Thus, while growth and division are under common controls, these controls are flexible. Perhaps some key compound must reach a critical level before division occurs, and the normal rate of production of this component is such that it usually accumulates to the appropriate level as the cell reaches a certain size. Or perhaps a much more subtle or complex control is responsible (Section 4.2.3). One experimental approach to this question is the creation of new combinations of nucleus and cytoplasm by transplantation or other procedures; such experiments will be discussed in Section 4.2.3.

3.3.4 Basal Bodies, Centrioles, Cilia, and Flagella

In ciliated protozoa such as *Paramecium,* the basal bodies of the cilia are arranged in a precise geometrical fashion (Fig. III–15). In one series of investigations, portions of the cortex (the cell surface plus the cytoplasm just below) were experimentally reoriented so that the basal body pattern was reversed. Such changed orientations persist through repeated cell duplications; some have been followed for 700 generations. This is one of a number of experiments lending weight to the idea that much of the surface pattern of *Paramecium* depends for its reproduction on local determinants rather than on the nucleus. Electron microscopy of duplicating protozoa shows that new basal bodies form in close association with old ones, suggesting some mode of self-duplication (see Section 2.10.7).

The basal bodies of ciliated protozoans show clearly the presence of a "cartwheel" structure within the 9 + 0 tubule patterns (see Fig. II–79). This is restricted to the end of the basal body opposite the end attached to the cilium or flagellum. Such observations on protozoa and other cells have raised the possibility that this structural polarity reflects a functional polarity, with one end of the organelle involved in producing cilia or flagella and the other in basal body duplication or growth. (As mentioned in Section 2.10.7, procentrioles form near the cartwheel end of the old body. Cartwheels are present early in procentriole development, with the forming organelles seeming to elongate from a cartwheel-containing base.)

Coordination and Control of Cilia- and Flagella-Based Motility
Organisms that move by means of cilia and flagella, including protozoa, show chemotactic and other changes in direction and rate of movement. These

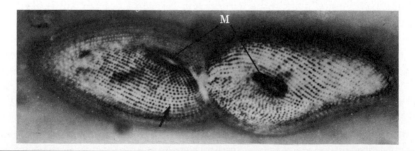

Figure III–15 *A dividing Paramecium* stained to show the basal bodies and related structures in the cell cortex (the cytoplasm just below the plasma membrane). The basal bodies are arranged in rows, which appear as lines of granules (arrow); the two daughter cells have identical patterns of basal bodies. The "mouths" of the daughters are indicated by *M.* (Courtesy of R.V. Dippel and T. Sonnenborn.)

changes stem from mechanisms controlling the frequency and form of the ciliary or flagellary beat. For example, ciliates such as *Paramecium* can alter their direction of motion when they encounter an obstacle or when they are placed in an orienting environment (such as an electrical field). They can even reverse the direction of the effective ciliary stroke (Fig. III–16) in order to swim "backward" away from an object blocking their path. Experimentally, the rate and direction of the ciliary beat are found to alter with changed ionic environments, especially changes in Ca^{2+} concentration. A convincing hypothesis derived from such observations suggests that alterations in electrical potential and in ionic permeabilities of the plasma membrane (see Section 3.7.3) also occur in protozoa under the impetus of mechanical stimuli such as an encounter with an obstacle. The result is a change in cytoplasmic Ca^{2+} concentration, mediated perhaps by channels in the ciliary membrane (Section 2.10.2) and generating, in turn (by way of calmodulin?), changes in ciliary motion. For further analysis of these phenomena, use is being made of mutant strains of *Paramecium* that show abnormalities in their patterns of motion ("behavioral" mutants).

That cilia can change the orientation of their beat is taken by many investigators as a manifestation of subtleties in the control of doublet sliding that are fundamental to movement of cilia and flagella. The cycle of sliding on which this movement is based (Section 2.10.3) must be timed so that not all doublets are doing the same thing at any given instant; if, for example, all were simultaneously being propelled toward the ciliary tip, it is difficult to see how bending could ensue. Perhaps the changes in orientation considered in this section reflect, in part, changes in the timing cycle.

Phototactic responses (orientation with respect to light) in *Euglena,* a flagellated chloroplast-possessing protist, are believed to depend on changes in permeability of the plasma membrane that occur in response to variations in intensity of light (Section 3.8.2).

Metachronal Waves The cilia on some cells beat in unison. This is true, for example, in the simple invertebrate organisms known as *ctenophores,* which swim by beating of rows of cilia; the cilia in a row exhibit synchronous effective and recovering phases (see Fig. II–81). Such *isochronal* beating is maintained, in this case, through linking of the individual cilia to one another by membrane associations. Many cells show a more complex *metachronal* wave of ciliary movement (Fig. III–16), in which steady swimming (or a steady flow of fluid past a stationary cell or cell region) results from a wavelike progression of activity through the population of cilia. At any time, some cilia are in the effective, power phase of the beat cycle, while others are recovering. Many hypothetical mechanisms have been advanced to account for this pattern, including waves of change in ionic permeability with associated electrical activity, and coordination through the systems of microtubules and cytoplasmic fibers associated with basal bodies, which can form a network linking the basal bodies together. In a few microsurgical studies aspects of coordination have been affected when

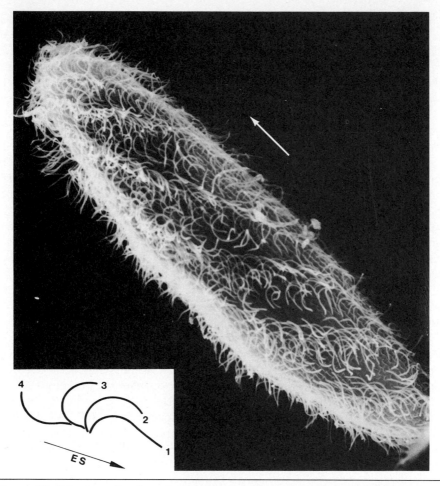

Figure III–16 *Coordination of cilia.* Micrograph of the surface of a Paramecium taken with a scanning electron microscope. The protozoan was fixed while swimming in the direction indicated by the white arrow. As the micrograph shows, the movement of the numerous cilia is coordinated so that they beat in waves (called *metachronal* waves); at a given time, some cilia are in phases of the effective stroke, and others are recovering (see Fig. II–81). The inset diagram shows the movement pattern of an individual cilium, seen from above; the arrow indicates the direction of the effective stroke, occurring in a plane perpendicular to the page, and the numerals show successive stages in recovery. × 1000. (From Tamm, S.L., T.M. Sonneborn, and R.V. Dippell, *J. Cell Biol.* **64:**107, 1975. Copyright Rockefeller University Press.)

the plasma membrane or the microtubule bundles were interrupted, but the results are inconsistent and hard to interpret. Interest has been aroused, therefore, in the possibility that metachronal coordination results in part from physical and mechanical effects, such as "hydrodynamic" interactions produced by local movements of water. Motile structures moving in close proximity to one another can fall into rhythms on the basis of effects of these sorts, without special coordinating devices. With cilia this could account for the fact that experimental changes in the viscosity (resistance to flow) of the medium in which ciliates are placed can markedly alter the pattern of ciliary waves. Hydrodynamic interactions could also explain findings made on the flagellate *Myxotricia,* which possesses, in addition to its flagella, a symbiotic population of spirochetes (Section 3.2.4) associated with its surface: Despite the absence of an obvious coordinating device such as a common plasma membrane or interconnected basal structure, these spirochetes, whose intrinsic motion is somewhat akin to that of a flagellum, beat in waves. It even appears that motion of the attached spirochetes can produce motion of the protozoan itself.

3.3.5 Endosymbionts

Many protozoa maintain long-term or permanent symbiotic relationships with other cells. *Myxotricia* with its spirochetes was just mentioned. Particularly interesting from the viewpoint of cellular evolution are the *endosymbiotic relationships* in which one cell lives inside another. Entry most often is by endocytic mechanisms; the entrant is presumably protected by mechanisms such as the ones discussed in Section 2.8.2. Both blue-green algae and eucaryotic algae have been found thriving inside species of protozoa, apparently "contributing" photosynthetic products and O_2 to the host and "receiving" the respiratory product CO_2 for use in photosynthesis, among other benefits. Bacteria also live inside protozoa: In one classic example, bacteria were originally detected in *Paramecium* through findings that some *Paramecium* strains could release particles capable of killing other strains. The particles turned out to be small bacteria normally harbored in vacuoles within the "killer" strains. Another case is the ameba *Pelomyxa,* in which no recognizable mitochondria are present but in which there are populations of endosymbiotic bacteria perhaps capable of providing the ameba with respiratory capacities.

As with other symbioses, the relative balances of benefits between the partners vary considerably, and the situation can approach parasitism. In many cases, one or the other partner can survive without the symbiotic association. However, a majority of students of cellular evolution believe that certain endosymbiotic relationships have evolved into virtually absolute interdependencies and that of these, some ancient ones account for the origin of intracellular organelles, especially chloroplasts and mitochondria. This belief will be pursued further in Section 4.5.2.

Chapter 3.4 Eucaryotic Plant Cells

Until recently, discussions of eucaryotic plant cells were frequently introduced with explanations about the slow progress of plant cell biology. Plants do pose special problems in preservation for microscopy and in biochemical analysis owing in large measure to their walls, which, for example, complicate the homogenization steps for cell fractionation. Many of these problems have been overcome, however, and progress in plant cell biology is accelerating rapidly. This is important both for the basic scientific information being gained and for practical matters of agriculture and ecology.*

Algae other than the blue-greens and all higher plants are eucaryotes. With very few exceptions their cells contain the organelles and exhibit the metabolic capacities described in Part 2. Also described in Part 2 was one of the characteristic cell organelles limited to plants, the plastids (Chap. 2.7). However, in addition to their photosynthetic capacities, plants show important differences from animals that extend to the cellular level: For example, the vascular system of higher plants carries water, minerals absorbed by the roots, and carbohydrates being transported from sites of synthesis in green tissues to sites of use or storage. A variety of small molecules are also translocated, including hormones. Unlike in higher animals, the extracellular fluids of the vascular system are not rich in macromolecules, nor is there an abundant population of circulating cells. Paralleling this, plant cells are surrounded by walls that permit passage of gases, water, inorganic ions, and other small molecules into and out of the cells but severely restrict access of larger molecules to the cell surface and also limit the interactions of adjacent cells with one another. As with bacteria, the walls provide mechanical support so that plants can grow in fresh water, and multicellular forms can circulate a relatively dilute solution; the cells can resist osmotically induced pressures without bursting. These characteristics, plus the presence in the plant cell of vacuoles controlling ion distributions within the cell, produce balances of ion and water movements that can be quite different from the ones usually found in animal cells.

What do these factors imply for the plant cell plasma membrane? It seems likely that the receptors, ion pumps, and other plasma membrane molecules are somewhat different from those in animals. There is fragmentary information suggesting that this is the case—for instance, as will be seen in subsequent sections, plant scientists have stressed transport of K^+ and H^+ across the plasma membrane more than the Na^+ transport of such strong interest to investigators of animal cells (Section 2.1.3). However, even such major relevant issues as the roles of cyclic nucleotides in plant cells are still quite unsettled. One question of interest is whether plants possess plasma membrane receptors that bind macromolecules and whether they engage in much endocytosis. At first glance neither endocytosis nor macromolecular binding would seem likely to be a major normal event for the plant cell surface given the restrictions imposed by the wall. But this probably is an oversimplified view. Excellent photomicrographs have been published showing coated vesicles apparently budding from the cell surface of algae, and when cell walls are exper-

---

*Readers with a special interest in plant science should refer to the index listing "Plant cell(s)" for material covered in this text, and to the Further Reading section at the end of this part.

imentally removed and macromolecular tracers are introduced, some plant cells seem able to take up the tracers. At the very least this might reflect cell surface turnover phenomena of the sorts discussed in Sections 2.1.6 and 2.8.4. Lectins capable of recognizing and binding specific carbohydrate configurations (Section 2.1.7) also are present at plant cell surfaces.

Nitrogen Fixation Another kind of evidence for endocytosis-like capacities, and perhaps for cell surface recognition systems that cooperate with intracellular receptors, comes from the responses of plant cells to invasive organisms— pathogens, parasites, and symbionts. For instance, the root nodules found on major crop plants like peas and beans are sites of interactions between Gram-negative bacteria of the genus *Rhizobium* and the plants' root cells. Through growth of infective threadlike structures that penetrate the roots, enzymatic activities and production of hormone-like signaling molecules by the plant host and by the invading bacteria, a sequence of events is orchestrated whereby the bacteria come eventually to reside in the host cell cytoplasm. They live there within cell surface invaginations and membrane-bounded vacuoles. Among other processes, the bacterial entry requires localized dismantling of the plant cell wall and induces a pattern of growth and division of the host cells that generates the characteristic lumplike, bacteria-containing nodules in the roots. Only certain species of host plants and specific strains of *Rhizobium* can establish the relationships described; this specificity argues for recognition devices, some of which may reside in the cells' walls or membranes. (Host cell lectins [Section 2.1.7] recognizing bacterial cell surface polysaccharides are under suspicion.)

Once inside, the *Rhizobium* cells ("bacteriods") use carbohydrates derived from the host plant's metabolism to support the energy production required for the fixation of atmospheric nitrogen. The plant is provided with a much richer nitrogen supply than that available from the nitrates and other soil sources. (This, of course, is also the basis for crop rotation in which soybeans, peas, or other *legumes* are planted periodically to re-enrich the soil with usable nitrogen.) The host cell apparently aids in maintaining an environment conducive to the functioning of the bacteriods it contains: A protein related to hemoglobin [leghemoglobin] is synthesized by the host. Along with the walls of the cells in the nodules, this substance limits and controls the amount of free oxygen that reaches the bacteria, permitting adequate amounts to support bacterial metabolism without inactivating nitrogen fixation (Chap. 3.2B). The derivatives of the host cells' plasma membranes, which surround the spaces in which the *Rhizobium* cells live, probably participate actively in the exchanges between host and bacterium.

In addition to the symbiosis-like relations just outlined, which apply only to certain plants, the nitrogen metabolism of plants in general involves other specialized features. For example, nitrates absorbed from the soil are first reduced to nitrites by a cytoplasmic (cytosolic?) enzyme system (nitrate reductase). The nitrites are then converted to ammonia and its derivatives by nitrite reductases and other systems in the chloroplasts that use the photosynthetic apparatus as the source of reducing capacities. Peroxisomal uric acid oxidase (Section 2.9.2) may also take part in the metabolic conversions by which nitrogen is made available for amino acid synthesis and other pathways.

Chapter 3.4A Algae

Algae are plants—some unicellular (such as *Chlamydomonas;* Figs. II–59 and II–61), others multicellular (some such as the giant kelps are quite large). They are classified partly on the basis of their color, which is determined largely by the nature of the pigments present in addition to the chlorophyll. For example, in red algae a major additional pigment is similar to the red pigment of blue-green algae (Chap. 3.2C).

As Figure III–17 indicates, the chloroplast membranes of algae are usually less elaborately developed than in higher plants (see also Fig. II–59). The plastids are membrane-bounded, and the internal sacs or thylakoids are arranged as long parallel sheets, with several adjacent thylakoids often associating more or less closely for substantial distances. (See Section 2.7.1. These groups are sometimes called "lamellae.") Most algae lack the stacked-coin-like grana characteristic of higher plants (see Figs. II–59 and II–60). The plastids contain ribosomes and fibrils of DNA. A *pyrenoid,* where starch is stored and probably synthesized, and a *pyrenoid sac,* which also stores polysaccharides, are part of the plastid.

The mitochondria may be like those in Figure III–17, or the cristae may be tubular as in many protozoa. The ER is relatively scanty; as usual, it is continuous with the nuclear envelope. The Golgi apparatus often lies near the nucleus in algae, and it seems likely that membrane from the nuclear envelope feeds into the outer surface of the Golgi stack through vesicles, as portrayed in the diagram. Lysosomes and peroxisomes have not been isolated from algae, but both organelles have been reported by electron microscopists. Autophagic vacuoles have been seen to increase in number in starved *Euglena,* the chloroplast-containing protist closely related to algae. In *Euglena,* too, the number of peroxisomes has been shown to increase greatly when the algae are shifted from a glucose medium to one with acetate or alcohol, suggesting to the investigators that the peroxisomes may function in lipid and carbohydrate metabolism as they do in seeds (Section 2.9.2). In contrast to most higher plant cells, paired centrioles are present in many algae.

Many algae have prominent cell walls based on cellulose, as in higher plants, but some do not. The wall surrounding *Chlamydomonas,* for example, is rich in glycoproteins and lacks cellulose; it more resembles an animal cell's coat than a typical wall of a higher plant. In certain algae the cell walls show specialized plaquelike scales. Small vacuoles, derived from the Golgi apparatus, transport the scales and other material to the forming cell walls. In diatoms, the walls are made largely of silica.

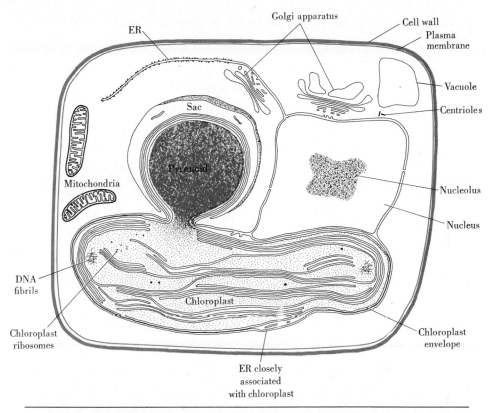

Figure III–17 *Relationships among organelles in a hypothetical brown alga.*
Not included in this schematic diagram are free polysomes and several types of
cytoplasmic granules. Note that the flat sacs within the plastid are arranged in
extended parallel arrays which tend to associate in groups of three; the groups
are referred to as *lamellae* by some authors but this term is used in other ways as
well. (Compare with Figs. II–59, II–60, and II–61). (After G.B. Bouck.)

With a large algal cell like *Nitella* (millimeters to centimeters in length), it
is possible to suck out and thus to analyze directly the contents of the central
vacuole that occupies much of the cell (as in Fig. II–96; Section 3.4.2). Much
of the conception of vacuole properties outlined below comes from such work.
These cells also lend themselves well to the investigation of cytoplasmic motility; their size facilitates observation of the cytoplasmic streaming around the
central vacuole. *Nitella* was one of the first plant cells in which the organization
of microfilaments described in Section 2.11.4 was elucidated; heavy meromyosin was used to demonstrate that the filaments are of actin.

Some unicellular algae have a single chloroplast that divides in synchrony
with the cell, consistent with the "self-duplication" of chloroplasts (Section
2.7.4). We have already outlined the proposal that the chloroplast-associated

ER prominent in some algae contributes to chloroplast growth and maintenance (Section 2.7.4).

Chapter 3.4B Higher Plants

Higher plants show a host of intriguing specializations of cell structure, chemistry, and physiology, some obvious and some subtle. For example, the varying properties of woods and plant-derived fibers such as cotton depend on the details of molecular arrangements in the cell walls of the tissues of origin. Section 2.7.5 describes and Figures II–66 and III–18 illustrate features of the cellular organization of leaves, as well as metabolic and structural specializations of the C_4 plants. This chapter presents some additional examples and deals with several major features of cells of higher plants. Figure III–19 illustrates the dominant structural features characteristic of many such cells—the presence of a prominent cell wall and of a large central vacuole. The Further Reading List should be consulted for discussions of the diversity of plant cells.

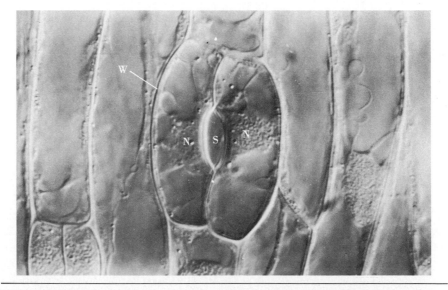

Figure III–18 *Stomatal complex* of an onion seedling photographed through a Nomarski interference microscope (Fig. I–12). The two guard cells (their nuclei are evident at N) control the extent of opening of the stoma, the space through which gas exchanges occur between the underlying tissues and the outside environment. W indicates the thin cell wall of one of the guard cells. Figure II–66 shows transverse sections through stomata. × 1500. (From Palevitz, B., and P. Hepler, *Chromosoma* 46:311, 1974.) *S* indicates the stoma.

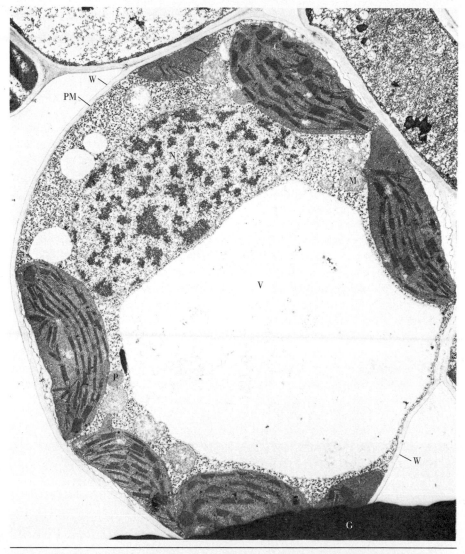

Figure III–19 *Cell of a higher plant.* This mesophyll cell (Section 2.7.5) from a leaf of *Phleum pratense* shows the major features of many mature plant cells. The vacuole *(V)* occupies much of the volume; the nucleus *(N)* and organelles such as chloroplasts *(C)*, mitochondria *(M)*, and peroxisomes *(P)* are found in a thin layer of cytoplasm at the cell periphery. The cell is separated from its neighbors by cell walls *(W)*. × 10,000. (Courtesy of M. Ledbetter.) G: grid edge (Fig. I-10.)

3.4.1 Cell Walls

Plant cell walls almost all contain cellulose, a polysaccharide made of glucose units. The cellulose molecules in the wall are in the form of multimolecular bundles or fibrils up to 25 nm thick in some algae, but generally 5 to 10 nm or less in higher plants (Fig. III–20). The fibrils can be several micrometers long. Within a fibril, the individual polysaccharide chains run parallel to each other (Fig. III–21). Approximately 30 to 75 chains are aligned in each fibril of a higher plant. The chains are associated by numerous hydrogen bonds, creating a very cohesive structure difficult to disrupt even by enzymes capable of breaking the covalent links between the glucoses. (Organisms that live on cellulose digest it slowly. Still, pathogens and symbionts can penetrate cell walls, using *cellulases* and other enzymes to do so. See Introduction to Chap. 3.3.) The point was made in Section 1.4.3 that the unbranched nature of the cellulose chains facilitates their close packing, as relatively flat structures, into compact fibrils. Figure II–44 summarizes terminology used in Figure III–21 and elsewhere in this chapter.

A matrix of other materials surrounds the cellulose; this matrix is rich in polysaccharides, mostly more complex in composition than cellulose, and involving a variety of sugars and related saccharides (Fig. III–21D). Early biochemical analyses of matrices led to the discrimination of two major categories of polysaccharide components, the *hemicelluloses* and the *pectins* ("pectic polysaccharides"), but each of these classes is itself complex. Work is still in progress on the analysis of the molecular structure of both the hemicelluloses and the pectins: The diagrams in Figure III–21D show present conceptions of the principal components of walls of tissue-cultured cells from a sycamore tree; details vary in different plants. Figure III–21 illustrates the most popular current model for the arrangement of fundamental wall components. The proponents

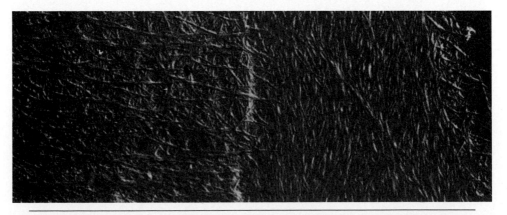

Figure III–20 *Cellulose fibrils in a cell wall.* A portion of the wall surrounding a cell of a grasslike plant, the rush *Juncus,* prepared for electron microscopy by shadowing technique (Fig. I–11). Many *cellulose fibrils* in the layer of the wall seen at the left of the micrograph run roughly perpendicular to the direction of the fibrils in the adjacent wall layer, seen at the right. × 38,000. (Courtesy of A.L. Houwink and P.A. Roelfson.)

of this model maintain that hemicelluloses are associated directly with the cellulose fibrils, linking to them by hydrogen bonds. Each hemicellulose molecule is estimated to be made of about 50 saccharide units. The pectins run between the fibrils: They have the form of quite large, multiply branched structures. In the example illustrated, the rhamnogalacturonan backbone would be hundreds, or perhaps thousands, of units long and would additionally be attached, covalently, to many elongate arabinose- and galactose-containing pectins. It is believed that the hemicelluloses are linked (perhaps to an extent covalently) to the pectins and that the pectic polysaccharides form various associations with one another covalently, noncovalently, and possibly by interactions involving Ca^{2+} (see Section 4.1.1). Thus, although a single covalently linked network is not present (as it is in bacteria), a strong, potentially quite stable network-like wall structure is formed around a plant cell.

The pectins are charged (the COOH in glucuronic acid tends to dissociate to the COO^- form), and both they and the hemicelluloses are hydrophilic. Thus the network of matrix components can take up ("imbibe") and bind water, thereby forming a gel. Pectins, especially, show marked tendencies to do this when obtained in purified form (hence their uses in food preparation). The denseness and consistency of matrix gels affect the mechanical properties of the wall, the ability of molecules to penetrate to the cell's plasma membrane, and the movement of water within the plant. Estimates of the size of molecules that *can* penetrate the matrix vary, but several studies suggest that structures larger than about 30 to 40 Å in diameter and proteins having molecular weights greater than 15,000 to 20,000 are very much impeded. In special circumstances, when plants secrete larger molecules to the space outside their walls, special channels or other devices come into play.

The interlinking of pectins and other components into three-dimensional networks can also help to bind adjacent cells together; pectins are abundant in the *middle lamella,* a layer found between adjacent cells in various higher plants (Fig. III–21). Removal of the pectins by enzymatic or other means permits cells with their walls to fall apart from one another, although the walls maintain their shape.

Matrix compounds such as *lignin,* a complex polymer that imparts strength and rigidity to the cell wall, are prominent in woody plants. In many plants, waxy substances in the cell wall help to protect cells from drying out; these substances and *polyesters* such as *suberin* and *cutin* also help the wall to resist invasion by fungi and bacteria. In plants living on land or otherwise possessing surfaces exposed directly to air, leaves and other organs are covered by a distinct protective extracellular layer *(cuticle)* made of such substances. The cuticle is interrupted at somatal openings (Figs. II–66 and III–18), permitting gases including water vapor to circulate into the underlying tissues.

Proteins can account for up to 10 percent of the cell wall's dry weight. Most probably are linked extensively to polysaccharides in the wall. One of the proteins identified in cell walls resembles the animal protein of connective tissue, *collagen* (see Fig. III–35), in being rich in the modified amino acid *hy-*

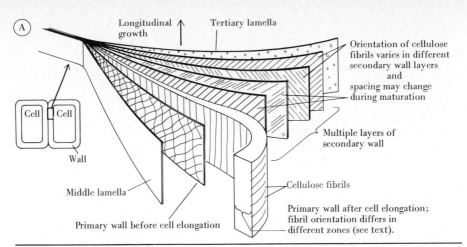

A

Longitudinal growth

Tertiary lamella

Orientation of cellulose fibrils varies in different secondary wall layers and spacing may change during maturation

Cell | Cell

Wall

Multiple layers of secondary wall

Middle lamella

Cellulose fibrils

Primary wall before cell elongation

Primary wall after cell elongation; fibril orientation differs in different zones (see text).

B Carbohydrates important in cell walls; Figs. I-26 and II-44 show glucose, galactose and mannose, and summarize terminology. (The carbon atoms of the ring structures illustrated here are indicated by the vertexes in the ring, to which other groups are joined, rather than by the letter *C*; this is a convention widely used to simplify diagrams. Sometimes, as in part C of this figure, the hydrogens attached to these carbons also are omitted.)

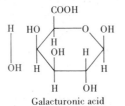

Galacturonic acid

(Glucuronic acid differs in the positions of the H and OH on 4th carbon as shown in red

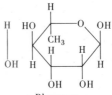

Rhamnose

(Fucose differs in the positions of the H and OH on the 4th carbon as shown in red.

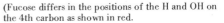

Arabinose
(5 carbons)

Xylose
(5 carbons)

C Two of the hormones that affect cell growth and influence patterns of cell wall organization. (See part B for notes on the illustration of the ring structures of the molecules.)

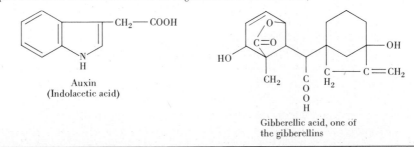

Auxin
(Indolacetic acid)

Gibberellic acid, one of the gibberellins

Figure III–21 *Plant cell walls.* **A. Schematic representation of the various layers of the cell wall of a higher plant. The chief orientation of the cellulose fibrils in the various layers is indicated. The orientation and spacing of fibrils may change during the growth of the plant (Section 3.4.1). (After K. Muhlethaler and others.) B. Saccharides (sugars and related molecules) present in plant cell walls.**

D Model of the primary cell wall of a dicotyledenous plant such as sycamore. Based chiefly on the views of P. Albersheim as summarized in Scientific American 232 #4 pp. 80-95 (1975) and in The Biochemistry of Plants (Ed. N.E. Tolbert), New York, Academic Press 1980; Vol. 1, pp. 91-162, Nomenclature of polysaccharides is based on the principal sugars present and the linkages (see Fig. II-44). Hence, for instance, *glucans* like each cellulose chain are polymers of *glucose*. For pectins especially, the structure shown is tentative. The primary cell wall of cultured sycamore cells is 23% cellulose, 34% pectin, 24% hemicellulose and 19% hydroxyproline-rich glycoprotein by dry weight.

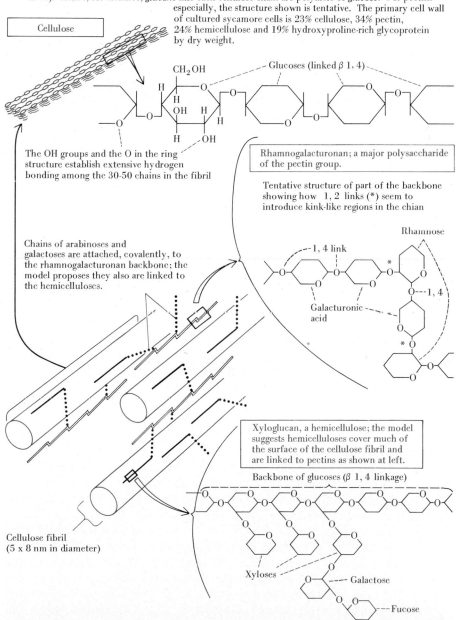

Cellulose

Glucoses (linked β 1, 4)

CH_2OH

The OH groups and the O in the ring structure establish extensive hydrogen bonding among the 30-50 chains in the fibril

Chains of arabinoses and galactoses are attached, covalently, to the rhamnogalacturonan backbone; the model proposes they also are linked to the hemicelluloses.

Rhamnogalacturonan; a major polysaccharide of the pectin group.

Tentative structure of part of the backbone showing how 1, 2 links (*) seem to introduce kink-like regions in the chian

Rhamnose

1, 4 link

Galacturonic acid

Cellulose fibril (5 x 8 nm in diameter)

Xyloglucan, a hemicellulose; the model suggests hemicelluloses cover much of the surface of the cellulose fibril and are linked to pectins as shown at left.

Backbone of glucoses (β 1, 4 linkage)

Xyloses

Galactose

Fucose

C. Two plant hormones that affect cell growth and elongation and influence other phenomena important in determining wall morphology (Section 3.4.5). D. Organization of a cellulose fibril and linking of fibrils by other components of the primary wall. The flatness of cellulose chains reflects the β 1, 4 links.

droxyproline, but these two ''structural'' proteins differ in numerous other features.

Wall morphology and chemistry vary greatly among the different plants and plant tissues and are responsible for many major features of the tissues. As we shall see in Section 3.4.5, openings present in the walls of many cells of multicellular plants permit passage of fluids. The fluid-conducting vascular system of higher plants is made partly of vessels composed of intercommunicating spaces bounded by complex cell walls left by dead cells. Cork owes its buoyancy to the closed and watertight compartments bounded by the cell walls, which enclose air-filled spaces left when the cells die. (Cell walls were the structures seen by Robert Hooke when, in 1665, he used the term *cell* in reporting results of his investigations of cork and piths from several plants.) Plant scientists also are coming to believe that the walls of growing seedlings are organized so as to transmit light to tissues of portions of the plant that are still underground; the organization calls to mind artificial fiber-optic systems.

Cell Wall Formation and Layers In higher plants, cell division is restricted largely to specialized growing zones or *meristems* present at the tips of roots and shoots, and in other organs. Some of the cells produced in the meristems continue to divide, providing for further growth while some differentiate into nondividing types of various tissues.

The wall initially laid down by a growing plant cell (Fig. III–21) is called the *primary wall.* This wall is capable of enlargement and changes in shape as the cell grows to reach its final form. As the cell matures it may produce an elaborate and rigid *secondary* wall. Secondary walls differ characteristically in different plant tissues. They often contain distinctive zones such as alternating thick ridges and thinner portions. Frequently they show distinct layers. In the walls along the sides of elongated cells, the individual cellulose fibrils in each layer are oriented parallel to one another, but the orientation differs among different layers, with the fibrils of adjacent layers sometimes even lying at right angles with respect to one another (Figs. III–20 and III–21). This layering contributes considerable strength to the wall.

The extension of the primary wall during growth requires explanation. Some facets of the process seem clear—for example, the wall as laid down is relatively plastic, and the force generating shape change often seems to derive from osmotic movements of water into the cell's vacuole as outlined in Section 3.4.2. However, precisely how the polymers in the wall change their relations during growth is still somewhat in dispute. In shoots and roots of higher plants, tissue differentiation includes substantial elongation of the cells. The wall lengthens and thickens during elongation. The overall distribution of cellulose fibrils in the elongating wall may look something like a virtually random network, as it actually is in the case of meristematic cells that maintain only a thin primary wall and do not elongate. For elongating cells, however, the apparent overall ''disorder'' masks more orderly processes; some of these processes continue in modified form during completion of the secondary wall by mature cells. Specifically, the prevailing opinion is that at each timepoint during elongation, the cellulose fibrils being newly added to the wall are oriented parallel to one another, as in secondary walls. Most investigators contend that the predominant orientation of new fibrils in primary walls is usually at right angles to the long axis of the cell so that the cell is encircled by its most recently made cellulose bundles (Fig. III–21). Although not everyone is convinced of such an orientation, it would help to provide resistance to expansion of the cell's width and thus encourage selective growth in length rather than expansion in all directions. (The fibrils at the cell's ends seem less ordered than those along its length, often appearing as a random network.)

As elongation continues, previously deposited cellulose fibrils are "buried" within the wall by deposition of newer cell wall materials, and the older fibrils become reoriented so that by the time elongation is complete, many of those in the outermost layers (the oldest fibrils) lie parallel to the cell's long axis (Fig. III–21). It is this reorientation that supposedly generates the multiple directions of fibrils seen in the primary wall as a whole—fibrils in the middle layers having orientations intermediate between those of the inner layers and those of the outer layers. If the model of wall structure as outlined in Figure III–21 is correct, such reorientation probably requires extensive breaking of bonds and establishment of new ones among the elements of the wall so that structures can slide past one another. Lytic enzymes secreted by the cells are thought to participate, perhaps under the control of and in conjunction with hormonally induced ionic changes (Section 3.4.5).

Involvement of the Golgi Apparatus, Plasma Membrane, and Microtubules The Golgi apparatus plays a central role in the secretion of cell wall polysaccharides. Figure III–22 illustrates the apparent movement of the Golgi-derived vacuoles to the cell surface where their contents are emptied by exocytosis. This release occurs when new cell walls form during cell division (see Fig. IV–28) and at other times as well. Autoradiographs of elongating cells show that the new polysaccharides being continually added to the wall are present along the whole length of the wall very soon after their release from the cell. Apparently they are added to the wall at many points rather than in a few specialized growing zones. (In a few special cases, such as the pollen tubes whose growth into the female parts of a flower carries the male gametes to the ova, there is selective addition of new material to the growing tip.)

The polysaccharides synthesized and secreted by means of the Golgi apparatus include the hemicelluloses and pectins, among others. Some of the steps completing their synthesis and linking them into the cell wall networks take place after they are released from the cell. In those algae whose walls are made of platelike cellulose-containing scales, cellulose fibrils and the scales themselves form in the Golgi apparatus as well. But in most plants the cellulose fibrils form outside the cell, at or near the surface. It is becoming increasingly accepted that the enzyme systems responsible for fibril formation (especially

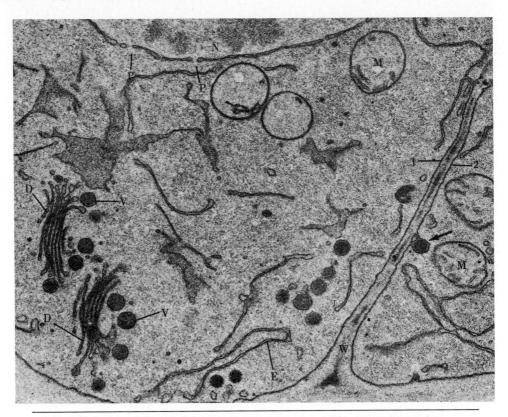

Figure III–22 *Golgi apparatus contributing to the cell wall.* Portions of two cells of corn root. The plasma membranes are seen at *1* and *2*. *N* indicates a part of the nucleus; *P*, nuclear pores. A mitochondrion is seen at *M* and endoplasmic reticulum at *E*. The Golgi apparatus *(D)* consists of numerous stacks of sacs (each stack is a "dictyosome"). The large Golgi vacuoles *(V)* contain an electron-dense material; they migrate to the cell surface, fuse with the plasma membrane (arrow), and contribute their content to the cell wall *(W)*. (Courtesy of W.G. Whaley, J.A. Kephart, and M. Dauwalder.)

cellulose synthetase) reside within the plasma membrane and move within the membrane as they "spin" out microfibrils of cellulose. This view arises from observations that cellulose fibrils often seem to form in association with rosettes of distinctive-looking intramembrane particles in the plasma membrane. Whether the Golgi apparatus contributes to these phenomena by providing precursors and perhaps the specialized membrane itself is an open question. The few suggestive reports of particle rosettes in Golgi-derived vesicles could be interpreted as demonstrating Golgi participation in passage of the synthetase system to the cell surface. This possibility, however, as well as, more generally, the involvement of the plasma membrane in cellulose synthesis, is still subject to several interpretations.

Elongating cells actively forming walls show highly oriented bands of microtubules located in the cytoplasm just below the cell surface (Fig. III–23). The cellulose fibrils forming outside the cell frequently have their long axes oriented parallel to that of the microtubules, and treatment of the cells with colchicine (Section 2.10.5) disrupts the pattern of cellulose deposition. These findings could indicate that the microtubules actually anchor or orient the cellulose-forming systems. Alternatively, they could reflect a less direct relationship, with both tubules and fibrils responding separately to common cues: The tubules for instance, may help to maintain cell shape by orienting cytoplasmic elongation until the wall has become rigid enough to take over a "skeletal" function.

Other pertinent control mechanisms need further study. For instance, the initial cellulose molecules made by the cells that produce cotton fibers are of varying length; they contain up to 5000 linked glucose molecules. When secondary wall formation begins, the number of glucose molecules per chain increases to almost 15,000, and the lengths of the newly made cellulose chains become more uniform.

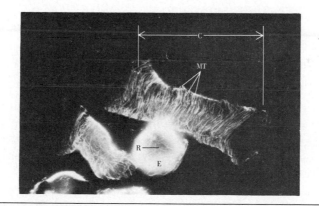

Figure III–23 *Oriented microtubules in a plant cell.* Preparation showing the region near the cell surface of an elongated cell *(C)* from onion *(Allium)* root tip. The cell was stained with fluorescent antibodies (Section 1.2.3 and Fig. II–86) specific for the microtubule protein, tubulin, and was photographed through a fluorescence microscope. This permits visualization of the numerous microtubules *(MT)*—each a bright line at this magnification—that run just below the cell surface; these cortical microtubules are seen to be oriented circumferentially, running perpendicular to the longitudinal axis of the cell. The bright object at *E* is an end-on view of another cell whose diffuse brightness is due to the overlapping of the images of many microtubules that occurs in such views. At *R*, however, the cortical cytoplasm that underlies the end wall of the cell is visible; here the microtubules assume a more random orientation than they do along the sides. As described in the text, there are correlations between the orientation of cortical microtubules and that of cellulose fibrils in the cell wall. × 640. (From Wick, S.M., R.W. Seagull, M. Osborn, K. Wever, and B.E.S. Gunning, *J. Cell Biol.* **89**:685–690, 1981. Copyright Rockefeller University Press.)

3.4.2 Vacuoles

The number and size of vacuoles varies in different cells and during development. In mature cells, a single vacuole is often present (Fig. III–19) and may occupy up to 80 percent or more of the cell volume. At earlier developmental stages, vacuoles usually occupy little of the cell volume, and several small ones may be present. The vacuole is delimited by a membrane, the tonoplast. As in algae, the cytoplasmic rim surrounding the vacuole of a cell such as that in Figure III–19 can show extensive cyclosis (rotational streaming; see Fig. II–96). The vacuole contents move in directions paralleling those of the streaming in the adjacent cytoplasm; the velocity of this motion diminishes with distance from the tonoplast, so that the fluid in central regions of the vacuole moves much more slowly than that near the periphery. Large vacuoles sometimes are traversed by strands of cytoplasm continuous with that at the vacuole rim and exhibiting streaming.

The vacuoles of many plants contain pigments such as the anthocyanins. The striking colors of petals, leaves, and fruits are due to such pigments (plus plastid pigments); these colors are important in attracting insects and other organisms that accomplish pollination and seed dispersal.

Cell Elongation; Turgor; Stomata   The concentration of salts, sugars, and other dissolved materials within the vacuole can be very high. When it is, there is a resulting osmotic entry of water through the tonoplast into the vacuole. This influx generates a pressure balanced by the mechanical resistance of the cell wall so that the cytoplasm is pushed firmly against the wall. (This pressure is usually referred to as *turgor pressure*.) The cell wall elongation that usually accompanies plant growth, described in the preceding section, is powered largely by the pressure generated by water uptake in the vacuole; the walls "stretch" under the force generated. Elongation may be very rapid (cells can elongate at rates of 20 to 75 μm per hour), but since it is primarily the volume of the vacuole that is increasing, the cell's synthetic apparatus need not generate massive amounts of new cytoplasm to keep up with the overall volume change.

Opening and closing of stomata also depends on the water balances of the cell. When the *guard cells* (see Fig. III–18) contain much water they press against their thin walls, distorting the walls' shape and opening the aperture between them. This mechanism probably is governed by osmotic phenomena reflecting the rates at which guard cells convert polymerized stores of sugar into separate sugar molecules and also the effects of ion (K^+) pumps and permeabilities of the cell surface. There is evidence for hormonal regulation (Section 3.4.5). Under excessively hot, dry conditions, when evaporation of water is rapid, water loss from the guard cells tends to close the stomata, retarding loss from deeper tissues.

Relations to Lysosomes; Storage and Degradative Functions   Through cytochemical techniques and through isolation from tissue homogenates, vacuoles of a variety of plant cells have been shown to contain acid hydrolases and thus to qualify as lysosomes. It has been suggested that vacuoles can engage in a type of autophagy in which a bud of cytoplasm protrudes into the vacuole and then pinches off and undergoes degradation. Recently some investigators have proposed that vacuole enlargement during cell differentiation and maturation may also involve autophagy; they suggest that small areas of cytoplasm are enveloped by membranes to form autophagic vacuoles, which become small, clear-looking vacuoles through degradation of their contents. The small vacuoles merge to produce larger ones; these eventually fuse to generate the large central vacuoles. The membrane systems involved in this autophagy may be similar to the Golgi-associated, lysosome-related GERL (Section 2.5.3) of animal cells. In general the ER and the Golgi apparatus have been suggested as the likely source of vacuoles, perhaps through budding of small "provacuoles" as well as through autophagy.

Degradation of materials within the vacuole produces some of the low–molecular-weight solutes that it contains. One role of the organelles thus may be in storage of small molecules—sugars and amino acids, for example—for eventual use in the cytoplasm. The tonoplast apparently contains transport systems that move materials into and out of the vacuoles. It is only recently that tonoplast purification has become practical; thus, not too much is known directly about its capacities. The nature of the vacuole contents suggests that the tonoplast probably has carriers of various kinds and ion pumps, presumably including ATP-dependent ones. Since vacuole pH can be much lower than that of the cytoplasm outside, H^+ transport systems are among those being sought in the tonoplast. The low pH may also be maintained by the presence of small acidic molecules that are among the products of vacuolar function.

In seeds, modified vacuoles store proteins and other nutrients. In addition, food reserves for seeds are often stored in *endosperm* tissues, as polysaccharides or other molecules. These tissues can be formed by endosperm cells as extracellular wall materials, or they can be produced intracellularly by cells that subsequently die. In grasses, the developing embryo mobilizes these reserves by secreting hydrolases from a specialized tissue known as the *aleurone* layer. The relations of these hydrolases to the ones in the cells' vacuoles require further study.

When plant cells die, the tonoplast becomes interrupted. Vacuole hydrolases leak out into the adjacent cytoplasm. There they participate in dissolution of the cell (autolysis). These events contribute, for instance, to formation of the fluid-conduction system: The cell walls are resistant to degradation and thus remain as hollow compartments (Section 3.4.4). In addition, autolysis is prominent during the withering and shedding ("senescence" and "abscission") of leaves and flowers; in this case it provides a sort of conservation mechanism, whereby useful soluble digestion products are withdrawn back into surviving portions of the plant prior to loss of the withered organs.

3.4.3 Plasmodesmata

Holes about 50 to 100 nm in diameter are usually found in the cell walls between adjacent cells in higher plants and many algae. Within these holes, extensions of cytoplasm are present; these are called *plasmodesmata*. Favorable views of plasmodesmata, as in Figure III–24B, show that the plasma membranes of the adjacent cells are continuous through the cell wall. Endoplasmic reticulum is often closely associated with these configurations, and extensions from the ER, of a still-debated nature, pass from one cell to the next. The frequency of plasmodesmata varies greatly among plant cell types, ranging from less than one to several dozen per square micrometer of cell surface. At the upper end of the range, the total plasma membrane surface represented

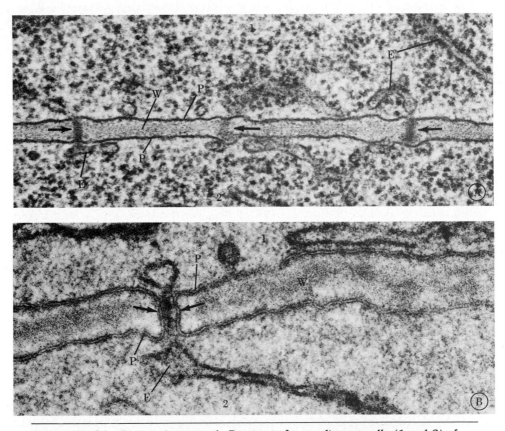

Figure III–24 *Plasmodesmata.* A. Portions of two adjacent cells *(1* and *2)* of *Arabidopsis,* a plant of the mustard family. The cell wall is seen at *W* and the plasma membrane at *P.* Arrows indicate plasmodesmata, which, in this preparation, appear as dense regions that traverse the wall as if connecting the cytoplasm of the two cells. Endoplasmic reticulum *(E)* is associated with the cell surfaces at the regions where the plasmodesmata are found. × 63,000. (Courtesy of M. Ledbetter.) B. A higher magnification view of a similar region, showing the details of a plasmodesma connecting two cells of corn. At the arrows it can be seen that the plasma membranes of the two cells are continuous, forming a channel connecting the cytoplasm of one cell with that of the other. The plasma membranes exhibit unit-membrane structure. × 130,000 (approximately). (Courtesy of H. Mollenhauer.)

by plasmodesmata can be more than one half of the total cell surface area; a large cell can be connected to its neighbors by many thousands of plasmodesmata. Exceptional cell types, such as germ cells, lack plasmodesmata.

The simplest explanation for the origin of most plasmodesmata is by the persistence of spots of continuity between dividing cells. When plant cells divide (see Fig. IV–28), cytoplasmic separation is accomplished by a system of microtubules and Golgi-derived vesicles, with the vesicles fusing to establish the boundaries—cell wall and plasma membrane—between daughter cells. Subsequently, the wall between daughter cells thickens (Section 3.4.1), and cell–cell continuities that survived through the earlier stages become distinct plasmodesmata. Although this seems to be their primary mode of origin in many cells, the possibility that plasmodesmata may form in other ways as well is still being explored.

In a sense, plasmodesmata represent a plant cell counterpart of the gap junctions of animal cells (Section 2.1.5), whose formation in plants seems precluded by the presence of cell walls. Plasmodesmata provide a pathway for the movement of a variety of materials—water, ions, nutrients, hormones—from cell to cell. In this way extensive circulation of materials within a network of interconnected cells (a *symplast*) can take place by what are essentially intracellular routes. Such an arrangement would, for example, facilitate the cell cooperation on which C_4 metabolism is based. The following section will describe the relevance of symplast organization to the functioning of the vascular system.

Still uncertain is whether large molecules generally can cross plasmodesmata. Organelles such as mitochondria or plastids do not, but some smaller structures, such as infecting viruses, apparently do (the viruses may alter the plasmodesmata structure as a consequence of or prelude to their passage).

3.4.4 The Vascular System of Higher Plants

The vascular system transports water, "minerals" (inorganic ions such as K^+, HPO_4^{2-}, Ca^{2+}), products of photosynthesis, and other components including regulatory molecules (discussed in the following section). Our interests are in the basic features of the system at the level of cells and cell walls. To bring out broad principles and issues, we will focus on vascular organization and functioning in flowering plants (angiosperms).

Apoplast and Symplast Movements of water and of solutes in plants occur both extracellularly and from cell to cell. The cell walls and the extracellular compartments they define within a given region of a plant constitute collectively the *apoplast* of that region. This term is employed to emphasize the fact that extensive networks of extracellular spaces plus the walls themselves frequently are readily accessible to water and other small components that diffuse within plant tissues. The symplast (Section 3.4.3), by contrast, is a network of *intracellular* media (''symplasm'')—the cytoplasm of cells connected by plasmodesmata.

The directions and rates of movement of water, and of solutes such as ions and sugars within a plant, are governed and mediated both by apoplastic phenomena and by symplastic phenomena. Barriers at specific extracellular locations are important, and active transport mechanisms at plasma membranes and tonoplasts are involved along with osmotic movements of water. Short-range movements of a few micrometers or tens of micrometers can occur by diffusion—''aided,'' perhaps, by the cytoplasmic streaming (Section 2.11.4) that moves materials rapidly in bulk within a cell. Long-range movements, over distances of tens of meters in large plants, occur within specialized vascular conducting and translocating tissues: *xylem* and *phloem.*

Xylem Water and minerals absorbed from the soil, as well as NO_3^- or other nitrogen compounds acquired or produced by the roots, move up through the plants by way of conducting structures in the xylem. Water and inorganic ions penetrate from the soil or other medium in which the plant is growing into the apoplast of the outer (cortical) regions of the root. Direct movement through extracellular routes into the vascular system at the core of the root is prevented by the *endodermis,* a layer of cells surrounding the vascular tissues. The endodermis is characterized by special regions of association between adjacent cell walls, rich in suberin and lignin. These *Casparian* strips (Fig. III–25) are barriers to diffusion; movement across the endodermis into the vascular system thus requires passage through the cells, an arrangement permitting selectivity and active processes of accumulation. Inorganic ions are absorbed by the cortical (and endodermal) cells through mechanisms including active transport; once in the cortical cells' cytoplasm, the ions can move through plasmodesmata to cross the endodermis along a symplastic route. Water enters the cells too, moving osmotically, since the total concentration of solutes within the cortical and endodermal cells is generally substantially higher than that in the medium in which the plant is growing. Inside the boundaries defined by the endodermal barrier, ions and water are released from the cells and pass into the xylem. The continual movement of the ''sap'' (water plus solutes) formed in this way, away from the roots through the xylem, maintains the concentration gradients needed to sustain the passage of water and minerals into the vascular system of the roots.

The units of the conducting system of the xylem (''tracheary'' or conduit elements) are called *tracheids* and *vessel elements.* The word ''element'' is used in place of cell since the units comprise cell wall–enclosed spaces (Figs. III–25 and III–26), the cells that produced them having died and disintegrated. The units individually measure a few hundred to a few thousand μm in length but are arranged in end-to-end and overlapping files and other arrays that can be tens of meters long in a tall tree. The cell walls are relatively thick and—in tracheids, for example—rich in lignin. The walls are, however, extensively perforated by openings of various kinds (Fig. III–26) that permit ready communi-

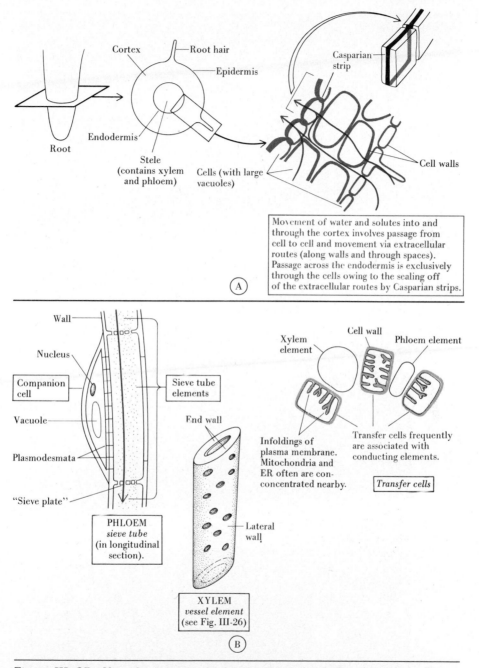

Cortex

Root hair

Casparian strip

Epidermis

Endodermis

Root

Stele
(contains xylem
and phloem)

Cells (with large
vacuoles)

Cell walls

Movement of water and solutes into and
through the cortex involves passage from
cell to cell and movement via extracellular
routes (along walls and through spaces).
Passage across the endodermis is exclusively
through the cells owing to the sealing off
of the extracellular routes by Casparian strips.

(A)

Wall

Nucleus

Companion
cell

Vacuole

Plasmodesmata

"Sieve plate"

Sieve tube
elements

End wall

PHLOEM
sieve tube
(in longitudinal
section).

Lateral
wall

XYLEM
vessel element
(see Fig. III-26)

Xylem
element

Cell wall

Phloem element

Infoldings of
plasma membrane.
Mitochondria and
ER often are con-
concentrated nearby.

Transfer cells frequently
are associated with
conducting elements.

Transfer cells

(B)

Figure III–25 *Vascular tissue and related cells of plants.* The diagram at the
upper right corner illustrates the three-dimensional organization of the Casparian
strip, which forms a band surrounding each endodermal cell at the surfaces
where the cell abuts on the neighboring endodermal cells. (After D.A. Baker,
B.E.S. Gunning, and many others.)

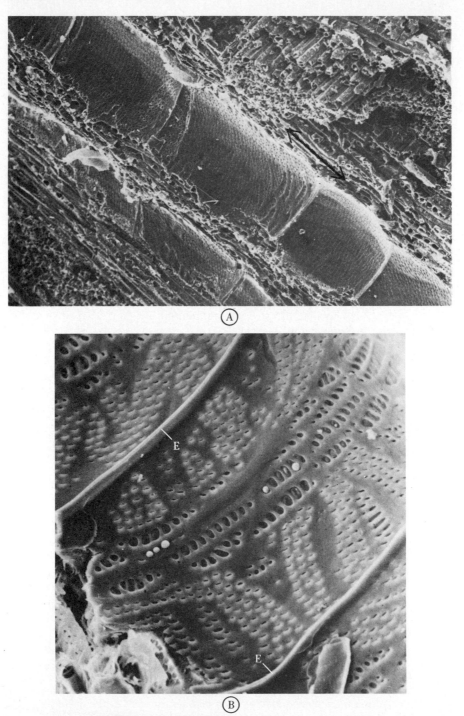

◀ **Figure III–26** *Xylem.* **Scanning electron micrographs (Section 1.2.1) of preparations of xylem elements (Fig. III–25) from oak wood *(Quercus).* Panel A, × 1,250; Panel B, × 2,750. The double-headed arrow in panel A indicates the longitudinal direction of a xylem conducting vessel that has been split open to expose its interior. In panel B the two end walls of one element are indicated by *E.* Note the numerous openings in the side walls. (Micrographs courtesy of M.C. Ledbetter.)**

cation between aligned units. These openings are produced and enlarged as the cells differentiate and die, through processes of controlled deposition of secondary wall components and degradation of portions of the wall. Tracheids intercommunicate through "pits," abundant at their end walls, in which the only barrier—a slight one—is a relatively loose meshwork of modified primary wall. Vessel elements in many plant types lose virtually all of their end walls as they differentiate and die, so that vertical files of such elements communicate directly through broad, open spaces, producing water-filled columns up to many meters long. The movement of the water, plus the solutes it carries, in such columns constitutes conduction by the xylem.

Under some conditions, the actual movement is pushed, to an extent by pressure arising through the osmotic events in the roots. A widely accepted model proposes that, in effect, a difference in osmotic pressure across the endodermis cells' membranes exists between the apoplast outside the endodermis and that enclosed by it. This gradient arises in part since solutes are continually being *absorbed* by the cortical cells and being *released* into the spaces around and in the vascular system; since the endodermal barrier prevents solutes from simply diffusing back out, a net osmotic influx of water results. However, most of the force that permits sap in the xylem to ascend many meters seems to be exerted from above. The notion is that loss of water through transpiration (evaporation through the stomata) tends to raise the concentration of solutes in the cells of aerial portions of the plant, such as the mesophyll cells of the leaves (Fig. II–66). The net result is osmotic movement of water from the vascular system into the cells. This movement in turn exerts a "pull" on the water further down in the vascular system. In this model, the water in the cells, that in the walls, and that in the conduction system is visualized as a single continuous, cohesive phase maintained by the interlinking of water molecules through hydrogen bonds; Section 1.4.1). Furthermore, the xylem structures in which water moves are relatively narrow (10 to a few hundred μm in diameter), so that adhesion of water molecules to the cell walls—capillarity—can help to sustain and to move the column of sap in the xylem.

In addition to the structures responsible for conduction of sap, the xylem contains supportive elements and living parenchymal cells. The latter function, for example, in the storage of carbohydrates and other materials. Modified parenchymal cells known as *transfer cells* show extensive infoldings of their plasma membranes and walls at the surfaces where they abut on conducting structures of the xylem. Evidently, this feature provides a large surface area through which materials such as nitrogen compounds can be absorbed from the sap for storage or subsequent distribution through the symplasts in which these cells are involved.

The wood of "woody" plants, including the annual rings of trees, is characterized chiefly by the arrangement of xylem. Often at the center—the oldest portion—of a woody stem, the xylem contains few if any living cells and no longer functions in conduction and transport. This "heartwood" still serves, however, as supportive tissue.

Phloem Phloem is the principal tissue in which carbohydrates produced by photosynthesis are conveyed from producing organs—primarily the leaves—to sites of utilization and storage such as developing fruit. The carbohydrate is translocated in the form of sugar molecules, primarily sucrose. Translocation is through sieve tubes (Fig. III–25) made up of sieve cells ("sieve elements") associated end to end. Unlike the conducting elements in the xylem, the sieve cells retain living cytoplasm; they do, however, lose their nuclei as they mature, and there is loss or modification of most other organelles, such as the Golgi apparatus and ER. Within the cells, considerable deposits of "P protein" (phloem protein) accumulate. This is not a single species of molecule but, rather, refers to proteinaceous structures, often filamentous, that abound in the cytoplasm.

The cells of the sieve tube typically are on the order of 100 to 500 μm long by 25 μm in diameter. They communicate directly with one another at their ends by modified cell walls—"sieve plates." Cytoplasm-to-cytoplasm continuity is maintained across these plates through enlarged and altered plasmodesmata. These pass through regions where the cell walls, at their ends, are correspondingly interrupted by numerous "pores." The pores range from less than 1 μm in diameter to more than 5 μm in different plant species and collectively occupy up to 50 percent of the end wall area. Their formation requires selective degradation of portions of the wall as the cells mature. The end wall is further modified by deposition of an impermeable glucose polysaccharide, *callose*, in the zones between the pores.

Associated with the cells of the sieve tube proper are *companion cells*, which maintain numerous connections, through plasmodesmata, with the sieve cells. The companion cells do have nuclei and "ordinary" cytoplasmic structure and thus are presumed to supply the sieve system cytoplasm with components that require nuclear participation for their synthesis. In many plants, sieve cells and associated companion cells have limited life spans, surviving for one or a few years. There are plants, however, in which the phloem cells live for many years (in palm trees, 50 or more). Patterns of phloem replacement, like those of xylem growth, vary, depending on the overall design and growth pattern of the plant species. Phloem also contains transfer cells and other parenchymal cell types.

The content of sucrose in the fluid within the sieve tubes can be as high as 10 to 25 percent. Movement of the sugar molecules is many thousands of times faster than rates accounted for by diffusion; estimated rates are 0.1 to

0.5 m per hour. Such rates permit, for instance, a growing potato to accumulate 50 to 100 g of starch in less than 100 days through a connection to the stem and leaves, whose phloem has a cross-sectional area of less than 0.5 mm^2.

Mechanisms of translocation within the phloem are incompletely understood. Movement is highly directional but can be in either direction; for example, sugars can be carried from the leaves to storage sites in a given season and then moved back out of storage tissues in another season when the stores are mobilized for use. It seems clear that translocation occurs primarily within the sieve tubes and depends on the extensive continuities between the longitudinally arrayed files of cells. There is no universal agreement, however, even about whether the pores are open channels for free flow. Many investigators are coming to believe they are, but when examined in the electron microscope the pores tend to be plugged by P protein or other material, perhaps as an artifact arising during tissue preparation for microscopy. (One possibility is that the pores can be reversibly opened and blocked as a mechanism regulating transport.)

Increasingly well established is that the "loading" of sucrose into the sieve tubes from a "source" such as the mesophyll tissues of a leaf is an energy-dependent process. Sucrose produced through photosynthesis is released by the mesophyll cells into the extracellular environment (the apoplast). The companion cells of the phloem, and perhaps the sieve cells themselves, utilize ATP to take the sugar from here into their cytoplasm. At present it is thought that the ATP directly powers an active extrusion of H^+ from the cytoplasm, coupled to uptake of K^+. This pump thus maintains a proton gradient across the plasma membrane. This gradient in turn drives uptake of sucrose, probably through a carrier mechanism or other system in which the sugar enters in concert with movement of H^+ down its concentration gradient (see Section 2.1.3).

The active concentrating of sucrose within the sieve tubes near "sources" results in osmotic movement of water into the tubes. Correspondingly, at the "sinks"—the sites where sucrose is being used or stored and thus is being removed from the phloem—water will tend to move, osmotically, out of the tubes. The "pressure-flow" hypothesis argues that these osmotic effects at sources and sinks generate pressures that account for bulk translocation of water and the solutes it carries within the sieve tubes. If the sieve plates are in fact largely open to bulk flow of materials, this hypothesis is the simplest explanation for movement in the phloem. But some investigators still adhere to competing hypotheses attributing active roles to P protein or to other components of the sieve tube cells in promoting longitudinal movement of phloem contents.

3.4.5 Hormones and Growth Regulators

Despite the absence both of a nervous system and of specialized endocrine glands, multicellular plants show a variety of coordinated responses to their environment that require capacities to sense key properties of the environment, and to communicate among distant cell groups. Such responses are particularly evident during plant growth and development when shoots, roots, and other specialized organs grow in directions strongly influenced by light (phototropism), gravity (geotropism), water, and other environmental stimuli. Plant hormones that stimulate or inhibit the division and growth of cells mediate such responses. Most are small molecules, ranging in size from ethylene (C_2H_4), through auxin (Fig. III–21), which is roughly the size of the amino acid *tryptophan* from which it is produced, to gibberellic acid, a carbon-ring structure slightly smaller than a steroid (Fig. III–21). Each hormone has multiple effects. Different cell groups respond differently to the same hormone, and the interaction of two or more hormones frequently is involved in determining the course of events.

The hormones are produced by a number of tissues, including especially the rapidly growing meristematic tissues at the tips of roots and shoots. Their distribution and concentration in different plant regions determine growth patterns. The movement of these "growth substances" within the plant can occur through the vascular system, through plasmodesmata, and by other routes. It is still not certain, however, how gradients and local concentrations are built up at sites distant from those of hormone production. Some of the transport of auxin is energy-dependent. One hypothesis suggests that the cells transporting auxin are somehow polarized to produce differences in the balance between auxin entry and auxin release at opposite ends of the cell. If even a slightly higher proportion of auxin is released at one end than at the other, and if this is repeated through a chain of cells, each passing auxin to the next, by the end of the chain a very substantial net directionality will have been imposed on the transport. A variant of this concept proposes that pH gradients control the distribution of hormones between the cell and extracellular spaces or between communicating cells (as with weak bases equilibrating between lysosomes and the media outside; Section 2.8.4). A difference in such gradients at the two ends of the cell could favor net uptake at one end and release at the other.

Hormones control the dormancy and senescence of plant tissues that is a prominent part of the annual cycle—the relative inactivity in cold seasons and the withering and shedding of leaves mentioned in Section 3.4.2. They also stimulate the local growth necessary to heal wounds and thus minimize entry of potential pathogens. On the other hand, it is suspected that auxin secreted by invading *Rhizobium* bacteria helps to evoke some of the responses in the plant host that lead to nodule formation (see introduction to Chap. 3.4), and that fungi which normally live symbiotically in the roots of trees also secrete auxins to induce a flow of nutrients into the regions they inhabit.

Major Plant Hormones The *auxins* (indoleacetic acid [Fig. III–21] is the chief natural one known; others have been synthesized in the laboratory) control the elongation of growing cells in regions such as stems or roots. They are therefore crucial in many of the tropisms as well as in determining features such as the branching patterns of stems. Auxin can both stimulate and inhibit elongation depending on its concentration and the sensitivities of the responding cells. Some of the apparently opposite effects of auxin on different tissues may result from its stimulation of production of other hormones such as ethylene. This type of situation is common in plants and animals: The effects of a given hormone may be modulated or extended by its evoking production of other hormones, whose influences include some feedback effects opposed to the direct effects of the first hormone.

A much-debated model proposed to account for auxin stimulation of cell elongation suggests that it engenders altered bonding among cell wall elements, making it easier for the wall to be changed in shape and size. This altered bonding could follow auxin-induced increases in secretion of H^+ ions by the responding cells; the resulting drop in pH in the wall matrix would loosen hydrogen bonds and perhaps stimulate lytic enzymes. Both of these last effects could result in a more pliable wall structure. Several facts indicate that this is a plausible hypothesis: Auxin-induced elongation of cells is accompanied by release of acid; adding acid to growth media can speed elongation; and some wall-associated enzymes work best at low pH.

Auxin is destroyed by an enzyme of the peroxidase type called *indoleacetic acid oxidase*. This enzyme is no doubt important in control of the hormone's levels. The peroxidase from horseradish—much employed as a cytochemical tracer (Fig. III–31)—may serve this function, among others, in its plant of origin.

The *cytokinins* stimulate cell division. In tissue cultures this is enhanced when auxin is also included in the growth medium, suggesting that the two hormones cooperate. It is striking that alteration of the relative amounts of auxins and cytokinins included in tissue culture media favors formation either of rootlike or of shootlike structures, through different balances among cell division, cell elongation, and growth of cells in width. In other words, through overlapping sets of effects—some cooperative, some antagonistic—the relative levels of different hormones can help to determine the directions of tissue formation.

The *gibberellins,* of which there are a variety, also can stimulate cell elongation, sometimes doing so at developmental stages different from those controlled by auxin. Gibberellins seem not to show the highly localized transport patterns of auxin and thus tend to have general effects on growth, in contrast to the often highly localized auxin effects that can produce directional growth as in the tropisms. Gibberellins also control the production of enzymes such as *amylase* in plant embryos; the enzymes release sugars from storage forms. The gibberellin-evoked production of amylase requires synthesis of new mRNAs (as do some effects of auxin): Changes in transcription of RNAs is one of the mechanisms by which a variety of hormones, plant and animal, are thought to act (see also Section 4.4.6).

Ethylene stimulates the sequence of biochemical changes responsible for the ripening of fruit, such as changes in respiratory rates and the hydrolysis of storage materials to liberate sugars. Together with other hormones, including auxin, it controls the changes leading to leaf withering and abscission (falling off).

Abscisic acid is often described as a "growth inhibitor" whose effects tend to counteract those of other hormones. It is one of the major hormones that produces the metabolic inhibitions and other changes involved in "dormancy," the inactive, nongrowing state characteristic of many seeds and plants under conditions of rigor, such as winter. One interesting physiological effect of administration of this hormone is on the stomata; these are closed, evidently by abscisic acid operating through influences on ion (K^+) distribution and thus on osmotic "turgor" of the guard cells (Section 3.4.2).

For the most part, the modes of action of these various hormones at the cellular level are yet to be elucidated in detail. There are many changes in metabolic patterns in the tissues that respond, and it is too often not clear which are primary and which secondary. Some of the effects occur within seconds after hormones are applied to experimental preparations, some take minutes, and some take hours or even days. The hormones often are effective at very low concentration, so it is presumed that they frequently act by means of receptors that can amplify their effects through second messengers such as ions, through changes in permeability, through interactions with the genetic apparatus, and so forth.

By analogy to animal systems it has been postulated that auxin effects and transport depend on the hormone's binding to receptor proteins in membranes and elsewhere. Proteins and membranes that can bind auxin have been prepared from plant tissues; their nature and roles are still being sought. For most researchers, it is not yet certain that the ability of these preparations to bind auxin is biologically significant.

Sensing; Telling Time   Organisms respond to forces such as those of gravity or acceleration (and sometimes of magnetism) through events thought to be triggered by particles or fluids whose orientation or motion responds to the forces. These particles or fluids often are extracellular but sometimes are intracellular (Fig. III–10). In these connections and others the sensory apparatus of plants has only begun to be understood. For example, the annual cycle of plants—growth, flowering, dormancy, and so forth—is closely timed to the seasons. This cycle involves responses to temperature, but it also depends on capacities of plants to detect the relative length of daily dark and light periods, which varies with time of year. The length of the night seems an especially important determinant.

From studies of the effects of different wavelengths of light upon flowering and other plant cycles, it has been concluded that one participant in the system that permits plant responses is *phytochrome*. Phytochrome is a protein that absorbs red light. (Another plant sensory protein, less thoroughly analyzed, absorbs blue light.) On absorption of red light (wavelengths of about 660 nm), phytochrome is converted into a form that absorbs maximally at 730 nm (the "far red" part of the spectrum). In the dark, or upon illumination at 730 nm, the far-red-absorbing form converts back to the red-absorbing form. (Thus, phytochrome-mediated effects of illumination at 660 nm can be reversed by illumination at the longer wavelength.) Plants growing in sunlight have much of their phytochrome in the far-red form during the day; reversion to the red form takes place at night. By the state of its phytochrome, a plant can tell whether it is day or night (the same system enables it to detect the extent of shading by other plants, which can also lead to adjustments in growth pattern.) As yet the "clock" that permits the plant to tell how *long* the day or night is has eluded detection. Phytochrome itself is largely localized in membranes; it appears that some of the effects of the transition to the far-red form on plant physiology come from changes in permeability or ion pumps affecting distributions of ions such as K^+, and of water. Many interesting cellular effects have been observed: For example, in algae, illumination at red wavelengths leads to orientation of chloroplasts with their broad surface perpendicular to incoming light; in contrast, illumination at longer wavelengths leads the chloroplast to orient with its long axis parallel to the incoming light (a position in which the narrow edge of the disc faces the light, minimizing light absorption).

Phytochrome mediates a very large repertoire of responses throughout the life cycle of many plants. Many of its effects involve the hormones discussed earlier in this section.

Chapter 3.5 Tissues; Absorptive Cells; More on Junctions

3.5.1 Tissues: The Small Intestine

Figure III–27 diagrams a cross section of the vertebrate small intestine; this organ, like many others, contains the four major tissue types formed in higher animals: epithelia, connective tissues, muscle, and nerve. This section serves to introduce these tissue types. Subsequent chapters will deal in more detail with specific components and features of particular interest to cell biologists.

Epithelia Epithelia serve as covering and lining tissues, at absorptive surfaces, in ducts, in skin, and elsewhere. They are continuous sheets of cells. Where the tissue plays primarily a protective or barrier role, as in skin, the epithelium is often several cells thick (stratified). At absorptive surfaces, where molecules are selectively passed across the epithelium, the sheet usually is one cell thick. The absorptive surface of the intestine is thrown into a series of folds (Fig. III–27) that provide a greatly increased absorptive area in comparison to that of a simple smooth-walled tube. (A much greater surface increase results at the cellular level, from the presence of microvilli; see Section 3.5.3.) The intestine is lined by a single layer of epithelial cells that selectively absorbs material from the *lumen,* the space enclosed by the organ. The cells are re-

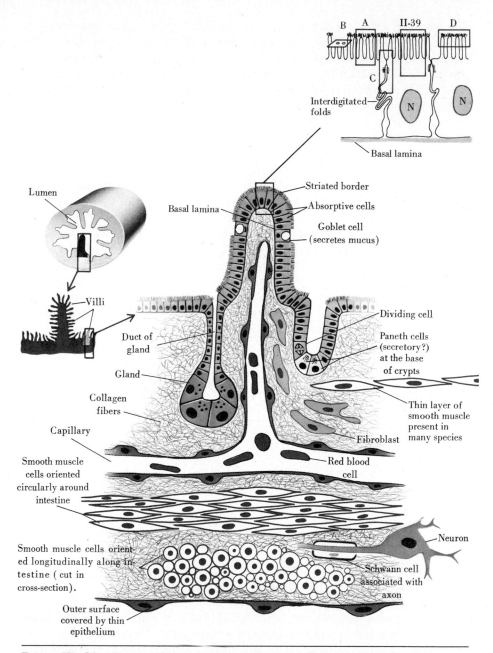

Figure III–27 *Tissues of the intestine.* Schematic diagram of major cell types and tissues of the mammalian intestine. Figures II–39 and III–28 A–D are electron micrographs showing the regions indicated by the outlines in the diagram of the epithelium.

ferred to as *columnar* because they are taller than wide. The epithelium rests on a *basal lamina,* a mat of extracellular fibers and other material that, in part, provides structural support and anchorage. The cells at the tips of the villi are continually sloughed off into the lumen and replaced by others that migrate from further down. For many mammals the average stay of a cell at the surface of a villus is about 2 days. The population is maintained by the division of cells in the indentations *(crypts)* found at the base of the villi.

In the skin the cells at the outer surface of the epithelium fill with a fibrous protein, *keratin,* providing a tough, impermeable layer. As the cells keratinize they die and eventually slough off from the surface, to be replaced from a dividing population at the base of the stratified cell layers.

Gland cells also are classified as epithelial, in part because during embryonic development glands arise from layers of epithelial cells. The intestine contains three types of gland cells. *Goblet* cells scattered among the absorptive cells secrete mucus that lubricates and protects the lining. Other cells thought to be secretory (some controversy persists) are present at the base of the villi and in the crypts (Fig. III–27). Multicellular glands are present well below the absorptive surface in the first part of the intestine, near the stomach. Their secretion passes into the lumen through ducts; it is alkaline, and probably, among other functions, neutralizes the acid in the stomach contents that enter the intestine.

By means of ducts, the intestine also receives secretions from epithelial cells of the pancreas (Fig. II–33) and of the liver. The pancreas contributes enzymes that digest macromolecules; the liver secretes bile (Chap. 1.1), which helps to break up fat particles into small aggregates ("emulsification"), making them more susceptible to digestion.

A terminological convention should be noted: In absorptive epithelia the *base* of the cell is the region facing the capillaries, and the *free surface* or *apical region* is the opposite pole where absorption occurs. A comparable convention is used for exocrine secretory cells (Chap. 3.6A); the apex of such cells is the zone where secretions are stored and released.

Connective Tissues   Connective tissues serve to bind other tissues together, providing support and a framework within which blood vessels, lymphatic vessels, and nerves course. They are composed of cells surrounded by abundant extracellular materials, usually produced by the cells with which they are associated. In fibrous connective tissues, widespread in the body, *fibroblasts* produce extracellular fibers, consisting mostly of the protein *collagen*. The fibers are surrounded by a matrix (to be discussed in Section 3.6.6); the matrix contains proteins and such polysaccharides as *hyaluronic acid*. Elastic connective tissues containing fibers of the protein *elastin* are found in the walls of arteries and elsewhere; such tissue provides resiliency to organs. Cartilage consists of cells surrounded by a stiff matrix of polysaccharides and proteins; bone cells are surrounded by calcium salts in an organic matrix. Blood also is often classified as a connective tissue; large numbers of red blood cells and fewer white blood cells are carried in the protein-rich *plasma*.

Adipose tissue, where fat is stored, is also regarded as a form of connective tissue. Fat is deposited as large intracellular globules within *adipocytes,* often filling most of the cytoplasm. Groups of such cells, surrounded by more fibrous connective tissue, make up the fat deposits present in various places in the body.

The connective tissue of the intestine is largely of the fibrous kind. In it, blood capillaries and lymphatics are present. These transport oxygen to the epithelium and other layers and remove nutrients absorbed from the lumen, as well as CO_2 and other waste products. The lymphatic vessels are particularly concerned with transport of fat. In addition, macrophages and other cells of the body's defensive system are present in the connective tissue of the intestinal wall to ward off possible entry of the microorganisms present in the lumen of the digestive system, and to deal with other potentially toxic or injurious materials.

The outer layer of the intestine consists of a thin connective tissue layer within a covering of *squamous* (flat) epithelial cells. This outer epithelial layer is continuous with the *mesentery,* a thin sheet of epithelial plus connective tissue which attaches the intestine to the body wall.

Muscle   There are three major types of muscle tissue: the voluntary *skeletal* muscle (Section 2.11.1), the involuntary *cardiac* (heart) muscle, and *smooth* muscle. The first two types show cross striations. In the intestine a thin layer of smooth muscle cells underlies the surface epithelium and influences the overall shape of the epithelial layer. Separated from this by a thick region of connective tissues are two additional smooth muscle layers. In the inner one, the cells are oriented circularly around the intestine. In the outer, the cells have their long axes oriented longitudinally along the intestine. The presence of two outer layers of muscle cells with the cells in one layer oriented perpendicular to the cells in the other makes possible peristalsis and other complex movements of the intestine.

Nervous Tissue   Nervous tissue consists of the impulse-conducting nerve cells *(neurons)* and certain associated cells called the *neuroglia* and *Schwann cells.* Some of the associated cells cover much of the surface of neurons and probably serve to modify neuronal activities.

One of the prominent groups of nerve cells in the intestine is disposed as a layer between the two outer muscle layers. These cells help to integrate the functions of the intestine with those of the rest of the body.

3.5.2 Junctional Structures

Cells of epithelia are held together in sheets or clusters by special junctional structures. There are several types of such structures; the particular ones present vary in different epithelia. *Gap junctions* have already been described (Section 2.1.5).

Tight Junctions; Desmosomes   In the absorptive epithelium lining the intestine, there are regions near the lumen where the outermost layers of plasma membranes of adjacent cells touch focally, obliterating the intercellular space that usually separates adjacent cells (Figs. III–28 and III–29). These regions are known as *tight junctions* and occur in a continuous band around each cell. This means, of course, that a given cell forms tight junctions with each of the several neighbors that abut on it.

Some distance below the tight junctions are *desmosomes* (Figs. III–28 and III–29). In contrast to the tight junctions, desmosomes occur in localized areas, or patches, rather than as a continuous band. A given cell forms several with each of its neighbors. In desmosomes the plasma membranes of the two cells are separated by an intercellular space of 20 to 30 nm. The extracellular material within the space may appear slightly denser than elsewhere, and sometimes it seems to be organized in closely associated layers. A current interpretation (Fig. III–29) suggests that the layered appearance reflects ordered systems of fine linking structures that run from one plasma membrane to the other. The adjacent cytoplasm shows a meshwork of filamentous materials, including microfilaments. Additional filaments from other portions of the cytoplasm, notably the intermediate-type "tonofilaments" described in Section 2.11.2, also can converge upon desmosomes. They associate with an electron-dense plaque found on the cytoplasmic side of the membrane (Figs. III–28 and III–29). This association of filaments with desmosomes is particularly evident in cells subject to much mechanical stress and is thought to reflect mutual anchoring of cytoskeletal systems and the desmosomes.

The Junctional Complex   In the intestine and other tissues, the junctional structures near the lumen are known collectively as the *junctional complex* (Fig. III–28C). The tight junction is sometimes called a *zonula occludens;* the desmosome, a *macula adherens.* The region between desmosome and tight junction includes a *zonula adherens* ("belt desmosome," "adhering junction," or "intermediate junction"). Its structure generally is not as distinctive as the other regions but often somewhat resembles the structure of desmosomes except that in place of a series of localized patches, it forms a "belt" around each cell. The two plasma membranes at belt desmosomes are separated by about 20 nm, and a band of filamentous material, microfilaments, and some intermediate filaments encircles each cell, lying just below the plasma membrane. This band merges with the "terminal web" (Fig. III–28A) of filaments at the cell apex and with the desmosomal filaments below.

Gap junctions in the intestinal epithelium occur below (basal to) the junctional complex, establishing cell-to-cell communication (Section 2.1.5). Additional desmosomes and interdigitations of folded regions of adjacent cell

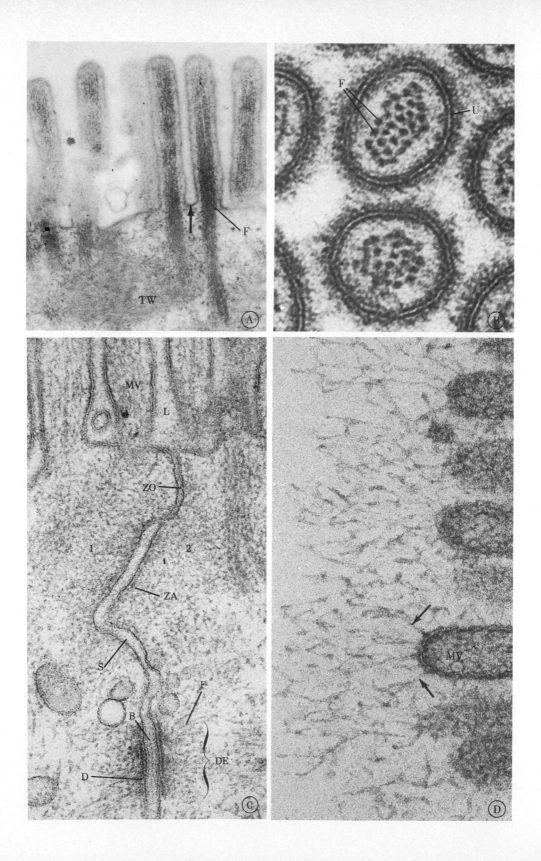

◄ **Figure III–28** *Intestinal epithelial cells.* Electron micrographs of the regions of intestinal epithelial cells indicated on the diagram in Figure III–27 (see also Fig. II–39).

A. The microvilli are fingerlike extensions of the plasma membrane (arrow); each contains a core of microfilaments *(F)* that merges into a zone of filaments and amorphous material (the terminal web, *TW*) in the cytoplasm below the microvilli. × 50,000. (Courtesy of J.D. McNabb.)

B. Cross section of two microvilli. The three-layered ("unit") plasma membrane is seen at *U*. The core filaments are seen transversely sectioned at *F*. × 275,000. (Courtesy of T.M. Mukherjee.)

C. Junctional complex between two absorptive cells *(1* and *2). MV* is a microvillus on one cell; *L* is part of the lumen of the intestine. The desmosome *(DE)* and tight junction (zonula occludens, *ZO*) are seen. *ZA* indicates a "zonula adherens" (see text). At the desmosome, cytoplasmic filaments *(F)* radiate from a dense line *(D)* adjacent to the plasma membranes (the three-layered membrane appearance is barely evident at this magnification). An additional faint line bisects the intercellular space between the cells *(B)*. × 96,000. (Courtesy of M. Farquhar and G.E. Palade.)

D. Tips of microvilli from a section specially prepared to show the polysaccharide-rich surface coat. The coat consists of fine filaments, some of which may be seen attached to the plasma membrane (arrow). × 160,000. (Courtesy of S. Ito.)

---

surfaces (Fig. III–27) also are present along the lateral cell surfaces; these contribute to the mechanical integrity of the epithelium.

Roles of Desmosomes and of Tight Junctions Desmosomes participate in anchoring cells together in epithelia. Under experimental conditions (for example, shrinkage) in which cells tend to pull apart from one another, they often remain attached to one another at desmosomes (see also Section 3.9.2). Since the "cytoskeleton" may also be anchored at desmosomes, these junctions seem to accomplish a kind of structural integration of both the surfaces and the cell interiors of multicellular arrays. "Half-desmosomes" sometimes form where a cell abuts on extracellular surfaces such as the basal lamina; the cell evidently can use the specialized membrane, extracellular material, and filament arrangements it establishes at desmosomes to help it to adhere to such surfaces.

Tight junctions are also widely found in epithelia; they may, to a limited extent, help to anchor cells together, but they also have a quite different function from that of desmosomes. It has been shown that tracer molecules—metals dispersed as fine particles and proteins such as peroxidase (Section 2.1.6)—do not penetrate into tight junctions, although they ordinarily pass readily through the spaces between cells, even at gap junctions (Section 2.1.5). For this reason it is believed that tight junctions can seal a surface by preventing material from penetrating between the cells. Thus, in the intestine, kidney, and other absorptive sites, movement through the cells themselves is the only means of passage across the epithelium (Section 3.5.3). Movement *through* cells is a more selective mechanism than movement *between* cells. In the latter, the size of a molecule as compared to the dimensions of the intercellular space is the chief limiting factor; in the former, many controlling devices determine movement (see Chap. 2.1).

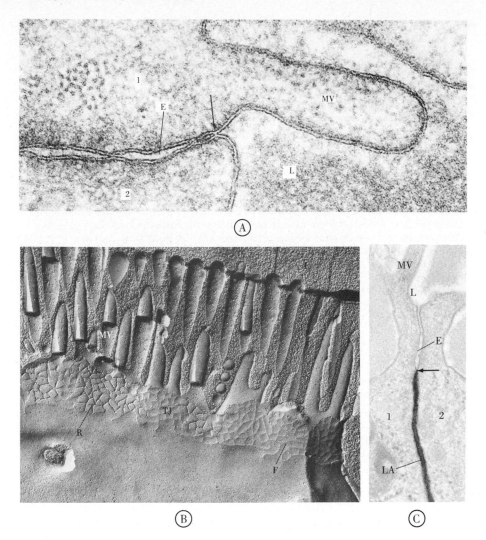

For secretory cells (Fig. III–29C) tight junctions can prevent secretions released selectively at a cell pole bordering on a secretory lumen (*see* Fig. II–33) or capillary (Fig. II–70) from leaking back between the gland cells. Less certain is whether they also limit diffusion in the plane of the plasma membrane. It is frequently asserted that when a tight junction is present at the border between two domains of a given cell's surface, such as between the apical and lateral surfaces of an intestinal absorptive cell, membrane molecules from one domain cannot diffuse into the adjacent one. This, then, may be one device whereby cells can maintain specialized regions of their plasma membrane. The weight of opinion still strongly favors such a viewpoint. However, insofar as proteins are concerned, the evidence is not always unambiguous since anchoring of proteins to cytoskeletal

◀ **Figure III–29** Part I. *Tight junctions; electron micrographs.*

A. Region of a tight junction between two adjacent cells (*1* and *2*) of rat intestinal epithelium. *MV* indicates a microvillus; *L,* the lumen of the intestine; and *E,* the extracellular space between the cells (see Fig. III–27). The plasma membranes bordering the cells show the usual unit-membrane appearance. At several points the adjacent plasma membranes of the two cells are closely apposed to one another; at one of these points (arrow), the membranes touch "focal fusion"), completely obliterating the extracellular space *(E).* × 150,000. (Courtesy of G.E. Palade.)

B. Freeze-fracture preparation of rat intestinal epithelium. *MV* indicates microvilli; *L,* the lumen. The fracture plane has passed through a region of tight junctions between adjacent cells (TJ), providing a face view of membrane structure at these junctions. Where thin sections show focal fusion (see part A), freeze-fracture preparations show ridges *(R)* and furrows *(F)* in the membranes. (The ridges and furrows represent the complementary membrane faces mentioned in Figure II–6 legend; the course of the fracture plane through the region is complex so that different membrane faces are seen at different points. The ridges are in one of the membrane's P faces and the furrows in an E face, but there is controversy as to which of the two plasma membranes is the source of the P and E faces seen.) × 35,000. (From Friend, D.S., and N.B. Gilula, *J. Cell Biol.* **53:**758, 1972. Copyright Rockefeller University Press.)

C. Tight junction (arrow) between two adjacent cells (*1* and *2*) of the rat pancreas. *MV* indicates a microvillus; *L,* the lumen into which secretions are released (see Fig. II–33). Electron-opaque lanthanum salts have been introduced into the extracellular spaces surrounding the secretory units (acini; Fig. II–33). Much lanthanum has penetrated into the space between the cells *(LA),* but its progress has been blocked by the tight junction; the extracellular space above the junction *(E)* is free of the tracer. × 35,000. (From Friend, D.S., and N.B. Gilula, *J. Cell Biol.* **53:**758, 1972.)

Part II (Segments D and E) on page 394.

---

elements may limit diffusion even when tight junctions are disrupted. It appears also that some lipid molecules in plasma membranes may be able to traverse the boundaries, diffusing, in the cytoplasmic half of the bilayer, through the tight junctional region. Still, it would be expected that where a membrane participates in establishing configurations like the ones shown in Figure III–29D, lateral movement of macromolecules of that membrane, through the zone involved in the configurations, would be impeded by comparison with movement elsewhere in the membrane.

When intestinal cells slough off from the villi, junctions must be broken between the departing cells and those that remain behind, and the continuity of the epithelium must be reestablished rapidly. This process has only begun to be studied. The observation (Section 2.1.5) that gap junctions functionally close when Ca^{2+} levels rise may be germane. The concentration of calcium free in the cytoplasm normally is kept low through energy-dependent mechanisms. When a cell dies, however, the energy supply or other conditions needed for those mechanisms may fail, leading to a rise in Ca^{2+} concentration and a consequent closing of gap junctions. This mechanism could permit the epithelium to seal off what were channels of communication between healthy cells and a dead or dying cell and thus protect the surviving tissue from loss of

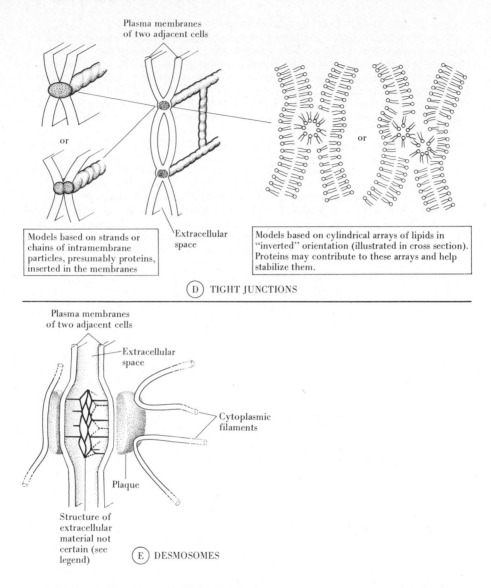

Plasma membranes
of two adjacent cells

or

Models based on strands or chains of intramembrane particles, presumably proteins, inserted in the membranes

Extracellular space

Models based on cylindrical arrays of lipids in "inverted" orientation (illustrated in cross section). Proteins may contribute to these arrays and help stabilize them.

or

(D) TIGHT JUNCTIONS

Plasma membranes
of two adjacent cells

Extracellular space

Cytoplasmic filaments

Plaque

Structure of extracellular material not certain (see legend)

(E) DESMOSOMES

metabolites or unbalanced influx of materials from the dead cell or the extracellular environment.

First Steps in Molecular Analysis   It is self-evident that the properties of the various junctional types depend on the specific proteins, lipids, and other molecules from which they are assembled. The characteristic freeze-fracture particles of gap junctions (Fig. II–10) are believed, reasonably, to be principally of protein molecules. Progress has been made in isolating and identifying pro-

◀ **Figure III–29** Part II. *Models of tight junctions and desmosomes.*

D. Proposed structures of tight junctions. Most models (left) have suggested that the junctions correspond to systems of "strands," usually presumed to be chains of transmembrane proteins, inserted in the membranes: the two plasma membranes might share a single strand or each might have its own, with the two strands touching as illustrated. Alternative models (right) have recently been proposed in which in place of chains of particles the strands are thought to represent cylindrical arrays of lipids in "inverted" orientation (that is, with their "polar heads" [Fig. II–3]) clustered and their hydrophobic tails associated with the hydrophobic central zone of the plasma membrane bilayers). (After Pinto de Silva, P., and B. Kachar, *Cell* 28:441–450, 1982; Kachar, B., and T.S. Reese, *Nature* 296:466, 1982; Staehelin, L.A., and B.E. Hull, *Scientific American* 238 (5):140–152, 1978.)

E. Proposed structure of a desmosome shown as if cut approximately in half. For an electron micrograph see Figure III–28. The structure and nature of material in the extracellular space is still being investigated: there does appear to be an organized layer midway between the cells but more must be learned of the structure of this layer and of the linking elements that attach the cell to this layer. As noted in the text, cytoplasmic filaments seem to be associated with the desmosome plaques as if anchored there; the filaments extend considerable distances from the desmosome. (Based largely on the Staehelin and Hull reference in part D.)

---

teins from gap junctions, but it is too early to relate properties of these proteins to junctional properties. Desmosomes also show freeze-fracture particles that some authors interpret as views of the linking structures that may cross the space between the cells and insert in the two plasma membranes. The extracellular space at desmosomes contains carbohydrates, presumably linked to glycoprotein proteins in the membrane but perhaps including some truly extracellular materials as well.

The strandlike structures in the membranes at tight junctions were initially assumed to represent primarily aligned protein molecules, but recently an alternative hypothesis suggesting they are organized largely of lipids has been advanced (Fig. III–29). Perhaps further investigation will reveal that both lipids and proteins are major structural elements of these strands.

Septate Junctions In insects and other invertebrates, *septate junctions* are seen at which the plasma membranes of adjacent cells are widely separated, but where fine "bridges" *(septa)* run from one to the other (Fig. III–30). (Somewhat similar-looking junctions have also been seen, but only rarely, in vertebrates.) These probably serve in cell-to-cell adhesion, though other roles—especially in limiting diffusion through intercellular spaces—have also been proposed and are now being evaluated.

Capillary Walls The junctions between the *endothelial* cells lining capillaries influence the passage to and from the blood stream of molecules such as sugars or proteins, which cannot diffuse readily through the cells as water or gases

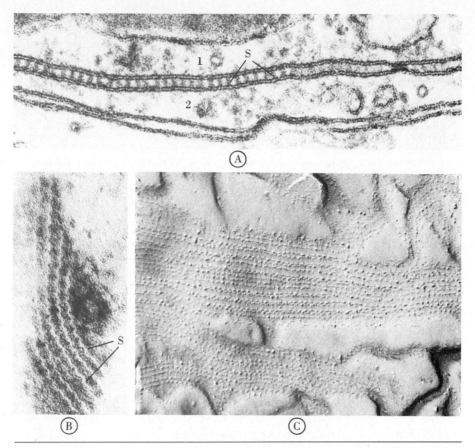

Figure III–30 *Electron microscopic views of septate junctions.* A. In sectioned preparations the plasma membranes of the two cells *(1,2)* at such a junction show thin "bridges"—septa *(S)* crossing the extracellular space between. × 150,000. B. A suspension of electron-dense lanthanum salts introduced extracellularly penetrates between the septa, accounting for the electron-dense (dark) deposits in this micrograph. The micrograph provides a face view of the intercellular space separating two plasma membranes; the area shown is equivalent to that between the two membranes in part A. The septa *(S)* appear light since they are outlined by the lanthanum; they are seen, in this junction, to have the form of narrow pleated ribbons. The "bridges" in part A are actually transverse sections of such ribbons. The ability of lanthanum to penetrate septate junctions suggests the junctions do not seal off the space between the cells (compare with Fig. III–29C). × 175,000. C. Freeze-fracture microscopy reveals particles arranged in rows within the P faces (Fig. II–6) of the plasma membranes; each row of particles in the membrane is probably associated with one of the septa in the intercellular space. × 100,000. (Micrographs courtesy of Nancy J. Lane.)

can. In some tissues, such as brain, adjacent endothelial cells are associated by tight junctions; no macromolecules move between blood and brain tissue, and exchanges of many other molecules depend on specialized local portions of the circulatory system (such as "choroid plexuses") that control movements between brain and blood. In other tissues, various less restrictive arrangements are found, and there are corresponding variations in the movement of material. For example, in the liver most molecules can pass readily through relatively large gaps in the endothelial lining of the modified capillaries called *sinusoids* (see Figs. I–1 and I–2). This facilitates the extensive exchanges that take place between the hepatocytes and blood stream. (An example of such exchange is the passage of albumin, a major protein component of blood plasma, which is synthesized and secreted into the blood by hepatocytes.) Along with their other roles, the phagocytic Kupffer cells (macrophages) found just outside the sinusoids probably serve to filter out potentially harmful materials that "escape" from the blood through the permissively structured sinusoid walls. In many endocrine glands, small "fenestrations" (Fig III–31) 100 nm or more in diameter are seen in capillary walls. These openings may have a fine nonmembranous layer across them, but this, it seems, does not prevent the passage of the secretions into the blood. Many of the capillaries in the intestinal villi are fenestrated. Most muscle capillaries lack such openings, presumably reflecting the absence of major macromolecular traffic from muscle cells to blood.

Capillaries also show varying numbers of cytoplasmic vesicles that can acquire microscopic tracers introduced in the circulation (Fig. III–31). These provide a transport route across the capillary wall, supplementing movement through the cytoplasm proper and the often restricted movement between the cells. The vesicles may form at one surface of the capillary and then pass to the opposite surface of the endothelial cell, where they release their contents by exocytosis-like fusion. But recent studies suggest they also can fuse with one another and with vesicles still continuous with the plasma membranes at the two cell surfaces, thus forming a chain of fused vesicles that opens a transitory channel across the capillary wall (Fig. III–31). In other words, to participate in movement of materials from one side of the endothelium to the other, a given vesicle need not itself move the entire distance and may not even have to separate from the cell surface.

3.5.3 Absorption in the Kidney and Intestine

The kidney and intestine well illustrate major features of absorptive processes in multicellular organisms: They provide examples of the use of cell surface and extracellular enzymes to prepare materials for absorption, employment of the several types of transport mechanisms available to the plasma membrane for moving materials into the cell, and subsequent intracellular metabolism of the internalized materials.

Kidney Tubule Cells   As Figure III–32 schematizes, a filtrate of blood enters the kidney tubules ("nephrons") from the modified capillaries that constitute the *glomerulus*. The capillary walls are fenestrated, showing many porelike openings a bit less than 100 nm across. Interposed between the capillary and the tubule lumen is a thin epithelial "capsule." The cells ("podocytes") of the capsule extend numerous slender processes that cover the surface of the cap-

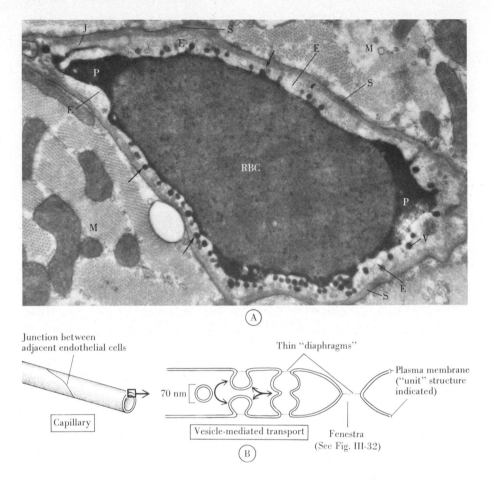

illary; these processes interdigitate snugly and extensively with one another but remain separated from one another by narrow slitlike extracellular spaces through which material from the capillaries can pass into the tubule lumen. Impelled by blood pressure, water, ions, and molecules up to the size of small proteins move through the fenestrations and slits and enter the kidney tubules from the blood; cells and most of the proteins normally present in blood are retained in the capillaries because their size and other properties preclude passage through the endothelium and capsule. The meshwork of extracellular (basal lamina; Section 3.6.7) material surrounding the capillary serves to impede passage of larger molecules and structures out of the capillary.

The tubule cells modify the filtrate they receive into urine, which then passes through ducts to the bladder. The tubules' activities make essential contributions to the mechanisms by which vertebrates control the volume of fluid and the concentrations of solutes—salts, glucose, urea, and so forth—in their blood and in the tissue fluids that exchange materials with the blood.

◀ **Figure III–31** *Capillaries.* A. A capillary (cut transversely) in the heart muscle of a rat that had been injected intravenously with the enzyme *peroxidase* 5 minutes prior to fixation. The tissue was reacted cytochemically so that sites reached by the peroxidase are marked by a dense reaction product. Most of the space within the capillary is occupied by a red blood cell *(RBC)*. Surrounding this is a peroxidase-filled region, the plasma *(P)* suspending the blood cells. The capillary wall consists of thin endothelial cells *(E)* outside of which a space *(S)* separates the vessel from the muscle cells *(M)*. Many peroxidase-containing vesicles *(V)* are present in the endothelial cells; some (arrows) are seen still to be connected to the cell surface (see also Part B). At *J* peroxidase has penetrated part way into the space between adjacent endothelial cells but is prevented from passing all the way across because a tight junction is present. × 25,000. (Courtesy of M.J. Karnovsky.)

   B. Diagram of a portion of an endothelial cell showing a fenestra, such as are present in some capillaries, and the vesicles present in virtually all. Vesicle-mediated transport across the capillary wall involves both movement of vesicles across the cytoplasm and the transient formation of channels across the endothelial by fusions of vesicles. The "diaphragms" across the openings of vesicles seem to be rich in carbohydrates, probably of glycoprotein side chains. The diaphragms on the fenestrae bear negative charge, thought to be contributed by glycosaminoglycans like heparan sulfate (Fig. III–37). These differences in composition between the diaphragms at different sites are an interesting example of local specialization of cell surface material. It seems likely that the differences help influence the routes taken across the capillary wall: the diaphragms do permit passage of a variety of molecules, including macromolecules, but negatively charged molecules would be expected to be discriminated against by the negatively charged fenestral diaphragms due to the repulsion of like charges. (After Simeonescu, M., N. Simeonescu, and G.E. Palade, *J. Cell Biol.* **94**:406.)

The tubule cells reabsorb salts, sugars, water, and other components, passing much of the reabsorbed material back into the blood in the capillaries that run near the tubules. The rates of reabsorption are adjusted by a variety of hormonal and other regulatory devices, so that when concentrations of particular solutes in the blood are abnormally high, increasing amounts pass through the kidney and are excreted in the urine. Reabsorption through the plasma membranes of the tubule cells is accomplished by passive diffusion or by active and facilitated transport, depending on the component. Associated with movement of solutes are osmotic movements of water in directions controlled by the solute concentrations in the tubule lumen and other extracellular spaces, and by the permeability properties of the cells' plasma membranes.

For example, the cells of the portion of the loop of Henle that ascends from the tip toward the collecting tubule (Fig. III–32) actively pump NaCl out of the lumen, into the spaces surrounding the tubules. (Most physiologists believe that it is Na^+ that is actively transported, with Cl^- following passively, but a persistent group argues that Cl^- is the actively transported ion.) This portion of the loop is impermeable to water, which otherwise would pass osmotically out of the lumen. The portion of the loop that descends from the glomerulus toward the tip is permeable to water and to NaCl. The net effect is

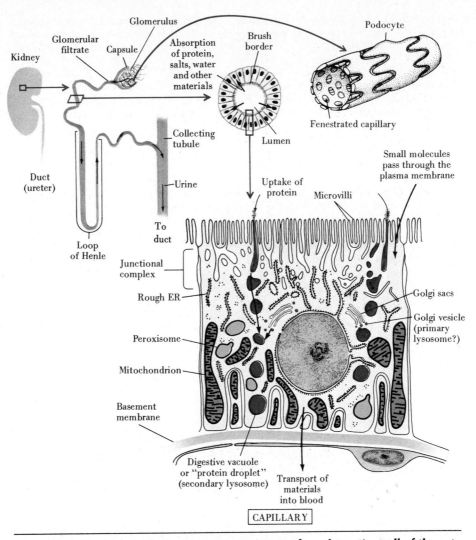

Glomerulus

Glomerular filtrate

Kidney

Capsule

Absorption of protein, salts, water and other materials

Brush border

Podocyte

Fenestrated capillary

Collecting tubule

Lumen

Duct (ureter)

Urine

Small molecules pass through the plasma membrane

Uptake of protein

Microvilli

To duct

Loop of Henle

Junctional complex

Rough ER

Golgi sacs

Golgi vesicle (primary lysosome?)

Peroxisome

Mitochondrion

Basement membrane

Digestive vacuole or "protein droplet" (secondary lysosome)

Transport of materials into blood

CAPILLARY

Figure III–32 *Kidney.* Schematic representation of an absorptive cell of the rat kidney. The fenestrated capillaries of the glomerulus and the slitlike spaces between adjacent podocytes in the capsule act as a filter (see text). Through selective absorption of some water, salts, proteins, and other components the filtrate of blood that enters the tubule from the glomerulus is modified into urine. Many of the salts and small molecules absorbed by the tubule cells are returned to the blood in the capillaries at the base of the cells. The diagram also illustrates the uptake and fate of protein, which is digested within the tubule cells.

that overall the loop of Henle functions as a so-called countercurrent system that produces a gradient in concentration of Na$^+$ and Cl$^-$ between different regions of the spaces *outside* the tubules. For our present concerns, the important consequence is that quite high concentrations accumulate in some of these spaces, such as near the tips of the loops of Henle. The collecting tubules run through these spaces; under physiological conditions in which the walls of the collecting tubules are permeable to it, water flows osmotically from the collecting tubule lumen into the extratubular zones where the ion concentrations are maintained high. From these regions outside the tubules, the water can be returned to the blood stream. These phenomena therefore conserve water for the organism by minimizing the amounts excreted. As water exits, the solutes remaining in the fluid within the collecting tubule, including waste products such as urea, become more concentrated; the urine produced passes to the bladder for subsequent excretion.

The permeability of the collecting tubule wall to water is under the control of *antidiuretic hormone* (ADH) released by the pituitary gland; ADH increases permeability. ADH release is controlled, in turn, by sensory cells in the brain, blood vessels, and elsewhere that monitor solute concentrations in tissue fluids and the pressure in the circulatory system; this pressure changes when, for example, blood volume is altered. By determining the levels of ADH release the sensory cells can adjust water balances in the body. In addition, the steroid hormone *aldosterone* produced by cells in the outer zone (cortex) of the adrenal gland, helps to regulate Na^+ reabsorption. Release of this hormone varies in response to changes in blood pressure and ionic concentrations monitored by cells in the kidney and elsewhere.

The amounts of protein that filter into the kidney tubules from the blood stream vary in different animals and under different physiological and pathological conditions. Normally the amounts are quite limited. In some animals, such as rats, a variety of proteins introduced experimentally into the blood stream do enter the tubules in appreciable quantity. Study of proteins detectable in the microscope indicates that in the initial portion of the tubule (the "proximal convolution"), proteins are absorbed by the cells so that except in pathological conditions, few of the molecules are lost in the urine. The proteins enter the cells in pinocytic vesicles that form at the ends of infolded tubular ("canalicular") structures projecting into the cell between the bases of the microvilli present at the luminal surface (Fig. III–32). These vacuoles merge to form larger *apical vacuoles* that move toward the base of the cell and acquire lysosomal hydrolases, probably by fusion with preexisting lysosomes. Proteins are broken down within the resulting secondary lysosomes, or *protein droplets* as they often are called. The soluble digestive products are probably utilized by the kidney cells.

Microvilli Both for the kidney tubule cells and for the cells lining the small intestine, the absorptive area at the lumen surface is greatly increased by enormous numbers of microvilli; these are arranged sometimes in what appear as precise geometric arrays and collectively constitute the *brush border* or *striated border* (Figs. II–39, III–28, and III–32). In the human intestine there are up to 200,000 1-μm-long microvilli per square millimeter of intestinal surface, so that the overall membrane surface of the intestinal lining reaches 30 square meters. For the intestine, it is considered that enzymes, built into and attached to the extracellular surface of the brush border, such as *disaccharidases,* hydrolyze various carbohydrates and other molecules and in so doing participate actively in digestion. The surface is coated with filaments 25 to 50 Å thick that are especially prominent in humans, cats, and bats. They are attached to the outer surface of the plasma membrane and are rich in carbohydrates, including carbohydrate chains linked to membrane glycoproteins. The chief role suggested for the filaments is as a filter that keeps large particles from approaching the plasma membrane. However, some enzymes of the brush border may also be associated with the filaments. Other enzymes are inserted, as integral proteins, in the microvillar plasma membrane. Both filament materials and the cell-associated digestive enzymes continue to be synthesized and moved to the brush-border surface as the cells migrate up the villi (Section 3.5.1). This accounts in part for the prominent ER and Golgi apparatus present in the cells and permits replacement of components lost or inactivated by the wear-and-tear of luminal events.

The occurrence of microvilli illustrates one evolutionary "solution" to the surface–volume "problem" raised in Section 3.3.3. The effective surface area is greatly increased by specialized cell shapes. Spherical shapes have minimal surface; alterations from the spherical increase the surface-to-volume ratio.

Many cell types not specialized for absorption nonetheless form microvilli or other surface projections. Those of the intestine are, however, unusually abundant and stable. The stability probably derives in part from the ordered microfilament core. This core merges at its base with a filamentous zone, the terminal web, which has the form of a layer located beneath the apical surface of the cell (Fig. III–28). The web blends at its borders with filaments associated with the junctional complex. The interlinking of filaments in these zones very likely provides architectural support. Whether the filament systems also mediate movements, or participate actively in separating groups of cells from the epithelium as they slough off (Section 3.5.1), requires further study (see Section 2.11.4).

Intestinal Triglyceride Absorption; Chylomicra and Other Lipoproteins Triglycerides are absorbed by the cells lining the small intestine. The few pinocytic vacuoles that are seen at the base of the microvilli (Fig. III–28) play little, if any, role in the absorption of triglycerides. (Tracer experiments show that the vesicles take up proteins and other macromolecules and deliver them to lysosomes, but since the intestinal lumen and brush border contain an abundance of digestive enzymes, such uptake probably is generally of minor importance for digestion of foods.) Triglycerides are digested in the intestinal lumen *outside* the cell and enter the cell, at the microvillar surfaces, in the form of small molecules: monoglycerides, fatty acids, and glycerol. These move into the ER where they are used in resynthesizing triglycerides and other lipids (see Fig. II–39). The lipids combine with specific proteins, producing lipoproteins within the ER. As outlined in Chapter 2.4, triglycerides, phospholipids, and steroids are thought to be made by ER, and protein is made by ribosomes attached to rough ER. In the apical region of the cell there are many continuities between rough and smooth ER, so that materials can move readily between the two. An extensive network of ER is present a short distance below the microvilli. Lipoproteins are transported by the ER toward the Golgi apparatus and possibly toward the lateral cell borders. Most of the lipoprotein is converted to droplets called *chylomicra* (or chylomicrons). The chylomicra apparently empty from the intestinal cells into extracellular spaces at the base of the epithelium by fusion of the membranes that enclose them (Golgi vacuoles or vesicles budding from smooth ER) with the plasma membrane at the lateral and basal regions of the cell. From the extracellular spaces, the droplets enter the lymphatic system through which they are carried to the blood stream for transport to the liver, *adipose* (fat-storing) tissue, and elsewhere.

It is potentially instructive that after a fatty meal has been eaten, when a great deal of fat enters the cell, smooth ER becomes more abundant, possibly by loss of ribosomes from rough ER and by conversion of the cisternae and tubules into an interconnected meshwork and vesicles. After a fat-rich meal, the Golgi apparatus also contains lipoprotein resembling that seen in the ER. Most likely this reflects addition of carbohydrate components to the chylomicra by the Golgi apparatus, as part of packaging processes like those described in Section 2.5.2.

Chylomicra are one of several types of lipoprotein particles present in the blood. Other types include the VLDL particles synthesized and secreted by the liver (Section 2.4.4 and Fig. II–47). The surface of these particles is a layer of molecules with hydrophilic zones—including proteins and phospholipids. This enables the particles to remain in suspension in the blood despite the abundance of hydrophobic molecules such as triglycerides and cholesterol derivatives (cholesterol esters) present in their cores. The proteins also mediate specific interactions of particular lipoprotein types with particular cells such as the binding to receptors that leads to endocytosis (Section 2.1.6 and Fig. II–11). The different types of circulating particles interact with one another, exchanging components and undergoing transformations as a result. Some, such as the chylomicra, are subject to hydrolysis by lipases built into the surfaces of capillary endothelial cells; this makes the triglyceride components, carried by the chylomicra, available for metabolism by the adjacent tissues (and converts the particles themselves into a type of lipoprotein known as *remnant* particles, which are then further metabolized by the liver). These various processes distribute lipids to storage tissues and to sites of use, and regulate their concentrations in the blood. A number of important human disorders are attributable to defects in the functioning of one or another aspect of lipoprotein metabolism (Section 2.8.5 and Chap. 5.2).

Distribution of Mitochondria In the kidney tubule cell (Fig. III–32) many elongate mitochondria are vertically aligned, in close relation to the plasma membrane at the base of the cell, near the capillaries. The plasma membrane shows deep infoldings. In addition, the lateral surfaces of adjacent cells fit together by protrusions from one cell fitting into indentations of the next; this interdigitation extends to the bases of the cells, so that a thin section of one cell shows mitochondria apparently in membrane-bounded compartments that are actually extensions of neighboring cells not included in the section. The mitochondria have a great many cristae and high levels of oxidative enzymes; thus they have the capacity to produce much ATP. The abundance of mitochondria and the complex folding of the plasma membrane are devices that make energy available to a large area of cell surface that is actively transporting ions and other substances (reabsorbed from the tubules) from the cell to the blood.

In the intestinal cell there are no basal interdigitations, and the mitochondria show no special orientation at the cell base. The *apical* mitochondria are more striking: They are very elongated and are oriented lengthwise in the cell, parallel to the ER strands that are also concentrated in the apical cytoplasm. Here, presumably, the mitochondria provide energy for the active processes involved in absorption of various molecules and, perhaps, in lipid synthesis.

Chapter 3.6   Secretory Cells, Not All in Glands

Chapter 3.6A Varieties of Secretions

We have encountered several secretory cells in the chapters on ER and Golgi apparatus (Chaps. 2.4 and 2.5). In pancreas (Figs. II–33), pituitary (Fig. II–46), and intestinal glands (Fig. II–42), as well as in others, proteins are manufactured on the polysomes and are transported by means of the ER to structures associated with the Golgi apparatus; here they are "condensed" into granules or viscous fluid as they are "packaged" into membrane-delimited vacuoles. These vacuoles subsequently open to the surface by fusion of the vacuole membrane and the plasma membrane to discharge the secretion (exocytosis). We further mentioned the likelihood that extracellular coats lining cell surfaces are also secreted by exocytosis.

Cells that secrete protein have a well-developed ER and Golgi apparatus. The same is true of many cells that secrete polysaccharide-rich material, such as the mucus-secreting intestinal goblet cells and *chondroblasts.* The latter synthesize and secrete the polysaccharide chondroitin sulfate, a major component of cartilage (see Chap. 3.6B). In Chapter 2.5 we outlined evidence that steps in polysaccharide synthesis and "packaging" take place in the Golgi apparatus, and we pointed out that in many (perhaps most) cells the secretion bodies formed by Golgi apparatus are mixtures in which both carbohydrates and proteins are present in varying proportions and associations. Cells secreting steroids have well-developed smooth ER (Fig. II–38). Those secreting lipoprotein particles often have substantial amounts of both types of ER (Chap. 2.12).

There are some obvious differences among secretory cells. *Exocrine* glands release their secretions into special duct systems (Fig. II–33); the secretions of *endocrine* glands (hormones) directly enter the blood stream (Fig. II–70). Not all secretions are released by simple processes of membrane fusion; for instance, in the *sebaceous glands* (which secrete the oils that coat the skin and hair), the cells fill with secretion and then die and disintegrate, releasing their content. Section 2.8.2 referred to the complex path taken by the secretions of the thyroid gland. The gland first releases *thyroglobulin* to an extracellular storage lumen and then takes it back into the thyroid cells, digests it, and releases the digestion products to the blood stream. Thyroglobulin is a glycoprotein to which iodines are attached through the action of a peroxidase enzyme (Fig. II–35). An interesting possibility is that this enzyme is itself secreted into the storage lumen directly from the ER or Golgi apparatus, through transport in small vesicles that fuse with the plasma membrane; in the lumen it can act to iodinate the thyroglobulin molecules. How steroid hormones are secreted is still a matter of disagreement: Some investigators believe that most of these hormones move in complexes with proteins through the Golgi apparatus, as their parent molecule, cholesterol, probably can; others think it possible that some steroids exit directly from the ER, perhaps crossing the ER and plasma membranes as individual molecules rather than being transported in vesicles.

Secretory cells need not form distinctive-looking large secretory granules. Gland cells generally do, probably since they retain appreciable stores of secretion, releasing them intermittently upon appropriate stimulation. Other cells seem to rely on rather ordinary-looking small Golgi vesicles for transporting and releasing their products. This is true, for example, of the plasma cells that secrete antibodies (Chap. 3.6C). These cells release their products more or less continuously and thus seem not to retain stores additional to the material in their ER and Golgi systems (see Fig. III–40).

3.6.1 Post-Golgi Processing and Accumulation

In discussing the synthesis of insulin we noted that several secretory proteins are synthesized in "pro" forms that, while still in the cell, are modified through proteolysis to produce the functional protein (Section 2.4.3). In a sense, the production and subsequent digestion of thyroglobulin represents an extreme and circuitous form of this type of process. A number of the pancreatic digestive enzymes as released from the gland are in inactive, "zymogen" forms; these are activated by enzymes in the lumen of intestine. Activation involves the clipping-off of a chain of several amino acids, from the end of the polypeptide chain in the case of the conversion of *trypsinogen* to *trypsin,* or from the middle, as when *chymotrypsinogen* is activated to *chymotrypsin.* Trypsinogen is converted by *enterokinase,* a protease secreted by intestinal glands, and chymotrypsinogen is converted by trypsin. (Both trypsin and chymotrypsin are themselves proteases.)

Many Hormones from a Single Precursor   Recently it has been discovered that a number of different hormones released by the pituitary arise from the same precursor protein (or a family of very closely related proteins). The details are not all known yet, but it seems that the cells produce a large molecule that can be cleaved proteolytically into several different polypeptide fragments, each active as a hormone, including ACTH (adrenocorticotropic hormone, which stimulates steroid release from the adrenal gland), β-lipotropin (which increases breakdown of stored fat in adipocytes), β-endorphin (which reduces pain sensations), and several others. Fragmentation of a common precursor into more than one biologically active polypeptide product may occur in other cases as well. Possibly, this sort of mechanism has evolved into processes that enable different cell types to generate different spectra of final products from the same precursor protein.

Catecholamine Accumulation   The "catecholamine" hormones *epinephrine* (Section 2.1.4) and *norepinephrine* (adrenaline and noradrenaline) are molecules (Fig. II–9), about the size of the amino acid *tyrosine* (Fig. I–28), from which they are derived. They are synthesized by cells in the inner zones (medulla) of the adrenal gland (different cells make the two hormones). These cells use their Golgi apparatus and ER to create an immature secretion granule lacking the hormones but possessing machinery for synthesizing and accumulating them. The granules contain certain of the enzymes of hormone synthesis *(dopamine β-hydroxylase)* plus transport systems that carry the hormone or intermediates in its synthesis across the granule membrane. The transport systems make possible a collaborative effort in which certain of the steps of synthesis occur in the cytoplasm outside the granule and others occur inside. How elevated concentrations of the hormone molecules are maintained differentially inside the granules even though the molecules can be carried across the granule membrane is gradually being clarified. An increasingly popular hypothesis suggests that the granule produces a pH gradient across its membrane, and that the hormones, being weak bases, accumulate, much as weak bases do in lysosomes (Section 2.8.4). Another possibility, not necessarily in conflict with the first, is that the hormone molecules form large multimolecular complexes inside the granule by associating with proteins, Ca^{2+} ions, and ATP molecules, which also are present.

3.6.2 Secretion of Ions

The *parietal cell* (oxyntic cell) of the vertebrate stomach is an important example of a cell specialized to secrete inorganic ions. This cell type secretes HCl into the stomach, where the acid has a number of roles in digestion including activation of the digestive enzyme pepsin, a protease secreted as an inactive zymogen (Section 3.6.1) by glands in the stomach wall. The cell surface exposed to the lumen is enormously enlarged by a great many microvilli and by deep infoldings into the cell *(canaliculi)* (Fig. III–33). *Active transport,* in which the cell expends energy (Section 2.1.3), is involved in moving hydrogen ions out of the cell into the lumen. It hardly surprising, therefore, that parietal cells contain many stout mitochondria with numerous cristae and high levels of ox-

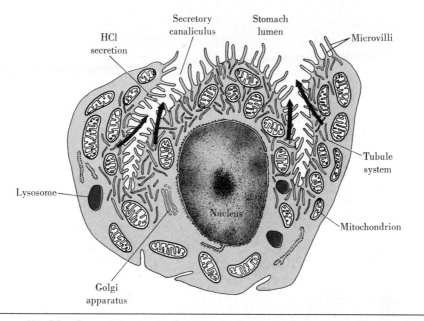

Figure III–33   *Secretion of acid.* Schematic diagram of a *parietal ("oxyntic")*
cell from the mammalian stomach. (After S. Ito and R.C. Winchester.)

idative enzymes. Within the cytoplasm, numerous tubules delimited by smooth
membranes are present. These are distinct from both the ER and the Golgi
apparatus, neither of which are extensive. Early literature, especially that concerning the frog, suggested that the tubules could establish continuities with the
plasma membrane, through fusions akin to exocytosis; these fusions supposedly released ions stored in the tubules bound, perhaps, to charged macromolecules. Recent studies have cast doubt on these concepts, and the situation
is presently uncertain. It has been established that with the stimulation of hydrogen ion secretion, the tubular system decreases in amount, and the microvilli of the secretory canaliculi become longer. When ion secretion is decreased
or inhibited, the reverse occurs. Therefore it does seem likely that the proposed
fusions occur. How they relate to ion transport, however, is less certain, since
there is no clear evidence yet that ions are actually stored in the tubules. Perhaps it is the tubule membrane rather than the contents that is important for
the secretion of acid into the stomach. It could be, for example, that this membrane possesses "pumps" or permeability properties permitting it to redistribute
H^+ and Cl^- ions across the cell surface after it becomes inserted in the cell
surface. Proposals of a comparable sort have been made to explain the increase in water permeability induced by ADH (Section 3.5.3) in the lining of
the urinary bladder of the toad—an aspect of water conservation in these organisms. ADH, probably operating via cyclic AMP, brings about reversible

structural changes in the surfaces of the bladder's lining cells, including exocytosis-like insertion of the membrane bounding cytoplasmic granules or tubules. It is postulated that this membrane is or becomes more permeable to water than the resting cell's plasma membrane, so that its insertion increases the cell surface permeability.

Marine birds (and some turtles) possess salt glands that function in salt excretion. The cells show numerous mitochondria and deeply infolded surfaces, providing a large area for ion transport. The folding of the surface of transfer cells in the xylem (Section 3.4.4) and that of the basal portion of the kidney tubule cell (Section 3.5.3) probably have the same significance.

3.6.3 Cnidoblasts in Hydra

An interesting modified secretory mechanism has been described in the primitive animal *Hydra,* a coelenterate. The organism is made essentially of two layers of epithelium. In the outer layer there are special cells, called *cnidoblasts,* that produce small projectiles, *nematocysts,* that are shot out to paralyze, penetrate, or entangle prey.

The cnidoblasts develop from undifferentiated precursor cells, the interstitial cells. The ER, sparse in the primitive cell, becomes extensively developed in the maturing cnidoblast. Ribosomes (presumably in polysomal form) usually lie free in the cytoplasm in the primitive cell but are arranged on the ER membranes during maturation (see Section 2.4.1). The Golgi apparatus also becomes highly developed as the cell begins to secrete the proteins that are stored in the *nematocyst.* The latter apparently begins as a Golgi vacuole, small at first and then enlarging greatly. Innumerable small vesicles develop from the much enlarged Golgi saccules and fuse with the nematocyst. As the nematocyst enlarges, its content becomes more electron dense (Fig. III–34). (A great many microtubules appear outside the nematocyst; they probably give rigidity and shape to the area in which the secretory vacuole is rapidly enlarging.) When the nematocyst attains its maximum size, the ER and Golgi apparatus regress, breaking into vesicles that progressively diminish in number. Upon triggering, the nematocyst releases its contents to the extracellular environment (lower left diagram in Fig. III–34A).

The striking development of the ER and Golgi apparatus, followed by their virtual disappearance, is but one of the interesting features of cnidoblasts. Another is the dramatic structural differentiation that occurs inside the nematocyst, *without apparent connections to other cell organelles.* (Speculation on how this might occur is found in Section 4.1.4.) Thus far, little attention has been given to the manner by which this differentiation occurs. Yet it seems likely to be a genetically determined process, involving a great many secretory proteins. In Hydra there are four distinct types of nematocysts. In 100 species of coelenterates related to Hydra, 17 types of nematocysts have been described. Each type is characterized by a structure that is characteristic for the particular species. In one of the abundant nematocysts in Hydra, the complex structure includes a capsule (which apparently contains a variety of phosphatase enzymes), a lid, a coiled thread, large stylets, and smaller barbs of specific shapes (Fig. III–34).

3.6.4 Secretory Controls

The rates of secretion from cells are under precise controls. In Hydra, a *cnidocil* protrudes from the cnidoblast surface; this is in part a modified cilium. Nematocyst release occurs when the cell is stimulated by the appropriate chemical and mechanical stimuli such as may result from the presence of the small organisms used as food. The cnidocil is probably a mechanical receptor; additional chemical receptors are thought to be built into the plasma membrane. Release is also influenced by the primitive nervous system found in Hydra.

In higher organisms, nervous and hormonal stimulation governs the rates of release of secretions from glands and from organs such as the pancreas. As Section 2.1.4 outlined, stimulation of secretion by neurotransmitters and hormones frequently involves increases in cytoplasmic Ca^{2+} concentrations due to changed plasma membrane permeability or to release of Ca^{2+} from internal stores. Cyclic nucleotides, microtubules, and microfilaments may also participate in the intracellular mechanisms governing movements of secretion granules to the cell surface and the interactions of membranes involved in release of secretions. Feedback mechanisms of various kinds operate to balance overall activities. A well-known example is the mutual interaction of the pituitary gland and the thyroid: The pituitary releases *thyrotropic* hormone or *thyrotropin* which stimulates release of thyroid hormone from the thyroid gland. Thyroid hormone, in turn, inhibits release of thyrotropic hormone so that the pituitary's further stimulation of the thyroid is turned down, until circulating levels of the thyroid's product drop and increased stimulation is again required.

Mast Cells Release of secretions from the variety of nonglandular secretory cells is also under precise controls involving cell surface receptors, modulating agents such as hormones, prostaglandins (Section 2.1.4), intracellular calcium, and so forth. *Mast cells,* for example, are relatives of white blood cells that are scattered in connective tissues in many parts of the body. Damage to the tissues, immune responses (Chap. 3.6C), and other events lead to the production or release of factors that stimulate exocytosis of the granules that abound in the mast cell cytoplasm. Among other components, these granules contain *histamine,* which alters the permeability of blood capillaries. This effect of histamine accounts for some of the swelling of tissues characteristic of the common cold.

Polarization An interesting matter deserving further study is the precise polarization of many secretory cells: Exocrine glands release their secretions only at the cell surfaces that are directly accessible to their ducts; bile is released only into bile canaliculi (Fig. I–1); and endocrine glands release their

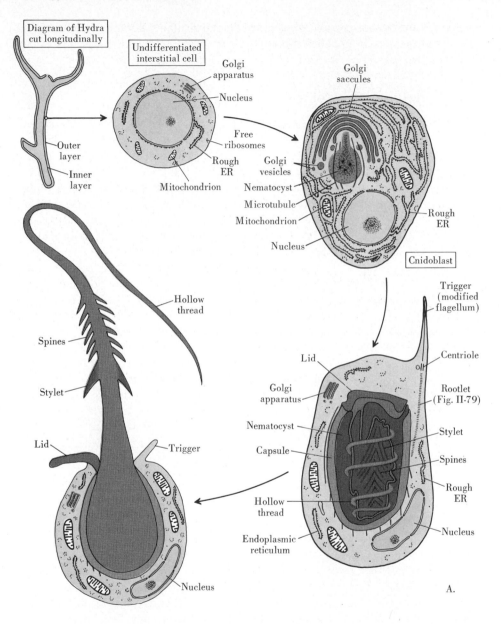

Diagram of Hydra cut longitudinally

Outer layer

Inner layer

Undifferentiated interstitial cell

Golgi apparatus

Nucleus

Free ribosomes

Rough ER

Mitochondrion

Golgi saccules

Golgi vesicles

Nematocyst

Microtubule

Mitochondrion

Nucleus

Rough ER

Cnidoblast

Trigger (modified flagellum)

Centriole

Rootlet (Fig. II-79)

Stylet

Spines

Rough ER

Nucleus

Lid

Golgi apparatus

Nematocyst

Capsule

Hollow thread

Endoplasmic reticulum

Hollow thread

Spines

Stylet

Lid

Trigger

Nucleus

A.

secretions at the cellular poles facing capillaries. Junctions between the cells help to define the local regions where secretions are released (Section 3.5.2) and prevent leakage of released materials into inappropriate spaces. "Cytoskeletal elements" (Chaps. 2.10 and 2.11) may move secretory bodies into the proper cell regions. But how are the membrane fusions and other intercellular events regulated, for example, so as to produce exocytosis only at the proper

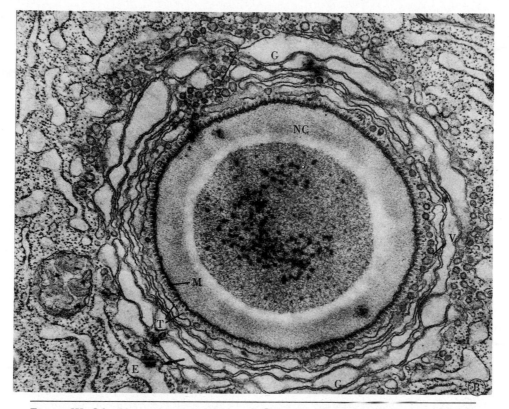

Figure III–34 *Nematocysts in Hydra.* A. Diagram illustrating the maturation of a nematocyst within a "cnidoblast." As the nematocyst increases in size there is a great increase in extent of rough ER and size of Golgi saccules and vacuoles. It is after the recession in development of ER and Golgi apparatus that the nematocyst undergoes the major part of its development of complex structure. (Based on studies of *Hydra* and other coelenterates by Slautterback, Westfall, Picken, Skaer, and Lentz.) B. Developing nematocyst *(NC)* in a cnidoblast of *Hydra.* Oriented microtubules surround the nematocyst; in this section most are sectioned transversely *(T).* As indicated in A, the nematocyst forms in association with the Golgi apparatus. Golgi saccules *(G)* and vesicles *(V)* are abundant. Much rough ER also is present *(E).* The arrow indicates a vesicle probably budding from the ER and contributing to the Golgi apparatus (or developing nematocyst; see Fig. II–33). × 30,000. (Courtesy of D.B. Slautterback.)

place? In a few cases there are specialized structural arrangements that may function as "recognition" or "guiding" devices controlling membrane fusions (Sections 3.3.1 and 3.7.5). In others it is proposed that the permeability changes leading, for instance, to increased Ca^{2+} levels occur very locally and lead to a highly localized, transient change in ion concentration that promotes the exocytosis only near the sites of permeability change (Section 3.7.5). Other proposed steps in the release of secretions such as specific enzymatic events (the production of prostaglandins or the addition of methyl groups to lipids) or interactions of proteins (Section 2.5.2) have also been postulated to occur only at particular zones of the cell surface.

Chapter 3.6B Extracellular Materials in Animals

Animal cells make and secrete a variety of "structural" materials. These range from the sometimes elaborate exoskeletons or cuticles surrounding invertebrates to the thin layers of connective tissue separating liver lobules (Fig. I–1) from one another or permeating the core of intestinal villi (Fig. III–27) or the spaces between pancreatic acini (Fig. II–33).

The protective outer layers of the bodies of vertebrates are formed in part of keratin (Section 3.5.1) in skin, feathers, and scales, and in part of secretions from glands and other cells near the surface. The cuticles of some invertebrates are of collagen and other materials like those of connective tissue discussed below. Insects and crustaceans, on the other hand, construct a cuticle of proteins, lipids, and filamentous structures rich in the polysaccharide *chitin*. These latter filaments are in layered arrays and bundles; sometimes each layer has parallel filaments, and successive layers show different filament orientations, as in cell walls of plants. The exoskeleton is secreted by the epidermal cells at the insect's surface and must be dismantled—part discarded through molting, part resorbed—and reconstructed when the animal grows too large.

For internal connective tissues, the basic molecular building blocks are proteins and carbohydrates (plus inorganic salts in the case of bone). Structurally, protein fibers, elongate polysaccharides, and large branched and meshlike molecular aggregates are intermingled in varying relationships. They can form harnesses capable of transmitting substantial forces; tough, protective covering layers; loose water-filled networks that permit ready diffusion of many materials within but also help to maintain tissue architecture; or firm supportive structures.

3.6.5 Fibers

Collagen, the major fiber-forming protein of connective tissues, is the most abundant protein in the mammalian body, accounting for as much as 25 percent or more of the total protein. It is secreted by a number of cell types, of which *fibroblasts* produce the most prominent collagen fibers. Several types of collagen are produced by different cells and at different times—they vary in details of molecular structure and in the fibers and other aggregates they form. In vertebrate organisms, there may be as many as nine distinct but closely related collagen polypeptide chains, different ones of which form different combinations in the five classes of collagen known. The account that follows is specifically concerned with the most abundant class, Type I collagen.

Collagen molecules are glycoproteins, especially rich in the amino acids *glycine* and *proline* and in amino acids modified by the addition of hydroxyl (OH) groups—*hydroxyproline* and *hydroxylysine.* This amino acid composition and proteolytic and other alterations of the molecules occurring during their life history are responsible for the forms assumed by the molecules (Fig. III–35). Collagen is made in the rough ER. The hydroxyl groups of the modified amino acids are added after translation, by ER enzymes known as *prolyl* and *lysyl hydroxylases.* It is to the hydroxylysines that the sugars are linked— generally in the form of disaccharides of which varying numbers are present in different collagens. In the ER, collagen is in the *procollagen* form of three elongate polypeptide chains—two chains of one type, α1(I), and one of another, α2. Each chain is a helix more than 1000 amino acids long; the three helices coil around one and bond strongly to one another by numerous hydrogen bonds.

Note that despite the use of the prefix α, the helices formed by the α-chains of collagen (Fig. III–35) are individually, much less tightly coiled than are the α-helices described in Figure I–28. Figure III–35 also points out that the ability of the collagen chains to coil together into a triple helix depends on unusual features of the amino acid sequence. Particularly striking is the fact that over stretches of the polypeptide chain, many hundreds of amino acids long, glycine appears in every third position (. . . glycine–other amino acid– other amino acid–glyceine . . .). Because glycine is the smallest of the amino acids, this arrangement allows the three chains to fit snugly together, with the glyceines located in the core of the triple helix.

Procollagen moves through the Golgi apparatus and is secreted by exocytosis. Outside the cell its ends are removed by proteolytic enzymes (peptidases) converting it to *tropocollagen,* a rodlike molecule about 0.3 μm by 1.5 nm. It is this step that controls the formation of collagen fibers, since unlike procollagen, tropocollagen can assemble into the characteristic banded fibers, the most prominent form in which collagen occurs (Fig. III–36). The assembly as such is apparently spontaneous and is based initially on noncovalent bonds—that is, it is a self-assembly in the sense discussed in Section 1.4.3. Subsequently, however, enzymes form covalent bonds that strengthen the links between tropocollagens (Fig. III–35). As a result collagen fibers are both insoluble and quite resistant to stretch—which is why, for example, bundles of them can serve in the tendons that link muscles to bones. Specific *collagenases* are required to degrade collagen; this degradation occurs normally during tissue remodeling in development, and in fact, collagenases were first detected in the tails of frog tadpoles during the tail resorption involved in metamorphosis into a frog.

Collagen fibers form thick bundles in some places such as tendons, flat sheets in others, and less regular arrays in the sorts of "ordinary" connective tissue found in places such as the intestinal wall and villi. The fibers associate

Peptide bond to adjacent amino acids
(see Fig. II-23)

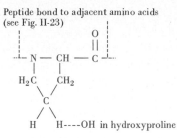

Proline: The ring structure and related
features of this amino acid strongly affect the
conformation of the polypeptide chains of
collagen, in which it is unusally abundant.
The collagen chains form helices that are much
less compact than the α-helices illustrated in
Fig. I-28.

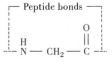

Glycine: The simplest and smallest amino
acid. Approximately every third amino acid
in the collagen polypeptide chain is glycine;
because glycine has no side group that would
protrude from the polypeptide, collagen's "α-
chains" can pack closely together in the triple
helix they form.

(A) Fundamentals of collagen structure

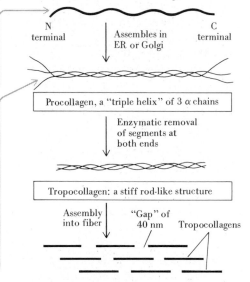

The individual polypeptide chains (α-chains)
are about 1000 amino acids long.

N terminal   Assembles in   C terminal
                              ER or Golgi

Procollagen, a "triple helix" of 3 α chains

Enzymatic removal
of segments at
both ends

Tropocollagen: a stiff rod-like structure

Assembly   "Gap" of
into fiber   40 nm   Tropocollagens

A collagen fiber (Fig. II-36) forms as a "staggered"
array of tropocollagens. The arrays are stabilized
by cross links established enzymatically between
the tropocollagen units as in C.

in varying patterns with the other components present in the extracellular environment. Thus, for instance, the calcium phosphate crystals in bone are
thought to grow in association with specific points along collagen fibers (the
"gaps" between successive tropocollagens in Fig. III–35). These sites may also
be preferred regions for interactions of collagen with other glycoproteins and
with polysaccharides.

Fibrin; Blood Clots Assembly of a different sort of extracellular fiber is
important in responses to tissue damage or injury. The protein *fibrinogen* is
secreted by the liver and circulates in soluble form in the blood stream. Tissue
damage leads to the release of factors that convert another circulating protein,
prothrombin, into an active proteolytic enzyme, *thrombin,* which specifically
cleaves fibrinogen, altering it into *fibrin.* Fibrin spontaneously aggregates into

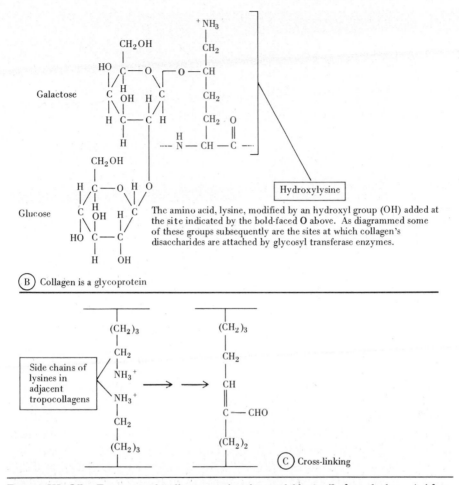

The amino acid, lysine, modified by an hydroxyl group (OH) added at the site indicated by the bold-faced O above. As diagrammed some of these groups subsequently are the sites at which collagen's disaccharides are attached by glycosyl transferase enzymes.

(B) Collagen is a glycoprotein

(C) Cross-linking

Figure III–35 *Features of collagen molecules and fibers.* (Left and above.) After J. Gross, E. Hay, L. Stryer and others. (For a clear, more detailed account *see* L. Stryer, *Biochemistry,* 2nd ed. San Francisco: W.H. Freeman, 1981, Chapter 9.)

the fibers that interweave to form a clot. Covalent cross-links subsequently stabilize the fibers.

Clotting at inappropriate places, as *within* blood vessels, is potentially quite dangerous for the organism; therefore, controls of the sites where clots form are essential. Clotting is kept localized to some extent by the fact that the activating factors are released locally where damage occurs, and also by protective systems that prevent potentially dangerous spreading. The blood contains a number of proteins that complex with proteases, inactivating them; these ensure that when thrombin spreads from sites where it is activated, it is rendered ineffective in promoting clotting. Heparin, a polysaccharide released

Figure III–36 *Collagen fibers* shadowed with evaporated metal atoms (Fig. I–11) to bring out their three-dimensional appearance. The prominent pattern of cross-banding is evident; the banding is based on the regular arrangement of the tropocollagen units in the fiber (Fig. III–35). × 50,000. (Courtesy of J. Gross.)

from mast cells, seems to cooperate in this inactivation. Dissolution of fibrin clots depends on enzymes such as *plasmin,* which also circulates in inactive form (plasminogen).

Circulating cell fragments, *platelets,* participate in blood clotting. When they escape from the blood stream through wounds they adhere to collagen fibers in the connective tissue. This sets off a chain of events in the platelet membrane, mediated perhaps by arachidonic acid (Section 2.1.4), which lead to secretion by the platelets of ADP, plus a hormone-like molecule called serotonin, and a protein called *platelet growth factor.* These secretions stimulate aggregation of more platelets with one another and with the forming fibrin network, helping to establish the clot. They also evoke narrowing of blood vessels, which reduces blood loss, and initiation of growth and division of cells in the blood vessel wall, which eventually helps to reestablish the integrity of the vessels. (Platelets, like other blood cells, are produced in the bone marrow. They arise by fragmentation of large cells, megakaryocytes; if not "used" they survive about 10 days in humans.)

3.6.6 Connective Tissue Matrix Materials

Glycosaminoglycans The matrix of materials surrounding the collagen fibers in connective tissues is rich in carbohydrates. Most of the carbohydrate is present as polysaccharides made of repeating disaccharide units of the sort shown in Figure III–37. These glycosaminoglycans ("GAGs," "acid mucopolysaccharides") are generally negatively charged owing to the presence of sulfate, or carboxyl, groups. They bind both appropriately charged ions and water. Different ones predominate in different tissues; chondroitin sulfate is the major one in cartilage and hyaluronic acid in ordinary fibrous connective tissue.

However, there is appreciable overlap, and mixtures or aggregates of different types are found in various tissues.

Proteoglycans The glycosaminoglycans in connective tissues frequently are linked into larger complexes by attachment to proteins. Each polypeptide chain generally has multiple glycosaminoglycans attached: By weight the saccharides can constitute more than 90 percent of the complex. The polypeptide is referred to as a "core protein"; the assembly as a whole is a "proteoglycan."

Proteoglycans in turn can be linked to elongate glycosaminoglycans forming very large multiply-branched arrays of the sort shown in Figure III–38. Various proportions and combinations of glycosaminoglycans, proteoglycans, and collagen can form: viscous fluids, as in joints, in which elongate hyaluronic acid molecules contribute much of the viscosity; loose gel-like meshworks, within which water and small solutes can move readily, as in ordinary connective tissues; or hard dense structures, as in cartilage. The molecules are produced and secreted by fibroblasts and by the chondroblasts of cartilage, the osteoblasts of bone, and a few other related cell types. The glycosaminoglycans are synthesized in the Golgi apparatus, and initial steps in linkage to protein probably occur there as well. Some of the subsequent linkages—especially those producing the very large aggregates—may, it is theorized, occur extracellularly.

3.6.7 Binding of Cells to Extracellular Materials; the Basal Lamina (Basement Membrane)

The term "basement membrane" was long used to refer to the mat of extracellular fibers and matrix materials that underlies epithelia. It is being replaced by "basal lamina," as the term "membrane" is now generally reserved for the familiar lipoprotein structures. A similar layer that surrounds capillaries, muscle fibers, and some other cells is often called the "external lamina."

These layers serve as anchors for groups of cells (Section 3.5.1), as barriers to passage of cells and large molecules across tissue boundaries, and in specialized roles such as in the kidney filtration described in Section 3.5.3. They are produced through secretion by the cells with which they are associated but may grade into adjacent connective tissue formed by fibroblasts. The collagen in basal laminae is present as a meshwork of fine fibrils, rather than the large fibers found elsewhere. It is of the type IV class—the molecules are of three polypeptide chains of the $\alpha1$(IV) variety. Other proteins (laminins) are present as well, and at some developmental stages at least, glycosaminoglycans are also found.

Fibronectin The adherence of cells to the basal lamina or to collagen fibers and other connective tissue constituents depends in part on components such as Ca^{2+} but also on proteins, the best-known of which is a large glycoprotein, made of two polypeptide chains, called *fibronectin*. (Other comparable pro-

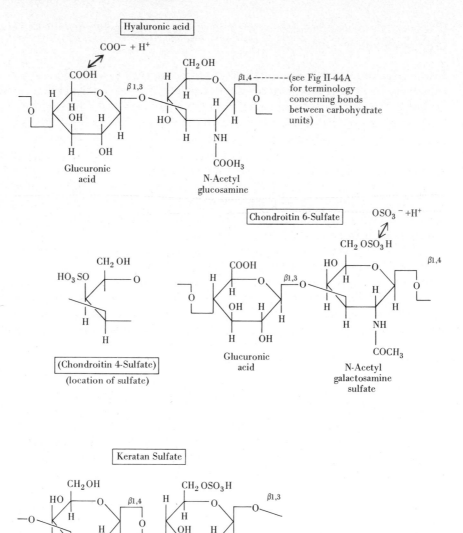

teins, such as *chondronectin* in cartilage, are now being discovered, and the laminins just mentioned may play similar roles.) Fibronectin is produced and secreted by fibroblasts and by a number of other cell types. Considerable evidence suggests it possesses a specific region ("domain") toward the middle of the elongate molecule, permitting it to bind noncovalently to cell surface com-

◀ **Figure III–37** *Molecular structure of three of the principal glycosaminoglycans.* Each is a polysaccharide made of repeating disaccharide units linked end to end. (See Fig. II–44 for terminology applied to the linkages. For simplicity, and according to convention, the carbons of the rings are indicated in the present diagram simply by vertices rather than by Cs). The saccharides of the units are modified forms of glucose and galactose. In each case one of the two has a group derived from an NH_2 (hence the "amino" in the generic name). The sulfates and carboxyls are acidic groups so these molecules were once called *acid mucopolysaccharides*. As illustrated, H^+ ions dissociate from the acidic groups, leaving negatively charged sulfates and carboxyls attached to the saccharides. Chondroitin sulfate occurs in two forms: chondroitin 6-sulfate, and chondroitin 4-sulfate, which differs from the chondroitin 6-sulfate as illustrated at the left.

The number of saccharides linked in a glycosaminoglycan chain varies in different types and tissues. The hyaluronic acid chain in Figure III–38 is several thousand saccharides in length, which is toward the upper end of the length spectrum. (After E. Hay's diagrams summarizing work by many others. [*J. Cell Biol.* **91**:205S, 1981.])

ponents, whereas toward one end it has a domain by which it associates with collagen. A glycosaminoglycan-binding domain is present near the opposite end. Such characteristics help to account for fibronectin's ability to promote adhesion of cells to extracellular materials. They also make for ambiguity in classification—is it properly an extracellular protein, a cell-surface protein, or a connective-tissue protein? This is a superficial aspect of a series of more exciting questions about how cells are integrated into tissues, and about the relations of the cell surface to materials on both of its sides—cytoplasmic and extracellular. Thus, fibronectin is considered by some workers as an agent through which the cell's cytoskeleton could interact, across the plasma membrane, with extracellular materials. That is, at sites along the cell surface where fibronectin molecules attach the cell surface to extracellular materials by binding simultaneously to both, cytoskeletal elements might establish special relations with the cytoplasmic side of the plasma membrane. Conceivably, the cytoskeleton can link to cytoplasmic portions of the same transmembrane proteins thought to bind to fibronectin on the extracellular side of the membrane.

Chapter 3.6C Antibodies and Immune Phenomena

In their immune responses, vertebrates utilize numerous secreted materials. In our consideration of these responses here, we shall see that immune phenomena depend on cellular activities of diverse kinds—not just on secretion—and that interactions of several cell types are required for the system to function in a discriminating and regulated manner.

The most clearly understood phase of the functioning of the immune system is the production of antibodies—glycoproteins of the *immunoglobulin* category. There are several different types of immunoglobulins, some of which circulate in the blood stream and some of which are associated with the plasma

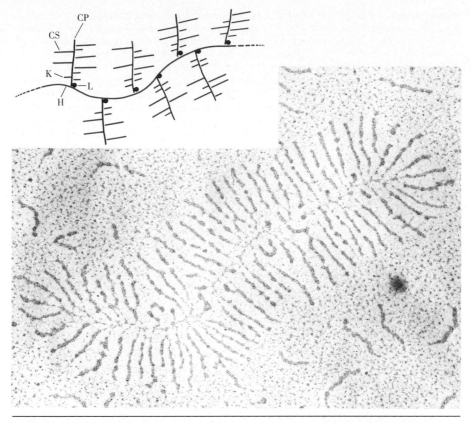

Figure III–38 *Proteoglycan aggregate* prepared from cartilage (bovine fetal epiphysis), spread on an electron microscope grid and shadowed with metal (Fig. I–11). The aggregate is 2 to 3 μm in length. As the diagram indicates, the backbone is an elongate molecule of the glycosaminoglycan *hyaluronic acid (H)* to which are attached polypeptide chains ("core proteins; *CP*); each core protein chain has attached to it several dozen additional glycosaminoglycan chains (keratan sulfates, *K* and chondroitin sulfates, *CS*). The keratan sulfates and chondroitin sulfates are linked to the amino acids *serine* and *threonine* in the core protein by O linkages like that in Figure II–44. The core proteins are bound noncovalently to the hyaluronic acid; this binding is stabilized by small linking proteins *(L)*. Consult Figure III–37 for details of the glycosaminoglycans' structures.

During preparation for microscopic examination the glycosaminoglycan chains collapse around the core proteins producing the thickened "clublike" appearance seen here. Note that the clubs are thinner near their base, reflecting the fact that in this region keratan sulfate chains predominate; these chains are markedly shorter than those of chondroitin sulfate. The aggregate as a whole has been flattened during preparation; in intact cartilage the proteoglycans—core proteins plus attached glycosaminoglycans—would protrude from the backbone in a three-dimensional array like a test tube brush, and each proteoglycan itself would have a brushlike form. Thus, the aggregate would constitute an extended hydrophilic mesh with many negatively charged groups. × 40,000. (Courtesy of J.A. Buckwalter and L. Rosenberg.)

membranes of the cells that produce them. Figure III–39A diagrams the structure of the *IgG* type, the major circulating variety and the most thoroughly studied. Antibodies of this sort are employed to recognize invading microorganisms and to initiate their sequestration and destruction.

Other classes of antibodies are found at various locales. There are five major types constructed on basically similar principles to those of IgG. Different ones have different biological activities. IgE, for example, bound to the surface of mast cells, mediates some features of allergic reactions through its triggering of the release of histamine (Section 3.6.4).

The major cell types that participate in immune responses are *lympho-cytes* and their derivatives and *phagocytes* (Section 2.8.2), such as the polymorphonuclear leukocytes and macrophages. These cells originate in the bone marrow and spleen, travel in the circulation, and take up residence in specialized sites such as the lymph nodes as well as in many other tissues.

Recall that an *antigen* is any molecule that evokes production of an antibody that binds specifically to it. It should also be noted that, quite often, proteins are identified and studied through immunological techniques, such as immunohistochemical methods, before their functions and other properties are understood. Thus, for example, when certain viruses induce cancers, they generate proteins in the responding cells known by names such as "T antigen" (Section 3.12.3); the name "antigen" here has no special immunological connotation, being simply a convenient label reflecting the procedures by which the proteins were first found.

3.6.8 Cell-mediated Immune Phenomena; T-lymphocytes

Different types of immune processes depend on somewhat different mechanisms. Rejection of tissue grafts or transplanted organs result from responses known as "cell-mediated" reactions, since the antigen-specific molecules that participate in the recognition of the foreign tissue are of types that remain associated with cells rather than being secreted. (The molecules are not ordinary immunoglobulins but may be related to them.) A category of cells called *T-lymphocytes* (because they spend a crucial period of their developmental history in the *thymus*) are instrumental in the recognition and destruction of the foreign tissues. Antigen-specific molecules in their plasma membranes enable these cells to identify foreign cells.

Histocompatibility; the Recognition of "Self" In the case of grafts and tissue transplants, the antigens on the transferred cells to which the recipient's T-cells respond are primarily cell surface glycoproteins. Even closely related individuals differ genetically in these components, accounting for the difficulties in achieving successful transplants. Collectively the major group of glycoproteins that determine whether a transplanted tissue will be accepted or rejected is called the set of *major histocompatibility antigens*. The group of genes that specify them is referred to as the MHC, the *major histocompatibility complex*.

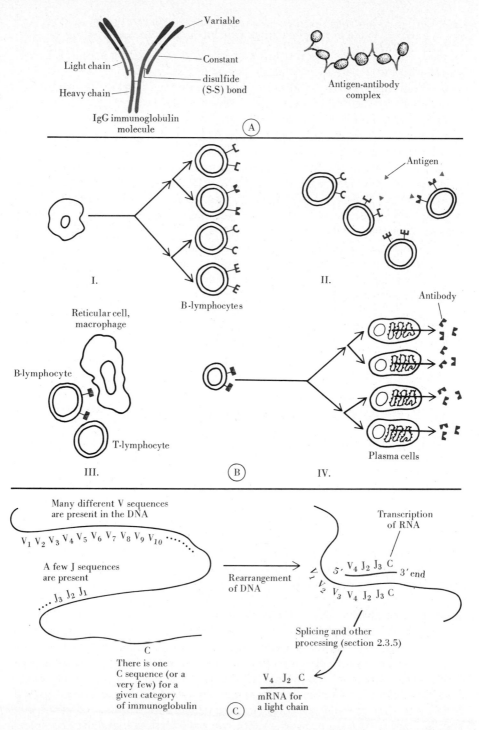

Variable

Light chain

Constant

disulfide
(S-S) bond

Heavy chain

Antigen-antibody
complex

IgG immunoglobulin
molecule

A

I.

B-lymphocytes

Antigen

II.

Reticular cell,
macrophage

B-lymphocyte

T-lymphocyte

III.

B

IV.

Antibody

Plasma cells

Many different V sequences
are present in the DNA

V_1 V_2 V_3 V_4 V_5 V_6 V_7 V_8 V_9 V_{10} ......

A few J sequences
are present

J_3 J_2 J_1

Rearrangement
of DNA

Transcription
of RNA

5′ V_4 J_2 J_3 C ——— 3′ end

V_2 V_3 V_4 J_2 J_3 C

Splicing and other
processing (section 2.3.5)

C

There is one
C sequence (or a
very few) for a
given category
of immunoglobulin

V_4 J_2 C

mRNA for
a light chain

C

◀ **Figure III–39** *Antibodies. Upper panel:* IgG molecules are the predominant type of antibody found in the circulation. One IgG molecule consists of a pair of light chains (each with approximately 215 amino acids [214 in humans]) and a pair of heavy chains (approximately 440 amino acids each [446 in humans]). The amino acid sequence in the variable regions (red) differs among different IgGs, accounting for their abilities to bind to different antigens. The constant regions vary little among immunoglobulins of a given class, such as the IgGs, but do differ among the immunoglobulin classes.

Under some conditions, antigen-antibody complexes form quite extensive aggregates because each IgG is "bivalent" (it has two identical variable regions) and therefore can bind to two antigen molecules; moreover, the antibodies produced by an organism responding to a complex antigen such as a microorganism or protein generally are a mixture of molecular species (a "polyclonal" mixture; see Section 3.11.2) that bind to different portions ("determinants") of the antigen.

Middle panel: Production of antibodies. *I,* B-lymphocytes ("B-cells") arise by division of precursor cells, whose origin in mammals is still uncertain. Different B-lymphocytes have cell surface receptors with different specificities; these receptors are immunoglobulins. *II,* An antigen encounters B-lymphocytes at a wound, irritation or site of infection or in spleen or lymph nodes. *III,* The antigen interacts with a B-lymphocyte carrying receptors that can bind the particular antigen. Other cells play roles in this interaction, but the natures of these roles have not yet been elucidated. *IV.* The lymphocyte, stimulated by the antigen binding and consequent events of unknown sorts, divides and differentiates into plasma cell. These plasma cells secrete antibodies, the antigen-binding specificity of the secreted antibodies is the same as that of the surface receptors on the lymphocyte. (Figs. II–8 and III–40 are micrographs of lymphocytes and plasma cells.)

Lower panel: Working hypothesis about the *genetic phenomena underlying production of a light chain.* Repositioning of DNA sequences generating combinations of different Vs and Js with the C in different lymphocyte clones occurs during development. Note however, that as illustrated, the production of the mRNAs specifying a particular antibody also depends on initiation of transcription at the proper place (presumably a promoter; Section 4.4.6), and on considerable processing of the transcript. (See Fig. V–1 for one of the several contending models for how the DNA translocations might come about.)

(The HLA complex in humans and "H-2" locus in mice are the MHCs that have been most extensively studied.) The genes form a cluster in a single region of one of the chromosomes. Certain of the proteins they specify occur as integral membrane proteins in most differentiated cell types. Others are more restricted; they are found on cells of the immune system and a few other cell types. The roles of the proteins in cellular activities are not well understood, but some are important for regulating interactions of lymphocytes with one another. A widespread view is that the primary functions of the MHC proteins are in permitting the organism to recognize its own cells. Cellular recognition phenomena are important in normal development (Section 3.11.2) and many other processes, both normal and pathological. Determining their molecular bases is a major item on the current research agenda in biology.

When the T-cells responsible for cell-mediated immunity encounter foreign tissue they secrete "factors," including small proteins called *lymphokines,* that attract cells such as the circulating macrophage precursors, *monocytes,* and stimulate their "cytotoxic" (cell-killing) activities. The destruction of foreign tissue by macrophages "activated" in this way involves release of toxic agents (one of which may be H_2O_2 produced at the cell surface [Section 2.9.2]) and phagocytosis. T-cells themselves can kill some of the cells they recognize as foreign by binding to them, somehow lysing them, and then detaching to seek another target.

The lack of response by the immune system to an individual organism's own tissue depends in part on another category of T-lymphocytes, the "suppressor T-cells," which inhibit the sequence of events just described. Although "self-tolerance" is still poorly understood, it seems to rely in part on processes early in development through which the organism "learns" not to respond "aggressively" to its own molecules. The pertinent T-cells may arise as an outcome of such processes.

3.6.9 Circulating Antibodies; B-lymphocytes; Plasma Cells

Cell-mediated immune events dependent on T-cell recognition, lymphokines, and macrophages also take part in the body's defenses against invading microorganisms. However, more is known about those responses to microorganisms that depend on the production of secreted antibodies, such as IgGs, that enter the blood stream; these antibodies are referred to as "circulating" or "humoral antibodies." Antibodies of this sort also are evoked when foreign molecules are injected; they can then be obtained from the blood serum (the fluid surrounding the cells) for various experimental uses (Section 1.2.3). Immunoglobulins in the blood serum or in other extracellular fluids can bind to the corresponding antigens producing *antigen–antibody complexes* referred to as *immune complexes.* It is the "variable" region of the immunoglobulin molecule that enables it to recognize the antigen (Fig. III–39). Binding of the variable region to the antigen results in conformational changes in the remainder of the molecule. When so altered, the antibody is also bound avidly by receptors on the plasma membranes of phagocytes (such as the "F_c" receptor of macrophages), leading to endocytosis of the molecule or particle bearing the antigen. The promotion of endocytosis by binding of a molecule such as an antibody to the material that is taken up is called "opsonization." Opsonization is a primary means by which the circulating-antibody branch of the immune system deals with antigens.

IgGs are "bivalent," meaning that each can bind to two identical antigenic sites (Fig. III–39). Consequently, when the concentrations of antibodies and antigens are correctly adjusted, large multimolecular complexes that will precipitate from solution can form. This property permits the use of antibodies in the laboratory for identification and purification of particular antigens (specific proteins for example) in complex mixtures.

Complement As they form, *immune complexes* (antigen–antibody complexes) interact with a set of circulating proteins known collectively as *complement*. Certain of these are secreted by the liver, some by the intestinal epithelium, and some by macrophages. Interaction of complement with immune complexes is initiated by binding of particular complement proteins to corresponding sites on immunoglobulins in the complexes. Alteration of various complement proteins ensues through a cascade of enzymatic and other events. The resultant products have a variety of biological effects. One of the products aids in opsonization. Others are chemoattractants for phagocytes (Section 2.11.4); because immune complexes form where antigens are present and since this generates the complement reactions locally, phagocytes are attracted to the sites where they are needed. Another set of complement proteins produces cytotoxic effects; the proteins insert in the plasma membranes of foreign cells, opening holes, with catastrophic effects on the cells' permeability controls.

B-lymphocytes; Plasma Cells IgGs are secreted by *plasma cells,* which are numerous in the spleen and lymph nodes but are found also in other tissues where responses to foreign materials are occurring. The plasma cells arise by maturation of *B-lymphocytes,* so called because they were first analyzed in the *bursa of Fabricius,* an immunological organ of birds whose counterpart in mammals is still a subject of controversy. (It may be the bone marrow.) As Figure III–40 illustrates dramatically, it is in the ER of the plasma cells that the antibodies destined for secretion are synthesized.

Clonal Selection How does antibody *specificity* arise? This issue has been investigated particularly extensively in mice since they can be studied through genetic approaches and experimental manipulation of the immune system coupled to biochemical and cell culture procedures; it is this constellation of techniques that has been required for progress. The picture of the genetic underpinning of the immune system that has developed in the last few years centers on the concept that the normal development of an individual organism's immune system *pre-equips* the system with the capacity to make hundreds of thousands of different antibody molecules. Each of these is characterized by a variable region (Fig. III–39A) capable of binding to a distinct molecular conformation. Antigens *evoke* the expression of the corresponding capacity (Fig. III–39B).

B-lymphocytes are the carriers of the capacities to make different antibodies. During development the progenitors of the B-cells diverge from one another in that each comes to have the genetic capability (DNA sequence) to make a different immunoglobulin from that made by the others. Each then gives rise to a clone of identical B-lymphocytes, which may include thousands of cells located in the immune tissues of different parts of the body. When they enter the body, potentially antigenic materials encounter these B-cells. When the encounter is with a cell that can produce a specific antibody capable of

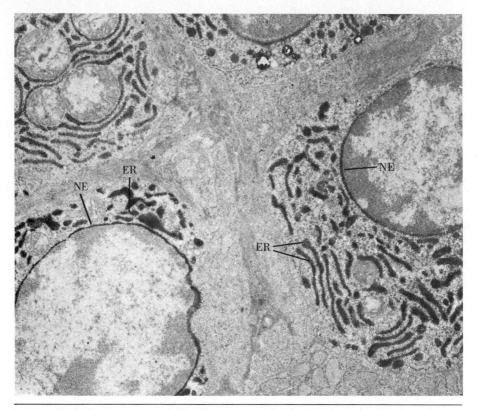

Figure III–40 *Plasma cells synthesizing antibodies in their rough ER.* Portions of several developing plasma cells from a rabbit injected with horseradish peroxidase (HRP) so as to induce formation of antibodies to HRP. Once antibody production was well under way (this usually is at least several days after injection), spleen tissue from the rabbit was fixed and cut into sections thin enough to expose the interior of many of the cells. Note that at this point the antibody-inducing HRP was no longer present in the cells; this is relevant because a second, different treatment with HRP was used to detect the antibodies. In the detection procedure, the fixed tissue was soaked in a solution of HRP (along with similarly prepared "control" tissue—spleen from a rabbit that had *not* been injected earlier with HRP and hence should not be producing antibodies). Next, the preparations were incubated with a cytochemical method that reveals sites of peroxidase enzyme activity (as in Fig. III–31) and then prepared for electron microscopy. The micrograph shows electron-dense reaction product, indicative of HRP enzyme activity, in the ER, including the nuclear envelope *(NE)*. This activity reflects the synthesis of antiHRP antibodies by the plasma cells since no such reaction product is observed in the control material. The reaction product seen in the micrograph is an indicator of the presence of the specific antibodies at the sites where the product is present because the reaction product is due to presence of HRP, and the HRP is there because it was bound by HRP-specific antibodies made by the cells. × 7500. (Courtesy of E. Leduc.)

binding to that antigen, an immune response is initiated. The B-cell recognizes the corresponding antigen by virtue of immunoglobulins with that B-cell's particular specificity, built into the cell's plasma membrane. The result is stimulation of cell division and of differentiation so that the B-cell gives rise to a set of plasma cells that produce and secrete the specific immunoglobulin. (Some of the products of the B-cell's division do not differentiate into plasma cells but persist as a clone of "memory" cells. These permit the organism to mount a more rapid response if the same antigen is encountered again, later in the individual's life. Plasma cells survive a few weeks; memory cells can survive for many years.)

Effective interaction of antigen and B-cell generating these responses frequently requires participation of other cells—specifically, T-lymphocytes called "helper cells" and macrophages or macrophage relatives such as the "reticular" cells of the spleen. The latter cells seem to "present" the antigen to lymphocytes in some appropriate form. (Perhaps this depends on orientation of the antigen on a membrane, its enzymatic modification, its linkage to another molecule, or its accumulation at a critical concentration.) Helper T-cells have been found to secrete factors that stimulate B-cell division and differentiation.

This analysis implies that the diversity of antibodies a given organism can make—although very large—is not limitless. Instead it is likely to be related, in ways now only dimly perceived, to the range of antigens encountered in the evolutionary history of the species and during the development of the individual. In fact, although an individual can make many thousands of different antibodies, within this set there are overlaps in specificity and also gaps, expressed as poor immunological responses to certain potential antigens.

Pathogens also evolve in part by selection for traits that permit them to evade or to overcome host organisms' immune systems. Life within phagocytes (Section 2.8.2) is one of the relevant mechanisms.

Vesicle-mediated Transfer of Maternal Antibodies; Secreted IgAs At birth, the immune system of many mammals is immature and affords little protection. For a time the newborn relies on antibodies acquired from its mother's milk, which pass into the offspring's bloodstream. This passage requires movement across the intestinal epithelium, a process carried out evidently by endocytic vesicles that employ specific plasma membrane–associated receptors to acquire the antibodies at the intestinal lumen and subsequently to deposit them, by exocytosis-like release, at the basal or lateral surface of the cell. From here the antibodies can enter the circulation. Such transepithelial transport in different mammals ceases after the first few days or weeks of postnatal life. A model made plausible by recent microscopic findings suggests that the vesicles responsible for antibody transfer fuse first with lysosomes in the epithelial cells, depositing the materials that are free within the vesicle interior; the antibodies to be transferred, however, remain bound to the receptors in the vesicle membrane. Next, the vesicle pinches off from the lysosome and moves to the basal or lateral cell surface with which it fuses. Now being tested is the hypothesis that the receptors involved bind antibodies avidly at low pH, such as that in the intestinal lumen or lysosomes, but release them at higher pHs, such as that of the extracellular spaces surrounding the cells.

Vesicle-mediated transport in the opposite direction occurs with a type of immunoglobulin known as *IgA*. IgAs are secreted into the lumina of various organs by the corresponding epithelia; they are found, for instance, in the protective mucus layers lining the intestine or respiratory tract, in saliva and in the colostrum and milk produced during the early stages of lactation in mammals. The IgAs are made by plasma cells, including ones present near the basal zones of the epithelia across which they are transported. Current hypotheses suggest that the epithelial cells synthesize receptor proteins (called *secretory component*) and insert them in their plasma membranes; these proteins bind IgAs, promote their endocytic uptake, and remain attached to the IgAs as the endocytic vesicles transit to the cell's luminal surface, where the vesicles fuse with plasma membrane. As released to the luminal content the IgAs still have portions of secretory component molecules attached; it is believed that this reflects release of the IgA from the membrane by a proteolytic cleavage that splits off part of the secretory component, freeing it plus the IgA.

3.6.10 Genetic Basis of Antibody Diversity

How do the DNA sequences specifying different IgGs or other antibodies arise? The process is based on the shuffling of modules to produce diverse combinations (Fig. III–39C). Antibody molecules such as IgGs are each coded for by a cluster of DNA sequences that start out, at the beginning of an organism's development, widely separated from one another. There is one or a very few different sequences capable of generating the constant region of the heavy chain of IgG (C-region sequences for the H chain), a similarly small number of C-region sequences for the L chain, but several hundred sequences potentially specifying H-chain variable regions and several hundred potential L-chain V-region sequences. In the progenitor of each clone of B-cells, a particular H-chain V-region sequence becomes associated with an H-chain C-region sequence, and the same occurs for L-chain V- and C-region sequences. This "association" refers to physical rearrangement of the DNA so that a C-region and a V-region sequence are moved ("translocated") into the proximity required for an RNA polymerase to produce an RNA that, after splicing and processing, can specify a corresponding polypeptide. In other words, a complete gene is put together from parts.

The two L chains of an immunoglobulin molecule are identical to one another, as are the two H chains, but with few restrictions, any L-chain gene generated by the mechanism just described can come to be expressed (transcribed and translated), together with any H-chain gene—each combination characterizing a particular B-lymphocyte line. The fact that tens of thousands of different combinations are possible among the hundreds of L chains and hundreds of H chains accounts for much of the diversity of antibody specificity.

Additional diversity results from the fact that the DNA stretches specifying a given variable region are actually constructed through translocation of several, initially separate, genetic "fragments." The genetic information for each heavy-chain variable region is assembled by associating a V sequence plus one of the several varieties of D sequence and one of the J_H sequences that exist in the progenitor cells. Each light-chain variable region involves a J_L sequence as well as its V. The J and D sequences are considerably shorter than the V sequences and seem to occur in fewer variations, perhaps less than a dozen each, but they multiply the diversity of possible amino acid sequences in immunoglobulins.

Still more diversity stems from changes in the DNA sequences that occur, probably during the shuffling of DNAs into the combinations required for expression—a sort of mutational process occurring in the progenitor cell.

The several classes of immunoglobulins differ in the C regions of the H chain but have variable regions in common. This reflects the origin of the corresponding genes through translocation involving different C-region sequences brought together, in different cells, with the same set of variable-region genes.

Progress is also being made in determining how a given cell inserts an immunoglobulin of particular specificity in its plasma membrane at one stage in its life history, whereas its offspring later *secrete* an immunoglobulin with the same specificity. Comparison of the alternate forms of the proteins and corresponding genetic analysis suggest this is a phenomenon of mRNA production, perhaps of different splicing patterns. That is, the same DNA sequence is used to specify both the membrane-inserted and the secreted form of the protein, but mRNAs are made that differ in the regions specifying the C-terminal region (see Fig. II–23) of the protein; evidently it is the differences in C-terminals that result in one version of the protein "sticking" in the membrane while the other passes entirely through.

3.6.11 Autoimmune Disorders

In a number of diseases, the restraints that prevent the immune system from generating antibodies to "self" break down, so that an individual organism produces "autoantibodies," ones that bind to its own cells and molecules. In some cases, this phenomenon is initiated when antibodies evoked by particular foreign organisms or molecules turn out to bind also to the responding organism's own molecules; some bacterial cell walls, for example, seem to have configurations resembling portions of molecules in the hosts they invade. (This resemblance could be the result of the pathogen's evolutionary "effort" to avoid immune responses by "masquerading" its molecules as host cell molecules.) In other cases, tissue damage may reveal or create antigens not normally encountered by the immune system.

When large amounts of immune complexes form in the blood stream, they can accumulate in critical places, such as the glomerular filter in the kidney (Section 3.5.3), where they pile up as layers along the capillaries and basal laminae. Phagocytes will also accumulate and attempt to engulf the complexes. Ordinarily, where numerous phagocytes are active, there is a certain amount of release of lysosomal hydrolases to extracellular spaces as a result of death of some of the phagocytes and also through accidental release when lysosomes fuse with forming phagocytic vacuoles before the vacuoles have sealed off completely from the outside. When large layers of immune complexes are present, this effect is probably exacerbated; under experimental conditions, when phagocytes attempt to engulf material too large for them to handle, they eventually release their hydrolases to the surface of the material, and the enzymes leak out into adjacent spaces. Some of the tissue damage in kidney disorders and in certain forms of arthritis may result from phenomena of these sorts.

Investigators are finding some autoantibodies quite useful. In *myasthenia gravis,* a neuromuscular disorder, antibodies to the receptor for the neurotransmitter *acetylcholine* (Fig. II–9) are produced. Employed in immunohistochemical studies, these antibodies are useful for discovering the locations of such receptors in tissues difficult to study with the physiological and biochemical methods usually utilized in searching for receptors. Analogous work is being done with antibodies that bind to components associated with centrioles or with kinetochores. How antibodies against these organelles come to be made is unknown, but they are proving helpful both for microscopic work and in beginning to investigate the chemistry of the bodies they can bind to. In the disorder *systemic lupus erythematosus,* autoantibodies that bind to nucleoproteins are formed; these are being used to isolate and analyze some of the lesser-known RNA–protein complexes of the cell such as the small nuclear RNAs (Chap. 5.1).

Discussion of antibody uses will continue when we consider "monoclonal" antibody production and diagnostic approaches (Section 3.11.2).

Chapter 3.7 Nerve Cells

Neurons include the longest cells in the body, ranging to several feet in length. They are the coordinating elements of an elaborately interconnected network. The network consists of sensory receptor cells, enormous numbers of connecting and integrating neuronal circuits, and neurons leading to effector organs such as muscle. A given neuron can interact, more or less directly, with hundreds of other cells.

At the level of individual cells, the developmental processes responsible for the establishment of specific neuronal geometry and connections depend upon many factors. Considerable research is under way on the developmental

timing with which particular cells appear, grow, migrate, and contact one another; studies focus also on the death of certain neurons as a facet of normal
development in nervous tissue, and on cell-surface "recognition" phenomena.
It is found, for example, that growing neuronal processes can be guided toward
their destinations by "mechanical" influences (Section 2.11.4) such as the organization and distribution of extracellular materials and of nonneuronal cells
(Section 3.7.4), and by chemotactic interactions. Major tasks for modern biologists are to determine how such factors are integrated to produce a functioning nervous system and to dissect the cellular and molecular features of the
system that alter with experience in the phenomena of learning and memory.

Figure III–41 shows two of the major morphological types of neurons,
unipolar and *multipolar*. Many unipolar neurons, such as vertebrate sensory
neurons, are characterized by a single process that divides into two branches.
One branch receives input from a sensory receptor, and the other connects to
other neurons by means of *synapses*. Multipolar neurons such as the motor
neurons of the spinal cord have numerous receptor processes *(dendrites)*,
which receive impulses at synapses with other neurons, and a single transmitter
process (the *axon*), which carries impulses from the cell body to effectors. Both
types of neurons are fundamentally similar in other respects.

3.7.1 Perikaryon

The neuronal cell body, or *perikaryon,* contains the cell nucleus and cytoplasmic organelles distributed around the nucleus in an arrangement that is
roughly symmetrical (Fig. III–41). The Golgi apparatus is a well-developed network (Fig. I–18). Rough ER is extensive; it is concentrated in the so-called *Nissl
substance*—"basophilic" (Section 2.2.3) patches (Fig. I–15) showing parallel
rough cisternae plus many free polysomes. Numerous lysosomes are present
and tend to be concentrated in the Golgi zone. Mitochondria are abundant
throughout the cell. In contrast, peroxisomes generally are sparse, and in some
neurons, only a very few are found with the techniques now available. In general, the cytological characteristics suggest extensive macromolecular synthesis:
large nucleolus, many ribosomes, extensive rough ER, and large Golgi apparatus. These features are in keeping with the role of the perikaryon as a synthetic center that supplies macromolecules and other materials to the rest of
the neuron. In extreme cases, such as the motor neurons of the spinal cord,
the total neuronal volume is thousands of times the volume of the perikaryon,
so that the task of supplying these materials is a considerable one.

As in other cells, autoradiographic studies indicate that protein made in
the rough ER passes into the Golgi apparatus of neurons (see Fig. II–43). In
neurosecretory neurons, found in the pituitary gland and elsewhere, the Golgi
apparatus packages hormone granules; these travel down the axons, and their
contents are released by exocytosis at the axon endings to enter extracellular
spaces or the blood stream. The "osmoregulatory" hormone ADH (Section
3.5.3) is an *octapeptide* (chain of eight amino acids) that is secreted in this

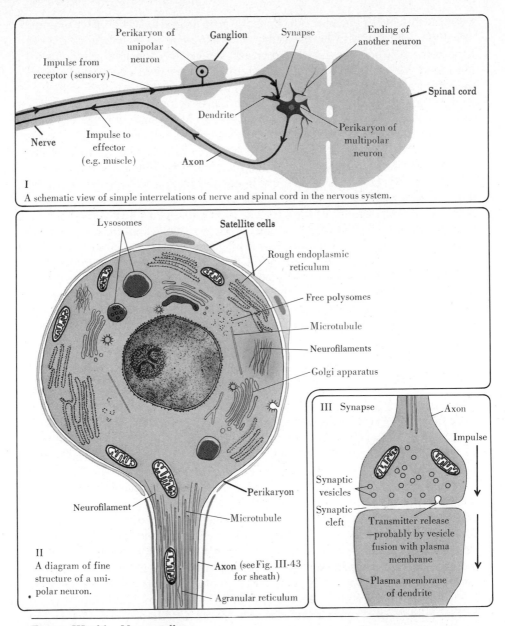

I

A schematic view of simple interrelations of nerve and spinal cord in the nervous system.

II
A diagram of fine structure of a unipolar neuron.

III Synapse

Figure III—41 *Nerve cells.*

way. In most other neurons the specific roles of the Golgi apparatus are less clear. Increasing evidence does suggest that many of the proteins passing through the apparatus are later transported down the axon (Section 3.7.2). Moreover, certain of the substances involved in the transmission of nerve impulses from one cell to another appear to be packaged in membrane-delimited structures, which may originate from Golgi-associated sacs. (As discussed below, transmission is based on phenomena akin to secretion that occur at axon endings.) The Golgi apparatus and associated structures also seem to participate in forming the numerous lysosomes.

The abundance of lysosomes is of uncertain significance. Some probably take part in the extensive turnover of membranes that accompanies neuronal function (Section 3.7.5). Many of the lysosomes of neurons are residual bodies. Neurons do not divide, and many live as long as the organism, slowly accumulating residual bodies. As mentioned in Section 2.8.4, with time, lipofuscin ("aging pigment") accumulates within residual bodies; neurons of aging animals often have a substantial lipofuscin content.

Since neurons do not divide and no reservoir of undifferentiated "precursor" cells exists, if the perikaryon is destroyed, the neuron is not replaced. However, if the axons are cut, only the portion no longer attached to the perikaryon degenerates; the remainder is capable of regenerating. Under appropriate conditions, neurons of the peripheral nervous system (nerves and ganglia) can reestablish some of the original connections and so restore neuron function. Proper conditions for full regeneration and reestablishment of the connections of neurons of the mammalian central nervous system (brain and spinal cord) have not been found.

3.7.2 Axons

In contrast to the cell body, the axon contains neither Golgi apparatus nor rough ER. There are few ribosomes. Most axons are largely inactive in the synthesis of protein or other macromolecules. On the other hand, along the entire length of most axons there are microtubules and filaments—mainly *neurofilaments,* a type of intermediate filament, but some microfilaments too; these elements are oriented parallel to the long axis of the axon. Studies are underway to determine if any of the individual microtubules or filaments are long enough to stretch from one end of the axon to the other. The limited information available suggests that most of the individual tubules and filaments are much shorter than this but overlap extensively at their ends. The axon contains some smooth-surfaced tubules and vesicles, including structures of endocytic origin, Golgi-derived structures, and smooth endoplasmic reticulum. Collectively these are sometimes called *agranular reticulum,* but this term does not imply that they constitute a single functional or structural system. In some types of neurons the smooth ER forms a network that can be continuous from one end of the axon to the other. The smooth ER or other elements of the agranular reticulum may help to regulate the levels of Ca^{2+} ions free in the axonal cytoplasm (see Section 3.9.1).

Axonal Transport Microscopic studies of living nerves indicate that there is a constant traffic of structures in both directions along the axon. This has been confirmed by biochemical investigations demonstrating a net movement of macromolecules—free and membrane-associated—down the axon (in the "orthograde" or "anterograde" direction); this transport supplies the axon itself and its terminals with essential components. Movement in the opposite direction ("retrograde transport") carries materials back to the perikaryon for reuse or degradation in the lysosomes as part of intracellular turnover. It is often speculated that in addition to serving supply and turnover, axonal transport in both directions also carries "informational" molecules. These are hypothetical signals the perikaryon might receive or send, permitting it to communicate with distant portions of the cytoplasm and cell surface in ways other than the conducting of impulses. In addition, exchanges between cells of materials transported along axons and dendrites could be important for cellular interactions.

The mechanisms of axonal transport are still to be definitively described. At one time it was thought to depend on a kind of peristalsis-like bulk flow of axoplasm within the axon, but this is now known to be inaccurate. Different materials can move at quite different rates along the same axon: Many membrane-bounded structures seem to move rapidly, at rates up to hundreds of millimeters per day; other material, such as the components of the microtubules and filaments and presumed cytosolic molecules, move slowly, at one or a few millimeters a day. Mitochondria show complex behavior interpreted as including periods when they move rapidly down the axon, periods when they reverse direction, and periods when they remain stationary. Corresponding to this variety of rates, several different transport routes and mechanisms could participate in generating movements in one or both directions. Processes mediated by microtubules or filaments (or the postulated microtrabecular system), such as those discussed in Section 2.11.4, are obvious possibilities. Many investigators think such processes propel vesicles, endocytic structures, and other membrane-bounded elements. Flow within the smooth endoplasmic reticulum and movements of molecules within the plane of the plasma membrane or ER membrane are also possible mechanisms of axonal transport.

3.7.3 The Nerve Impulse

The most readily observable feature of the passage of a nerve impulse along an axon is a change in the "potential" (the difference in electrical potential) that exists across the plasma membrane; this change (the *action potential*) moves like a wave along the axon surface. Current theory, strongly based, holds that the underlying mechanism depends on the movement of ions accompanying a wave of permeability changes that passes down the axon membrane.

Membrane Potential; Resting Potential   In resting state, K^+ concentration is maintained high and Na^+ concentration low in the axon (as compared with the extracellular space) by an energy-requiring active transport mechanism (Section 2.1.3). A somewhat oversimplified but useful view of the consequences of this asymmetric distribution of ions follows: It can be shown in model experimental systems and by theoretical treatment that when such asymmetries are established across a membrane, a difference in electrical potential aross the membrane *(membrane potential)* will result if some of the ions can diffuse through the membrane more rapidly than others. K^+ can pass across the axon membrane much more rapidly than Na^+. (What is being considered here is the passive movement down concentration gradients that continually tends to restore equal concentrations on both sides of the membrane.) The tendency for K^+ to leave the axon more rapidly than Na^+ enters produces, at equilibrium, a net relative deficiency of positive charges *locally,* on the inside of the membrane. In consequence, the inside of the cell membrane is at an electrically negative potential as compared with the outside; this can be measured as a voltage with suitable electrical devices, and it is referred to as the "resting membrane potential" or, more often, as the *resting potential.* The potential reaches a magnitude of the order of 50 to 75 mV (millivolts; 1 mV = 1/1000 volt) in the nerve cells most studied, such as the giant axons of the squid. This magnitude expresses the equilibrium situation in which the tendency of K^+ to exit down its concentration gradient is just balanced by the potential that results: The negative potential retards the further exit of positive ions. At equilibrium there is no *net* loss of K^+ although individual K^+ ions continue to move in both directions.

Action Potential Impulse conduction by a given region of the axon is based on the following sequence (Fig. III–42):

1. A localized permeability change permits the rapid influx of Na^+. This influx leads to changes in the potential across the membrane; eventually the inside becomes positive with respect to the outside.
2. A second set of permeability changes restricts the movement of Na^+ but permits the rapid efflux of K^+. This is associated with eventual restoration of the original negativity of the inside.
3. Propagation of the impulse to adjacent axon regions occurs. This results from the fact that the local changes of steps 1 and 2 produce an axon region that differs in ion concentration and electrical potential from adjacent regions. A flow of ionic current takes place between this region and the adjacent regions. This *electrotonic* flow takes place both within the axon and outside it. (The direction of flow outside is opposite that inside, so that overall the current flows in a complete circuit.) The longitudinal current results in initiation of the cycle of permeability changes (steps 1 and 2) in membrane regions adjacent to the region we have been considering. Thus the impulse moves along the membrane.
4. The intracellular low Na^+–high K^+ condition is restored, relatively slowly, by a "sodium pump" that actively moves Na^+ out, with a reciprocal influx of K^+. This can occur relatively slowly, since only a small percentage of the ions move in a single impulse. The very large initial asymmetry in Na^+ and

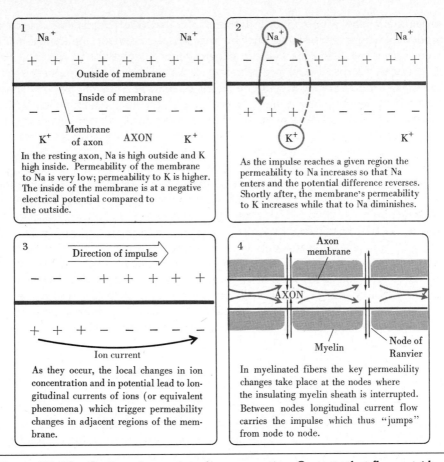

1

Na$^+$ Na$^+$

$+$ $+$ $+$ $+$ $+$ $+$ $+$ $+$
Outside of membrane

Inside of membrane

$-$ $-$ $-$ $-$ $-$ $-$ $-$

Membrane
of axon   AXON
K$^+$ K$^+$

In the resting axon, Na is high outside and K
high inside. Permeability of the membrane
to Na is very low; permeability to K is higher.
The inside of the membrane is at a negative
electrical potential compared to
the outside.

2

Na$^+$ Na$^+$

$-$ $-$ $-$ $+$ $+$ $+$ $+$ $+$

$+$ $+$ $+$ $-$ $-$ $-$ $-$

K$^+$ K$^+$

As the impulse reaches a given region the
permeability to Na increases so that Na
enters and the potential difference reverses.
Shortly after, the membrane's permeability
to K increases while that to Na diminishes.

3

Direction of impulse

$-$ $-$ $-$ $+$ $+$ $+$ $+$ $+$

$+$ $+$ $+$ $-$ $-$ $-$ $-$ $-$

Ion current

As they occur, the local changes in ion
concentration and in potential lead to longitudinal currents of ions (or equivalent
phenomena) which trigger permeability
changes in adjacent regions of the membrane.

4

Axon
membrane

AXON

Myelin   Node of
Ranvier

In myelinated fibers the key permeability
changes take place at the nodes where
the insulating myelin sheath is interrupted.
Between nodes longitudinal current flow
carries the impulse which thus "jumps"
from node to node.

Figure III–42 *Outline of nerve impulse propagation.* **Currents that flow outside
the axon accompanying the currents inside are not shown.**

K$^+$ distributions can permit the passage of many impulses before pumping
becomes absolutely essential to further functioning. Normally the cycle of
changes in permeability at a given axon region is completed in a very few
thousandths of a second (milliseconds). The impulse passes down the axons
of different nerves at rates of tenths to tens of meters per second.

The axon is capable of conducting impulses in both directions along its
length. Normally impulses move in one direction because they are initiated at
one end; furthermore, as an impulse passes a given region, the changes that
occur make that region unable to conduct another impulse for a time (the
refractory period).

Only in the last step related to impulse propagation (step 4 above) does
the cell actively expend chemical energy. In the living cell the mitochondria

provide energy for the sodium pump, but the axon membrane itself (perhaps together with some closely associated material) seems to be solely responsible for conduction of impulses. This has been nicely demonstrated by using the giant axons, 500 μm in diameter, found in squid. Virtually all of the intracellular organelles and other cytoplasm can be removed from such axons and replaced by suitable solutions of ions in which the Na^+ concentration is kept low and the K^+ concentration high; the preparations still conduct impulses quite efficiently. As expected from the analysis developed above, the *resting* potential sustained by these preparations depends on the K^+ concentrations used, whereas the *action* potential requires the presence of Na^+ (or of certain substitute ions). (Section 2.1.3 discusses active transport of Na^+.)

Gates and Channels The nature of the changes in the axonal plasma membrane that are responsible for conduction of impulses is a major unresolved issue. The models usually advanced propose that the membrane contains channels for K^+ and Na^+ and that these channels are opened and closed by "gates." This process can be visualized, for example, as a change in protein conformation, leading to a localized movement that reversibly unblocks a selective "hole" in the membrane (see Fig. II–4). The longitudinal currents of ions that carry the impulse do so by changing the membrane potential. Specifically, the currents originating when a given axon region conducts an action potential reduce the difference in potential across the plasma membrane of adjacent regions. When in consequence the membrane potential in these adjacent regions passes a threshold level, the Na^+ "gates" open, and the cycle of events of the action potential commences. If the "gates" represent charged molecules or parts of molecules, their movements could be explained by the fact that charged groups move in response to changes in electrical field such as those resulting from changes in the membrane potential.

Evidence that K^+ and Na^+ cross the axon membrane through separate pathways comes from observations with naturally occurring toxins and with molecules synthesized in the laboratory: Some of these, such as *tetrodotoxin,* obtained from puffer fish, selectively block Na^+ permeability; others, such as tetraethylammonium ions, selectively block permeability to K^+. That Ca^{2+} may participate in events of conduction is suggested by numerous observations showing effects of changed Ca^{2+} levels on thresholds and other characteristics of relevant membrane phenomena.

3.7.4 Myelin; Glial Cells

In vertebrate peripheral nerves two major types of axons are found, *myelinated* and *unmyelinated*. In both, *Schwann cells* are closely associated with the axon. In the unmyelinated types the axon occupies a pocket formed by the indentation of Schwann cells (Fig. III–43). The Schwann cells associated with the myelinated axons wrap repeatedly around the axon, surrounding it with many lay-

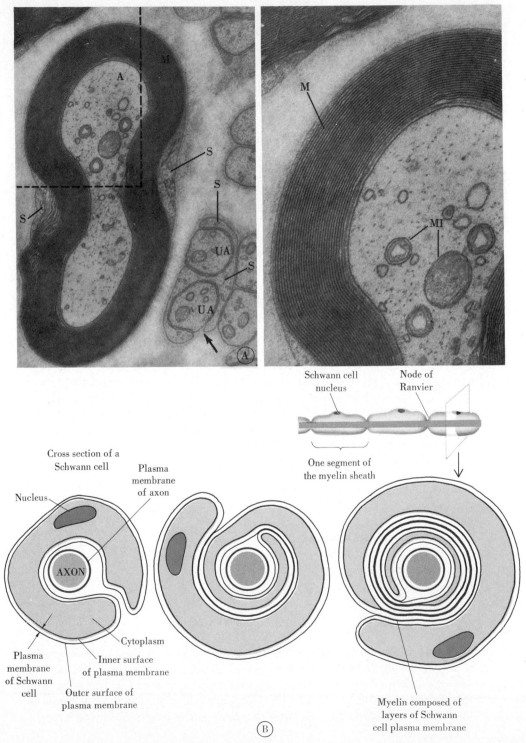

Cross section of a Schwann cell

Schwann cell nucleus

Node of Ranvier

One segment of the myelin sheath

Nucleus

Plasma membrane of axon

AXON

Plasma membrane of Schwann cell

Outer surface of plasma membrane

Inner surface of plasma membrane

Cytoplasm

Myelin composed of layers of Schwann cell plasma membrane

◀ **Figure III–43** *Myelin.* A. Electron micrographs of a portion of sciatic nerve (guinea pig) sectioned transversely. A myelinated axon *(A)* and several unmyelinated axons *(UA)* are shown. *S* indicates Schwann cell cytoplasm. A narrow space continuous with the extracellular space outside the Schwann cell (arrow) separates unmyelinated axons from the Schwann cells that surround them. The arrangement often resembles that shown in the lower left diagram. Myelin *(M;* enlarged at right) is of multiple membrane layers. Mitochondria in the axon are seen at *MI.* Left, × 20,000; right, × 50,000. (Courtesy of H. Webster.) B. Schematic diagram of myelin formation by the establishment of a multilayered spiral-like pattern of Schwann cell plasma membrane. The cytoplasmic space and extracellular space originally separating membrane layers are obliterated so that the final pattern consists of layers where the original external surfaces of the membrane are closely apposed, alternating with layers where the original cytoplasmic surfaces are closely apposed. In mammals the myelin segment contributed by a single Schwann cell (this corresponds to the distance between successive nodes) can cover a length of several hundred micrometers of axon.

ers of their plasma membranes. The multilayered membrane system constitutes myelin (Fig. III–43); it is part of the Schwann cell. The rest of the cytoplasm and the nucleus remain at the outer surface of the multiple layers of plasma membrane.

Each Schwann cell contributes one segment of myelin to the sheath that covers most of the length of the axon. The segments of sheath contributed by two Schwann cells abut, but do not fuse, at *nodes of Ranvier.* The lipoprotein of which myelin is made is a good electrical insulator; unlike most membranes, there is considerably more lipid than protein, by weight. Thus, a myelinated fiber is surrounded by a layer of insulating material that is interrupted at the nodes. This arrangement modifies the conduction of the nerve impulse. Current tends to flow preferentially along paths of low resistance, and in myelinated axons the nodes define such a path. The changes in membrane permeability responsible for conduction take place chiefly at the nodes (Fig. III–42). This kind of propagation, by "jumping" ("saltation") from node to node, is called *saltatory conduction;* each node may be thought of as reamplifying the impulse and passing it down the axon to the next node. The permeability change at one node results in local changes in concentration and potential; these changes set off an electrotonic flow of ionic current within the axon, which in turn triggers the permeability change at the next node. The "jumping" is responsible for the more rapid impulse conduction by myelinated nerves, as compared with that of unmyelinated nerves; the latter lack the extensive insulation that makes saltatory transmission possible.

The myelin of central nervous tissue is laid down by *oligodendrocytes,* one of the *glial* cell types (see next paragraph). Many of the unmyelinated axons of central nervous tissue lack an individual nonneuronal sheath; they do, however, form groups and bundles of various sorts, which are penetrated and segregated from one another by processes from glial cells, especially astrocytes.

Schwann Cells; Glial and Satellite Cells   In addition to the Schwann cells and oligodendrocytes, both central and peripheral tissues contain other unique nonneuronal cell types. The term *glia* (or neuroglia) is sometimes used inclusively to refer to all of these nervous system–specific cells. But in embryological origin and some other features, those of the central nervous system differ from the Schwann cells and other nonneuronal cells of the peripheral system; therefore, we shall restrict use of *glia* to refer collectively to the central nervous system types. In some brain regions there are far more glial cells than there are neurons.

Study of the various nonneuronal cells has been much less intensive than the study of neurons, in part since the former do not conduct nerve impulses and thus are not directly involved in the flow of information. Precisely what they do (aside from the types that form myelin) has yet to be discerned; there is a general prejudice toward believing that they serve as structural supports and linking elements that help to maintain overall tissue architecture. They may also provide some molecules to neurons, establish barriers that limit access of materials to the neuronal surface, and otherwise affect pathways for diffusion in the extracellular space. In certain cases, they are thought to "mop up" neurotransmitters released at nerve terminals by transporting the transmitters into their cytoplasm (Section 3.7.5).

When axons of peripheral nerves are interrupted, the Schwann cells surrounding the portions no longer continuous with the perikarya first engage in activities such as phagocytosis that help to dispose of the degenerating axon (they also break down myelin sheaths they hitherto had maintained). Then they proliferate while remaining near their original locales and can subsequently help to guide the regenerating axon back to the sites it initially occupied. In addition, at least for some of the giant nonmyelinated axons of invertebrates, molecules made in the Schwann cells, perhaps including proteins, can be transferred to the interior of the axon. Certain of the glial cells may help to guide the migrations of neurons or growing axons during development of central nervous tissue.

The cell bodies of peripheral ganglion cells (Fig. III–41) are invested with a thin sheath of *satellite* cells. These, it has been postulated, serve nutritive roles for the neurons and screen out inappropriate components in extracellular fluids. The satellite cells show a number of interesting but little-explored cytological characteristics, such as the presence of numerous peroxisomes—an abundance unusual for nervous tissue.

Astrocytes are one of the two major glial cell types of the central nervous system (the oligodendrocytes are the other). These cells extend elaborate processes among the neurons; some of the processes end on capillaries, and some seem to segregate surfaces of adjacent neurons or bundles of axons from one another. This separation may, for example, prevent the spread, to other neurons, of neurotransmitters released at the synapses on a given neuron; such spread could otherwise trigger inappropriate responses. In one major class of astrocytes, prominent bundles of a unique variety of intermediate filaments are present, evidently serving supportive roles. Under a number of circumstances glycogen accumulates in astrocytes; this finding has led to proposals that the cells store nutrients destined ultimately for neurons.

It should be recalled that the capillaries of the brain are unusually restrictive about what can cross the "blood–brain barrier" (Section 3.5.2). In fact, the extracellular fluid of the central nervous system (the "cerebrospinal fluid") communicates with the blood stream by a number of special tissues (such as the choroid plexuses) whose effect is to buffer the brain's neurons against too dramatic change in the composition of the fluids that surround the cells and against potentially noxious materials. Glial cell activities may also contribute to this protection.

3.7.5 Synapses

Neurons communicate with one another, and with effector organs such as muscles or glands, through *synapses.*

Electrotonic Junctions In a limited number of cases transmission from neuron to neuron depends on gap junctions. These form a low-resistance pathway (an "electrotonic junction" or "electrical synapse") through which ions can flow from cell to cell, coupling the changes in potential in one cell directly to those in the other. Such junctions provide rapid, reliable transmission under some circumstances in which secure coordination is called for. For example, certain fish can discharge considerable electric currents through action of their "electric organs," specially modified muscles. In South American electric fish, these organs are controlled by a group of nerve cells in the spinal cord that fire simultaneously, leading to coordinated discharge by the electric organs. The nerve cells are linked to one another by gap junctions so that stimulation of one of them causes all to initiate action potentials. (The nerve cells communicate with the electric organ by chemical synapses like those described in the following paragraphs.) Some of the very rapid muscle movements by which organisms escape from predators are also coordinated by neurons associated by gap junctions.

Communication by gap junctions between muscle cells will be described in Section 3.9.2.

Chemical Synapses; Neurotransmitters; Acetylcholine In most cases, electrical transmission is precluded by the absence of the requisite special junctions. Instead, chemical transmission takes place. *Acetylcholine* and *norepinephrine* are among the known chemical transmitters (Fig. II–9). Different classes of neurons use different transmitters. Most of the known or suspected ones are the size of amino acids or slightly larger. Some *are* amino acids, such as *glycine* and *glutamic acid.* In addition, a growing variety of small peptides has been found to be released from different nerve terminals.

One synapse is diagrammed in Figure III–41; others are shown in Figures III–44 and III–47. The vesicles hold neurotransmitter molecules. When the impulse reaches the synapse, the vesicles fuse with the membrane and release the transmitter into the extracellular "synaptic space." Involved in initiating this process is an influx of Ca^{2+} ions into the synapse. At the large neuromuscular junctions by which motor neurons control striated muscles (Fig. III–47), synaptic vesicles fuse with the axon membrane at localized "active" zones. These are characterized by rows of prominent intramembrane particles seen in freeze-fracture preparations. An attractive though still unproved hypothesis is that the particles are gated Ca^{2+} channels that respond to passage of an action potential into the nerve terminal by opening, permitting a very localized influx of Ca^{2+}. By the mechanisms discussed earlier (Section 2.1.4), the rise in Ca^{2+} is kept transitory, so that a single action potential results in release of a limited amount of transmitter.

Electron-dense material is often aggregated in bands, ribbons, or other arrangements along the axonal plasma membrane in the local regions where vesicle release occurs. Narrow bands of such materials occur, for example, at the active zone of the neuromuscular junction. In synaptic terminals of photoreceptors, vesicles near certain of the sites of transmitter release associate with synaptic "ribbons," elongate aggregations of electron-dense material that protrude from the plasma membrane into the cytoplasm (see Fig. III–44). Though relatively undistinguished-looking in ultrastructure, such bands and ribbons may reflect special organization of the cytoplasm that "guides" the vesicles in their interactions with the membrane, perhaps promoting fusion at the proper places.

The transmitter released from a nerve terminal interacts with specific receptor sites on the plasma membrane of the "receiving" cell (the *postsynaptic* plasma membrane). The effect varies, depending on the transmitter and receptor. Receptors on many of the neurons or muscle cells that respond to acetylcholine mediate excitation that can initiate an impulse in the responding cell (Sections 3.7.6 and 3.9.1). This excitation is based on changes in membrane potential (depolarization; see following section) reflecting altered membrane permeability: The interaction of acetylcholine with its receptors on muscle cells or on the electric tissues of electric fish (Fig. II–9) transiently opens channels through which both Na^+ and K^+ can pass. The interaction is reversible: The acetylcholine molecules bind to specific groups in the receptor but subsequently dissociate from these groups.

Hydrolysis and Reabsorption of Transmitters   An enzyme, *acetylcholinesterase,* is present in the extracellular space at synapses of neurons transmitting by acetylcholine. This enzyme can inactivate acetylcholine, by splitting it into acetate and choline (Fig. II–9). Consequently, of the several thousand acetylcholine molecules released from each synaptic vesicle, relatively few survive to interact with receptors. Moreover, once an acetylcholine molecule that has bound to a receptor dissociates from it, chances are that the acetylcholine will be dismantled by the cholinesterase before it can bind again. Therefore, the enzyme gives an important measure of control to the system. If the acetylcholine remained intact in the synapse, it could continue to reassociate with receptors and to excite the postsynaptic cell: A nerve impulse might set off a long series of twitches in a muscle rather than the single one that is usually obtained. Axon endings that release norepinephrine can also take up this compound from the extracellular space. This is one of the mechanisms that control the persistence of the transmitter extracellularly and also conserve it for reuse. Glial cells similarly absorb certain transmitters.

Formation and Filling of Synaptic Vesicles   It is too early to generalize from the few synapses studied in detail and to assert that exocytosis necessarily is the only important way in which chemical transmitters are released from nerve terminals. Although most investigators agree that the synaptic vesicles are central agents in transmitter release, there is still much to learn. For example, unlike macromolecular synthesis, which is confined largely to the perikaryon, extensive synthesis of neurotransmitter molecules such as acetylcholine or norepinephrine occurs in axon terminals at synapses. Certain of the enzymes for such synthesis are associated with the vesicles, but some are outside. Storage of transmitters may therefore involve events like those described in Section 3.6.1: cooperative efforts of vesicular and nonvesicular enzyme systems, transport across the vesicle membrane, and binding inside the vesicle. Details of where transmitters in the terminals are made and how they are stored in the vesicles are in dispute.

   Similarly, although it is reasonably well established that the macromolecules of synaptic vesicles are produced chiefly in the perikaryon and transported to the synapses, and that vesicles can fill with transmitters while in the terminals, the life history of the vesicles is incompletely understood. Some seem to be formed in the perikaryon in the Golgi zone, but there may be additional sites of vesicle formation such as the agranular reticulum along the axon and in terminals. Subsequent to fusion with the plasma membrane during neurotransmission, vesicle membrane is "retrieved" from the cell surface by endocytosis-like processes (Section 2.5.4). The membrane (or its macromolecules) seemingly can be reused repeatedly to form functional synaptic vesicles, cycling between the interior of the terminal and the surface many times in the course of its existence. Eventually the membrane is degraded, perhaps by transport to the perikaryon and incorporation in lysosomes.

3.7.6 Neuronal Integration

A given neuron affects the cells it controls by the *frequency* with which it fires action potentials and the "strength" of its synaptic effects—the extent and duration of the changes evoked in the recipient cell. Each action potential conducted by that neuron's axon is essentially the same as the others and has essentially the same effect at a given synapse. (This last sentence overstates the simplicity of the situation, but the qualifications called for fall beyond the scope of our concerns here.) Most sensory systems also code information about the intensity of stimuli in terms of the frequency of action potentials and consequent synaptic activity (see Section 3.8.1). Some combinations of sensory cells and the neurons they connect to produce more action potentials when exposed to stimuli of greater intensity; others (Section 3.8.1) produce fewer. Much progress has been made in unraveling the mechanisms by which the nervous system is able to use such coding to bring about or modulate complex behaviors. Crucial to these mechanisms are the patterns of the synaptic connections among the cells involved and the details of interactions of the different synaptic inputs a cell may receive.

Excitation and Inhibition; Depolarization and Hyperpolarization
Quite often a given neuron receives synapses from several or many other neurons. Some of these may inhibit the recipient, and some excite it. In the cases studied in most detail, excitation amounts to *depolarization*—a rise in the potential across the membrane that, when large enough to move the potential past a threshold level, can trigger an action potential, as outlined in Section 3.7.3. Recall that at rest the inside of the neuron is at a negative potential with respect to the outside: A *rise* in the membrane potential means a positive change—in other words, change of the negative potential toward zero. Since the potential difference across the membrane becomes *less,* the term "depolarization" is used. Inhibition usually involves a change in ionic permeability that tends to stabilize the membrane potential near the resting potential or actually to *hyperpolarize* the cell (drive the membrane potential to a more negative level, further from the threshold at which an action potential can be initiated); increases in permeability to K^+ or Cl^- are commonly implicated. Which occurs at a particular synapse—excitation or inhibition—depends both on the transmitter released and on the receptors present: A given transmitter can mediate excitatory effects at some synapses and inhibitory effects at others.

Summation The strength of effect—the magnitude and duration of the potential change induced—also differs at different synapses. Some, for example, release the contents of relatively few synaptic vesicles per impulse; others, the contents of many. Moreover, the impact of synaptic activity can depend in part on the distribution of the synapses on the recipient cell. Thus, many neurons can initiate action potentials only at limited portions of their surface, often the base of the axon—the "axon hillock." Spread of potential changes from synapses elsewhere—on the dendrites, for instance—is by current flow akin to the longitudinal, electrotonic flow along an axon. This flow attenuates with distance from the synapse that initiated it. A synapse releasing only a small amount of transmitter or one located far from the axon hillock may not produce a great enough effort to trigger an action potential. One located at or near the hillock can have a more immediate and greater effect.

It is common for the effects of many of the excitatory synapses on a given recipient neuron, such as a motor neuron, to be insufficient individually to trigger an action potential in the recipient. This is a key factor in permitting a neuron to integrate the many inputs that may impinge upon it. The *pattern* of synapses present on a given cell arises developmentally and often is quite precisely determined. The *timing* of activation of the synapses depends on the responses of the input neurons to the physiological factors that activate or inhibit them. At any given moment, the state of a neuron receiving inputs at a variety of synapses will depend on the summation of excitatory and inhibitory inputs—their strength and distribution. Each synapse brings about a transient change in the membrane potential of the recipient neuron: The effect of a given synapse dies away in a relatively short time. The net effect of several synapses therefore depends on the timing with which they are triggered and the pattern of spread of the resultant potential changes along the recipient's membrane. If several synapses act to change the membrane potential in the same direction during a short enough interval and with an appropriate distribution, their total effect can push the recipient past its threshold; this effect leads to an action potential. If contrary actions occur at different synapses transmitting during a given period—some excitatory and some inhibitory—the net effect will depend on the timing and the relative strength of the various inputs: The inhibitory ones can counteract the excitatory ones.

We have been considering primarily neuron-to-neuron synapses of the sort that occur in the brain, for example, and account for much of the subtler modulation of physiological activities in higher animals. There also can be both excitatory and inhibitory inputs, nervous and hormonal, to effector tissues such as muscle or endocrine glands.

Storage of Experience   Less progress has been made in understanding how the nervous system acquires and stores experience over prolonged periods—learning and memory. For a while popular theories attributed memory to the synthesis of particular information-carrying molecules such as RNAs. But although macromolecular synthesis is likely to be involved in memory storage, it now seems probable that this involvement is of a general sort—serving cellular growth, or changes in synapses rather than the creation of specific "memory molecules." Most present theories center on changes in the connections among cells: the activation or stabilization of synapses or alterations making synapses that have fired a number of times more likely to fire again. There are plausible proposals for underlying molecular events relating to membrane properties, Ca^{2+} influx, cyclic nucleotides and other constituents, and features of synapses and axons. However, no consensus has yet developed about which of the many competing models applies best, and it may well be that different events occur in different organisms or in different learning phenomena.

Chapter 3.8 Sensory Cells

Different sorts of sensory receptor cells occur throughout the body. Each is specialized so that it responds to a given category of stimulus, usually by changing its membrane potential. Some types of sensory cells respond to mechanical pressure; others, to specific classes of chemicals. One interesting type responds to light. This is the *retinal rod cell,* capable of mediating vision at low illumination levels. Another similar set of retinal cells, the *cones,* functions chiefly in high-intensity light and permits color vision. These photoreceptive cells line the retina, and they make synaptic contact with connecting cells that synapse with neurons of the optic nerve.

3.8.1 The Retinal Rod

Structure; Disc Renewal   The rod cell (Fig. III–44) consists of a cell body with a nucleus, numerous mitochondria, and other organelles, connected by a cytoplasmic bridge or stalk to an *outer segment* specialized for light reception. At the base of the connection, a basal body with centriole-like structure is found, and within the cytoplasmic stalk, an arrangement of microtubules characteristic of cilia is present. From this appearance and from developmental studies, it has been concluded that the outer segment of the rod is a greatly modified derivative of a cilium. Other cilia of the sensory type are known, but often the modifications are not as great as in the rod. Sensory cilia generally lack the central pair of microtubules characteristic of most cilia and flagella (see Section 3.6.4 for another example). They are spoken of as 9 + 0 cilia, in contrast to the 9 + 2 cilia discussed earlier (Chap. 2.10).

The rod outer segment is a cylindrical body, delimited by a plasma membrane, containing several hundred flattened sacs (discs) stacked on top of one another (Fig. III–44). The membrane-bounded discs contain molecules of the visual pigment *rhodopsin.* This is a glycoprotein, *opsin,* with which is associated *retinal,* a derivative of vitamin A. Opsin is synthesized on the rough ER and moves through the Golgi apparatus on its way to the discs. Before becoming incorporated in the discs it may be inserted in the plasma membrane since the discs originate as plasma membrane foldings that pinch off into the outer segments. (One chief morphological difference between cone cells and rods in many organisms is the persistence of continuity of the stacked sacs with the cone cells' plasma membranes.) If a "pulse" (Fig. I–23) of radioactive amino acids is administered and then the label is traced by autoradiographic techniques, a band of radioactivity is observed to move up the rod outer segment. This finding reflects the fact that the discs are continually being replaced: Each disc gradually passes up the outer segment, reaching the tip in one to several weeks, depending on the species. At the tip the discs are shed from the rods and are phagocytized and degraded in lysosomes by cells of the *pigment epithelium* that surround the tips of the photoreceptors.

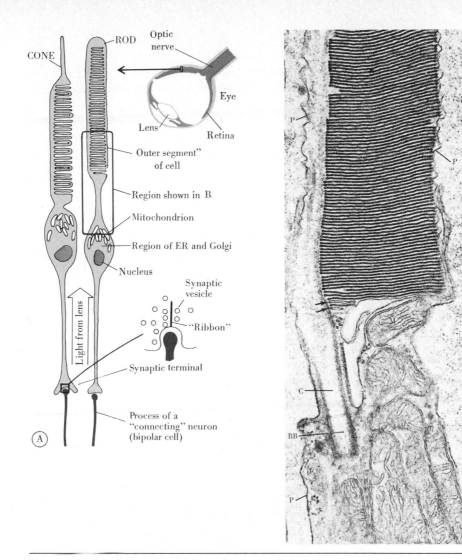

Figure III–44 *Retinal receptor cells of vertebrates.*

A. Schematic diagram of a portion of the retina. The receptor cells have their light-sensitive portions directed toward the back of the eye. As shown, in many organisms the stacked membrane systems in the cones, unlike those in the rods, remain connected to the plasma membrane. At certain of the principal sites of transmitter release, the synaptic terminals of the photoreceptors contain distinctive looking "ribbons" of unknown composition, oriented perpendicular to the plasma membrane.

B. An electron micrograph of a portion of a rod cell like that indicated by the rectangle in part A. The basal body *(BB)* and cilium *(C)* that connect the inner and outer segments of the cell are seen in longitudinal section. Many mitochondria *(M)* are present near this connection. *P* indicates the plasma membrane. The outer segment contains the stacked membrane systems *(discs)* in which the photoreceptive pigments are located. These discs form as sacs (arrows) that eventually flatten, obliterating the space within. (Courtesy of D.W. Fawcett.)

Rhythmic Shedding and Elongation When animals of various species are maintained under lighting conditions approximating the normal day–night cycle, such as alternating 12-hour periods of light and darkness, shedding of *rod* discs occurs chiefly during a period shortly after the onset of light. This, then, is one of the vast variety of cellular and organismal processes that occur with a 24-hour cyclical rhythm. In the fewer species where shedding of *cone* discs has been studied, such shedding seems also to occur rhythmically, but the peak period is shortly after the onset of *darkness;* this difference from the rods may be related to the fact that the cones are used most for vision in bright light and the rods in dim light. The genesis of these rhythms is not known. They continue for some time even when the animals are maintained for several days in the dark. Thus, they resemble many of the other *circadian* or *diurnal rhythms* exhibited by cells and organisms. These are rhythms found in many physiological processes, based on a roughly 24-hour cycle, and produced in part by what are conventionally called "biological clocks" or "oscillators." The nature of endogenous biological clocks—those that function autonomously, responding slowly, if at all, to environmental changes—is poorly understood. Some clocks can, however, be "reset" by alteration in light–dark periodicity or other cues; sleep patterns are an obvious example. From such cases it is clear that "ordinary" physiological and biochemical mechanisms, such as periodically fluctuating levels of enzyme activities or balances among different hormones, help to mediate the expressions of the clocks (see Section 3.11.2 for an example). What are not yet understood are those underlying timekeeping mechanisms that seem significantly independent of known environmental cues.

In fish and other lower vertebrates, in addition to the rhythmic shedding, the photoreceptors cyclically elongate and contract so that the rods are at maximal length in bright light and the cones, in the dark. When they are elongated, the cells' photoreceptive outer segments are shielded by the pigment epithelium at the back of the retina (which is rich in light-absorbing dark pigment granules); this cycle in effect exposes the rods for use in dim light and the cones for use in bright light. The contractions and elongations are controlled in part by illumination conditions and in part by an endogenous circadian rhythm. Cyclic AMP and the microtubule and filament systems longitudinally disposed in the cells participate in the length changes.

Functioning The arrangement of the photoreactive material in stacked membranes provides a large surface where light is absorbed. Studies with polarizing microscopes (Section 1.2.2) have shown that the rhodopsin molecules within the membrane exist in orientations that provide high efficiency of light absorption. When it absorbs light, the rhodopsin molecule changes conformation, which leads to the release of the light-absorbing *retinal* portion from *opsin,* the protein. The appearance of one or a few such altered rhodopsin molecules changes the bioelectrical behavior of the cell through mechanisms that are still being sought. Competing theories ascribe primary importance to alterations involving GTP, cyclic nucleotides, or Ca^{2+}. What does seem clear is that the cells' Na^+ channels are closed by the effects of light. The consequence, in electrical terms, is that the cell hyperpolarizes; such a stimulus-driven change in membrane potential is called a "receptor potential." These effects spread directly to the terminal with no action potential required. The outcome is an alteration in the activity of the synaptic end of the cell. Present evidence is interpreted as indicating that photoreceptor synapses actively release neurotransmitters in the dark and that the hyperpolarization induced by light reduces the rate of such activity. Different light intensities produce different levels of change in the cells' membrane potential; these are reflected in different rates of transmitter release, which in turn affects transmission through a sequence of other retinal neurons. Eventually, the information about which photoreceptors received light and about the extent of the stimulus is sent to brain in the form of a pattern of action potentials carried along the optic nerve by the axons of retinal neurons called "ganglion cells." That is, through intermediary, connecting neurons ("interneurons"), the photoreceptors responding to light determine which of the ganglion cells fire action potentials and the rate and timing of this firing.

Rhodopsin accounts for approximately 50 percent of the mass of the disc membranes; most of the remainder is lipid. Thus, the numerous particles (50 Å in diameter) seen in the membranes with freeze-fracture techniques almost certainly are rhodopsin molecules, or at least structures dependent upon the presence of rhodopsin. Studies now under way are aimed at determining whether the distribution or other features of the particles change during response to light, in ways that might help to explain the functioning of the membranes.

There also is much interest in recent observations that light-activated rhodopsin interacts with a set of other proteins in the outer segment, with the net effect of activating a *phosphodiesterase* enzyme (Section 2.1.4). This enzyme converts cylic GMP into the inactive, noncyclic nucleotide, so it could play a pivotal role in connecting light reception to the subsequent cellular phenomena. A single activated rhodopsin can activate hundreds of molecules of the phosphodiesterase, thus amplifying the effects of light reception.

3.8.2 Light Receptors in Invertebrates

The structure of eyes differs widely among organisms. In contrast to the single light-sensitive retina of the vertebrate eye, many invertebrates (for example, crustaceans and insects) have *compound eyes* made of repeating photoreceptive units (ommatidia), each with its own set of cells. In these units, light reception centers at the *rhabdomeres,* which are densely packed aggregates of microvilli protruding from the surfaces of the light-receptive cells (Fig. III–45). The microvilli, like the retinal rod sacs, are oriented with their long axes perpendicular to the surface from which light enters. They provide a large area of light-receptive material. Leading hypotheses suggest that light-induced changes in the molecules of the rhabdomere membrane result directly in changes of

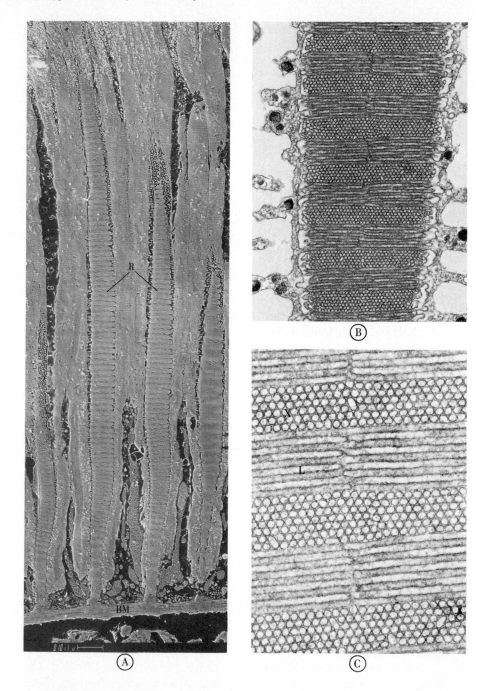

permeability to inorganic ions. Since the microvilli are part of the cell surface, the permeability changes could result directly in changes in the potential across the plasma membrane.

Like the rod cell discs, the rhabdomere membrane also undergoes periodic turnover on daily schedules. In numerous species, at dusk the microvillar
system undergoes expansion through addition of membrane. At dawn, membrane is withdrawn back into the cell; in a few species, some is also shed to be
phagocytized by other cells. This cycle relates, presumably, to the nocturnal
habits of many of the organisms and the need for a greater expanse of photoreceptive membrane when light intensities are low. The routes by which
membrane is added to the surface are unknown. Withdrawal involves endocytic processes with at least some of the membrane winding up in lysosomes,
where it is degraded.

Certain invertebrates possess light-sensitive cells outside of their eyes. For
example, crayfish have light receptors on their abdomens that apparently serve
as protective devices against predators; the response to altered illumination
leads to rapid movement of the organism. One of the cell types that has been
tentatively identified as an abdominal light receptor is shown in Figure III–46.
The cell contains many membranes arranged in a whorl, which presumably
provides *extensive* membrane surfaces comparable to those found in other
light receptors.

Unicellular Organisms   Many unicellular organisms possess special light-
absorbing *eyespot* or *stigma* regions. These include elements in *phototaxis* systems for orienting the organisms with respect to light. Such orientation is
especially important in photosynthetic forms. Sometimes the eyespots are discrete zones within chloroplasts (see Fig. II–59). In other cases they are separate
structures. Often they are found near the basal body of a flagellum. The pigment in some eyespot regions is in the form of large spherical bodies.

The role of light-absorbing pigments in eyespots is quite different from
that played by photosensitive pigments in receptors of higher organisms. For

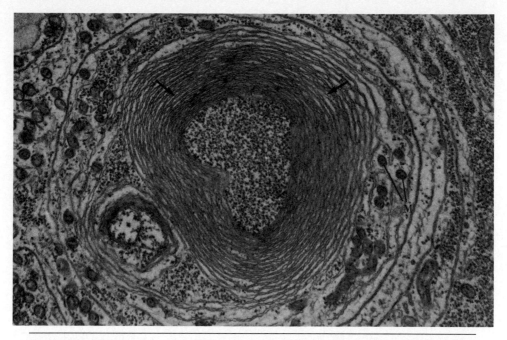

Figure III—46 *Portion of a cell thought to be an abdominal light receptor* in *the crayfish.* The extensive system of concentric membranes found in the cytoplasm is indicated by arrows. Mitochondria are seen at **M. × 15,000. (Courtesy of A.J.D. DeLorenzo.)**

example, an interesting explanation has been put forth for the light responses of the flagellated, chloroplast-containing protist *Euglena.* It has been suggested that the small pigmented eyespot near the flagella is so positioned that when the organism is oriented at certain angles to the light, a shadow of the eyespot is cast on a special swelling located at the base of one flagellum. The swelling is presumed to be sensitive to light, with its responses generating permeability changes in the cell surface at or near the flagella. Alterations in light intensity at the swelling resulting from different shadow positions eventuate in alterations in flagellar motion, with consequent effects on the movement of the cell.

Note that this relatively primitive sensory system may be evolutionarily related to the vertebrate photoreceptor, since both are based on light-sensitive membranes associated with a cilium or flagellum. Is it conceivable that distant evolutionary relationships also exist between these systems and the rather different "sensory" use of flagella in mating by *Chlamydomonas* (Section 2.10.2)?

Hair Cells An evolutionary "use" of microvilli for sensory activities quite different from those in the invertebrate eye is represented by the *hair cells* of the vertebrate ear. These cells have a single 9 + 2 cilium plus rows of microvilli of varying length protruding from the surface of each cell. Each microvillus is stiffened by a core of closely arranged microfilaments. Slight movements of the microvilli in response to the vibrations transmitted within the ear set off changes in membrane potential that initiate the neural mechanisms responsible for hearing and other responses to sound.

Chapter 3.9 Muscle Cells

Skeletal muscle is made of bundles of large *fibers* (see Fig. II–89). Each fiber is a "syncytium": As a result of cell fusions during development (Section 3.11.2), many nuclei are present in a common cytoplasm. Both smooth and cardiac muscles are made of separate uninucleate cells; smooth muscle is composed of bundles or sheets of cells (Fig. III–27), cardiac muscle of a branched network of cells joined end to end.

The contraction of all three major types of muscles—skeletal, cardiac, and smooth—is dependent upon the presence of the proteins actin and myosin. As discussed in Section 2.11.1, it is the regular arrangement of actin- and myosin-containing filaments that is responsible for the *striations* of cardiac and skeletal muscle. The sliding of filaments past one another is the basis of contraction of both muscles.

3.9.1 Structure and Function of the Sarcoplasmic Reticulum

A skeletal muscle fiber (Figs. II–89 and III–47) is bounded by a *sarcolemma*, the plasma membrane with an overlying external lamina (Section 3.6.7). Numerous peripherally located nuclei are present in each fiber. Golgi apparatus, rough ER, and ribosomes are scanty and are concentrated near the nuclei. Myofibrils occupy most of the cytoplasm. Closely adjacent to the myofibrils are mitochondria; their close topographic relation to the contractile material results in the efficient transfer of ATP. Glycogen, in the form of particles 200 to 300 Å (20 to 30 nm) in diameter, is scattered through the cytoplasm, providing a storage supply of carbohydrates.

Neuromuscular Junction In vertebrates, the contraction of a skeletal muscle fiber is evoked by a large synaptic terminal of a motor neuron from the spinal cord or brain (Fig. III–47). Through branching of its axon, a single motor neuron can control many fibers; the neuron and the fibers whose simultaneous contraction it stimulates is a "motor unit." At the neuromuscular junction the muscle surface is thrown into a series of folds whose tips are very rich in intramembrane particles roughly 10 nm in diameter. Many of these almost certainly are acetylcholine receptors basically similar to the one shown in Figure II–9. The active zones of the nerve terminal (Section 3.7.5 and Fig. III–47) are aligned along the mouths of the folds. The release of acetylcholine from the

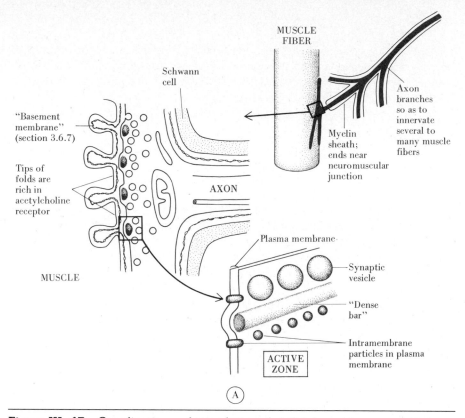

MUSCLE FIBER

Schwann cell

"Basement membrane" (section 3.6.7)

Tips of folds are rich in acetylcholine receptor

MUSCLE

AXON

Axon branches so as to innervate several to many muscle fibers

Myelin sheath; ends near neuromuscular junction

Plasma membrane

Synaptic vesicle

"Dense bar"

Intramembrane particles in plasma membrane

ACTIVE ZONE

A

Figure III–47 *Coordination and stimulation of skeletal muscle; sarcoplasmic reticulum.*

A. In higher vertebrates, a given axon of a motor neuron controlling skeletal muscle can branch so as to establish synaptic endings *(neuromuscular junctions)* on several to many fibers in a muscle. Such organization facilitates coordinated control of the muscle. The group of fibers innervated by a single neuron is referred to as a *motor unit.* The diagram illustrates a portion of a neuromuscular junction; the nerve terminal extends within an indentation in the muscle fiber surface. The terminal shows "active zones" thought to be the actual sites of release of neurotransmitters. At the zones, synaptic vesicles are aligned along electron-dense "bars," and the plasma membrane exhibits aligned rows of the intramembrane particles visualized by freeze-fracture procedures (Fig. II–6). The plasma membrane of the muscle at the junction is thrown into a series of folds whose tips are rich in receptors for acetylcholine, the transmitter released by the nerve terminal. The enzyme *acetylcholinesterase* (Section 3.7.5) is concentrated in the material present in the space between the nerve and muscle. (After concepts developed by R. Couteaux, J. Heuser, T.S. Reese, K. Pfenninger, and others).

B. (On facing page.) The T system and sarcoplasmic reticulum of a frog muscle. The transmitter molecules (acetylcholine) released by the neuromuscular junction controlling the fiber initiate an action-potential-like impulse at the fiber surface. The conduction of this impulse by the plasma membrane depends on mechanisms comparable to those occurring at the axon membrane (Section 3.7.3); this probably is true within the T system as well. The mechanisms of coupling between events in the T system and those in the sarcoplasmic reticulum are incompletely understood. (Diagram modified from L.D. Peachey and W. Bloom and D.W. Fawcett.)

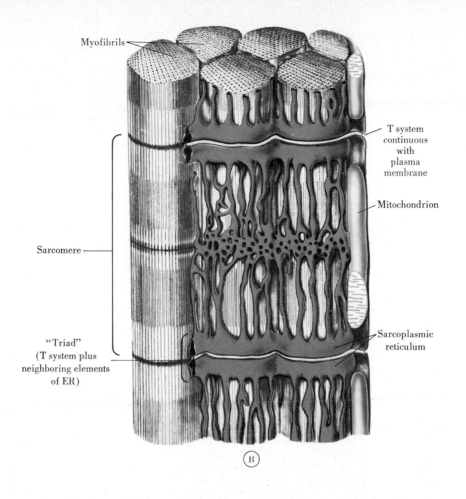

Myofibrils

T system
continuous
with
plasma
membrane

Mitochondrion

Sarcomere

Sarcoplasmic
reticulum

"Triad"
(T system plus
neighboring elements
of ER)

B

neuromuscular junction locally depolarizes the plasma membrane of the muscle and initiates an action potential–like impulse (Section 3.7.3) that spreads along the muscle fiber. Thanks to the T-tubule system, the impulse can travel deep within the fiber.

T-Tubules; Sarcoplasmic Reticulum   A membrane-bounded tubule is associated with each sarcomere of a skeletal muscle. The tubules constitute the *T system*. Tracer macromolecules such as ferritin and peroxidase can be seen to pass rapidly from the extracellular medium into the T system. This confirms the finding by electron microscopy that the tubule system is continuous with the plasma membrane and is therefore open to extracellular fluids. In essence, then, every sarcomere lies close to an extension of the plasma membrane.

The smooth ER of striated muscle is called the *sarcoplasmic reticulum*. The sarcoplasmic reticulum is closely associated with each T-system tubule and forms a network between the tubules (Fig. III–47). Although the sarcoplasmic reticulum comes into close contact with the tubules, the reticulum and T system are not continuous with one another; extracellular peroxidase and other markers do not pass into the reticulum. Muscles of different organisms often contain somewhat different arrangements of the T system and sarcoplasmic reticulum, but the basic function is the same in all striated muscles.

Release and Reabsorption of Calcium   The elaborate arrangement of the T system and sarcoplasmic reticulum membranes coordinates the contraction of the fiber. Coupling of the contraction of the muscle fibrils to the electrical impulse carried by the T system results from the fact that the interaction of the T system and the sarcoplasmic reticulum eventuates in a release of calcium ions from the sarcoplasmic reticulum into the myofibrils. The Ca^{2+} ions initiate contraction (Section 2.11.1). As the calcium is reabsorbed by the sarcoplasmic reticulum the muscle relaxes.

The conclusion that Ca^{2+} controls muscle contraction is based in part on observations that injection or localized application of Ca^{2+} to muscle fibers can cause contraction. Additionally, direct evidence for intracellular release of calcium associated with contraction has been obtained through the use of *aequorin,* a protein obtained from luminescent jellyfish that emits light (fluoresces) in the presence of calcium ions. When aequorin is injected into barnacle muscles and the muscles are stimulated, a burst of fluorescence is noted before contraction.

Microsome fractions (Chap. 1.2B) containing portions of the sarcoplasmic reticulum can concentrate Ca^{2+} from the medium through active transport dependent on a membrane-associated Ca^{2+}-ATPase. This ATPase activity and also calcium-binding proteins (for example, *calsequestrin*) are found in such microsome fractions and can also be demonstrated within muscle fibers by cytochemical procedures. The reticulum can reabsorb Ca^{2+} rapidly and store considerable concentrations with its lumen.

In the absence of something like the T system, it is difficult to see how an external plasma membrane change triggered at the neuromuscular junction could produce virtually simultaneous responses of all the sarcomeres of all the myofibrils throughout the mass of a muscle fiber many micrometers in diameter. If "trigger" substances had to diffuse in from the outer surface of the fiber, the outer fibrils would contract far in advance of the ones located at the center of the fiber. The T system penetrates throughout the fiber, so that even the sarcomeres of fibrils at the center can be affected directly by the impulse at the plasma membrane. Impulse conduction by the membrane is a much more rapid process than diffusion: Molecules move by diffusion at rates measured in micrometers per second; impulses are transmitted at rates of meters per second. Similarly the close association of sarcoplasmic reticulum with each sarcomere permits the rapid, local changes in calcium concentration needed for precise control of contraction. The question of how the T system produces a sudden release of Ca^{2+} from the sarcoplasmic reticulum is the major unknown at present.

The rise in Ca^{2+} concentration on stimulation not only prompts the contraction of muscle but also activates the enzyme *phosphorylase kinase.* As with cyclic AMP–mediated hormonal activation (Section 2.1.4), the result is phosphorylation of *glycogen synthase* and *glycogen phosphorylase,* leading to increased availability of glucose for production of the ATP needed by the muscle. (The effect is eventually turned off, once stimulation stops, by phosphatases that dephosphorylate the synthase and kinase.)

3.9.2 Features of Smooth and Cardiac Muscle

Coordination of Cardiac Muscle Sarcoplasmic reticulum and a T system are also found associated with the myofibrils of cardiac muscle. The contraction of this type of muscle is based on calcium-controlled sliding of filaments. However, the coordination of the individual units of cardiac muscle, so that the heart regions beat in proper rhythm, depends on special features.

In higher vertebrates, the rhythm itself is generated by a pacemaker system of cardiac muscle fibers that spread a rhythmic change in membrane potential to the various regions of the heart. This system is subject to influence by neuronal and hormonal stimuli coming from outside, but its basic rhythm is self-generated and arises from cyclic variations in the Na^+ and K^+ permeability of the plasma membranes of some of the fibers. The individual cells of which the heart muscle is made are coordinated with the pacemaker system and with one another by electrical (ionic) communication through the extensive gap junctions that exist between the cells.

End-to-end linkage of the cardiac muscle cells into networks depends on extensive interdigitating of their plasma membranes, which form large adhesive junctions reminiscent in many respects of desmosomes (Section 3.5.2) and referred to as "intercalated discs."

Smooth Muscle Smooth muscle cells form sheets and bundles (see Fig. III–27) in which the contraction of individual cells is coordinated. Here again gap junctions serve to couple potential changes in the separate cells. How smooth muscles' actin and myosin produce contractions is incompletely understood. Actin microfilaments abound in the cytoplasm, and some thick, myosin-containing filaments are present as well, but these are not arrayed in the regular overlapping arrangements of striated muscle. The myosin filaments of smooth muscle seem much more susceptible to alteration by the techniques used to prepare tissues for electron microscopy than are those of striated muscle, one of a number of observations suggesting there are differences in organization. Scattered in the cytoplasm of the cells are electron-dense patches (dense bodies) that contain α-actinin (Sections 2.11.1 and 2.11.2) and thus were originally thought to anchor the actin filaments, as do the Z-discs of striated muscle. Many investigators still emphasize this possibility, but a few have swung recently toward stressing the observed dense-body linkages of *intermediate filaments,* which might create a kind of cytoskeletal harness for the oriented transmission of contractile forces. The contractile filaments, particularly the actin filaments, may be anchored at the plasma membrane instead of, or in addition to, anchoring at the dense bodies.

Some smooth muscle arrays undergo fairly rapid cycles of contraction and relaxation, but in many of their roles, such as controlling the size of blood vessels, the cells engage in slow sustained cycles; some of the cells in the group are always active so that muscle *tone* is maintained. Smooth muscle cells are controlled by numerous hormones, as well as by neurons and the cells' own direct responses to mechanical stimuli that result from stretching of tissues the muscles surround. For instance, histamine (Section 3.6.4) provokes contraction of the smooth muscles around small veins; the resultant pressures within capillaries force fluids out into the tissues accounting for some of the symptoms of allergic reactions. Ca^{2+} is a likely participant in the intracellular processes controlling smooth muscle contraction. Plasma membrane permeability and transport and the activities of an ER system less extensive and less elaborately organized than that of striated muscle govern Ca^{2+} levels in the cytoplasm. The intermediary in Ca^{2+} effects on the muscle filaments appears to be calmodulin (Section 2.1.4) rather than troponin (Section 2.11.1) as in striated muscles. A favored proposal is that when complexed with Ca^{2+}, calmodulin stimulates *kinase* enzymes to add phosphates directly to myosin in smooth muscle; this phosphorylation activates myosin's ATPase capacities and related interactions with actin.

Chapter 3.10 Gametes

The details of cellular mechanisms related to sexual reproduction vary considerably among different organisms. However, in virtually all sexually reproducing eucaryotes, two features are constant. First, at some point in the life cycle, two special cell divisions occur; together the two divisions constitute *meiosis*. The meiotic divisions result in the halving of the number of chromosomes and, thus, halving of the amount of DNA per nucleus. Second, at a subsequent stage two cells with the halved chromosome numbers, resulting from meiosis, fuse to form a zygote. The nuclei also fuse to produce a single nucleus combining the chromosomes from the two cells. The details of chromosome behavior and the genetic consequences of these processes will be considered in Chapter 4.3. For the moment it should be noted that zygote formation usually involves the fusion of cells and combination of chromosomes from two different parent organisms. This results in a zygote nucleus with a new genetic combination.

The cells that fuse and contribute the chromosomes of zygotes are generally referred to as *gametes*. Sometimes, as in many unicellular organisms and in some lower plants and animals, male and female gametes do not differ much in structure. In some cases fusion is partial and temporary. Thus during conjugation in *Paramecium,* two micronuclei (Section 3.3.2) are present in each cell; one from each cell passes into the other cell through a cytoplasmic bridge that forms as a temporary structure. In *Chlamydomonas,* on the other hand, the cell associations initiated by the flagella (Section 2.10.2) culminate in complete fusion of the cells.

In most higher organisms, male and female gametes are distinctly different, and gametes fuse completely (fertilization). Usually, though not invariably, the male gametes are smaller and motile; cilia or flagella are present on most male gametes of multicellular animals and lower plants, but some species have ameboid gametes. In flowering plants, pollen grains germinate to form a pollen tube that grows through the female structures of the flower and ultimately brings the male gametes into the ovary where the egg is located. In these plants "double fertilization" occurs: The egg cell is located in the embryo sac or female gametophyte (Section 4.3.1). One sperm cell fertilizes the egg cell to produce a zygote. A second sperm cell fertilizes another cell of the embryo sac; the products of this fusion give rise to nutritive "endosperm" tissue.

In this chapter we shall consider the gametes of multicellular animals that exemplify a high degree of gamete specialization. Those of organisms such as fish and water-dwelling invertebrates have been studied most. They can be obtained in large quantities, and since fertilization normally occurs outside the organism, experimental investigation of this process is facilitated.

3.10.1 Nuclei

The differences between sperm and egg are related to their different roles in zygote formation. Transmission of chromosomes from the male parent depends on specialized features of sperm cells. Egg cells carry the chromosomes from the female parent and also contain the reserve material used in the early development of the embryo. Gametes form from the *gonial* cells (spermatogonia or oögonia). These divide to produce a population of *spermatocytes* or *oöcytes.* Each spermatocyte undergoes the two cell divisions of meiosis (Section 4.3.1) to produce four *spermatids,* which start off as more or less round cells but differentiate into mature elongate *sperm* (see Figs. III–48 and III–50). Oöcytes usually undergo extensive growth before or during meiosis; in many animals the meiotic divisions are not completed until after fertilization, and usually only one of the four division products gives rise to a mature *ovum* (the others degenerate).

The *nuclei of mature sperm* are small, and the chromatin is densely packed and entirely inactive in RNA synthesis. In some animals (for example, fish) the histones (Section 2.2.3) are replaced by *protamines,* very basic proteins unusually rich in the amino acid *arginine.* Frequently unusual lamellar or tubular patterns of chromatin distribution are seen (Fig. III–48). As in many higher animals, the chromatin may be so densely packed that it appears to be a homogeneous mass. Pores in the nuclear envelope are sparse or absent.

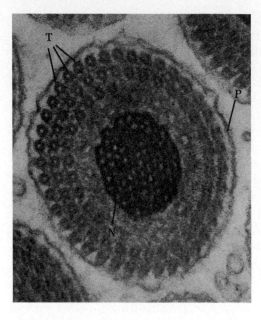

Figure III—48 *A sperm head.* Cross section of the head of a sperm from the insect *Steatococcus.* The head has an elongate cylindrical shape. The many microtubules oriented longitudinally *(T)* are presumed to contribute to the establishment or maintenance of this shape. In the nucleus *(N)* the chromatin is arranged as dense material surrounding less dense tubular regions, which appear in section as light circles. *P* indicates the plasma membrane. × 100,000. (Courtesy of M. Moses.)

Some sperm nuclei are elongate; they may be more than ten times longer than wide. In such nuclei the chromosomes are often lined up as coiled fibers, with long axes running along the length of the head.

The nuclei of growing oöcytes and ova are quite large, and the chromatin is dispersed within to such an extent that often it cannot readily be seen in the microscope, even after use of highly specific stains for DNA. The chromosomes may assume a special "lampbrush" configuration (Section 4.4.5), and sometimes a great many nucleoli are present (Section 2.3.3). The large oöcyte nuclei engage in intensive RNA production, supporting their own cytoplasmic growth and providing RNA that is stored and used in later embryonic development; the special features of their nuclei are related to this RNA production.

3.10.2 Cytoplasm

Sperm   Sperm of higher animals are elongated, highly modified cells (Figs. III—48 and III—50); widths of a few micrometers and lengths of 50 to 100 μm are not uncommon. Most of the length is accounted for by the flagellum. The *head,* which contains the nucleus, may make up only 5 to 10 percent of the length. The front of the head contains an *acrosome.* This large membrane-bounded structure is produced by the Golgi apparatus as the spermatid matures. Lysosomal enzymes have recently been shown to be present in the acrosome of some sperm, leading to the proposition that acrosomes are highly modified lysosomes.

It is also during spermatid maturation that the flagella grow in association with the centrioles. As the flagella form, the centrioles become situated at the base of the nucleus. The flagella show the usual 9 + 2 tubule pattern, often accompanied by additional fibrous structures that run alongside the tubules (see Fig. III–50). In the *midpiece*, the region just behind the head of the sperm, the mitochondria are aggregated near the flagellum. Quite often the mitochondria are in ordered arrays; sometimes they form a spiral ribbon that twists around the flagellum for quite some distance. Most likely this facilitates transmission of ATP used in sperm movement. The flagellum projects beyond the midpiece as the *tail*. In mammals the individual mitochondria of a given spermatid remain structurally separate, but in some other vertebrates and in various invertebrates they fuse with one another, yielding a large composite body. Sometimes, as in insects, the internal structure of the mitochondria is substantially modified, with crystal-like accumulations displacing the cristae.

The entire sperm is covered by a plasma membrane, which shows a number of distinct regions specialized in structure and function, but aside from the acrosome, mitochondria, centrioles, and flagellum, little cytoplasm is present. Golgi apparatus, ER, ribosomes, and other cytoplasmic components are sloughed in a cytoplasmic bud that forms as the spermatid matures into a sperm and eventually separates from the sperm and disintegrates. (In many species it is phagocytized by the *Sertoli cells* of the testis.)

Ova   In contrast to sperm, ova are often enormous cells with diameters ranging from roughly 100 μm in many mammals to more than 1 mm in some invertebrates and amphibians; occasionally they are even larger, as in reptiles and birds. Their abundant cytoplasm accumulates during oöcyte growth, which may require days or even months. The cytoplasm contains the usual organelles, plus distinctive *yolk bodies* that may occupy much or most of the cytoplasmic volume (Fig. III–49). Yolk contains varying proportions of stored lipids, carbohydrates, and proteins, used later in development. Numerous ribosomes—some bound, many free—are present in egg cytoplasm. These are used by cells of the early embryo. In some organisms new ribosomes are not synthesized by the embryo until relatively late in development (Chap. 4.4). The ribosomes of mature but unfertilized eggs are inactive in protein synthesis. They are activated after fertilization.

The details of yolk formation in the oöcyte vary in different species. Yolk may accumulate in Golgi vacuoles, in the ER (Fig. III–49), or occasionally in mitochondria. In many species considerable amounts of material are also taken up by pinocytosis and contribute to the yolk (Fig. III–49). Some of this material comes from the follicle cells or nurse cells that surround many growing oöcytes.

Nurse cells and oöcytes in some organisms, such as the fly *Drosophila*, are connected by bridges of cytoplasm. These represent continuities persisting after division of a common progenitor cell. The bridges permit direct transfers of even relatively large molecules and structures, including RNAs.

Much material taken up by the oöcytes of many species comes from the blood. In amphibians, yolk is made largely of complexes of proteins and lipids derived from a large precursor molecule, *vitellogenin,* secreted by the liver.

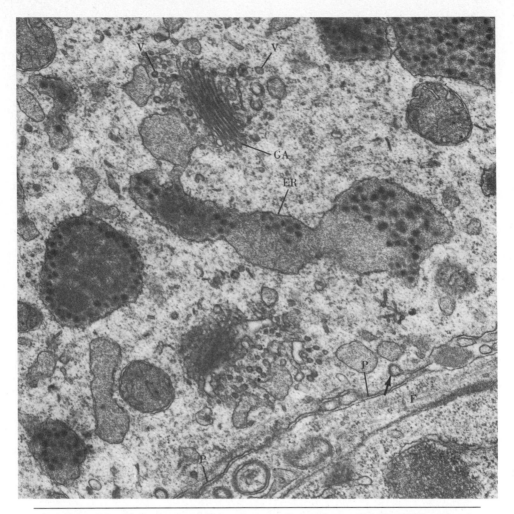

Figure III–49 *Yolk.* Portion of an oöcyte in the spider crab *Libinia.* Part of one follicle cell in the follicle cell layer surrounding the developing egg is seen at *F.* Yolk accumulates as electron-dense material in cisternae of the endoplasmic reticulum *(ER);* dilated regions appear to separate from the ER as large yolk spheres. Vesicles, some coated (arrow), form at the plasma membrane. These coated pinocytosis vesicles and other vesicles *(V)* associated with the Golgi apparatus *(GA)* are thought to add material to the yolk; the vesicle membranes fuse with the ER-derived membrane of the yolk sphere. × 50,000. (Courtesy of G. Hinsch and V. Cone.)

The liver is induced to synthesize and release quantities of this molecule—a phosphorylated protein associated with lipids—by steroid hormones, *estrogens.* The oöcyte plasma membrane has vitellogenin receptors that permit it to endocytize vitellogenin reaching the cells through the blood stream. Endocytic vesicles deliver the protein molecules to yolk bodies, and it is probably there that they are cleaved, proteolytically, into the two smaller molecules that are the actual storage forms: *lipovitellin* and *phosvitin*. How incoming vitellogenin evades lysosomal degradation in the oöcytes, which do have some lysosomes, is an interesting question. Perhaps the endocytic vesicles are routed so that they do not fuse with lysosomes; perhaps some do fuse, but there are too few lysosomes to make much of an impact on the extensive importation; or perhaps lysosomal enzymes are present even in the yolk bodies but are inactivated in some way. Digestion of yolk during early development of the egg may be carried out in part by lysosomal enzymes, though the evidence for this is still equivocal.

The cytoplasm just below the egg surface is often organized as a gel-like "cortex." In many species special cortical granules or fluid-containing vacuoles accumulate there. The role of these granules in fertilization is discussed below.

Eggs of various organisms show asymmetric distribution of microscopically visible cytoplasmic components. Of particular interest are the special cytoplasmic regions in some eggs that contain substances involved in differentiation of specific cell types during development (Section 4.4.3).

Protective extracellular coats, sometimes in multiple layers and usually rich in polysaccharides, surround the eggs of many species. In some cases a thin surface layer may be formed by the egg itself, but additional thick coats or shells are contributed by ducts down which the eggs pass after release from the ovary, or by follicle cells as in the case of the *chorion* surrounding insect eggs (see Section 4.4.6).

3.10.3 Fertilization

The details of fertilization differ from species to species, but there are common underlying themes. The egg and sperm must recognize one another as a compatible pair; the sperm must penetrate the coats surrounding the egg; plasma membrane fusion between egg and sperm must occur, followed by more extensive fusion and nuclear migrations; and once one sperm has been successful, fertilization by additional ones must be inhibited. The details outlined in this section derive principally from studies of a few invertebrate species, notably sea urchins and starfish, but versions of the same sorts of events apply to most animals.

The Acrosome Reaction The acrosome of the sperm generally plays a central role in sperm penetration through the exterior coats that surround the egg. The egg may release chemoattractants that bring numerous sperm into its vicinity. On contact with the outer coats, some of these sperm release their acrosomal contents by exocytic fusion with the plasma membrane (Fig. III–50). Acrosomal enzymes, such as *polysaccharidases* (such as *hyaluronidases;* Section 3.6.6) and *proteases,* digest portions of the coat, permitting a closer approach by the sperm to the surface of the egg cell. The acrosome membrane, now continuous with the sperm's plasma membrane, extends as one or more

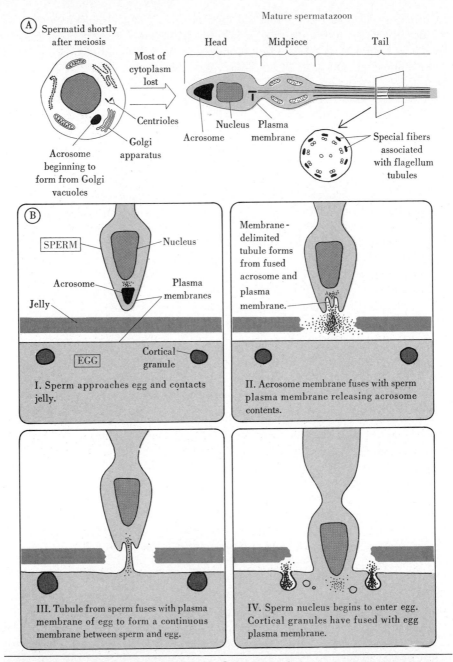

Figure III–50 *Sperm; fertilization.* A. Schematic diagram comparing a *sperma-tozoon* (of a higher animal) with the *spermatid* from which it differentiates. The arrangement of mitochondria, the acrosome, centrioles, and other organelles varies in detail in different animals. B. Outline of fertilization based primarily on studies of invertebrates. (After L. Colwin and A. Colwin.)

tubular projections. In the few species that have been studied thoroughly this extension seems to be powered by cytoplasmic filaments. Section 2.11.4 described one way in which this can happen, by new assembly of filaments. In other cases, rearrangements of preexisting filaments are responsible for pushing the projections out. Still not certainly identified are the sources of the membrane that must be added as the acrosomal processes grow; the nuclear envelope, the plasma membrane, and intracellular stores of lipids are among the candidates being investigated in sperm of different organisms.

The acrosome tip seems to contain "recognition" molecules (one. a protein called "bindin," has been tentatively identified in sea urchin sperm). Discriminating receptors at the egg cell surface, probably glycoproteins in the innermost of the extracellular coats (the "vitelline layer"), bind these molecules and thus initiate firm attachment between sperm and egg. After binding, plasma membrane fusion takes place. Note that the surfaces involved are shielded before the acrosome reaction occurs—the sperm membrane is within the sperm cell, as a part of the acrosome, and the egg surface is covered by the coats that are digested by the acrosomal enzymes. The uncovering of the membrane regions that subsequently fuse, by the acrosome reaction and related events, contributes to controlling the timing and specificity of fertilization. Soon after fusion the sperm contents including the nucleus pass into the egg by formation of a conelike protuberance from the egg surface that seemingly sucks them into the egg's cytoplasm.

The Egg's Responses As sperm and egg fuse, the egg undergoes changes. One rapid alteration prevents fertilization by additional sperm. The character of this alteration is complex but seems to involve a rapid change in membrane potential somewhat akin to a nerve impulse. This rapidly erected "barrier" is soon reinforced, in many eggs, by the formation of a *fertilization membrane* that surrounds the egg. This is not a true membrane but is rather a thick layer of proteins and polysaccharides. It includes what was the egg's vitelline layer plus contributions from the cortical granules. These undergo exocytosis (Fig. III–50), releasing their contents. The granule membranes thereby become incorporated in the plasma membrane, leading eventually to lengthening of the microvilli present at the egg surface. (Actin microfilaments may take part in the lengthening.) Enzymes released from the cortical granules have several effects. Some cross-link the extracellular molecules, toughening the coat.

Fusion of the cortical granules with the plasma membrane occurs in a wave that proceeds around the egg from the point of sperm entry, starting at about 30 seconds after the sperm contacts the egg's plasma membrane. This wave is provoked by a preceding wave of increase in local Ca^{2+} concentration that also moves through the egg and can be visualized by use of aequorin (Section 3.9.1). This transient rise in Ca^{2+} concentration, and other ionic changes outlined below, may be due in part to altered cell surface permeability and in part to altered distribution of ions among intracellular compartments. Changes in ionic concentrations contribute also to the sperm's reactions: A Ca^{2+} influx seems to trigger the acrosome fusion with the plasma membrane, and in those cases where new microfilaments are assembled to extend the acrosome process, decreases in H^+ concentration are believed responsible for freeing actin from proteins that prevent its polymerization.

Next Steps; Changes in pH; Centrioles Changes in hydrogen ion concentrations also occur in the egg cytoplasm soon after sperm fusion. The pH may rise as much as a few tenths of a unit. (Remember that pH $= -\log [H^+$ concentration], so a change of pH by one full unit amounts to a tenfold change in concentration.) The loss of H^+ may be accomplished through a plasma membrane pump, perhaps one that exchanges this ion for Na^+; it occurs somewhat later than the increase in Ca^{2+} described above but is observable within a minute or two after sperm entry in sea urchin eggs. (The calcium increase occurs within seconds of sperm contact, and the sperm contents enter the egg within a half-minute.)

There is much enthusiasm about the observations of changes in Ca^{2+} and pH in fertilized eggs, because they hold promise for explaining crucial steps in the onset of the development of the egg. Acting through calmodulin or other intermediaries, calcium could affect enzyme activities and the cytoskeletal systems. Changes in pH also can affect activities of proteins by altering conformations or by dissociating inhibitors. Increases in the respiratory activity that occur soon after fertilization and the later turning on of DNA synthesis and of protein translation (Chap. 4.4) are among the major processes now being investigated from these perspectives.

The roles of the sperm in bringing about these changes in the egg need further study. It is obvious that under normal conditions sperm do *trigger* them. Even before the sperm nucleus enters the egg, a small amount of material from the region just behind the acrosome can enter. It has been speculated that this material, the acrosome membrane, or perhaps even Ca^{2+} ions entering the egg from the sperm are the normal triggering agent. In a few species such as the brine shrimp (end of Section 4.3.4), the egg develops parthenogenetically (without sperm); eggs of other species can be induced to do so by pricking with needles or by treatment with certain chemicals. The egg, apparently, is partly "preprogrammed" so that the sperm need not carry a whole battery of specific molecules, each of which activates a different phase of egg response. (Of course, the *initial* stages of sperm–egg interaction are quite specific, with respect to species and other features, as is biologically required, and in nature the egg rarely encounters effective triggers other than the appropriate sperm.) Steps in egg activation and sometimes substantial parthenogenetic development can be induced by ionophores promoting increases in cytoplasmic Ca^{2+}, and some of the increases in synthetic activities of fertilized eggs are mimicked in eggs exposed to ammonia, which increases the cytoplasmic pH.

In many species, not only the sperm nucleus but also the mitochondria, centrioles, and flagellum enter the egg. Entry of the centrioles has traditionally been thought to be of great importance since it was assumed that the egg lacked functional centrioles and thus could not organize the microtubule-based division apparatus. Something like this may be true in some species. It does appear that the sperm basal bodies (more likely, the associated satellites; Section 2.10.6) often organize microtubules in the fertilized egg; a radiating set of such tubules (an "aster") can appear around them soon after fertilization. The general situation remains quite cloudy, however; at least some eggs may in fact have centrioles, and associated materials, or relevant precursors, that function as microtubule-organizing centers but require activation to do so. Whatever the case, fertilization leads to the presence of active microtubule-organizing centers needed for cell division and other processes. One of the early things the microtubules so engendered could do is to take part in bringing the egg and sperm nuclei together; in sea urchins this occurs about a quarter of an hour after fertilization and is followed by intensive DNA synthesis and cell division.

There are species in which sperm and egg nuclei remain separate until development begins: maternal and paternal chromosomes behave in parallel and coordinated fashion during the first division of the fertilized egg, and a single "mixed" nucleus is formed in both daughter cells.

The fate of the sperm's mitochondria is not clear. In the eggs of some animals they are observed to degenerate, as does the flagellum (see also Section 4.3.6).

Chapter 3.11 Cells in Culture

Early in the century it was shown that many embryonic cells and some cells removed from adult tissues of animals could grow and divide outside the animal if adequate numbers were present to begin with and if proper nutrients were added to the culture medium. About the middle of the century it was demonstrated that, if properly fed, some single cells of mammalian tissues would divide to form *clonal* lines—ongoing populations descended from the single parent. This could be done, for example, by growing the parent cells in company with a "feeder" layer of other cells, still metabolically active but prevented from dividing by irradiation or other means. Such a support population establishes extracellular fibers and matrices and exchanges nutrients and other materials among its own members and with the cloned cells. The need for feeder layers in the early work emphasizes the dependency of cultured cells on an appropriate local environment, and the occurrence of significant cell–cell interactions in culture.

The advent of antibiotics that could be added to the growth medium permitted control of bacterial contamination and thus made large-scale cell culture for long periods of time a much simpler procedure. Human cells were first systematically used for cell culture in 1952.

Many cell types can now be grown on plastic or glass surfaces in thin layers, which are well suited for microscopic study of living cells (Figs. I–13, II–92, and II–93). Some can also be grown suspended in large volumes of growth medium to produce huge numbers of cells suitable for biochemical work and other uses.

3.11.1 Culture Conditions and Types

Cell Lines; Immortality; Crisis Certain strains of cultured cells appear to be immortal, especially those from malignant tumors, or cells that become "transformed," spontaneously or following addition of tumor viruses or other agents to the cultures (Chap. 3.12). If the cell population is subdivided and transferred to fresh medium after some days (the interval in days depends on the cell type and its rate of growth), the cells grow and divide as often as once every day or two for indefinite periods. The same is true of a few seemingly normal cells, that is, cells that do not form tumors if injected into the species of origin and that maintain their normal set of chromosomes in culture (over the long run, most cultured cells that survive prolonged culturing do not—they become "aneuploid"; Section 4.5.1). However, most normal cells divide for a limited period and then die even if culture conditions are carefully adjusted. For example, human "fibroblasts" that are apparently normal and maintain their ordinary, *diploid* (Section 4.2.1) chromosome number will divide only 40 to 50 times. Some investigators believe that study of this cell "senescence" will provide important information regarding the aging process in humans and other organisms (Section 4.4.7). Others point out that it is quite often the case in normal development for cells to divide for a time and then to differentiate into nondividing but still long-lived cells (neurons are an obvious example); they wonder whether the cells in culture may be "attempting" this differentiation but instead are dying since they have not been provided with the culture conditions that would be optimal for their differentiated state rather than their proliferative state. The two views converge, to an extent, in the proposition that cell death itself may be a normal, terminal state of differentiation.

In cultures started with normal cells from a number of species, such as mice, one or more of the cells in the population often survive the "crisis"—the period when most of the population dies. The survivors, which can go on to divide indefinitely, show properties that depend on the growth conditions under which the population was maintained prior to crisis. When the cells were grown always in crowded, densely populated conditions the survivors frequently show distinctly abnormal properties, including transformation into tumor-producing lines. If grown under less demanding conditions the survivors often are closer to normal and rarely are tumor-producers. It appears that the changes permitting survival through the crisis period arise spontaneously in cells of the original population during its life span rather than being induced by the conditions at the crisis period. The character of the survivors reflects selective pressures in the population, like the natural selection that operates in evolving populations of organisms; the population that results is based on the reproductive success of cells that do best under prevailing growth conditions.

Cell lines, or more precisely, *continuous cell lines,* are those strains of cultured cells whose cultures can be maintained indefinitely. (Sometimes the term "cell lines" is used more loosely to refer to any specific strains of cultured cells.)

"Fibroblasts"; "Epithelial" Cultures   Many common cultures are referred to as "fibroblasts," "fibroblast-like," or "fibroblastic." Even though quotation marks are generally omitted, this does not mean that the cells simply are ones that were destined to serve as fibroblasts in the animal from which they were obtained. Cultures often are started by explanting complex mixtures of cells—fragments of tissue, for example—and the cells that survive are those that grow best under the conditions used. Frequently the cells that do so look like fibroblasts—they are spindle-shaped, something like the smooth muscle cells in Figure II–27 (see also Fig. III–52B) and can make products that fibroblasts make. The actual origin of such cells is often not certain, however; their products, such as glycosaminoglycans and collagen, are not unique to fibroblasts. Some cultures, for example, originate with "stem" cells, whose developmental fate was not yet fully decided at the time of explantation; they are capable of differentiating along several paths (Section 4.4.1). Such cultures can evolve in complicated or varying ways, depending on growth conditions.

According to one view, many of the cultures with "epithelial" characteristics—the cells in these grow to form flat sheets quite different from the "fibroblastic" cultures—are related to the endothelial cells that ordinarily line capillaries and can proliferate and grow within the body, for example, when tissues are healing after wounds. (These endothelial cells are, in fact, related in embryonic origin to true fibroblasts.)

The point to bear in mind is that many of the properties of cells in culture, including their shape and biochemistry, are strongly affected by the conditions used to establish and maintain the culture. The long-established cell lines that have been in wide use for decades mostly are of uncertain ancestry. Detailed study of a cell line's history and metabolic capacities is required to ascertain its precise nature.

(As an unpleasant extension of this last point, it is becoming increasingly well appreciated that when different cell cultures are maintained in the same laboratory, cross-contamination can occur unless precautions are thorough. What is thought to be a culture of a particular cell type may wind up as a culture of quite a different cell type, sometimes even from a different species of organism. Embarrassing experimental artifacts can ensue.)

These problems aside, much still can be learned by paying attention to the shape that cells adopt in a given culture. This is an easy characteristic to observe rapidly, and it serves as a rough monitor of the state of the "cytoskeleton" and of properties of the plasma membrane such as its ability to interact with other cells in the culture and with the substrate (the surface on which the cells are growing).

Culture Conditions For culturing cells from solid tissues, the tissues must be dispersed into individual cells or small groups. This generally is accomplished by treatments with enzymes, such as the protease *trypsin,* which disrupts intercellular cementing materials. Often, agents that reduce Ca^{2+} concentrations, such as EDTA (Section 2.4.3), are also required. During the pioneering phase of culture work it was tacitly assumed that these treatments simply separated the cells from one another, leaving them otherwise unchanged. With knowledge gained about the importance of Ca^{2+} for cell behavior and the potential effects of proteases on cell surfaces (Section 3.12.1), this assumption is actively being re-examined as will be seen shortly.

Growth of cells in culture requires media with suitable concentrations of salts, glucose or another sugar, amino acids, vitamins, lipids, and "growth factors." These last factors were originally provided by adding the sera from the blood of horses, fetal calves, or other sources (or coconut milk for plant cells). The specific components necessary to sustain cell division and metabolism had not been identified, but sera and other natural media that normally sustain the growth, division, and metabolism of cells proved to be a rich source of the necessary materials. Early cultures were in fact grown in blood clots, providing both serum factors and surfaces suitable for cell adhesion (see below). The factors required for growth of specific cell types are now being identified as specific hormones and related molecules, circulating proteins involved in transport of various blood-borne materials, and trace metals and other materials. Insulin or one of a number of hormones closely related to it, such as *somatomedin,* is very often required, as well as proteins needed for maintenance or proliferation of specific cell types ("fibroblast growth factor," "epithelial growth factor," and so forth). Thus, precisely specified growth media can be used for an increasing variety of cell types.

For many normal cell types the surface provided for them to grow on has a strong influence on their behavior in culture. In order to thrive, they must be able to adhere well to the substrate, and to spread out. Frequently it has proved advantageous to coat culture surfaces with collagen and other cell products. Components such as fibronectin (Section 3.6.7) are believed to participate in the attachment of the cells to the substrates on which they grow. Most of the cell lines that do not require effective adhesion to a surface ("anchorage") to thrive are those of tumor cells. Normal lymphocytes and other blood cells can be experimentally stimulated to divide without adhesion (Section 4.2.3), probably reflecting the fact that they are suspended in the blood or tissue fluids for a significant part of their life.

Cells can be stored, apparently indefinitely, in liquid nitrogen at $-196°$ C. When frozen carefully, in order to limit the formation of ice crystals (sometimes "cryoprotectants" such as glyerol or dimethylsulfoxide [DMSO] are added), and then carefully thawed, cells (not necessarily all of them) will resume growth and division. Some embryos at early developmental stages can also be preserved in this way, but most complex tissues have thus far proved refractory to such handling.

Inhibition of Growth and of Movement   Normal cells in culture generally show limitations on the extent of their growth and of their movement on surfaces. These effects are potentially instructive ones because they point to reciprocal interactions of cells in populations. From early work with "epithelial cells"—those that spread out to form epithelia-like layers in culture—it was presumed that growth and movement generally stopped at the point when a *confluent monolayer* formed: a layer one cell thick that covers the entire surface. Under optimal conditions, however, many "fibroblastic" cell types can form layers a bit thicker. Nevertheless, there are what appear to be intrinsic limits on growth of cells attached to substrates.

If, after reaching the inhibited state, the culture is "wounded" by scraping a zone clear of cells, or if the cells are separated and placed in sparser numbers on a fresh surface, they resume active movement and proliferation, indicating that no permanent change has occurred.

Presumably, in the organism, the limitations on growth and movement reflected in these observations help to maintain normal tissue architecture while permitting healing of damage. Initially the limitations on movement, growth, and division were regarded as manifestations of the cells' contacting one another under crowded conditions and were collectively referred to as "contact inhibition." This term has been replaced by "density-dependent inhibition" as certainty about the causes has waned; moreover, although the effects on movement and those on growth and cell division may share some common origins, most investigators believe they are not indissolubly linked. "Density" refers to the extent of packing together of the cells in the population.

There are many hypotheses but little general agreement about factors that are responsible for the density-dependent effects. It is frequently asserted that the inhibition of cell movement reflects the fact that to move, a cell in a continuous layer of cells must crawl across or under its neighbors and thus encounters problems in forming the adhesions required for crawling (Section 2.11.4). Relevant evidence is that even under nonconfluent conditions, density-responsive cell types moving in tissue culture do not crawl across one another but change direction of motion when they encounter one another.

Changed conditions of cellular adhesion and direct cell surface interactions may contribute to density-dependent inhibition of growth and division as well. This possibility is suggested both by observations that some factors stimulating cell division act at the cell surface (Section 4.2.3) and by experimental findings such as the observation that partially purified plasma membrane preparations added to cell cultures can sometimes mimic certain effects of crowding. The latter result could be due to membrane interactions like the ones that occur among cells, although whether it shows the specificities and other features of cell–cell interactions has yet to be seen. Other lines of investigation stress the possibility that the cells influence one another less directly, by altering the extracellular environment in the immediate vicinity of the cell layers in some crucial fashion that may be quite localized. Perhaps elevated concentrations of inhibitory substances accumulate at rates too fast for them to dissipate adequately into the growth medium, or the growth medium is depleted of molecules needed for continued proliferation. Possibilities of this type have been put forth to explain experiments indicating that the final density attainable by a growing cell population can be increased by continually agitating the culture dish in order to promote rapid mixing of the medium near the cells with the bulk of the medium; stimulation of growth and division can also be achieved by repeatedly replacing the growth medium with fresh medium, or by adding more serum to the medium, or even by careful choice and maintenance of the pH of the medium (most cells grow best at pHs a bit above 7). Still, on the one hand, simply replacing or stirring the medium does not permit cells to grow without limit, and on the other, changes at the genetic level can transform cells into states in which their ordinary growth limitations fail to operate (Sections 3.12.2 and 3.12.3). Therefore, growth control cannot be solely a matter of the state of the extracellular environment.

Differentiated Cells Many of the long-established cell lines are regarded as relatively undifferentiated in structure and function. That is, although they can secrete molecules such as collagen, they do not produce highly specialized products of the sort that uniquely characterize particular tissues or organs. There are numerous exceptions, however, and they often prove very interesting. For example, about 20 years ago, a clonal cell line was established from a mouse *neuroblastoma* (a tumor of the cells that differentiate into neurons) whose cells under proper conditions (e.g., the addition of cyclic AMP) form long cytoplasmic processes resembling those of neurons, and also are neuron-like in structure and in electrical and enzymatic activities. Molecular biologists, biochemists, and neurobiologists are vigorously studying this line for clues to nerve growth and function. *Erythroleukemic cells* grown in culture have been used to study the biology of the red blood cell progenitors from which the tumor line arose. The next chapter will describe some hepatocyte-like *hepatoma* cells that have been maintained as cultures. There is a school of thought that views even the "undifferentiated" cell lines as specialized: The cells' properties do reflect characteristics adapted to growth under the particular culture conditions used to select them; furthermore, discrimination of *specialized* from *nonspecialized* depends on arbitrary recognition criteria.

Even when differentiated cell *lines* have not been established, much information is obtainable from so-called *primary cultures*. These are tissues removed from the animal (embryonic or fetal material is generally best) and grown for a limited period as dispersed cells, or in some cases (such as fetal liver tissue) as small pieces of tissue. Such preparations often maintain many of the special structural and biochemical features of the corresponding cells and tissues of the intact animal. Fetal nervous tissue explanted into cultures and suitably nourished ("nerve growth factor" often helps) can establish and maintain complex functions for periods of weeks or months.

For study of mature multicellular tissues, "organ" cultures are often used: Whole organs or large pieces can be explanted into suitable media and maintained for a while. (The cells do not divide, nor do the organs grow, but important physiological and biochemical properties are often maintained intact in such organ cultures for many hours or even days.)

3.11.2 Some Uses and Manipulations of Cell Cultures

The experiment illustrated in Figure II–28 was done with HeLa cells, one of the most widely used of mammalian cultured cell lines (the line originated from a human cervical cancer). The experiment required use of a large cell population, accessible to labeled precursors and to drugs, and uniform enough that representative samples could be taken at intervals. Doing such experiments with, for example, the livers of intact animals requires waiting until injected materials reach the liver through the blood stream. The materials are greatly diluted in the blood, and large proportions enter organs other than the liver. There may be serious problems in obtaining successive samples. If, as is customary, several animals are used for different time periods, it often is impossible to be sure that the animals are sufficiently similar to one another in nutrition, genetics, and other properties to permit reliable comparison. If the experiments require exposure to drugs, hormones, toxins, or similar compounds, the direct effects on the liver sometimes are difficult to distinguish from secondary effects in which the agent has acted upon another organ, such as an endocrine gland, which in turn has acted upon the liver. In addition, liver contains several cell types, a complicating factor for interpretation of biochemical analyses.

Efforts to distinguish the various components of biological timekeeping mechanisms are being pursued by culturing cells that might participate. This facilitates strict control of the environment and isolates the cells from potentially complicating interactions with other cells. Recently, for instance, it has been found that organ cultures of chick pineal gland, and possibly even separated cells from the gland, maintain a circadian rhythm (Section 3.8.1) in the activity of the enzyme N-acetyltransferase for several days or more. This enzyme is a key participant in the synthesis of the hormone *melatonin* through which the pineal gland controls a number of biological processes. Evidently, in birds, there is an endogenous, pineal gland–dependent rhythm in production and release of the hormone, that is based partly on an intrinsic rhythm in enzyme activity in the cells. This rhythm is, however, not entirely autonomous since it can be modified by exposure of the bird, or of the organ-cultured pineal gland, to appropriate cycles of light and dark. Studies of pineal gland cells under varying conditions in culture should help greatly in working out the mechanisms by which the cells and the gland maintain their rhythms and respond to environmental change.

Developmental Interactions of Cells and Tissues For reasons of experimental simplicity similar to those just described, cultured preparations are very useful for studying various interactions on which embryonic development depends. For example, when cells destined to develop into different sorts of tissues are prepared as suspensions of separated cells and mixed together, they can reaggregate in culture and "sort out": Cells of like origin associate with one another, and the aggregates can go on to form arrays quite reminiscent of normal tissues—liver, cartilage, neural tissue, and so forth. With usual trypsin-EDTA isolation procedures, the cell surface is depleted of some of its coat, and if replenishment of the surface molecules is interfered with through use of suitable metabolic inhibitors, sorting out of like cells does not occur. This phenomenon provides one of the key experimental starting points for the investigation of the molecules—glycoproteins and others—responsible for the differential adhesion and recognition phenomena on which the sorting-out process apparently depends.

The presence of one tissue type or its products in a culture often strikingly affects the differentiation of another. For instance, epithelial structures—glands, absorptive epithelia, the lining of the lung—normally develop from a layer of cells in the embryo known as the "endoderm." They do so in conjunction with development of underlying connective tissues from another embryonic cell group, the "mesodermal" cells. In many cases when endoderm from a given region of the early embryo is cultured together with mesoderm from other regions, the epithelia that form have the characteristics of the region from which the mesoderm was taken. Correspondingly, some epithelial structures will not form in culture unless mesoderm from the *same* region of the embryo is included in the culture. Others, such as pancreatic tissue, are more permissive in terms of the source of mesoderm that will support their development. Such studies, as well as ones in which partially purified materials extracted from cells or tissues are substituted for one or another of the partners in mixed cultures, indicate the following conclusions about cell and tissue interactions: The interactions often are mutual—each partner influences the other; some interactions are "instructive"—they impose a direction of differentiation—and some "permissive" (presence of the second tissue is required to permit the first to express a capacity it already has); and many of the interactions are mediated by materials released to the extracellular environment. Certain of these materials are molecules that can diffuse from one tissue to another, even if the tissues are separated by porous barriers such as filters that prevent direct cell-to-cell contact. Most likely the responses to these diffusible materials depend on receptors of the sorts discussed in Section 2.1.4. Other extracellular materials influencing development seem to be components of the organized fibers and matrices of connective tissues—collagen, glycosaminoglycans, and basal lamina materials. The architecture of tissues depends in appreciable degree on the mutual arrangements of the extracellular substances and the cells. Their most obvious influences are mediated by phenomena of selective adhesion and the like, affecting cell shape and distribution, but molecules of the extracellular matrix may also affect intracellular phenomena through cell-surface receptors or "intermediary" molecules such as fibronectin. Therefore, specific influences on enzyme or gene activity are conceivable.

Single uninucleate embryonic muscle cells will divide repeatedly in culture; then they will cease dividing and fuse to form multinucleate skeletal muscle fibers. This finding was important evidence in settling a decades-long dispute between researchers who believed that the numerous nuclei in one skeletal muscle fiber result from fusion of separate cells and those who maintained that they arise by repeated nuclear division without division of cytoplasm. It also provides a means for obtaining long-term cultures of functional muscle fibers, which is otherwise difficult to do since mature muscle fibers do not divide and are very difficult to separate in intact form and maintain. A similar approach, the culturing of stem cells capable of division and differentiation in culture, can be used to study some other cells that do not divide when mature, such as those of the blood.

Vaccines; Amniocentesis   Among the practical uses of cultures, probably the first major application was their employment for growth of large quantities of viruses that require specific eucaryotic host cells for replication. It is possible in this way to obtain material such as inactivated viruses, or genetically "attenuated" strains, that can be injected without causing infections but still evoke antibodies active against related, pathogenic viruses.

In the diagnostic procedure known as *amniocentesis,* small numbers of fetal cells taken from the fluid surrounding a developing fetus are grown in culture into large enough populations for biochemical analysis and cytological study. This procedure permits detection of severe developmental abnormalities in the fetus, especially those of genetic origin.

Somatic Hybrids; Heterokaryons   Techniques of cell fusion applied to cultures have proved very useful. Fusions occur spontaneously with low frequency, but if certain agents, most often specific viruses or glycoproteins from their envelopes, are added to the growth medium, fusion of cells with each other occurs on a large scale. One popular tool is *Sendai virus* made noninfective by exposure to ultraviolet light. The virus binds to the cell surface (Section 3.1.1); two cells adhering to the same virus particles can eventually fuse with one another. Chemicals used to promote fusion include polyethylene glycol. In early experiments demonstrating cell fusions, human cells (HeLa cells) were fused with a mouse tumor line that was initiated by Paul Ehrlich early in the century. If the DNA of the HeLa cell is prelabeled with ^{3}H-thymidine and if the Ehrlich tumor cells are unlabeled, the fused cells can be shown by autoradiographic techniques to contain both types of nuclei, labeled and unlabeled. Fused cells containing separate nuclei of different cell types are called *heterokaryons.* Section 4.4.3 discusses the activation of the ordinarily inactive nuclei of red blood cells of birds that occurs in red blood cell–HeLa heterokaryons. Such experiments are providing insights into interactions of the nucleus and the cytoplasm.

If, as the fused cell divides, the chromosomes of the two "parents" mingle to form a single joint nucleus, the product is a *somatic hybrid.* In many cases such cells can divide and be perpetuated in culture as more or less stable types. Hybrids have been produced by fusion of the cultured pigment-producing cells from hamsters with the nonpigmented cells of mice. That the hybrids do not synthesize pigment (melanin; Fig. II–49) suggests that some mouse genes can cause the synthesis of molecules able to repress the hamster genes responsible for melanin synthesis. Experiments of this sort opened approaches to the identification of regulatory molecules, which are now being pursued further through the DNA transfer techniques outlined below.

Monoclonal Antibody Production   Somatic hybridization is being extensively exploited for the production of antibodies. Ordinarily, antibodies for experimental work, such as immunohistochemical studies, or for therapeutic uses are obtained by injecting the antigen of interest into a rabbit or other animal. Later, blood is obtained from the injected animal, and the blood cells are removed, leaving an *antiserum,* serum containing the evoked antibodies among its immunoglobulins. Typically, a given antigen evokes a spectrum of antibodies that bind to different portions of the antigenic molecule (to different "antigenic determinants"): In other words, several different B-cell clones (Section 3.6.9) respond to different parts of the molecules and contribute to the antiserum. Not infrequently the predominant species in the evoked mixture binds to an antigenic determinant other than the one of primary interest. Although the immunoglobulins can be isolated as a group from serum, it is generally quite difficult to separate these overlapping antibodies from one another. In addition, a given animal can produce only limited quantities of blood, imposing an inherent limitation on the quantities of antibodies that can be obtained.

If, however, after injection of an antigen, responding antibody-producing *cells* are isolated from the animal and are grown as separate clones in culture, an antibody specific for one antigenic determinant—a *monoclonal antibody*—can be obtained from each clone. In principle, an endless amount of antibody should be obtainable from a tissue culture so long as the cells remain alive. The B-cells themselves, however, will not grow into large-enough proliferating populations. Cell fusion techniques can be used to prolong greatly the life span of antibody-producing cultures and to facilitate their growth into very large populations. In practice this is accomplished by provoking fusions of the cells present in a responding tissue of the antigen-injected animal (spleen cells of mice are most convenient to use) with *myeloma* cells. These are tumor cells arising from B-lymphocytes or plasma cells. The *hybridomas* resulting from the fusions divide for indefinitely long periods (some may be practically immortal), and for a prolonged time at least, they continue to produce the antibodies characteristic of the normal B-cell partners in the fusions. Clonal lines can be established from the mixture of spleen cell–derived hybridomas. Each clone will produce an antibody specific for a particular determinant and secrete the immunoglobulin molecules into the culture medium, from which the molecules can readily be obtained.

Cultured Plant Cells and Tissues   Masses of unspecialized plant cells (callus) can be grown in culture and used for developmental and other studies. Structures resembling roots or buds can be induced to form in cultured callus tissue by treatments with appropriate levels of specific plant hormones (Section 3.4.5). Work with cultures facilitates study of the actions and interactions of the hormones: For example, with some cells in culture, addition of cytokinins influences gene expression so that the *kinds* of peroxidase "isozymes" produced are affected. (Isozymes are distinguishable forms of a given type of enzyme that have similar activity but differ in details of molecular structure, such as in amino acid sequence or oligosaccharide side chains.) Auxin addition will then modulate the *amounts* of isozymes synthesized, increasing some and decreasing others. It is expected that such studies in cultured cells will elucidate hormonal actions and interactions in the living cell.

Cells from mature carrots or potato leaves and from other plants such as sycamore tree seedlings and soybean plants have been grown in culture for prolonged periods. They divide at rates that double the cell number every three or four days. Small clusters of such cells can produce complete new plants by growth under proper conditions, including the presence of appropriate hormones. Apparently a single cultured cell can give rise to an entire plant. The conclusion from such experiments is that many plant cells are not irreversibly fixed in differentiation. Normally the cells of the mature carrot from which the culture was begun would have restricted functions based on differentiation controlled by hormones and other vehicles of cellular interplay; however, after growth and division as separated cells in culture they can produce all the cell types of a plant. The ability to grow entire plants from cultured cells may have practical importance. The initiating cells can be in "protoplast" form (stripped of their walls), rendering much easier their experimental modification. In the hopes of producing cell lines that will generate plants with desirable agricultural properties, plant scientists are using a range of cell-culture techniques to select variants and to alter the genetic constitution of cells (see below) and to fuse different cells.

3.11.3 Genetic Studies With Cultured Cells

Mutation and Selection   The availability of mutants in bacteria is one of the major reasons so much is known about procaryotic genetics and biochemistry. Comparable momentum is now building in work with eucaryotes. The usual strategy for obtaining mutant eucaryotic cells in culture is the devising of a selective culture medium formulated so that mutant cells of the desired type survive while other cells do not. As in all cell populations, mutants arise spontaneously in cultures; the frequency of mutation can be increased by exposure to agents affecting DNA such as radiation or chemical mutagens.

The most widely used selective media contain components that are altered into cytotoxic agents by cellular enzymes. For example, the enzyme *hypoxanthine-guanine phosphoribosyltransferase* (HGPRT) normally permits cells to use *hypoxanthine*, provided in the culture media, to synthesize adenine and guanine nucleotides for its nucleic acids. But if cells are grown in a medium containing *azaguanine,* or *thioguanine,* abnormal analogues of the nucleic acid bases, the presence of the HGPRT enzyme leads to the entry of the analogue into the cell's nucleic acids with lethal effects. Thus, only mutant cells lacking functional HGPRT survive in media to which the analogue is added.

Introduction of DNA   Some of the techniques for genetic analysis developed for bacteria can be modified for use with eucaryotes. For example, with cultured cells it is now possible to introduce DNA purified from one cell into another by adding the DNA to the growth medium under proper conditions. As with bacterial transformation (Section 3.2.6), the DNA can be integrated into the recipient cell's chromosomes, permitting study of a variety of genetic and metabolic regulatory processes. The availability of well-defined mutations is an essential aid for this work. To evaluate whether cells have incorporated DNA to which they were exposed, experimenters can use a recipient cell line lacking a genetic capacity or carrying a mutant gene and, with suitable selective media, can determine whether corresponding normal genes from the DNA donor have become incorporated in the recipient's chromosomes. Genes specifying *thymidine kinase,* an enzyme of nucleic acid metabolism, have been employed effectively in this way; selection schemes have been devised in which the presence of functional enzyme is required for cell proliferation even though under other experimental conditions its presence can be fatal.

It is expected that such eucaryotic cell transformation and also introduction of foreign DNAs through viruses (Sections 3.12.3 and 4.3.3) will permit use of eucaryotic cell cultures for recombinant DNA approaches of the sort now employed primarily with bacterial systems (Section 3.2.6). This could have significant benefits for study of eucaryotic genes and for genetic engineering; as Section 3.2.6 pointed out, bacteria probably are not capable of some of the control mechanisms that govern gene expression in eucaryotes, and there may be important processing steps required for production of functional molecules that normally take place only in eucaryotes.

Strains of plants useful for agriculture might be produced by creating new genetic combinations in cultured cells and then by producing plants from the cultures, as outlined above. The idea that generally springs to mind first is the incorporation of genes conferring nitrogen-fixing ability, from procaryotes, directly into cells of a eucaryotic crop plant, thus reducing needs for nitrogen-containing fertilizers or other nitrogen sources. This idea may prove feasible eventually but faces at least two difficulties: First, as just mentioned, mechanisms controlling gene expression differ significantly between procaryotes and eucaryotes. Second, the *nitrogenase* enzyme is quite sensitive to oxygen (Chap. 3.2B) and probably would be rapidly inactivated in ordinary plant cells.

Mass Enucleation   The drug *cytochalasin B* (Section 2.11.2) causes a proportion of the cells in a culture to extrude their nuclei. This effect has led to the development of a simple technique for obtaining large numbers of enucleated cells. Cells are cultured as monolayers on coverslips. If, before reaching the confluent distribution, the cells are placed face down in a medium containing cytochalasin B and are gently centrifuged, the nuclei are extruded and go to the bottom of the centrifuge tube. The coverslip is left with cells, almost all of which are enucleated; these are still viable for a while if returned to fresh medium without the drug. Such preparations can be used directly to study the metabolic capacities of cytoplasm, or further experimental manipulations such as the introduction of a foreign nucleus can be carried out. In one series of experiments, enucleated preparations were made of cultured mouse "fibroblasts" that were *resistant* to the drug *chloramphenicol,* which ordinarily affects mitochondrial protein synthesis (Section 2.6.3). They were then fused with intact (nucleated) mouse fibroblasts of the "wild type" (sensitive to chloramphenicol). Some of the products gave rise to cell lines that were resistant to the drug. Since the resistant cells used for the fusions lacked nuclei, the likely explanation is that in this experiment chloramphenicol resistance is due to a mutation inherited through the cytoplasm, probably through mitochondrial DNA (Section 2.6.3).

Gene Amplification   Interesting observations have been made on a number of mouse cell lines exposed to the drug *methotrexate,* which interferes with metabolism of *folic acid,* an essential nutrient that is an important participant in a variety of pathways. Some cultured cell populations become relatively resistant to the drug through synthesis of enhanced levels of the enzyme *dihydroxyfolate reductase.* This increase, in turn, depends on the presence in the cells of tens to hundreds of extra copies of the genes coding for the enzyme. Such selective "gene amplification" in cultures should prove a useful model system for analyzing mechanisms that could contribute to amplifications known to occur under more "natural" conditions (Section 4.4.6).

Chromosome Mapping   The traditional means for systematic mapping of the locations of genes on chromosomes depend upon selective breeding of organisms possessing different forms of given genes (Section 4.3.3). This is obviously an inappropriate technique for studying human genetics, except for analysis of naturally occurring genetic combinations, and it is slow and laborious for vertebrates in general since their breeding cycles are much more lengthy than those of the favorite organisms of traditional genetics, such as fruit flies or yeast. Much progress in mapping has been made possible by cell-culture procedures, complemented more recently by enhanced capacities to examine DNA molecules directly (Section 4.2.6).

For human cells, somatic hybridization and related culture techniques have dramatically expanded knowledge of which genes are on which chromosomes. As interspecies hybrids, such as mouse cell–human cell hybrids, are continued in culture, chromosomes are eliminated by abnormal divisions or other events. Human chromosomes are eliminated more often than mouse chromosomes, but the elimination is partially random so that different hybrid clones from the culture, containing different combinations of human and mouse chromosomes, can be prepared. The enzymes and other proteins synthesized by clones with different chromosome constitutions can be compared. In this way it is possible to determine which chromosomes carry genes for particular proteins of interest. New methods for microscopic identification of chromosomes (Section 4.2.1) have made such work markedly easier.

Under some conditions the nuclei of cultured cells treated with colchicine fragment into "micronuclei," each with one or a few of the cell's chromosomes. These micronuclei can be isolated from the cells by treatment with cytochalasin B, as above, and then fused to intact cells from a different species. In this way experimenters can produce clones of cells, each containing specific foreign chromosomes whose genetic content and behavior can then be followed. Both mapping and investigation of gene regulation are being pursued with this approach.

Chapter 3.12 Cancer Cells

Part 3 concludes with consideration of cancer, a most important form of cell and tissue abnormality. The basis (or bases) of cancer remains unknown. Optimism regarding its conquest, however, is generated by current progress toward understanding the genetic machinery of cells and the mechanisms by which the body can defend itself.

Permanent changes in the heredity of cells can upset the mechanisms that restrict their division and keep them in their normal places. The cells accumulate in abnormal large masses called *tumors*. *Benign* tumors remain as discrete masses. *Malignant* tumors (cancers) generally *metastasize*—cells spread to distant parts of the body through blood or lymph, and there they take root. As they do, they produce factors that stimulate the ingrowth of capillaries to provide a blood supply, disrupt the local tissue arrangements, and grow into new tumors. Cancers originating in epithelial tissues are called *carcinomas*. Connective tissue cancers are referred to as *sarcomas*. Cancers originating in blood and immune system tissue include the *leukemias, lymphomas,* and *myelomas,* among other types. Often, when particular known cell types are involved, the cancers are designated accordingly: *Melanomas* arise from pigment-producing *melanocytes,* and *neuroblastomas* (Section 3.11.1), from *neuroblasts.*

In recent years, tissue-culture growth of cells with the potential for producing malignant tumors ("tumorigenic" cell lines) has proved of great value for the study of cancer, especially of its molecular biology. Traditionally the chief sources of cancer cells for morphological and biochemical studies had been the transplantable tumors of mammals. In transplanting tumors, a small piece or a suspension of tumor cells is injected under the skin or into the leg muscle or body cavity. Before the tumor has grown to a size that kills the animal (from several days to several months, depending upon the growth rate of the tumor) a bit is again transplanted. Some transplantable tumors are obtained in *ascitic* form. These grow in the abdominal cavity where they elicit the production of fluid; the cell-laden "ascitic fluid" can be removed from the body cavity as a convenient source of malignant cells.

Different animal and human cancers vary enormously in etiology (developmental history), histology, and biology. Historically, cancer researchers have oscillated between the belief that this diversity reflects significant differences in the basic nature of the transformation to the malignant (tumorigenic) state and the belief that this transformation is essentially of the same sort in many and perhaps all cancers. Presently, the ultimate aim of basic research on cancer focuses on understanding the nature and cause of the changed heredity. These efforts are hampered by lack of adequate understanding of the normal mechanisms that regulate cell division and growth (Section 4.2.3); correspondingly, progress in cancer research may help to elucidate these normal mechanisms as it reveals their perturbations.

3.12.1 Comparative Studies

Hepatomas; Primary vs. Secondary Changes For decades biologists have been comparing the cytology and biochemistry of normal and malignant cells in the hope of detecting changes common to several or many cancers that might provide clues to the fundamental alterations. Cancers of many organs have been studied, but none more extensively than *hepatomas,* cancers of the liver. This is because the liver has been so thoroughly investigated biochemically and morphologically, and because a broad spectrum of hepatomas varying in growth rate is available for investigation.

Both rapidly and slowly growing hepatomas possess organelles like those of normal cells, although the number and arrangement of the organelles may differ somewhat from normal (Fig. III–51). Cells of the slowly growing *Morris hepatoma* look much like hepatocytes, though their microvilli may be somewhat less regularly arranged and there are some differences in the distribution of cell surface enzymes. The rapidly growing *Novikoff hepatoma* differs from the Morris hepatoma in possessing much less endoplasmic reticulum; in having many more free polysomes, a smaller Golgi apparatus, and smaller mitochondria; and in lacking nucleoid-containing peroxisomes (it possesses only small microperoxisomes). From these differences, it seems likely that the Novikoff hepatoma originated from a different type of liver cell than that for the Morris hepatoma. This situation illustrates one of the problems faced by cancer biologists in their analysis of the origins of particular cancers: Cancers arise by proliferation of single altered cells, and most organs contain several or many cell types that can potentially become malignant. Moreover, a tumor in a given organ can arise from metastatic cells originating elsewhere; particular cancers show characteristic patterns of spread to other organs. In addition, populations of cancer cells probably undergo appreciable evolution as they encounter the organism's defenses and as they spread and establish themselves. Thus the cells that eventually characterize a particular tumor may differ significantly from

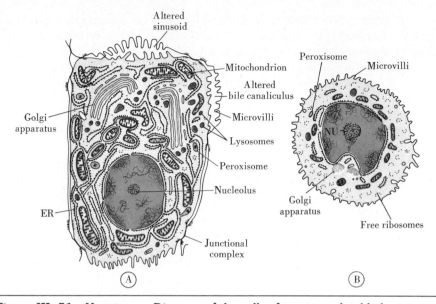

Figure III–51 *Hepatomas.* Diagram of the cells of two transplatable hepatomas of rat. A. A slowly growing (differentiated) Morris hepatoma. B. A rapidly growing (undifferentiated) ascites Novikoff hepatoma. Comparison with Figure I–5 will confirm that the differentiated hepatoma cells resemble hepatocytes in many respects far more closely than do the undifferentiated hepatoma cells. Yet even the differentiated type differs from hepatocytes in cell-surface features, in amounts of ER, and in the size and nature of the lysosomes.

their progenitors that started the process of tumor growth. For such reasons, not infrequently morphological studies must be supplemented by detailed biochemical studies or by detection of antigens (Chap. 3.6C) characteristic of specific cell types in order to determine the origins of a tumor.

More severe difficulties arise in deciding which of the changes from normal observed in a population of cancer cells are fundamentally related to the underlying transformation from normalcy to malignancy. Many of the more obvious changes are secondary manifestations of, for example, alterations in rate of division and growth. A given cell type may differ markedly, when growing and dividing rapidly, from the same cell type under conditions favoring slower growth and more expression of specialized characteristics. Cells of some slowly growing cancers, like the Morris hepatoma cells, can look strikingly like their normal counterparts.

One firmly established feature of malignant cells had been their maintenance of a high rate of glycolysis (Secion 1.3.1) even in the presence of oxygen ("aerobic glycolysis"). Analysis of slowly growing hepatomas, however, has shown some to have low rates of aerobic glycolysis, similar to that of normal liver. The point is that it is proving exceedingly difficult to find any individual metabolic or structural feature that universally distinguishes tumor cells from normal ones.

Even the rate of cell division is not always accelerated in cancers. We tend to assume it is because some cancers arise from cell types that normally divide infrequently, such as liver cells. For cancers arising from cell types that ordinarily do divide rapidly, rates of cell duplication may not be strikingly elevated. Thus, for the rapidly dividing stem cell populations that normally serve to replace blood cells or cells of the skin and other epithelia, the key controls that go awry appear to be those that balance division and differentiation. In the cancers of blood and epithelia, the dividing cell populations no longer are restricted to specific sites in the body's organs: Unlike the normal situation in which daughter cells enter a nondividing, differentiated state as they move away from their place of origin, leaving behind the population that continues dividing (Section 3.5.1), most or all of the cancer cells' offspring retain the ability to divide.

It is tempting to conclude that the cellular changes generating malignancy are quite different in different cancers. However, genetic changes affecting regulatory molecules or processes can have very different effects on different cells, depending on the overall sets of metabolic and structural control mechanisms operating in particular cell types. DNA alterations of one or a few types could therefore be reflected in varied patterns of change in cell characteristics.

Cell Surface and Cell Architecture; Density-dependent Inhibition; Protease Secretion Though not applying universally to all cancers, features commonly observed in tumorigenic cells in culture may help to explain some of the biology of malignant tumors. Most of the pertinent observations have been made on cells "transformed" to malignancy in culture through use of viruses or other agents, as will be discussed shortly. Many such transformed cells release unusually high levels of proteolytic enzymes into the media in which they are cultured. These high levels were detected by studying the cells' ability to activate the serum enzyme *plasminogen,* which, upon action of proteases like those secreted by the cells, becomes capable of degrading fibrin, the protein of blood clots (Section 3.6.5). (Under normal conditions, activation of plasminogen permits dissolution of clots as tissues heal after injury.) The release of proteases by cancer cells may help to explain their detachment from sites at which they normally are anchored, or how they can penetrate some of the barriers that normally maintain tissue architecture such as extracellular matrices and basal laminae. In addition, the proteases may act on proteins of blood serum and other extracellular fluids, or even directly on cell surfaces, altering or removing molecules and thus generating effects that may account for some of the changes seen in tumors. When proteases are added to the medium surrounding cultured nonmalignant cells, some of the properties of malignancy are transiently mimicked: Alterations in adhesiveness, motility, and even division frequency can be produced.

Owing in part to changes in their surfaces, many types of cancer cells, when placed in culture, show diminished susceptibility to density-dependent inhibition of division and movement. Often the cells can grow in multiple layers to high density under conditions where nontumorigenic cells are much more

restricted (see Fig. III–52). As mentioned in Chapter 3.11, some lines of malignant cells can even be grown in suspension; they no longer require anchorage
to a substrate for growth and division. Such characteristics seem to parallel
important properties of cancer cells in the body, most notably their "escape"
from controls that normally limit cell movement and keep cells in place, and
the modifications in the controls governing cell growth and division. There
seem to be contributory changes in the amounts of particular molecules present at the cell surface, and in the mobility or other characteristics of membrane
molecules. Thus, some cultured lines of cancer cells surround themselves with
substantially less fibronectin than that produced by their normal counterparts.
This decrease would affect their adhesion to extracellular surfaces and perhaps
to one another. Experimental binding of lectins (Section 2.1.7) to malignant
cells produces clumping (agglutination) of the cells at much lower lectin concentrations than those needed to clump normal cells, suggesting differences in
the plasma membrane glycoprotein or glycolipid population. It is also striking
that some lines of cultured malignant cells show much less cell-to-cell communication by gap junctions than that typical of similar "normal" cells.

Changes in the cell surface may contribute to the altered requirements
for particular components of growth media exhibited by cultured cancer cells.
Some are able to thrive in lower concentrations of serum factors (Section
3.11.1) and of ions such as Ca^{2+} than those needed for comparable nonmalignant cells. Perhaps this diminished requirement is attributable to alterations
in the state of receptors, permeability channels, ion pumps, carriers for nutrients, and other plasma membrane molecules. Such alterations could lead to
modified balances of intracellular metabolism and could affect cellular control
systems.

The arrangement and abundance of cytoplasmic structures such as microtubules or microfilament bundles often differs markedly between cancer
cells studied in culture and the nonmalignant cells from which they derive.
These alterations could directly affect the cells' movements and other properties: For instance, the numerous possible interactions between the cell surface
and the cytoskeletal and cytoplasmic mobility systems could account for various observed characteristics of malignant cells.

Note, however, that as with other cellular characteristics, none of the
deviations from normal observed with cancer cells in culture is a perfect pre-
dictor of tumorigenic capacity. That is, there are nontumorigenic cells that
show altered density-dependent effects, diminished fibronectin production, or
others of the traits just outlined, at least under some growth conditions.

"Reversal" of Malignancy in Plants and Animals Plants also develop
tumors. For example, if a tobacco plant is treated with extracts of a particular
strain of bacterium *(Agrobacterium tumefaciens),* within 3 days a tumor called
a *crown gall* forms. (The responsible agent appears to be a plasmid transferred
from the bacterium; this agent is now receiving much attention as a potential
tool for genetic engineering of plants.) As with normal plant cells, single tumor

cells can be isolated in culture, and they will grow to form multicellular masses. When masses of tumor-derived cultured cells are grafted to normal tobacco plants, they form normal tissues. In some cases, merely growing the tumor cells in culture will yield cells capable of generating normal plants. These observations imply that the expression of malignancy can alter with the cells' developmental history and environment.

For animal tumors, a possibly analogous entraining of malignant cells to normal behavior is seen with *teratocarcinoma* cells. Teratocarcinomas are tumors originating in the tissue that normally gives rise to germ-line cells (primarily gametes); they have the interesting capacity to generate tumor cells resembling several to many different tissues. When such cancer cells are incorporated into developing mouse embryos using the embryo construction procedures that have recently been perfected (Section 4.4.1), their progeny contribute to the apparently normal tissues of the mouse that develops; under these circumstances the cells do not generate tumors.

It is still uncertain whether the initial alterations generating malignant behavior in the crown galls and teratocarcinomas are similar to those characterizing other cancers. Some investigators feel they reflect regulatory imbalances rather than permanent genetic changes. Thus, although the observations that the expression of malignancy can be reversed are encouraging, the breadth of their implications for the control of cancer needs further exploration.

3.12.2 Carcinogenesis in Animals

Agents that produce cancers are known as *carcinogens*. A great many chemical carcinogens are known in animals and man. There are also physical carcinogens, such as X-rays and ultraviolet radiation.

Initiators and Promoters Carcinogenesis frequently involves cooperative action of more than one agent and sometimes seems to occur in several stages or steps. For instance, skin cancers result in mice if they are fed small amounts of coal tar ingredients such as dimethylbenzanthracene and then the skin is painted with irritants extracted from the plant derivative known as croton oil. Cancers will still appear without the croton oil exposure but with a lower frequency and greater delay. Croton oil is said to "promote" the cancers: It does not itself induce them, but it does hasten their appearance after exposure to cancer "initiators" such as coal tar. Exposure to promoters can be delayed for long periods—sometimes years—and still be effective. This suggests that the initiators cause long-lasting or permanent effects, probably on the genetic machinery. The promoters speed the expression of these effects. The mechanisms are still obscure, although most investigators suspect that the actions of certain promoters are tied to their ability to stimulate cell proliferation through wounding, irritation, or other effects. One possibility is that they alter the abundance or responsiveness of the receptors through which cells respond to the hormone-like serum growth factors and other regulatory molecules described in

Sections 3.11.1 and 4.2.3. This explanation ties in with current proposals that abnormal responsiveness to these factors is a distinguishing feature of tumorigenic cells or that some tumor cells overproduce such factors.

Some of the chemicals known to have carcinogenic effects are themselves innocuous but become modified into carcinogens by the exposed organism. Ironically, several coal components are altered into carcinogenic forms by the enzyme systems of hepatocytes responsible for detoxification of drugs and some other potentially harmful agents (Section 2.4.4).

There is good reason to believe that the eventual effects of most agents that initiate cancer are on DNA. Various chemical carcinogens can bind to DNA and damage it or affect its replication. Radiation can break nucleic acid strands or induce covalent cross-links between thymine bases on the two DNA strands; the cell has mechanisms for repairing such damage but in doing so sometimes makes errors that lead to mutations. As outlined in the following section, viruses that transform normal cells to malignant ones affect cellular nucleic acids. In recognition of the likely involvement of DNA alterations in carcinogenesis, one of the tests widely used to determine whether a given compound, such as a new industrial product, is a potential carcinogen is the determination of whether it causes mutations in bacteria. The correlations are not perfect since, for example, metabolic modifications of the type required to generate a carcinogenic derivative may not occur in bacteria. Moreover, it is not yet established that all carcinogens are mutagens. However, for initial, rapid screening of a large variety of products the approach is proving helpful.

In general, the fundamental genetic changes in cancer cells are likely to be much more subtle than the abnormalities in numbers and shapes of chromosomes seen in many but not all cancers. These gross chromosome abnormalities are most often considered effects rather than causes of the malignant transformation. Some probably reflect cell division "errors" engendered by sustained high rates of proliferation. On the other hand, in one type of chronic leukemia in humans, particular chromosomes change consistently (absence or reduction in size of one or two of the chromosomes); in this case there might be a causal connection between the chromosome abnormality and the malignancy. An extra copy of a particular chromosome is seen in many mouse leukemias, and a number of other cancers show characteristic chromosome rearrangements (translocations; Section 4.5.1). Such chromosome modifications could affect cellular metabolism by changing balances and altering regulation of the genes that come to be present in abnormal numbers or positions. These possibilities will be explored further in Section 3.12.3, where we will also have more to say about effects of mutations.

Immune Surveillance? Unanswered questions with potentially important therapeutic implications have to do with the immunological status of cancer cells. A still controversial hypothesis suggests that throughout the life of higher animals, potential cancer cells arise continuously, are recognized as altered by the immune system, and consequently are destroyed. Those cancers that do develop, it is speculated, do so because their properties permit them to evade this *immune surveillance* or *screening*. The evidence for this is highly equivocal. There are findings suggesting that when an animal's defensive systems are stimulated, as occurs in infections, proliferation of tumor cells may sometimes be slowed, and some types of cancer cells can induce immune responses when transplanted from one animal to another. However, the sorts of cancer found most frequently in humans do not occur with dramatically increased frequency in individuals whose immune systems are defective.

A related matter being studied is the use of immunological methods for early diagnosis of cancers. Since some cancer cells do seem to have unusual surface molecules that can act as antigens, it may be possible to identify them soon after they first arise with the very sensitive methods immunologists have developed for detecting small amounts of antigens. Attempts also are under way to use highly specific monoclonal antibodies (Section 3.11.2) in anticancer therapy, to destroy, selectively, tumor cell populations.

3.12.3 Virally Transformed Cells

Oncogenic viruses—those that transform cells to a potentially tumorigenic state—have proved to be exceptionally useful experimental tools. This is so because some of the viruses are exceedingly simple, because mutant viruses can be produced, because the viruses can transform cells in tissue culture, and because the large fund of information on the molecular biology of viral infection aids in interpretation of findings. Theories that infectious viruses or long-term "proviral" inhabitants of cells (Section 3.1.2) are responsible for many naturally occurring cancers have waxed and waned in popularity in recent years. Most investigators have adopted a wait-and-see attitude, recognizing that only a very few natural cancers have thus far been found to have such an origin, but convinced of the experimental utility of viral oncogenesis.

Viral Nucleic Acids; Reverse Transcriptase   *Polyoma virus* is one of the simplest known viruses: Its DNA consists of about 5000 nucleotide pairs (5224 in the most studied strain). It produces cancers when inoculated into newborn mice, rats, or hamsters. A similar virus, SV40 (SV = simian virus), has been obtained from cells of monkeys. When cultured rodent fibroblasts are infected with such viruses, some of the cells are transformed (Fig. III–52). They show the alterations in growth requirements and behavior outlined in the preceding section, and if introduced into an animal, they will multiply and produce cancers.

Since polyoma and SV40 are very simple it appears that very few genes need be involved in a cell's transformation to a malignant state. Transformed cells do not contain viruses as such, but when fusions are induced between certain transformed cells, the virus "reappears"; that is, separate viruses are again detectable. Evidently, as is true with certain bacterial viruses, the oncogenic viruses can insert their nucleic acids into the host cell chromosomes,

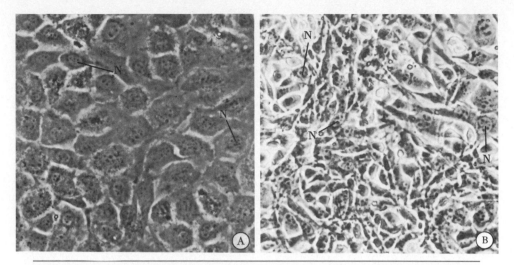

Figure III–52 *Oncogenic transformation.* Cells of a strain of cultured mouse "fibroblasts" photographed through a phase-contrast microscope. Nuclei of a few of the cells are indicated by N. The cells in part A are normal cultured cells that have formed a single layer (monolayer) of flattened cells on the glass surface. The cells in part B have been infected with polyoma virus and thereby "transformed." They no longer form a monolayer but overlap in multiple layers, clumps, and irregular arrays. (Courtesy of R. Dulbecco.)

where they replicate and are transmitted essentially like host cell DNA. That the host cell chromosome contains DNA encoding viral information is known also from the observation that purified nucleic acids from the viruses will hybridize with DNA from the transformed cells' chromosomes. Once a cell is transformed, its DNA can be used to transform other cells by the experimental techniques outlined in Section 3.11.3.

These findings caused some conceptual difficulty initially, because many of the known oncogenic viruses have RNA rather than DNA as their genetic material, and it was unclear how RNA could insert into the host cell genome. In 1970, however, an RNA-directed DNA polymerase was discovered in a virus that causes a chicken tumor (*Rous sarcoma* virus) and in another virus that causes a mouse leukemia (*Rauscher leukemia* virus). The enzyme, now known to be generally characteristic of oncogenic RNA viruses, is known as *reverse transcriptase* since it reverses the usual flow of information, which is from DNA to RNA. The viruses that employ this enzyme are called *retroviruses*. DNA copies of the viral RNA are made when the virus enters suitable hosts; these copies insert in the host chromosome. Retroviruses are of the membrane-enveloped type (Section 3.1.1) and are unusual in containing two copies of the viral nucleic acid in each virus.

Virally Encoded Proteins Transformed cells produce proteins coded for by the virally derived nucleic acid; these proteins are frequently referred to as transplantation antigens (T or t antigens) since many were first detected with immunochemical procedures. A given virus may produce several such proteins, and different ones accumulate in different parts of the cell. Intensive study of these proteins promises much information about how viruses transform cells.

Much attention is presently focused on one of the products of the Rous sarcoma virus. A viral gene referred to by the shorthand designation *src* (or *sarc*) specifies a protein (called *pp60src*) that localizes in the cytoplasm of the host cell; among other sites, immunohistochemical methods reveal it at or near the plasma membrane, and in regions where the plasma membrane and cytoplasmic filament and microtubule systems show close associations such as at adhesion plaques (Section 2.11.4). The *src* gene product is thought to be an enzyme of the kinase type—that is, one that regulates the activity of other proteins through phosphorylation (Sections 2.1.4 and 3.9.2). Unlike the majority of cellular kinases, which add phosphates to the amino acids *serine* or *threonine,* the *src* kinase phosphorylates the amino acid *tyrosine* (see Fig. I–28). Since kinases most often are regulatory enzymes and since some act on more than one target, the presence of such an enzyme among the viral products could help to explain the several interlinked cellular changes involved in transformation. This would be particularly easy to imagine if it acts directly on the plasma membrane or on components of the cytoskeletal and motility system. Indeed, one of the proteins phosphorylated by *pp60src* is *vinculin,* whose roles seemingly include involvement in interactions of cytoskeletal with membrane proteins (Section 2.11.2) and perhaps, through interactions with fibronectin, involvement also in anchoring of cells to extracellular surfaces.

Thus, it is possible to begin to develop testable hypotheses about how effects of transformation on cell shape and adhesion might come about. A number of retroviruses other than the Rous sarcoma virus may carry "oncogenes" similar to *src*. (*Oncogene* is the name proposed for genes responsible for tumorigenicity.) There is also initial evidence that *epithelial growth factor,* one of the serum factors normally controlling cell growth, affects phosphorylation of tyrosine. In brief, the findings on *src* are providing clues applicable to several, perhaps many, cancers, and also to normal processes of growth control.

As might be anticipated from our discussion of bacterial viruses in Sections 3.1.1 and 3.1.2, SV40 can multiply in cells and lyse them as well as transforming certain cells. Lytic infections permit study of the mechanism by which the virus controls its own replication and may therefore provide clues to how its products could affect normal cellular processes. The initial viral product made in an SV40 infection is a T antigen that accumulates in the host cell nucleus. This protein consists of about 300 amino acids and can bind to specific sites on SV40's DNA. These sites include the place where duplication of viral DNA begins (the "origin of replication") and nearby regions that serve as promoters (Section 3.2.5), regulating transcription of the mRNAs made in the early stages of viral infection. The binding of T antigen, it is thought, helps to trigger viral DNA synthesis; the binding also represses further production of mRNAs coding for the T antigen protein itself, without affecting transcription of the mRNAs needed for later stages in viral maturation (such as those coding for the coat protein). Comparable effects of the T antigen protein on host cell DNA in a transformed cell might well upset the normal controls of nucleic acid replication or transcription.

Viral Genes, Host Genes, and Carcinogenesis One tantalizing finding concerning the *src* gene product is the discovery that normal, nontransformed cells make a similar protein. Among several possible explanations, the one currently most appealing proposes that as the virus evolved it acquired its *src* gene from a host cell genome. Several other viral oncogenes may also have host cell counterparts; over a dozen different groups of genes are now under investigation in this connection.

A related hypothesis explains viral transformation by arguing that when the viral gene is introduced into the cell, it is not subject to the same cellular mechanisms that ordinarily control the expression of the cell's own copies of the similar genes; perhaps mutations have occurred in the course of the virus' evolution that alter responses to regulatory agents, or perhaps it is simply that the gene lacks the proper regulatory sequences or is in the wrong position or is present in too many copies for the cell's controls to be effective. The consequence might be that the gene product is produced in abnormal amounts (''dosage'') or with abnormal timing. By extension it has been suggested that in cases where oncogenic viruses do not themselves introduce oncogenes, viral nucleic acids may be positioned so that they ''turn on'' host cell genes or at least alter their regulation. Is it possible also that some cancers of nonviral origin result from rearrangements or mutations of the cell's own DNA that alter the regulation of potential oncogenes? Such ideas open perspectives, some of which are now being tested experimentally. For instance, they suggest that one way in which malignant cells arise is through imbalances in what are essentially normal processes. Correspondingly, through use of techniques for manipulating DNA and cells such as those outlined in Sections 3.2.6 and 3.11.3, the cellular counterparts of certain viral oncogenes can be isolated, coupled to regulatory DNAs (promoters; Section 3.2.5) from a virus, and reintroduced into cells. The cells are thereby transformed, evidently since the genes, though themselves normal, are ''freed'' by the viral promoter from proper regulation and thus make more of their products than normal. Irrespective of the validity of this interpretation, such findings are exciting because they signal the opening of an era of more direct investigation of fundamental processes of carcinogenesis.

Genes that seem to be involved in producing a malignant state have now been identified in cancerous human bladder cells; through comparison of the nucleotide sequences of these genes with corresponding DNA of nonmalignant cells (Section 4.2.6), mutational changes in sequence were detected, and these are now being studied intensively as possible specific alterations responsible for oncogenic effects. Evidently, in one line of bladder carcinoma cells, a single base pair in the gene coding for a protein has changed from G-C to T-A, leading to the substitution of the amino acid *valine* for one of the *glycines* normally present in the protein. Presumably, the consequent changes in the protein's biological properties crucially alter some aspects of cell structure or metabolism. This finding is especially important since it establishes a specific "qualitative" change in a cell's own genes as central to a particular malignancy. The studies on viruses have tended to concentrate on "quantitative" effects— altered amounts of gene products.

There may also be situations in which viral products, such as the SV40 T antigen, interact with otherwise normal host cell proteins, altering the proteins' properties.

Even when there is good reason to believe that a virus is involved in the genesis of a particular type of tumor, as with mouse leukemia, carcinogenesis is not simply a question of the presence of the virus. Genes of the mice influence the incidence of leukemia even in the presence of virus. Mice have been inbred so that in one strain, 80 to 90 percent of the animals develop leukemia. In another inbred strain fewer than 5 percent do so, and still other strains show intermediate incidences. Furthermore, mouse strains that ordinarily would show a high frequency of leukemia show much lower incidence if kept on a diet restricted to essential nutrients. Expression of mouse mammary tumor virus genes in host cells is strongly influenced by steroid hormones that normally mediate a number of regulatory responses by mammalian organisms. For humans, the once-popular theory that a significant proportion of natural cancers actually reflect expression of viral material incorporated in the genome cannot, by itself, account for epidemiological and experimental findings associating heightened frequencies of particular cancers with environmental or occupational variables, such as exposure to carcinogens, or with personal behavior such as smoking.

Further Reading

Viruses and the Like
Diener, T. O. Viroids. *Scientific American* 244 (1): 66–73, 1981.

Luria, S. E., J. E. Darnell, D. Baltimore, and A. Campbell. *General Virology,* 3rd ed. New York: John Wiley & Sons, 1978, 578 pp. *A comprehensive introduction.*

Spector, D. H., and D. Baltimore. The molecular biology of poliovirus. *Scientific American* 232 (5): 25, 1975.

Procaryotes
Adler, J. The sensing of chemicals by bacteria. *Scientific American* 234 (4): 40–47, 1976.

Barile, M. F., and S. Razin (eds.). *The Mycoplasmas.* New York: Academic Press. *Projected as a multiple volume series; the first volume appeared in 1979.*

Berg, H. How bacteria swim. *Scientific American* 232 (2): 36–44, 1975.

Cohen, S. N. The manipulation of genes. *Scientific American* 233 (1): 24, 1975.

Koshland, D. E. Biochemistry of sensing and adapting. *Trends Biochem. Sci.* 5: 297–302, 1980.

Novick, R. P. Plasmids. *Scientific American* 243 (6): 102–127, 1980.

Sharon, N. The bacterial cell wall. *Scientific American* 220 (5): 92–98, 1969.

Stanier, R. Y., E. A. Adelberg, J. L. Ingraham, et al. *The Microbial World.* Englewood Cliffs, N.J.: Prentice-Hall. *An excellent introduction to procaryotic cells. (Revised at intervals; the latest (fifth) edition was published in 1980.)*

Stoeckenius, W. The purple membrane of salt-loving bacteria. *Scientific American* 234 (6): 38–46, 1976.

Watson, J. D. *The Molecular Biology of the Gene,* 3rd ed. New York: Benjamin, 1976, 739 pp. *An excellent introduction to* E. coli.

Woese, C. R. Archaebacteria. *Scientific American* 244 (6): 98–122, 1981.

Algae, Protozoa, and Other "Lower" Eucaryotes

Excellent, brief introductions to protists and other unicellular organisms are included in many introductory biology books. A fine, readily available standard book of this type is Keeton's Biological Sciences *(3rd ed., New York: Norton Co., 1980).*

Beckett, A., I. B. Heath, and D. J. McLaughlin. *An Atlas of Fungal Ultrastructure.* London: Longmans, Green & Co., 1974, 221 pp. *A collection of micrographs of yeast and other fungi showing their basic structure and discussing matters such as growth and reproduction.*

Bold, H. C., and M. J. Wynn. *Introduction to the Algae.* Englewood Cliffs, N.J.: Prentice-Hall, 1978, 706 pp. *One of the standard introductory texts with much information on algal structure.*

Doughty, M. J. and S. Dryl. Control of ciliary activity in *Paramecium:* an analysis of chemosensory transduction in a eucaryotic, unicellular organism. *Prog. Neurobiol.* 16 (1): 1–115, 1981.

Patterson, D. J. Contractile vacuoles and associated structures: their organization and function. *Biol. Rev.* 55: 1–46, 1980.

Pickett-Heaps, J. D. *Green Algae.* Sunderland, Mass.: Sinauer, 1975, 606 pp. *A strong presentation of the ultrastructure of this group of algae.*

Satir, B. The final step in secretion. *Scientific American* 233 (4): 28, 1975. *Describes secretory phenomena in the protozoan* Tetrahymena.

Sleigh, M. A. *The Biology of Protozoa.* Baltimore: University Park Press, 1980, 324 pp. *An introduction to the protozoa, with emphasis on ultrastructure.*

Animal Cell Types (see also the specific types below)

Kessel, R. G., and R. H. Kardon. *Tissues and Organs: A Text Atlas of Scanning Electron Microscopy.* San Francisco: W. H. Freeman, 1979, 317 pp. *Good illustrations of cellular arrangements in tissues.*

Lentz, T. L. *Cell Fine Structure.* Philadelphia: W. B. Saunders, 1971, 437 pp. *An atlas of schematic drawings of many cell types of mammals with explanatory material.*

Wilson, J. A. *Principles of Animal Physiology,* 2nd ed. New York: Macmillan, 1979, 890 pp. *One of a large number of good physiology texts discussing the context in which cell functions occur in multicellular, higher animals.*

Histology texts: Many excellent ones have been published, and new editions of most of these appear at regular intervals. Among those most useful from the perspectives of cells and organelles are those by Bloom and Fawcett (Histology, *Philadelphia: W. B. Saunders, 1975), Ham and Cormack* (Histology, *New York: Harper & Row, 1979), Weiss and Greep* (Histology, *New York: McGraw-Hill, 1977), and Rhodin* (Histology, *New York: Oxford University Press, 1974). The book* Comparative Animal Cytology and Histology *by U. Welsch and V. Storch (Seattle: University of Washington Press, 1976) is an unusually broad comparative survey of cell and tissue structure in the animal kingdom.*

Physiological Reviews *is a periodical devoted to extensive reviews of animal cell physiology, often with good details on cellular behavior.*

Higher Plant Cells

Consult the Annual Reviews of Plant Physiology *for reviews of recent research.*

Albersheim, P. The wall of growing plant cells. *Scientific American* 232 (4): 80–95, 1975.

Baker, D. A. *Transport Phenomena in Plants.* New York: Halsted Press, 1978, 80 pp. *A brief, useful book.*

Bauer, W. D. Infection of legumes by rhizobia. *Ann. Rev. Plant Physiol.* 32: 407–449, 1981.

Cocking, E. C., M. R. Davey, D. Pental, and J. B. Power. Aspects of plant genetic manipulation. *Nature* 293: 265–270, 1981.

Cronshaw, J. Phloem structure and function. *Ann. Rev. Plant Physiol.* 32: 465–484, 1981.

Encyclopedia of Plant Physiology. New York: Springer-Verlag. A multivolume series, begun in 1975, dealing with many aspects of plant structure and function including vascular tissue, hormones, C_4 metabolism, and other topics discussed in the present text.

Guerrero, M. G., J. M. Vega, and M. Losada. The assimilatory nitrate-reducing system and its regulation. *Ann. Rev. Plant Physiol.* 32: 169–204, 1981.

Gunning, B. E. S., and A. W. Robards (eds.). *Intercellular Communication in Plants: Studies on Plasmodesmata.* Berlin: Springer-Verlag, 1976, 387 pp. *A collection of research articles.*

Gunning, B. E. S., and M. W. Steer. *Ultrastructure and Biology of Plant Cells* and *Plant Cell Biology, An Ultrastructural Approach.* London: Edward Arnold & Co., 1975. *These books discuss plant cell structure and function from the viewpoint of electron microscopists. The second is an atlas of micrographs taken from the first, which is a good introduction to plant cell biology.*

Jensen, W. A. Fertilization in flowering plants. *Bioscience* 23:21, Jan. 1973.

Ledbetter, M. C., and K. R. Porter. *Introduction to the Fine Structure of Plant Cells.* New York: Springer-Verlag, 1970, 188 pp. *An atlas of micrographs with explanatory material.*

Moore, T.C. *Biochemistry and Physiology of Plant Hormones.* New York: Springer-Verlag, 1979, 274 pp. *A relatively brief but extensive introduction.*

Raschke, K. Stomatal action. *Ann. Rev. Plant Physiol.* 26: 309, 1975.

Raven, P. H., and R. F. Evert. *Biology of Plants.* San Francisco: Worth, 1981, 686 pp. *A basic, introductory text.*

Stumpf, P. K., and E. F. Conn (eds.). *The Biochemistry of Plants: A Comprehensive Treatise.* New York: Academic Press. *A multivolume series covering many aspects of plant structure and function. The first volume (1980, edited by N. E. Tolbert) reviews organelles and cell structure.*

Epithelia, Junctions, and the Like

Pinto da Silva, P., and B. Kachar. On tight junction structure. *Cell* 28: 441–450, 1982. *Reviews basics of structure and proposes a still-controversial model.*

Staehelin, L. A., and B. E. Hull. Junctions between living cells. *Scientific American* 238 (5): 140–152, 1978.

Moog, F. The lining of the small intestine. *Scientific American* 245 (5): 154–179, 1981.

Connective Tissues, Extracellular Fibers, and So Forth

Doolittle, R. F. Fibrinogen and fibrin. *Scientific American* 245 (6): 126–136, 1981.

Hay, E. Extracellular matrix. *J. Cell Biol.* 91 (3, Part 2): 205S–223S, 1981.

Ross, R. Wound healing. *Scientific American* 220 (6): 40, 1969.

Zucker, M. B. The functioning of blood platelets. *Scientific American* 242 (6): 86–103, 1980.

The Immune System

Hood, L. E., I. L. Weissman, and W. B. Wood. *Immunology.* Menlo Park, Cal.: Benjamin, 1978, 467 pp. *An introductory text that covers the essentials in an unusual, "problem-approach" format. Inevitably, somewhat out of date for the intimate details of nucleic acid changes.*

Leder, P. The genetics of antibody diversity. *Scientific American* 246 (5): 102–115, 1982.

Rose, N. R. Autoimmune diseases. *Scientific American* 244 (2): 80–103, 1981.

Scharff, M. D., S. Roberts, and P. Thammana. Hybridomas as a source of antibodies. *Hosp. Pract.* 16 (1): 61–66, Jan. 1981.

Silverstein, S. The militant macrophage. *The Sciences* 21 (6): 18–22, July/Aug. 1981.

Nerve, Muscle, and Receptor Cells

The Handbook of Physiology, *a multivolume series published by the American Physiological Society, has many useful review articles.*

Binkley, S. A timekeeping enzyme in the pineal gland. *Scientific American* 240 (4): 66–71, 1979.

Hudspeth, A. J. The hair cells of the inner ear. *Scientific American* 248 (1): 54–64, 1983.

Fine, A., and E. Z. Szuts. *Photoreceptors: Their Role in Vision.* Cambridge: Cambridge University, 1982, 212 pp. *A first-rate brief introduction.*

Kandel, E.R. and Schwartz, J. H. (compilers and part authors). *Principles of Neural Science.* New York: Elsevier/North Holland, 1981, 731 pp. *A fine introductory text ranging from the cellular to the behavioral.*

Keynes, R. D. Ion channels in the nerve cell membrane. *Scientific American* 240 (3): 126–135, 1979.

Kuffler, S. W., and J. G. Nicholls. *From Neuron to Brain.* Sunderland, Mass.: Sinauer, 1976, 486 pp. *A well-written introduction to the nervous system, stressing physiology.*

Murray, J. M., and A. Weber. The cooperative action of muscle proteins. *Scientific American* 230 (2): 58–71, 1974.

Peters, A., S. L. Palay, and H. deF. Webster. *The Fine Structure of the Nervous System.* Philadelphia: W. B. Saunders, 1976, 406 pp. *A beautifully illustrated discussion of the principal cell types of mammalian nervous tissue.*

Squire, J. *The Structural Basis of Muscle Contraction.* New York: Plenum Press, 1981, 698 pp. *Ultrastructure and function of muscle.*

Gametes
See the developmental biology texts listed at the end of Part 4 for detailed treatments.
Epel, D. The program of fertilization. *Scientific American* 237 (5): 128–138, 1977.

Tissue Culture; Cancer
Bishop, J. M. Oncogenes. *Scientific American* 246 (3): 80–93, 1982.
Cairns, J. *Cancer: Science and Society.* San Francisco: W. H. Freeman, 1978, 199 pp. *Though its molecular biology is inevitably becoming outdated, this brief book is a good introduction to basic perspectives.*
Cooper, G. M. Cellular transforming genes. *Science* 217: 801–806, 1982.
Fuchs, F. Genetic amniocentesis. *Scientific American* 242 (6): 47–53, 1980.
Kasal, C., and J. R. Perez-Polo. Circadian rhythms in vitro. *Trends Biochem. Sci.* 7: 59–61, 1982.
Nicholson, G. L. Cancer metastasis. *Scientific American* 240 (3): 66–76, 1979.
Marx, J. L. Change in cancer gene pinpointed. *Science* 218: 667, 1982.
Pollack, R. (ed.). *Readings in Mammalian Cell Culture,* 2nd ed. New York: Cold Spring Harbor Press, 1981, 718 pp. *A comprehensive collection of classic and recent research articles on properties of cells in culture, including oncogenic transformation, with thoughtful introductory segments by Pollack.*
Ruddle, F. H., and R. S. Kucherlapati. Hybrid cells and human genes. *Scientific American* 231 (1): 36–44, 1974. *Use of tissue culture methods for chromosome mapping.*
Shepard, J. F. The regeneration of potato plants from leaf-cell protoplasts. *Scientific American* 246 (5): 154–166, 1982.
Varmus, H. E. Form and function of retroviral proviruses. *Science* 216: 812–820, 1982.

PART 4

Duplication and Divergence: Constancy and Change

Duplication of even the simplest procaryote involves far more than replication of DNA. Cell growth and division are integrated processes based on coordinated activities of virtually all components. Mechanisms operate for duplication of all cell structures and for separation of the cell into two daughters, each with its share (in general, approximately equal) of the structures and macromolecules of the parent.

In bacteria such as *E. coli,* and probably in most other procaryotes, only a single chromosome is present per nuclear region. For some bacteria, it is proposed that special regions of the plasma membrane function in separating the two daughter chromosomes produced by DNA replication (Section 3.2.4).

In eucaryotic cells there is much more DNA than in procaryotic cells, and it is "packaged" in several or in many chromosomes. The chromosomes, unlike those of bacteria, contain much protein and other non-DNA material. In almost all eucaryotic cells the chromosomes are duplicated and separated into daughter cells by the process of *mitosis,* which involves the formation of a temporary intracellular structure, *the spindle.*

Self-assembly and Organelle Biogenesis Mechanisms for synthesizing nucleic acids, proteins, and other macromolecules are now reasonably well analyzed. Organelle biogenesis has only recently begun to be understood. In

496

Section 1.4.3 we introduced the presumption that self-assembly is central to such biogenesis. That is, some macromolecules possess built-in features that automatically produce a correct association with other macromolecules, so that a given three-dimensional structure results spontaneously from properties inherent in the molecules of which it is composed. No special, external, "structure-specifying" system ("template") is needed to impose three-dimensional ordering on groups of macromolecules in a manner analogous to the way in which mRNA controls the amino acid sequence of proteins. If such structure-specifying systems are involved at one level or another, they themselves must ultimately derive their structure from a self-duplicating system (presumably one containing nucleic acids) that essentially specifies its own structure. Otherwise an endless chain of systems specifying the structure of other systems must be postulated.

The presumption of a central role for self-assembly in the biogenesis of cell structures is borne out by reconstitution experiments of the types we have discussed for a number of the organelles (Sections 2.3.2, 2.6.1, and 2.10.5). For some of the cell's structures, such as multienzyme complexes or even ribosomes, assembly in the cell may well resemble that in the test tube reasonably closely. Even for these cases, however, it is important to bear in mind that the cell must control the timing of appearance, the amounts, and the concentrations of the assembling molecules. In addition, the molecules' environment must be controlled: For example, the pH and other ionic concentrations must be such as to favor assembly.

For complex organelles, such as chloroplasts or mitochondria, it is improbable that a *complete* organelle could self assemble even if the constituent molecules were mixed in the correct proportions in a medium of proper composition. Mitochondria and plastids grow and divide, and it seems likely that as an organelle grows, the previously existing structure aids in orienting newly added molecules. In other words, as the structural complexity of the cell has evolved, modified assembly mechanisms permitting duplication of structures that cannot simply self-assemble from separate molecules have developed. Experiments pertaining to these mechanisms will be outlined in the following chapter. The experiments provide hope that the duplication of organelles will soon be describable in molecular terms.

Change: Physiological, Developmental, and Evolutionary Duplication leads to similar cells, but cells also undergo change. Change occurs on three time scales: the very rapid physiological changes that have been discussed at numerous points in the preceding parts, cell differentiation during embryonic development, and the slower processes of evolutionary change.

The cell diversification that occurs in embryonic development results from complex patterns of interaction between nucleus and cytoplasm and between different groups of cells. When an egg is fertilized, a precise program of events is set in motion, a program that depends on factors of which many remain to be appreciated. An important part of the story is the formation of different messenger RNAs in different cells as different genes are activated or freed from the repression that keeps them inactive. Interesting changes in chromosome structure are associated with this gene activation.

It is self-evident that basic cellular processes, such as photosynthesis or mitosis, have an evolutionary history. At some point in evolution they arose from simpler processes. Their history is known only in vaguest outline, but some research and much speculation have been devoted to its elucidation.

Evolutionary changes are based on the natural selection of variant organisms resulting from mutations of DNA. Mutated DNA is often expressed as altered proteins that engender changes in cellular metabolism. A change in any major step of metabolism tends to affect many distantly connected metabolic processes owing to the interweaving of pathways. Variation is greatly increased by sexual reproduction, which generates new combinations of genes. In most eucaryotic organisms, sexual reproduction depends on *meiosis,* a special pair of cell divisions.

The evolution of cells involves the evolution of chromosomes, as evidenced by differences in the number and shape of chromosomes in different species. Cytoplasmic factors also are involved; some cytoplasmic organelles possess genetic information that is expressed partly independently of the nuclear genes. Mutations of cytoplasmic genetic factors can occur and can result in permanent alterations of cell characteristics.

Chapter 4.1 Macromolecules and Microscopic Structure

This chapter will investigate ways in which self-assembly processes are employed and modified in the formation of cellular structures. We shall outline some general points that emerge from and extend the earlier discussions of the biogenesis of particular organelles. The chapter also introduces ongoing studies on particularly useful experimental material, such as viruses.

4.1.1 Time and Place of Assembly

Time is an important factor in the formation of biological structures. Even basic processes such as the folding of a protein molecule can be affected by temporal aspects of the molecule's history: As the parts of the protein made first (the N-terminal end; Section 2.3.1) leave the ribosome or cross the ER membrane, they can begin to fold up independently of the parts yet to be completed, so that the final conformation of the protein may be different from that attained when the entire molecule is experimentally unfolded and encouraged to refold.

One obvious way in which the cell determines where and when particular structures form or grow is by controlling the timing of synthesis of the structures' constituents and by accumulating these constituents in the appropriate cell regions. A host of known or suspected mechanisms contribute to this regulation, ranging from those that govern mRNA production (Sections 2.3.5 and 3.2.5 and Chap. 4.4) and enzyme activity (Sections 1.4.2, 2.1.4, 3.2.5, and 3.6.1) to those that determine transport of macromolecules into specific cytoplasmic compartments (Sections 2.4.1, 2.4.3, 2.5.2, 2.6.3, 2.7.4, 2.8.4, and 2.9.3).

Inorganic Ions   Cellular controls over concentrations of inorganic ions also influence assembly. Most attention has been paid to the "divalent" ions (those with two units of charge), because such ions, especially calcium, have so many effects both on living cells and on test-tube-assembly systems. We have seen, for example, that microtubules will not assemble if too much Ca^{2+} is present (Section 2.10.5) and that ribosomes fall apart into their two subunits if Mg^{2+} concentrations are reduced (Sections 2.3.2 and 2.4.3). Calcium also can change the packing of lipids in membranes, induce membrane fusions, and alter the conformation of proteins so that groups essential for the proteins' activities and interactions are exposed or are buried. One of the ways in which ions like calcium can have such effects is by binding simultaneously to two separate negatively charged groups, thus bringing molecules or parts of molecules into close association with one another or helping to stabilize previously existing associations.

Divalent ions are, however, by no means the only ones important for assembly processes. Section 2.5.2 took up roles of large, charged organic molecules in the condensation of secretion granules; many of the properties of connective tissue matrices also depend on interactions of macromolecules bearing ionizing groups, such as sulfates (Fig. III–37). At the other extreme of ionic size, in Sections 2.11.4 and 3.10.3 we outlined proposals that changes in hydrogen ion concentration (pH) affect the assembly of actin filaments by controlling the association of assembly-inhibiting protein molecules with actin molecules. It is not only the formation of distinct ionic bonds that contributes to such effects. For example, the degree to which particular macromolecules associate with one another and the conformation of elongate multiply charged molecules like the polynucleotide strands of DNA or RNA are influenced strongly by the "ionic strength" of the medium—the overall concentration of ions. The negatively charged phosphates of a polynucleotide strand repel one another, but if there are many positively charged small ions present in the solution, these tend to associate with the phosphates and "mask" their charge; this "masking" allows the strand to coil into a more compact array than that permitted by the repulsions when few such ions are around. Similar "masking" can change the conformation of proteins or facilitate their association with one another. Physical-chemical theory indicates that the inorganic ions involved need not form stable one-to-one associations with the charged groups of the macromolecules. Rather, the larger molecules may be surrounded by a dynamic "cloud" of constantly moving inorganic ions, and the masking just described can be thought of in statistical terms as reflecting a tendency for appropriately charged ions to spend more time in the vicinity of the macromolecules' charged groups. In similar fashion, "layers" of inorganic ions appear to associate with extended charged surfaces such as the external faces of phospholipid bilayers. These affect the passage of materials to and from the bilayer surfaces and the interactions of separate membranes with each other.

The overall concentrations of inorganic ions in the cytoplasm are controlled by the permeability properties of the plasma membrane; by the transport and storage mechanisms discussed in Sections 2.1.3, 3.6.4, and 3.9.1; and by binding of the ions to cellular molecules. Comparable mechanisms seem to maintain the special local ionic environments found in some of the organelles (Sections 2.8.4 and 3.6.1). In addition, by extension of the information about Ca^{2+} (Sections 2.1.4 and 3.9.1), it is surmised that localized and temporally controlled changes of ion concentration in particular cytoplasmic regions occur as a result of local effects on ion influx through the plasma membrane or release from intracellular compartments, binding to cytoplasmic proteins, and transport into storage compartments or out of the cell.

Proteolytic and Other Posttranslational Modifications of Proteins Our consideration of extracellular materials (Section 3.6.1 and Chap. 3.6B) revealed several cases in which the timing and location of assembly is controlled by modifications of proteins occurring some time after their synthesis. The clearest examples are the formation of fibrin and collagen fibers from molecules whose assembly is promoted by proteolytic cleavages. The molecular segments removed by these cleavages apparently inhibit spontaneous assembly of procollagen molecules or fibrinogen molecules into fibers. This effect permits the organism to limit fiber formation to the extracellular sites where appropriate proteases are present or are activated, as in the cascade of events leading to blood clotting.

The proteolytic processing of insulin occurs inside the cell after the proinsulin molecule has folded into an appropriate conformation (Section 3.6.1). The two chains that result from the processing are difficult to reassociate correctly if experimentally separated. In other words, proinsulin contains assembly information that is lost once "correct" assembly of insulin is achieved. The biological significance of this fact is still being debated (perhaps it facilitates eventual inactivation of the hormone), but it serves as a dramatic demonstration that the cell must control the timing and location of intracellular, posttranslational modifications of its proteins. The locations and specificities of the enzymes that accomplish particular processing steps are essential factors in the control. Recall that for insulin, as for other proteins, there is first a proteolytic removal of the "signal sequence," which takes place in the ER (Section 2.4.3). Then, after the resultant "pro-" form folds and is transported to the Golgi apparatus and secretory granules, the second proteolytic modification is carried out by an enzyme quite different from the ER one that removes the "signal sequence." This second enzyme splits the proinsulin chain into three segments: the two that constitute insulin, and a third, which having contributed to the folding of proinsulin, plays no further known part in insulin's structure or function.

Less is known about the impact of posttranslational modifications other than proteolysis. Many investigators believe that the carbohydrate chains attached to glycoproteins or glycolipids affect molecular conformation, interaction among molecules, or anchoring of proteins and lipids in membranes. Probably modifications like the kinase-mediated phosphorylations that regulate various protein activities (Sections 2.1.4 and 3.9.2) also contribute to governing assembly.

Assembly Initiators Microtubule-organizing centers (Section 2.10.5) are responsible for determining where at least some of the cell's microtubules form. Such control is especially evident in the participation of basal bodies in the genesis of cilia and flagella. Beginnings have been made toward experimental analysis of the timing of microtubule-organizing activity. It appears that duplication of organizing centers occurs in association with preparations for cell division (Section 4.2.7) and that previously inactive centers can be "turned on" at appropriate times (Section 4.2.7). Future work should clarify the underlying mechanisms and reveal their interplay with the mechanisms regulating tubulin synthesis (Section 2.10.6) and those that create intracellular environments favorable to microtubule assembly.

Oriented Growth from Nucleating "Seeds"; Actin Meshes and Bundles; Bacterial Flagella The extent to which the non-self-duplicating organelles other than microtubules rely on "organizing" centers to control their distribution is uncertain. It is known that quite short stretches of polymerized actin can lengthen rapidly by addition of actin molecules (Section 2.11.2), the initial polymers serving to "nucleate" or "seed" growth. Perhaps actin networks sometimes form in the cell, when conditions favor polymerization, by a *relatively* slow, spontaneous assembly of such seeds, followed by their rapid elongation; the elongation of each individual filament would be highly oriented, but since the seeds themselves would form at random, filaments running in many directions would appear. Assembly of highly oriented bundles of microfilaments might require an additional orienting device analogous to a microtubule-organizing center. The search is on for structures that might serve such a function: As mentioned in Section 2.11.4, one suspect has been encountered associated with the membranes that participate in elongation of acrosome processes by rapid actin assembly.

Bacterial flagella can be dispersed into flagellin molecules that, by themselves, reconstitute flagella only slowly. Reconstitution of flagella-like fibers is much more rapid if short pieces of broken flagella are present. The fragments serve as "seeds" that initiate growth, and they do so by establishing an oriented pattern of flagellin addition. The flagella are relatively straight in some bacterial strains, and wavier in others. The differences in shape reflect genetically based differences in the flagellin molecules themselves. Nevertheless, if long fragments are used to seed growth, then fragments from "wavy" flagella combined with flagellin from "straight" ones yield wavy flagella. If both fragments and flagellin molecules are from straight flagella, the products are straight. Apparently, in the test tube at least, the bonding of flagellins to preexisting structures can override their own "predispositions."

4.1.2 Size Control; Molecular Cooperation

Some biological structures are of *self-limited* size. They possess closed surfaces: Once the structure is complete, no additional room is left for adding more subunits (Fig. IV–1). The simpler "spherical" viruses with polyhedral coats (Section 3.1.1) show such assembly; their proteins and nucleic acids can be separated and then recombined in the test tube to reconstitute complete, infectious viruses.

Elongate structures such as microtubules or filaments have no such intrinsic limitations (Fig. IV–1). Correspondingly, test-tube assembly produces varying lengths of tubules or filaments, depending on conditions. Yet, the cell can exert precise controls on the length of such structures, as is most evident from the regular arrays of filaments in striated muscle or the differences in lengths of flagella or cilia formed by different types of cells. In some cases, cellular boundaries such as the plasma membrane may be termination points for growth, but often no such obvious limiting devices are seen at the ends of filaments or microtubules.

Control of the length of microtubules and microfilaments might depend in part on balances among competing processes of assembly, disassembly, growth of preexisting structures, and initiation of new ones. When, for example, high concentrations of unpolymerized actin or tubulin are present and when conditions favor assembly, long filaments or tubules tend to form. Per-

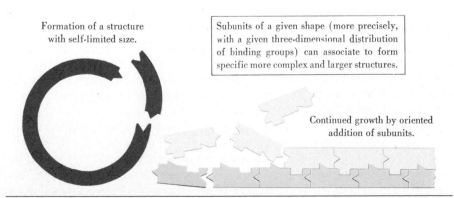

Formation of a structure with self-limited size.

Subunits of a given shape (more precisely, with a given three-dimensional distribution of binding groups) can associate to form specific more complex and larger structures.

Continued growth by oriented addition of subunits.

Figure IV–1 *Self-assembly* **of a self-limited structure and of a structure without an intrinsic size control.**

haps, however, if there also are many active initiation sites or many preexisting structures that can add more subunits to their ends (see Fig. IV–1), assembly of a larger number of shorter structures can be favored. More stringent controls probably require additional participants such as components of the membranes with which cytoskeletal structures associate, microtubule-associated proteins (Sections 2.10.1 and 2.10.5), and various of the proteins known to interact with actin. In the test tube, different proteins seem able to block growth of actin filaments at one or the other end (Section 2.11.2); the length of filaments formed in a given test-tube mixture depends on the particular proteins present along with actin, as well as on the concentration of actin molecules, ATP, and inorganic ions.

Tobacco Mosaic Virus (TMV)   Tobacco mosaic virus is a cylindrical structure of RNA and protein (Figs. III–2 and IV–2A). The protein subunits can be separated from the RNA molecule and made to assemble by themselves into cylindrical structures in the test tube. Under some conditions the cylinders formed are of a helical array of proteins resembling that in the intact virus, although of more variable length. Significantly, however, under conditions of pH and other ion concentrations like those found in the host tobacco cell, the proteins assemble to form a different array—flat discs that can then stack into short cylinders (Fig. IV–2). Under these same conditions, with both RNA and protein present, the native viral structure forms.

Detailed studies of viral assembly indicate that, early in the process, a disc of protein molecules binds to a specific region of the viral RNA (a looped zone stabilized by H-bonded base pairs; see Fig. IV–2). This binding results in a realignment of the proteins into a short helix; growth of the coat then continues by binding of additional discs to the RNA and by their realignment into helical form until the RNA has been covered with protein, whereupon addition of protein ceases.

These experiments show that TMV's RNA not only carries the genetic information specifying the structure of its coat protein molecules but also interacts with the protein molecules to assemble the viral structure. In a sense, then, the RNA both brings about its own duplication and specifies the three-dimensional ordering of itself and of the proteins for which it codes. The length of the assembled virus, for example, depends on the length of the RNA, which is an intrinsic, self-perpetuating property of the nucleic acid molecule. A better summary, perhaps, is that the RNA carries the underlying genetic information expressed in the assembled virus, while the assembly depends on mutual interaction or "cooperation" of the protein molecules and the RNA: The RNA helps to align the proteins, and the proteins both "recognize" the correct RNAs (thus, they do not make virus-like structures with other, cellular RNAs) and help the RNAs to orient into the helical strand found in the virus.

The important general conclusion is that cooperative interactions among different macromolecules can modify and control assembly, influencing overall length of structures, among other properties.

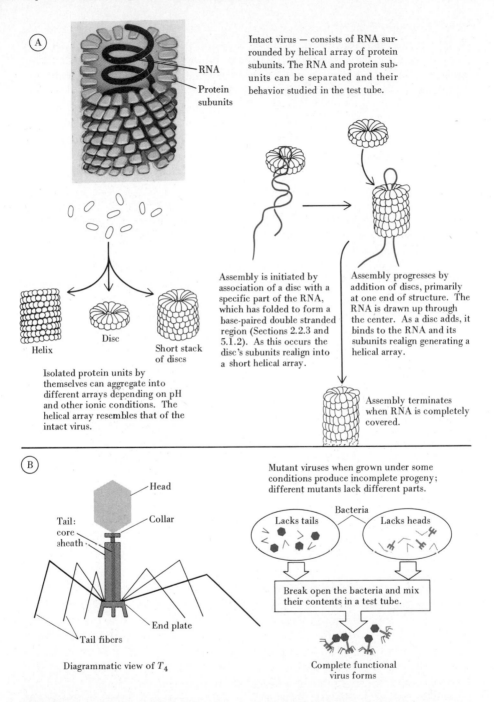

(A) Intact virus — consists of RNA surrounded by helical array of protein subunits. The RNA and protein subunits can be separated and their behavior studied in the test tube.

RNA
Protein subunits

Helix

Disc

Short stack of discs

Isolated protein units by themselves can aggregate into different arrays depending on pH and other ionic conditions. The helical array resembles that of the intact virus.

Assembly is initiated by association of a disc with a specific part of the RNA, which has folded to form a base-paired double stranded region (Sections 2.2.3 and 5.1.2). As this occurs the disc's subunits realign into a short helical array.

Assembly progresses by addition of discs, primarily at one end of structure. The RNA is drawn up through the center. As a disc adds, it binds to the RNA and its subunits realign generating a helical array.

Assembly terminates when RNA is completely covered.

(B)

Head
Collar
Tail: core sheath
End plate
Tail fibers

Diagrammatic view of T_4

Mutant viruses when grown under some conditions produce incomplete progeny; different mutants lack different parts.

Bacteria

Lacks tails

Lacks heads

Break open the bacteria and mix their contents in a test tube.

Complete functional virus forms

Scaffolds The protein arrangements of the heads of various bacteriophages (Figs. III–2 and IV–2) seem not to form by simple, spontaneous self-assembly. Rather, their formation depends on cooperative interactions of the head pro-

◀ **Figure IV–2** *Viral reassembly in the test tube* (see Fig. III–2 for electron micrographs of the viruses). A. Experiments with tobacco mosaic virus (Section 4.1.2). Under conditions like those prevailing in host cells the viral proteins assemble into discs but do not form long helical rods unless the RNA is also present. This, in effect avoids wasting the proteins through assembly into structures that cannot propagate. B. Experiments with T_4 bacteriophage (Section 4.1.3). (From the work of Frankel-Conrat, Kellenberger, Wood, Edgar, Klug and others. See Butler, P.J.G., and A. Klug, *Scientific American* **239**(5):62, 1978.)

teins with other protein molecules that are not included in the finished head. Once the proper structure has formed, the latter ("scaffold") proteins dissociate from it; in some viral strains, they are reused to form new "scaffolds" for additional heads, whereas in others, they probably are degraded.

The assembly of viral head structures also can depend on proteolytic cleavages rendering previously inactive proteins capable of proper association. As with the fibers discussed above, the cleavages presumably engender changes in conformation or removal of inhibitory segments.

4.1.3 Sequential Assembly

The Assembly of T_4; Temperature-sensitive Morphogenetic Mutants The assembly of the bacteriophage called T_4, which infects *E. coli,* illustrates how the processes discussed so far can contribute to formation of a complex structure. T_4 (Fig. IV–2B) possesses over 40 genes that are known to affect the structure of the virus or of viral parts. It is probably the most complex biological structure whose formation is understood in reasonable detail—in large measure because numerous mutants have been isolated, including a set of "temperature-sensitive" mutants. These grow normally at some temperatures but produce defective virus, which cannot propagate, when grown at a different (usually higher) temperature. The temperature sensitivity is quite useful since it is possible to find temperatures at which to grow large populations of viruses altered genetically in ways that are interesting but potentially fatal. The sensitivity arises from the fact that temperature changes can alter protein conformation with resultant magnification of defects in protein structure.

Some of the structure-affecting ("morphogenetic") temperature-sensitive mutants of T_4 fail to make heads. Others cannot make tails or parts of tails. If test-tube mixtures of incomplete viruses are put together so that all parts are present, complete, infectious viruses can form (Fig. IV–2). However, there are restrictions on test-tube assembly: For example, the collar, and usually the head, must be attached to the tail before the fibers will attach to the other end of the tail. (There seem to be elongate, fine, whisker-like extensions of the collar that help to align or to attach the fibers.)

From such experiments it has been concluded that in the normal course of events, heads, tails, and tail fibers assemble from their components as separate structures through processes of self-assembly, proteolytic modifications, and molecular cooperation, including scaffold formation. These processes and subsequent ones occur in sequences that lead to orderly, step-by-step assembly of the virus; the sequences are based on temporal controls of synthesis of viral components (Section 3.1.3) and on assembly restrictions like the ones revealed in the test-tube reconstitutions. The tail, for example, forms by binding of its outer proteins to a narrow cylindrical core, which assembles first, in association with a previously made baseplate (Fig. IV–2). Completed heads, tails, and collars associate spontaneously. Once this association has taken place, tail fibers can attach.

Still unsolved is how the DNA, which is many micrometers long, becomes packaged within a head with a diameter of only a fraction of a micrometer. For a number of viruses, including T_4, the DNA is synthesized in the form of very long molecules, each composed of multiple, repeated copies of the viral sequences. The DNA for an individual head is enzymatically cleaved from such a molecule. Apparently this cleavage takes place just before or simultaneously with the entry of the DNA into the head; the DNA then probably associates with head proteins that "collapse" it into a compact arrangement that can fit inside.

Some of the proteins specified by T_4's DNA affect assembly without either participating in the final structure or serving obvious scaffolding roles. Investigators of this situation suspect that enzymes or enzyme-like proteins help to produce specific alignment of parts and perhaps to catalyze formation of specific bonds that attach parts to one another. One such protein may link the fibers to the tails. If "morphogenetic" enzymes do exist, they add an additional level of complexity and of potential control that can help to explain how structures even more elaborate than T_4 may assemble.

Ribosomes Ribosomes probably also assemble through an obligatory sequence of steps (Sections 2.3.2 and 2.3.4). Test-tube reassembly experiments suggest that the RNAs and certain of the proteins associate first to form a "core" structure; the molecular arrangements established in this structure create sites for the specific addition of other proteins.

4.1.4 Membrane-bounded Organelles

As we have already done in Sections 2.5.4, 2.6.3, 2.7.4, and 2.8.4, we can make plausible guesses about major features of the assembly of membranes and of organelles delimited by membranes. A host of mechanisms are available, at least for conceptual constructs. Both free and bound polysomes can contribute, along with lipid-synthesizing enzymes in the ER, mitochondria, and plastids, saccharide-adding and remodeling systems in the ER and Golgi apparatus, and targeting and recognition devices that determine where particular molecules end up. Membrane movements and transformations almost certainly are involved, as are differential movements of particular molecules in the plane of the membrane and transport of lipid molecules through the cytosol by carrier proteins. Interactions of membrane components with one another and with adjacent cytoskeletal or extracellular materials help to determine mutual locations of molecules. Under some conditions at least, complex membranes can be assembled step by step (Sections 2.5.4 and 2.7.4), suggesting the operation of assembly sequences in which parts of structures (such as the F_1 particles of mitochondria or plastids?) form through partly independent processes and then link together.

The fact that membranes arise from preexisting membranes explains certain features of their assembly that are difficult to duplicate in the test tube. Consider, for example, the asymmetric distribution of membrane proteins and lipids (Section 2.1.2). When membranes are reconstituted in the test tube, both surfaces of the bilayer tend to be of similar composition, since there is nothing to guide molecules differentially to one side or the other; for similar reasons, it is often difficult to reconstitute the specific orientations of protein molecules characteristic of natural membranes. In the cell, of course, components are added in oriented fashion—for example, from ribosomes or enzymes restricted to one side or the other of the membrane. This is one of the ways in which preexisting structures can strongly affect their own growth. Another likely one is by the provision of specific sites to which new molecules can add: The preexisting molecules serve as a framework orienting the integration of incoming ones.

Three-dimensional Geometry We still have only shadowy notions about how the elaborate three-dimensional structure of a complete organelle like a mitochondrion or chloroplast arises and is maintained. Why are the mitochondrial cristae tubular in some tissues and flat, shelflike structures in others? What underlies the differences in thylakoid arrangement between algae and higher plants or between mesophyll and bundle sheath cells in C_4 plants (Section 2.7.5)?

Experimental attacks on such questions have begun. For instance, through manipulation of the concentration of Ca^{2+} and other ions, isolated chloroplast thylakoids can be made to associate into closely stacked arrays reminiscent of those in the cell or to dissociate from such arrays. Thylakoid stacking in the intact chloroplast is affected by mutations or other changes that alter the population of proteins and intramembrane particles present in the membranes. Thus, in the reasonably near future molecular analysis of at least some major features of chloroplast geometry should be forthcoming.

Rough ER generally has the form of flattened sacs, whereas smooth ER often is substantially more tubular in outline. Could these differences reflect the impact on membrane morphology of the ER proteins that participate in the binding of ribosomes?

Nematocysts of *Hydra* In Section 3.6.3 it was shown that the nematocysts of *Hydra* have a complicated structure that arises inside a vacuole originating from the Golgi saccules. A wide assortment of proteins and other substances is probably present inside the vacuole. From these, the capsule, lid, coiled thread, stylets, and barbs are fashioned. Each of these structured elements has a morphology that is specific for a given *Hydra* species; therefore they must be under genetic control. The structures develop at a time when the endoplasmic reticulum and Golgi apparatus are rapidly regressing, suggesting that protein synthesis has been turned off and that little, if anything, is being added to the vacuole.

An interesting possibility is that nematocyst structures within the vacuole form, like the T_4 virus, by the separate assembly of individual parts, with the presence of binding sites on each part for the attachment to other parts and by the sequential addition of parts to form a whole structure. Perhaps enzymes present in the vacuole are involved. This kind of assembly process has not been demonstrated for nematocysts, and the possibility cannot be excluded that more complex processes are involved. However, it would be interesting to attempt isolation of individual nematocysts at a time when protein synthesis by the cnidoblast is turned off and when visible differentiation of nematocyst structures has not yet progressed. If the notion of T_4-like assembly within a vacuole containing all necessary components has any merit, differentiation of structure might well proceed in isolated nematocysts. The next experimental steps would be the extraction of nematocyst proteins and attempts to reconstitute the complex structures in the test tube.

Chapter 4.2 Cell Division in Eucaryotes; Chromosomes

4.2.1 Chromosome Constancy and Mitosis: A Brief Review

That genes are arranged in linear order on chromosomes was one of the great biological discoveries of this century. It opened the route to a series of studies that have culminated in recent years in the analysis of heredity in terms of DNA base sequences. Some major conclusions emerging from these studies are especially relevant to the discussions in subsequent chapters.

1. In dividing and nondividing cells the nuclear DNA is located in the chromosomes. When division is not occurring, the chromosomes are usually unfolded, and the DNA participates in controlling cell structure and function. During division the chromosomes fold ("condense") into compact structures whose behavior accounts for transmission of DNA to daughter cells. Chromosomes of a dividing plant cell are shown in Figure IV–3. Note that the chromosomes are of different lengths and that recognizable specialized regions are visible. *Centromeres* (also called *primary constrictions*) are present in all chromosomes; a specialized structure, the *kinetochore* present within the centromere region (Fig. IV–21), functions in chromosome movement during cell division. (Note: The terms *centromere* and *kinetochore* often have been used interchangeably since both refer to the zone of the chromosome that associates with the spindle.) *Secondary con-*

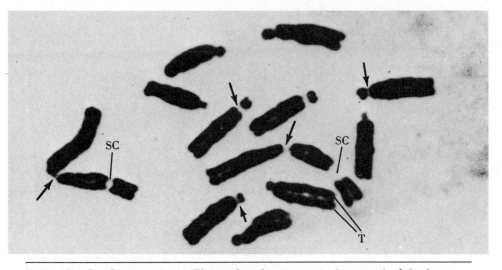

Figure IV–3 *Chromosomes.* The twelve chromosomes (six pairs) of the bean *Vicia* during mitotic metaphase (*see* Fig. IV–6). Each consists of two chromatids *(T)*, and each has a centromere (primary constriction) seen as a depression or lightly stained region of the chromosome (arrows). The chromosomes differ in length and position of the centromere. Two of the chromosomes, members of one pair, show a second "gap" (lightly stained region) along their lengths, the secondary constrictions *(SC)*. These are the nucleolar organizer sites, the places where the nucleolus is attached when it forms (*see* Fig. II–27). × 2000. (Courtesy of J.E. Trosko and S. Wolff.)

strictions are present only on a few chromosomes; they include the sites of nucleolar organizers. Each chromosome is divided longitudinally into two identical *chromatids;* this is characteristic of cells soon to enter division.

2. Figure IV–4 shows the human *karyotype* (chromosome complement). The chromosomes of one cell have been arranged to demonstrate the fact that they occur in pairs of similar *(homologous)* chromosomes, referred to as *homologous pairs.* Cells in which the chromosomes are present in such pairs are referred to as *diploid.* The pairs may differ considerably from one another in sizes of chromosomes and in positions of centromeres. Although unicellular organisms, algae and fungi differ from this pattern (see Section 4.3.1), usual multicellular animals or higher plants contain identical diploid sets of chromosomes in most of their cells. The major exceptions are the mature sex cells which are *haploid;* only one chromosome of each pair is present. Also, especially in plants, some cells may contain twice (or greater) the normal total number of chromosomes, with corresponding numbers of extra identical copies of each chromosome (polyploidy; Section 4.5.1).

3. Most organisms of a given species have the same karyotype. Different species differ in chromosome number, which varies from two to several

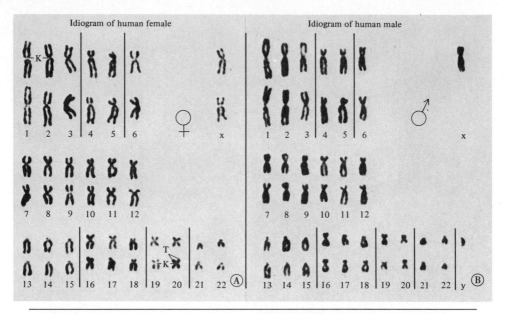

Figure IV–4 *Metaphase chromosomes (Fig. IV–6) from human cells.* The cells have been spread on a slide, stained, and photographed. The chromosomes have been cut from the photographs and arranged as an "idiogram" showing the 22 pairs of "autosomes" and the two "sex chromosomes" (in females, two X chromosomes; in males, one X and a smaller Y chromosome). *K* indicates centromeres, and *T,* chromatids. × 2000. (Courtesy of T. Puck and J.H. Tijo.)

hundred, and in chromosome morphology (lengths, positions of centromeres, and so forth).

4. A particular chromosome in a given species carries a particular set of genes (see Section 2.2.1) controlling the characteristics of that species. Nonhomologous chromosomes carry different genes. (Differences between homologous chromosomes will be discussed in Section 4.3.1.)

The constancy of chromosome type and number among the different cells within an organism is dependent on mitosis (see Fig. IV–6; Figs. IV–23, IV–24, and IV–29 are micrographs of cells in mitosis). The two chromatids of each chromosome are the result of chromosome duplication prior to division. Chromosome duplication includes replication of DNA and results in two chromatids containing identical copies of DNA. The chromatids are separated to opposite poles of the spindle; in consequence, both daughter cells possess one daughter chromosome derived from each chromosome of the parent cell. Both thus contain identical genetic information.

The constancy of chromosomal type and number within a species is based on reproductive mechanisms resulting in transmission of chromosomes from parent organism to offspring. For some species, mainly those of unicellular organisms, mitosis is the chief mode of reproduction. The special chro-

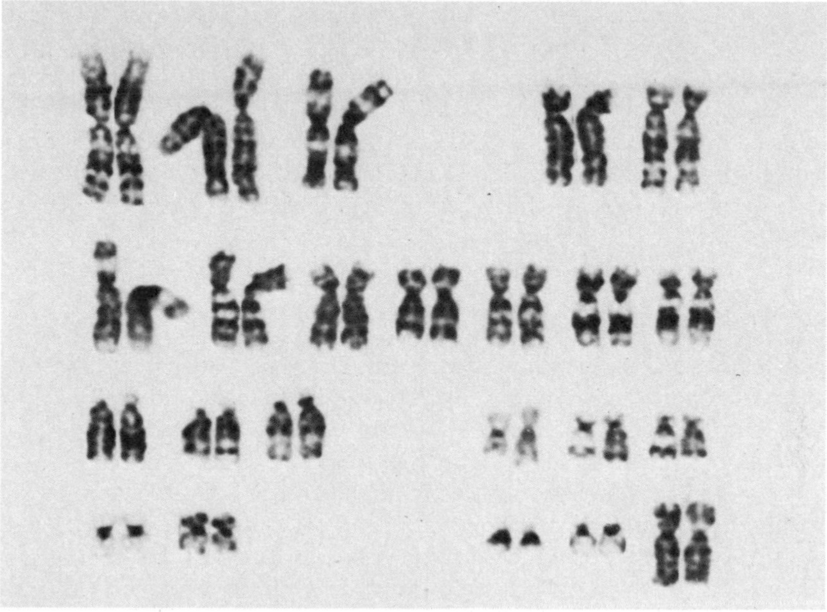

Figure IV–5 *Banding.* Human karyotype from a preparation stained by a method that produces a banded appearance of the chromosomes. The cells used were from a female; the two X chromosomes are at the lower right corner. The method used employs a mixture of acidic and basic dyes (Section 2.2.3) known as *Giemsa's mixture;* thus the banding is sometimes referred to as "G-banding." (Courtesy of T.C. Hsu.) With other staining mixtures and procedures the same chromosomes show different banding patterns: "Q-banding" is the pattern obtained with a category of fluorescent dyes called *quinacrines.* "C-banding" refers to methods that stain heterochromatin flanking the centromeres (Section 4.2.6). Hypotheses about the functional significance of banding are outlined in the text: proposals such as that a given G-band groups replicons that replicate in synchrony (Section 4.2.4) are now being tested.

mosome behavior involved in sexual reproduction will be discussed in Chapter 4.3.

Recently developed staining procedures utilizing colored dyes or fluorescent dyes result in characteristic banded appearances of the chromosomes (Fig. IV–5). The underlying mechanisms are incompletely understood. Banding demonstrated with some procedures seems to mirror patterns of heterochromatin and especially sites where very highly repetitive DNAs (Section 4.2.6) are concentrated. Other techniques, like the one used for Figure IV–5, may accentuate subtle structural differences that exist along the chromosomes, perhaps related to longitudinally arranged coiled regions called "chromomeres" that sometimes are observed (Fig. II–27 and Section 4.4.5). Alternatively, aspects of banding may be based on differences in the relative abundance of A-T as opposed to G-C base pairs, or in protein–DNA relations in different regions of the chromosome. The legend to Figure IV–5 gives some additional details.

Despite the uncertainties about the staining mechanisms, the techniques have proved very valuable for chromosome identification. The banding pattern for a given chromosome in the karyotype is constant from cell to cell, whereas chromosomes that may differ little in size or centromere position frequently are readily distinguishable on the basis of their banding (compare Figs. IV–4 and IV–5).

4.2.2 The Cell Life Cycle

The life cycle of dividing cells is usually divided into (1) *interphase* and (2) the mitotic division stages: *prophase, metaphase, anaphase,* and *telophase* (Fig. IV–6). Interphase was originally so named because it is the stage between successive mitotic divisions and usually shows few dramatic chromosomal changes that are readily recognizable in the microscope. In the earliest periods of microscopic work on cell division, interphase was regarded as a resting stage. However, since then it has become obvious that interphase is the period of intense cellular metabolic activity. Most cells spend most of their time in interphase.

DNA Synthesis; (G_1, S, and G_2) Cytochemical and autoradiographic evidence indicates that DNA synthesis (replication) in preparation for division occurs during the interphase prior to division. In interphase a period of DNA synthesis called S, a pre-S "gap" period called G_1, and a post-S predivision period called G_2 (Figs. IV–7 and IV–8) are recognized; these are distinguishable in experiments like the following.

A culture of cells of a given type is exposed briefly (for a few minutes to an hour) to tritiated thymidine (^{3}H-thymidine; Section 1.2.4) and then grown in nonradioactive medium. Only those cells actively synthesizing DNA during the brief exposure will incorporate label. If samples from the population are studied by autoradiograpic techniques at intervals thereafter, it is observed that the first cells to enter the stages of actual division—prophase and later mitotic stages—show *no* label in their chromosomes (Fig. IV–7). These cells were in G_2 at the time label was given; they had already completed DNA synthesis and so incorporated no radioactive thymidine. After a time the cells entering the

---

Figure IV–6 *Mitosis in an animal cell.* **The behavior of chromosomes is similar in virtually all eucaryotes, plants, and animals. Some of the other organelles vary in behavior as outlined in the text. See Figure IV–28 for one of the distinctive features of plant divisions. Once the two chromatids have separated after metaphase, it is a matter of terminological convenience as to when they should be referred to as daughter** *chromosomes.*

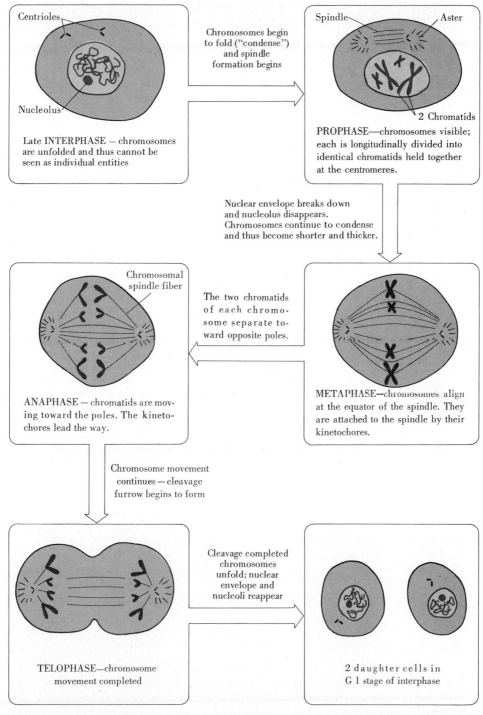

Centrioles

Nucleolus

Late INTERPHASE — chromosomes are unfolded and thus cannot be seen as individual entities

Chromosomes begin to fold ("condense") and spindle formation begins

Spindle

Aster

2 Chromatids

PROPHASE—chromosomes visible; each is longitudinally divided into identical chromatids held together at the centromeres.

Nuclear envelope breaks down and nucleolus disappears. Chromosomes continue to condense and thus become shorter and thicker.

Chromosomal spindle fiber

The two chromatids of each chromosome separate toward opposite poles.

ANAPHASE — chromatids are moving toward the poles. The kinetochores lead the way.

METAPHASE—chromosomes align at the equator of the spindle. They are attached to the spindle by their kinetochores.

Chromosome movement continues — cleavage furrow begins to form

Cleavage completed chromosomes unfold; nuclear envelope and nucleoli reappear

TELOPHASE—chromosome movement completed

2 daughter cells in G 1 stage of interphase

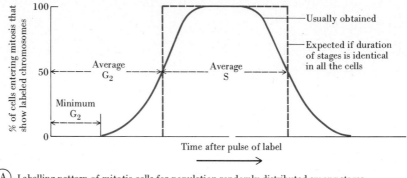

(A) Labelling pattern of mitotic cells for population randomly distributed among stages

Figure IV–7 *Analysis of cell division cycles.*

A. Curves like the one shown as a solid line are obtained when a population of dividing cells is studied by autoradiography at intervals after a relatively brief exposure to a radioactive label that enters their DNA. In the analysis, the cells are assumed to be all of the same type but randomly distributed in the different stages of the cell cycle. Note that what are counted in the autoradiographs are the cells that actually have *entered the stages of mitosis:* the curve shows the *proportion* of such cells *whose chromosomes are radioactively labeled* observed in samples taken at successive intervals. The *total number* of cells in mitosis is the *same* at all the time points.

The fact that for an initial time period all the cells in mitosis are *unlabeled* demonstrates the existence of a G_2 period: the first cells to divide are not labeled since they had already completed DNA synthesis but had not yet divided when the label was administered. The first *labeled* cells to divide are ones that had almost completed S when label was presented: thus the interval before they appear is a measure of the time it takes to pass from the end of S to the stages of actual division, in other words, a measure of G_2. If all cells in a population showed precisely the same length for each stage of the cycle, the curve expected would be that shown in dotted lines. The quite different curve shape actually obtained (solid line) demonstrates that the cells in the population vary somewhat in the duration of G_2, S and G_1. The time of appearance of the *first* radioactive cells in division stages is a measure of the *minimal* length of G_2; the time when 50% of the dividing cells are labeled is a measure of the *average* G_2 (remember that at the various points on the upward arm of the curve the *labeled cells* are those that *were near the end of* S at the time of exposure to label while the *unlabeled ones were in* G_2). As time progresses, a period is reached during which all the dividing cells are labeled since all were in S during the period of exposure to label. Then unlabeled cells again appear (the curve falls) as *cells that were in* G_1, pass through S and G_2 and begin to divide. They are unlabeled since the label was no longer available for incorporation into DNA by the time they entered S. (Part B on next page).

mitotic stages will begin to show label, as the cells that were synthesizing DNA when label was given pass through G_2 and go on to divide. The time between exposure to tritiated thymidine and the *first* appearance of labeled cells in division is a measure of G_2, since it represents the time taken by cells almost finished synthesizing DNA to proceed through the subsequent G_2 and reach the mitotic stages.

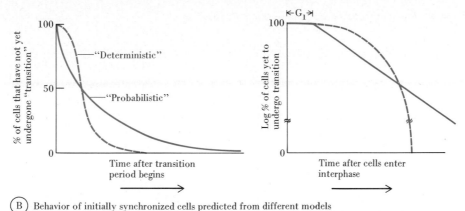

Figure IV–7B *Analysis of cell division cycles (continued).* Expectations derived from different models of the cell cycle, especially the "deterministic and probabilistic" ones outlined in Section 4.2.3. We include the curves not so much for the details but as an example of work in progress: the essential point is simply that different models of the cell cycle lead to different predicted behavior of cell populations that are potentially experimentally evaluable. The curves illustrate the predicted behavior of a cell population, all of whose members are *initially synchronized* at the same point in the cell cycle, as the cells undergo a major cell cycle change such as the transition from G_1 to S or from noncycling to cycling states. The curves at the left would be obtained by evaluating the percent of cells in the population that have *yet to undergo* the transition, at successive intervals, after the time when the period of transition begins. The dotted line would be expected if the cells are all moving through the cycle on the basis of a determining chain of events but vary somewhat in timing, more or less as in part A (previous page). The solid line would be expected from a probabilistic model in which the population behaves like a population of radioactive atoms undergoing radioactive decay; the cells make the transition as if governed by a random factor such that the *probability* of a given *proportion* of the population undergoing the change during a given time interval remains constant. Note from the curve that this latter model predicts, for example, that there should be some cells that are delayed almost indefinitely in making the transition. The "probability model" curve is in fact a simple exponential curve which, mathematically, never reaches 0.

The right-hand curves concern the same two models but graph the expected data in a different way—actual studies of cells tend to use this mode of analysis since the differences in expectations are particularly simple to visualize clearly and to test. The percentage of cells yet to undergo transition is indicated in logarithmic terms, with the starting point for the measurement being the *beginning of interphase.* The initial flat part of each curve describes the time before any cells undergo transition (for example, the G_1 period preceding the G_1-to-S transition). The declining portions illustrate the same type of data as the left-hand curves, but because the data are expressed here in terms of *logarithms,* the curves are shaped differently which is the point the illustration is intended to convey; note that the probability model predicts a straight-line decline while the deterministic model predicts a curved decline. (After Smith, J.A., and L. Martin, *Proc. Nat. Acad. Sci.* USA **70**:1263–1267, 1973; Smith and Martin favor probabilistic models on the basis of their studies of dividing tissue culture cell populations.) Solid line; probabilistic model; dashed: deterministic.

Comparable analyses are being carried out for a variety of population phenomena other than cell division. Section 2.8.3, for example, mentions that steady-state turnover of many macromolecules follows a time course like the probability model curves (though without the initial flat "lag" portion in the logarithmic version of the curve, since turnover is a continuous process).

This time is the *minimal* G_2 characteristic, of those members of the population that passed through this period most rapidly. However, even cells of a single type vary somewhat in the duration of the different stages of the cell cycle. The *average* G_2 period is measured by the time it takes after exposure to label for 50 percent of the cells entering division to show label in their chromosomes (Fig. IV–7).

Similar approaches are used to determine the duration of the other interphase stages. For example, under some circumstances the percentage of cells found in autoradiographic studies to incorporate ^{3}H-thymidine during a very brief exposure gives a rough measure of the length of the S phase as compared to the other cycle stages. This is so because the probability of any given cell's being in S when the label is presented depends directly on the percentage of the total cycle represented by S. (Note that these techniques assume that the population being studied is reasonably homogeneous in terms of the duration of the different stages of the cell cycle, and that the cells are distributed at random among the stages at any given moment—some are in mitosis, some in G_2, some in S, and some in G_1. Such assumptions generally hold best for unicellular organisms and tissue-culture populations.)

Other Aspects of the Cycle Not only DNA synthesis, and related processes such as the accumulation of nucleic acid precursors, but also many other cellular activities show regular variations during the different stages of the cycle. For example, RNA synthesis and protein synthesis decline sharply during mitosis; by metaphase, very little RNA is being made, and rates of protein synthesis are dramatically reduced. These changes correlate with structural changes—especially the folding of the chromosomes, the dispersal of the nucleolus, and evidently, the dissociation of many polysomes, both free and bound, into separate mRNAs and ribosomes.

Note that RNA synthesis continues at high rates throughout S. The actual duplication of a given region of DNA requires only a brief period during S (Section 4.2.5), and only during this time is this portion of the DNA effectively precluded from the possibility of being transcribed (see also Section 3.3.2). (Of course, whether or not a given stretch of DNA actually is transcribed depends on the regulatory mechanisms governing gene expression [Section 4.4.6].)

The synthesis of proteins important for division, like the tubulins and histones, shows cyclical variations during the cell cycle. Histones, for example, are made primarily during the S period, although some limited synthesis occurs at other times in certain cell types and circumstances. Cyclical changes in regulatory molecules such as cyclic AMP and cyclic GMP are also observed, and corresponding variations in phosphorylation of tubulins, histones, and other proteins are found; the details differ somewhat for different cell types and have yet to be definitively interpreted.

Noncycling Cells Most cells and cell types that stop dividing, permanently or temporarily, as in density-dependent inhibition (Section 3.11.1), do so when

	Interphase	Mitosis (M)
Sea urchin egg cleavage		
at 2 cell stage	0.25-0.5 hours	0.5 hours
at 200 cell stage	2	0.5
Plant root meristems	16-30	1-3
Cultured animal cells	12-24	0.5-2
Mouse intestinal epithelium	12-24	1

	G_1	S	G_2
Mouse intestinal epithelium	9	7	1-5
Cultured mouse fibroblasts	6	8	5
Bean root meristems	9-12	6-8	4-8
Ameba proteus	undetectable probably no G_1	3-6	30

Figure IV–8 *Approximate rates of division and durations of cell cycle stages* **for some rapidly dividing cells. For a given cell type, G_1 is the most variable period. (Data from the original observations of Howard and Pelc and from discussions from Mazia and others.)**

diploid, before entering S. (There are some exceptions: See Sections 4.4.4 and 4.5.1). This led to the concept of a special G_0 state that cells supposedly entered as an alternative to G_1. It is more fashionable now to speak of nondividing cells as "noncycling" or as "arrested" in G_1, since it is by no means clear that there is a unique state into which nondividing cells enter. There may instead be several alternative ways in which cells can move out of and into the division cycle.

More generally, the division cycle involves so many coordinated cell activities that cell biologists have yet to accomplish the experimental "teasing out" of the mechanisms regulating any one of these: the initiation of DNA synthesis, changes in the coiling of the chromosomes, the formation of the spindle, and so forth.

4.2.3   Rates and Controls of Division

Rates of division vary considerably among cell types (Fig. IV–8). The rates are dependent on conditions of growth. For example, *Ameba proteus* at 23° C divides every 36 to 40 hours, but if the temperature is lowered to 17° C, this rate slows to one division every 48 to 55 hours. Few eucaryotic cells divide as rapidly as bacterial cells; a bacterial population may double in cell number and mass every 15 to 20 minutes. Yeast cells can divide every 2 hours or so under optimal growth conditions. Early cleavages of developing eggs are relatively rapid because no growth occurs between divisions; the egg is separated into smaller and smaller cells (see Fig. IV–37). Such cleavages can occur at rates of one per hour to several per hour; the G_1 period of the cell cycle is short or even absent, and DNA synthesis is unusually rapid. Very fast divisions are found in some insect eggs where only the nuclei divide in the earliest stages of development (see Section 4.4.3 and Fig. IV–10). Cells of plant root tips or mammalian cells in culture can double in mass and divide every 12 to 24 hours. Most other cells divide much less frequently, and some highly specialized cells, such as mature neurons, do not divide at all.

The time it takes for a dividing population of unicellular organisms or tissue-culture cells to double in number is called the "doubling time" or, more often, the "generation time" for that cell type. By extension, the terms are sometimes used to refer to the average time required for a cell type to complete the cell cycle.

Wound Healing: Regeneration　If 70 percent of a rat's liver is removed, the remaining cells, which normally divide very slowly, begin to divide at rates exceeding cancers of rat liver. When the original mass is regained, cell division slows to the original rate. Similarly, when the skin is wounded, epithelial cells, fibroblasts, and endothelial cells of capillaries begin proliferation, providing the cell populations needed to heal the wounds. Along with the experiments already described for tissue-culture cells (Section 3.11.1), such observations make it clear that cell division is subject to controls exerted by the cell's "environment." On the other hand, both the fact that cells do have characteristic division rates and observations that division behavior can be altered at the genetic level (Chap. 3.12) dramatize the operation of control mechanisms internal to each cell. In some manner, the interplay of internal and external cues determines how a particular cell behaves.

Division and Differentiation　It is often said that the more complex or specialized a cell becomes in its morphology and metabolism, the more unlikely it is that the cell will divide; similarly, cells preparing to divide supposedly will not make specialized products. Such statements are oversimplifications but do point to interesting facts. We have already described cases in which organs maintain stem populations of rapidly dividing cells, some of whose daughters go on to differentiate into highly specialized cell types that divide infrequently, if at all, while others remain behind as reservoirs for future divisions. This phenomenon is observed in epithelial tissues such as the intestine or skin (Section 3.5.1), in plant shoots and roots, where division is confined to meristems (Section 3.4.1), and in the bone marrow and spleen, where blood cell precursors are produced by division of less specialized stem cells.

When embryonic blood-forming tissue, muscle cells, or cartilage cells are placed in tissue culture under suitable conditions, they often show an initial stage of rapid proliferation. This stage is followed by a period when multicellular aggregates form (the muscle cells fuse; see Section 3.11.2). In these aggregates cell division slows or stops. It is only then that large amounts of specialized tissue products are synthesized: hemoglobin, muscle proteins, or cartilage matrix. These observations are among a set suggesting that the program of events that leads to differentiation includes cell division, in the sense that cells subjected to influences that commit them to differentiate along particular lines still tend to or must undergo some additional divisions before such commitment is expressed as the production of specialized molecules.

The equivocal language used in the last sentence emphasizes that there are no hard and fast rules summarizing convincingly the relations of differentiation and division. Some complex and specialized cells divide rarely; neurons and red blood cells are the extreme examples. It makes evolutionary "sense" that mature neurons not divide, since they must maintain precise patterns of synaptic connections to other cells. However, protozoa also are complex and can have very specialized morphology, yet they divide rapidly. In addition, there are tumors and other proliferating cell lines that make specialized products (Section 3.11.1). The essential conclusion from the mass of information summarized here and in Section 3.11.2 and Chapter 4.4 is that cell growth, division, and specialization are under complex coordinated controls whose elucidation will demand detailed information about genetic and other regulatory mechanisms. The controls produce different behavior for different cell types under similar conditions and for a given cell type under different conditions. They result, for example, in the fact that organs and tissues show the same organization and size range in different individuals of a given species and in the fact that many cells that undergo repeated divisions almost precisely double their mass during the interphase prior to division. (As mentioned above, however, cells in early cleavages of many developing embryos do not grow at all between divisions.) The controls permit some normally quiescent cells to be "activated" when needed for repair of tissue damage—an important feature. Perhaps the key point to bear in mind is that words like *specialized* or *differentiated* are not very precise; they actually are shorthand designations for clusters of characteristics of which only some are well defined. Certainly, in common-sense terms, cells that produce particular products, such as hemoglobin, or assume extraordinary architecture, such as neurons, are obviously specialized. Nonetheless, although cell division is not a similarly unique property, it is hardly absurd to cite plant meristems or the crypts of the intestine as examples of tissues specialized for continued rapid division.

For some time developmental biologists have been intrigued by the finding that growth of cultured muscle or cartilage cells in the compound *bromouracil deoxyriboside* (BUdR) can sometimes inhibit the cells from synthesizing their characteristic products (cartilage matrix or abundant contractile proteins) while permitting them to continue making those molecules needed for survival and division. BUdR is incorporated in replicating DNA in place of nucleotide *thymidine,* but beyond this the basis of its seeming ability selectively to shut off expression of "differentiated" cell characteristics is not known.

We shall have more to say about differentiation in Chapter 4.4.

External Controls; "Mitogens" Experimentally, it is easy to demonstrate that various cell types can be stimulated to divide by agents in their surroundings. The most obvious example is the fertilization of an egg cell (Section 3.10.3). Plant hormones (Section 3.4.5) and "factors," primarily polypeptide hormones, in blood serum (Section 3.11.1) have "mitogenic" effects on suitable "target" cell populations in tissue culture; they can induce division of previously nondividing cells. Similar effects occur in the living organism. In their normal life history, B-lymphocytes are stimulated to divide by interactions with antigens and T-cells, initiated at the cell surface (Section 3.6.9). The proliferation of fibroblasts, endothelial cells, and other cell types seen in wound healing is stimulated in part by hormone-like molecules released from damaged tissues or platelets. It cannot, however, be assumed that all experimental mitogens have mitogenic functions in nature. Thus, certain of the lectins (Section 2.1.7) isolated from plant tissues can stimulate division of specific animal cells even though this seems unlikely to be their natural role. T-lymphocytes, for example, which normally are mitotically quiescent when obtained from the blood stream, can be stimulated to synthesize DNA and divide when exposed to the lectin *concanavalin A,* which binds to mannose units (Figs. II–44 and II–45) in saccharide side chains of glycoproteins or glycolipids; the *phytohemagglutinin,* from beans, whose name derives from its ability to bind to red blood cell surfaces and thus produce cell clumping ("agglutination") has similar effects.

As with these lectins, the effects of many mitogens depend on interactions of the stimulating agents with particular cell-surface molecules, such as receptors specific for given hormones. This could tie in with the fact that cell-surface changes often are detectable in cells showing altered growth control (Section 3.12.1). The division cycle itself involves cyclical alterations in cell-surface properties and also in cell shape. The most obvious of the latter changes is the rounding up of cells, even those normally flattened against the surface of a tissue-culture vessel, which takes place during mitosis. The changes in shape reflect the reorganization of the cell's cytoskeletal and intracellular motility system in preparation for chromosome movement and cytoplasmic division (Sections 4.2.7 and 4.2.9), plus changes in adhesion to extracellular materials.

In most cases mitogens probably act by means of "second" messengers (Section 2.1.4) or other intermediaries. Changes in the distribution of receptors, such as capping (Figure II–8), internalization of mitogenic molecules by endocytosis, alterations in Ca^{2+} or in cyclic AMP and in cyclic GMP, and a wealth of other potentially significant phenomena may ensue when mitogenic agents bind to cell surfaces. For example, in a fair number of cell types in culture, though not in all, effective mitogens provoke a selective, temporary rise in levels of cyclic GMP while inhibiting production of cyclic AMP. (Later on, cyclic AMP levels rise again just before or just as the cell enters the S period.) Section 3.10.3 mentioned the changes in cytoplasmic pH that precede the onset of cleavage in newly fertilized eggs. No doubt, changes of these types do control phases of the preparation of the cell for division such as cytoskeletal reorganization and the activation of enzyme activities and synthetic pathways. However, the details have yet to be traced out, and most importantly, which one or which set of changes triggers the crucial events leading to DNA synthesis and, later, to mitosis is not known definitively for any cell type. It remains quite possible that the effective changes differ among cell types and that searching for the universal trigger will be fruitless.

In fact, although we have stressed *stimulation* or *triggering* of cell division in previously quiescent cells, an alternative viewpoint formulates matters in terms of the release of *inhibition* or the reversal of "arrest." Some workers have elaborated theories that argue that cell division in mitotically inactive multicellular organs and tissues is normally inhibited by the presence of specific—mostly hypothetical—agents, sometimes called "chalones." Aspects of the cell proliferation that results from wounding, for example, could reflect a "disinhibition" resulting either from the release of a counteracting agent by the wounded tissue or from wound-induced declines in levels of the inhibitors. Although the specifics of this viewpoint remain controversial, it should be recalled that inhibitory effects on cell growth and proliferation are readily observed in tissue culture (Section 3.11.1) and have yet to be adequately explained. Furthermore, yeast cells release mating hormones of which one effect is the arrest of the division cycle in cells of the opposite "sex" at a point prior to the replication of DNA; this arrest ensures that upon cell fusion in sexual reproduction, the nuclei of the two partners are both in the pre-DNA duplication state appropriate for zygote formation and the mitotic divisions that follow.

Whatever the links are between mitogens arriving at the cell surface and the subsequent onset of DNA synthesis and division, they require time and synthetic activities by the cell. At least several hours usually pass before mitogen-induced DNA synthesis begins, and during this period synthesis of both RNA and protein generally takes place. As might be expected, the "re-entry into the proliferation cycle," or the "reversal of arrest in G_1 or G_0," to use current parlance, requires substantial reorganization of the cell's activities.

Transitions and Clocks: Deterministic or Probabilistic? Although external agents can markedly affect rates of cell duplication, they must operate in a context of intracellular controls. The latter presumably account for many of the characteristic differences in timing, rates, and responses among different cell types and for the fact that genetic changes alter division behavior as described in Chapter 3.12 and later in this section. A related conceptual issue of much concern at present is whether the transitions from one stage to another in the cell cycle, and among proliferating, nonproliferating, and "differentiated" states, are best portrayed in the traditional "deterministic" terms. That is, we tend to think of cells as being locked into series of events that, in the absence of outside interference, inevitably propel them from one stage or state to the next. Alternative theoretical positions emerge in part from frustration with efforts to produce adequate models for the internal "clocks" governing cell division rates. Their proponents emphasize that even with the most homogeneous cell populations, such as continuous cell lines (Section 3.11.1), there is variation among the cells in generation time and in the duration of stages such as G_1. They do accept that some transitions are more or less rigidly determined—generally, once cells enter S, only major disruptions prevent them from going on to complete S, G_2, and mitosis. Other transitions, however, especially the one from G_1 to S, may be better described in "probabilistic" terms. In fact, the equations and curves that best fit the behavior of some cell populations passing through these transitions resemble the sort called "first order," like the ones applicable to radioactive decay (see Fig. IV–7). The population as a whole shows characteristic timing, different from other cell types, and the rates of division do respond to "environmental" triggers such as growth factors. Within this overall framework, however, individual cells seem governed by apparently random probability factors. (For the mathematically inclined: One way of interpreting the data is to regard the probabilities as being reflected in the *proportion* of cells undergoing the transition of interest in a given time interval. During the overall period when the population is undergoing the transition, this proportion [probability] remains more or less constant. The result, however, is that the *number* of cells undergoing the change per time interval declines, exponentially: As more and more cells complete the transition, there are fewer left to do so; hence, a constant proportionality results in a declining absolute number.)

The crux of the problem is that in many cases it is only for large populations that some of the details of the cell cycle can be predicted with reasonable accuracy. The duration of G_1 is particularly variable within a given population. Different individual cells or small subpopulations show variation in G_1 without any obvious external cause. Does this mean that there are truly random internal factors influencing how particular cells behave, or merely that their behavior really is fully determined by well-defined chains of causes and effects but that we lack the methods to detect this? Such questions are familiar ones to students of populations of all sorts—sociologists and atomic physicists as well as cell biologists. Much more analysis of the governance of cell division is needed before even a tentative answer can be made.

Different kinds of models for "clocks" governing cell division pertain to different positions in these disputes. As with much other research on cellular "time-telling" (Sections 3.1.3, 3.4.5, and 3.8.1), the models are entirely hypothetical and are intended mainly to focus thought rather than as literal proposals for real phenomena. A "deterministic model" might view control of the timing of the cell cycle as governed by the transcription of particular species of RNAs from corresponding DNAs. The rate of transcription is fixed by the properties of RNA polymerases: In eucaryotes, growing RNAs elongate at rates of a few tens of nucleotides per second. For a cell type with a generation time of 24 hours, there would be a controlling DNA sequence, a bit more than 4 million base pairs long, which is transcribed into RNA at about 50 nucleotides per second. At the start of each timing cycle, transcription begins at one end of the DNA sequence and progresses toward the other end, initiating production of a series of RNA transcripts that serve to control key events in the nucleus or cytoplasm. The sequential production of these RNAs and of the proteins for which they code leads the cell to pass through the several stages of the cell cycle (as with viral replication? [see Section 3.12.3]). Once transcription is completed, 24 hours after it began, it is either reinitiated or not, depending on extracellular and intracellular cues that determine whether the cell should again prepare to divide. Cells with different generation times would differ in the length of the controlling DNA sequences.

"Probabilistic" models might instead attribute passage through a key transition, such as past an "arrest" point in G_1, to the achievement of threshold levels of cytoplasmic regulatory molecules. The amounts of such molecules might fluctuate continually on the basis of competing processes occurring at varying rates, some tending to increase the amounts and some to decrease them. Thus, at any given time cells at comparable points in the cell cycle might differ substantially in the levels of the controlling molecules present, depending on complex details of their recent metabolic history. Some cells will have reached the threshold needed to move on; in others the levels will be increasing or decreasing; and still others may fall "victim" to influences that move them out of the cycle altogether before the threshold is reached.

Cell division cycles in a number of unicellular organisms are roughly circadian (Section 3.8.1), but this clearly is not the case in many of the cells we have been discussing. This does not mean, however, that the "clocks" governing the division cycle and those governing circadian properties of cells are unrelated; so little is known of the actual timing mechanisms that no possibility can yet be ruled out.

Coordination; Synchronization; Mutation; Nucleocytoplasmic Interaction The control of cell division and the coordination of its many steps and facets has been studied in a variety of ways and in a corresponding variety of cells. Section 3.3.3 outlined the pioneering, though inconclusive, investigations on amebae concerning the relations of surface-to-volume ratios and nucleocytoplasmic ratios to the triggering of division. Egg cells have proved useful for work on many aspects of cell division, since they are large and relatively easy to manipulate experimentally, and because they undergo precisely timed series of divisions that can be initiated at will (by fertilization or parthenogenetic activation). Recently, for example, developing amphibian embryos have been found to contain cytoplasmic molecules that induce the dissolution of the nuclear envelope when injected into oöcytes. This activity, demonstrable in cytoplasmic extracts, varies cyclically during the cleavages of the egg, indicating that the molecules may be responsible for controlling the dissolution of nuclear envelopes that normally takes place in prophase of cell division (Section 4.2.10). With tissue-culture cells subjected to metabolic inhibitors it can be shown that if synthesis of RNA or of protein is prevented, the cell cannot pass from G_1 to S, suggesting that specific new macromolecules are required for this transition.

Division of cells in culture and of unicellular organisms can be reversibly prevented by exposure to abrupt temperature changes, or to certain metabolic inhibitors interfering with nucleic acid metabolism, or to colchicine (Section 2.10.5), which disrupts the mitotic spindle. These treatments prevent cells from passing through one or another critical stage in the cycle necessary for initiation or completion of mitosis. If a population of cells is treated for a time period long enough to permit most cells to reach the same stage and to stop there, and if the conditions preventing division are then reversed, synchronous mitosis often occurs, and synchrony persists for a few cycles. Over 90 percent of the cells may pass through the several stages of the cell cycle in synchrony and divide at the same time. In the absence of external synchronizing factors, dividing cell populations tend to be asynchronous; at any given moment they contain cells in all stages of the cycle (early cleavages of a developing embryo being among the few exceptions). The availability of methods of artificial synchronization is an important asset for biochemical studies of division stages, since these techniques provide homogeneous populations for analysis. It should be possible to investigate the accumulation and utilization during the division cycle of components such as cyclic nucleotides, Ca^{2+}, ATP, specific enzymes, or spindle proteins, and thus eventually to identify key control points.

Another promising approach is the study of the interaction of nucleus and cytoplasm by the direct experimental alteration of the cytoplasmic environment surrounding a given nucleus. The hope in such studies is to identify critical activating or inhibiting agents or balances of metabolites responsible for the remarkable coordination of nucleus and cytoplasm in the division cycle. In several different situations, DNA synthesis is induced in nuclei that ordinarily do not synthesize DNA, by exposure to the cytoplasm of cells preparing for division (this is true with adult toad brain nuclei transplanted into toad eggs, nuclei transplanted from amebae in G_2 to amebae in S, and hybrid cells such as the ones formed by fusing HeLa cells with chicken red blood cells [Section 3.11.3]). The folding of interphase chromosomes into more condensed forms like those of mitosis can be induced in S-phase nuclei by fusion with cells in mitosis. As with the cleaving amphibian eggs mentioned above, such observations demonstrate the existence of cytoplasmic molecules that are capable of moving to the nucleus and exerting profound effects there, and that show cyclical variations during the cell cycle. Efforts are under way to identify the specific factors that are involved. Most likely they include enzymes, gene-activating proteins, and other proteins transferred from the cytoplasm to the nucleus.

Mutations disrupting cell division are difficult to study because ordinarily they prevent the obtaining of large populations of affected cells. However, analysis of cell division is now being aided by "temperature-sensitive" mutants (Section 4.1.3) blocked at one stage or another of the cell cycle. With bacteria, study of such mutants has led to the conclusion that specific proteins must be synthesized to initiate DNA synthesis. The mechanism regulating the actual division of the bacterial cell into two normally seems tied closely to, or even triggered by, the completion of chromosome duplication, but the ties are disrupted in some mutants, so that offspring "cells" lacking chromosomes may be formed. For eucaryotes, studies have begun with yeast and with algae. Unicellular yeasts of the genus *Saccharomyces* normally duplicate by mitosis, coupled with the budding of a small daughter cell. Like cells of most eucaryotes, they tend to divide at a characteristic size and thus slow their division rate when growth is slowed by limitation of nutrients. There are "wee" mutants, however, that divide and function at smaller sizes than normal; these are being studied for hints about the mechanisms that normally coordinate growth and division. In other yeast mutants, much of the process of bud formation (but not final separation of the bud) occurs despite the absence of DNA replication and related steps in nuclear duplication. Conversely, there are mutants that duplicate their nuclei without forming buds. It may be inferred that nuclear duplication and the cytoplasmic events of bud formation occur through independent sequences of processes, the two sequences being coordinated at their beginnings by controlling steps on which both depend (one of these may be tied to the replication of the nuclear envelope plaques responsible for spindle formation; see Section 4.2.7) and at their ends by controls that ensure that the cytoplasm does not complete its division if the nucleus has not divided.

For cells having well-defined centrosomal regions, such as cells with centrioles, the cycles of centriolar and centrosomal duplication and migration are among the ones now being looked at anew, particularly intensively. Because centrosomes and centrioles and related microtubules can participate in many cellular mechanisms and do take part in organizing the apparatus for cell division, investigators hope they may prove to have experimentally accessible regulatory or coordinating functions in the cell cycle.

Chapter 2.2 mentioned that even if deprived of their nucleus, some eggs can be induced to cleave a number of times. This observation as well as others suggests a certain degree of autonomy of cytoplasmic cycles from nuclear ones in early embryonic development (see also Section 4.4.3). Bear in mind, however, that the situation is a "simplified" one in that no cytoplasmic growth occurs between cleavages.

4.2.4 Replication of Chromosomal DNA

Basic Events A *molecule* or *duplex* of DNA is a *double helix*, composed of two *polynucleotide strands*. DNA duplicates by a "semiconservative" mechanism (Figs. II–15 and IV–9). The double helix separates into its two polynucleotide strands. Each strand remains intact and acquires a new complementary partner that is formed by the sequential alignment of nucleotides along the old strand by base-pairing and by the enzymatic linking of the nucleotides into a polynucleotide chain. Figure IV–10A and B are micrographs showing replicating DNA.

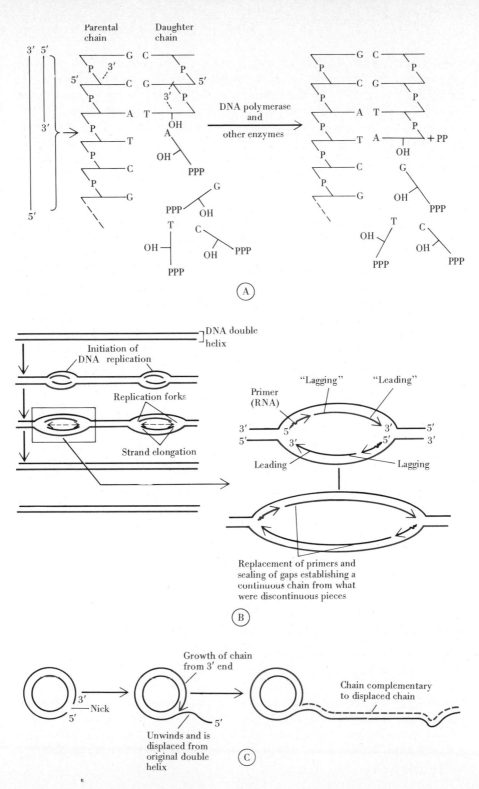

◄**Figure IV–9** *Replication of DNA.*

A. This set of diagrams illustrates a new DNA chain growing as the complementary partner of a parental chain.

Nucleotides are schematized here by a conventional skeletal structure (compare with Fig. II–15) in which the base is indicated by a letter, the 5-carbon backbone of the sugar by a straight line (the positions of carbons 5' and 3' are shown in red in the diagram), and the phosphodiester bonds that connect successive nucleotides by -P-. The free nucleotides employed for incorporation into DNA are in triphosphate form, hence the *PPP* in the diagram. All DNA chains grow in the 5'→3' (5' to 3'; 5'—3') direction, that is, as if they start at the 5' end (Section 2.2.1) and add nucleotides sequentially to their 3' end. As nucleotides are added, the most recently added one has a hydroxyl (OH) group available on its 3' carbon. As the next nucleotide to add aligns by base pairing with the parental chain, it reacts with this group, releasing two phosphates (still linked together, as *pyrophosphate*) and establishing a 3'—5' phosphodiester bond; the OH group on the 3' carbon of the newly added nucleotide is now available for a similar reaction with the next nucleotide.

The base sequence of nucleotides in the growing chain depends, of course, on base pairing so that the parental chain specifies the sequence of the daughter. The directionality of growth is imposed by the properties of the DNA polymerase enzymes involved, which cannot add nucleotides in the 3' to 5' direction.

B. *Replication of DNA in procaryotes and eucaryotes is bidirectional* and, in eucaryotes, can be initiated simultaneously at many points along a given double helix. At each replication fork, both daughter chains grow in the 5' to 3' direction as described in the text (Section 4.2.4). Note that one end of a given growing DNA chain is the "lagging" strand at one replication fork whereas the other end of the same chain is the "leading" strand at the other replication fork of the same replicating unit.

C. A *"rolling circle"* mechanism thought to apply to replication of certain DNA molecules. When a nick is introduced into one chain of a circular duplex (double helix), nucleotides can add to the 3' end of the nicked chain as in the ordinary replication outlined above. Such growth produces a new partner complementary to the intact, parental chain of the original duplex; this involves the unwinding of the growing chain at its 5' end and the progressive displacement of that end from the circle. Continued growth and displacement can separate the original nicked parent completely from its old partner and can continue on to generate a "tail" containing multiple copies of the original DNA sequence linked end-to-end in "tandem" (head to tail) array. (The displaced strand of DNA eventually forms a complementary partner and thus generates a double helix). It is as if the intact strand of the original duplex rotates ("rolls") during its replication with each complete revolution producing a complementary partner that will be displaced into the extending tail during the next revolution.

Progress is being made in understanding of the enzymatic details of DNA replication, but important gaps remain. Especially because of the availability of temperature-sensitive mutants, in which one or another of the many participating enzymes and other proteins is abnormal, more is known for procaryotes and viruses than for eucaryotes.

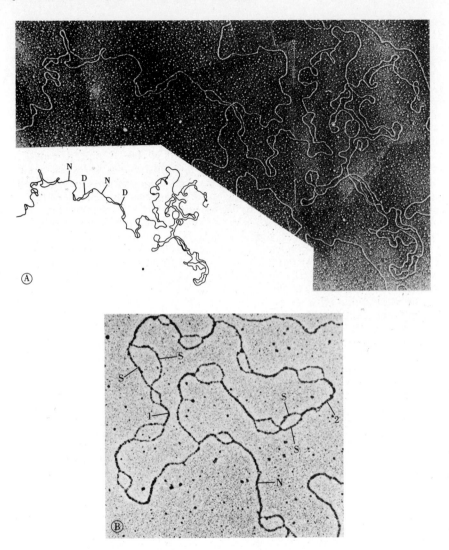

As presently conceived, the duplication of a DNA duplex can be summarized as follows:

1. Replication begins when *binding proteins* bind to each of the polynucleotide strands and *unwinding* enzymes (some have been called "helicases") commence to separate the strands through ATP-dependent processes.

2. Since the two strands of the duplex are mutually coiled around one another, the separation of the two requires that they be able to swivel around each other; otherwise the entire double helix would have to rotate repeatedly or, for circular DNAs, twist up into an increasingly tight series of "su-

◀**Figure IV–10** *Replicating DNA* isolated early in the cleavage stage of *Drosophila* embryos, prepared by shadowing techniques (Fig. I–11). The cleavages are very rapid (Section 4.2.3; interphase may last less than 5 minutes), and the corresponding rapid duplication of DNA facilitates study of replication.

A. Replicating units of DNA arranged as would be predicted from biochemical studies of replication (see Fig. IV–9). Stretches of replicated DNA *(D)* alternate with stretches not yet replicated *(N)*. (The sketch at the lower left is a tracing of the photograph.)

B. One of the already replicated regions from a molecule like that in A. *N* is a portion of double helix not yet replicated; *1* and *2* are the two daughter double helices in a region similar to one of the D regions in panel A. This molecule has been *partially denatured* by treatment with formamide. As in the present case, it is observed that some treatments can bring about local separation of the double helix into its two constituent strands (S), while adjacent regions of the helix retain their double-stranded configuration. The regions that separate are those relatively rich in A-T base pairs (see Fig. II–15); A and T bind to one another by two hydrogen bonds, in contrast to G and C, which form three such bonds and are thus more strongly associated. In this experiment, partial denaturation confirms the expected presence of two polynucleotide strands in the replicated DNA. In other cases this procedure is being used to map DNA molecules in terms of the locations of G-C–rich and A-T–rich regions. (*Note:* A and B are printed in opposite photographic contrast.) (Courtesy of H.J. Kriegstein and D.S. Hogness.)

pertwists" (Fig. III–11 shows the more limited supercoiling of circular DNA found normally). The amount of unwinding necessary can be appreciated from the fact that the double helix makes one turn every 10 nucleotides, and even the simplest viruses have several thousand nucleotide pairs. The perturbations resulting from local uncoiling of the double helix are minimized through action of enzymes known as DNA *topoisomerases* (some are called *gyrases*), which temporarily cut ("nick") the DNA sugar-phosphate backbone near the zones where the two strands are separating. This effect permits the strands to rotate locally around one another and possibly also can enable one segment of duplex to pass through a transient gap in another. The enzymes then seal the nicks, restoring the integrity of the DNA.

3. The unwinding, nicking, and sealing progress along the double helix, creating twin *replication forks* that move in the two directions away from the site of initiation of replication. That is to say, replication is *bidirectional* (Figs. IV–9 and IV–11). At each fork both strands of DNA form complementary partners using nucleotides in the "energized," triphosphate form (comparable to ATP). As each nucleotide is linked into the growing chain by *DNA polymerases,* a pair of phosphates is liberated and energy is made available for the polymerization.

4. The DNA polymerases "edit" or "proofread" as they go—that is, if a nucleotide makes an incorrect association with a parental DNA strand, the polymerases (or other enzymes) are able to remove it even if it has accidentally already been incorporated at the end of a growing chain. (The probability of such incorporation is itself low because the polymerases rarely catalyze the necessary bonds for mismatched bases.) This correction mech-

Figure IV–11 *Bidirectional replication of DNA.* Autoradiographs of DNA from a cultured mouse cell (L cell) that was exposed first to highly radioactive thymidine for 30 minutes and then to much less radioactive thymidine for an additional 30 minutes before the cells were broken open and their contents spread on a slide. The DNA molecules themselves cannot be seen, but the portions that have replicated during the labeling period are radioactive and produce lines or tracks of grains in the autoradiograph, seen as dark zones in the micrograph (the magnification is too low for individual grains to be visible). Each track corresponds to a replicating unit of DNA (see Figs. IV–9 and IV–10; also Section 4.2.4). The unit at b commenced replication while the highly radioactive thymidine was present, and it was still replicating when the cells were switched to the less radioactive medium. Since replication begins at the middle of a unit and proceeds in both directions from there (see Fig. IV–9), the central portion of the track shows heavier radioactivity (denser accumulation of grains) than that apparent in the more lateral portions (these correspond to DNA made in the less radioactive medium). The unit at a had started replicating before label was present; thus it shows an unlabeled central region (arrow) flanked on either side by densely labeled DNA (made after the cells were placed in the highly radioactive thymidine) and then by less heavily labeled DNA. Determinations of the length of tracks produced with different labeling periods indicate that an end of a single growing DNA strand adds nucleotides at a rate of approximately 1 to 3 thousand per minute. × 200. (Courtesy of R. Hand.)

anism helps to ensure the accuracy of replication. It is one of a number of repair and error correction devices the cell can use for its nucleic acids (see also Section 3.12.2). In procaryotes the mechanism is based on the ability of the polymerase itself to remove nucleotides (an "exonuclease" activity of the polymerase). The equivalent mechanism for eucaryotes is not known. It is interesting that RNA molecules seem not to be similarly "edited" as they are made; presumably this difference relates to the fact that the cell makes many copies of RNAs and thus can tolerate an occasional error better than it can the mutations that result from errors in DNA synthesis.

Discontinuous and Continuous Synthesis; RNA, Primers; Rates The formation of the new DNA strands is not as straightforward as the account thus far implies. At least two complications are involved:

First, DNA polymerases can only catalyze elongation of chains in the 5'–3' direction; the 5' carbon of the incoming sugar becomes linked, by means of a *phosphodiester bond*, to the 3' carbon of the sugar in the nucleotide added just before (Figs. II–15 and IV–9). One of the newly forming DNA strands at each replication fork could grow directly in this *continuous* way; by convention, this strand is called the "leading" (or "forward") strand. The other ("lagging" or "retrograde") DNA strand is oriented in the wrong direction for such direct elongation. It grows instead by the formation of a series of short chains (Fig. IV–9) that *do* polymerize in the 5'–3' direction; these chains are then attached to the growing strand, by enzymes known as *ligases*. In other words, the lagging strand is produced in initially discontinuous stretches (named "Okazaki fragments") thought to be a few hundred nucleotides long in eucaryotes and five or ten times longer in procaryotes.

Opinion about whether the "leading" strand actually is made in similar discontinuous fragments has vacillated in recent years. The present majority favors continuous synthesis of this strand but the issue is not yet closed.

Second, each piece of DNA, including each discontinuous segment of the "lagging" strand, is begun by a short chain of *RNA*, 10 or more nucleotides long, and complementary to the corresponding stretch of parental DNA. These RNA pieces are synthesized by enzymes called *primases*, which may resemble RNA polymerases. Later the RNA segments are removed enzymatically. The gaps left by removal of the RNA "primers" are filled in with appropriate DNA nucleotides before the ligases seal the pieces of DNA to the growing chain. The "filling in" is done by a type of DNA polymerase (or more than one sort of enzyme) different from those responsible for the fundamental growth of the chains; some investigators believe that in procaryotes at least, this same enzyme (or enzymes) is responsible for removal of the primers. (Both procaryotes and eucaryotes possess at least three distinguishable varieties of DNA polymerases: Polymerase α—"III," in procaryotes—catalyzes most of the chain growth; polymerase β—"I" in procaryotes—controls the filling in of gaps. The roles of the third enzyme are not fully understood, though in eucaryotes, polymerase γ probably replicates mitochondrial DNA. Polymerase α can be inhibited, selectively, by a fungal toxin, *aphidicolin,* which therefore prevents growth of eucaryotic cells, but not of procaryotes.)

In procaryotes like *E. coli,* a given DNA strand can grow at 500 to 1000 nucleotides per second. (Eucaryotes are five- to tenfold slower.) Thus, the discontinuous segments exist as such only for very brief periods; the gaps are rapidly sealed over, and the segments are linked to growing strands.

Rolling Circles Some of the circular DNA molecules in biological systems are replicated by mechanisms differing from those just described in details, though not in such essentials as the specification of sequence by base-pairing, or in the use of polymerizing enzymes to link nucleotides. Replication in the so-called "rolling circle" manner can produce copies of circular DNAs, in which the DNA sequence is repeated as illustrated in Figure IV–9. The copies can assume a linear form, or they can form a circle. A rolling circle mechanism seems to be one of the ways in which the larger precursor DNAs from which a viral "headful" is cut (Section 4.1.3) are generated. A similar mechanism may be responsible for producing the "extra" copies of the nucleolar organizer DNA present in the oöcyte nuclei of Section 2.3.4.

Mitochondrial DNA   Replication of the small circular DNAs of the mitochondria of vertebrates can involve types of modified mechanisms that still are being studied. As a working hypothesis, many investigators have adopted the proposition that replication of the two strands is substantially asynchronous and that both may be duplicated by a continuous mechanism akin to that of the "leading" strand of "ordinary" replication. Replication begins at a specific point ("origin of replication") but for some time only one fork is involved. This fork moves unidirectionally, duplicating one of the strands, while the other member of the parental duplex is left as a single strand that loops out from the circle. Eventually, however, this strand also forms a replication fork and generates a duplex. Still unknown events permit the two daughter circular duplexes to come apart, rather than interlocking like links in a chain, and other processes supercoil the circles into their mature supercoiled form (Fig. II–57).

Other features of DNA replication may also differ in detail from the general picture drawn above. For instance, in certain viruses, such as the adenoviruses that infect mammalian cells, protein molecules covalently linked to the ends of DNA chains may play essential parts in initiation of DNA replication.

Initiation; Origins; Replicons   In *E. coli* and other bacteria the chromosome is a single circular molecule of DNA. Replication begins at a specific point of the circle (the "origin of replication"—in *E. coli*, a base sequence of about 250 nucleotides) and proceeds in both directions from that point. Most viral DNAs also have a specific origin of replication. Progress in purifying the proteins that recognize and interact with the *E. coli* origin of replication to initiate DNA synthesis, has begun with successful preparation of "cytoplasmic" extracts that carry out initiation of replication in the test tube. Including the components of these extracts, 20 or more different enzymes and other proteins are required for DNA replication in *E. coli*. Section 5.1.2 considers features of nucleic acid structure that may relate to the interactions of these proteins with the origin of replication.

The chromosomes of eucaryotes are larger, more complex, and more numerous than those of procaryotes or viruses. Each consists of amounts of DNA that often are much greater than the total DNA content of a bacterium, plus many protein molecules. Duplication of eucaryotic chromosomes is based on semiconservative replication of DNA as described above, but factors additional to those already discussed come into play. These permit the large amounts of DNA to be replicated in reasonably short times, and to be distributed to daughter chromosomes in orderly fashion.

When eucaryotic cells are exposed for a few minutes to tritiated thymidine and then grown in a nonradioactive medium and studied by autoradiographic techniques, not all the chromosomes of a given cell are invariably labeled. Those that are labeled show radioactivity at multiple discrete, and often quite distant, points along their length. These findings indicate (1) that different portions of chromosomes duplicate at different times or at very different rates during interphase (only those that have duplicated significant amounts of DNA during the brief exposure to label become radioactive) and (2) that several regions replicate simultaneously on a given chromosome (if only one replication point per chromosome were operating, only one site would be labeled). Of interest is the fact that the schedule of DNA replication in different chromosomes and chromosome portions appears to be the same in successive divisions of a given cell type. Quite often this schedule involves duplication of the DNA in heterochromatic (Sections 2.2.5 and 4.4.6) chromosome regions relatively late in S, or very slowly.

The conclusion from such observations is that in its replication, the DNA of a eucaryotic chromosome behaves as if composed of several hundred to a few thousand replicating units (replicons) whose sizes range from roughly 10 μm to a hundred μm or more in different cells and circumstances (see Figs. IV–8, IV–9, and IV–10). Initiation of DNA replication in a replicon results in a pair of replicating forks that progress outward from its center at 0.5 to 2 μm per minute. The mechanisms coordinating the initiation of replication in each unit with that in the others are presently unknown, though it does seem that adjacent replicons tend to replicate as synchronous groups.

It might be imagined that replicons each contain a special controlling DNA sequence where replication begins. This is still being studied, for example, through efforts to understand the fact that the very rapid replication of DNA during early embryonic cleavages is based on simultaneous initiation of DNA replication at far more sites than those operating simultaneously later on in development (Fig. IV–10). From the available, incomplete, information it appears that this is not simply a matter of scheduling: The embryonic cells seem to have a larger number of shorter replicating units than are found later in development. (From present knowledge this appears somewhat paradoxical because, overall, the DNA of embryonic cells and that of the "adult" cells to which they give rise is presumed to be essentially identical [Chap. 4.4].) Differences in the duration of S phase among cell types of a given organism, as well as variations in rates of replication of different chromosome regions, are also considered to reflect, in part, differences in size and abundance of replicons. In addition, the subdivision of DNA molecules into segments capable of independent replication may help to explain how some cells can replicate portions of their chromosomes selectively (Sections 4.4.4 and 4.4.6).

In bacteria, a new round of DNA replication can be initiated at the origin of replication well before the replication forks have completed their circuit—in other words, before the prior round of replication has been completed and before the cells have divided. (As Section 3.2.3 points out, under some growth conditions bacteria contain more than one copy of the chromosome per cell.) In eucaryotes, renewed initiation of replication of a chromosome generally must await completion of replication already underway, even though there are cases where chromosome replication need not be followed by division of the cell (Sections 4.4.4, 4.4.6, and 4.5.1).

Behavior of the Chromosome as a Single DNA Duplex Once the DNA of a eucaryotic chromosome is replicated, how is it apportioned between daughter chromosomes? An experiment designed to answer this question is outlined in Figure IV–12 (see also Fig. IV–13). The key finding is that the distribution of label in chromosomes after the second duplication is different from that at the first duplication following label administration. Frequently this experiment is interpreted as showing simply that chromosomal DNA replicates semiconservatively, as expected. In fact, the implications go significantly beyond this. At the time the experiment was done there was no way of knowing whether the DNA of a single eucaryotic chromosome was in the form of several or even many separate molecules, as the existence of replicons might conceivably imply, or whether it actually was one, continuous molecule that somehow can replicate in segments. The experiment showed that the DNA behaves as if there were only one DNA double helix per chromosome before duplication and per chromatid after duplication. It was over a decade later before techniques progressed to the point where more direct evidence sustaining this conclusion could be gathered (Sections 4.2.5 and 4.4.5).

Membranes; Matrices Many investigators are uncomfortable with the idea that all these initiations, enzyme activities, movements of replication forks, and so forth take place on DNA floating free in the "nuclear sap" (Section 2.2.5). There would seem to be too many possibilities for errors and inefficiencies and for tangling of the extremely elongate molecules. For procaryotes, there is evidence that during replication, the chromosome is anchored to the plasma membrane (Section 3.2.4). Some believe that initiation of DNA replication or the enzymatic steps of replication itself occur at the points of anchoring, although this is still the subject of much contention. By extension, it has sometimes been proposed that attachment of eucaryotic chromosomes to the nu-

---

Figure IV–12 *Chromosomal replication.* J.H. Taylor's experiment on chromosome duplication. Taylor used the drug colchicine to suppress cell division (Section 2.10.5) without preventing chromosome duplication. Chromatids separate, but in the absence of a spindle, they remain in the same cell. This provides a convenient method for determining the number of times the chromosome set being studied has duplicated during the experiment. Each duplication results in doubling of the number of chromosomes per cell. Red strands are radioactive. See also Figure IV–13.

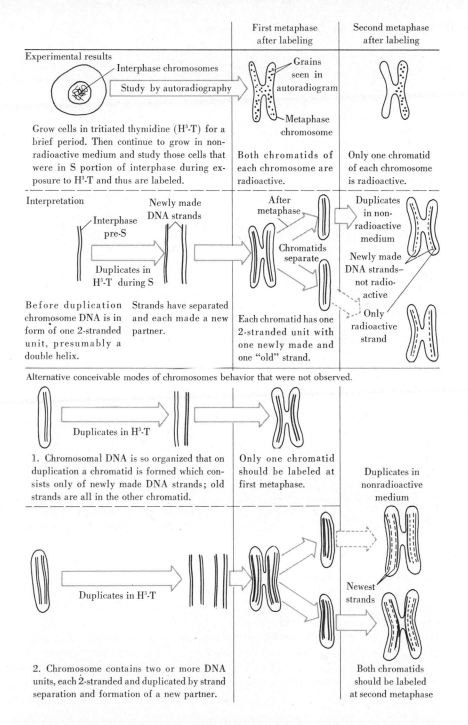

First metaphase after labeling

Second metaphase after labeling

Experimental results

Interphase chromosomes

Study by autoradiography

Grains seen in autoradiogram

Metaphase chromosome

Grow cells in tritiated thymidine (H³-T) for a brief period. Then continue to grow in non-radioactive medium and study those cells that were in S portion of interphase during exposure to H³-T and thus are labeled.

Both chromatids of each chromosome are radioactive.

Only one chromatid of each chromosome is radioactive.

Interpretation

Interphase pre-S

Newly made DNA strands

Duplicates in H³-T during S

After metaphase

Chromatids separate

Duplicates in non-radioactive medium

Newly made DNA strands— not radioactive

Only radioactive strand

Before duplication chromosome DNA is in form of one 2-stranded unit, presumably a double helix.

Strands have separated and each made a new partner.

Each chromatid has one 2-stranded unit with one newly made and one "old" strand.

Alternative conceivable modes of chromosomes behavior that were not observed.

Duplicates in H³-T

1. Chromosomal DNA is so organized that on duplication a chromatid is formed which consists only of newly made DNA strands; old strands are all in the other chromatid.

Only one chromatid should be labeled at first metaphase.

Duplicates in nonradioactive medium

Duplicates in H³-T

Newest strands

2. Chromosome contains two or more DNA units, each 2-stranded and duplicated by strand separation and formation of a new partner.

Both chromatids should be labeled at second metaphase

clear envelope serves a similar organizing function for eucaryotic DNA replication. This concept has been eclipsed recently by the alternative proposition that the structural matrix hypothesized to pervade the eucaryotic nucleus (Section

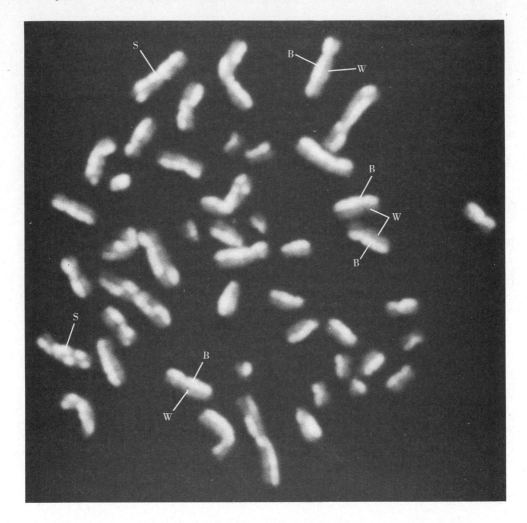

2.2.5) incorporates the enzyme systems needed for DNA replication, providing an organized array of anchoring and replication locales from which the DNA loops out.

Whatever the eventual fate of these proposals, numerous observations indicate that even when maximally unfolded in interphase, chromosomes do not invariably fall into random patterns within the nucleus. In dividing cells of onion, for example, when the chromosomes "reappear" in prophase they seem to be in the same relative positions they occupied when they began to unfold at the end of the prior division. Studies on a few other cells also suggest that specific chromosome regions, such as heterochromatic portions or centromeres, are grouped or clustered together in the interphase nucleus, but no general pattern characteristic of most cell types has yet been demonstrated

◀ **Figure IV–13** *Study of DNA replication by fluorescence microscopy.* These methods may eventually evolve into procedures for observation of replication in living cells. Here, they have been used for an experiment like the one in Figure IV–12. Human lymphocytes were grown in BUdR which substitutes for thymidine (BUdR = Bromo-uracil-deoxyriboside; it is also referred to as BrdU; see also Section 4.2.3). The growth period in BUdR was such that the cells went through two interphases. The cells were then fixed and stained with a dye (33258 Hoechst) that binds to DNA and fluoresces when irradiated with ultraviolet light, as in a fluorescence microscope. The fluorescence of this dye decreases when the DNA contains BUdR. By the same logic used in Figure IV–12, one would expect that after two rounds of replication in the presence of BUdR, each metaphase chromosome will have one chromatid in which both strands of the DNA double helix contain BUdR and one in which only one strand contains BUdR. This figure shows one piece of evidence that confirms this: each chromosome has one chromatid whose fluorescence is very weak *(W)*, presumably with 2 BUdR strands, and one with stronger fluorescence *(B)*, presumably with only one BUdR strand. In some cases "sister chromatid" *(S)* exchanges seem to have occurred; two chromatids of a chromosome have apparently exchanged portions with one another. (From S. Latt, *J. Histochem. Cytochem.* **22**:478, 1975. Copyright 1975, The Histochemical Society, Inc.)

convincingly. Additional examples of nonrandom arrangements are discussed in Sections 4.2.7, 4.3.4, and 4.4.4.

4.2.5 Chromosome Structure; Histones and Nucleosomes; Duplication

The architecture of the eucaryotic chromosome is one of the outstanding unsolved problems of cell structure. Recent progress has been rapid, and models of chromosome structure are now evolving that should provide a more adequate basis for understanding the arrangement and control of transcription and duplication, the condensation (folding) cycle, and other major unresolved issues.

Fundamentals of Chromosome Architecture: Folding; Scaffolds (?); Uninemy A chromosome that measures at mitotic metaphase, 5 to 10 μm in length and a micrometer or less in width, can contain an amount of DNA that would be 10,000 times longer—several centimeters in length—if stretched out in one straight double helix (Section 3.1.3). "Packaging" of this DNA in chromosomes must involve extensive coiling and other folding. This was already evident through light microscopy; chromosomes during cell division are seen to consist of coiled and folded fibers (Fig. IV–14) whose coils and other folds relax during interphase. The condensation of a chromosome into the compact array of mitosis must be governed by reasonably precise rules because the morphology of the folded structure is reproduced at successive cell divisions. Chromosome length and overall shape, the positions of longitudinal differentiations such as kinetochores, secondary constrictions (Section 4.2.1), banding patterns (Fig. IV–5), and patterns of local folding ("chromomeres";

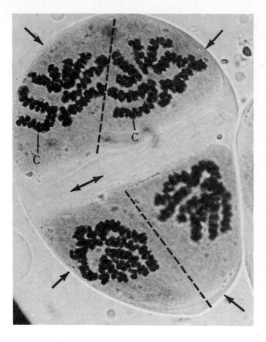

Figure IV–14 *Coiling and folding of chromosomes.* A microsporocyte of the plant *Trillium*. It has been fixed at the end of anaphase (second meiotic division) as the chromosomes *(C)* begin to unfold. The two meiotic divisions (Fig. IV–30) produce four cells that give rise to male gamete nuclei. In the present case the first division took place along the plane of the double-headed arrow; the second would have been along the dotted lines. The single-headed arrows indicate cell borders. × 1000. (Courtesy of A. Sparrow.)

Section 4.4.5 and Fig. II–27) are the same cycle after cycle, and in different cells of a tissue or organism.

To some investigators the ability of the chromosome to fold up repeatedly into the same mitotic configuration is best accounted for by the assembly of an organizing "scaffold." This structure is presumed to be of chromosomal proteins that link temporarily to one another as the chromosome folds, creating a longitudinal backbone from which loops of DNA extend (Fig. IV–15). A structure of this sort with DNA loops 10 to 30 μm long (up to 100,000 nucleotide pairs) is seen when isolated metaphase chromosomes are suitably mistreated to remove the histones and spread out their DNA. Whether this treatment simply *reveals* the backbone or artifactually *creates* it is in dispute; the reality of the scaffold in the intact cell is still being investigated. Of great potential interest is the possibility that the scaffold proteins derive from the nuclear "matrix" (Sections 2.2.5 and 4.2.4), which could be a powerful clue to how chromosome structure ties in with DNA replication and other processes.

None of this, however, implies that the basic chromosome structure is of a protein backbone to which DNA is attached at intervals, as was sometimes suggested during earlier periods of cytological work on chromosomes (Fig. IV–15). Since chromosomes are found to fragment into shorter pieces when treated with the enzyme DNase but do not do so when digested with proteases (Section 4.4.5), it seems clear that the DNA itself is responsible for the longitudinal (end-to-end) continuity of the chromosome.

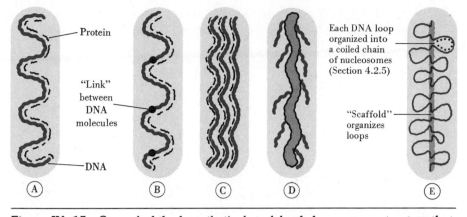

Figure IV–15 *Several of the hypothetical models of chromosome structure* that have been advanced at various times. The models drawn are intended as schematic views of metaphase chromatids but presumably would apply, with suitable unfolding, to interphase as well. A. One elongate coiled DNA molecule (double helix) with protein and other material closely associated. B. Several DNA molecules linked together end to end by unknown materials. C. Several to many DNA molecules arranged longitudinally in parallel strands or, more likely, coiled together as in a rope. D. A non-DNA backbone (for example, of protein) to which DNA molecules are attached at intervals. E. One elongate DNA molecule organized into loops by a "scaffold" of protein molecules (some think RNA may also be part of the scaffold). In all cases, the material in the actual chromosomes would be far more extensively folded and coiled; for example, all the loops in Part E are thought to be in the "string-of-beads" (nucleosomal) form described in Section 4.2.5 and schematized here for one loop.

For several decades, extending into the 1970s, arguments raged about whether the structure of an interphase chromosome and of a mitotic chromatid is based on a single threadlike, coiled and folded DNA-containing unit or whether there are several or many units packed in parallel bundles or coiled together as in a cable (Fig. IV–15). The arguments were fueled by occasional observations in the light microscope of what could be interpreted as longitudinal subdivisions of chromatids into two or more "sub-chromatids" and by observations that some special chromosomes indubitably are made of multiple parallel units (Section 4.4.4). Electron microscopy of chromosomes in the usual thin sections provided important details about chromosome structure, such as the absence of a delimiting membrane and the presence of specialized organization at regions like the kinetochore (see Fig. IV–21). Work with conventional preparations, however, failed to resolve the issue of the number of longitudinal threads present. Chromosomes cut into the thin sections generally used for electron microscopy show numerous fine fibers (fibrils) grouped in a variety of bundles. In diameter they are as small as 20 to 30 Å, the dimensions of a DNA molecule or DNA–protein complex; bundles 10 nm (100 Å) thick are common. Only a short length of a given fibril is included in a section (as in

Fig. IV–21), so that it is impossible to tell whether the chromosome is of one or a few very long coiled fibrils, of many independent fibrils, or even of a complex interconnected network. Attempts at serial-section reconstruction (Section 1.2.1) have been of limited value owing to the small dimensions of the fibrils and to their high degree of folding.

A further complication was the fact that when DNA is isolated from eucaryotic nuclei, the molecules obtained usually are too short for there to be only one of them per chromosome. This could be explained on the basis of breakage of much longer molecules during their isolation, a phenomenon common with DNA molecules, which fragment under the impact of even relatively mild mechanical stresses. However, taken together with the finding that eucaryotic chromosomes replicate as if made of multiple DNA units (Section 4.2.4), it provided ammunition for those who argued that chromosomes are basically "multinemic." The suffix -*nemic* denotes "threadlike" and in this context refers to the hypothetical basic unit of chromosome structure much larger and more complex than the two strands of a DNA double helix. A multinemic chromosome would have more than one such unit.

At present, however, a near-consensus seems to be developing around "uninemic" models for interphase chromosomes and metaphase chromatids. A uninemic mode, in which the basic thread is equivalent to a duplex of DNA, accords most simply with genetic evidence and with the findings on DNA duplication described in the preceding section. Moreover, when metaphase chromosomes are isolated carefully from cells and examined as "whole mounts" (without sectioning), they appear as masses of long fibers, 20 to 30 nm in diameter, that show extensive looping, as would be expected if there were actually one very long thread folded into a compact chromosome (Fig. IV–16). Few free ends are seen, demonstrating that the chromosome is not a cable of many fibers, although the complexity of the structure as well as breakage and other alterations that occur during preparation preclude the unequivocal demonstration in this way that only *one* fiber is present.

Fortunately it now is possible to demonstrate the likely existence of DNA molecules long enough for one molecule to account for all the DNA of a chromosome. This has been done for organisms such as yeast and *Drosophila,* whose chromosomes contain *relatively* small amounts of DNA. (The amount of DNA in different yeast chromosomes ranges from about 150 to about 2,000 "kilobases" [Section 2.2.1], with an average of about 750, which amounts to a length of roughly 250 μm [Section 2.2.1]. By contrast, individual human chromosomes contain 10 to 100 times as much DNA.) Preparation of the DNA was accomplished by carefully breaking open (lysing) cells in a solution of detergents, proteolytic enzymes, and other components. This mixture opens the cell very gently, digests proteins, including those of the chromosomes, and simultaneously inhibits cellular DNases that would otherwise degrade the DNA. The lysis was carried out within the chamber of a device used to measure *viscosity* properties of a solution, so that the size of the DNA could be studied without transfer from one piece of apparatus to another, thus minimizing the

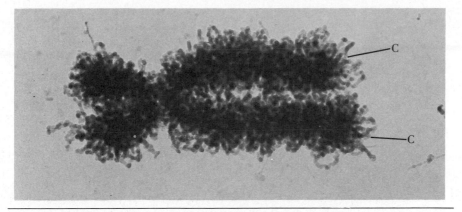

Figure IV–16 *A human chromosome* isolated at the metaphase stage of mitosis and viewed without sectioning. The two chromatids are readily visible *(C)*; they are attached to one another at the centromeric region. Since very few ends of fibers are seen, it has been argued that each chromatid consists of one long looped and coiled fiber. Approximately × 50,000. (Courtesy of E.J. DuPraw.)

opportunity for breakage. Viscosity measurements examine the rate of flow and related hydrodynamic and elasticity features of solutions; the measurements can be interpreted in terms of the size and geometry of dissolved molecules. Strictly speaking, the findings show only that DNA-containing elongate threads are present that are of the appropriate length. They do not completely rule out, for instance, the possibility that these strands are of several duplexes linked end to end by unknown non-DNA materials that somehow resist the proteases and other components of the lysing medium. Convincing evidence favoring such linkage is not at hand, however. Furthermore, electron micrographs of isolated yeast chromosomal DNAs show apparently continuous molecules with no striking interruptions by material of different appearance. Therefore, while some skeptics still support more complex arrangements of the DNA, most investigators have tentatively drawn the simplest conclusion—that the basic strand of a chromosome is a single duplex.

Nucleosomes   How is the DNA duplex, 20 Å in diameter, associated with proteins to construct the thicker fibers, 10 to 30 nm across, seen in interphase and metaphase chromosomes? Until recently it was generally believed that the proteins surrounded the DNA either coating the DNA molecules or running alongside (Fig. IV–15). This view has been changed radically on the basis both of microscopic and of biochemical evidence.

To examine interphase nuclei without sectioning, the nuclei are carefully lysed under conditions that extract some proteins and disperse the nuclear contents to a limited extent (nuclei can, for example, be gently lysed in suitable salt solutions and briefly fixed in formaldehyde; Section 1.2.1); the contents are then spread on the surface of a suitable salt solution and picked up on an

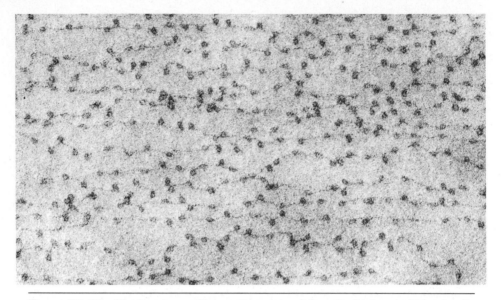

Figure IV–17 *Nucleosomes.* Chromatin prepared from a chicken red blood cell nucleus by a procedure involving swelling of the nuclei, fixation in formaldehyde, and negative staining. Note the numerous small globules (10 nm diameter) connected to each other by a fine fiber. These globules are nucleosomes; normally they are packed more closely together as in Figure IV–18B, but the preparation procedure has spread them apart, revealing the stretch of DNA that links each nucleosome to the next. × 170,000. (Courtesy of D.E. Olins and A.L. Olins.)

electron microscope grid. Interphase chromatin so prepared has the appearance illustrated in Figure IV–17. Each bead is approximately 10 nm in diameter. When chromatin is gently and briefly digested with certain types of DNases, usually obtained from bacteria ("staphylococcal" or "micrococcal" DNases), it is cut into a series of fragments. Each fragment contains a DNA piece approximately 200 nucleotide pairs long or a simple multiple thereof. Longer incubation reduces the DNA pieces to about 146 nucleotide pairs. The enzymes that generate these fragments are of the type known as *endonucleases,* signifying that they can break the DNA by cutting directly across the molecule (rather than chewing away only at the ends, as "exonucleases" do). When DNA, purified from chromatin by removal of proteins, is exposed to the same type of enzyme it is rapidly broken down into a set of very small fragments of random size. The inference is that the regular pattern of fragments obtained with chromatin is due to the presence of chromosomal proteins. This is confirmed and extended by experiments showing that when purified DNA is mixed with histones before exposure to micrococcal DNase, the 200-nucleotide–pair pattern is again obtained, as well as the microscopic appearance shown in Figure IV–17. Nonhistone proteins are not required for this reconstitution.

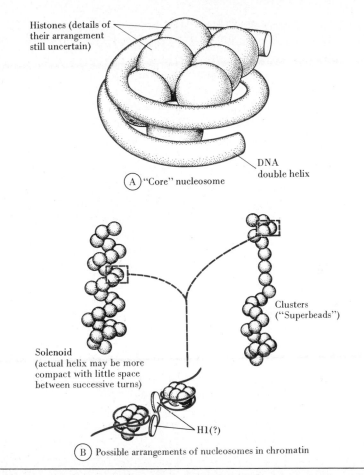

Histones (details of their arrangement still uncertain)

DNA double helix

(A) "Core" nucleosome

Clusters ("Superbeads")

Solenoid (actual helix may be more compact with little space between successive turns)

H1(?)

(B) Possible arrangements of nucleosomes in chromatin

Figure IV–18 *Models of nucleosomes.* A. Model of the core structure of a nucleosome based largely on proposals and diagrams by R.D. Kornberg, A. Klug, and G. Felsenfeld. The DNA double helix is in a "supercoil" (Section 5.1.2) that makes 1¾ turns around the histone octamer. The precise arrangement of the eight histone molecules is not yet known. B. Proposals as to how nucleosomes are arranged in a fiber of chromatin; both types of arrangement may occur in nature. Note that the association of H1 with the core nucleosome is thought to permit the DNA to make a full two turns around the histone octamer.

It has been concluded that the basic longitudinal unit of the chromosome has the form of a string of beads; each "bead" is now called a *nucleosome.* Each nucleosome has a core structure consisting of a stretch of DNA about 146 nucleotide pairs long and coiled *around* a cluster of histone molecules (Fig. IV–18). A stretch of DNA about 50 base pairs long links each core nucleosome to the next. The association of DNA with histones affords some protection against DNase digestion, though this is only partial and can be overcome by prolonged exposure to the enzyme. The linking DNA stretch is more susceptible to attack by DNase than is the core structure, owing probably to differences in association with histone (see below). Thus, initially, DNase cuts across the links between core nucleosomes, separating the DNA into the 200-nucleotide-pair-long fragments; then the linking DNA becomes more extensively degraded, initially leaving fragments about 160 nucleotide pairs long and then (once histone H1 has dissociated from the particles—see below) leaving the 146-nucleotide-pair DNA of the core nucleosome structure.

Note that the concepts of "core DNA" and linker portions are structural notions relating to nucleosome organization. They bear no obvious functional significance in terms, say, of information content or gene activity (Section 4.4.6). Note also that a given gene occupies a stretch of DNA that will form several to many nucleosomes (Sections 3.1.3 and 4.2.6).

Chromatin is organized as nucleosomes both in dividing and in nondividing cells and throughout the division cycle. Some fine details, such as the lengths of the linking DNA segments, differ in different organisms, but the basic arrangement is essentially ubiquitous for eucaryotes. It is found even on the DNAs of certain of the viruses infecting eucaryotic cells (such as SV40; see Section 3.12.3).

Nucleosomes are not found on procaryotic chromosomes or in mitochondria and plastids. (There are sporadic reports of limited amounts of histone-like basic proteins associated with bacterial chromosomes and of "beaded" structures on bacterial chromosomes isolated by some procedures, but these have yet to be definitively interpreted or fully understood.) However, in procaryotes and in organelles, the DNA is "supercoiled" (Section 3.2.3 and Figs. II–57 and III–11). Section 5.1.2 will outline the proposition that such twisting of DNA, like the coiled arrangement in nucleosomes, has important effects; it may, for example, make it easier for proteins to separate the two strands in transcription into RNA and in DNA replication.

Histones   The histones of virtually all animal and plant species, multicellular and unicellular, are of five distinct types. The molecules are small proteins. They are named according to the relative proportions of the basic amino acids (Section 2.2.3) *arginine* and *lysine,* and by numbers deriving historically from the procedures used to isolate them. Arginine and lysine account for roughly 25 percent of the total amino acids in each. The "very-lysine-rich" or H1 histone is composed of about 215 amino acids (molecular weight 21,000) and has 20 times as many lysines as arginines. The other four types are made of 100 to 130 amino acids (molecular weights of 11,000 to 15,000). They include two "lysine-rich" types: H2A, which has 1.25 times as many lysines as arginines, and H2B, with double this ratio. The two "arginine-rich" histones have about 1.3 times as many arginines as lysines; they are called H3 and H4.

The histones are similar in different cell types and even in different species—for H4 there are only two differences in the overall sequence of amino acids between peas and calves, and H3 shows only four differences. However, although these similarities are striking, histones are not totally invariant. In some organisms at least, changes in the histones present are observed during development. H1 is the most variable among species and cell types; for example, in red blood cells that retain their nuclei such as those of birds, it is replaced by the closely related *H5*. In sperm of various species, the histones are replaced by even more basic proteins, the arginine-rich *protamines* (Section 3.10.1). (See Section 4.4.6 for some of the modifications of histones possibly involved in control of genetic activity.)

Histone and DNA Arrangement in Nucleosomes   To reconstitute nucleosomes from purified DNA, H2A, H2B, H3, and H4 are required. Careful analysis of both reconstituted and native nucleosomes indicates that there are two of each of these molecules per nucleosome—hence, an ''octamer'' (or ''octet'') of histones is associated with each 200 ''base pairs'' (nucleotide pairs) of DNA. Among the pieces of evidence substantiating this is the observation that chromatin exposed to cross-linking agents (Section 2.3.2) yields clusters of cross-linked histones involving up to eight molecules, two of each kind. The presently favored (though still developing) model illustrated in Figure IV–18 suggests that the DNA molecule makes 1¾ turns around the histone octamer, producing a disc-shaped ''core'' structure about 11 nm in diameter—the dimensions of nucleosomes seen in the microscope; this arrangement requires the 146-nucleotide-pair length of double helix. Binding of H1 to the core structure results in association of an additional 20 nucleotide pairs of DNA with the nucleosome, completing the particle. The remainder of the DNA links adjacent nucleosomes. There is one H1 molecule per nucleosome core particle. This molecule seems to be bound to the outside of the DNA and the core particle and may, for example, help to link nucleosomes to one another. There is more uncertainty about H1's locations and roles than about those of the other histones but it is found that when H1 is selectively removed in the test tube, the nucleosomes spread apart more (or cluster less tightly).

The coiling of the DNA double helix around the histone octamer is a form of ''supercoiling'' comparable to that in Figures II–57 and III–11 (see Section 4.2.4) although, of course, the DNA is not a closed circle.

Generally, it is presumed that the linkage of histones to DNA depends on electrostatic attactions between the positively charged basic amino acids and the negatively charged DNA phosphates. The four histone types that participate in the core of the nucleosome each have the majority of their basic amino acids located toward the N-terminal (Section 2.3.1) end of the molecule, especially in the initial one third of the polypeptide chain. This organization might mean that binding of histones to DNA depends on the N-terminal portion of the proteins. Octamer assembly in the test tube, as outlined later in this section, shows strong dependence on the less basic two thirds of the molecules; these portions of the histone molecules, where apolar (Section 1.4.1) amino acids are relatively abundant, could mediate interaction of the histones with one another and perhaps with additional, nonhistone proteins (Section 4.4.6). Some investigators, however, emphasize the lack of definite evidence that the histone regions especially rich in basic amino acids are the exclusive or even the primary sites of association with DNA. Several other roles of these regions, including in interactions with other proteins, are still conceivable.

H1 has clusters of basic amino acids at both ends, probably relating to its arrangement along the DNA.

The location of the DNA duplex at the surface of nucleosomes (Fig. IV–18) implies that some of the phosphate groups are not likely to be bound strongly to oppositely charged groups on histones. This may account for the fact that "basic" dye molecules can link to DNA in fixed nuclei (Section 2.2.3); "acid" dyes are impeded from reaching the histones in the nucleus, since the histones are clustered, and surrounded in part by DNA.

The Chromosome Fiber The existence of nucleosomes roughly 10 nm in diameter spaced at close intervals along the DNA provides a ready explanation for observations of a 10-nm fiber in chromosomes prepared for microscopy in an appropriate manner. Many models assume that H1 and chromosomal proteins additional to the histones help to pack the nucleosomes in orderly arrangements within this fiber, but no general model has won universal acceptance.

The nonhistone proteins associated with chromosomes include enzymes of DNA replication and transcription as well as other proteins thought to modify or to control genetic activities. RNAs and their associated proteins are also present on chromosomes, at least as transitory components (see Fig. IV–40). How nonhistone proteins and RNAs associate with the nucleosome structure, however, is still to be detailed. Part of the problem is in evaluating which of the nonhistone proteins actually are part of the chromosome proper and which are appropriately regarded as nonchromosomal; the isolated chromosomes and interphase chromatin prepared by available procedures are extensively contaminated with other nuclear contents, and there are few unequivocal experimental criteria for deciding whether a particular protein is a contaminant, a true chromosomal component, a transient occupant of chromosomal sites, or a molecule that truly belongs both to the chromosomes and to the extrachromosomal compartment of the nucleus.

At the structural level, a key question is how the 10-nm string of beads is coiled or otherwise folded into the fiber 20 to 30 nm wide seen in whole mounts of chromosomes (Fig. IV–16). Since this latter fiber often shows distinct nonuniformities in diameter along its length, one view has been that the nucleosomes are grouped or clustered into "superbeads" whose overall diameter can reach 30 nm; superbeads may be linked together by thinner regions with smaller clusters of nucleosomes. An alternative "solenoidal" model envisages the coiling of the nucleosome chain into a helical structure (Fig. IV–18) with the beads arranged in a manner somewhat reminiscent of the monomeric tubulin units in the microtubule wall shown in Figure II–77. Compromise models and ones involving mixtures of different types of packing of the nucleosomes are also under consideration. One hypothesis, for example, is that the "solenoid" arrangement is maintained by H1-mediated linking of adjacent nucleosomes and that this linking is loosened, leading to a more flexible, less regular nucleosome arrangement in chromosome regions undergoing transcription or duplication (see below and Section 4.4.6).

Duplication of Chromosomes   Until recently, accounts of the way in which chromosomes duplicated could go very little beyond the replication of DNA because there were so many ambiguities about the fundamentals of chromosome structure. This situation is improving rapidly, although many questions have yet to be answered.

The chromosomal proteins are made in the cytoplasm and move into the nucleus. The histones, at least, are small enough that they probably can enter readily through the pores in the nuclear envelope.

In continually dividing cell populations the total content of protein in the nucleus doubles during interphase as the nuclear volume doubles. Although overall increases in nuclear protein are observed throughout interphase, the nuclear histone content doubles during S, in parallel with the DNA. The weight of opinion is that mRNAs for histones are made selectively during S and not at other stages, accounting for the timing of histone synthesis. This has, however, occasionally been challenged by investigators working with certain cell types, who report that some histone synthesis occurs at periods other than S; limited amounts of H1 especially, may be made during the G_1 period. Although they agree that the *nuclear* content of histones doubles in S, they believe that for some cell types, the timing of synthesis may be less restricted than is generally assumed.

For several cell types it is claimed that overall increases in nuclear volume and nuclear protein content are particularly marked late in S and early in G_2. How general this is has yet to be evaluated.

Since nucleosomes can be reconstituted by mixing DNA and histones, it is likely that they form in the cell by modified self-assembly mechanisms like those discussed in Chapter 4.1. Nonhistone nuclear proteins that may speed this assembly have been identified in oöcytes and are being sought elsewhere. Some investigators think that DNA topoisomerase enzymes (Section 4.2.4) also participate, but such enzymes have been little studied in eucaryotes. The histones by themselves interact with one another in ways that probably are important for the generation of nucleosome structure. Mixtures of H3 and H4, for instance, produce "tetramers" containing two molecules of each; perhaps in the cell such tetramers form first and then associate with H2A and H2B to yield octamers. Correct folding of histone molecules and histone–histone associations depend largely on the two thirds of each molecule furthest from the N-terminal end—that is, proper associations still occur after experimental removal of the region richest in basic amino acids described above. This finding hints at an interesting intramolecular division of labor, which is now under intensive investigation.

As DNA replication proceeds, the newly made strands are rapidly incorporated into nucleosomes, and the DNA awaiting replication remains in nucleosomal form. Dissociation of the histones and DNA during replication could be postulated, but if this process occurs at all, it takes place very locally and is very transitory. Probably "release" of DNA strands from histones is limited to the vicinity of the replication fork. If the histones synthesized during one S period are labeled with radioactive and density (see Section 4.3.4 and Fig. IV–

34) labels and the cells are then grown in label-free media so that older histones are labeled and newer histones are unlabeled, it is found that the members of an octamer stay together in succeeding divisions; labeled octamers remain fully labeled rather than becoming "diluted" by separation of labeled members and acquisition of unlabeled new ones. The conclusion is that once they have formed, histone octamers remain intact; new octamers are assembled entirely from new histones. In this limited sense, then, unlike the situation with DNA duplexes duplication of chromosome structure is "conservative."

This pattern implies, however, that as the DNA in each nucleosome replicates, one of the two parent strands and its new partner must acquire a newly assembled histone octamer. Whether it is the same strand for all the nucleosomes of a given region of DNA is still unclear. At the level of the whole *chromosome* it seems quite unlikely that the old nucleosomes all go with one parental DNA strand and that the new ones go with the other; this arrangement would produce a pair of chromatids of which one has entirely new histones and the other entirely old histones, and there is no evidence that this is the case. (The situation would be very much analogous to the first of the two "alternative conceivable modes" in Fig. IV–12.) However, there may be a tendency for immediately adjacent nucleosomes to behave in similar manner in terms of which parental DNA strand retains the preexisting octamers. A suggestion that was widely adhered to for some time was that at each replication fork, the newly made octamers are associated with the forming DNA double helix that contains the "lagging" strand of newly synthesized DNA (the strand known to be synthesized in discontinuous fashion; Section 4.2.4). Note in Figure IV–9 that at the opposite replication forks in each replicating unit, it is the opposite parental strand that is associated with the "lagging" newly made one. This would mean, in other words, that at one replication fork, one parental strand would stay with the old histones, while at the other fork, the other strand would stay with the old histones. Consequently the two duplexes produced by each replicon would have, overall, a mixure of associations and so, therefore, would the chromatids. This model is currently being challenged by investigators who believe that even for adjacent nucleosomes, old and new histone octamers become associated with either DNA strand essentially at random.

There is only very preliminary information about how the structure of the chromosome as a whole reorganizes during duplication to produce a pair of chromatids linked (until anaphase) in the centromere region but separated elsewhere. Several sorts of findings suggest that structural duplication starts early, well before DNA replication is completed. Thus, when interphase chromosomes are made to condense prematurely by fusing interphase cells with mitotic cells (Section 4.2.3), chromosomes in G_1 are seen to be of a single chromatid-like structure, and those in G_2, of two chromatids, as might have been anticipated; in contrast, S-phase chromosomes condense into fragmented or "pulverized"-appearing structures, as if they were caught in the process of some major structural reorganization that had altered the parental arrangement but had not yet progressed to the point where two well-defined, continuous chromatids had formed. (This interpretation depends, of course, on the assumption that the cell fusions simply cause the chromosomes to fold up, permitting convenient microscopic viewing, and do not themselves disrupt the chromosomes.)

Another important matter requiring more study is the degree to which chromosomal proteins turn over (introduction to Part 2). In rapidly dividing cells the histones seem stable for at least several cell generations; they do not seem to undergo much breakdown. However, there have been few adequate studies over long periods of time either for dividing cells or for ones that no longer are dividing. Findings obtained before the histones were well understood as a class of proteins suggest that there may be some replacement over the long run; efforts to confirm this are under way. For nonhistone proteins there is little doubt that appreciable turnover takes place, but this class of chromosomal proteins is quite heterogeneous, and detailed studies of different subtypes have not been carried out. These matters are particularly important because the nuclear proteins are thought to participate centrally in regulating genetic activity (Section 4.4.6), and the degree to which particular molecules retain stable associations with DNA and in the ways which local patterns of association may be reestablished during or after replication could be crucial for the relevant mechanisms.

4.2.6 DNA and Chromosomes: Work in Progress

Experimental Probes; "cDNAs" The next major task in the correlation of microscopic and molecular portraits of the chromosome is a detailed description of the arrangement of particular types of DNA sequences—those coding for proteins, those coding for nontranslated RNAs such as rRNAs or tRNAs, and those serving regulatory or structural functions. Important advances have already been made, particularly with respect to the nucleolar organizers (Section 2.3.4). Comparable information is badly needed about other microscopically distinctive chromosome regions such as the centromeres, and about the bulk of the chromosome fiber, which has only a few longitudinal landmarks (pp. 537, 613, and 616).

At first glance the problems seem overwhelming, since any particular DNA sequence makes up only a tiny fraction of the total present in the nucleus. One approach to overcoming this difficulty is the exploitation of special forms of chromosomes. As Section 4.4.4 will describe, some of these are large enough to permit even the dissection by hand, aided by light microscopy, of particular regions, which can then be studied biochemically. For more general work, a battery of relevant molecular biological techniques is evolving. We have already described a number of the most useful procedures, including the hybridization of RNAs with DNAs (Section 2.2.2), the production of cloned copies of DNA sequences by "recombinant DNA technology" (Section 3.2.6), and the transfer of DNA from one cell to another (Sections 3.2.6 and 3.11.3).

Once-formidable problems in determining the base sequences of long stretches of nucleic acids have been largely overcome so that investigators now know the precise sequences of many DNA regions and RNAs of interest. For several organisms, collections ("libraries") of cloned DNA sequences have been established covering the entire chromosome complement. Increasingly large regions of each chromosome are being mapped in detail by starting with cloned fragments containing DNAs representing genes whose locations are known from more "classical" techniques (Sections 3.11.3, 4.3.3, and 4.4.4) and using these as reference points for step-by-step mapping of the locations of other DNA sequences. First, fragments whose ends overlap with those of the known ones are identified in the "library"—the sharing of sequences indicates that these come from chromosome regions adjacent to the previously identified gene locales. Once this is determined, the newly mapped fragments can be used for establishing the locations of other fragments with which they overlap; this can be repeated, permitting investigators to move further and further from the original known genes.

Also helpful is the synthesis of nucleic acid "probes" for use in *in situ* hybridization (Section 2.3.4 and Fig. II–30) and for purification of desired DNAs and RNAs from complex mixtures. Radioactive RNA transcripts of purified DNAs can be made by incubating the DNAs with RNA polymerases plus radioactive precursors (triphosphate forms of the nucleotides). DNA copies of RNAs can also be made by using *reverse transcriptases* (Section 3.12.3); it often is easier to purify particular mRNAs or other RNA types than to isolate the corresponding DNAs directly from the nucleus. Once made, such "cDNA" (complementary DNA) copies of the RNAs can be produced in quantity by the techniques described in Section 3.2.6 and can be used for study and isolation of the corresponding natural DNA sequences and also in studies of the RNA species to which they are complementary. Recall, for example, that nucleic acids can be purified from mixtures by attaching a molecule with complementary base sequence to an otherwise inert surface such as a filter or a column of small plastic beads (Section 2.3.5). Thus, whether particular cells make a particular mRNA under given circumstances can be determined by preparing the appropriate cDNA and then using it to analyze the RNA species produced under the conditions of interest. This can sometimes be done even by *in situ* hybridization using radioactive cDNAs to detect whether cells contain corresponding RNAs, thus reversing the more conventional procedure in which RNAs are used to detect DNAs (Fig. II–30).

Since they are made by copying mature mRNAs, cDNAs can differ from the corresponding natural genes in lacking features such as intervening sequences (Section 2.3.5). Therefore, it is hoped, for example, that cDNAs will prove useful in achieving expression of eucaryotic genetic information in bacterial systems (Section 3.2.6) that cannot carry out the splicing ordinarily required for production of eucaryotic mRNAs.

Why So Much DNA? (The "C-value" Paradox) Genes are usually detected by mutation. Their presence is revealed by the discovery of inheritable changes in an organism's appearance, chemistry, or other characteristics. From the number of genes thus detected, from estimates of the numbers that code for rRNAs, tRNAs, and comparable nucleic acids, and from studies on special forms of chromosomes particularly amenable to correlated cytological and genetic analysis (Section 4.4.4), it would appear that well-studied organisms such as fruit flies have on the order of 5000 genes; mice, 10,000 to 20,000; and humans, perhaps 25,000 to 50,000.

When such calculations were first done the results seemed quite paradoxical, since they suggested that most eucaryotes have far more DNA than that needed to account for the genes coding for proteins, rRNAs, tRNAs, and so forth. This excess is still notable even with the highest probable estimate of 100,000 genes for mammals, calculated on the basis that roughly 100,000 different species of RNA are produced overall, by the various cells of active tissues, such as brain. (These numbers emerge from hybridization studies aimed at determining how many different species of RNA are present in cells of active tissues; see Chap. 4.4. For higher plants similar studies suggest that roughly 25,000 mRNAs are made.) Even the intuitively "obvious" explanations—that "advanced" organisms require much more DNA than simpler ones—are ruled out by observations that organisms such as salamanders and lilies contain far more DNA per cell than that in apes or humans, although they do not make a correspondingly larger variety of RNAs or proteins.

Much of the apparent DNA "excess" can now be "explained" in the sense that we can categorize the DNAs involved and describe major features of their locations and behavior. A summary of the essential findings follows below. The point is that much of the deviation from the expected relations between complexity of organism, on the one hand, and DNA content, on the other, is accounted for by repetition of DNA sequences and by the existence of DNA sequences that do not code for proteins, or for rRNAs, tRNAs, and the like. In other words, *total* DNA content is often an inaccurate measure of the diversity of genetic information available to and expressed in the organism. Progress in understanding the *significance* of the several DNA categories found in nuclei has been slow: Some are of the types expected from classical genetic analysis; some seem chiefly to be devices by which the cell is able to produce particular types of RNAs at very rapid rates. For the rest, speculation has centered on suggestions that certain of the DNAs reflect evolutionary phenomena (Section 4.5.1), that some regulate the activities of genes or the replication of DNA (Section 4.4.6), and that some are important for special features of chromosome structure (see below). These suggestions are not mutually exclusive, and all could be found to have elements of truth.

Intervening Sequences ("Introns") One major category of DNA not expected from classical analysis is that of intervening sequences (Section 2.3.5). For many of the genes studied so far, such DNA is present in amounts several- to many-fold greater than the DNA of the expressed sequences of the same genes. In other words, a protein of 350 amino acids can be specified by a sequence tens of thousands of nucleotide pairs long, rather than only the thou-

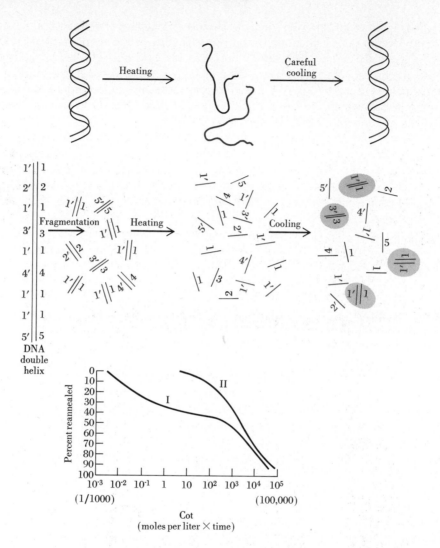

sand or so pairs needed to code for its amino acid chain. The notion that genes generally are made up of separate segments has spurred conceptual progress on many fronts. The most dramatic advance so far is the analysis of immunoglobulins in terms of separate coding sequences that are reassorted during development (Chap. 3.6C). Section 4.5.2 will discuss some evolutionary considerations. Still unclear, however, is whether intervening sequences serve positive functions beyond separating expressed sequences—in gene regulation, perhaps, or in the structural organization of chromosomes. Bits and pieces of evidence suggest they do: For example, through differential processing of RNAs, the same DNA sequence can code for distinct proteins or forms of pro-

◄ **Figure IV–19** *DNA reannealing.* Reannealing is also called *reassociation* or *renaturation.*

(Top), The reestablishment of a double helix by two DNA strands separated by heating.

(Middle), Reannealing of DNA, containing repetitive and nonrepetitive sequences, that has been fragmented by ultrasonic vibrations or other means. The annealing has reached an intermediate stage; at a later time, the remaining single-stranded DNA will also return to the double-helical form. *1* and *1'*, *2* and *2'*, and so on, indicate complementary sequences present on the two DNA strands. The double-stranded 1–1' fragments form more rapidly during reannealing because there are several 1' fragments with which a given 1 fragment can reassociate; it need not "find" its original partner. For the nonrepetitive sequences *(2, 3, 4, 5)* there is but one partner.

(Bottom), Reannealing time course for eucaryote DNA. Curve I is the type of pattern generally obtained. If few or no repetitive sequences were present, a curve like II would be obtained. Each point on the curves represents the percent reannealing for a given *"Cot"* value. *Cot* is *t* (the time elapsed since reannealing started) multiplied by *Co* (the total concentration of DNA in the solution used for the experiment). The use of *Cot* facilitates combining and comparing data ob tained using different *Co*s. Reannealing is faster at higher DNA concentrations. Thus for convenience and accuracy in constructing curves like the present ones, the relatively slow reannealing of the unique sequences (right-hand portion of curve I) is often studied with high *Co*s, while the very rapid reannealing of the highly repetitive sequences (left-hand portion of the curve) is studied at low *Co*s. (Based primarily on work by R.J. Britten and D.E. Koehne.)

teins (Section 3.6.10). In addition, for yeast mitochondria, genetic evidence has been interpreted tentatively as meaning that some of the introns specify RNA sequences for portions of enzymes ("maturases") that process RNAs. It is interesting that one of the maturases may catalyze splicing reactions thought to destroy the mRNA that codes for its own production, and by so doing regulate its own synthesis.

Repetitive DNAs   When purified DNA duplexes are separated into their complementary strands, the strands can reconstitute double helices in the test tube by base-pairing between complementary sequences (Fig. IV–19). The rate at which such *reannealing* occurs depends upon the frequency with which appropriate strands with complementary base sequences encounter one another during the random collisions occurring in solution. Especially if the DNA is fragmented into pieces a few hundred base pairs long before the strands are separated, reannealing may have a complex time-course (Fig. IV–19). This complexity is observed with the DNAs from the nuclei of virtually all eucaryotes, whereas most viruses and procaryotes show a simpler reannealing pattern (like curve II in Fig. IV–19). The eucaryotic "Cot" curve (see Fig. IV–19, legend) is explained by the coexistence of populations of DNA sequences with very different frequencies in the nucleus. When the DNA is fragmented so that previously linked sequences no longer influence one another's behavior, these populations reanneal more or less independently, at rates reflecting the frequencies with which they are repeated in the nuclei of the cells from which the DNA is obtained.

A proportion of the DNA, generally ranging from 40 to 75 percent for different organisms but sometimes less abundant, is "unique sequence" ("nonrepetitive," "single-copy") DNA: The sequences are present once or a very few times per haploid chromosome complement ("per genome"). This finding is what was expected from genetic studies for genes of the "traditional" sort—those coding for enzymes and most of the other proteins whose effects are detected in usual mutational analysis. Confirming such expectations, mRNAs for a number of proteins have been found to hybridize with DNAs that reanneal at the leisurely rates characteristic of unique-sequence DNAs. This is true even for proteins that can be made in very large quantities such as ovalbumin (Section 2.3.5).

The remainder of the DNA falls into various "repetitive" ("reiterated") classes: Sequences may be present in tens or hundreds of copies up to tens of thousands to millions of identical or similar copies per genome. The sequences are generally classified as "middle repetitive" if present in tens to tens of thousands of copies; as "highly repetitive" when tens of thousands to hundreds of thousands of copies are found; and as "very highly repetitive" if present in hundreds of thousands to millions of copies. Different investigators, however, use somewhat different boundaries and terminology. The repetitive DNAs account for much of the apparent DNA excess not represented by intervening sequences. Their special abundance in some plants and amphibia is largely responsible for the unexpectedly large amounts of DNA per nucleus found in these organisms. However, the cells of certain salamanders contain 30 times as much DNA as those of humans. Only 80 percent of this is repetitive (versus 30 percent in humans); thus, there still is a disparity between salamanders and higher mammals requiring elucidation.

Procaryotic chromosomes are almost entirely of unique-sequence DNA, though as mentioned in Section 2.3.4 a limited amount of repetition occurs for sequences such as the rRNAs.

"Simple-sequence" Repetitive DNAs   As far as is known, the most highly repetitive DNAs do not code for proteins or for distinctive RNAs and are transcribed in the living cell only under quite exceptional circumstances if ever. (The very active nuclei of certain amphibian oöcytes may be one such exceptional case.) For a number of organisms, some very highly repetitive sequences have been purified by centrifugation, on the basis of their density. DNA density increases with increased proportions of GC base pairs (Fig. II–15), and thus DNA species particularly poor or rich in guanines and cytosines can sometimes be separated from the bulk of the DNA as "satellite" bands in density gradients (Chap. 1.2B). When tested through hybridization, none of the cell's RNAs are found to be complementary to such very highly repetitive DNAs. However, radioactive RNA copies of the DNA can be synthesized, as outlined at the beginning of this section, and these copies can be used for *in situ* hybridization.

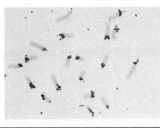

Figure IV–20 *Very highly repetitive DNA localized near the centromere.* This is an *in situ* hybridization preparation (see Fig. II–30) of mouse chromosomes, fixed at mitotic metaphase. After denaturation of their DNA, the chromosomes were incubated with highly radioactive RNA. The latter was prepared in the test tube by adding radioactive RNA precursors plus RNA polymerases to a very highly repetitive "satellite" DNA fraction isolated by centrifugation from mouse nuclei (Section 4.2.6). Thus the RNA is complementary to the very highly repetitive DNA. The DNA in question is a sequence 240 base pairs long repeated 10^6 times per genome. Clusters of grains are found close to the centromere (in mice, the centromere is located near the end of each chromosome), indicating that it is here that the very highly repetitive DNA is concentrated. × 750. (From Pardue, M.L., *Science* **168**:1356–1358, 1970. Copyright 1970 by the American Academy for the Advancement of Science.)

For a variety of species, such approaches show that certain heterochromatic regions of the chromosome, particularly those located near the centromere (*centric* or *centromeric* heterochromatin), are rich in very highly repetitive DNA sequences (Fig. IV–20).

One of the very highly repetitive DNAs of *Drosophila* has the sequence ACAACT arranged head to tail in repeating stretches ("tandem arrays") hundreds of base pairs long. (ACAACT is the sequence on one DNA strand; the other strand, of course, has the complementary sequence.) Other *Drosophila* sequences differ, and very highly repetitive DNAs in other species frequently have more complex sequences. In general, however, the sequences are relatively simple when compared with those that code for RNAs or proteins.

Plausible guesses can be made about how the staining behavior, folding patterns, or the timing of replication of chromosome regions rich in such sequences might come to differ from the average, as is observed to be the case for heterochromatin (Sections 2.2.5, 4.2.1, and 4.2.4). More mysterious is whether the clustering of the very highly repetitive sequences is significant for functioning of the centromere or the kinetochore at mitosis: The centromere region is the last portion of the chromosome to separate structurally at mitosis. It may be duplicated by G_2 (Section 4.2.7), but chromatids remain linked at their centromeres until anaphase. The centromere, or, when present, its kinetochore, is the place at which the chromosome attaches to the spindle (Fig. IV–21). Molecular explanations for these distinctive features are badly needed, and it is hoped that analysis of highly repetitive DNAs may help to provide them (see also Section 4.2.7).

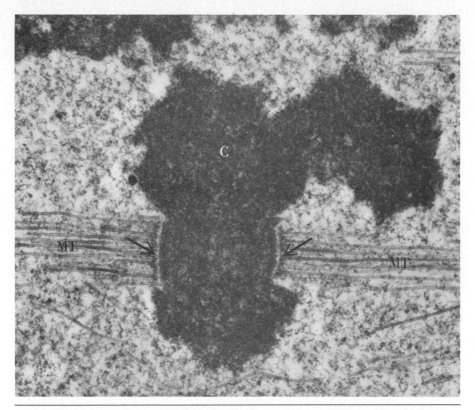

Figure IV–21 *Kinetochores.* Metaphase chromosome *(C)* in the green alga *Oedogonium.* The arrows indicate the kinetochores of what will be two daughter chromosomes once the sister chromatids separate at anaphase. Spindle microtubules are evident at *MT;* the tubules appear to be attached to the kinetochores. The kinetochores themselves exhibit a distinctive layered structure; the molecular organization of the layers is yet to be established although some investigators have put forth the still controversial proposal that the outermost zone is organized as a diffuse "corona" whose structure includes chromatin with which the spindle microtubules interact. × 35,000. (Courtesy of M.J. Schibler.)

Also to be explained is how very highly repetitive DNAs, and others of the repetitive categories, are kept reasonably homogeneous in sequence. For some of the DNAs, the uniformity is precise, with all copies in a cell or organism being identical or close to identical. For others, the homogeneity is less exact. The sequences do evolve so that some comparable ones may differ even between closely related species. This finding indicates that, as with all other DNAs, mutations can occur. Theoretical considerations indicate that without mechanisms to maintain the similarities among the members of a particular category of repetitive DNAs, mutation would produce substantially more divergence within a given organism than that actually found. One proposal envisages the operation of correction mechanisms through which sequences of the

DNAs of a particular type are somehow compared with one another and divergences repaired. Another suggests that repetitive sequences undergo regular cycles in which their number is diminished and then reestablished by selective replication of one or a few "master copies." The latter process could resemble gene amplification mechanisms (Section 4.4.6). Most popular in recent years are suggestions that repetitive DNAs arise somehow through complex crossing-over events analogous to the ones that can generate gene duplications (Section 4.5.1); Older views postulated "rolling circle" DNA replication (Fig. IV–19) or other amplification mechanisms. More recent speculation focuses on movable DNA (see below). All of these potential mechanisms for the evolutionary origins of repetitive DNAs have been invoked in one or another proposal for how many copies could be kept similar in sequence.

Repetitive DNAs of Known Functions   The DNAs present in several copies per genome, to hundreds or thousands of copies per genome, include types we have already discussed, such as the sequences coding for rRNA. Those coding for tRNA and those coding for histones also are of this category. That repetitive sequences for the rRNAs and tRNAs arose in evolution probably reflects the fact that many cells must make large numbers of these RNAs rapidly, and all cell types need them during much of their life history.

For the most part, when large quantities of *proteins* are required, transcription of many mRNAs from the same very few DNA sequences provides them. The repetitiveness of histone genes is attributable partly to the fact that virtually all dividing eucaryotic cells must make large numbers of histone molecules each time they divide. Even for other proteins, however, the situation can be more "interesting" than "a single DNA sequence per protein type per genome." For a number of types of protein there are several genes that code for similar though often distinguishable forms of the protein. The actins provide one example (Section 2.11.2). A more extensive group, with as many as several dozen members, is the family of keratin genes coding for feather proteins in birds; several keratins, distinct but related in amino acid sequence, are coded for by the genes in this group. A large family of genes codes for the proteins of the chorion of insect eggs—a tough protective covering secreted by follicle cells (Section 3.10.2). In some of these families, particular genes are present in a few apparently identical copies, perhaps facilitating rapid production of the corresponding proteins. The *divergences* within the families can reflect the evolutionary origins of different genes and therefore of different proteins, from common ancestors (Sections 4.5.1 and 4.5.2). Small differences among proteins can have effects—subtle or substantial—on their functioning and control. The variable regions of immunoglobulins (Section 3.6.10) provide an example of an extensive family of related protein sequences with varying functional capacities. More speculative is the possibility that even when the proteins specified by a family of genes are functionally identical, different genes in the family are associated with varied regulatory DNA segments (Section 4.4.6), so that different cell types can use the proteins differently; this might be a device for coordinating the production of these proteins with the synthesis of other proteins, in varied timing or patterns appropriate to different contexts. Commonly, different genes within a family are expressed (transcribed and translated) at different times in development. During the development of humans or other mammals, for example, different, closely related genes are used at different stages for production of one pair (the β-chains) of the two pairs of globin chains that constitute hemoglobin molecules. The hemoglobins differ correspondingly; those in the fetus have a higher affinity for oxygen, permitting efficient gas exchange across the placenta.

The genes within these various families may differ somewhat in nucleotide sequence, but when they are sufficiently similar, their DNA can behave like repetitive sequences in experiments of the sort illustrated in Figure IV–19. To reanneal, two DNA polynucleotide strands need only have long stretches of complementary base sequences. Perfect matching along the entire length is not required, although perfect matching can be distinguished from imperfect matching because of effects on the stability and other properties of the duplexes formed.

There are, as well, some rare, intriguing situations in which cells selectively produce extra copies of what are normally unique-sequence DNAs, "converting" them to middle repetitive types. As outlined in Section 3.11.3, certain lines of tissue-culture cells exposed to potentially toxic drugs can, in this way, amplify genes needed to produce detoxifying enzymes, hinting at control mechanisms now being actively sought. Note that this situation is quite different from the ones in which the entire genome or large parts of it are amplified, as in the formation of macronuclei in ciliated protozoa (Section 3.3.2); it resembles, more, the production of extra nucleoli in oöcytes (Section 2.3.4), though with at least some of the drug resistance genes, the "extra" copies remain in the chromosomes rather than separating as with the nucleoli.

Amplification of much of the genome by repeated chromosome replication *is* seen in certain tissues of insects such as the silk glands of silkworms gearing up to produce massive amounts of the silk protein *fibroin,* or the salivary glands of several insect genera (Section 4.4.4).

As is true for the DNAs coding for rRNAs, the other repetitive DNAs of known function sometimes are clustered together in discrete regions of the chromosome. In animals such as *Xenopus,* those coding for 5S RNA are in clusters at sites found at the ends of many of the metaphase chromosomes (see also Section 4.5.1). In humans, on the other hand, most 5S RNA genes are part of a single cluster. In some types of organisms the genes for histones are highly clustered, while in others they may be somewhat dispersed, with individual copies or small groups separated from the rest. The DNAs coding for tRNAs also vary in distribution: In yeast, for example, few if any large clusters are found, whereas in some higher organisms many of the DNA copies coding for a particular species of tRNA are grouped together, with spacer DNAs separating the individual copies in a manner reminiscent of the rDNAs (Section 2.3.4). The genes in the multiprotein families are found in a variety of arrangements; clustering is common but not universal.

Interspersed Repetitive DNAs The remainder of the repetitive DNAs are sometimes lumped as "middle repetitive," but they are heterogeneous in terms of size, complexity of sequence, and copy frequency (they range from tens of copies to hundreds of thousands per genome). Insufficient order has emerged from this diversity to permit confident generalization. Very likely, several different types of DNAs coexist in this grouping.

Although the boundaries between the categories may not be sharp, most of the repetitive DNAs in question seem to be more complex in sequence than the simple highly repetitive sequences described above. Yet, they appear not to carry information for proteins or for the well-known RNA species like the rRNAs or tRNAs. (The necessity for the use of qualifying and tentative language here reflects the fluidity of the field; very little has been established beyond doubt, though matters are advancing quite rapidly.) Some are transcribed and thus are represented in the nuclear RNAs and to a markedly lesser extent in cytoplasmic RNAs. Nucleotide sequences corresponding to "middle repetitive DNAs" are, for example, common in hnRNAs but are absent or markedly reduced in size and abundance in the mRNAs to which the hnRNAs give rise. Apparently they are among the sequences that are selectively excised during splicing and other processing steps; accordingly, the mature mRNAs are chiefly derived from unique-sequence DNAs even though the RNA molecules originally transcribed contained unique sequences and middle repetitive sequences linked to one another. Section 5.1.2 will describe recently discovered species of small RNAs that may be related to middle repetitive DNAs.

Often these DNAs are not all clustered together; rather, individual copies or pairs or small groups are scattered widely among the unique-sequence DNAs—hence the name "interspersed" or "dispersed" repetitive DNAs. Their distribution seems to differ somewhat in different organisms, but in many, the DNA, when carefully fragmented, yields fairly small pieces in which unique-sequence DNA is present along with one of the repetitive sequences. This proximity ties in with the ability of RNA polymerases in the cell to transcribe RNA molecules containing information from both types of DNA.

Based on such findings, an idealized model applicable to several types of organisms suggests that the interspersed sequences can be thought of as collections of families. The sequences within each family range up to a few hundred nucleotides in length (or, in some cases, up to a few thousand) and are very similar to one another, though not necessarily strictly identical. Members of a given family are associated with different unique-sequence DNAs, being found adjacent to genes or within intervening sequences. In *Xenopus*, there are 10,000 to 20,000 distinguishable families of DNA sequences repeated hundreds of times; these may fall into larger groupings. The "Alu" family, present as the major interspersed repetitive DNA in human cells, as well as related sequences in various mammalian cell lines, is made up of several subfamilies and totals about 500,000 copies of similar sequences roughly 300 nucleotide pairs long. This implies that there is one "Alu" sequence for every few thousand pairs of other nucleotides in the genome and that Alu sequences account for approximately 5 percent of the DNA in the nucleus of a human cell. hnRNAs contain sequences transcribed from members of this family. (Section 5.1.2 describes some other tantalizing features of the sequences and of RNA molecules that may be related to them. The name "Alu" refers to one of the "restriction enzymes" [Section 3.2.6] used to characterize the DNAs.)

The model of repetitive DNAs as interspersed among "ordinary" genes, which stresses similarities rather than the differences within families, was developed to help to conceptualize possible functions of the DNAs. Different genes might be expected to resemble each other in sequences coding for RNA regions subject to common processing steps or to be similar in "promoter" sequences, "origins of replication," and other DNA sequences recognized by enzymes and other proteins responsible for DNA transcription or replication (Sections 4.2.4 and 4.4.6). Some such sequences might be represented in mature mRNAs, especially in the "flanking regions" (Section 4.4.6), those adjacent to the portion of the molecule that actually encodes information for amino acid sequences. Others might be removed during processing of precursor RNAs, accounting for the observations on hnRNAs mentioned earlier; perhaps these repetitive sequences are recognized by enzymes that process the RNAs. It is also conceivable that coordinated control over different genes could be exerted through influences on DNA neighbors they have in common. Different interspersed families, then, might regulate the transcription of different collections of genes. Alternatively, they might control chromosome folding, providing recognition devices whereby previously distant segments of the chromosome fiber could reproducibly be brought into close proximity. Precise folding is required for the gene-to-gene pairing phenomena of chromosomes in meiosis (Chap. 4.3).

Movable DNAs   Attractive as such schemes for possible functions of repetitive DNAs are, they remain hypotheses with little direct experimental backing. Before the proposals can be evaluated adequately, account must be taken of an extraordinary complication that has become evident only recently. Detailed study of the distributions of DNA sequences in bacteria, yeast, and *Drosophila* has shown that some DNA sequences are located at different places in different members of the same species. Comparable sequences seem to exist in corn where, in fact, the first clear hints of movable genes were obtained through genetic studies done before the roles of DNA in heredity were appreciated. Such sequences are strongly suspected to occur among the interspersed repetitive sequences of many organisms and cell types, including mammalian cell lines. The most detailed information has been gathered in bacteria, for the usual reasons (Chap. 3.2B).

These sequences with varying location are sometimes called "insertion elements," "transposable elements," or "transposons," depending on details of their characteristics. They range in length from a few hundred nucleotides to several thousand. Each is characterized by specific short DNA sequences at its ends, which appear to be crucial for the ability to move. (Different movable elements differ in the sequences, but in each the same sequences are present at the two ends [as "inverted repeats" and "direct repeats"; Section 5.1.2], which might permit the enzymes and other proteins, largely unknown, that accomplish the "movement" to recognize the movable elements and to interact with them in similar manner at both their ends.) One set of movable sequences in yeast is present in 30 copies per genome. A number of the families of sequences in higher eucaryotes present in thousands to hundreds of thousands of copies may also be movable.

In bacteria and plasmids, genes present in movable elements confer resistance to antibiotics. Direct genetic activity of movable elements in eucaryotes has not yet been detected, but as is true also in bacteria, some may be linked to ordinary genes that can be carried along when the movable elements move. Several are thought to be transcribed in the cell.

Movable elements can move both from one part of a chromosome to another and from chromosome to chromosome. In some cases they insert within genes, interrupting the normal DNA sequence. In others, they insert next to genes. Evidently, their movement is different from the insertion and excision of episomes (Section 3.2.6) in that the elements do not maintain an existence independent of the chromosomes. For bacteria at least, the movement does not *require* physical departure of the element from its original site; rather, a transposable element on one chromosome can remain in place and somehow generate a copy in another chromosome, probably through mechanisms related to genetic recombination (Section 4.3.4). However, the elements *can* physically leave chromosomal sites and in this sense are truly mobile. Various lines of speculation suggest that episomes, some viruses, and movable elements are different evolutionary states of similar types of nucleic acid sequences whose capacities for independent existence depend on whether or not they have incorporated the appropriate genes from their hosts at some point in their history. It is also speculated that DNA sometimes may "move" by still-to-be-discovered processes of reversed transcription (see Section 3.12.3) in which RNA copies of a gene serve as templates for production of DNA at a new chromosomal site.

The movement of a movable DNA sequence into the vicinity of an ordinary gene can alter the expression of that gene, making it subject, for example, to new regulatory controls, or turning on or off the expression of the gene. As they move, the elements also promote large-scale chromosomal changes, including rearrangements of large portions of the chromosome fiber (see Section 4.5.1). Their discovery is too recent for consensus to have developed about their roles or significance, but their existence confirms the suspicion, already justified by the DNA rearrangements found in antibody-producing cells, that cellular DNA is not quite as staid a molecule as was often assumed. Its plasticity has only begun to be factored into our understanding of chromosome structure and behavior in development and evolution.

4.2.7 Spindles; Kinetochores; Centrioles

Light Microscopic Findings   The precision of chromosome behavior in mitosis reflects the operation of the spindle. This highly organized, fairly rigid, gel-like region of cytoplasm has a fibrous appearance in the light microscope.

Three types of spindle "fibers" seem to be distinguishable by light microscopy: *continuous* ones that pass from pole to pole, *chromosomal* or *kinetochore* fibers that attach to the chromosome kinetochores, and *interzonal* fibers that are present between two groups of chromatids as they separate at anaphase (Figs. IV–6, IV–24, and IV–26). The interzonal fibers probably include the continuous fibers that persist after the chromatids move away from the spindle equator in anaphase and telophase.

In some organisms, the "continuous" (pole-to-pole) fibers of the spindle are concentrated in the center, with the kinetochore fibers and their attached chromosomes arranged around the periphery. In others, the two types of fibers are more extensively intermingled.

Demonstration of spindle fibers in living cells using polarization microscopy (see Section 1.2.2 and Fig. IV–22) resolved a dispute lasting many decades about the reality of the spindle's fibrous organization.

In animal cells, spindles generally form in association with the centrioles (Fig. IV–6). More precisely put, the fibers extend from the centrosomal region surrounding the centrioles (Section 2.10.6). At each pole, an aster is present (see Figs. IV–6 and IV–24). In the light microscope asters appear as roughly spherical arrays of fibers oriented radially around the centrioles and merging with the spindle. Centrioles and asters are not seen in many sorts of plant cells.

Electron Microscopic Findings   Electron microscopic study shows that spindles (and asters) contain numerous microtubules (Fig. IV–21) as well as ribosomes and vesicles, many of which probably derive from the endoplasmic

Figure IV–22   *The spindle of a living egg* of the worm *Chaetopterus,* photographed through a polarizing microscope (Section 1.2.2). The equator where the metaphase chromosomes (not visible in this photograph) are located, is indicated by *E*. Clear indication of the fibrous organization of the spindle may be seen at *F*. The plasma membrane is present at *P*. × 1500. (Courtesy of S. Inoue.) See also Figure IV–24.

Figure IV–23 *Microtubules in the spindle.* Cell from an onion root tip fixed at metaphase of mitosis and stained immunohistochemically to demonstrate tubulin (Section 1.2.3 and Fig. II–86). The "fibers" of the spindle stain strongly, reflecting the presence of microtubules. The arrows indicate the two ends of the spindle. × 850. (From Wick, S.M., R.W. Seagull, M. Osborn, K. Weber, and B.E.S. Gunning, *J. Cell Biol.* **89**:685–690, 1981. Copyright Rockefeller University Press.)

reticulum. The fibrous organization observed by light microscopy is found to be due chiefly to the presence of bundles of microtubules (Figs. IV–23). However, the distribution of the individual microtubules is not as neatly describable as the light microscopic picture of continuous and kinetochore fibers would lead one to expect. Microtubules do terminate at the poles and at the kinetochores (Fig. IV–21), and there are individual ones that run from pole to pole and from near the poles to the kinetochore. As with many other features of the spindle, however, details of this sort vary considerably among different organisms. In many organisms, microtubules of such "simple" types are a minority. Instead, numerous tubules extend only part way from the kinetochore to the poles, or part way from one pole to the other; some are "free," that is, ending neither at a pole nor at a kinetochore. The bundles constituting the light microscope "fibers" evidently contain microtubules of varying lengths.

Some electron microscopists regard the spindle as being chiefly composed of two categories of microtubules: kinetochore tubules, responsible for the association of chromosomes with the spindle, and "framework" tubules of several sorts, which overlap to establish collectively the pole-to-pole integrity of the spindle. Others think the spindle is better described as composed of two half-spindles (see Fig. IV–26), each consisting of some microtubules that connect to kinetochores and some that extend toward the opposite pole. Whether one view or the other is more appropriate, or whether both are of equal use and validity, will be clear only with a better understanding of the actual functioning of the spindle. There is need also for more thorough evaluation of possible sources of distortion in the microscopic image, such as alterations in microtubules that might occur during preparation of specimens for microscopy.

Composition of the "Mitotic Apparatus" The *mitotic apparatus* (spindles, centrioles, asters, and chromosomes) can be isolated as a unit from developing eggs (Fig. IV–24). The apparatus retains much of its original morphology, and chromosomes remain attached to it. However, it has not yet been possible to induce normal movement of chromosomes on isolated spindles, although as will be seen, progress in this direction is being made with

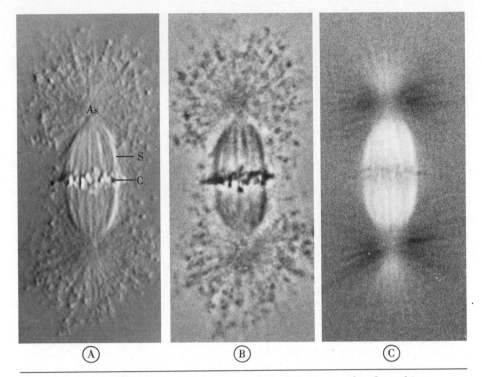

Figure IV–24 *A spindle isolated from an egg* of the sea urchin *Lytechinus* as seen with three different light microscopic optical systems (Section 1.2.2): A. Differential interference contrast (Fig. I–12); B. Phase contrast; C. Polarization. In panel A the position of the chromosomes is indicated by the letter *C*, that of the asters by *As* and that of the spindle itself by *S*. The fibrous organization of the spindle is readily apparent. Approximately × 1,000. From Salmon, E.D., and R.R. Segall, *J. Cell Biol* **86**:355, 1980. Copyright Rockefeller University Press.)

"permeabilized" (introduction to Chap. 2.11) or partially "lysed" cell preparations. Much material is lost from the spindle during isolation, and there may be other alterations rendering it nonfunctional. As expected, some of the proteins that can be obtained from isolated spindles are tubulins, like those of other microtubule-rich structures. Nevertheless, microtubules as such may account for less than 25 percent of the total mass of isolated spindle preparations. Little is known with certainty about the remainder. Immunohistochemical procedures show tubulin (Fig. IV–23) and microtubule-associated proteins in the spindle. Actin and some myosin have been detected in this way as well; correspondingly, microfilaments can be seen in suitably prepared electron microscopic material. Certain of the regulatory proteins, notably calmodulin, also are present.

Assembly   Spindle proteins such as tubulins are present in the cell in large amounts long before the spindle forms, and spindle formation involves primarily assembly of previously made subunits. Eggs and early embryos have large pools of spindle proteins to sustain their rapid divisions. As Section 4.2.3 mentioned, other cells often change shape markedly before undergoing cell division—animal cells frequently "round up" and in so doing loosen their associations with adjacent cells. Dividing tissue-culture cells decrease their contact with the substrate on which they are growing. These shape changes reflect reorganization of the cytoskeleton, including the recruitment of tubulin molecules for construction of spindle microtubules by depolymerization of microtubules of other cell areas.

Presumably, control of spindle assembly depends on cellular regulation of levels of available tubulin and of Ca^{2+} and other components, as well as on activation and deactivation of organizing centers. The assembly itself is based chiefly on hydrophobic interactions and other noncovalent bonds (Section 1.4.1).

Centrosomes; Centrioles   In most cells with centrioles, the spindle forms between the centriole pairs as they separate (in prophase; Fig. IV–6). Centriole duplication occurs well in advance of mitosis and may begin as early as telophase of the prior division. In interphase, as DNA synthesis progresses, each centriole of the initial pair is accompanied by a new daughter procentriole that will mature into a centriole oriented at right angles to the parent (Section 2.10.7. (Centriole duplication continues even if DNA synthesis is experimentally inhibited, but not if the nucleus is removed from the cell.) Generally the nuclear envelope breaks down in prophase as the chromosomes become associated with the spindle.

Experiments of the sort illustrated in Figure IV–25 suggest that as the centrioles duplicate, there is a corresponding duplication of the centrosome-associated microtubule-organizing capacity. Other studies show that under appropriate conditions of ion concentration and temperature, aster-like groupings of microtubules will assemble in homogenates of clams' eggs activated to begin cleavage, but no such asters form before activation. The interpretation is that shortly before the onset of cleavage, microtubule-organizing capacities of the centrosome region are somehow "turned on." Comparable though more hypothetical changes have been invoked to explain the cyclical formation and depolymerization of spindle microtubules during the cell cycle of cells in general; initial findings hint, for example, that the microtubule-organizing capacities of the centrosomal region increase in early mitotic stages and then decline at anaphase.

Centromeres; Kinetochores   Less is known about centromere and kinetochore behavior than about centrioles. (Recall that as used here and by many other authors, the centromere is the *region* by which the chromosomes attach to the spindle, and the kinetochore, a specialized structure present within this

Figure IV–25 *Microtubule organizing centers (MTOCs;* Section 2.10.5) *in cul-*tured mouse cells (3T3 line). The MTOCs were visualized by depolymerizing the cells' microtubules (the drug Colcemid was employed to do this) and then observing where new microtubules form when the cells are exposed to purified tubulin. This exposure is carried out in the presence of a detergent (Triton X100) that makes the cells quite permeable to large molecules and under conditions that promote microtubule assembly. The field shows the microtubules that have assembled in two cells, as visualized immunohistochemically (Section 1.2.3 and Fig. II–86) with fluorescent antitubulin antibodies. The cells' nuclei are evident as large pale globular shapes. The microtubules are arranged radially around the MTOCs that initiated their assembly (arrows). One of the cells has a single MTOC, the other, two. This difference reflects the fact that the cells are at different stages in the cell cycle; in the 3T3 cells duplication of MTOCs, in the functional terms revealed here, occurs in the middle of S. Electron microscopy indicates the MTOCs are associated with the centrioles. Approximately × 1000. From Brinkley, B.R., S.M. Cox, D.A. Pepper, L. Wible, S.L. Brenner, and R.L. Pardue; *J. Cell Biol* **90**:559, 1981. Copyright Rockefeller University Press.)

region, in many organisms, that seems to be involved in attaching the chromosome to spindle microtubules; see Figs. IV–3 and IV–21 and Section 4.2.1.) A discrete, recognizable kinetochore is not necessary for attachment to the spindle: Organisms such as yeast lack such a structure, seeming to show, instead, direct association of chromosome regions with microtubules. Even in certain of the organisms that do have kinetochores, some spindle microtubules may interact with other portions of the centromere as well. Recently the specific DNA sequences present at the centromere regions of yeast have been isolated and cloned. They show some unusual features, such as a base composition rich in As and Ts; there are also indications that the nucleosome structure may be somewhat different from the average. Investigators are following these leads in the hope of determining, for example, whether they signify special capacities for interacting with spindle proteins. Section 4.2.6 pointed out that in many organisms very-simple-sequence, highly repetitive DNA segments are concentrated near the centromere.

Kinetochores are not invariant either. Most animal cell types and many plants show a well-defined disc with layered structure (Fig. IV–21 shows one kind of arrangement) that is located in a pinched-in appearing centromere *(primary constriction)* of metaphase chromosomes (Fig. IV–3). Layering is less evident in the kinetochores of some plant cells. Meiotic chromosomes of *Drosophila* have kinetochores that lack the innermost layer seen in Figure IV–21. Certain insects have "diffuse kinetochores"—a band of electron-dense material comparable to that seen at conventional kinetochores runs along much of the length of the surface of the metaphase chromatids, and association with the spindle involves this extended surface rather than the limited zone seen as in most cases.

In organisms possessing well-defined kinetochores, by the time the chromatids become visible in prophase, each has its own kinetochore (Fig. IV–21). Consequently, even though chromatids remain attached to one another in the region of their kinetochores, each can associate with the spindle by its own set of kinetochore microtubules. Very recently, antibodies that bind specifically to still-unknown components of the kinetochore have been found in the blood serum of some humans with autoimmune disorders (Section 3.6.11). When fluorescent derivatives of these antibodies are used to stain dividing tissue-culture cells, twice as many stainable structures are seen in nuclei of cells in G_2 as those present earlier in interphase. This suggests that the kinetochore becomes structurally duplicated by G_2. Kinetochores can act as microtubule-organizing centers, at least in experimental situations such as when isolated chromosomes are incubated in a tubulin-rich solution under conditions favoring microtubule assembly (Section 2.10.5).

Microtubule Polarity   Techniques like the ones illustrated in Figure II–84 have been used to evaluate the polarity of spindle microtubules. Those microtubules radiating from the spindle poles are oriented with their (+) ends (the ends characterized by more rapid assembly and disassembly; Section 2.10.5) directed away from the poles. Their (−) ends are at the poles, as might be expected if growth were initiated there. There is still disagreement about the kinetochore tubules: When microtubules are caused to grow in association with chromosomes under test-tube conditions, their (+) ends may point away from the kinetochore; but an opposite orientation, in which the (+) ends are at the kinetochore, has been reported for at least some of the kinetochore microtubules present normally in a few cell types. It is expected that the polarity data will eventually be very useful in determining how the spindle forms and functions (see below). Remember that microtubules can "polymerize" or "depolymerize" from either end, depending on concentrations of tubulins and on factors such as whether one or the other end is blocked, but that intrinsic rates of addition and loss of tubulin differ at the two ends. Remember also that special organizing sites may not always be needed to initiate tubulin polymerization. Experiments now being done should soon demonstrate if the oriented growth of microtubules from the spindle poles reflects simply initiation there, and subsequent stabilization of the $(-)$ end by the microtubule-organizing material. Additional experiments should clarify whether kinetochores normally initiate growth of the microtubules with which they associate, whether they attach to preexisting tubules, or whether both phenomena occur (see below).

Diverse Modes of Spindle Formation Spindles in some organisms are made of a few microtubules; in others there are thousands. Cases are known in unicellular organisms, especially certain yeasts, in which only a single microtubule constitutes an entire "kinetochore fiber," in contrast to the situation in Figure IV–21, where the fiber is of a bundle of microtubules. We have already mentioned that the arrangement of the tubules within the several categories of "fiber" also varies. Not surprisingly, such variation extends to details of spindle formation. For instance, the many types of plant cells that lack structures obviously corresponding to centrioles form spindles that separate the chromosomes in essentially the same manner as in centriole-containing cells. Clearly, then, centrioles as such are not required for spindle assembly or functioning. It has even been proposed that the location of centrioles, when present, at the poles of the spindle is a device insuring orderly segregation of the centrioles to daughter cells rather than having anything to do directly with spindle formation.

In some protozoa and in fungi and algae, the spindle forms within the nucleus, and the nuclear envelope does not break down. In other cells, even though the envelope remains in place, the chromosomes attach to an extranuclear spindle in a manner not understood (Section 4.2.10). Yeast cells and some other fungi form an intranuclear spindle between plaques of electron-dense material (called *spindle pole bodies* or *nucleus-associated organelles*) associated with the nuclear envelope. Before division, the single plaque initially present duplicates, and the resulting "daughter" plaques migrate around the surface of the nucleus with microtubules forming between them; this migration continues until the two spindle pole bodies come to lie at opposite poles of the nucleus, with the spindle running between them.

In some protozoa, even if the nucleus is removed from the cell, a spindle of continuous fibers forms between the centrioles. On the other hand, in some insect and plant cells, spindle fibers are seen to form initially in association with the kinetochores of the chromosomes. Accordingly, various compromise schemes propose that kinetochore fibers are a mixture of microtubules nucleated at kinetochores and microtubules nucleated at the poles. Note, however, that few of the cases where spindle fibers seem to form, under normal conditions, initially associated with kinetochores, have been studied by electron microscopy. In one case that has, an insect, the "fibers" turned out to be membranous structures that prefigured the distribution of microtubules but did not include the microtubules until later in the process of spindle formation. Thus, even though kinetochores can organize microtubules under test tube circumstances, their capacities to do so in normal cells remain matters of controversy.

Overall, the available information does not yet permit convincing generalization about how the spindle forms or about how the centrosomal region and centromeres are involved. It is still difficult to reconcile the variety of observations on spindle formation in different cell types with the apparent near-uniformity of the end result of spindle functioning in cell division.

Questions About the Movements of Chromosomes in Prophase and About the Cytoskeleton Among questions to which answers are still being sought are the following:

1. How do the chromosomes attach to the spindle so that the two mitotic chromatids separate one to each pole, rather than both moving to the same pole? Is this simply a matter of the location of kinetochores on opposite surfaces of the chromatids (Fig. IV–21)?
2. How is the precise alignment of chromosomes at metaphase brought about? Is the notion that the alignment results from balanced pulling from both poles valid? (That much subtler mechanisms may also contribute is suggested by circumstances under which abnormal spindles form with only a single pole [*monopolar spindles*] but where the chromosomes nevertheless accumulate at what would be the spindle equator were the second pole present.)
3. Do special associations exist, during interphase, among chromatin strands, centrioles, and the nuclear envelope? Do such associations account for the behavior of chromosomes during duplication and early stages of division or for the fact that chromosomes in certain cells move little during interphase, from the positions they had at the prior telophase (Section 4.2.4)?

In some cells the chromosomes move within the prophase nucleus along a path paralleling the movement of the centrioles outside the nucleus. In fact, during interphase and prophase the centrosomal region sometimes seems anchored to the nuclear envelope, remaining attached even if the cell is experimentally disrupted. In several unicellular organisms chromosome associations with the envelope are integral to the mechanisms by which the chromosomes are segregated into daughter nuclei (Section 4.2.10). A few investigators are convinced that kinetochores tend to lie close to the nuclear envelope during interphase of "higher" eucaryotes as well, but studies with the fluorescent antibodies that stain kinetochore regions have so far failed to support this as a general phenomenon. On the other hand, association of chromosome *ends* with the nuclear envelope is common in prophase stages of *meiosis* (Section 4.3.4).

The cytoskeletal and motility-generating components most studied in dividing cells are the ones that comprise the spindle proper, especially the microtubules. However, studies on tissue-culture cells indicate that actin-rich micro-filament bundles become much less prominent as a cell prepares for mitosis and that intermediate filaments change their distribution as well. In other words, division profoundly alters much of the cell's "skeleton." The intermediate filaments generally are excluded from the spindle but lie in groups or bundles in the surrounding cytoplasm. Sometimes they appear almost like a basketwork system around the spindle. Might such arrangements provide mechanical support that aids in spindle functioning? Comparable proposals have been made for astral microtubules, especially since in various cell types these can extend out to the plasma membrane where they could anchor.

4.2.8 Chromosome Movement

In separating at anaphase, the daughter chromosomes move at rates in the range of 1 to 10 μm per minute. This is not very fast by comparison with other intracellular movements (Chaps. 2.10 and 2.11); evidently not much force is required. The movement often involves two components (Fig. IV–26). In most familiar cell types and organisms, separating chromosomes approach the spindle poles ("anaphase A" movement). In many cells, the spindle as a whole *also* elongates, so the ends move apart ("anaphase B"). In a few cases chromatid distance from the poles changes little, and the movement apart of the chromosomes depends primarily on such anaphase B separation of the spindle poles.

When the kinetochore is near the middle of a chromatid, a characteristic V shape is seen, with the kinetochore in the lead as the chromosome moves

Figure IV–26 *Chromosome movement during cell division.* A. A diagram of a hypothetical animal cell showing the separation of sister chromatids. The relative contributions of spindle elongation and of movement toward the poles to chromosome movement vary in different cell types. B. The shortening of the chromosomal fibers evident by light microscopy almost certainly involves loss of subunits (tubulin) from microtubules; less certain is whether this occurs exclusively at the spindle poles, as most hypotheses have suggested. C. Various hypotheses propose that kinetochores actively move along the spindle fibers; or that microtubules from the kinetochores slide along other microtubules pulling the kinetochores along; or that elastic or contractile material of still unknown nature moves the kinetochores in directions oriented by spindle microtubules. D. and E. Present evidence suggests that elongation of the spindle may occur by somewhat different mechanisms in different cell types and organisms; two possibilities are illustrated here. In both cases, chromosomes attached to the spindle would be moved apart by the changes diagrammed. Model D assumes that the two half spindles "slide" apart; the microtubules of one half spindle are presumed to be of opposite "polarity" (Fig. II–84) to those in the other, a presumption which seems justified from work with the techniques described in Figure II–84. (After discussions and proposals by S. Inoue, A. Forer, R. McIntosh, Z. Cande, D. Mazia, F. Schrader, A. Salmon, and many others.)

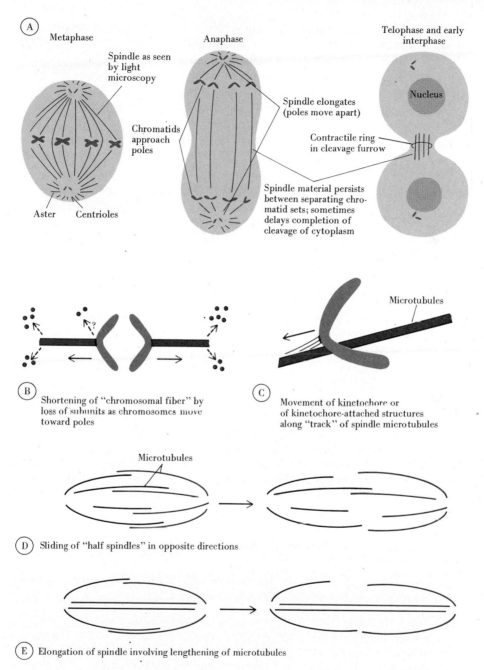

(A)

Metaphase

Spindle as seen
by light
microscopy

Chromatids
approach
poles

Aster   Centrioles

Anaphase

Spindle elongates
(poles move apart)

Spindle material persists
between separating chro-
matid sets; sometimes
delays completion of
cleavage of cytoplasm

Telophase and early
interphase

Nucleus

Contractile ring
in cleavage furrow

(B) Shortening of "chromosomal fiber" by
loss of subunits as chromosomes move
toward poles

Microtubules

(C) Movement of kinetochore or
of kinetochore-attached structures
along "track" of spindle microtubules

Microtubules

(D) Sliding of "half spindles" in opposite directions

(E) Elongation of spindle involving lengthening of microtubules

in anaphase (Fig. IV–6). Chromosome fragments without kinetochores can be
produced by treating cells with radiation or with certain chemicals. Such fragments do not attach to spindle fibers and do not show oriented movement.

(The fragments can, however, simply separate into two corresponding chromatid fragments, indicating that this initial step of chromosome segregation does not require pulling by the spindle.)

The most straightforward conclusion is that the spindle moves the chromatids at anaphase by exerting force on the kinetochores. An alternative view, with a few adherents, is that the kinetochore is an active agent, "crawling," "sliding," or otherwise moving along the microtubules and pulling the remainder of the chromosome along. In either event, the nature of the forces remains the focus of intense uncertainty.

Mechanisms: First Thoughts At first glance, "simple" mechanisms for chromosome movement come to mind: Chromosomal "fibers" could shorten and pull the chromosomes to the poles. Alternatively, the chromosomal fibers with their attached chromosomes could be moved along the continuous fibers. The spindle might elongate in response to pushes exerted on the poles by the middle region of the spindle itself, or to pulls exerted by the asters, when present. Since sliding of microtubules with respect to one another can exert force (Section 2.10.3), sliding seems an excellent candidate for the mechanism underlying chromosome motion. On the other hand, addition of tubulin to one or the other ends of spindle microtubules would permit the spindle to elongate simply by microtubule growth. *Disassembly* of microtubules could account for striking facts about the functioning of the spindle: Studies with ordinary light microscopes or with polarizing microscopes show that as the chromosomes move towards the poles, the chromosomal fibers to which they are attached shorten, but do not get thicker, even when fiber shortening is to 25 percent or less of their original length, as frequently is the case. Nor can thickening or comparable changes be seen in the spindle's microtubules when preparations are examined by electron microscopy. Thus, unlike a muscle, or a stretched rubber band, chromosomal fibers seem to lose part of their mass as they shorten; depolymerization of the microtubules—loss of tubulins—is an obvious way in which this could take place. (The loss is most readily envisaged as occurring at the poleward ends of the kinetochore-attached microtubules, but this certainly is not an established fact.)

The Spindle as a System in Dynamic Equilibrium Other important observations relate to the loss of mass during shortening:

1. Chromosomes can be detached from the spindle by fine glass needles. They are able to reattach and to move in normal fashion during division.
2. Exposure of cells to cold or to high pressure dissolves the spindle and its microtubules; the tubules and spindle reappear upon return to normal conditions.
3. Treatment of cells with a number of agents, including D_2O ("heavy water"; D is deuterium, an isotope of hydrogen), "freezes" the spindle, increasing the amount of ordered material detected by the polarizing microscope. The

spindle remains intact and sometimes actually is larger than normal, but chromosome motion is slowed or prevented. Similar effects are noted when cells are gently "permeabilized" by detergent treatment and then exposed to high concentrations of tubulin or to drugs such as *taxol*—conditions that promote the assembly of microtubules and stabilize them against influences favoring disassembly (Section 2.10.5).

4. Irradiation of a local region of the spindle with a narrow beam of ultraviolet light abolishes the organization of the irradiated portion of the spindle. With the polarization microscope a region of the spindle disorganized in this way, at metaphase or anaphase, can be observed to move toward the pole and to disappear even if chromosome motion has not yet begun. Other observations (Section 4.2.10) also suggest that even at metaphase, when the chromosomes are not moving, there is some sort of movement of material from the equator toward the pole.

The main point that emerges from these considerations (Fig. IV–26) is that the spindle is a system capable of rapid, oriented, and controlled assembly and disassembly. Indeed, many investigators are convinced that the spindle is a structure in dynamic equilibrium with its surroundings and that its functioning depends on the state of this equilibrium. In this view, subunits, of which tubulin dimers almost certainly are centrally important, are continually entering and leaving spindle "fibers" even when the spindle as a whole is not changing in size or shape. During prophase, as the spindle forms, the gain of subunits outweighs the loss. During anaphase, the situation is reversed, and the loss of subunits outweighs the gain, at least for the microtubules attached to the kinetochore. This loss is required for chromosomes to approach the spindle poles. At the same time some of the other microtubules might gain tubulins as the spindle elongates. The polymerization and depolymerization of microtubules occurs at the ends of the structures. Since under appropriate conditions either end of a microtubule is capable of net growth or net loss of tubulin, determining the polarity of spindle microtubules (Section 4.2.7) and establishing how their association with kinetochores or structures at the pole affects assembly and disassembly should be especially illuminating. It is possible, for instance, to imagine various relations of the postulated equilibrium to the "treadmilling" (Section 2.10.5) of tubulins that seems to occur, at least in the test tube.

What Generates the Forces Underlying Movement?   One form or another of the theory that spindle functioning normally requires controlled microtubule assembly and disassembly is now widely accepted. The theory leaves open, however, the question of how the forces producing movement arise. At present the best guess is that different cell types knit together combinations of mechanisms in somewhat different ways. For that matter, even within a given cell, movement of the chromosome toward the poles could well be based on different mechanisms from those that move the poles apart. This is suggested by such observations as the finding that incubation of permeabilized cells in excess tubulin, which seems to retard microtubule disassembly, inhibits the poleward movement much more than the elongation.

Cells permeabilized by the methods mentioned in Chapter 2.11 continue to move their chromosomes if incubated in the proper medium containing ATP; elongation of the spindle is the process most clearly shown to be dependent on ATP. With such preparations, the levels of Ca^{2+} in the medium markedly affect chromosome motion. Together with circumstantial evidence, such as the presence of calmodulin, this finding suggests that Ca^{2+} ions may participate in governing the motion. It is reasonable, therefore, to posit mechanisms similar to those producing oriented movements of other kinds (Chaps. 2.10 and 2.11)—sliding or other lateral interactions between microtubules, participation of microfilaments, or perhaps involvement of the postulated "microtrabecular" latticework. The membranous sacs and vesicles associated with the spindle, some of which are ER-related, could function in manners analogous to the sarcoplasmic reticulum (Section 3.9.1), controlling local calcium distributions. Tests of such proposals are actively under way. A brief summary of important findings follows.

In some organisms, patterns of change in the distribution of microtubules in the spindle during elongation suggest the mutual sliding of the two "half-spindles"; the degree of microtubule overlap diminishes, just as expected if tubules associated with the two poles were sliding past one another (Fig. IV–26). The problem is that in other organisms, such changes are not evident—in some cases, for example, if the tubules are sliding past one another during elongation of the spindle, they must be simultaneously growing longer by adding tubulins, because the extent of overlap is not markedly diminished. There do seem to be organisms, such as some of those with an intranuclear spindle, in which elongation of the spindle clearly does involve growth of microtubules, some of which seem to be continuous from one pole to the other. Note also that if tubules both slide and grow simultaneously, they can maintain overlapping portions, through which they might interact, even after what were their original ends would have moved past one another. In other words, tubule growth could permit lateral interaction between microtubules to elongate the spindle to a greater extent than that possible with the lengths of tubules present at the beginning of elongation.

These considerations imply to many investigators that some spindles probably elongate by pushing on their poles through growth, sliding, or some combination of the two. Matters may not always be this simple. For instance, in recent studies on a fungus it was found that laser irradiation of the midplane of the elongating spindle—the zone between the separating sets of chromosomes—*increases* the rate at which the chromosomes are moved apart. The radiation interrupts the set of microtubules that runs between the two spindle pole bodies. A possible implication is that some of the force producing spindle elongation in this organism arises from pulls exerted on the spindle pole bodies and that the spindle actually provides some resistance to this force (perhaps the force is exerted by asters anchored in the plasma membrane or pulled on by some other cytoplasmic system). Whether this notion of "resistance" proves true or not, the observations bring to mind the possibility that some of the spindle's microtubules may function largely to orient or control motion generated from some other source, or to transmit rather than produce force. Furthermore, a framework like that represented by interdigitating half-spindles might be required in part simply to hold the poles apart so that the force responsible for chromosome movement toward the poles does not instead pull the poles in toward the chromosome.

Some investigators who attribute spindle elongation to active sliding of two half-spindles are encouraged, whereas others are mildly perplexed, by the fact that the microtubule polarity in one half-spindle is opposite that in the other—in each, the (+) ends of the tubules point away from the corresponding pole. In other words, the spindle tubules thought to interact laterally, exerting mutual pushes, have opposite polarity with respect to one another. In the axoneme, the only system for which sliding is well established, all of the peripheral doublets—the axonemal elements that slide with respect to one another—seemingly have the *same* polarity of growth (Sections 2.10.3 and 2.10.6). What such considerations might mean in terms of interactions mediated by dynein (Section 2.10.4) or comparable energy-providing proteins is uncertain.

Chromosome motion is not inhibited by injection of antibodies that bind to myosin, into cleaving eggs, or by exposure of permeabilized cells to other agents that should link to components of the actin–myosin system and block their activities. These observations pose a challenge for those who think actin filaments play key roles in spindle movement; still, with the variety of ways in which cytoskeletal and motility-generating systems can operate (Section 2.11.4), roles of the filaments are by no means ruled out. Motion in permeabilized preparations *is* diminished or prevented by vanadate and other agents at concentrations where dynein is inhibited (Section 2.10.4). The effects of the various inhibitory agents used to study the spindle are complex and not entirely understood; thus, such observations do not constitute definite proof. However, taken together with microscope observations of fine bridgelike connections between adjacent microtubules, they do suggest that cross-bridging between tubules, perhaps of the sort mediated by dynein in flagella, could be essential for aspects of spindle functioning such as elongation. That the postulated dynein-like molecule may not be identical to the flagellary protein is suggested by observations on mutant *Chlamydomonas* that lack functional flagellary dynein but nonetheless divide normally.

Nor has it been ruled out that the *disassembly* of microtubules attached to the kinetochores results, more or less directly, in force exerted on the chromosomes. Carefully graded application of colchicine to dividing cells at concentrations that do not dissolve the spindle, or careful application of hydrostatic pressure or low temperature, can produce or speed chromosome motion, as if the promotion of disassembly directly enhances the poleward-directed forces or at least permits them to "express" themselves as chromosome motion. It will be a difficult matter to determine whether the process of microtubule depolymerization itself is somehow harnessed to generate movement or whether depolymerization, for example, allows some elastic, contractile, or sliding system to move the kinetochore or the entire fiber attached to the kinetochore. In general, movement of the chromosomes toward the pole remains more mysterious than the elongation of the spindle, although obviously neither is well understood at present.

Controls If calcium ions do help to control spindle movement, they could do so by effects on ATPases or other enzymes or by influencing the balances between microtubule assembly and disassembly. Injection of Ca^{2+} into cleaving eggs can produce localized but transitory disruption of spindle regions. This finding confirms the expectation from what is known of microtubules (Section 2.10.5) that the state of assembly of the spindle in the cell should be sensitive to Ca^{2+}. (The short-term and localized nature of the effects also indicates that there probably are local mechanisms within the spindle—most likely in the membrane systems present—that adjust calcium concentration). The calmodulin found in the spindle in immunocytochemical studies is located between the chromosomes and the poles; little is present in the zone between the separating chromosome groups at anaphase. The same is true of actin visualized with fluorescent antibodies. There are lingering doubts about the reality of these localizations, owing to the possibilities for artifactual relocation of molecules during the rather harsh treatments required to prepare the tissues for examination. (The plasma membrane must be disrupted and some soluble macromolecules lost to permit the labeled antibodies to penetrate.) Nevertheless, such observations represent first steps toward a more adequate dissection of the molecular machinery operating in the spindle.

Eventually in explaining the control of spindle function we must account for a host of phenomena observed over generations by students of cell division. Importantly, chromosome behavior in mitosis and meiosis shows intriguing combinations of independent movement by individual chromosomes with precise coordination among all the chromosomes. For one thing, anaphase almost never starts until all the chromosomes have moved to their equatorial metaphase position ("metaphase plate"), but individual chromosomes often arrive at this position at different times; therefore, some must await the arrival of the rest. For another, although most chromosomes subsequently move more or less in synchrony away from the metaphase plate, there are numerous exceptions, especially in meiosis, where particular chromosomes move with their own timing. Cases also occur in which specific chromosomes fail to attach to the spindle at preprogrammed stages in an organism's development, leading to selective chromosome loss (Sections 4.3.4 and 4.4.3).

A "Thermodynamic Digression" with Practical Ramifications The fact that cold depolymerizes spindle microtubules and other microtubules is fascinating. Since raising the temperature tends to disrupt noncovalent bonds (Section 1.4.1), cold is usually associated with stabilization of complex structures, and higher temperatures with tendencies toward their disruption. The formal thermodynamic equation governing this sort of consideration is one that defines the "free energy" of the system: $\Delta G = \Delta H - T\Delta S$. The "free energy change" (ΔG) characterizing a phenomenon depends on the amount of energy that is released or absorbed (ΔH), on the temperature (T), and on the change in entropy: ΔS. The ΔS can be regarded as a measure of the overall change in total *order* (or randomness) of the components involved. Increased entropy implies increased "disorder." By convention, when energy is released, ΔH is given a *negative* value, and when entropy is increased, ΔS is given a *positive* value. As used by physical chemists, the equation is a mathematical summary of the conclusion that chemical reactions that can occur spontaneously under given circumstances are those that result *overall* in a combination of energy gain or loss and entropy changes yielding a negative ΔG—a "release" of "free energy." Thus, in simple cases, chemical reactions involving changes in bonds that result in substantial release of energy are favored, as are those that increase disorder, such as the hydrolysis of a large molecule into many smaller ones. For any given reaction, however, the overall balance between energy and entropy effects must be considered; a release of energy, such as occurs on hydrolysis of ATP, can be used to power a process involving a *decrease* in entropy, such as the formation of macromolecules from smaller precursors.

Examined in this light, the effects of temperature on microtubules are explained by the fact that as T is diminished, a set of bonds other than those linking tubulin dimers to one another are thermodynamically favored. To associate with one another, tubulins must break some of the noncovalent bonds through which they associate with water molecules when floating free in solution. When the tubulin dimers dissociate, they reestablish these bonds; the loss of order among the tubulins is "compensated" by the increased ordering of water molecules resulting from the association of more of them with tubulins.

These abstract considerations have practical impact. It turns out that the most convenient way of purifying microtubule proteins from tubule-rich structures, such as brain, is to take advantage of the unusual "cold lability" of microtubules. The tissue is homogenized at low temperature (0 to 4° C), promoting dissociation of the tubulins but stabilizing most other structures, for which the thermodynamic balances are different. On centrifuging using speeds at which most cell structures will sediment as a pellet (Chap. 1.2B), a supernatant is obtained, containing dissolved microtubules plus the normally soluble molecules of the cell. If this supernatant is now adjusted in terms of ionic ingredients and other components that control microtubule polymerization (Section 2.10.5) and the resulting solution is warmed up, microtubules form while the other soluble molecules remain separate. Since they are now large structures, the tubules can be separated as a centrifugal pellet and then redissolved by lowering the temperature, yielding a solution of purified microtubule components.

4.2.9 Cytoplasmic Division ("Cytokinesis")

In general, cells at telophase divide along the plane of the spindle equator (where the metaphase chromosomes had been located), and two cells of equal size are produced. The two daughter cells usually receive similar shares of cytoplasmic organelles, although there are some dramatic exceptions (such as the divisions of a maturing animal oöcyte into a large future ovum and small polar bodies destined to degenerate; Section 4.3.1).

Timing of increase and duplication varies among different cytoplasmic organelles and cell types. Section 4.2.7 outlined the duplication cycle for centrioles, which generally is coordinated with the overall division cycle. Enucleation of cells prevents the completion of centriole duplication but inhibition of DNA synthesis does not, in the sense that the appearance of a second centriole in company with its "parent" (which normally occurs in S) still takes place. In the relatively few cases studied in detail, replication of organellar DNAs such as those in mitochondria shows no universal timing pattern; it is not, for example, closely synchronized with replication of chromosomal DNA, though there is an overall doubling of total mitochondrial DNA content during the course of the cell cycle. The timing of increases in mass of organelles such as mitochondria during interphase differs in different cells and circumstances; but, for instance, mitochondrial growth seems not to be limited to well-defined periods of nuclear change such as S. In some algae, where a single plastid that occupies much of the cytoplasm is present (Fig. II–61), the actual division of the plastid occurs in synchrony with that of the rest of the cell.

In cells with large numbers of mitochondria or plastids, these organelles are often distributed apparently at random in the cytoplasm, so that roughly equal numbers of organelles are contained in the two daughter cells. Sometimes equal distribution results from a special arrangement of organelles. Thus, lysosomes often cluster near the spindle poles. In the dividing spermatocytes of some insects, the mitochondria group around the spindle to form an aggregate of elongate mitochondria; this is cut in half as the cell divides.

The actual processes of division of organelles such as mitochondria or plastids are not well understood. Some microscopic observations suggest they pinch in two. In other cases, the process resembles, more, the growth of a partition or septum that eventually crosses the width of the organelle and separates it into two daughters.

Animal Cells In animal cells, a *cleavage furrow* "pinches" the cell in two by moving in perpendicularly to the long axis of the spindle (Fig. IV–27). Many hypotheses have been advanced over the years for the mechanism of furrowing. The best-supported one proposes that a ring of "contractile material" pinches the cell in two. This contractile material is in fact a belt of microfilaments generally about 10 μm wide by 0.25 or less μm thick. A ring of this sort has been observed just below the plasma membrane at the cleavage furrow in many animal cell types and organisms; cleaving eggs have been most thoroughly studied. In addition to the actin of the microfilaments, myosin is asso-

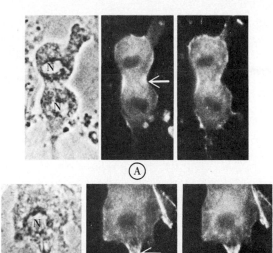

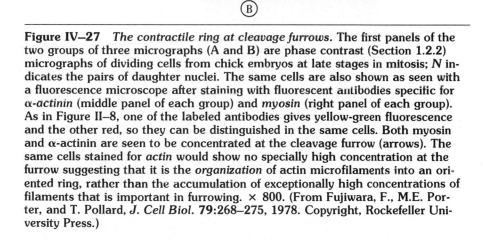

Figure IV–27 *The contractile ring at cleavage furrows.* The first panels of the two groups of three micrographs (A and B) are phase contrast (Section 1.2.2) micrographs of dividing cells from chick embryos at late stages in mitosis; *N* indicates the pairs of daughter nuclei. The same cells are also shown as seen with a fluorescence microscope after staining with fluorescent antibodies specific for α-*actinin* (middle panel of each group) and *myosin* (right panel of each group). As in Figure II–8, one of the labeled antibodies gives yellow-green fluorescence and the other red, so they can be distinguished in the same cells. Both myosin and α-actinin are seen to be concentrated at the cleavage furrow (arrows). The same cells stained for *actin* would show no specially high concentration at the furrow suggesting that it is the *organization* of actin microfilaments into an oriented ring, rather than the accumulation of exceptionally high concentrations of filaments that is important in furrowing. × 800. (From Fujiwara, F., M.E. Porter, and T. Pollard, *J. Cell Biol.* **79:**268–275, 1978. Copyright, Rockefeller University Press.)

ciated with the ring, and α-actinin is also present. Although no distinct set of thick filaments has yet been described, the composition of the ring has fueled speculation that some kind of actin–myosin sliding underlies the contractile processes through which the ring narrows to divide the cell. The bundle of filaments comprising the ring does not get thicker or wider as the ring narrows, suggesting that disassembly or disaggregation of filaments accompanies their rearrangement. The ring as a whole has a relatively brief life span—assembling, functioning, and disappearing in a few minutes.

Agents that inhibit actin–myosin interactions, such as antibodies that bind to myosin, also prevent furrowing when injected into dividing cells. So does exposure to cytochalasin B (Section 2.11.2). Thus, the evidence supporting a filament-based cleavage mechanism is strong. Microtubules seem not to be involved directly, as judged from the lack of effect of colchicine. Ca^{2+} ions appear to play controlling roles similar to those observed with other motility phenomena.

Plant Cells   In higher plants, cytoplasmic division occurs by the formation of a *cell* plate. Vesicles, many derived from the Golgi apparatus, accumulate in the middle of the cell, flatten, and fuse to establish the new cell boundaries (Fig. IV–28). This process begins at the *phragmoplast,* a region of vesicles that accumulate near microtubules persisting at the spindle equator for a time after the chromosomes have separated (Fig. IV–29).

Location of the Division Plane   If the mitotic apparatus is experimentally shifted during prophase or metaphase in a fertilized egg cell or tissue-culture cell, the plane of the cleavage furrow is shifted accordingly. If the mitotic apparatus is shifted after metaphase, or even removed, there is no effect upon the cleavage furrow. Some interaction between the spindle and the rest of the cell appears to "set" the cleavage plane; once this has occurred, the spindle is no longer required. One view suggests that attachments sometimes found between asters and the cell surface contribute to setting the cleavage plane. (See also the studies on nucleocytoplasmic interaction and coordination in Section 4.2.3.)

In higher plants a special arrangement of microtubules, evidently related to the future orientation of cytokinesis, appears transiently during the early phases of cell division. This ringlike, circumferential "preprophase band," located in the peripheral cytoplasm, encircles the cell near the zones where the new cell plate will later join the original walls of the parent cell.

Interesting observations have been made on the division of the "guard mother cell" that produces the guard cells bordering the stomata of onions. The mother cell is elongate, and the initial movement of the chromosomes is toward opposite ends of the cell (the orientation actually is somewhat oblique). Late in division, however, the spindle and phragmoplast rotate in the cell, so that the final separation into daughters occurs along a plane parallel to the mother cell's long axis, rather than the usual transverse plane (Fig. IV–28). Thus, two elongate guard cells result (Fig. III–18). Preceding this division, a preprophase band is present in the appropriate position, correctly "predicting" the division plane.

Many comparable cases are known in which specially oriented cell division contributes to the geometry of developing tissues in plants and animals or produces daughter cells with very different amounts of yolk, mitochondria, or other inclusions (Section 4.4.3). With eggs of the alga (seaweed) *Fucus,* for example, the first cleavage separates the cell into two daughters with distinctly different developmental fates: One will generate structures by which the plant attaches to surfaces; the other, the remainder of the plant. The plane of this

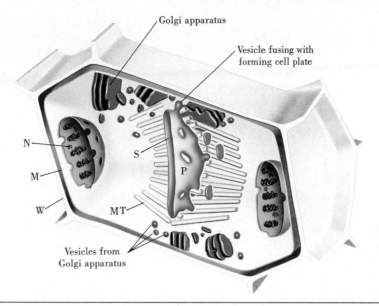

Golgi apparatus

Vesicle fusing with forming cell plate

N

S

M

P

W

MT

Vesicles from Golgi apparatus

Figure IV–28 *Cytokinesis in higher plants.* **Schematic representation of a cell of a higher plant as seen at telophase in mitosis. The cell wall is indicated by *(W)*, the plasma membrane by *(M)*, and one of the two daughter nuclei by *(N)*. In the phragmoplast region, a region of membranes and microtubules *(MT)*, a cell plate forms *(P)* and grows until it separates the cytoplasm into two daughter cells. The cell plate develops as a membrane-delimited structure enclosing a space *(S)* in which a new cell wall will form. The Golgi apparatus contributes many vesicles to the phragmoplast membrane. The vesicle membranes are incorporated into the membrane of the cell plate, and the vesicle contents enter the forming cell wall. (Modified from M. Ledbetter.) See also Figure IV–29.**

division can be set by environmental signals such as the orientation with which light is shone on the eggs. The response to such signals is the establishment of zones of different ionic permeability at different regions of the egg's plasma membrane, and this in turn leads to an oriented flow of ionic current within the egg. Investigators are now attempting to determine how this could influence the plane of cleavage; the involvement of Ca^{2+} ions is strongly suspected.

4.2.10 Nuclear Envelope and Nucleolus

Nuclear Envelope Segregation of daughter chromosomes in procaryotes is thought to depend on their anchoring to the plasma membrane (p. 330). In some lower eucaryotes the nuclear envelope seemingly plays a similar role or at least one that stems from common evolutionary antecedents. Dinoflagellates are primitive alga-like eucaryotes. Their chromosomes differ from most other eucaryotes in lacking histones. During mitosis the chromosomes remain inside the nuclear envelope, associated closely with the inner membrane. Bundles of

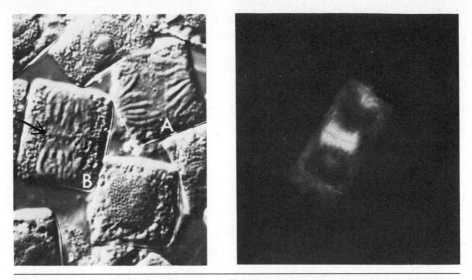

Figure IV–29 *Dividing plant cells.* (Left), Dividing cells from an onion root tip, that were fixed, isolated from the root and then viewed in a differential contrast interference microscope (Fig. I–12). Cell A is in anaphase and B in late telephase; the chromosomes are readily seen. The arrow indicates the site of the forming cell plate (Fig. IV–28). × 800. The right panel shows a cell like B in the left panel, but stained immunohistochemically to demonstrate sites of tubulin. The phragmoplast's system of microtubules associated with the forming cell plate is clearly evident. × 900. (From S. Wick et al.; reference in Fig. III–23.)

microtubules run in tunnels through the nucleus but remain separated from the chromosomes by the nuclear envelope. The chromosomes are segregated to daughter nuclei as though moved by the envelope along paths established by the microtubules. Could it be that in these organisms it is the nuclear envelope that achieves chromosome segregation, with the microtubules serving simply as guides? Alternatively, at least in some dinoflagellate species, the tubules might actually generate the motion through mechanisms and attachments yet to be fully detailed. In either event, it is possible to imagine evolutionary sequences from procaryotic division mechanisms to the processes just described or to the ones outlined in earlier sections for certain fungi, in which nuclear envelope–associated spindle pole bodies participate in organizing a spindle. There also are organisms that divide by mechanisms that could be evolutionary intermediates between the ones in which the nuclear envelope participates and those in which the envelope breaks down; in some, for instance, the kinetochores associate with the nuclear envelope.

The disappearance of the nuclear envelope in prophase, as occurs in many ''higher'' eucaryotes, results from fragmentation of the envelope into numerous sacs and vesicles. Little is known of the signals governing this fragmentation beyond the observations that cytoplasmic factors probably are involved (Section 4.2.3). Studies by fluorescent antibody staining suggest that the proteins of the inner fibrous lamina (Section 2.2.6) become dispersed in the cytoplasm as the envelope becomes fragmented. They reaccumulate around the forming nucleus in telophase when the new nuclear envelopes of the daughter cells form. Reestablishment of the envelope itself appears in the electron microscope as a process of fusion of sacs and vesicles that accumulate around the chromosomes. Some investigators believe that these are membranous elements of the old envelope that have persisted in scattered form during division. However, it is not easy to distinguish them from other ER structures. The formation of pores in the new envelope occurs rapidly: this process and the subsequent increase in pore number that can take place as the nucleus enlarges during interphase have been little studied. Formation of pores in telophase, and in early G_1, seems to continue even if protein synthesis is inhibited, suggesting that preexisting proteins suffice. On occasion configurations resembling pore complexes have been seen on the chromosomes at mitotic stages or in sacs in the cytoplasm during mitosis, but at present there is no compelling evidence that pores as such generally maintain their architecture through the cycle of envelope disintegration and reestablishment. Perhaps, as it reassembles, the inner fibrous lamina helps to reestablish the distribution of pores and other features of the envelope.

In those cells where the nuclear envelope does not disappear, the two separating sets of chromosomes accumulate at opposite ends of the nucleus and become surrounded by a portion of the old envelope. The necessary reorganizations of the envelope's membranes and pore structures have been little studied.

Nucleolus The nucleolus, as an organized entity, generally disappears during cell division. However, in at least some cell types nucleolar components probably are preserved during division and participate in forming a new nucleolus. Microscopic and autoradiographic studies suggest that during division, some materials from the nucleolus become distributed along the chromosomes, while some pass into the cytoplasm. Much of this dispersed material returns to the new nucleolus formed after division. This is one facet of what some believe to be a broader phenomenon of movement of nuclear RNAs and some proteins into the cytoplasm when the nuclear envelope breaks down and their return after the subsequent telophase. If this really occurs it indicates the operation of still unknown recognition and migration mechanisms.

In a few cell types nucleoli persist as intact bodies throughout the division cycle. Sometimes they pass into daughter cells along with the chromosomes to which they are attached, but in some plant cells, portions of nucleoli, associated with the chromosomes until metaphase, can move to the spindle poles before chromosome movement begins. Apparently the nucleoli are carried passively by some sort of poleward flow of material within the spindle. Thus, they are transported independently of the chromosomes.

Chapter 4.3 Genetic Basis of Cell Diversity

Thus far, discussion has included mechanisms that tend to keep structures and cells *constant*. The present chapter considers the origins of *diversity* through *genetic mechanisms,* diversity that is the basis of evolution. It is not the genes themselves but rather the genetically determined characteristics of organisms that are exposed to the environment and thus to natural selection (although there are some special cases of fairly direct interaction of gene and environment, such as enzyme induction; see Section 3.2.5). More precisely, an organism's characteristics (its *phenotype*) are an expression of its genetic constitution (its *genotype*) but generally depend on the interplay of many genes with one another and on the developmental and environmental context in which the genes are expressed. Those organisms with characteristics that prove advantageous in the particular environment produce, on the average, more offspring than those produced by organisms with fewer advantageous characteristics. They make a larger contribution of genes to the next generation of a population. With time, populations undergo evolutionary change.

4.3.1 Meiosis: A Brief Review

Mutation and Alleles; Zygotes   Mutations in DNA result in altered base sequences, which ultimately are translated into proteins with altered amino acid sequences. Sometimes the change of just one amino acid in a protein has profound effects. For example, in *sickle cell anemia,* the human inheritable disease, a single amino acid in the hemoglobin molecule is replaced by another. This results in drastically altered structure and function of red blood cells. (The cells are fragile, and under certain conditions they tend to assume a shriveled sickle shape, in contrast to the normal disc shape.) See also p. 491.

Mutations give rise to *alleles,* which are alternate forms of a gene. (For example, one allele of an *eye color* gene may produce blue eyes, whereas brown eyes are produced by another of the several alleles of the same gene.) For the most part, inheritable differences among individuals of a given species are attributable to the fact that different individuals carry different alleles.

All sexually reproducing eucaryotic organisms (that is, most unicellular organisms and virtually all multicellular organisms) go through a *zygote* stage or its equivalent. Usually a single nucleus is formed, containing a mixture of alleles from both parents. As outlined in Chapter 3.10, the zygote nucleus may result from the fusion of nuclei carried by distinctive gametes, as in most higher organisms. In many unicellular organisms and in some lower plants, nuclei contributing to the zygote are transferred between two cells without the formation of obviously specialized gametes. The transfer involves processes such as partial and temporary fusion *(conjugation)* of ciliated protozoans, producing a cytoplasmic bridge through which each sends a micronucleus to the other, or the complete fusion of two cells in *Chlamydomonas* or yeast.

Meiotic Stages Meiosis is complementary to zygote formation in the life cycle of organisms. It accomplishes the segregation of alleles. This results from the fact that the two members of each homologous pair of chromosomes (Section 4.2.1) are separated by meiosis into different cells. Usually the members of a homologous pair both carry the same genes arranged in the same sequence along their lengths; however, the particular alleles present on each homologue may differ. Meiosis involves two rounds of cell division, as outlined in Figure IV–30. The figure legend summarizes the several distinctive stages of the first meiotic prophase; Figures II–27, IV–14, IV–32, IV–33, and IV–39 are micrographs illustrating the unique chromosome configurations of this prophase. Meiosis is preceded by DNA replication during interphase. S may be relatively prolonged in comparison with that for mitotic cells of the same organism. The divisions themselves are different from mitosis in two key respects: First, in the first meiotic prophase, homologues pair gene for gene; they come together and align so that the site *(locus)* of a given gene on one homologue lies next to the same gene locus on the other homologue (Figs. II–27, IV–32, and IV–33). No such pairing occurs in mitotic prophase. Second, the chromosomes remain paired until metaphase. Then one *chromosome* of each pair goes to each pole (Fig. IV–30). Unlike mitosis, the two *chromatids* of each chromosome remain together. Thus, the number of chromosomes per nucleus is halved. The *second* meiotic division resembles mitosis in that no pairing occurs and chromatids separate from one another. However, the second division is not preceded by DNA replication or chromosome duplication; the chromatids that separate are the ones already present during the first division. The net result of the two divisions is four haploid cells (Figs. IV–14 and IV–30), each having half the DNA and half the number of chromosomes (one of each pair) of the parent diploid cell.

Life Cycles The subsequent behavior of the meiotic products varies considerably in different organisms. Many algae and protozoa are normally haploid; the meiotic products each divide by mitosis to establish a new cell line. The members of a given line produced in this way are genetically identical, except for mutations occurring subsequent to the meiotic divisions that produced the first cell of the line. Lines differ from one another in alleles, since they have resulted from products of meiosis that differ in alleles (see Fig. IV–30 for an explanation of such differences). Sexual reproduction is an occasional process; it occurs, for example, in *Paramecia* under conditions of low food supply).

Yeast can divide mitotically either as diploid or as haploid cells. They remain haploid until opposite mating types encounter one another and fuse. The diploid cells resulting from fusion persist as diploids until subjected to starvation, whereupon they form thick-walled inactive spores. When nutrients are again available these spores germinate by dividing meiotically, producing four haploid cells that go on to propagate by mitosis. The association of meiosis with survival through hard times as in *Paramecium* and yeast reflects the evolutionary importance of sexual reproduction.

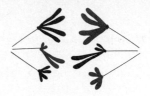

(A) Overall behavior of chromosomes

At metaphase the chromosomes line up in pairs at the spindle's equator.

At mid-prophase of the first of the two meiotic divisions homologous chromosomes are paired; each chromosome is of two chromatids.

At anaphase homologous chromosomes separate to opposite poles while the two chromatids of each chromosome stay together.

Daughter nuclei from the first division contain one chromosome from each homologous pair. No DNA synthesis or chromosome duplication precedes the subsequent division

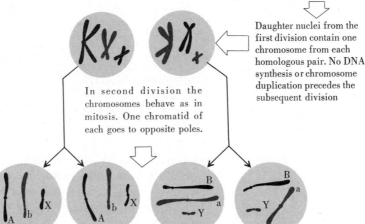

In second division the chromosomes behave as in mitosis. One chromatid of each goes to opposite poles.

Thus, each cell entering meiosis produces four haploid daughters. Each homologous pair segregates independently of the others. For example, A enters the same daughter as B or b with equal frequency. Thus, another (Aa, Bb, XY) cell entering meiosis will produce daughters abY, ABY, aBx or AbY.

(B) Crossing over

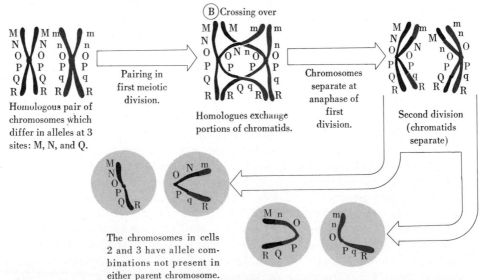

Homologous pair of chromosomes which differ in alleles at 3 sites: M, N, and Q.

Pairing in first meiotic division.

Homologues exchange portions of chromatids.

Chromosomes separate at anaphase of first division.

Second division (chromatids separate)

The chromosomes in cells 2 and 3 have allele combinations not present in either parent chromosome.

◄**Figure IV–30** *Chromosome behavior in meiosis.* The first meiotic prophase is usually thought of as having five major stages: (1) *Leptotene,* when the chromosomes have coiled and folded from their interphase state to the point where they are visible as discrete slender threads; although their DNA has duplicated, separation into two chromatids cannot be seen. (2) *Zygotene,* when homologues begin visible pairing ("synapsis"). (3) *Pachytene,* when gene-to-gene pairing is completed (see Fig. II–27); each chromosome pair is referred to as a "bivalent." (4) *Diplotene,* when the two chromatids of each chromosome become clearly visible. (Since four chromatids are present per pair, the array is called a "tetrad.") This is the stage when RNA synthesis related to oöcyte growth is maximal; the chromosomes may assume special lampbrush forms as in Figure IV–39. (5) *Diakinesis,* when the chromosomes coil to reach maximum thickness and, although they remain associated, relax pairing so that homologues are associated only at some points, the *chiasmata* (Fig. IV–32). The X and Y chromosomes drawn in the diagram represent the sex-determining chromosomes seen in many organisms (Fig. IV–30 and Section 4.3.3). The four products of the two meiotic divisions are illustrated in Figure IV–14.

In multicellular lower plants, such as most seaweeds (multicellular algae), the haploid cells resulting from meiosis are released as spores. These divide by mitosis to form a gamete-producing plant that is haploid and multicellular *(gametophyte).* Gametes are produced by differentiation of some of the gametophyte cells. A diploid zygote is formed by the fusion of gametes. The zygote develops by mitosis into a diploid multicellular *sporophyte.* In the sporophyte some cells undergo meiosis to produce the haploid spores. In some species such as ferns and other lower plants, alternating *sporophyte* and *gametophyte* generations are separate plants; in others, one generation is reduced in size and appears as a special portion of the more highly developed generation. In higher plants such as the flowering plants the gametophytes are microscopic. They are parts of the flower that form gametes—male gametes in the pollen and egg cells in the base of the flower.

Thus, in higher plants and almost all multicellular animals, most cells are diploid—the products of mitotic divisions of the zygote. Each individual organism starts with an essentially unique set of alleles, except for the asexual reproductive phenomena that occur in various plants and for the occasional occurrences in animals when a single zygote produces two (identical twins) or more organisms. Meiosis occurs in specialized reproductive cells and results in gametes (see Chap. 3.10). In males all four meiotic products form gametes. In females three of the cells resulting from meiosis generally degenerate, and only one functions in reproduction.

The cells resulting from meiosis in animals differentiate directly into sperm and ova. Meiosis itself can, however, be a quite prolonged process. Oöcytes of many organisms remain in the first meiotic prophase for weeks or months, during which they grow extensively, accumulating the stores of RNA, yolk, and other materials needed for embryonic development (Chap. 4.4). In *Xenopus,* the duration of this period is controlled hormonally. The cells are stimulated to complete the first meiotic division by steroid hormones *(progesterone)* secreted by the follicle cells associated with them in the ovary. The oöcytes arrest in metaphase of the second meiotic division until fertilization.

In mammals such as humans, the ovary's oöcytes reach first meiotic prophase during fetal development and arrest there, awaiting hormonal signals that may come years later, during sexual maturation and the ovulation cycles.

In flowering plants, the nucleus surviving meiosis in females divides mitotically to produce eight nuclei distributed in seven cells, which together constitute the gametophyte. The two nuclei that occupy a common cell fuse with one another and subsequently fuse with one of the two haploid male cells delivered by the pollen tube (Chap. 3.10); the cell that results is triploid and will divide to generate nutritive, endosperm tissue. The zygote forms by fusion of another of the female gametophyte cells with the other male cell.

4.3.2 The Transmission of Hereditary Information; an Example

The diversity in detail of gamete and zygote behavior is important for understanding the hereditary patterns of different organisms. But the variety of life cycles all involve two constant features, the reduction of diploid cells to haploid (meiosis) and the formation of new diploid cells with new combinations of alleles generated by fusion of haploid nuclei (sexual reproduction). This constancy produces predictable patterns of transmission of chromosomes and genes from one generation to the next, making genetic analysis possible. For example, earlier (Section 2.3.4) it was mentioned that matings of certain individuals of the toad *Xenopus* produce offspring of which one quarter are inviable due to a chromosomal defect affecting ribosome and nucleolus formation. This pattern of inheritance is explainable by the proposal that the parents in the matings each contain one normal *(N)* and one abnormal *(No)* homologue in the chromosome pair responsible for formation of nucleoli. Since a normal chromosome is present, ribosome synthesis can take place. However, meiosis results in gametes of which half contain the *N*, and half the *No*, homologue. If "m" represents the homologue contributed by the male parent and "f" the homologue contributed by the female parent, then equal numbers of the following types of zygotes will formed; mNfNo, mNofN, mNfN, mNofNo. Thus, one quarter of the zygotes are *NN*, one quarter *NoNo*, and one half *NNo*. The *NoNo* category are the ones that die, since they have no normal homologue. Their survival through the early stages of development is based on ribosomes stored in the oöcyte before the completion of meiosis. At this time a normal homologue is still present in the eggs that *eventually* contain only the *No* homologue, since the oöcytes, like the other *diploid* cells of the parents, are all *NNo*.

This explanation predicts that there should be a class of surviving offspring that contains one *N* and one *No* homologue and another class with two *N* homologues. There are two lines of evidence that this is so. Among the viable offspring of the matings under discussion, some have many cells in which two nucleoli are present (in some cells the two fuse to form a single organelle), while in the cells of other individuals only one nucleolus is ever present. Also, DNA can be extracted from embryos prior to the death of the *NoNo* class and its ability to form molecular hybrids with purified rRNA determined. The expected three categories of individuals are found: Some have essentially no DNA sites that bind rRNA, and, of the rest, one group has twice as many sites as the other. This last finding indicates that the abnormality in *No* chromosomes involves the loss (*deletion;* Section 4.5.1) of the rRNA-producing DNA segments.

4.3.3 Genetic Diversity; Crossing-over

Sexual reproduction and meiosis continually generate new genetic combinations. In Figure IV–31 it is shown that a hypothetical organism with six chromosomes (three pairs) can produce gametes with eight different combinations of chromosomes. Figure IV–31 diagrams the ten different zygote combinations that can result from matings of two individuals, each having only four chromosomes (two pairs) with the same arrangement of alleles in both individuals.

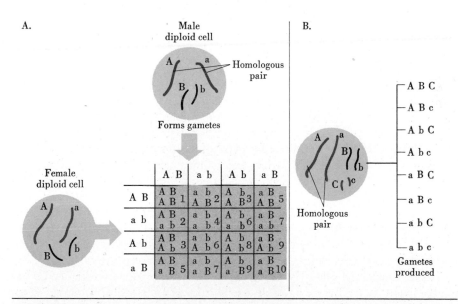

Figure IV–31 *Some genetic consequences of meiosis.* A. The results of a genetically simple mating. The diagram indicates the four different genetic types of gametes and 10 different genetic types of zygotes formed in a mating between genetically identical males and females possessing two pairs of chromosomes. From the viewpoint under consideration the zygote combination (AB$_{maternal}$; ab$_{paternal}$) can be considered to be the same as (AB$_{paternal}$; ab$_{maternal}$). B. The gametes that can be produced by an individual whose diploid cells contain three pairs of chromosomes.

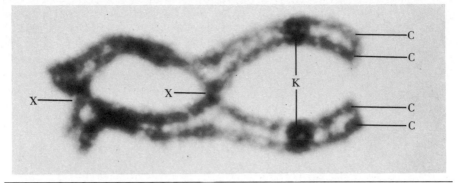

Fig. IV–32 A pair of homologues during late prophase (shortly before metaphase) of the first meiotic division of a salamander spermatocyte. The four chromatids of the two chromosomes are readily seen at C, and the two centromeres are indicated by K. At X chiasmata are present. That only one chromatid from each chromosome is involved can clearly be seen in the chiasma at the right. (Courtesy of J. Kezer.)

These are extremely simple examples. With large numbers of chromosomes and matings, as is usual between individuals carrying many different alleles, enormous numbers of combinations are possible.

Even greater diversity of gametes and zygotes results from *crossing-over (recombination)* between homologous chromosomes. This depends on the reciprocal exchange of portions of chromatids during the first meiotic division (Fig. IV–30). Crossing-over produces characteristic chromosome configurations known as *chiasmata* (Fig. IV–32), which are visible just before and during the first meiotic metaphase. It is not an occasional phenomenon but rather an almost invariable feature of the association of chromosomes in meiosis in most organisms studied. (There are a few exceptions, such as males of the fly *Drosophila*, in which a type of pairing occurs but crossing-over is rare so chiasmata do not form.) In any given crossover the two chromatids usually exchange comparable portions that may differ in alleles but that carry the same set of gene sites *(gene loci)*. However, when a particular homologous pair is compared in different cells undergoing meiosis, the chromosome portions involved in crossing-over vary almost at random, so that different sets of loci are exchanged in different cells. The frequency with which crossing-over occurs between given loci of a chromosome is the basis of classical techniques for mapping gene locations; the further apart two loci are, the more likely a crossover will occur between them. (For more recently developed methods, see Section 4.2.6.)

Sex Determination Sex determination can often be readily demonstrated to depend on meiosis. For example, in humans, males are usually XY and females XX, where X and Y refer to special homologous sex chromosomes that pair at meiosis (Fig. IV–4), although they differ in genes and in morphology. The segregation of these chromosomes in meiosis in males (Fig. IV–30) results in equal numbers of X and Y sperm. Since female gametes all are X, half the zygotes will be XX and half XY. In other organisms somewhat different situations are found. For example, in some insects males have one X and females two; no Y chromosome is involved, and males have one less chromosome in their complement than in that of females. In meiosis in these males, equal numbers of gametes have either one X or none; all female gametes have one X. In birds such as chickens, where the sex chromosomes are frequently designated Z and W, males are ZZ and females ZW.

There are organisms that differ radically from these patterns, such as *hermaphroditic* invertebrates, in which a given individual produces both types of gametes. In many of the *hymenopteran insects* (bees, ants, wasps), females arise from fertilized eggs, and males, parthenogenetically (Section 3.10.3), from eggs that are not fertilized. In reptiles, the temperature at which the eggs develop has strong influence on whether they differentiate into males or females.

Key genetic information responsible for sex differentiation is probably carried on the sex chromosomes, but the nature of the genes is still unknown. There do appear to be antigenic (Chap. 3.6C) differences between tissues of males and females that lead, for instance, to rejection of skin grafts from one sex by the other. These presumably are products of sex chromosome genes; accordingly, one of the cell surface–located antigens is called the *H-Y antigen*. However, it is not yet clear how or if these genes and their products relate to the differentiation of gonads into ovaries vs. testes and to the variety of hormonal and other controls responsible for generating male and female characteristics.

"Sexual" Reproduction in Procaryotes Meiosis does not occur in procaryotes. The few species of bacteria that have been adequately studied are haploid, in the sense that there is only a single copy of the chromosome per nuclear region. The number of nuclear regions per bacterial cell may vary under different conditions of growth, but each nuclear region apparently contains a copy of the same chromosome. The multinucleate condition represents a feature of cell growth and division rather than one of sexual reproduction.

Procaryotic cells do, however, exchange genes. *Genetic recombination* can be produced in the laboratory through transfer by viruses ("transduction") or by plasmids and through transformation by purified DNA (Section 3.2.6). Recombination by one or another of these mechanisms is thought to occur under appropriate circumstances in nature as well.

As mentioned in Section 3.2.6, when the sex factor (F) plasmid becomes integrated into a chromosome in *E. coli,* the bacterium becomes capable of transferring genes to a recipient bacterium lacking the sex factor. This reflects the fact that sex factor–containing (F$^+$) cells can produce *sex pili,* protein filaments that protrude from the cell surface (see Section 3.2.4). When F$^+$ cells encounter F^- cells of the same strain, they attach to them by the tip of a pilus, initiating an incompletely understood series of events by which a channel is opened between the cells; perhaps retraction of the pilus produces close contact between the surfaces of the two cells. DNA from the F^+ cell enters the F^- cell through the pilus-established contact. This DNA is a linear copy of the F^+ cell's chromosome produced by a "rolling circle" mechanism (Fig. IV–9). The end of the DNA molecule corresponding to the sex factor enters the recipient bacterium first, followed by part or, occasionally, all of the chromosome in which it is integrated. The recipient is thereby made partially diploid, but this is a temporary state: recombination, akin to crossing-over, takes place between the donated DNA and the recipient's chromosome, generating a recombinant, haploid nuclear region.

4.3.4 Problems of Meiosis

There are still many unanswered questions regarding meiosis. What makes spermatogonia and oögonia, after numerous mitotic divisions, switch their mode of chromosome behavior to meiosis? How are the "arrests", in first meiotic prophase and later, regulated? What determines the unequal division of many animal oöcytes into a small "polar body" (a cell that degenerates) and a much larger cell, the mature egg cell? (See also p. 587, bottom.)

Pairing; Spindle Attachment Homologous chromosomes pair gene for gene, probably with their centromeres and kinetochores in special state or arrangement that will permit chromatids to remain together and chromosomes to attach appropriately to the spindle. Is there something unique about the chemistry or structure of meiotic chromosomes that permits this discriminatory mode of association? Is there any relationship between pairing of chromosomes and base-pairing of nucleic acids? Are repetitive DNAs involved? Is the *somatic pairing* of homologues seen in the nonmeiotic cells of some insects (such as *Drosophila;* see Section 4.4.4) based on mechanisms similar to meiotic pairing? In a few instances, differences have been reported between histones of meiotic and those of mitotic cells, but the significance of this is unknown.

In grasshopper spermatocytes it is possible to dislodge paired homologues from the meiotic metaphase spindle with a microneedle and to turn the pair around so that the chromosome that was about to segregate to one spindle pole now faces the opposite pole. The paired chromosomes reattach to the spindle and separate normally, but they segregate to the poles opposite those they would have moved to without the experimental reorientation. This experiment strongly hints that the chromosomes themselves, rather than some external system, control attachment of homologues to the spindle in the manner necessary for meiotic separation. Presumably the kinetochores are the responsible parts.

In *Drosophila,* during late meiotic prophase, the kinetochore of a given homologue has the form of a hemispherical dome apparently shared by both chromatids. By late metaphase this transforms into a pair of more conventional discs, one per chromatid, but both oriented toward the same pole. For *Drosophila,* then, a meiotic chromosome as it begins to attach to the spindle seemingly has not yet completed structural duplication of its kinetochore. This could help to explain the subsequent behavior of the chromosomes and chromatids (see below).

The Synaptonemal Complex   In almost all types of eucaryotic organisms, electron microscopy of meiotic stages reveals a special structure, the *synaptonemal complex* (Fig. IV–33), formed between paired chromosomes. Further analysis of this structure may illuminate meiotic mechanisms. Each homologue

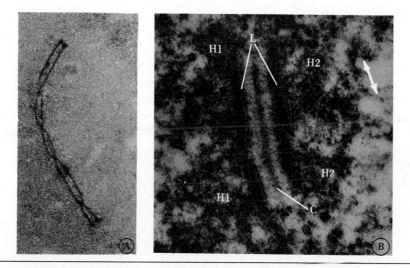

Figure IV–33   *The synaptonemal complex.* A. A pair of homologous chromosomes from human spermatocytes (pachytene stage of meiosis; see Figs. II–27 and IV–30). The cells have been disrupted by placing them at the interface between a salt solution and air, and then the chromosomes have been picked up on a thin plastic film for viewing in the electron microscope without sectioning. The paired structures seen in the micrograph are the synaptonemal complex at which the two homologues are associated. B. A thin section of part of a synaptonemal complex of a rooster spermatocyte. The two chromosomes are represented by the irregular fibrous regions at *H1* and *H2*. The long axes of the chromosomes are indicated by the double-headed arrow. The synaptonemal complex is seen at the surfaces where *H1* and *H2* are associated. The complex consists of the pair of dense bands (called lateral elements) indicated by *L* and the single central element *(C)* found between them. Faintly visible filamentous material runs from the central element to the lateral ones. It is the lateral elements that are visible in part A. The lateral elements are roughly 50 nm across; their width varies somewhat in different species. The space between them is about 80 to 100 nm. A, × 9,000; B, × 50,000. (From Moses, M., S.J. Counce, and D.F. Paulson, *Science* 187:363–365, 1975. Copyright © 1975 by the American Association for the Advancement of Science.)

contributes one of the ribbon-like lateral elements. These are seen first early in meiotic prophase (at leptotene as "axial elements" running the length of the chromosomes and located, some think, between the two chromatids). The central element appears when the two lateral ones become associated during pairing and disappears as pairing relaxes into relations like those in Figure IV–32; the lateral elements persist for a longer time. Sometimes distinct striations cross the central element or each of the lateral ones, resulting in ladder-like appearances of the elements. Both RNA and protein seem to be present in the synaptonemal complex, but little is known of the particular molecular species represented. Many hypotheses assume that DNA from the chromosomes extends into the complex. The complex itself, however, is generally thought of as a structure separate from the chromosomes proper, which assembles along chromosome surfaces in the first prophase of meiosis and disassembles once the paired chromosomes separate. This viewpoint derives partly from observations that stacks of synaptonemal complexes ("polycomplexes") appear in the nuclei of some organisms, separate from the chromosomes, late in the first meiotic prophase. It is not clear yet how the presumed cycle of assembly and disassembly relates to proposals that the two chromatids of each chromosome remain together during the entire first meiotic division partly as a result of the presence of persistent materials between the chromatids; these hypothetical materials are supposed to produce cohesion of the chromatids to one another and contribute as well to the axial elements visible at leptotene, and to the synaptonemal complex. The more usual explanation for the chromatids' remaining together is that the centromere of the parent chromosome is delayed in completing its duplication or its separation into two, until late in the first meiotic division as outlined above; this does seem to be the case, as far as can be judged from microscopically visible structure.

Local zones of thickening or enhanced electron density are present along the central element of the synatonemal complexes in a number of species. These have sometimes been called "recombination nodules" to emphasize the postulate that they are forerunners of the chiasmata seen later in prophase and that they may reflect events of genetic recombination. With the present uncertainties about recombination mechanisms in eucaryotes discussed below, this proposal cannot yet be evaluated decisively. It gains some support from the fact that the frequency of nodules correlates with the frequency of chiasmata.

Toward a Molecular Mechanism of Crossing-over   The molecular mechanisms of pairing and crossing-over in eucaryotes still are unknown. Leads should soon come from the rapid progress being made in analysis of genetic recombination in viruses and procaryotes.

Genetic recombination between viral "chromosomes" (DNA molecules) occurs when different viral strains infect the same cell. A single recombinant DNA molecule is formed that carries hereditary information contributed by both parents; this is comparable to a chromatid resulting from a crossover.

One set of theories (*copy-choice* theories) states that viral recombination results from phenomena of DNA replication. During its synthesis a strand of

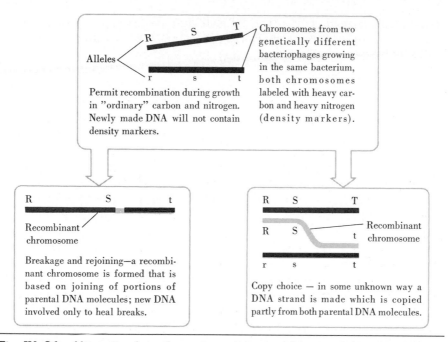

Fig. IV–34 *Alternative hypotheses to explain recombination in bacteriophages.* The finding that some of the chromosomes (DNA double helices) that can be shown to be genetically recombinant are essentially as dense as the parent molecules indicates that recombination *can* take place by breakage and rejoining; extensive DNA replication is not required. However, the experiment is not designed to establish that this is the only mechanism by which recombination is possible. (From the work of J.J. Weigle and M. Meselson.)

DNA might start to form by alignment of nucleotides along a strand of one parental DNA molecule. After partial completion it might somehow switch and finish growth by alignment of nucleotides along a strand of a second parental DNA molecule (Fig. IV–34). For part of its length the new strand would have information specified by the base sequences of one parental DNA; for the rest of its length the information would have been specified by the other parental DNA.

An alternative theory is based on "breakage and rejoining" ("breakage and exchange"); two parental DNA molecules might break, and portions of each join with portions of the other. An experiment was designed to determine whether or not this is a mechanism of recombination (Fig. IV–34). Viral DNA can be labeled with *density markers* (such as ^{13}C and ^{15}N, isotopes of carbon and nitrogen), which permit separation from nonlabeled viral material by density gradient centrifugation (Chap. 1.2B). The higher the proportion of markers in the DNA, as compared with ^{12}C and ^{14}N, which normally predominate, the greater the DNA density. When recombination takes place between two different strains of viruses, both equally labeled, some of the recombinant chromosomes with genetic information from both parents have DNA essentially as dense as the parental DNAs. This means that breakage and rejoining can produce recombinant DNA and that extensive new DNA synthesis is not necessary for recombination. Under the conditions of the experiment, newly synthesized DNA would contain no label, and its presence in recombinant chromosomes would lead to a decrease in density to levels below that of the parental chromosomes. It also is observed that many of the steps in recombination of viral DNAs can take place even when the bacterial hosts are subjected to conditions inhibiting DNA synthesis. However, this information does not imply that extensive DNA synthesis is never involved in recombination. Much new DNA can be made as part of the processes by which the segments of DNA molecules participating in recombination are joined together. The important point is that the initial events of recombination involve physical rearrangement of preexisting DNA (Fig. IV–35), rather than the synthesis of an entirely new strand.

Figure IV–36 shows a recombinant DNA molecule formed between bacteriophage DNAs. One point illustrated is that recombination involves *single* strands, one from each of the two DNA duplexes. This is a major link in the chain of evidence suggesting that the pairing up of participant DNAs required for recombination is founded on base-pairing between complementary sequences on DNAs (Fig. IV–35). Genetic evidence also indicates that *heteroduplex* DNA molecules can form as a result of recombination. These are molecules whose two strands have regions where bases are mismatched; complementarity in sequence is not perfect. This can readily be explained if the two DNA strands at a region of exchange between two duplexes came from parent molecules carrying different alleles. Heteroduplexes, and molecules like that in Figure IV–36 are,

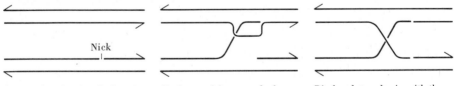

Two DNA molecules (duplexes) with similar base sequence. The arrowheads indicate the 5'-3' directionality of the strands. Recombination begins with the introduction of a nick or small gap in one strand.

Single strand from one duplex "invades" a homologous region of the other duplex by base pairing. This displaces corresponding strand of invaded duplex.

Displaced strand pairs with the single strand region complementary to invading strand; during this process it is also nicked.

Figure IV–35 *Initial steps in genetic recombination* as postulated from studies on bacteria and viruses and with bacterial enzyme systems. Several equivalent models have been proposed: the one presented is intended as an example of how base pairing and exchanges of single strands between duplexes (double helices) might occur. The subsequent events of recombination probably include additional reorganization of the molecules, healing of the nicks and gaps, and other processes, most of which have yet to be fully analyzed. (Based chiefly on the discussions by F.W. Stahl of proposals by R. Holliday, C.M. Radding, and others, in F.W. Stahl, *Genetic Recombination*, San Francisco: W.H. Freeman, 1979.) See also Figure IV–36 for a micrograph of a configuration of DNA molecules that probably arose by phenomena akin to those diagrammed here.

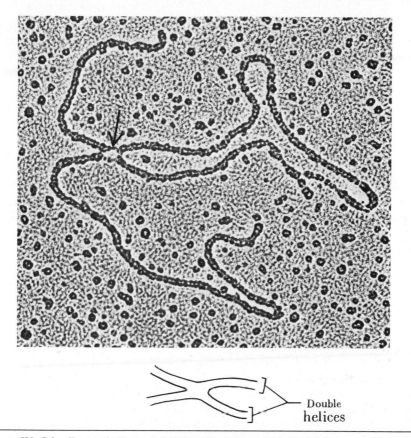

Double helices

Figure IV–36 *Recombination of DNA.* DNA from a bacterial virus (phage "M13 Gori 1") treated in the test tube with *E. coli* proteins thought to mediate early events in genetic recombination (*rec* proteins, Section 4.3.4). The sketch diagrams the region indicated by the arrow; in this region the two double helices in the field have undergone a rearrangement such that one strand from each molecule switches partners. See Figure IV–35 for one mechanism by which such switches might occur in the cell. The DNA was prepared for examination by a modified shadowing procedure (Fig. I–11). Approximately × 250,000. (From DasGupta, C., A. Wu, R.P. Cunningham, and C.M. Redding, *Cell* **25**:507, 1981. Courtesy of the authors and MIT Press.)

of course, temporary intermediates in the recombination process. That is, when they undergo DNA replication, each strand of such structures builds the usual perfectly complementary partner and thereby forms an "ordinary" DNA molecule.

There are several competing models of how the pairing and exchanges between DNAs might occur. Most current ones propose that a single strand from one duplex unwinds to a limited extent from the duplex and "invades" an adjacent duplex at a "region of homology"—that is, where it finds extensive base sequences complementary to its own. It competes with the DNA strand already there and displaces it (Fig. IV–35; Section 5.1.2 outlines some features of DNA that may be relevant). This is most readily imagined if the unwinding starts at a "nick" in the invading strand generating a free end (Fig. IV–35) but could occur by other means. The several models vary in their proposals for such details and for the details of the steps that follow, in which the enzymes responsible for recombination complete the exchanges, nibble away loose ends, and seal breaks. The models share a conceptual advantage in that they envisage the individual strands as being reshuffled and interrupted one at a time, and step by step. The process could begin at random along the length of the initiating duplex, but once it starts, base pairing ensures that the appropriate precise alignments and exchanges occur. It would be more difficult to see how recombination might work if the DNA had first to be broken across both strands simultaneously and then to be linked to another molecule in which breaks were somehow introduced independently at precisely the same points.

Increasingly successful efforts to understand the molecular machinery responsible for recombination stem from the availability of mutants such as those of the *rec* class in *E. coli*. These mutants are abnormal in recombination frequency, or in details of recombination, both for the bacterial DNAs and for phage and plasmid DNAs present in the cells. The effects can be traced to alterations in enzymes and other proteins that produce the nicking, unwinding, filling in of gaps, and other DNA modifications responsible for recombination. The "recA" protein, for example, can catalyze the interactions of a single strand from one DNA duplex with a complementary sequence in another duplex through events that require ATP. A number of the proteins involved in recombination also take part in "ordinary" DNA replication. The basic point is that the models proposed for recombination in procaryotes and bacterial viruses can be constructed from known properties of enzymes and other proteins.

Eucaryotes   Recombination in eucaryotes is still difficult to study at the level of DNA. Fragmentary evidence suggests that at least some of the major mechanisms resemble those just described. Detailed genetic studies of eucaryotes have uncovered phenomena most readily explained by such mechanisms. For example, there is evidence for the initial formation of short DNA regions where complementarity between the two strands is imperfect. Such regions "disappear" during subsequent DNA replication; when the strands separate during S, each will build a perfectly complementary partner, so that zones of mismatched bases will no longer be present. Along with some other phenomena predicted by the models described above, these events can lead to ratios of genetic traits among offspring different from the simple ones predicted by the mechanisms of Figure IV–31. Such "deviations" are rare, but they can be detected in suitable experimental material such as fungi, with which very large numbers of matings can be carried out and then the fate of each product of meiosis followed.

Several other observations also indicate that "breakage" of the chromosome fiber and exchange of chromosome portions are very likely to be involved in crossing-over in eucaryotes: In some strains of corn and other organisms, slight morphological differences occur between members of homologous pairs, and recombinant chromosomes are produced that are visibly combinations of both. In addition, if chromosomal DNA is labeled with tritiated thymidine before meiosis, autoradiographic techniques can subsequently be used to demonstrate exchanges of DNA between homologues.

Since crossing-over requires that parts of different chromosomes be close to one another, crossing-over and pairing probably are related processes. However, pairing of chromosomes as seen in the microscope bears an unknown relationship to the kind of DNA base-pairing found in viral recombination; viruses, after all, do not undergo meiosis or form synaptonemal complexes. There is, in fact, no unequivocal evidence that crossing-over must occur only *after* the chromosomes are visibly paired in prophase; it might be part of the pairing process. It even has been suggested that crossing-over actually takes place during the S period of premeiotic interphase, when most of the DNA replication occurs. (This timing might be expected if copy choice were the predominant mechanism.) However, several lines of investigation suggest that crossing-over occurs later than S. For example, in the life cycles of certain fungi and algae, meiosis, presumably including genetic recombination, takes place *immediately after* two parental nuclei are brought together in a zygote. In several of these species, premeiotic DNA replication occurs in the separate parental nuclei before the chromosomal complements come together to participate in meiotic events. In other words, the homologues are in separate nuclei during S.

Findings made initially on plant material and then extended to some animals (including mice) indicate that there may be one or more periods of limited DNA synthesis during meiotic prophase. Although only 1 percent or less of the total DNA is involved, this synthesis could be quite significant. For example, exposure of meiotic prophase cells to deoxyadenosine, an inhibitor of DNA synthesis, can result in fragmentation of chromosomes and interruption of pairing and continued meiosis. Is the DNA made during prophase important in joining separate DNA segments or in other events of pairing or crossing-over? Studies on DNA isolated from meiotic cells and on the enzymes present in such cells offer tentative support for the proposal that during meiotic prophase, DNA is nicked extensively, presumably to "permit" recombination, and that this nicking and the subsequent sealing of gaps in the DNA are carried out by enzymes similar to those implicated in bacterial genetic recombination. Interruption of protein synthesis during meiotic prophase also leads to abnormal chromosome behavior, one sign that the chemical events underlying visible pairing and crossing-over may involve coordinated production and interplay of proteins and DNA.

Pairing and exchange in meiotic prophase is between chromosomes that are appreciably folded. During stages of visible pairing the chromosomes are about 1000 times shorter than the length of DNA double helix they contain. Furthermore, pairing may be mediated or stabilized by a structure interposed between the chromosome, the synaptonemal complex described earlier. Thus, there seem to be structural constraints on the extent to which DNA sequences from homologous chromosomes can approach one another and interact. Some investigators speculate that the synaptonemal complex represents an organizing framework, composed of an array of enzymes and other factors, that participates in DNA recombination, for example, by fostering appropriate interactions between stretches of DNA. These stretches, it is posited, might penetrate from the lateral zones into the center of the complex where molecular interactions between the homologues could occur. Other proposals assert that repetitive DNAs strategically localized at intervals along the chromosome play special roles in organizing the events of pairing and crossing-over. It also is argued that the association of chromosome ends ("telomeres"; Section 4.5.1) with the nuclear envelope may help to set the stage properly for meiotic pairing. In some species, pairing of the homologues seems to begin near chromosome ends that are associated with the envelope and progress in zipper-like fashion from these points. However, this is certainly not invariably true, because in many species initial pairing is observed far from the ends. Not infrequently, the ends of all of the pairing homologues migrate so that they cluster near one another at one zone of the envelope, with the chromosomes extending into the nuclear interior in "bouquet"-like loops. (Often the pole of the nucleus where the chromosome ends accumulate is associated with the cytoplasmic centrosomal zone, containing the centrioles.)

Modifications of Meiosis Exceptions to the general rules of meiosis are known; the exceptions indicate the extent to which even fundamental cellular processes have undergone evolutionary change. During sperm formation in the insect *Sciara*, segregation occurs between *maternal* and *paternal* chromosomes, that is, between the chromosomes derived from those originally contributed by the male and female parents to the zygote that produced the *Sciara* individual under study. At the first meiotic division, the spindle that forms is *unipolar* (monopolar)—it is a conical structure equivalent to half of a conventional spindle. Maternal and paternal chromosomes both associate with this spindle, but subsequently only the maternal chromosomes move to the spindle pole; the paternal chromosomes stay at what would be the metaphase plate of the spindle were two poles present. Only cells receiving the maternal chromosomes form functional sperm. The paternal chromosomes are not transmitted to the next generation. Accordingly, the patterns of heredity are quite unlike those generally found. In most species maternal and paternal homologues separate at random to the poles of the first meiotic division, and therefore all gametes usually contain some chromosomes derived from those contributed by both parents. (Crossing-over additionally mixes maternal and paternal chromosomal material.) The recognition devices responsible for the unusual behavior of *Sciara* chromosomes are not known; a current hypothesis proposes that the kinetochores of the maternal and paternal chromosomes differ in their association with the spindle at meiotic prophase.

In the brine shrimp *Artemia*, development is by *parthenogenesis,* the development of an egg without fertilization (Section 3.10.3). The first division of oöcyte meiosis is normal, but in the second, the chromatids separate without cytoplasmic division, and a single nucleus eventually forms with both chromatids of each chromosome. Consequently, the egg is diploid, and although there is no contribution of chromosomes by sperm, the egg produces a diploid organism.

4.3.5 Some Consequences of Genetic Diversity

The variation in genetic constitution resulting from mutation, meiosis, and sexual reproduction is the raw material upon which selection operates in evolution. The pattern of interaction of gene and environment differs somewhat in organisms with different types of life cycles. For example, mutations in diploid organisms are less rapidly, or less directly, exposed to selective "testing" by the environment than those of haploid organisms. In a haploid organism a mutation that results in the production of a defective form of an important enzyme will result in an inviable cell. In diploid organisms this is not necessarily so, because each gene is represented at least twice in each cell. Thus, the presence of a newly mutated allele on one chromosome may be "masked" (the allele acting as a *recessive*) by the presence on the homologous chromosome of a different *(dominant)* allele that can support the synthesis of adequate amounts of "normal" enzyme. In addition, the presence of two different alleles *(heterozygosity)* in an individual organism may be more advantageous than the presence of the same allele on both homologues *(homozygosity)*. Also, the extent to which a particular allele is advantageous, of neutral advantage, or disadvantageous often depends on the environment. For instance, humans homozygous for the sickle cell allele usually die from severe anemia. Heterozygous persons with the allele for sickle cell anemia on one chromosome and an allele for normal hemoglobin on the other homologue may have little or no difficulty (beyond a slight anemia) at low altitudes in temperate climates. However, at high altitudes where the oxygen pressure is low, their red blood cells will assume the abnormal shriveled shape. This heterozygosity, however, also confers resistance to forms of malaria, a disease prevalent in the tropics.

In many cases, the effect of a given allele depends strongly on the genetic and developmental context—for example, on the particular set of alleles of other genes that are present—so that classification of an allele as "advantageous" or "disadvantageous" becomes difficult and of questionable utility.

The net result of factors such as these is that populations of diploid organisms tend to have a much more complex *gene pool* (the total number of genes and alleles in all individuals) than do populations of haploid organisms. The preservation, in diploid populations, of alleles that may be selectively disadvantageous under some conditions provides a source of evolutionary variability not available to haploids. If the environment changes, these alleles may become advantageous. On the other hand, the large populations and rapid division rates of many haploid unicellular organisms favor rapid evolution by environmental selection among mutants that continually arise at random within all populations.

4.3.6 Cytoplasmic Inheritance

In one type of cytoplasmic inheritance, *maternal inheritance,* characteristics of the offspring are determined solely by the female parent. As in cytoplasmic inheritance generally, the transmission patterns of the relevant hereditary information are not those expected for genes on chromosomes governed by meiosis and mitosis. In most cases DNA molecules in plastids or mitochondria seem to be the sites of the genetic information, and it is the pattern of their transmission that is reflected. In many species paternal cytoplasmic genetic factors are not passed on to the offspring. For higher organisms this may be due, in part, to the gross disparity between the sperm's and the egg's contribution of cytoplasm to the zygote. However, uniparental cytoplasmic inheritance also occurs in unicellular organisms, such as *Chlamydomonas,* where sexual reproduction involves fusion of two cells of similar size but opposite mating type. Apparently there are factors, still unknown, that differentially suppress the perpetuation of cytoplasmic genes from one parent cell. In some of the cases where paternal cytoplasmic organelles do enter the zygote, they soon show microscopic signs of deterioration; eventually, these cases should provide clues to the suppressive mechanisms at work. In *Chlamydomonas,* the chloroplasts (Fig. II–61) of the two parental cells are thought to fuse in the zygote, but the chloroplast DNA from one parent—that of the (−) mating type—is selectively degraded during the first few hours after zygote formation. The chloroplast DNA of the (+) mating–type parent persists as the hereditary material of the zygote's plastid.

Both spontaneously occurring and experimentally induced mutations of cytoplasmic genes have become important experimental tools. In yeast, "petite" mutants possess mitochondria that are abnormal in structure and function. Aerobic metabolism is disrupted, and the cells grow slowly since they must rely on the less efficient anaerobic metabolism (Section 1.3.1). Some "petite" yeast strains result from mutations in nuclear genes; others involve changes in mitochondrial DNA, including marked alterations in base composition. With the fungus *Neurospora,* cytoplasm from a strain with mutant mitochondria has been injected into a normal strain, and the successful "colonization" of the recipient by proliferating abnormal mitochondria has been observed. Such studies help to confirm the concept of semiautonomous mitochondrial duplication outlined earlier (Section 2.6.3).

Efforts are under way to detect and analyze mutations affecting centriole or basal body structure and behavior (Section 2.10.7). The findings may help to settle the controversy over the possible presence of nucleic acids in these organelles (Section 2.10.7), since it should be possible to demonstrate whether or not cytoplasmic mutants occur. We have already outlined evidence that the *patterns* of aligned basal bodies in protozoa can be inherited independently of the nucleus (Section 3.3.4). This reproduction of pattern, however, need not depend upon genetic autonomy of the basal bodies, because the pattern involves microtubules, cell-surface components, and other structures and molecules as well as basal bodies.

Cytoplasmic mutants of chloroplasts that affect plastid structure or function often yield plants with varying patterns of groups of cells that are green or white (absence of chlorophyll). The patterns depend upon the details of plastid transmission during sexual reproduction and during cell divisions in the embryo. Some ornamental plants owe their origin to such cytoplasmic inheritance, and there also are a few phenomena of this type that are important in agriculture.

Especially convenient for genetic analyses are plastid mutations leading to altered resistance to experimentally administered drugs. Through use of such mutants in *Chlamydomonas,* and of comparable mitochondrial mutants in yeast, it has been found that under some experimental conditions sexual reproduction can lead to patterns of inheritance suggesting genetic recombination between cytoplasmic DNAs from both parents. Do the DNAs interact when organelles contributed by the parents fuse? That this may be the case is suggested by microscopic observations indicating that in situations where organellar DNAs from both parental types survive in the zygote, the DNA-containing regions of fused organelles coalesce for a time. Later, the ordinary pattern of multiple separate DNA-containing regions per organelle is restored. Another line of investigation concerns the possibility that a given organelle possessing several DNA molecules may sometimes stably possess different information in the different DNAs (Sections 2.6.3 and 2.7.4). Thus far, the patterns of inheritance detected suggest that such heterogeneity is not very marked, if it occurs at all.

Chapter 4.4 Divergence of Cells: Embryonic Development and Differential Gene Expression

4.4.1 Development and Differentiation

Major Developmental Stages   Figure IV–37 illustrates the major stages in the development of eggs of higher animals. The fertilized egg divides by mitosis to produce the *blastula*. As mentioned earlier (Section 4.2.3), the divisions are referred to as *cleavages*. Little cell growth occurs; thus, sequential divisions separate the fertilized egg into successively smaller cells. In many cases, the cells produced are of different sizes, and the cleavage planes are oriented so that particular spatial arrangements of the daughters are generated (see Section 4.2.9); these patterns are important for subsequent development. Moreover, especially in certain invertebrates, eggs have special localized cytoplasmic regions that become included in only a few of the cells resulting from the cleavages. As will be described in Section 4.4.3, some of these cytoplasmic materials determine the developmental directions of the cells in which they are present. Such cases provide emphatic reminders that at the level of individual cells, it is the *interaction* of nucleus and cytoplasm that governs cell differentiation. Overall development cannot be understood by attending only to the nuclear genes,

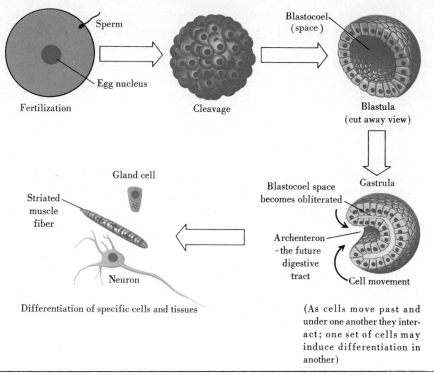

Figure IV–37   A schematic outline of early development in higher animals.

or only to cytoplasmic determinants (though in particular developmental programs of different organisms or tissues, either the nucleus or the cytoplasm may be of more decisive apparent influence in committing the cells to a specific fate).

Furthermore, only limited facets of development are comprehensible in terms of autonomous properties of cells. Thus, in many organisms, most of the cells resulting from early cleavages are not *intrinsically* restricted to differentiate into particular types later in development. Among the more recent of the numerous demonstrations of this are experiments in which cells obtained from different early mouse embryos are combined. These combinations can be implanted into the uterus of a suitably prepared ("pseudopregnant") mouse and will develop essentially like normal embryos. They produce "allophenic" mice, animals in which different cells come from what were originally different embryos. If the original cells are selected so that they differ in appropriate genes, such as one controlling visible pigments, their descendants in the mice can be identified. From examination of what tissues arise from particular cells in allophenic combinations it appears that the developmental fate of early cleavage cells can be changed in accordance with the cell's position in the embryo—the neighbors and environment with which it interacts. Later in development as well, cells' locations and the time they arrive there often are crucial for determining what becomes of them.

The blastula forms a *gastrula* by kinds of cell movement that vary according to the species; during this process, some cells are brought into new spatial relations with others. Subsequently, *differentiation* of cell types occurs, and the embryo increases in size and mass by cell growth and division. The juxtaposition of cells in new geometric relations during gastrulation is accompanied by the *induction* of differentiation in groups of cells through interaction with other cells that have come to lie close to or in contact with these groups. Cells acquire characteristics enabling them to "recognize" one another (Section 3.11.2). Additional patterns of cell growth and migration result in the establishment of other specific associations of different cell types, for example, the complex relationships among neurons, sensory receptors, and muscle cells. Selective death and destruction of cells and tissues also is part of normal development (Section 4.4.7).

Estimation of how many different types of cells a given organism forms is tricky since it requires arbitrary decisions about where to draw the boundaries. Figures commonly given are 100 to 1000 for humans and 30 to 50 or more in higher plants. These numbers contrast with the two stable cell types (spores and growing cells) formed by bacteria and the few forms yeast cells can assume. It should, however, be kept in mind that both bacteria and yeast can modify their metabolism rather considerably under varying environmental circumstances (Sections 2.6.3, 2.9.1, and 3.2.5); relations between such modulation and the differentiation of cell types in multicellular organisms may exist at the level of genetic control mechanism.

Some Perspectives from Cell Biology   Many of the methods and conceptual frameworks described in earlier chapters are being applied to analysis of development. It is believed, for example, that coordination of cells in very early stages involves gap junctions. Very likely, such junctions plus glycoproteins and other macromolecules at the cell surface also mediate many of the interactions among neighboring cells that occur later on. In addition, neighboring cell groups can affect one another's developmental fate through extracellular matrices; experiments in tissue culture that explore such effects were outlined in Section 3.11.2.

Other interactions, including ones between distant cell groups can be brought about through secreted molecules such as the hormones and growth factors we have described for animals and plants (Sections 2.1.4, 3.4.5, 3.11.1, and 4.3.1). For instance, vertebrates maintain "stem cell" populations (p. 518) throughout much of their life span. Those from which the various cell types of the blood differentiate are found in bone marrow, spleen, and other tissues. These cells constitute a multipotent "reserve"—they can differentiate into a variety of different blood cell types. Throughout the life of the organism, groups of them become restricted in their developmental potential and mature, eventually, into circulating blood cells. The initial steps of their commitment to particular directions of differentiation are assumed to depend on local cell interactions. The last stages are triggered by hormone-like agents such as *erythropoietin,* a glycoprotein released by the kidney in response to altered levels of oxygen; erythropoietin evokes or permits differentiation of those cells destined to form red blood cells.

Chapter 3.6C outlined some of the processes by which antigens provoke differentiation of B-cells into plasma cells.

At present, techniques for manipulation of cells and of DNA are, of course, being used on developing systems. Section 3.12.1 outlined the incorporation of cultured (teratocarcinoma) cells into allophenic embryos. Another example, perhaps foreshadowing "genetic engineering" of agriculturally important animals, is the introduction of foreign DNA into early developmental stages. Cloned (Section 3.2.6) rabbit genes and genes from viruses have been injected into fertilized mouse eggs or early cleavage stages of mouse embryos; the genes are found to integrate into the nuclei of the growing mice to the extent that they even can be transmitted to the next generation through normal sexual reproduction. Study of the expression of foreign genes in developing systems should give useful information about genetic regulation. It has also proved possible, by injecting suitable DNAs, to introduce and obtain expression of appropriately modified mouse genes in developing mouse eggs, and of chosen alleles of *Drosophila* genes in *Drosophila* embryos. This has permitted control of the eye color of the resulting adult flies, and of the levels of growth hormone—and the consequent sizes of the animals—in the mice.

Gene Activity   Our emphasis in this chapter is chiefly on the nuclear and chromosomal aspects of differentiation, especially as they relate to the specific activation or repression of genes. Strong evidence supports the proposition that in different cell types, different genes are switched on or off so that cells come to differ in their structural and chemical characteristics, although they possess the same genetic constitution. When different genes are switched on, different specific mRNAs are produced from different DNA templates. Consequently, cell-specific proteins (such as enzymes, secretory materials, or contractile proteins) are synthesized. There must, however, also be some DNA "templates" that function in virtually all cell types, such as the genes for ribosomal or transfer RNAs.

Attempts to estimate the extent to which different genes are expressed in different cells have been based on systematic RNA–DNA hybridization studies (Section 2.2.2) comparing the DNA sequences represented by nuclear and cytoplasmic RNAs in various cells and circumstances; supplementary information comes from analysis of the effects of mutations and increasingly accurate determinations of how many different polypeptide chains are present in particular cells. The complexities of cell populations available for study and remaining uncertainties about the organization of chromosomal DNA, and about the processing steps that intervene between transcription and maturation of RNA molecules, has impeded this work (see Section 4.2.6). Moreover, only a few cell types and organisms have been studied in detail. Fortunately these include material from some of the favorite objects of embryological study such as developing sea urchin eggs; these eggs and the eggs of amphibians are available in large quantities and, since they develop outside the mother's body, can readily be manipulated in the laboratory. A few generalizations have gradually gained tentative acceptance as conceptual platforms for future work (this is an area of strenuous effort, and more precise data should soon be available):

1. A given cell type transcribes a relatively small proportion of its DNA into RNAs with life spans long enough and abundance sufficient to be detectable with current methods. For a number of mouse tissues, less than 5 percent of the nonrepetitive or minimally repetitive DNA sequences are expressed as RNAs; in the brain, which possesses a variety of active cell types, the total for all of the cells is less than 20 percent. Even in yeast cells, no more than 20 to 25 percent of the unique sequences are represented by RNAs under given growth conditions. (Since only one of the two DNA strands in a given gene is transcribed [Section 2.2.2], these figures should be doubled to estimate the percentage of potentially active genes represented.)

2. The total number of distinguishable species of mRNAs produced in different cells ranges from a few thousand in the "simpler" eucaryotes (3000 to 4000 in yeast) or in some differentiated tissues, like the digestive tract of sea urchins, to perhaps 10,000 to 25,000 in the oviduct or liver of a chicken or in the leaves of plants. Particular mRNAs may be present in great abundance in particular cell types. In a chick oviduct stimulated to secrete ovalbumin by exposure to steroid hormones (estrogens), each cell contains up to 100,000 copies of the ovalbumin mRNA. Cells of other tissues or organisms also generally have several species of mRNAs present in many tens of thousands of copies. In addition, up to several hundred mRNAs may be present in hundreds or thousands of copies. In various cell types, however, the bulk of the mRNAs—well over half—is present in only a few copies per gene, often 1 to 10 per cell. Especially for the latter types it frequently is difficult to decide whether a given mRNA is present in all the cells in the population or whether the mRNAs vary somewhat among different cell types in a tissue or at different stages in a cell's history (for example, at different points in the cell cycle.)

3. As might be expected, some mRNAs are very abundant in one cell type and rare or absent in other cell types from the same organism. When tissues such as liver and kidney are compared, the most obvious differences are in the mRNAs present in large numbers of copies per cell. The differences in the other mRNAs are less striking than the similarities: Different cell types seem to overlap very considerably in the mRNAs they contain, especially the "scarce" ones, present in only a few copies per cell. The same is true when tissue-culture cell lines are compared in proliferating and nonproliferating states. The many thousands of genes seemingly expressed in common among different cell types must include "housekeeping" genes—those that

code for proteins needed by virtually all cells, such as ones involved in fundamental energy metabolism, ribosome structure, or plasma membrane functions. Of much more uncertain status is the possibility that some of the scarce species of mRNAs are not needed either for housekeeping or for specialized functions in some of the cells that synthesize them; they might arise as a result of "imperfections" or subtleties in genetic control mechanisms that make it impossible or inappropriate to turn off a gene totally.

In rough terms, then, the differences among cell types of a given organism appear to depend on differential expression of several hundred to a few thousand genes against an extensive background of genes that are expressed—at least in terms of RNA transcription (see also Section 4.4.6)—in most of the cells.

4.4.2 Differentiation in Embryos

The earlier stages of development—gamete formation and the programs of cell division and migration—do not simply provide raw materials for the later ones. They themselves represent specialized states of the cells. Differentiation depends on patterns of gene expression of cells in these states as well as on the subsequent changes in expression underlying production of tissue-specific products.

RNA Production   Oöcytes store substantial amounts of ribosomes and of mRNAs for use in the protein synthesis that starts soon after fertilization. Even some proteins are stored in quantity: Histones are stored in the oöcyte cytoplasm and employed for the rapid duplication of chromosomes during early divisions of the cleaving egg; tubulins to be used in forming spindles and structures such as the cilia of sea urchin blastulae also are stored. Experiments with sea urchin eggs have shown that cleavages up to the blastula stage can take place in the absence of a nucleus. The cells even form the numerous cilia characteristic of blastula cells. Further differentiation, however, depends on the presence of a nucleus and on the interaction of nucleus and cytoplasm. In amphibian eggs, nucleoli first become visible in the gastrula when rRNA production is dramatically stepped up.

From such observations it might be thought that the embryo's genome remains entirely inactive until after cleavage has generated a mass of cells, and that it turns on at about the gastrula stages in preparation for the differentiation of specialized tissues that follows. This clearly is not the case. Patterns of utilization of stored RNAs and of their breakdown and replacement by newly made RNAs from the embryo's nuclei vary considerably among different species of organisms and even among different genes in a given species. Some embryos, such as those of mammals, receive relatively small stores and rely on their own production (of rRNAs, for example) almost from the beginning of embryogenesis. Sea urchin oöcytes store histones, but during its early cleavages the embryo itself begins to make mRNAs for histones and to translate them.

Inventories like those outlined in the preceding section show that sea urchin oöcytes and early embryos (gastrula and pregastrula stages) contain a strikingly greater variety of mRNAs than is seen in differentiated tissues such as cells of the digestive tract. The oöcyte possesses the greatest variety, including all the types present in the gastrula. It is difficult to interpret these findings in detail, because the mRNA populations, especially at stages preceding the blastula, are mixtures of species made by the embryo and those stored in the oöcyte. In addition, the differences among different cells of the gastrula and earlier stages cannot be evaluated. Still, it does appear both that some of the mRNAs being translated into protein differ in different periods of early embryogenesis—even cells at the blastula stage make some mRNAs present only during this stage—and that there is considerable overlap among the RNAs used at different times. Perhaps most interesting, the sea urchin oöcyte and gastrula both seem to synthesize a considerably greater variety of RNAs than is made by the individual differentiated cell types of later developmental stages. Mature tissues do make some mRNAs not made in the gastrula; these probably specify specialized products. However, the overall pattern of mRNA changes during development quite possibly implies that, for some cell types, one aspect of differentiation is the diminution of the variety of genes transcribed into RNA. This could mean that many of the genes expressed in common by many different cell types are ones whose transcription commences very early in development and is not turned off later on. Products of other genes may be extensively required for early developmental processes or structures but later may be needed in lesser quantities (or not at all by certain cell types). Note also that *transcription* of a particular RNA, even a potential message, does not necessarily mean that the corresponding protein will be made; possible "posttranscriptional controls" will be outlined in Section 4.4.6.

Nuclear Transplantation Experiments Does the differentiation of the post-gastrula embryo depend on changes in the nucleus that are *irreversible?* This question is important because the answer will help to define better the kinds of molecular changes that might underlie differential gene expression. Selective loss or degradation of DNAs, for example, is more likely to be irreversible than is the binding of proteins to chromatin. In experimental terms, the question has been approached by efforts to determine whether nuclei taken from cells at progressively later stages in development can replace the initial zygote nucleus and support the differentiation of all the cell types of the organism.

The experiments in which whole plants are grown from the cloned offspring of a single differentiated cell (Section 3.11.2) suggest that for those plant tissues in which the cells remain alive, differentiation need not involve a permanent loss of genetic information or the inactivation of a portion of the genetic material in a way that makes reactivation impossible.

For animal cells the same general conclusion has been drawn from experiments on developing amphibian embryos. Nuclei are taken from cells at different developmental stages and substituted by microsurgical techniques for the nucleus of an unfertilized egg by procedures that also activate development of the egg (Section 3.10.3). With several species of amphibia, nuclei from blastula stages or earlier support normal development well. With *Xenopus,* apparently normal adults have also been grown from eggs with nuclei transplanted from intestine cells of tadpoles. Thus, the nuclei of at least some differentiated cells have the information needed to support the development of all the cell types. Nuclei from some tissues of adult *Xenopus,* after growth of the tissue for a while in culture, can also sustain development to reasonably advanced tadpole stages. However, even with *Xenopus,* only a small percentage of the transplants with nuclei from advanced stages go on to develop normally; many produce abnormal or incomplete development. With eggs of the frog genus *Rana,* nuclei taken from most kinds of cells at stages later than the blastula support only limited and abnormal development of embryos. In *Rana* transplants especially, abnormal chromosomes (for example, ring-shaped chromosomes) are seen with much higher frequency than is the case in ordinary embryos. Overall, most investigators conclude that the nuclei of some and perhaps many differentiated cells retain potential for differentiating along multiple lines *(pluripotency* or *multipotency).* There is more discord over the extent to which "totipotency"—unrestricted capacity—is maintained, as in the plant cell systems. Many think that even if the DNA itself does not change irreversibly, the biochemical machinery of the nucleus—the enzymes and other proteins needed for transcription or replication of DNA, for instance—may evolve as differentiation progresses, creating difficulties when the nucleus is forced, experimentally, to operate in accordance with the needs of earlier stages. Such changes could turn out to be as interesting as the possible genetic ones.

4.4.3 Interactions of Nucleus and Cytoplasm

In frogs, nuclei from one species that are capable of supporting normal development in eggs of that species cannot do so when transplanted into eggs of a different though related species. Some features of nucleocytoplasmic interactions are abnormal.

The normally inactive nuclei of chicken red blood cells can be induced to metabolic activity, such as RNA synthesis, when placed in the cytoplasm of mammalian tissue-culture cells. This may be done by fusion of the red blood cells with the HeLa strain of human tissue-culture cells using the methods outlined in Section 3.11.2. The red blood cell nuclei normally are small; their chromatin is densely packed and shows little or no incorporation of radioactive RNA precursors into RNA. In the fused cells, the chicken nuclei enlarge, the chromatin spreads out, and much RNA synthesis is detectable in autoradiographic studies. Apparently the HeLa cell cytoplasm has "activated" the red blood cell nucleus. DNA synthesis also may be initiated in the red blood cell nuclei. Normally, red blood cells do not divide and so do not synthesize DNA.

These experiments must be interpreted cautiously, since they involve abnormal conditions and drastic manipulation of cells. However, they provide direct evidence that the interaction of nucleus and cytoplasm is a reciprocal process. RNAs synthesized in the nucleus are the major agents through which the nucleus influences the cytoplasm. At least some of the influences in the opposite direction depend on movement of proteins from cytoplasm to nucleus. RNAs and proteins that can move in both directions and thus "shuttle" between nucleus and cytoplasm have been postulated to exist in protozoa and, by extension, in higher organisms as well. Whether these mediate reciprocal influences or whether instead they reflect transport processes for moving nuclear products into the cytoplasm, or cooperation of nucleus and cytoplasm in activities such as ribosome formation (Chap. 2.3), are matters for further study.

There are a number of organisms in which distinctive cytoplasmic regions present in egg cells have been shown clearly to have effects on nuclear behavior and differentiation. The specific molecules responsible have yet to be characterized: Most investigators assume they are RNAs, proteins, or complexes of RNAs and proteins. In the snail *Ilyanassa,* a temporary cytoplasmic protrusion called the "polar lobe" forms at one end of the fertilized egg just prior to the beginning of cleavage. The planes of the first and second cleavages are oriented so that the polar lobe cytoplasm becomes incorporated into only one of the four resulting cells. Embryos from which the polar lobe is microsurgically removed before cleavage develop into quite abnormal larvae. Key organs such as the heart, intestine, and eyes do not form. During early cleavages a sort of programmed "parceling out" of polar lobe materials into the cells of the developing embryo can be demonstrated, with specific cells apparently receiving cytoplasmic materials that enable them to differentiate along specific lines; selective destruction of individual cells that have received portions of the polar lobe yields embryos that fail to form particular polar lobe–associated organs such as eyes, or heart and intestine. Evidently, substances promoting particular lines of cellular development are distributed in the egg's cortex in patterns that lead to their preprogrammed segregation into particular cleavage products; the substances are probably bound firmly in the cortex, since they are not displaced by vigorous centrifugation of the egg.

A variety of insects have distinctive cytoplasm at one end of their eggs, the *germinal pole.* In *Drosophila,* this region exhibits microscopically recognizable features, including the presence of distinctive cytoplasmic granules rich in RNA. After fertilization, the nuclei of the developing insect egg undergo mitosis for a time, without cytoplasmic division, so that they become distributed through the egg's cytoplasm. Later, cytoplasmic division separates the egg into uninucleate cells. Those cells containing the germinal pole cytoplasm ("pole plasm") become the *germ-line* cells, the ones that eventually give rise to gametes. That the cytoplasm is the determining element can be demonstrated through transplantation experiments in which nuclei that ordinarily would not enter the polar cytoplasm are exposed to such cytoplasm and found to respond to it. In a few species of insects, the nuclei of the germ-line cells are the only ones that do *not* undergo an unusual series of mitoses, occurring early in embryonic development, in which a specific group of chromosomes fails to move on the spindle. These chromosomes are eliminated from the nuclei and disintegrate in the cytoplasm. In consequence, the germ-line cells have a different chromosomal complement from that of the other (somatic) cells of the insect.

The maturing egg cells of Xenopus provide an example of temporal changes in the cytoplasm that are of likely importance for reciprocal interactions with the nucleus. Nuclei from other tissues, inactive in replication of DNA or transcription of RNA, can be activated when transplanted into the cytoplasm of Xenopus oöcytes. When transplantation is done during the stages when the oöcytes are themselves storing RNA, activation of transcription, but not replication, is observed. The reverse pattern is found later on when the oöcytes are mature and awaiting fertilization (with the consequent onset of cleavage-associated DNA replication). In other words, at different stages in its development, the oöcyte's cytoplasm contains different components—probably enzymes and regulatory molecules—that influence essential features of nuclear metabolism.

4.4.4 Salivary Gland Chromosomes

Some tissues of the larvae of insects such as Drosophila and Chironomus (a midge) show interesting chromosomal behavior. This behavior has been studied most intensively in the salivary glands. The cells of these glands are extremely large, and their nuclei are correspondingly large. Although the cells are nondividing, the chromosomes are clearly visible. Measurements of the DNA content indicate that the nuclei may contain multiples of the normal amount of DNA as great as 1000 times or more. Cytochemical and autoradiographic studies of the cells as they enlarge indicate that this DNA accumulates by repeated replication of most of the chromosomal DNA. Some of the repetitive sequences seem not to take part in this replication or to take part to a lesser extent than the unique-sequence DNA; thus they are "under-represented" in the mature nucleus. Histones and some of the other nuclear proteins are present in elevated amounts, paralleling the DNA. The number of chromosomes is normal (four pairs in species of Drosophila), but homologous chromosomes are usually paired gene for gene. (In contrast to the paired chromosomes of meiosis [Sections 4.3.1 and 4.3.4], no synaptonemal complexes are present.) Each chromosome consists of a great many parallel fibers (such chromosomes are referred to as polytene). The chromosomes are several hundred micrometers in length and several micrometers thick, in contrast to the ordinary chromosomes of diploid cells of the insects, which are a few micrometers long and less than a micrometer thick. The most reasonable interpretation is that the chromosome has undergone uncoiling and has repeatedly replicated without separation of the daughters. It is thought by proponents of the "unineme" view of chromosome structure (Section 4.2.5) that one fiber of a salivary gland chromosome is comparable to the morphological and functional unit of an ordinary chromosome.

Bands; Chromomeres; Genes Study of these giant chromosomes has afforded a unique opportunity to correlate cytological, genetic, and developmental information. Figure IV–38 shows a polytene chromosome from *Chironomus,* an organism in which the polytene chromosomes in some cells contain more than 10,000 times the haploid DNA amount. The striking pattern of transverse bands is evident. Although DNA and protein are present along the entire length of the chromosome, the bands show a high concentration of DNA and histone. This results from the alignment in the bands of tightly folded or coiled regions *(chromomeres)* of the parallel fibers. The size, appearance, and arrangement of the bands along a given chromosome is identical in the two homologous chromosomes but different in nonhomologous chromosomes. Several tissues other than the salivary glands also have polytene chromosomes, and the banding pattern of a given chromosome is the same in all tissues. In other words, there is a parallelism between bands (or, more precisely, between the chromomeres comprising the bands) and genes. Genes also are linearly arranged along chromosomes, and they are the same in the

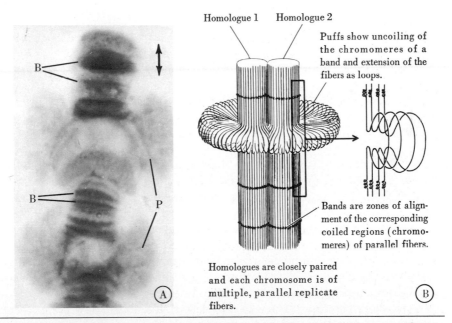

Figure IV–38 *Salivary gland chromosomes.* A. Portion of a chromosome from the salivary gland of the insect *Chironomus.* The double-headed arrow indicates the direction of the long axis of the chromosome. The cross-banding *(B)* characteristic of these chromosomes is readily visible as are two puffs *(P)* where the chromosome bands have been altered as indicated in part B of the diagram. × 1900. (Courtesy of U. Clever.) B. Schematic representation of a polytene chromosome like the one shown in A. What appears at first glance to be a single thick chromosome is actually two homologues closely paired, gene for gene, and each consisting of many parallel longitudinally arranged fibers.

two homologous chromosomes but different in nonhomologous chromosomes. Different cells of a given individual have identical endowment. By a variety of techniques it is possible to map the chromosomes, that is, to establish the location of specific genes controlling different characteristics (eye color, wing morphology, various enzymatic activities, and many others) at specific bands. The techniques used include observation of correlations between altered characteristics of the organism and loss (deletion; Section 4.5.1) of specific chromosome portions identifiable by examination of the salivary gland chromosomes.

The chromosome set in a nucleus of a *Drosophila* salivary gland cell contains on the order of 5000 distinguishable bands. In general a given band seems to be associated with a single genetic function, in the sense that each is associated with a particular inherited characteristic that can show mutations. This is one of the criteria on which *Drosophila* is estimated to possess roughly 5000 distinguishable "genes" (Section 4.2.6). However, matters are not quite this simple. The DNA present at a single band can contain several thousands to tens of thousands or more base pairs. From rough calculations of the quantities of DNA required to code for proteins (Section 3.1.3), it appears that each individual fiber at a band contributes several to many times more DNA than that needed to specify the amino acid sequences of typical proteins. As is broadly true for eucaryotes, not all of this excess is understood; some may represent repetitive information, and some, sequences involved in gene regulation or RNA processing (see Section 4.2.6).

Certain repetitive genes, such as those for histones, may be present at more than one band, and a few bands seem to be sites of more than one gene (see also the comments on heat shock responses below).

Puffs At certain times some bands show modifications called *puffs* (Fig. IV–38), which result in part from uncoiling of the chromomeres. These puffs are rich in RNA and are seen in autoradiographic studies to incorporate radioactive precursors of RNA rapidly. In some species there is a localized increase ("amplification") of the amount of DNA in certain puffs. Of great interest is the fact that different bands form puffs in different tissues; differences in location of the puffs are seen even among cells of the salivary glands that produce different secretions. Furthermore, specific puffs appear and disappear at specific times of development. From these observations it has been hypothesized that puffs are sites where particular genes are especially active. Puffs show the expected differences from cell to cell and also the high levels of RNA synthesis that would be expected of such gene sites. Polytene chromosome puffing is considered a morphological manifestation of a fundamental molecular and developmental process—the activation of specific genes. To be kept in mind, however, is that RNAs also are produced at bands that are not obviously puffed; the correlations between puffing and gene expression are not all-or-none.

Puffing patterns in salivary gland cells are influenced by materials originating elsewhere in the body. This is experimentally demonstrated by injection of *ecdysone,* a steroid hormone that plays a key role in the control of the developmental cycle of molting (shedding of the cuticle) and cocoon building. Injection of the hormone into young larvae induces a puffing pattern that is not normally seen in the salivary chromosomes until later in development, at the time when the organism's own ecdyson normally acts. Interruption of part of the circulatory system during normal development, so that ecdyson reaches only some of the salivary gland cells, will limit puffing to these cells; cells not reached by the hormone do not puff. Thus, intracellular events at the chromosomal level in the cells of an organ (the salivary gland) are controlled by hormones made by cells of another organ (the prothoracic gland). Such cell interactions during development are difficult to study in most organisms because they lack the convenient specializations, giant chromosomes.

One particular puff in *Chironomus* has been studied extensively by microscopy and by biochemical techniques. The puffs are large enough to be isolated by hand (microdissection) for detailed analysis. As it is synthesized, the puff RNA takes the form of ribonucleoprotein granules attached laterally to the DNA fibers of the puff (as in Figs. II–31 and IV–40). The transcript eventually is released as part of a ribonucleoprotein granule of the size and appearance of those shown traversing the nuclear envelope in Figure II–21. The RNA molecules that move to the cytoplasm are unusually large (75S; molecular weight at least 10 million) and include some sequences corresponding to repetitive DNAs. Little is yet known of the protein molecules for which they code, although one of the potential candidates could conceivably be 1000 to 3000 amino acids long, which is big enough to require a "giant" RNA for its message.

Heat Shock When *Drosophila* larvae, or tissues excised from them, are incubated at elevated temperatures (37° C), new puffs appear within a few minutes, and some previously existing ones regress. Correspondingly, the cells switch their pattern of protein synthesis, ceasing production of a number of proteins and commencing (or drastically increasing) production of a half-dozen or more new ones. Similar effects on protein synthesis are seen with tissue cultures of adult insect cells. The biological significance of these responses to *heat shock* is still being sought, but the effects are very useful for experimental purposes. They permit convenient evocation of discrete changes in gene expression that can be studied both in molecular and in cytological terms such as by *in situ* hybridization of heat shock–engendered RNAs with the polytene chromosomes; from work in progress it is anticipated that each of the RNAs and proteins corresponding to the responsive bands and puffs on the polytene chromosomes can be identified unequivocally and the factors controlling their synthesis analyzed. It appears that one type of "heat shock protein" is actually a "family" of proteins (Section 4.2.6) coded for by DNA sequences present in several similar copies per genome. A few such copies may be clustered at a single chromosomal locus (band), but copies of similar sequence may also be present at more than one band. Conversely, a locus represented by one band in the polytene chromosomes may contain DNA coding for more than one distinctive heat shock–induced RNA. Such observations should soon remove some of the remaining uncertainties about the relations of genes and bands alluded to above.

Heat shock effects comparable, in general terms, to those just described, have now been detected in cultured cells from diverse types of organisms.

4.4.5 Lampbrush Chromosomes

Meiotic Chromomeres In many species, the chromosomes of the initial stages of meiotic prophase show a type of chromomeric organization when viewed in the light microscope. That is, they show longitudinal patterns of darker and lighter regions, corresponding to tighter and looser folding and coiling (Fig. II–27). The chromomere pattern differs characteristically among the different chromosomes in a cell. Whether this is comparable in some way to the organization of polytene chromosome fibers is not yet clear. The number of readily visible chromomeres per meiotic chromosome is usually far too small—a few dozen, perhaps—for "one gene per one chromomere" hypotheses. Even when hundreds or thousands can be observed, as in the lampbrush chromosomes discussed below, there is markedly more variation among genetically related organisms and at different stages in the chromosome's life history than is observed for polytene chromosomes.

Lampbrush Chromosomes The meiotic chromosomes of oöcytes in amphibia have proved particularly interesting. During the later stages of first meiotic prophase the oöcytes deposit large amounts of yolk and grow enormously. At this time the chromosomes are up to 100-fold longer than they are at mitotic metaphase, and they possess numerous lateral loops; this feature gives them their "lampbrush" (or test-tube brush) appearance (Fig. IV–39). Each pair of homologous chromosomes has a characteristic pattern of loops; the loops differ in length, apparent thickness, and other morphological features. A total of several thousand loops is present in the chromosome complement. Similar organization, though often not as extreme or long-lasting, has been found in many nonamphibian species. The prominence of amphibian lampbrush chromosomes may relate in part to the unusually high DNA content in amphibians (Section 4.2.6); individual chromosomes in certain species have DNA that would be close to a meter long if unfolded into an elongate double helix.

When the ends of lampbrush chromosomes are pulled with fine needles, the loops pull out as if they were kinks in a continuous thread (Fig. IV–39). Thus, the loops are not separate structures attached to the chromosomes but are specialized regions of a continuous structure running the length of the chromosome. Much RNA is present on the loops; its removal by enzymatic (RNase) treatment thins the loops but does not disrupt the longitudinal structure of the chromosome (Fig. IV–39). Proteases have similar effects. On the other hand, DNase treatment quickly fragments the chromosome into short pieces, indicating that DNA maintains the continuity of the chromosome. This finding is im-

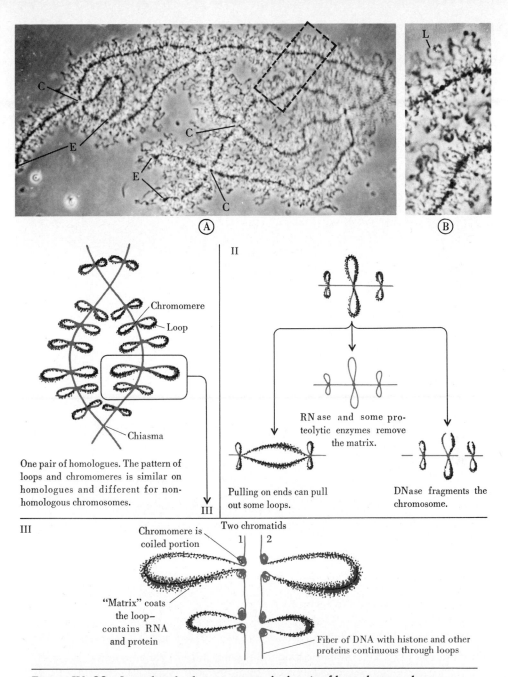

Figure IV–39 *Lampbrush chromosomes.* A. A pair of homologous chromosomes from an amphibian *(Triturus)* oöcyte nucleus (first meiotic prophase). The four ends of the two chromosomes are seen at *E;* chiasmata holding the homologues together are seen at *C.* Projections (loops) along the entire length of the chromosomes give them a brushlike appearance. The outlined area is enlarged in the photograph at the right to show the lateral loops *(L).* Approximately × 400; right × 800. (Courtesy of J. Gall.) B. Experiments and observations to reveal the structure of lampbrush chromosomes and the location of key macromolecules. (After the work of H.G. Callan, H.C. MacGregor, J. Gall, and others.) Panel III summarizes present viewpoints as to the organization of a region like the one indicated in panel I.

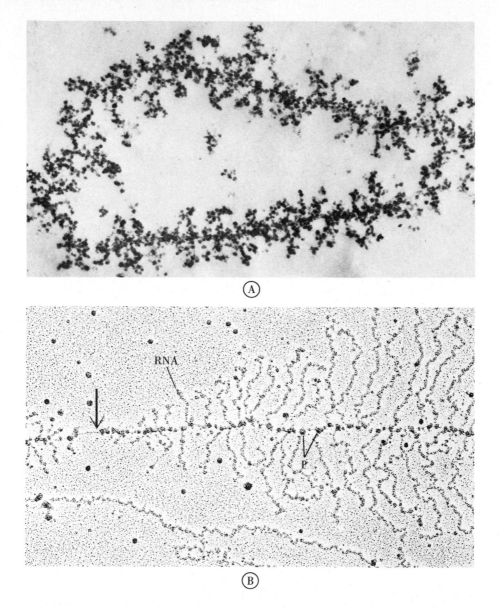

portant evidence that the basic longitudinal structure of chromosomes depends on DNA and does not involve detectable non-DNA linkers (Section 4.2.5). A related line of argument asserts that the loop structure readily seen in the lamp-brush state simply exaggerates the general involvement of loops in chromosomal structure, more or less like the model schematized in Figure IV–15E.

The lampbrush configuration apparently reflects a relatively unfolded ("decondensed") state of the chromosomes (corresponding mitotic chromosomes are 10 to 100 times shorter) and extensive activation of genes. At the

◀ **Figure IV–40** *RNA of lampbrush chromosome loops.* Electron micrographs of portions of loops of amphibian oöcyte lampbrush chromosomes. A. Loop of an isolated *Triturus* chromosome seen in a conventional thin section. Chains and clusters of small granules, 30 nm in diameter, extend from the loop axis; these granules are ribonucleoprotein particles, and each group of particles is thought to represent a large RNA molecule (like the ones in part B) that is undergoing transcription and the initial stages of processing. × 25,000. (Courtesy of H.G. Callan.)

   B. Part of a loop from a *Pleurodeles* chromosome prepared for examination by the spreading techniques of Miller (Fig. II–31); these methods remove some of the proteins and spread out the nucleic acids, which accounts for the fact that the RNA molecules in this part are not organized as chains of granules, as they are in part A. The micrograph shows a portion of the loop region containing a DNA sequence coding for a large species of RNA; the 5′ end of the transcriptional unit ("transcription unit"; Fig. II–31 and Section 3.2.5) is near the arrow; the sequence that is transcribed extends to the right. The DNA is visible at the arrow as a thin filament; over the remainder of its extent, it cannot be seen because of the numerous RNA polymerase molecules *(P)* that have associated with it. Each polymerase associates with the DNA in the region near the arrow and then moves along the DNA to the right, generating an RNA which remains associated with the polymerase as it elongates; thus, the further to the right in the figure, the older and longer the RNA, which accounts for the Christmas tree appearance. (Since the DNA is a double helix that unwinds locally during transcription, in the intact chromosome, RNAs probably protrude at every angle from the chromosome axis, and change their orientation as they elongate; the appearance of RNA molecules sticking only up and down from the chromosome axis results from the preparative method.) Remember that lampbrush chromosomes are unusually active in transcription; many genes prepared in this way from ordinary chromosomes show only one or two RNA polymerases and associated RNAs present along the length of the gene at a given time. × 60,000. (Courtesy of U. Scheer.) Micrographs modified from Callan, H.G., *Proc. R. Soc. Lond.* B214:417, 1982.

---

base of each loop the chromosome is tightly coiled (Fig. IV–39), whereas in the loop itself the chromosome is uncoiled.

Transcriptional Units   The enzymatic digestion studies imply that much RNA is concentrated at the loops. Autoradiographic studies show that this reflects local synthesis; radioactive precursors are rapidly incorporated into RNA at the loops. Electron microscopic and cytochemical studies indicate that each loop includes one or, in some cases, a few "transcriptional units." A unit is conceived of as the DNA responsible for production of a particular RNA species: The overall arrangement of the DNA and the associated growing RNA molecules in each functioning unit is described in the legend to Figure IV–40. Since each RNA molecule grows progressively longer as the polymerase synthesizing it progresses along the DNA, the matrix coating the loops often appears progressively more extensive from one end of a loop to the other. (As with other RNAs, the molecules transcribed on lampbrush chromosome loops associate rapidly with proteins as they grow. They are seen as elongate filaments in preparations such as those illustrated by Fig. IV–40B, owing to the techniques conventionally used to spread the chromosomes out in preparation for examination. In life, the molecules almost certainly are coiled into granule-like ribonucleoprotein structures, as in Fig. IV–40A.)

Still somewhat uncertain is the nature of the RNA produced by lamp-brush chromosomes. Most investigators believe it to be chiefly hnRNA—that is, mRNA precursors. In a "reversal" of usual *in situ hybridization* procedures (as in Section 4.2.6), radioactive *DNA* strands containing base sequences for histones have been obtained and allowed to bind to lampbrush chromosomes; the DNA associates with specific loops, apparently through hybridization with corresponding precursors of histone-specific mRNAs being transcribed there. This finding is one of the lines of evidence suggesting that the chromosomes are producing mRNAs both for use and for storage by the oöcyte. There is still much controversy, however, over the reliability of the methods used in such studies and over the detailed interpretation of the results. Various repetitive DNAs are included in the sequences transcribed in the oöcyte nucleus.

The polytene chromosomes of insects are found mainly in cells that are at terminal stages in differentiation; they are destined eventually to degenerate. The chromosomes do not return to "normal" or take part in mitosis. The situation is different with the lampbrush chromosomes. The oöcyte nuclei complete meiosis and do return to "normal"; the lampbrush chromosomes change back to ordinary chromosomes.

4.4.6 Control of Gene Expression

Variety Study of the mechanisms that contribute to the regulation of the expression of genetic information is progressing rapidly. Explanations are needed both for prolonged or permanent changes, such as those involved in differentiation, and for temporary ones through which cells can adjust their metabolism to short-term circumstances. Most attention has been paid to *transcriptional controls,* those that govern the rates of initiation of transcription and thus the rate at which particular DNA templates are used to generate RNAs; these controls are considered in the last two thirds of this section. Other types of controls known or strongly suspected to operate as well are presented first.

Sections 2.3.4, 3.3.2, 3.11.3, and 4.4.4 describe instances in which the *amounts of DNA* in the nucleus are increased with corresponding increases in pertinent RNAs. This amplification can involve many genes or a few specific ones and can be long-term (Sections 3.3.2 and 4.4.4) or transitory (Section 2.3.4). A variety of interesting biological phenomena, other than those already discussed, are involved. For instance, in certain insects such as *Drosophila,* which can produce eggs at extraordinarily rapid rates, intensive amplification of DNA occurs in the nurse cells that provide RNAs and other components to the developing oöcytes (Section 3.10.2). To support their intense activity during egg production, the follicle cells that manufacture the chorion ("eggshell") surrounding *Drosophila* eggs show general amplification involving most of the DNA, plus additional selective amplification of DNA sequences coding for chorion proteins. From present knowledge, however, there is little reason to believe that such effects (or the loss of some DNA; Section 4.4.3) play roles in most cell types; they have been detected so far only in a restricted range of organisms.

Careful comparison of the RNA molecules made in the nucleus with those that are transported to the cytoplasm suggests that a proportion of the hnRNA molecules transcribed from DNA are entirely degraded within the nucleus. Perhaps during the processing of hnRNAs to mRNAs, or the transport of mRNAs to the cytoplasm, transcripts of some genes can be excluded *selectively* from reaching the cytoplasm. Despite being transcribed, the genes will not be represented by protein molecules, and hence they will not affect cytoplasmic metabolism. Comparison of nuclear and cytoplasmic RNAs at different stages in embryonic development (as in Section 4.4.2) and comparisons between cells in growing and nongrowing states provide some evidence that different proportions of hnRNAs do in fact generate mRNAs at different times in a cell's history; this phenomenon does appear to involve "selection" of different transcripts at different times for processing into cytoplasmic messages. The nature, extent, mechanisms, and implications of such "posttranscriptional" selection have only begun to be studied, however. Some investigators believe it is a crucial contributor to the differences among cell types, at least in certain organisms such as the sea urchin. A few suggest that small RNA molecules may play major controlling roles (see Section 5.1.2). No consensus has yet developed concerning such matters, especially since additional complexities are becoming evident: For instance, in sea urchins, patterns of transcription of repetitive DNA sequences, and transport of such transcripts to the cytoplasm show interesting changes during development. The RNAs stored by the oöcyte, which have been thought to represent potential mRNAs, seem to contain more sequences corresponding to interspersed repetitive DNAs (Section 4.2.6) than is typical of the mRNAs translated during development; among many other possibilities, this difference could mean that additional processing is required before the stored RNAs can be used or even that the sequences from repetitive DNAs serve as controls governing the programmed use of the RNAs during development. Perhaps the key thing to keep in mind at present is simply that splicing and other posttranscriptional steps have already been shown to repre sent phenomena that the cell can vary (see Sections 3.1.2 and 3.6.1); thus these processes are possible regulatory points as well as mechanisms whereby different cell types may use the same stretch of DNA to generate different products. (See also Section 3.6.10.)

There may be several types of *translational controls* governing the rates at which information encoded in mRNA is translated into protein. Few unequivocal demonstrations of translational control have been accomplished as yet, but it is frequently included in hypotheses proposed to account for changes in gene expression. An "extreme" example, studied in most detail in sea urchins, is the storage of ribosomes and mRNA–protein complexes in separate, inactive form during oögenesis and their subsequent association to form active polysomes during development. Section 4.2.2 mentioned that as cells undergo mitosis, translation is markedly depressed; correspondingly, fewer polysomes are formed from the mRNAs and ribosomes present in the cytoplasm. In Section 3.2.5 we outlined a translational control mechanism governing ribosomal protein synthesis in bacteria; efforts are under way to see if similar controls regulate eucaryotic ribosome production. Speculative hypotheses propose that different mRNAs have different affinities for ribosomes, so that even if present in similar abundance, the mRNAs will differ in the efficiency with which they form polysomes and are translated.

The foregoing examples concern the initiation of protein synthesis. A different, well-studied case is the inhibition of the synthesis of globin by reticulocytes when there are no *heme* groups (Section 1.4.2) available to complex with the globin to form hemoglobin; in this situation the effect is mediated by a system of *kinases* (Section 2.1.4), which respond to the absence of heme by phosphorylating one of the *elongation* factors (Section 2.3.2) needed for protein synthesis, thereby inactivating it.

The life spans of molecules, including both RNAs and proteins, also are subject to regulation—they vary in different growth conditions and under the impact of hormones and other agents. rRNAs, for example, are degraded at a much slower rate in rapidly growing cells than the rate in the same cells under conditions where growth has ceased. During the differentiation of reticulocytes (Section 2.3.5), mRNAs specifying globin appear to be selectively spared from the degradation that drastically reduces the cells' content of many other species of mRNA. Other controls, still to be detailed, may affect posttranslational modifications and packaging of proteins or other processing steps required for macromolecular function. Of course, the *activities* of enzymes as well are affected by a variety of metabolic mechanisms, such as feedback inhibition (Section 3.2.5) which do not require altering the *amounts* of the enzymes present in the cell.

"Flanking" and "Internal" Control Sequences   Persisting uncertainties about the organization of DNA in eucaryotic chromosomes still hinder analysis of transcriptional controls, though this situation is improving rapidly. By analogy with regulation of procaryotic genes (Section 3.2.5), it is assumed that one aspect of such controls is interaction of other molecules—proteins, or perhaps even RNAs (Section 5.1.2)—with DNA sequences adjacent to ("flanking") the ones to be transcribed. In particular, promoter sequences (Section 3.2.5) might be expected to be present in the DNA flanking the "5′ end" of the transcribed region—the end specifying the beginning of the mRNA (Section 2.2.4). DNA in this zone—that is, flanking the 5′ end of the gene—is colloquially referred to as "upstream" from the gene.

The search for relevant sequences is on. The nucleotide sequences in DNA molecules corresponding to various genes have been compared; most share in common a flanking "TATA box"—the sequence TATAAAA, or a close relative rich in As and Ts. This sequence is present in the flanking region of the DNA; usually the sequence is located about 30 nucleotides "upstream" from the 5′ end of the gene. Though the specific sequences present differ from those of eucaryotes, the promoter regions of different genes in procaryotes show comparable similarities to one another (Section 3.2.5); this leads to the suggestion that the "TATA boxes" of eucaryotes are portions of promoters that help RNA polymerase molecules to associate with DNA or to assume the correct position for initiating transcription at the proper nucleotide.

One of the functions of the promoter, or of associated sequences, must be to specify which of the two DNA strands is to be transcribed. For a given gene, only one strand codes for the corresponding product, but for different genes—sometimes even adjacent ones along a given stretch of DNA—different strands carry the information and are transcribed. (Since the direction of transcription is always 5' to 3', and since the two DNA strands have opposite directionality in this respect [Sections 2.2.1 and 2.2.4] there are cases in which the RNA polymerases transcribing a given gene move in opposite directions along the DNA double helix, from the polymerases transcribing an adjacent gene.)

A fruitful experimental approach compares transcription of DNA molecules fragmented so as to include varying lengths of transcribed and flanking sequences. The aim is to detect the effects of the presence or absence of particular sequences. If they do contain the requisite set of sequences, the fragments can, in fact, be transcribed when introduced by microinjection into the nucleus of a *Xenopus* oöcyte, or when added to an extract produced from such nuclei or to certain other cell extracts. The results of such experiments suggest that there are additional nucleotide sequences important for initiation of transcription that are located further away from the 5' end of the transcribed sequences than the TATA box; some lie 100 nucleotides or more from the 5' end. Current hypotheses propose that these sequences also help RNA polymerase molecules, or regulatory factors that interact with the polymerase, to associate correctly with the DNA. Surprisingly, experiments of this type also have shown that, for 5S ribosomal RNAs and probably for tRNAs, sequences *within* the gene region that is transcribed into RNA participate in controlling the initiation of transcription. This finding was not foreshadowed recognizably by work on procaryotes and may reflect the fact that in eucaryotes, these RNAs are transcribed by a distinctive type of RNA polymerase (type III, as opposed to the polymerase II that is used for mRNAs; Section 2.2.4). Perhaps somewhat different control systems are employed to govern the functioning of the different types of RNA polymerases and the production of the different categories of RNA: Production of the nontranslated RNAs (rRNAs, tRNAs, and the like) synthesized in virtually all cells might be expected to be controlled differently from that of mRNAs, which vary from cell to cell. Such findings reemphasize the warning in Section 3.2.5 against too rapid an assumption of similarity between procaryotic mechanisms and those in eucaryotes. (Recall that procaryotes use the same type of RNA polymerase for all their RNAs.)

Chromosome Structure and Transcriptional Control; Heterochromatin The observations described in Sections 4.4.4 and 4.4.5 show that gene activation in eucaryotes can be accompanied by striking morphological changes in the chromosomes. Both in lampbrush chromosomes and in puffs of polytene chromosomes, very active transcription is accompanied by local unfolding or uncoiling ("decondensation") of the chromosomes, involving many genes in the lampbrush chromosomes and a few in the polytene chromosomes.

Several observations on cells with ordinary chromosomes point in a similar direction. Changes in chromosome morphology correlated with gene activation or inactivation are generally interpreted as demonstrating alternative states of chromatin: a *condensed* form that is inactive ("turned off") and an *extended* form that is active ("turned on").

1. In sperm cells, the chromosomes are densely packed, and little RNA synthesis occurs. During cell division the chromosomes are coiled into compact arrays, and again there is little RNA synthesis.
2. In some strains of mice and cats, when the two X chromosomes carry different alleles for coat color, females may show coats that are mosaics of the two colors. The explanation, often called the *Lyon hypothesis* after one of its formulators, is that during development, a given coat region is formed by the clonal progeny of a single cell; in the embryo one or the other of the two Xs is turned off in different progenitor cells, and this X remains the one that is inactive in the cell's offspring. Experiments sustaining this viewpoint and extending it to other types of genes demonstrate that when the Xs of a female carry alleles for two different forms of an enzyme, tissue-culture clones grown from different single cells of the individual animal manufacture either one or the other form of the enzyme but not both. In other words, no individual cell expresses the genes on both Xs, but in different cells in the same animal, different Xs are transcribed. These phenomena correlate with the condensation of one X chromosome (Fig. II–19) in many cell types of female mammals.
3. In white blood cells much chromatin is coiled into dense regions. Autoradiographic studies show these regions to synthesize much less RNA than that produced by adjacent extended chromatin. Most cells show at least some condensed chromatin in their interphase nuclei (Fig. II–18).
4. When inactive nuclei are activated, as in Section 4.4.3, their chromatin often "decondenses" (extends) as the nuclei swell.

As indicated in Section 2.2.5, the term *heterochromatin* is generally used to designate the coiled dark-staining condensed regions of interphase chromatin. The heterochromatin we are considering here is referred to as "facultative," to indicate that it varies from cell to cell and under different conditions. (The repetitive DNA–rich heterochromatin discussed in Section 4.2.6 is called "constitutive" heterochromatin since it is not known to vary in activity.) *Euchromatin* refers to the extended regions. It is conceivable but still speculative that the tight packing of condensed chromatin helps to maintain the inactive state, simply by impeding penetration of substrates and enzymes to the DNA. This might be one of the mechanisms through which agents controlling gene activity exert their effects. The nature of the underlying molecular events leading to condensation is not known, however. Furthermore, with the exception of salivary gland chromosomes, the evidence on chromosome condensation concerns inactivation of whole chromosomes or large portions of chromosomes with extensive blocks of genes. It is an open question whether this "coarse" control is an exaggeration of similar events at the level of individual genes. Different more subtle regulatory factors operating at the latter level seem necessary to explain, for example, why different genes that are active in a given cell can be transcribed at very different rates (see Section 3.2.5). Clearly, being included in a euchromatic region of a chromosome does not by itself mean that a gene will be particularly active. Indeed, the available evidence suggests that many genes in euchromatin are completely inactive or very nearly so.

Nucleosomes Since transcription requires access by RNA polymerase to the DNA, and local though temporary separation of the two DNA strands, changes in nucleosome structure might be expected to be involved in the relevant regulation. There are hints that such changes do occur. In general terms it is argued that the "decondensing" of chromatin associated with gene activation is based on a change in nucleosome packing from a tightly coiled "solenoid" or "superbead" configuration (Fig. IV–18) to a more stretched-out, flexible chain of beads. Actively transcribed genes are still organized as nucleosomes. However, where transcription is exceptionally rapid, with many closely spaced RNA polymerases functioning simultaneously (Fig. IV–40), the nucleosomes are sometimes very sparse and spaced quite far apart along the DNA. This appears to be true for sites of rRNA production, for some of the puffed regions of salivary gland chromosomes, and for some lampbrush loops. (The histones still seem to be present in these cases, but their state and distribution are unknown.) Less active genes show fewer and more widely dispersed RNA polymerases and transcripts (only one or two may be present at any given time) and a more ordinary nucleosome distribution. However, even for the latter types of genes the nucleosomes seem different from those of inactive genes, as judged by the response of isolated chromatin to digestion with certain types of DNases. *(DNase I,* obtainable from the pancreas and other sources, is most often used.) When chromatin from cells actively transcribing particular mRNAs (those for globin, for example) is gently digested, the DNA sequences coding for the mRNAs are rapidly and selectively degraded. One hypothesis suggests that this reflects an "opening up" of the nucleosome structure, making the DNA more accessible to enzymes—to DNase in the experimental case under consideration, and to RNA polymerases or other transcription proteins in the living cell.

It has also been found that transcription of genes is strongly correlated with the presence of sequences flanking ("upstream" from) their 5' end that are exceptionally susceptible to DNase degradation, even more so than are the transcribed sequences. The exciting possibility exists that this "hypersensitivity" reflects alterations directly involved in controlling the initiation of transcription.

There also are hints that in certain active genes, perhaps including those coding for tRNAs, the nucleosomes exhibit "phasing." That is, over a limited portion of the gene or flanking DNA the nucleosomes are specifically positioned so that particular DNA sequences are located at particular sites in the nucleosome structure, presumably with significant functional consequences. Inactive genes, it is thought, show more random association of DNA and histone octamers in the sense that no sequence-specific distributions are evident.

Histones   Proteins are prime suspects as agents that might control chromosome condensation or regulate genes at the molecular level. The histones are one obvious center of attention. There are only a few distinct types of histones per cell, and these generally vary little from cell to cell or during development (Section 4.2.5). Thus, by themselves they lack the specificity required for regulators of individual genes. Nevertheless, histones could respond to other, selective, regulatory agents and participate in bringing about alterations of chromosome structure or other changes of the sort we have been discussing.

In the test tube, with either isolated chromatin or mixtures of purified DNAs and proteins, histones inhibit transcription, as well as DNA replication. Removal of histones from DNA or conditions that weaken the binding between the two enhance these activities. Sperm of various organisms have protamines associated with their DNA (Section 3.10.1). These proteins are even more basic than the histones, and their tight binding to DNA may contribute to the inactivity of the sperm cell nucleus. On the other hand, histones are subject to enzymatically imposed modifications, in the course of a cell's history, that can affect the strength of their interaction with DNA; most weaken the binding of the histones to DNA. In dividing cells, for example, histone H1 shows a cycle of phosphorylation; the abundance of phosphates linked to the amino acids *serine* and *threonine* reaches a maximum early in mitosis and subsequently declines markedly. The reversible addition of acetate and methyl groups to the basic amino acids such as *lysine* in histones also takes place, and there are cases in which such modifications seem to be correlated with changes in gene activity. Even ADPs may become attached to histones or other chromosomal proteins by means of linkage through the ADPs' ribose sugars. Certain polypeptides may also become covalently attached to histones; one (highly controversial) series of investigations suggests that among these is the polypeptide *ubiquitin,* which has also been implicated (in equally controversial work) in regulating turnover of cytoplasmic proteins (Section 2.8.3).

Nonhistone Proteins   The nonhistone proteins of the chromosome are also receiving increasing attention. These proteins show much more variability in type and amount from tissue to tissue than that characteristic of the histones; changes in the nonhistone protein population are noted at different stages of cell life cycles and during development. Remember, however, that there are problems in deciding which of the proteins detectable in isolated chromatin or mitotic chromosomes truly are chromosomal and which have become associated, artifactually, during preparation of the material (Section 4.2.5). Moreover, some of the proteins associated with the chromosome are bound to the RNAs undergoing transcription rather than to the DNA. Others are enzymes involved in transcription or translation. Still, from admittedly fragmentary evidence, many investigators claim that underlying its apparent heterogeneity, the nonhistone chromosomal protein population contains a relatively well-defined set of molecules, a bit less abundant than the histones and of substantially greater variety, that participates in fundamental structural and regulatory phenomena.

There have been reports that when RNA polymerases are added to isolated chromatin, the nature of the RNAs produced depends on the nonhistone proteins present. When chromatin is artificially reconstituted by mixing histones, DNA, and nonhistone protein, the mRNAs transcribed correspond to the cell types from which the nonhistone proteins were obtained: Reticulocyte nonhistone proteins specify globin messages. However, furious debate continues about the reliability of the techniques used and about the interpretation of such experiments, and these findings cannot be taken conclusively to demonstrate specific effects. One problem is that most of the studies used RNA polymerases from bacteria, since those from eucaryotic cells have been unavailable in suitable form. Another is that it has not always been clear that under the experimental conditions used, *initiation* of new RNA molecules occurred rather than simply *completion* of RNA molecules already being transcribed when the chromatin was prepared from the cell.

What is especially needed to evaluate the roles of nonhistone proteins are clear demonstrations of effects of particular species of proteins, rather than of the heterogeneous and complex mixtures generally employed. A valuable ongoing line of investigation is focused on steroid hormones, such as estrogens (Sections 3.10.2, 3.12.3, and 4.3.1) or ecdysone (Section 4.4.4), which stimulate the synthesis of particular proteins by target cells. The hormones enter the cells and complex with receptor proteins in the cytoplasm; the complexes then enter the nucleus, where they stimulate specific transcription, apparently by interacting directly with the chromatin. (Presumably the complexes bind to specific regions of DNA or to configurations of DNA and chromosomal proteins.) Other studies have concentrated on a particular group of nonhistone proteins obtainable from chromatin of a variety of vertebrates; they are called the *high mobility group* (HMG) since they are relatively small and thus move rapidly on the gels used to separate proteins (as in Fig. II–37). The tantalizing finding has been made that removal of two particular types of HMG from chromatin abolishes the DNase I sensitivity of actively transcribed genes described above, and that this is reversed (sensitivity is restored, selectively) when the proteins are added back. Evidently these HMGs interact with the histones to help to alter nucleosome structure. The nature of the signal that determines which genes are to undergo such alteration is being sought.

Synthesis of the 5S ribosomal RNA in *Xenopus* seems to require the presence of a protein "transcription factor"; this binds to the DNA, described earlier in this section as constituting a region within the gene that is needed for initiation of transcription. The proposal has been advanced (and is gaining experimental support) that as 5S RNA accumulates in the cell, it complexes with this protein, competing with the DNA; the diminution in available transcription factor results in reduced rates of 5S RNA synthesis—yet another example of feedback control (Section 3.2.5).

Section 3.12.3 mentioned that the virus SV40, which infects eucaryotic cells, differentially regulates expression of its own genes through synthesis of a protein that can inhibit transcription of one set of mRNAs while not affecting transcription of another. If, as seems likely, this situation reflects differential recognition by the protein of specific features of the promoter regions of the genes involved, it could prove highly instructive. In general, because viruses "take over" and make use of host cell machinery for their transcription, they offer useful tools for investigating transcriptional controls (Section 3.1.3).

Modifications of DNA If chromosomal proteins are responsible for controlling transcription, how is the proper pattern of active and inactive genes perpetuated from one cell generation to the next? From our prior discussion (Sections 3.11.1, 3.11.2, and 4.2.3) it appears that the relations of cell division and differentiation are complex: Cells may become committed to particular lines of differentiation early on in development but undergo division, even repeatedly, before this commitment is fully expressed. Certain highly specialized cell types never divide, but there are cell types that continue dividing for a while after they have begun to make their specialized products. Differentiated plant cells allowed to divide in culture for a period seem able to return to a less specialized state. For a number of cell types, one daughter cell from a mitotic division differentiates into a nondividing, specialized state, while the other remains as an "undifferentiated" stem cell; in other cases both daughters have similar fates.

The chromosome duplication associated with cell division requires extensive reorganization of the chromosomes' proteins, including assembly of new histone octomers and association of new nonhistone proteins with the chromosome. Pathways can be imagined through which new or altered proteins might add to the chromosome as it duplicates and subsequently affect gene expression. It is less obvious how both daughter chromosomes could "remember" and reproduce the particular pattern of DNA–protein association characterizing their common parent. Sometimes it may simply be that the pattern is reestablished anew through the same regulatory system that produced it in the first place. An alternative possibility is that the DNA itself is modified as part of the control of gene expression and that such modifications are perpetuated. This line of thought was triggered by observations such as the fact that on the order of 5 percent of the cytosine bases in the DNA of many cell types have methyl groups (CH_3) attached. Methylation, which is accomplished by a specific methylating enzyme, can strongly affect interactions of enzymes and other proteins with DNA. Unlike the acetylations and phosphorylations of histones, methylation of DNA is stable over prolonged periods. Most of the methyl groups are located on cytosines directly adjacent to guanines; this finding is intriguing since a G-C on one DNA strand is always mirrored by a C-G on the complementary strand, and there is evidence that at such sites, when methyl groups are present, the Cs on both strands are methylated. If further investigation proves that methylation of a C on one strand can foster methylation at the corresponding spot on the other, there is at least the potential for perpetuating the same pattern of methylation for both daughter DNA duplexes. Such considerations could be particularly important in light of findings in many cases that the DNA of active genes shows levels of methylation markedly lower than those present in corresponding genes that are not being transcribed.

It would be very premature to conclude that selective methylation of DNA is a replicable signal that permits the cell to recognize those genes it has turned "off." Until much more evidence is available, the only point to be made is that there are possible ways in which the DNA itself might "record" the influences of regulatory systems and help to determine the stability of gene control and the patterns of perpetuation among a cell's offspring. (Another possible change in DNA will be outlined in Section 5.1.2.)

4.4.7 Aging

Theoretical Considerations Aging ("senescence") as a normal biological process is generally thought of as primarily an attribute of "higher," multicellular organisms. For such organisms, both death of individuals and reliance on sexual reproduction as the primary mode of hereditary transmission favor genetic diversity and evolutionary flexibility within the population. Unicellular organisms are usually regarded as potentially immortal in the sense that their descendants, produced directly by division, can continue to grow and divide without limit so long as nutrients and other conditions permit. However, there are exceptions: Some protozoa—*Euplotes* (Fig. III–14) or *Paramecium* (Figs. III–15 and III–16), for example—proliferate and survive only for a limited number of cell generations unless they undergo sexual reproduction. On the other hand, numerous multicellular plants can be propagated by asexual means for periods much longer than the life of the individual plant, and perhaps virtually without limit.

As with other phenomena in which cells or organisms "tell time" (Sections 3.1.3, 3.4.5, 3.8.1, and 4.2.3), there are many hypotheses about aging but only fragmentary evidence, especially at the cellular and organelle level. For instance, even for pathological situations it is often not clearly understood precisely why cells die (see below; Section 2.8.5 and Chaps. 3.6C and 5.2 discuss pathological changes in cells). Thus we include a brief segment on these topics chiefly to emphasize the need for more information. We do not mean to imply necessarily that it is aging of cells that initiates aging of organisms; more likely there is mutual interaction between the individual cells and the systems of which they are parts.

There are two major general classes of hypotheses: In the first, investigators propose that senescence is a programmed, *normal* process, controlled by mechanisms of gene expression, hormone balances, and other regulatory processes. Cell death stemming from such mechanisms is a feature of normal tissue modeling and remodeling during development. This is observable in most organisms but is especially notable in plants (Sections 3.4.2 and 3.4.4) and also in insects, where large-scale destruction of tissues and their replacement by new ones occurs during the successive metamorphoses of larvae into adults.

In the second class of hypotheses, aging is viewed as resulting from cumulative *stresses* or *"errors,"* arising from within the organisms or from without, that eventually overcome the organism or its cells. Speculation along these lines has focused on phenomena such as mutations in DNA due to errors in replication or to mutagenic agents, on deterioration of "defensive" systems such as the immune system, and on biochemical changes, such as the damage that might be done to lipids and proteins by peroxides. (See Section 2.9.2; peroxides can arise as a by-product of some enzymatic activities, as well as from extrinsic sources such as certain types of radiation.) This type of proposal has led to experiments in which the life spans of organisms or of tissue cultures are found to be reduced by exposure to low levels of radiation or other chromosome-damaging agents, or to chemicals that engender various biochemical abnormalities.

Both types of hypotheses have been put forth in various forms and combinations. Many authors point to the fact that signs of aging are seen in many different tissues of a multicellular organism and argue that aging is a widespread process intrinsic to many tissues and cell types or that it is a common response to a broadly targeted set of signals. Others advance the premise that there are "pacemaker" tissues whose deterioration controls the process—aging of other tissues is a secondary consequence or could be avoided were the pacemakers not to fail.

Cell Death With few exceptions (see Chap. 3.6C) we know little of the mechanisms by which the actual death of cells is engendered during embryonic development or later in the organism's life. Changes in plasma membranes, in lysosomes, and in other cytoplasmic organelles are detectable in cells that are in the process of dying, and as the cells die, these changes extend also to the nucleus. The chains of causation, however, are still to be unraveled by cell biologists; most of the alterations observed are likely to be secondary effects of fatal primary initiating factors yet unknown.

Senescence in Tissue Culture Section 3.11.1 mentioned tissue-culture experiments indicating that populations of normal cells from multicellular organisms seem to have limited life spans. In the sense discussed for the cell cycle, this may be a "probabilistic" phenomenon (Section 4.2.3) rather than one in which all the cells undergo essentially the same fixed number of divisions before the population ceases growth and dies. There is evidence in some cases that as they pass through succeeding divisions, cells "drop out" at random, slowing their proliferative rate and eventually ceasing to divide. Still, populations of cells of different types and from different organisms do have characteristic life spans, and there are important sets of cells within the organism—

those of the immune system (Chap. 3.6C) or the digestive tract epithelium (Section 3.5.1), for example—that must remain capable of dividing if the organism is to persist in good condition.

Many investigators are impressed by apparent correlations between the number of doublings of which a culture is capable before it fails and the normal life span of the species from which the culture was made, or the age of the individual organism from which the cells were initially obtained. These correlations, however, simply show that under the tissue-culture conditions devised thus far, normal cells have limited proliferative and functional life spans. They do not demonstrate necessarily that this is a *causal* phenomenon in the aging of organisms. It is unlikely that organisms die because they "run out of functioning cells." Some, at least, of the key cells and tissues in aged organisms seem capable of essentially normal functions. In addition, it is relatively easy to demonstrate that cells can proliferate, and organs like the kidney can function for periods a good deal longer than the characteristic life span of the organism from which they come. This can be the case when cells or organs are transplanted from animal to animal, rather than being explanted as cultures; it is assumed that transplants among immunologically compatible animals experience a less abnormal environment than that in cultured preparations.

Signs of Cellular Aging Within the Organism   Damaged or abnormal molecules or organelles may have little effect on rapidly growing and dividing cells since they can be diluted out through increasing dispersion among newly made molecules. Dilution cannot occur in nongrowing cells; therefore, the results of "errors" or abnormalities may persist and accumulate unless they are subject to "repair" or "editing" mechanisms (Section 4.2.4) or are capable of degradation by the cell's "turnover" processes. Slowly dividing and nondividing cells in normal organisms do change with age in detectable ways. We have alluded to two such changes already: Red blood cells in the circulation undergo alterations that probably affect their cell surface molecules and their mechanical properties; these changes promote the eventual destruction of the cells (Section 2.8.2). In addition, neurons, cardiac muscle, and hepatocytes accumulate "aging pigments" within their lysosomes (Section 2.8.4); these often are interpreted as reflecting the piling up of indigestible materials resulting from peroxidations and other slow but inexorable abnormal processes. Whether aging pigments have deleterious effects on cell function is still being examined.

Extracellular Materials   Extracellular components also undergo changes as organisms age. Striking examples are afforded by extracellular fibers, such as collagen, which continues to undergo cross-linking (Section 3.6.5) for prolonged periods after release from the cell. This has some obvious effects, including the readily observed changes in elasticity of the skin associated with aging in humans. Various investigators are exploring a host of less obvious possible impacts ranging from changes in elasticity of the walls of arteries to significant alterations in the microenvironment of cells that affect their adhesion, their access to nutrients, or their interactions.

Chapter 4.5 Divergence of Cells During Evolution

4.5.1 Evolutionary Changes in Chromosomes

Figure IV–41 is a diagram of the more common chromosomal changes in number or morphology that contribute to the differences in karyotype (Section 4.2.1) among species.

Changes in Chromosome Number; Polyploidy and Aneuploidy Many plant species appear to have evolved, in part, by *polyploidy,* the presence in the cell of extra *sets of chromosomes.* The likely evolutionary process is related partly to the manner by which polyploidy can overcome sterility of *species hybrids* (the products of matings between organisms of two different species). In most such hybrids, when the two species are only distantly related, the chromosomes contributed by one parent are not homologous to those from the other parent; they differ in number, gene arrangement, and so forth. Normal meiotic pairing cannot occur, and the gametes produced have variable numbers and abnormal combinations of chromosomes; zygotes rarely survive. If, however, the entire chromosome complement of the hybrid is doubled, or if the gametes that produce the hybrid are diploid rather than haploid, the hybrid will possess *pairs·*of chromosomes; normal meiosis and gamete formation can take place. Such doubling may occur naturally by accidental failure of cytoplasmic separation in a dividing cell. Plant breeders induce doubling by the use of agents, such as *colchicine,* that dissolve the spindle. Chromosomes duplicate to form chromatids, which eventually separate from one another (Section 4.2.8), but in the absence of a spindle, both chromatids end up in the same cell.

Polyploidy often results in larger cells, with nuclei enlarged in proportion to the number of extra chromosome sets. Some large polyploid cells are often present along with the diploid cells in tissues (liver, roots, and so forth) of multicellular organisms, but beyond the change in size, the significance of this is not clear. Presumably, in some cases the effect is similar to that of other forms of DNA amplification, such as polyteny (Section 4.4.4).

Several human disorders are related to *aneuploidy,* the presence of abnormal numbers of one or several chromosomes of the complement, rather than of entire sets of chromosomes as in polyploidy. Among these disorders is Down's syndrome, a form of mental retardation. This condition is associated with the presence of three copies of one chromosome (number 21; Figs. IV–4 and IV–5) rather than the normal two copies. Thus, a total of 47 chromosomes per cell is present, rather than the normal 46.

Aneuploidy commonly results from accidental failure of homologues to separate from one another at first meiotic division ("nondisjunction"); both chromosomes enter the same daughter cell. The fact that the presence of an extra chromosome can lead to abnormalities suggests that for some genes at least, normal balanced regulation and expression depend in part on the num-

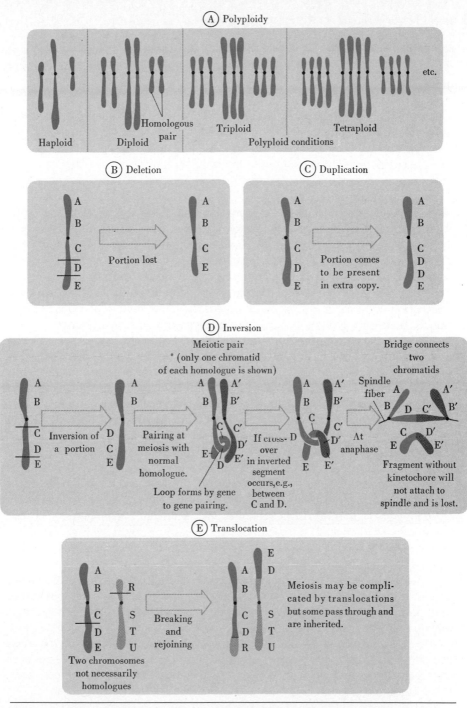

Figure IV—41 *Chromosome abnormalities.* The abnormalities illustrated occur occasionally in nature and may be induced experimentally by certain drugs, irradiation, or other means.

ber of copies present (gene "dosage"). The evolution of the chromosome and gene set of a species involves not only specification of the appropriate *kinds* of metabolism but also the balance and integration of metabolism.

Changes in Length and Gene Distribution; Unequal Crossing-over Rearrangements of chromosomes that alter their length and gene distribution must also have occurred during evolution since related species can differ in chromosome morphology and the arrangement of genes. Radiation and other forms of damage can break chromosomes and foster rearrangements through rejoining of the pieces in new patterns. Movable DNAs (Section 4.2.6) can also mediate such effects. It is interesting that in the course of these events, the *ends* of chromosomes rarely, if ever, become inserted within the interior of the rearranged structures; evidently there is something special about the organization of chromosome ends ("telomeres") that prohibits their fusing with other chromosome regions, so that even when shuffled about, they wind up located at the ends of the new chromosome combinations. The DNA regions corresponding to chromosome ends in yeast have recently been obtained and cloned, promising progress in molecular analysis of telomere behavior.

Phenomena occurring during crossing-over could also generate important rearrangements. As Figure IV–42 illustrates, especially where repetitive DNA sequences are involved, pairing and recombination events that result in unequal distribution of genetic material to the recombinant chromatids can be readily envisaged. When such unequal exchanges occur within clusters of identical or similar sequences, chromosomes with varying numbers of copies of the genes result (Fig. IV–42). Note, however, that the existence (Section 4.2.6) of similar stretches of highly repetitive DNA on different chromosomes and the presence of dispersed repetitive sequences and of movable DNA elements could, potentially, lead to even more dramatic effects. Such sequences might occasionally engender recombinations between otherwise nonhomologous chromosomes, fostering drastic changes in overall gene arrangement. These matters are still largely in the realm of hypothesis, but some such explanations are needed to account for the observed promotion of rearrangements by movable DNAs and for a variety of other phenomena.

Recombination-like exchanges between sister chromatids are observable, at least under experimental conditions (Fig. IV–13). This is another phenomenon whose potential impact on evolutionary and developmental genetic rearrangements needs to be studied now.

Additional clues to mechanisms by which chromosomes can change should come from further investigation of the ways in which DNA sequences specifying immunoglobulin molecules are put together from previously scattered segments (Section 3.6.10; see also Section 5.1.2).

Duplications; Deletions; Inversions; Translocations Cytogeneticists have detected a variety of chromosome alterations ("abnormalities"; Fig. IV–41) thought to spring from mechanisms like those just discussed, or from other

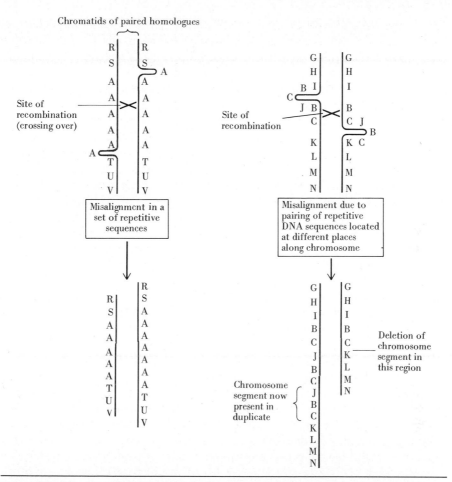

Figure IV–42 *Possible phenomena of genetic recombination (crossing over; Fig. IV–30) involving repetitive DNA. A, B,* and *C* represent repetitive sequences; the other letters represent unique sequence regions of the chromosome. The fundamental point is that gene-to-gene pairing in meiosis may sometimes misalign the chromosomes in regions including repetitive sequences (or even occasionally associate regions of nonhomologous chromosomes which have repetitive sequences in common). Crossing over might then produce a variety of effects, two of which are illustrated here. The diagram schematizes conceivable behavior of two chromatids, from two homologous chromosomes paired at first meiotic prophase. Similar pairing and exchange between nonhomologous chromosomes could generate translocations like those in Figure IV–41.

mechanisms such as the ones responsible for gene amplification (Section 4.4.6). Some have severe, negative impact and thus are unlikely to be perpetuated. Others seem to have contributed to evolution.

Deletions are usually harmful; for example, defective recessive genes on one homologue may be unmasked (Section 4.3.1) if the corresponding "normal" segment of the other homologue is deleted. If both homologues lack the same regions, functions vital to the cells may be absent. The abnormal nucleolar chromosomes in *Xenopus* (Sections 2.3.4 and 4.3.2) probably have undergone deletion of the DNA segments responsible for rRNA production.

Duplications can also sometimes result in abnormalities. Apparently, balances existing among the genes normally present in two copies are upset by the introduction of additional copies. Duplications are of interest from other viewpoints, however. For one thing, repetitive DNAs presumably arise in evolution by some type of duplication of preexisting sequences. For another, by providing "extra" copies of DNA sequences, duplication may open possibilities for change. If the extra sequences undergo mutation or enter into new combinations, the effects on the organism may be less dramatic than is ordinarily the case, because the cells still have unchanged copies of the same DNAs. In time, the extra segments can diverge quite widely from their original nucleotide sequence; this change can lead to the formation of proteins with new amino acid sequences, including, for example, enzymes with new specificities of action. In effect, then, new genes can arise from previously existing genes, while copies of the "old" genes continue to be present in the same cells. Evidence that this has occurred in evolution will be discussed in Section 4.5.2. Very similar considerations probably apply to polyploidy, which also provides "surplus" copies of genes.

Inversions have a different significance. Crossing-over constantly generates new combinations of alleles of different genes *within* chromosomes. Under some circumstances it is of selective advantage to the population that particular *combinations* of alleles present on one chromosome and favorable in a given environment not be disrupted. If a chromosome region becomes inverted, crossing-over with a noninverted chromosome is, in effect, suppressed. Close meiotic pairing may be prevented for the region. If pairing and crossing-over do take place in the chromosome region, the products are abnormal and usually will not be transmitted by viable offspring to the next generation. As seen in Figure IV–34, a crossover in the inverted region will produce chromosome fragments and also *bridges* that are connected to the kinetochores of both homologues; these bridges break in meiosis when the homologues separate. Only the chromatids in which no crossover has occurred in the inverted region will be normal, and usually only they will contribute to viable offspring. Thus, the combination of alleles within an inversion tends to be inherited as a block that is not changed from generation to generation by crossing-over. Different populations of a single species of the fly *Drosophila,* living at different altitudes or temperatures, show differences in the pattern of inversions on their chromosomes, an evolutionary result of such phenomena.

One finding of interest in organisms carrying *translocations* (Fig. IV–41) is that the expression of a given gene in terms of cell characteristics may differ depending on its neighbors. Gene *A* may behave differently if placed in chromosome region *B* than it does when placed in *C;* this effect also is observed with some inversions. Although the basis for these "position effects" is not completely understood, the findings emphasize that genes are not to be considered as totally separate and autonomous agents strung out along the chromosome. A plausible explanation for some aspects of position effects suggests that genes translocated to places near heterochromatin may be "turned off" as a result of inclusion in condensed chromosome regions (Section 4.4.6).

If by chance a translocation separates a nucleolar organizer region into two, nucleoli can be formed by both chromosomes that receive part of the organizer. This finding supports the other evidence that the genetic information in the nucleolar organizer is present in many copies (Section 2.3.4). The effect of the translocation is to move some of the copies elsewhere, so that two groups of copies are present rather than one.

4.5.2 Cellular Evolution: Some Facts and Some Hypotheses

Evolutionary theory was constructed initially on the basis of observations of existing plants and animals coupled to studies of fossils. Modern biology has made major strides in understanding the genetic processes that underlie evolution and is beginning now to develop molecular concepts that complement and powerfully extend the more classical views. In addition, initial experimental steps have been taken in understanding how the fundamental molecules of biological systems could originally have come into being. A variety of simple organic molecules, such as formaldehyde, are detectable in outer space even nowadays. Amino acids, components of nucleic acids, and other familiar molecules form spontaneously from such organic molecules when mixtures of components plausibly thought to simulate the Earth's early atmosphere and bodies of water are exposed to electrical discharges or other sources of energy of the sort that seem likely to have been present during the ancient history of the Earth. There obviously are severe limits to the certainty with which primitive conditions can confidently be reproduced in the laboratory, but these experiments do make clear that the fundamental biological molecular building blocks could have arisen through chemical and physical events of sorts amenable to direct study.

Similarly, reasonable conceptual and experimental models are being constructed for the origins of simple self-replicating molecular systems and of macromolecules, as well as for the transitions from molecules to primitive cells, from procaryotes to eucaryotes, from haploid to diploid, and from unicellular organisms to multicellular organisms. A good deal of guesswork is still required in work on such topics, but the goal is to develop ideas that can be tested through laboratory investigation or by observations in the field and fossil record. There is, for instance, experimental evidence that even relatively simple biological molecules tend spontaneously to form specific types of associations that foreshadow the more complex self-assembly properties of macromolecules—aggregates and complexes of several types arise under a variety of conditions. Fundamental facets of cellular evolution can be studied in laboratory populations of unicellular organisms or by following natural phenomena such as the arising of drug-resistant strains of microorganisms.

Evolution as a Historical and "Opportunistic" Process   An important point to keep in mind is that we have no reason to believe that evolution *had* to have taken precisely the course it did or that the biological molecules and living forms we see around us are the only ones that *could* have arisen on Earth (or elsewhere). As currently understood, evolution is an "historical" process in the sense that each step builds on the prior ones. Evolutionary directions taken by chance, as the result, say, of particular genetic mutations occurring during a particular time, strongly condition the range of subsequent possibilities. It is generally agreed, for instance, that oxygen became abundant in the atmosphere roughly 1.5 to 2 billion years ago as a result of the evolutionary appearance of photosynthetic plants that release O_2. As oxygen levels increased over a period of tens or hundreds of millions of years, many previously existing organisms were put at a disadvantage because they lacked protective systems permitting them to cope with the potentially toxic components, such as peroxides, that can arise from oxygen. However, the presence of oxygen also opened possibilities for aerobic metabolism, with consequent substantial increases in the energy available for cellular activities. It led as well to the establishment of the layer of ozone in the atmosphere that reduces penetration of ultraviolet light and some other forms of radiation to the earth's surface. Ultraviolet light may have been a source of energy for the early evolution of complex *molecules,* but it is potentially damaging to most known *cell* types. Its reduction by the ozone layer was a factor that appears to have facilitated the massive invasion of land surfaces by living organisms that previously "depended" on shielding by water. The presence of oxygen as the result of prior evolution thus exerted both positive and negative selective effects with dramatic impact on subsequent evolution. Some organisms became extinct, some evolved into oxygen-utilizing forms, and some survived by sheltering in persistent anaerobic environments. Note also that atmospheric oxygen is a product of living systems, exemplifying the fact that natural selection involves mutual interaction of organisms and their environment. Organisms condition one another's environment in a host of ways.

It is sometimes claimed that living cells are wildly improbable entities since the probability that the particular amino acid and nucleotide sequences found in the macromolecules of present-day forms could arise by shuffling the units (amino acids and nucleotides) around at random is very remote. Leaving aside the fact that over the Earth's 4.5-billion-year history, improbable events had a good deal of time to take place, such reasoning involves simple fallacies. The probability of *any* particular arrangement of large numbers of *any* sort of unit arising by chance is vanishingly small when viewed with no attention to the actual history of the arrangement. Even a square inch of beach appears preposterously improbable if the likelihood that the grains of sand fell into the arrangement they occupy all at once is simply calculated, ignoring the winds and tides and treating the arrangement actually found as the only arrangement of sand grains conceivable or possible for that square inch.

Outlines Information about tissue organization in multicellular plants and animals can be obtained by microscopic examination of fossils; bones and wood are, of course, better preserved than softer tissues. Aspects of primitive cell chemistry can be deduced from the analysis of organic deposits such as coal or certain petroleum oils, which are formed by the transformation of organisms and their products. Present methods, however, do not permit unambiguous deductions regarding fine structure or metabolic organization from fossils. Knowledge of cell evolution must, therefore, depend heavily on the comparative study of existing species, on extrapolation from known features of macromolecules and cells, and on postulated features of primitive environments. In the attempt to comprehend details of the evolution of metabolism, the reasonable assumption is made, for example, that some of the present-day anaerobic procaryotes occupying unusual habitats (described in Chap. 3.2B) are relics of earlier evolutionary stages.

A variety of early cells have been preserved in the fossil record (Fig. IV–43); objects resembling present-day procaryotes have been identified in rocks formed up to 3 billion years ago. Although it is not always possible with the very earliest deposits to determine whether these formations are truly the remains of cells or merely incidental mineral deposits, they include structures that

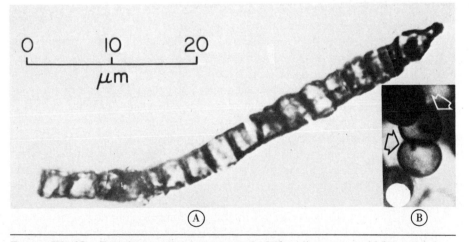

Figure IV–43 *Fossilized cells* approximately 850 million years old from the Bitter Springs geological formation in Australia. The structure in panel A resembles the many strains of present-day blue-green algae that are organized as multicellular filaments. Those in panel B resemble a variety of "modern" microorganisms. A, × 2,000. (Courtesy of J.W. Schopf. [See *J. Paleontology* **42**:651, 1968.]) B, × 1100. (Courtesy of E.S. Barghoorn. [See *Science* **190**:52, 1975.])

are very likely to represent forms of blue-green algae that appeared over 2 billion years ago. In addition, algae that probably were eucaryotic, as well as certain protozoa, can be identified in ancient fossil deposits from characteristics of the cell walls of the algae and of silicon- or calcium-containing "exoskeletal" structures of the protozoa. Eucaryotic cells first appeared at least 1 billion years ago.

As judged from their wide distribution among an immense diversity of organisms, many basic metabolic pathways apparently were laid down early and have persisted. This applies, for example, to much of carbohydrate metabolism—the fundamental enzymes of CO_2 fixation and the Calvin cycle, those of the Krebs cycle, and those of anaerobic sugar breakdown are thought to be of very ancient origin. That enzymes responsible for fixation of atmospheric N_2 are inhibited by oxygen suggests that they arose early (and emphasizes both that evolution cannot anticipate what is to come later and that it is not all change; long-term preservation is also prominent). The "problems" plants encounter in the present-day high O_2–low CO_2 atmosphere (Section 2.7.5) presumably reflect the origins of photosynthesis at a period when O_2 was scarce and CO_2 relatively more abundant. Features of various present-day procaryotes outlined in Chapter 3.2B, such as the use of a variety of hydrogen sources for photosynthesis, variations in the respiratory chains of O_2 metabolism, or the existence of relatively direct light-driven chemiosmotic systems for generating ATP, provide insight into the ways in which primitive organisms "exploited" the energy from sunlight and the other resources available to them. Through attention to such features we can trace the likely steps through which such fundamental metabolic machinery as electron transport or the employment of proton gradients across membranes could have arisen. Once the abundant and widely distributed molecules such as H_2O, and later O_2, could be utilized effectively by such machinery, metabolic energy was obtainable at sufficient levels and in diverse environments to permit expenditure for "luxuries" such as the replication and segregation of large amounts of DNA and the maintenance of large, multicellular aggregates. As cells evolved more complex structures, the selective advantages in metabolic efficiency, flexibility, or adaptation to different environments must have outweighed the problems of maintenance and reproduction of more complex organization.

Mitosis, a key mechanism in eucaryotes, probably arose early—it is responsible for division of virtually all present-day eucaryotic cells. The orderly segregation of large amounts of DNA by mitosis is a central foundation for eucaryotic genetics and for multicellular life. Steps in the evolution of cell-division mechanisms may be reflected in the mechanisms alternative to mitosis found in rare cases (the macronucleus of ciliated protozoa, for example, is simply pinched in two after its contents have duplicated) and in the broad variations in detail observed for mitosis in different unicellular organisms (Section 4.2.7). One pertinent issue still to be resolved is whether or not procaryotes can synthesize tubulin and form microtubules. Some investigators believe that certain spirochetes do possess these capacities, but more data are needed to evaluate their findings. The "problem" addressed in resolving this issue is, of course how the microtubule-based mechanisms characteristic, in one form or another, of essentially all eucaryotes might have evolved, given the very different division mechanisms utilized by procaryotes (see also Section 4.2.10 for comments about relevant evolutionary possibilities).

At present the origins of most of the highly diversified cells of present-day organisms remain a matter of conjecture. Yet, comparative cell biology studies provide some clues. We can already imagine in shadowy outline how, for example, the systems by which unicellular organisms respond to their environment might evolve into the networks of receptor cells, gland cells, and neurons and of hormones, transmitters, chemotactic agents, and receptor molecules, through which cells of multicellular organisms respond to one another and to the outside world (see, e.g., Sections 2.11.4 and 3.8.2). It is no doubt significant that a relatively small variety of regulatory agents, like Ca^{2+} or cyclic nucleotides, is used for such a diversity of controls in different cells and circumstances. Molecules that seem to be related to familiar hormones such as insulin, as well as receptors capable of responding to the neurotransmitters described in Section 3.7.5, have been detected in a large variety of organisms; in many of these organisms, such as protozoa, it is most unlikely that these substances are used as they are in specialized tissues of vertebrates. Future students of cellular evolution may well be able to make sense of these facts in terms of the origins of the multiplicity of present-day mechanisms from a relative handful of primitive cellular responses and from signaling and communication systems of ancient origin.

Molecules and Genes   Nucleic acids are the fundamental carriers of genetic information in all living organisms. A similar nucleotide genetic code is employed universally, though intriguing variations do occur (Section 2.6.3), and recognizably related RNAs and ribosomes are used to make proteins by all species studied—procaryotes and eucaryotes alike. Evidently, the nucleic acids and present-day mechanisms of replication, transcription, and translation arose very early in the evolution of life. Therefore, detailed comparisons of the macromolecules of present-day organisms, down to the level of the nucleotide sequences of comparable genes, afford a means for directly evaluating the genetic relationships among different species of animals, plants, and microorganisms and thus for reconstructing the history of evolutionary divergence. There is impressive agreement between the picture of evolution that emerges from such comparisons and the older proposals based on anatomical, physiological, and paleontological evidence.

Comparative studies of the amino acid sequences in the respiratory protein *cytochrome c* indicate that it arose early in the evolution of aerobic metabolism. The amino acid sequences of portions of the molecule are identical in the cytochrome *c* of yeast, invertebrates, mammals, and higher plants. Organisms as distantly related as yeast and mammals show the same amino acid sequence for almost half of the hundred or so amino acids of which the molecule is made. We have already mentioned that some of the histones of eucaryotes show even more impressive evolutionary conservation. Actin, tubulin, and a number of other proteins also are highly conserved. The basic explanation is that once these proteins evolved as effective components of complex systems, the great majority of mutations that changed the amino acid sequence of the portions of the molecules crucial to their functions, or to interactions with other proteins, would have been selectively disadvantageous because they adversely affected the functioning of the systems.

Part of cytochrome *c* has, however, changed in evolution, indicating that not *every* mutation is deleterious and that not *every* part of the protein is equally vital to its function. Still, there would be conceptual problems in understanding evolution if the only available process for genetic change were mutation. How could organisms maintain their preexisting functions and yet acquire new ones? In addition, rates of change based solely on simple mutation seem to be too slow to account for the estimated actual rates of evolutionary change.

Comparisons of amino acid sequences in proteins, and of nucleotide sequences in DNAs coding for proteins, indicate that different enzymes, hormones, and the like stem from common ancestors. The mammalian digestive enzymes *trypsin* and *chymotrypsin* catalyze different reactions in the breakdown of proteins, but they probably evolved from the same ancestral protein since they show considerable similarity in amino acid sequences. *Hemoglobin,* the iron-containing protein of red blood cells, and *myoglobin,* an iron-containing, oxygen-storing protein found in some muscles and a few other tissues probably also diverged from a common starting point. Hemoglobin contains four subunits, each a polypeptide chain about the size of the single chain of which myoglobin is constituted. Each hemoglobin chain has a three-dimensional conformation similar to myoglobin, and both proteins show a number of stretches of similar amino acid sequences including those in the region of the molecule responsible for binding the heme groups needed to bind oxygen (Section 1.4.2). Several different hormones of the pituitary gland also resemble one another in amino acid sequences. Troponin, a mediator of the effects of Ca^{2+} in some cell types (Section 2.11.1), shares amino acid sequences with calmodulin (Section 2.1.4), which mediates other effects of Ca^{2+}.

The beginnings of the explanation for such observations very likely reside in "duplications" of the sort discussed in Section 4.5.1: "Extra" copies of genes can diverge from their original functions and give rise to "new" proteins. An important extension of this concept has recently been proposed. Since the genes in eucaryotes are made up of separable DNA sequences coding for portions of the polypeptide chain (Section 2.3.5), new genes and corresponding proteins might arise by linking together previously unrelated DNA segments. We already know that different portions of a given protein can have different roles in the molecule's functions—some can bind to specific groups or catalyze enzymatic activity, others guide the protein into or across membranes, and still others control the protein's interactions with other proteins or with regulatory agents. A helpful way of envisaging the possibilities is to consider a protein molecule as being made of several different "domains," each a segment of the overall amino acid sequence that is responsible for a particular aspect of the molecule's function and structure. If some of the different domains are coded for by separate, discrete segments of DNA (separated by "introns"—intervening sequences; Section 2.3.5), then the evolution of a protein with a new ensemble of functional capabilities could be imagined to start by the association of "domains" belonging to one protein with those belonging to others. (Of course, it would be the corresponding DNA sequences—presumably extra copies—that actually underwent the evolutionary shuffling, perhaps through events comparable or analogous to the ones that generate the genes of the antibody system [Section 3.6.10].) The point is not that proteins are all quite so simply organized, or that such changes have definitely been proved to occur. It is just that as we have learned more about the details of gene organization and function, perspectives on mechanisms that may contribute to evolution are broadening dramatically. Indeed, for certain proteins, the underlying assumption made earlier, that different functional domains are coded for by DNA sequences separated by intervening sequences—has been borne out by detailed genetic analyses. For other proteins, however, this correspondence is not as obvious, which has led to suggestions that intron-separated modular units smaller than functional domains undergo evolutionary shuffling and that introns themselves are subject to various alterations and rearrangements.

Puzzling questions remain. If procaryotes and eucaryotes are of common ancestry, how did they come to differ in such a key respect as the presence of intervening sequences in eucaryotic genes and their absence in procaryotes? Does all the DNA present in an organism have a function, or could some of the apparent "excess" (Section 4.2.6) have little or neutral bearing on the organism's survival but be perpetuated simply because it can replicate and because the cell lacks effective means for recognizing and excising it? Could a portion of the cell's DNA be "debris" from evolutionary processes, such as duplications that failed to yield functional genes? (In evolutionary terms, such inefficiency might be the "price" paid for maintaining the existence of mechanisms that generate "new" genes. In fact, nonfunctional DNA sequences closely similar to functional ones are found in the genome; some are called "pseudogenes.") We know that chromosomal genes can be transferred from one procaryotic cell to another by plasmids or viruses. Do such transfers occur in nature, among eucaryotic organisms, and do they play roles in evolution? This process may seem implausible as a general phenomenon, but some such possibilities do exist in the genesis of cancer (Section 3.12.3), and there are some interesting unexplained facts that have been accounted for along such lines in avowedly speculative schemes. (One example, now the subject of intense interest, is the capacity of a few higher plants to make the protein *leghemoglobin* [introduction to Chap. 3.4], related to the hemoglobin of red blood cells of animals.)

Organelles The plasma membrane presumably is the evolutionary descendant of ancient spontaneously formed aggregates of lipids and other molecules.

The establishment of a membrane-like boundary was decisive in the evolutionary transformation of molecular aggregates to cells. A membrane-delimited compartment has advantages for the accumulation and maintenance of interacting collections of molecules and also provides surfaces of types that can facilitate important chemical processes. If current views of energy metabolism are correct (Sections 2.6.1, 2.7.3, and 3.2.4), membranes must have had central roles in such metabolism from early on.

Multienzyme complexes might have arisen when mutations resulted in the presence of groups, in previously separate enzymes, that promoted the binding of the molecules. When such associations, arising by chance, yielded enhanced efficiency, their perpetuation would have been favored.

Not all the organelles, however, need have arisen separately by such evolutionary "self-assembly." The "endosymbiosis theory," currently much in vogue, maintains that some organelles, especially plastids and mitochondria, evolved through symbiotic associations of unicellular organisms resembling associations of types still seen today (Section 3.3.5). One hypothetical sequence, necessarily oversimplified, suggests that a primitive nonmotile photosynthetic cell, such as an alga (see Chap. 3.2C), was taken up through endocytosis by a primitive ameba-like cell; for some reason it was not digested (see Section 2.8.2). The resulting combination of motility, photosynthetic capacity, and ability to endocytize nutrients from the environment would be advantageous. If the two cells multiplied synchronously the association might be stable and in time become necessary for the survival of the partners. The photosynthetic cell might, under these circumstances, evolve into a plastid. (Many parasitic organisms and some symbiotic ones survive very poorly if at all when separated from the "hosts" they normally occupy; this extends to some of the present-day cases in which unicellular organisms live within protozoa. What is considered by many though not all investigators to be an example of the latter is provided by the flagellated protist *Cyanophora paradoxa,* which contains "cyanelles," vacuole-enclosed structures that closely resemble blue-green algae; the cyanelles have yet to be successfully cultured independently, perhaps in part because they contain only 10 percent as much DNA as that in typical blue-green algae.)

Evolutionists are now seeking evidence to decide whether different lines of plant evolution have originated through different symbiotic relationships and whether the ancestral cells that yielded plastids might have been procaryotic in certain cases and eucaryotic in others. The plausibility of endosymbiosis-based schemes is strengthened by the partial self-duplicatory capacities of mitochondria and plastids and by the consideration that the protein-synthesizing machinery and genetic apparatus of both types of organelles resemble those of procaryotes to a degree, and differ from comparable eucaryotic systems (Sections 2.6.3 and 2.7.4). Especially for mitochondria, however, there are unique features of the "chromosomes" and protein-synthesizing apparatus (Section 2.6.3) that suggest that if these organelles did originate through symbiosis, either the ancestral organism was distinctly different from the organisms that have persisted in free-living form to present times, or that subsequent to the initiation of the symbiosis, a good deal of evolutionary change has occurred within the genetic system of organelles or of the nucleus. For that matter, the fact that both plastids and mitochondria depend very heavily on genetic information in the nucleus for their replication and maintenance warns against too rapid acceptance of a symbiotic origin. We might imagine that the initial partners in the symbiosis shared many genetic capacities and that "unnecessary" duplication was gradually eliminated. Traces of this process may be represented by the fact that at least one of the proteins of the mitochondrial "ATPase" (Section 2.6.1) is coded for in the nuclei of some organisms and in the mitochondria of others. Among students of cellular evolution, however, there remain a significant number who believe that mitochondria and plastids could have evolved step by step from simpler cellular membranes and enzyme systems, without symbiosis—that is, basically within the confines of a given cell lineage. From this viewpoint the genetic apparatus of cytoplasmic organelles might reflect an ancient parceling out of the cell's DNA, occurring perhaps when the eucaryotic nucleus itself arose. Alternatively, perhaps organellar DNAs originated from episomal or movable nucleic acids.

As yet, the evolutionary effects of mutations of genes in the cytoplasmic organelles have been little studied. Although such mutations represent a potential source for the variations observed in structure and chemistry in different organisms, the fact remains that most organelles of a given class are basically similar in a vast diversity of organisms. This similarity suggests that the fundamental features of the organelles were established relatively early and that once they had become knit into the intricate network of metabolic interactions in which they function, major changes tended to have deleterious effects and thus tended not to be perpetuated.

Further Reading

Topics of interest for this part of the book are covered in several of the general references listed at the end of Part 2, especially International Cell Biology, 1980–1981 *(edited by Schweiger) and* Discovery in Cell Biology *(edited by Gall, Porter, and Siekevitz). The* Annual Reviews of Biochemistry, *the* Cold Spring Harbor Symposia, *and the journal* Chromosoma *are also useful sources of recent information. The collection of articles,* Cell Biology, A Comprehensive Treatise *(see Further Reading, Part 2), is also a good source of information.*

Assembly
See Further Reading, Part 2, for sources discussing individual organelles, filaments, and so forth. The assembly of fibrin and collagen is discussed in standard biochemistry texts such as the one by Stryer listed at the end of Part 1.

Anderson, R. G. W. Assembly of biological structures. In Prescott, D., and L. G. Goldstein (eds.). *Cell Biology, A Comprehensive Treatise.* New York: Academic Press, 1980.

Butler, P. J. G., and A. Klug. The assembly of a virus. *Scientific American* 239 (5):62–69, 1978.

J. Supramolecular Struct. 2:2–34, 1974. *This is a collection of research articles on problems of self-assembly.*

Nomura, M. Ribosomes. *Scientific American* 221 (4):28, 1969.

Tanford, C. The hydrophobic effect and the organization of living matter. *Science* 200: 1012–1018, 1978.

Wood, W. B., and R. J. Edgar. Building a bacterial virus. *Scientific American* 217 (1):60, 1967.

Cell Division

The books listed under Tissue Culture and Cancer at the end of Part 3 contain much information about the control of cell division; see especially the one by Pollack. The International Cell Biology, 1980–1981 *and* Discovery in Cell Biology *collections (Part 2, General Reading List) contain surveys of information about cell division mechanisms that are up to date as of the time they were written; the ones by Inoué and McIntosh are good starting places.*

Heath, I. B. Variant mitoses in lower eucaryotes: indications of the evolution of mitosis. *Int. Rev. Cytol.* 64: 1–80, 1980.

Holley, R. W. Control of cell proliferation. *J. Supramolec. Struct.* 13:191–197, 1980.

Kubai, D. F. Evolution of the mitotic spindle. *Int. Rev. Cytol.* 43: 167–227, 1975.

Lloyd, D., R. K. Poole, and S. W. Edwards. *The Cell Division Cycle.* New York: Academic Press, 1982, 523 pp. *A reasonably compact survey of the cell cycle in eucaryotes and procaryotes that considers methodology, genetics, biochemistry, and ultrastructure.*

Mazia, D. The cell cycle. *Scientific American* 230 (1):54–64, 1974.

Pickett-Heaps, J. O., D. H. Tippit, and K. R. Porter. Rethinking mitosis. *Cell* 29: 729–744, 1982.

Zimmerman, A. M., and A. Forer (eds.). *Mitosis/Cytokinesis.* New York: Academic Press (in press).

Structure and Replication of Eucaryotic Chromosomes

See also the further reading for the Nucleus following Part 2.

Britten, R. J., and D. E. Kohne. Repeated segments of DNA. *Scientific American* 222 (4):24–31, 1970.

Kornberg, A. *DNA Replication.* San Francisco: W. H. Freeman, 1980. *A comprehensive survey dealing with procaryotes, eucaryotes, and organellar DNAs.*

Kornberg, R. D., and A. Klug. The nucleosome. *Scientific American* 244 (2):52–64, 1981.

Long, E. O., and I. B. Dawid. Repeated genes in eukaryotes. *Ann. Rev. Biochem.* 49: 727–764, 1980.

McGhee, J. D., and G. Felsenfeld. Nucleosome structure. *Ann. Rev. Biochem.* 49: 1115–1156, 1980.

Wang, J. C. DNA topoisomerases. *Scientific American* 247 (1):94–109, 1982.

Genetics

There is a wide variety of excellent genetics texts. One useful one, written by a person with strong background in cell biology, is Genetics (3rd ed.) *by U. Goodenough (Philadelphia: Saunders College Publishing, in press).*

Cohen, S. N., and J. A. Shapiro. Transposable genetic elements. *Scientific American* 242 (2): 40–49, 1980.

Dyer, T. A. Methylation of chloroplast DNA in *Chlamydomonas. Nature* 298: 422–423, 1982.

Grun, P. *Cytoplasmic Genetics and Evolution.* New York: Columbia University Press, 1976, 435 pp. *An introduction to cytoplasmic genetics.*

Stahl, F. *Genetic Recombination, Thinking About it in Phage and Fungi.* San Francisco: W. H. Freeman, 1979, 333 pp. *A useful, moderately advanced introduction to modern views of recombination mechanisms.*

Development; Aging

There are many good textbooks of developmental biology. Four of the most useful from the perspectives of the present book are Karp and Berrill's Development *(2nd ed., New York: McGraw-Hill, 1981); Ham and Veomett's* Mechanisms of Development *(St. Louis: C. V. Mosby, 1980); Browder's* Developmental Biology *(Philadelphia: W. B. Saunders, 1980), and Grant's* Biology of Developing Systems *(New York: Holt, Rinehart and Winston, 1977).*

Behnke, J. A., C. E. Finch, and G. B. Moment (eds.). *The Biology of Aging.* New York: Plenum Press, 1978, 388 pp. *Discussions of various theories.*

Davidson, E. H., B. R. Hough-Evans, and R. J. Britten. Molecular biology of the sea urchin embryo. *Science* 217: 17–26, 1982.

DeRobertis, E. M., and J. B. Gurdon. Gene transplantation and the analysis of development. *Scientific American* 241(6): 74–82, 1979.

Hayflick, L. The cell biology of human aging. *Scientific American* 242 (1):58–65, 1980.

Illmensee, K., and L. C. Stevens. Teratomas and chimeras. *Scientific American* 240 (4):120–133, 1979. *Experiments with allophenic mice.*

Gene Expression

Beerman, W., and U. Clever. Chromosome puffs. *Scientific American* 210 (4):150, 1964.

Brown, D. D. Gene expression in eukaryotes. *Science* 211: 667–674, 1981.

Brown, D. D. The isolation of genes. *Scientific American* 229 (2):20, 1973.

Callan, H. G. Lampbrush chromosomes. *Proc. R. Soc. Lond. [Biol.]* B214: 417–448, 1982.

Darnell, J. E. Variety in the level of gene control in eukaryotic cells. *Nature* 297: 365–371, 1982.

Kolata, G. B. Genes regulated through chromatin structure. *Science* 214: 775–776, 1981.

Lamb, M. M., and B. Daneholt. Characterization of active transcription units in Balbiani rings of *Chironomus tentans. Cell* 17: 835–848, 1979.

Lewin, B. *Gene Expression 2,* 2nd ed. New York: John Wiley & Sons, 1980, 1160 pp. *An on-the-whole successful attempt at a comprehensive review of recent research on eucaryotic chromosomes, especially the regulation of genetic activity. Provides very good background even though, inevitably, it is out of date in some areas.*

Kolta, G. Z-DNA: From the crystal to the fly. *Science* 214: 1108–1110, 1981.

Martin, G. R. X-chromosome inactivation in mammals. *Cell* 29: 721–724, 1982.

Miller, O. L. The visualization of genes in action. *Scientific American* 228 (3):34, 1973.

O'Malley, B. W., and W. T. Schrader. The receptors of steroid hormones. *Scientific American* 234 (2):32–43, 1976.

Schlesinger, M. J., G. Aliperti, and P. M. Kelley. The responses of cells to heat shock. *Trends Biochem. Sci.* 7: 222–225, 1982.

Weisbrod, S. Active chromatin. *Nature* 297: 289–295, 1982.

Evolution

The September 1978 issue of Scientific American (239 [3]) *is a set of articles on evolution, including good discussions of the origins and early history of cells.*

Dillon, L. S. *Ultrastructure, Macromolecules and Evolution.* New York: Plenum Press, 1981, 708 pp. *A discussion of organelles and biochemistry from the comparative viewpoint, aimed at evolutionary insight. Has many details of interest, although some interpretations are noncritical and others excessively idiosyncratic. Especially useful as an antidote against too rapid acceptance of endosymbiosis as the only viable theory of organelle evolution.*

Frederick, J. F. (ed.). Origins and Evolution of Eukaryotic Intracellular Organelles. *Ann. N.Y. Acad. Sci.* 361: 1–512, 1981. *A collection of articles, mostly supporting endosymbiotic viewpoints, but published along with some critical discussion.*

Gilbert, W. DNA sequencing and gene structure. *Science* 214: 1305–1312, 1981.

Lewin, R. Do jumping genes make evolutionary leaps? *Science* 213: 634–636, 1981.

Margulis, L. *Symbiosis in Cell Evolution.* San Francisco: W. H. Freeman, 1981, 419 pp. *A detailed and documented presentation of the endosymbiosis theory of cell evolution.*

PART 5

Toward A Molecular Cytology

Chapter 5.1 From Wilson to Watson

It is now apparent that E. B. Wilson's confidence (see beginning of Part 1) was well founded when he predicted more than 50 years ago that many "puzzles of the cell" would be solved as more powerful tools of analysis became available. None has yet been solved completely, but all are yielding to modern methods. Wilson's "puzzles" included the manner by which nuclear genes affect chemical reactions in the cytoplasm; the nature of cell differentiation; the continuity from one cell generation to the next of centrioles, mitochondria, and chloroplasts; and the influence of the cell's complex organization upon the behavior of macromolecules.

In Wilson's day, morphological description was the basic approach; chemical analysis was in its infancy. Today cell biologists have available a wide battery of methods, and qualitative description is now based on a variety of microscopes. Increasingly the problems defined by such descriptions are analyzed by precise quantitative measurements made with biochemical and biophysical procedures. Important strides are being made in manipulating cells in culture and in taking structures apart and then reconstituting them in the test tube. Although Wilson recognized that form and function were inseparable, it has taken the expanded knowledge of biochemical events and their intracellular localizations, coupled with the extension of microscopic observations into the molecular realm, to establish this for all organelles within the cell.

649

5.1.1 A Historical Example: Nucleoli

The complex path of progress in cell biology is well illustrated by the history of the study of nucleoli. Clues have come from many directions. Initially it was noted by light microscopists that nucleoli are found in virtually all eucaryotic cells, but that they are particularly prominent in active cells which synthesize much protein (Chapter 2.3). Later, cytochemists found much RNA in nucleoli and correlated this with the presence of abundant RNA in the cytoplasm; the presumption of transfer of RNA from nucleolus to cytoplasm could be supported by autoradiographic studies (see Fig. I–24). The discovery of genetically distinctive "anucleolate" strains of *Xenopus* (Sections 2.3.4 and 4.3.2) established the fact that the absence of nucleoli is accompanied by the absence of the formation of ribosomes. Molecular biological work explained this by showing that nucleolar RNA gives rise to ribosomal RNA (Section 2.3.4). Large precursor molecules are formed in the nucleolus and are modified then to produce the RNA molecules found in the subunits of the ribosomes (see Fig. II–28).

Nucleoli do not seem to be self-reproducing, however; they disappear as organized entities in many cells during division. How are they formed? Again, early light microscopy provided an important clue by showing that there is special chromatin associated with nucleoli during interphase and during early stages of cell division. In many organisms nucleoli are attached to special nucleolar-organizer regions of specific chromosomes (see Figs. II–27 and IV–3). When fusions between nucleoli are taken into account, the number of nucleoli parallels the number of organizer regions. RNA–DNA hybridization studies (Section 2.3.4) supported by other evidence (Section 4.3.2) have now shown that the nucleolar-organizer regions contain the genetic information for ribosomal RNA. They have revealed that this information is present in repetitive form; that is, up to many hundred copies of the information, apparently identical, may be present in each organizer. In amphibian oöcytes even more copies of the information are temporarily present under special circumstances of intensive rRNA production. Multiple replicates of the nucleolar-organizer region are produced, and each makes a nucleolus. This special situation has been used to advantage for isolation of the DNA responsible for RNA synthesis. Figure II–31 is an electron micrograph of the DNA "caught in action"; the interspersal of transcribed rDNA regions and nontranscribed spacers is clearly seen, as is the simultaneous use of a given rDNA segment for synthesis of many RNA molecules.

5.1.2 Structure and Function: Some Molecular Issues

Figure II–31 also illustrates an important technical aspect of modern cell biology, the utilization of microscopes to gain information about molecules.

Other examples are shown in Figures II–9, II–32, II–57, II–91, III–8, III–38, IV–10, IV–11, and IV–36. Such micrographs dramatize the arbitrariness of customary boundaries among "structural," "biochemical," and "functional" realms. To understand the functioning of biological molecules we must know how they are organized in the cell. Correspondingly, to understand the functioning of cell organelles, we must know how their component molecules are put together, and to comprehend the functioning of macromolecules, we must look to the three-dimensional conformation of the molecules.

Supercoiling and Other Features of DNA Conformation   For nucleic acids, most emphasis has been given to the linear arrangement of nucleotides along polynucleotide chains, but here too, three-dimensional structure is now receiving increased attention. For example, it has become apparent that in their native states, most DNA molecules are "supercoiled" ("superhelical," "supertwisted"; Figs. II–57 and III–11); this conformation can be produced through the operation of enzymes (Section 4.2.4) and through the influences of the histones of eucaryotic chromosomes. Supercoiling helps to explain details of the "packaging" of DNAs in biological structures. The enzymes capable of producing such altered geometry also function during DNA replication to twist the DNA duplexes so that the twisting effects intrinsic in strand separation (Section 4.2.4) are counteracted. Another potentially important consequence could be that supercoiling strains the DNA double helix in ways that favor separation of the two strands when enzymes and other proteins of DNA transcription and replication interact with DNA. Investigators even hypothesize that the double helix normally "breathes," in the sense that there are spontaneous, temporary, very local separations of the two strands. Phenomena of these types might help RNA and DNA polymerases to make their requisite associations with individual polynucleotide chains in DNA duplexes; perhaps they are involved as well in the strand "invasions" and exchanges of genetic recombination (Section 4.3.4). Strand separation might be expected to occur more readily in DNA regions rich in As and Ts (see Fig. IV–10B, legend), which may account in part for the occurrence of certain AT-rich DNA sequences near sites where strand separation is initiated by the cell (possibly including promoters [Sections 3.2.5 and 4.4.6]).

Even fundamental features of the double helix may be subject to functionally important structural change. Very recently it has been discovered that DNA in the test tube, and very likely in the cell as well, can sometimes assume a so-called "Z" conformation; the customary base-pairing relations are maintained, but the sugar-phosphate "backbone" of each chain spirals in zig-zag fashion to the left, rather than forming the smooth, right-handed helix of the more usual "B" form of DNA (Fig. II–15). An exciting line of imaginative ideas, supported at present only by bits and pieces of evidence, proposes that the transitions between these alternative DNA forms depend on methylation of cytosines (Section 4.4.6) related to gene activation. It will be interesting to determine how these changes could relate to the ones detected at the nucleosome level and in chromosome condensation that were discussed in Section 4.4.6.

Inverted and Direct Repeats Certain classes of nucleotide sequences in DNA and RNA can have structural effects whose possible functional ramifications are intriguing. *Inverted repeat* sequences (Fig. V–I), sometimes perfect, and sometimes with a few mismatched bases, are surprisingly common. As

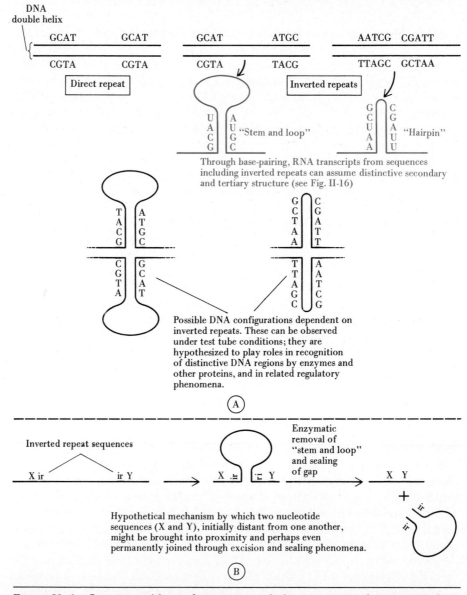

Through base-pairing, RNA transcripts from sequences including inverted repeats can assume distinctive secondary and tertiary structure (see Fig. II-16)

Possible DNA configurations dependent on inverted repeats. These can be observed under test tube conditions; they are hypothesized to play roles in recognition of distinctive DNA regions by enzymes and other proteins, and in related regulatory phenomena.

Ⓐ

Hypothetical mechanism by which two nucleotide sequences (X and Y), initially distant from one another, might be brought into proximity and perhaps even permanently joined through excision and sealing phenomena.

Ⓑ

Figure V–1 **Some possible configurations and phenomena involving inverted repeat sequences in DNA and in corresponding sequences in RNA transcripts. See also Figure III–11A.**

Figures II–16 and V–I illustrate, the presence of inverted repeat sequences permits RNA molecules and single strands of DNA to fold into various looped and hairpin structures whose configurations depend on the locations of the sequences. The three-dimensional conformation of RNA molecules is strongly affected by such loops (Chap. 2.2). Evidence is being sought now to support hypotheses proposing that configurations of these sorts serve as landmarks and recognition sites for enzymes that process nucleic acids in various ways and for regulatory molecules (see also Fig. IV–2). Inverted repeat sequences are present in the flanking regions responsible for control of the β-galactosidase operon in *E. coli* (Section 3.2.5), where they seem to represent sites to which regulatory proteins bind. Inverted repeat sequences found at the other (3'; Section 2.2.4) end of some bacterial genes are believed to be involved in termination of transcription, signaling where certain mRNAs end. Inverted repeats in DNA are one of the types of sequences being examined as possible origins of replication (Section 4.2.4). In certain situations, inverted repeats might also represent devices by which previously distant portions of nucleic acid strands are brought close to another in precise alignment, preparatory to rearrangement (Fig. V–I). Such concepts have been put forth and are being debated in connection with important phenomena such as the bringing together of the segments of genes specifying immunoglobulins (Section 3.6.10).

Direct repeat sequences are DNA sequences that are identical or nearly so, present with similar orientation at both ends of a gene or other stretch of DNA (Fig. V–1). Such sequences also are under suspicion as possible recognition devices for diverse phenomena.

Useful hints may come from the fact that various distinctive nucleic acids have a sequence at one end that is an inverted or a direct repeat of a sequence at the other end: This is true, for instance, of movable DNAs (Section 4.4.6), of retroviruses (Section 3.12.3), and probably of the short pieces produced by fragmentation of the chromosomes in macronuclei of ciliated protozoa (Section 3.3.2). In these cases the sequences are thought somehow to facilitate enzymatic recognition of the ends of distinctive regions of the DNA.

Repetitive DNAs; RNA Processing; Small Nuclear and Cytoplasmic RNAs At least as isolated from cells and studied in the test tube, hnRNAs show both considerable folding and association with proteins (Fig. IV–40). These features may be related to one another and to the extensive processing undergone by the molecules as they generate mRNAs (Sections 2.3.5 and 4.4.6). The folding derives, in part, from nucleotide sequences transcribed from interspersed repetitive DNAs (Section 4.2.6). Many of these sequences, such as those corresponding to the Alu family in human and other mammalian cells (Section 4.2.6), contain inverted repeats; Figure V–1 illustrates how the RNAs transcribed from DNAs with inverted repeats, might fold.

Recently it has become appreciated that the nucleus and cytoplasm each contain a number of species of RNAs additional to the ones whose roles in protein synthesis are understood. The several *small nuclear RNAs* (snRNAs) and *small cytoplasmic RNAs* range in length from about 100 to 300 nucleotides (roughly 4–7S) and occur in distinctive complexes with proteins. Strikingly, several of these small RNAs have nucleotide sequences corresponding to parts of interspersed repetitive DNAs, suggesting they are transcribed from such DNAs; certain ones seemingly come from particular members of the Alu family. Many conceivable roles are being examined for those RNAs and for the ribonucleoprotein particles in which they are present. One line of speculation focuses on their possible participation in the transport of other RNAs from nucleus to cytoplasm or in the transport of proteins across membranes (Section 2.4.3). Another possibility is that certain of the snRNAs take part in processing of hnRNAs. This at least could be an explanation for the facts that one of the snRNAs can form hydrogen-bonded, base-paired associations with hnRNAs and that there are sequences in this snRNA that seem related to the ones at the boundaries between intervening and expressed sequences in hnRNAs (Section 2.3.5). Such lines of reasoning are encouraged by the central roles of RNA–RNA interactions in aligning the participants in protein synthesis (an example is the interactions of tRNAs with mRNAs); there is also a (still scanty) body of observations suggesting the existence of (and roles in RNA processing for) enzymes that can differentially recognize and act on single-stranded or double-stranded regions of RNA molecules. Folding is one way of producing specific, localized regions of double-stranded RNA within a larger molecule; base pairing of a small RNA with a large one is another. It is considered possible that the elaborate but precise processing of hnRNAs, to generate mRNAs, depends in part on the processing enzymes recognizing structural "landmarks" established by the folding of hnRNAs described above, or created by base pairing of snRNAs with regions of hnRNAs to which the snRNAs' base sequences are complementary.

Still other proposals argue that the small RNAs (and regions of other larger RNAs) are important in producing conformations of the proteins with which they are associated, or in other ways contribute to the enzymatic activities of the structures in which the RNAs participate. This last proposition ties in with a still highly tentative set of findings suggesting that in certain processing steps, RNA molecules themselves may play catalytic roles in breaking and forming bonds—roles of a sort more usually ascribed to enzymes or other proteins. Situations—still quite rare—have turned up in which RNA molecules seem to catalyze RNA splicing.

Some investigators also suggest that association of snRNAs with DNA alters accessibility of DNA strands to polymerases or to other enzymes.

No doubt these matters will be considerably clearer soon, as the intensive research now under way bears fruit.

5.1.3 From Wilson to Watson and Back

Of central importance in elucidating cell metabolism and the genetic machinery that controls metabolism has been the use of the procaryotic bacteria and of the noncellular viruses. Study of bacteria and viruses has led to revolutionary changes in the way the cell is viewed and in the experimental questions being asked. These changes are dramatized in the classic text* by J. D. Watson, who with Francis Crick first unraveled the structure of DNA. For example, in Wilson's day and long after, a mutation was an event of unknown nature, affecting a gene of unknown chemistry, and resulting in a change detectable in the organism only at a level such as the color of the eye, far removed from the primary effect. Today, largely from work on *E. coli* and bacteriophages, many mutations are explained as a change in base sequence of DNA, resulting in a changed mRNA, leading in turn to a protein whose enzymatic or other properties are altered, ultimately producing the visible effect in the organism.

It is instructive to realize that much of this progress has come from study of experimental material whose structure and biochemistry were largely beyond the reach of the techniques of Wilson's day. The successes with bacteria and viruses demonstrated the power of analytical approaches based on detailed biochemical studies of relatively simple systems and on the use of mutations and other genetic manipulations to perturb biological function. This has prompted many investigators of eucaryotic cells to seek experimental material that can be exploited in comparable ways; the broad employment of tissue-culture preparations and work with yeast and other unicellular organisms reflect this effort. Many of the fundamental conceptual and technical lessons in molecular biology and biochemistry learned from procaryotes and viruses have proved applicable to eucaryotes as well. Even for such fundamental processes as the formation of ribosomes and messenger RNAs, however, eucaryotes have special features of their own (Sections 3.2.5 and 4.4.6). Most obviously, the presence of nuclear envelopes and nucleoli imposes both complexity and potential levels for regulation; even the proteins of eucaryotic ribosomes must traverse the nuclear envelope before assembling with the RNAs made in nucleoli; mRNA precursors in eucaryotes undergo much more elaborate processing than in procaryotes. Mitosis, embryonic development and differentiation, and a variety of physiological processes such as the functioning of the nervous system have, at best, distant parallels in the procaryotic world. The themes posed in the Introduction to this text—constancy, diversity, and the dependence of progress on techniques and on choice of organisms—are illustrated well by the historical shifting of investigational focus in cell biology from eucaryotes to procaryotes and viruses and now back to a more balanced approach in which studies on all sorts of organisms provide mutual reinforcement and stimulation in many aspects of the field.

A key lesson driven home as the analysis of eucaryotic cell structures and functions in molecular terms has deepened is appreciation of the fact that organelles and other intracellular structures constitute higher levels of integration than that of molecules. A mitochondrion is not a membrane-delimited sac in which respiratory enzymes, DNA, and other molecules are dissolved at random. It is a complex *organized* structure that couples phosphorylation and oxidation, transfers electrons in a highly efficient manner, and carries out other

---

*Watson, J. D. *The Molecular Biology of the Gene*, 3rd ed. New York: Benjamin, 1976.

integrated functions. Nuclear DNA in eucaryotes is not a naked template. It is part of an organized chromosome, capable of complex interactions with other chromosomes (for example, in meiotic pairing) and with other organelles (for example, the mitotic spindle). Light-absorbing proteins in the plasma membranes of some bacteria are oriented to produce *directed* movements of ions; this generates gradients that are harnessed by other plasma membrane–associated proteins, to make ATP (Section 3.2.4).

There is great intellectual excitement and far-reaching practical importance in molecular explanations of cytological events. But there is also profound drama in cell activities at higher levels of integration: the ceaseless beating of heart muscle cells for many decades, the spectacular specializations of structure and function within a single protozoon cell, or the beautifully synchronized and precisely balanced changes in the development of an animal or plant embryo.

Wilson's book,* decades after its publication, elicits astonishment at the variety, complexity, and beauty of cells. Watson's book engenders enthusiasm for the precision and power with which events at the molecular level can now be described.

Chapter 5.2 Cell Biology and Pathology

Even while the concept of the cell as the unit of form, function, and duplication in higher organisms was being established as a principle of biology, it was extended to the study of diseased tissue. The basis for cell pathology was laid in the mid-nineteenth century by Rudolph Virchow. Since that time, pathology, cell biology, and cytology have been interdependent. With the present focus in cell study upon organelles and molecules, cell pathology is naturally moving in similar directions. Explanations of abnormal cell functions, particularly their origins (pathogenesis), are being sought in terms of organelles and molecules. Tantalizing clues abound—witness our discussion of cancer in Chapter 3.12.

Because the cells and proteins are so convenient to study, the human diseases best understood in molecular and genetic terms are those affecting the blood, especially disorders involving hemoglobin. We have already mentioned sickle cell anemia, in which a single amino acid change in a protein, resulting from mutation, leads to grossly abnormal red blood cells. In a group of anemias called the *thalassemias,* hereditary changes lead to abnormal ratios in the abundance of different hemoglobin chains. Some result from gene deletions, whereas others appear to stem from "subtler" genetic changes affecting the processing or stability of mRNAs coding for globins, or steps in translation of the mRNAs by polysomes.

Like cancer, other major human diseases that hitherto have been mysterious are now beginning to be better understood through the emerging con-

*Wilson, E. B. The Cell in Development and Heredity, 3rd ed. New York: Macmillan, 1925.

cepts of cell biology. For instance, the work on lipoprotein receptors and endocytosis described in Sections 2.1.6 and 3.5.3 bears directly on cardiovascular disorders—heart disease and stroke—many of which involve defective control of levels of circulating lipids. Among the recent findings of interest are observations that some humans carry genetic mutations affecting the lipoprotein receptor systems and in consequence show abnormalities in their blood lipids. For several widespread diseases, such as arthritis, autoimmune phenomena are suspected to be important.

Various mammals are afflicted by abnormalities closely similar to those in humans. Sometimes this provides useful material for study. For example, the Chédiak-Higashi syndrome is a poorly understood, genetically transmitted, rare human disease in which a spectrum of abnormalities is encountered, ranging from unusual hair pigmentation to low resistance to infection. Very similar disorders are known to occur in mutant strains of mink, cattle, and mice. These animal strains can be maintained in the laboratory and studied conveniently with fewer ethical problems than those associated with work with human subjects. Studies on one such strain, the "beige" mutant of mouse, have been important in demonstrating that many cell types have unusually large lysosomes and related inclusions such as pigment granules. Moreover, the network of lysosome-related tubules and sacs (GERL; Section 2.5.3) in hepatocytes is remarkably enlarged in the beige mouse. Very likely, such observations point toward defects in intracellular transport and digestion, or in phenomena of membrane fusion, that are significant for the pathogenesis of the disease.

The changes in lysosomes and other cellular structures of the beige mouse exemplify the exaggeration of structure or function that often is encountered in pathological material. Such exaggeration can make evident subtleties difficult to detect in normal material—the abnormal teaches about the normal. The beige mouse is proving useful in working out the nature and roles of GERL. The manner of synthesis of membranes and enzymes of smooth ER in hepatocytes is being studied in animals treated with the drug phenobarbital, since such treatment leads to the manufacture of large amounts of smooth ER (Section 2.4.4). Drugs used to lower cholesterol levels in the blood (hypolipidemic drugs) have been found to increase the number and size of the peroxisomes in hepatocytes of some mammals (Section 2.9.2). The implications of this phenomenon for peroxisome formation and function are now being analyzed.

Parasitic infections are receiving renewed interest, spurred by hopes for dealing with widespread diseases as well by appreciation of the general insights to be gained from studying very highly specialized cells such as parasites. For instance, the protozoa (trypanosomes) that cause sleeping sickness, a disease common in Africa, seem able to evade destruction by the host immune system through changing a glycoprotein present in abundance at their extracellular surface. Change occurs repeatedly in the course of an infection so that successive generations of the protozoa proliferating in the blood stream show different amino acid sequences in the cell-surface glycoprotein. Thus, antibodies made against one generation do not adequately recognize later ones, delaying the mounting of an effective host defensive response. Elucidation of the genetic and regulatory mechanisms that enable the trypanosomes to change the glycoprotein could provide keys to more widely occurring controls. A favored current hypothesis is that genes specifying the different varieties of the cell-surface glycoprotein behave as movable DNAs (Section 4.2.6). According to this hypothesis, the genes depend for expression on their transposition from sites where they are inactive to a specific chromosomal site where they become active; different genes of the family undergo this transposition at different times in the course of the infection. By this mechanism, different genetic variants of the glycoprotein come to be made and shipped to the cell surface at different periods.

Diseased or otherwise abnormal tissue is often the source material for isolating or identifying biologically important substances. The demonstration that liver peroxisomes contain enzymes of fatty acid oxidation was greatly facilitated by the increase in peroxisome number and enzyme content evoked by hypolipidemic drugs. It was from pus cells that Miescher, almost 100 years ago, isolated "nuclein," later renamed DNA. Pus contains large numbers of white blood cells in easily obtainable form and already partially separated from the anucleate red blood cells.

Diagnosis and Therapy   Enzymatic changes in the so-called "inborn errors of metabolism" have been known for a relatively long time. These changes include the abnormal enzymes of amino acid metabolism in several rare mental disorders, the absence of an enzyme of pigment metabolism in certain forms of albinism, and defects in lysosomal enzymes in storage diseases. A growing variety of such abnormalities, and other important ones such as the aneuploid chromosome complements of Down's syndrome (Section 4.5.1), can be detected prenatally through amniocentesis and related diagnostic procedures (Section 3.11.2). In general, with the expansion of the armory of methods available comes progress in the ability to diagnose problems early and accurately. It is anticipated, for example, that antibodies produced on a mass scale with the recently developed monoclonal techniques (Section 3.11.2) will aid greatly in detecting abnormal cells and molecules. One goal is the improvement of early diagnostic capacities for cancer.

The ultimate aim of cell pathology is to obtain sufficient insight into·the basic causes and progressions of diseases so that cure or prevention is possible. With some of the inborn errors of metabolism detected at birth or before, relatively simple adjustments in diet can dramatically improve prospects if imposed early enough. For other diseases as well, changes in environmental factors or in life style hold promise for prevention. Where more direct curative or preventative intervention is required it is increasingly possible to design new therapies or to modify traditional ones based on detailed understanding of the pathologic process. Despite the limits of available information, what is known about the mechanisms and control of cell division has greatly aided rational development of a battery of chemotherapeutic agents that slow the progression of cancers. Agents that reduce or reverse the sickling phenomena in red blood cells are being sought.

Some of the lysosomal storage diseases should prove susceptible to therapy based on replacement of the missing enzymes. With tissue-culture cells, the abnormal intracellular stores of lipids or polysaccharides can sometimes be reduced by adding the enzyme that is missing to the growth medium or by co-culturing normal and abnormal cells. Exogenous enzymes can gain access to the defective lysosomes through endocytic entry into the cell and by fusion of the endocytic vesicles with the lysosomes. Introduction of missing enzymes into patients has been tried; it does have the expected effects on tissues that engage extensively in endocytosis. Unfortunately, tissues such as cardiac muscle seem inadequately active in endocytosis, and others such as central nervous tissue cells are not normally accessible to blood-borne macromolecules. Thus, full clinical cures of storage diseases have yet to be accomplished. Nonetheless, the effort represents a first step in a direction that is promising for many diseases other than the storage disorders—the therapeutic use of the variety of means by which particular types of molecules can be delivered to cells and organelles (Section 2.1.8). Many investigators share the hope that increased understanding of cell surfaces and of the interactions of antibodies and other molecules with cell surface receptors will permit the design of ameliorative agents that can be delivered efficiently and specifically to the appropriate cell types.

Some of the perspectives opened by present-day "DNA technology" are outlined in Sections 3.2.6 and 3.11.3.

Chapter 5.3 Epilogue

Biology in general is in the midst of a revolutionary period, and experiments inconceivable ten years ago are everyday exercises today. Within the career span of many researchers, we have gone from the point where genes were theoretical black boxes to the stage where determination of the base sequences of genes thousands of nucleotides long is a routine matter and where automated machinery that will synthesize nucleotide sequences on order is being produced.

At an accelerating pace, the national and international investments made in biology through agencies such as the National Institutes of Health and the National Sciences Foundations in the United States, and comparable bodies elsewhere, are yielding new concepts and techniques for medicine and agriculture. There are new issues and debates as well, ranging from ongoing public discussions of the safety and of the ethics of genetic engineering to questions about the proper relations between publically funded research and privately managed economic activities that depend on the research results. In other words, we are experiencing the kind of multileveled and exciting ferment that often has characterized periods of intense scientific and technical advance. The February 11, 1983 issue of *Science* (293:4585) is a dramatic survey of progress in "biotechnology" and, in 1983, a new periodical, *Biotechnology*, began publication under the auspices of the journal *Nature*.

One of the extraordinary aspects of science is that each generation learns enough to warrant optimism for the future and to arouse excitement in the succeeding generation. The prospects for cytology and cell biology have never been as bright as they are today. Roughly 40 years elapsed between the last edition of Wilson's book and the first edition of Watson's. Probably, 40 years from today, features of the cell will be described in terms of electrons and atomic nuclei. Already, impressive hypotheses for electron and proton transport in mitochondria, chloroplasts, and bacterial membranes are being constructed along these lines. More progress may be foreshadowed by the current use of techniques such as nuclear magnetic resonance to investigate cellular metabolism: It is anticipated that these methods, and their offspring, will eventually yield a detailed picture of the metabolic patterns of intact, functioning cells and tissues, that will complement the very detailed information we now have about the biochemistry of cell fractions and purified enzymes. Knowledge of the cell changes continuously. The excitement of studying cells derives partly from solving old problems, but each solution brings new questions that require answers.

Index

Numbers in italics refer to illustrations.
Numbers in boldface refer to definitions or principal treatments or summaries of material.